MESONS AND LIGHT NUCLEI

Related Titles from AIP Conference Proceedings

594 Hadrons and Nuclei: First International Symposium
Edited by Il-Tong Cheon, Taekeun Choi, Seung-Woo Hong, and Su Houng Lee,
November 2001, 0-7354-0037-7

549 Intersections of Particle and Nuclear Physics: 7th Conference, CIPANP2000
Edited by Zohreh Parsa and William J. Marciano, December 2000, 1-56396-978-5

540 Particle Physics and Cosmology: Second Tropical Workshop
Edited by José F. Nieves, October 2000, 1-56396-965-3

536 Instrumentation in Elementary Particle Physics: VIII ICFA School
Edited by Sehban Kartal, September 2000, 1-56396-960-2

512 Nuclear Physics at Storage Rings: Fourth International Conference: STORI99
Edited by Hans-Otto Meyer and Peter Schwandt, June 2000, 1-56396-928-9

508 Hadron Physics: Effective Theories of Low Energy QCD
Edited by A. H. Blin, B. Hiller, M. C. Ruivo, C. A. Sousa, and E. van Beveren,
March 2000, 1-56396-927-0

496 Workshop on Instabilities of High Intensity Hadron Beams in Rings
Edited by T. Roser and S. Y. Zhang, December 1999, 1-56396-910-6

494 New Directions in Quantum Chromodynamics
Edited by Chueng-Ryong Ji and Dong-Pil Min, November 1999, 1-56396-908-4

484 Trends in Theoretical Physics II
Edited by Horacio Falomir, Ricardo E. Gamboa Saraví, and Fidel A. Schaposnik,
July 1999, 1-56396-894-0

448 Workshop on Space Charge Physics in High Intensity Hadron Rings
Edited by A. U. Luccio and W. T. Weng, October 1998, 1-56396-824-X

432 Hadron Spectroscopy: Seventh International Conference
Edited by Suh-Urk Chung and Hans J. Willutzki, June 1998, 1-56396-765-0

To learn more about these titles, or the AIP Conference Proceedings Series, please visit
the webpage **http://proceedings.aip.org**

MESONS AND LIGHT NUCLEI

8th Conference

Prague, Czech Republic   2–6 July 2001

EDITORS
J. Adam
P. Bydžovský
J. Mareš
Nuclear Physics Institute, Řež

Melville, New York, 2001
AIP CONFERENCE PROCEEDINGS ■ VOLUME 603

Editors:

J. Adam
P. Bydžovský
J. Mareš

Department of Theoretical Physics
Nuclear Physics Institute
250 68 Rez, Prague
CZECH REPUBLIC

E-mail: adam@ujf.cas.cz
          bydz@ujf.cas.cz
          mares@ujf.cas.cz

Authorization to photocopy items for internal or personal use, beyond the free copying permitted under the 1978 U.S. Copyright Law (see statement below), is granted by the American Institute of Physics for users registered with the Copyright Clearance Center (CCC) Transactional Reporting Service, provided that the base fee of $18.00 per copy is paid directly to CCC, 222 Rosewood Drive, Danvers, MA 01923. For those organizations that have been granted a photocopy license by CCC, a separate system of payment has been arranged. The fee code for users of the Transactional Reporting Service is: 0-7354-0047-4/01/$18.00.

© 2001 American Institute of Physics

Individual readers of this volume and nonprofit libraries, acting for them, are permitted to make fair use of the material in it, such as copying an article for use in teaching or research. Permission is granted to quote from this volume in scientific work with the customary acknowledgment of the source. To reprint a figure, table, or other excerpt requires the consent of one of the original authors and notification to AIP. Republication or systematic or multiple reproduction of any material in this volume is permitted only under license from AIP. Address inquiries to Office of Rights and Permissions, Suite 1NO1, 2 Huntington Quadrangle, Melville, N.Y. 11747-4502; phone: 516-576-2268; fax: 516-576-2450; e-mail: rights@aip.org.

L.C. Catalog Card No. 2001097920
ISBN 0-7354-0047-4
ISSN 0094-243X
Printed in the United States of America

LOCAL ORGANIZING COMMITTEE

J. Adam *(chair)*, P. Bydžovský, J. Dobeš, J. Mareš, M. Sotona
Nuclear Physics Institute, Řež

D. Chrien, B. Sopko
Faculty of Mechanical Engineering, Czech Technical University, Prague

INTERNATIONAL ADVISORY COMMITTEE

V. Burkert (JLab), T.W. Donnelly (MIT), A. Gal (Jerusalem),
F.L. Gross (W.&M., JLab), O. Hashimoto (Tohoku),
T. Johansson (Uppsala), D. Kaplan (INT), J.-M. Laget (Saclay),
H.O. Meyer (IUCF), T. Motoba (Osaka), E. Oset (Valencia),
D.O. Riska (Helsinki), W. Plessas (Graz), C. Schaerf (Roma),
A. Stadler (Evora), G. Wagner (Tuebingen)

SUPPORTED BY

Czech Power Company ČEZ,
Thomas Jefferson National Accelerator Facility
Czech Committee for Cooperation with JINR Dubna

CONTENTS

Preface xv
Opening Address xvii

INVITED TALKS

Nucleon Structure in the Resonance Region 3
    V. D. Burkert
Chiral Dynamics in Few-Nucleon Systems 17
    E. Epelbaum, U.-G. Meißner, W. Glöckle, C. Elster, H. Kamada, A. Nogga, and H. Witała
Pion Pairs in Nuclear Matter 29
    E. Fragiacomo (for the CHAOS Collaboration)
An Introduction to Few Nucleon Systems in Effective Field Theory 41
    H. W. Grießhammer
The Deuteron: A Mini-Review 55
    F. Gross and R. Gilman
Recent Progress of Spectroscopy of Light Hypernuclei 69
    K. Imai
Point Form Analysis of Deuteron and Nucleon Form Factors 79
    W. H. Klink
Electroweak Structure of Few-Body Nuclei 89
    L. E. Marcucci
Recent Progress in Light Front Dynamics 101
    J.-F. Mathiot
Search for a Three-Nucleon Force 113
    H. O. Meyer
Recent Topics in Hypernuclear Spectroscopy 125
    T. Motoba
Probing the ηN Interactions with η Production Reactions 137
    M. T. Peña
Probing the Effectiveness: Chiral Perturbation Theory Calculations of Low-Energy Electromagnetic Reactions on Deuterium 149
    D. R. Phillips
Hard Exclusive Processes at HERMES 161
    G. van der Steenhoven (for the HERMES Collaboration)
First Experiment on Spectroscopy of Λ-Hypernuclei by Electroproduction at JLab 173
    L. Tang, T. Miyoshi, M. Sarsour, L. Yuan, X. Zhu, A. Ahmidouch,
    P. Ambrozewicz, D. Androic, T. Angelescu, R. Asaturyan, S. Avery,
    O. K. Baker, I. Bertovic, H. Breuer, R. Carlini, J. Cha, R. Chrien,
    M. Christy, L. Cole, S. Danagoulian, D. Dehnhard, M. Elaasar, A. Empl,
    R. Ent, H. Fenker, Y. Fujii, M. Furic, L. Gan, K. Garrow, A. Gasparian,
    P. Gueye, M. Harvey, O. Hashimoto, W. Hinton, B. Hu, E. Hungerford,

Italicized name indicates the author who presented the paper.

C. Jackson, K. Johnston, H. Juengst, C. Keppel, K. Lan, Y. Liang,
V. P. Likhachev, J. H. Liu, D. Mack, K. Maeda, A. Margaryan,
P. Markowitz, J. Martoff, H. Mkrtchyan, T. Petkovic, J. Reinhold, J. Roche,
Y. Sato, R. Sawafta, N. Simicevic, G. Smith, S. Stepanyan, V. Tadevosyan,
T. Takahashi, H. Tamura, K. Tanida, M. Ukai, A. Uzzle, W. Vulcan,
S. Wells, S. Wood, G. Xu, Y. Yamaguchi, and C. Yan

Two-Nucleon Interaction and Chiral Symmetry 187
R. G. E. *Timmermans*

Quark Degrees of Freedom in Hadronic Systems 199
V. *Vento*

Pion Production in pN Collisions near Threshold 211
J. *Złomańczuk*, R. Bilger, W. Brodowski, H. Calén, H. Clement,
C. Ekström, K. Fransson, G. Fäldt, L. Gustafsson, B. Höistad, J. Johanson,
A. Johansson, T. Johansson, K. Kilian, S. Kullander, A. Kupść,
B. Morosov, W. Oelert, R. J. M. Y. Ruber, J. Stepaniak, A. Sukhanov,
P. Sundberg, A. Turowiecki, G. J. Wagner, Z. Wilhelmi, C. Wilkin,
J. Zabierowski, and A. Zernov

CONTRIBUTED PAPERS

EFFECTIVE FIELD THEORY APPROACH

**A Renormalisation-Group Approach to Two-Body Scattering with
Long-Range Forces** 229
T. Barford and *M. C. Birse*

**Doublet Channel Neutron-Deuteron Scattering in Leading Order
Effective Field Theory** 233
B. Blankleider and *J. Gegelia*

The Rho Meson in Nuclear Matter—A Chiral Unitary Approach 237
D. *Cabrera*

Effective Chiral Theory for Pseudoscalar and Vector Mesons 241
E. Gedalin, *A. Moalem,* and L. Razdolskaya

Pseudoscalar Meson Mixing in Effective Field Theory 245
E. Gedalin, A. Moalem, and L. Razdolskaya

**On Convergence of the χPT HFF Expansion for One Loop
Contribution to Meson Production in NN Collisions** 247
E. Gedalin, A. Moalem, and L. Razdolskaya

Compton Scattering from the Proton at NLO in the Chiral Expansion 249
J. A. *McGovern*

DIRAC Experiment: Measurement of Pionium Lifetime 251
J. *Smolik* (for DIRAC Collaboration)

Italicized name indicates the author who presented the paper.

FEW BODY SYSTEMS AND RELATIVISTIC TECHNIQUES

Relativistic Covariant Model for Proton-Proton Bremsstrahlung 257
    M. D. Cozma, O. Scholten, R. G. E. Timmermans, and J. A. Tjon

Quark Model Study of the $NN^*(1440)$ Components of the Deuteron 263
    B. Juliá-Díaz, D. R. Entem, F. Fernández, and A. Valcarce

Quark Model Study of the Triton Bound State 267
    B. Juliá-Díaz, F. Fernández, A. Valcarce, and J. Haidenbauer

Stability of Bound States in the Light-Front Yukawa Model 271
    V. A. Karmanov, J. Carbonell, and M. Mangin-Brinet

New Polarization Test for Delta-Isobar Admixture in ^{3}He 275
    V. M. Kolybasov

Asymptotic Structure of the Three-Body Coulomb Green's Function
for the Case of Two Charged Particles 279
    S. B. Levin, S. L. Yakovlev, and N. Elander

Phase-Shift Analyses of p-^{3}He Scattering at $T_L = 4.0-200$ MeV 283
    M. Matsuda, V. Limkaisang, J. Nagata, and H. Yoshino

Meson Dynamics and the Resulting "3-Nucleon-Force" Diagrams:
Results from a Simplified Test Case 287
    L. Canton, T. Melde, and J. P. Svenne

Retardation Effects from Quark Confinement on Low-Energy
Hadron Dynamics 291
    R. K. Gainutdinov and A. A. Mutygullina

Quark-Gluon Retardation Effects and an Anomalous Off-Shell
Behavior of the Two-Nucleon Amplitudes 295
    R. K. Gainutdinov and A. A. Mutygullina

The Two-Nucleon System in the Δ Region including
Full Meson Retardation 299
    M. Schwamb and H. Arenhövel

Neutron-Proton and Neutron-Neutron Quasi-Free Scattering in the
nd Breakup Reaction at 26 MeV 303
    A. Siepe, W. Glöckle, V. Huhn, L. Wätzold, C. Weber, H. Witała, and
    W. von Witsch

Recent Measurements with the Out-of-Plane Spectrometer System at
MIT-Bates 307
    S. Širca (for the OOPS Collaboration)

Electromagnetic Isoscalar $\rho\pi\gamma$ Exchange Current and the
Anomalous Action 311
    E. Truhlík, J. Smejkal, and F. C. Khanna

Nuclear Forces and Quark Degrees of Freedom 315
    M. Lacombe, B. Loiseau, R. Vinh Mau, P. Demetriou, J. P. B. C. de Melo,
    and C. Semay

Covariant Electromagnetic and Axial Form Factors of the Nucleons
from a Chiral Quark Model 319
    R. F. Wagenbrunn, S. Boffi, L. Y. Glozman, W. Klink, W. Plessas,
    and M. Radici

Italicized name indicates the author who presented the paper.

Trinucleon Two-Body Photo Disintegration with Δ-Isobar Excitation 323
    L. P. Yuan, K. Chmielewski, M. Oelsner, P. U. Sauer, J. Adam, Jr.,
    and A. C. Fonseca
Trinucleon Magnetic Form Factors 327
    L. P. Yuan, J. Adam, Jr., K. Chmielewski, H. Henning, S. Nemoto,
    M. Oelsner, and P. U. Sauer

HADRON STRUCTURE

Measurement of the Electric Neutron Form Factor G_{en} in the Exclusive Reaction ^{3}He($\vec{e}$, e' n) at Momentum Transfer Q^2=0.67 (GeV/c)^2 331
    J. Bermuth (for the A1 Collaboration at MAMI)
Investigation of Δ Medium Effects Using the $^4He(\gamma,\pi^+n)$ Reaction 335
    D. Branford (for the Edinburgh, Glasgow, Tübingen, Mainz PiP/TOF
    Collaboration
Landau Parameters and Collective Modes in Nuclear Matter with Relativistic Nonlinear Models 339
    J. C. Caillon, P. Gabinski, and J. Labarsouque
Mesons on a Transverse Lattice 343
    S. Dalley
Heavy Quark-Antiquark Admixture in the Proton and Its Test in Exclusive Hadronic Process 347
    M. Dillig, G. F. Marranghello, S. S. Rocha, and C. A. Z. Vasconcellos
The GDH Sum Rule and Pion Photoproduction on the Nucleon 353
    H. Holvoet (for the GDH/A2 Collaboration)
Photoproduction of Heavy Quarkonia 357
    B. Jäger and W. Schweiger
Separation of Meson and Quark-Gluon Degrees of Freedom in the Hadron Scattering Reactions up to 1 GeV Energy Region 361
    A. I. Machavariani
Single and Multi-Nucleon Antiproton-^{4}He Annihilation at Rest 365
    P. Montagna (for the OBELIX Collaboration)
Deuteron and Dibaryons in the Skyrme Model 369
    E. Norvaišas, A. Acus, T. Krupovnickas, and D. O. Riska
Study of Charm Production in Neutrino Interactions 373
    O. Sato (for the CHORUS Collaboration)
Polarizabilities of Proton and Neutron Investigated by Compton Scattering on the Proton and Light Nuclei 377
    M. Camen, K. Kossert, M. Schumacher, and F. Wissmann
Two-Quark Correlations in the Hard Electromagnetic Nucleon Form Factors 381
    M. Schwärz, and W. Schweiger
The GDH-Experiment at ELSA 385
    T. Speckner (for the GDH-Collaboration)

Italicized name indicates the author who presented the paper.

Critical Behavior of the Characteristics of Hadron-Nucleus and
Nucleus-Nucleus Interactions Depending on the Centrality Degree of
Collisions 389
    M. K. Suleymanov, M. Šumbera, A. S. Vodopianov, and I. Zborovský

Virtual Compton Scattering: First Results from JLab 391
    R. Van de Vyver (for the JLab-Hall A/VCS Collaboration)

PHYSICS OF STRANGENESS

Non-Mesonic Weak Decay of Λ-Hypernuclei and Nuclear Structure 397
    M. F. Aristizabal, H. C. Wu, and W. A. Ponce

Low-Energy K^- Optical Potentials: Deep or Shallow? 401
    A. Cieplý, E. Friedman, A. Gal, and J. Mareš

Few-Body $\Lambda\Lambda$ Hypernuclei and the Onset of Stability for
$\Lambda\Xi$ Hypernuclei 405
    I. N. Filikhin, and A. Gal

Optical Potentials in Kaonic Atoms 417
    C. García-Recio, J. Nieves, E. Oset, and A. Ramos

A Meson Exchange Model for the YN Interaction 421
    J. Haidenbauer, W. Melnitchouk, and J. Speth

Four-Body Calculations of $^4_\Lambda$H and $^4_\Lambda$He with Realistic YN and
NN Interactions 425
    E. Hiyama, M. Kamimura, T. Motoba, T. Yamada, and Y. Yamamoto

How Does Nucleus Shrink by Participation of Λ Hyperon? 429
    E. Hiyama, M. Kamimura, T. Motoba, T. Yamada, and Y. Yamamoto

Single Particle Properties of Λ Hypernuclei in the Density Dependent
Relativistic Hadron Field Theory 433
    C. Keil and H. Lenske

Near Threshold K^+K^- Meson-Pair Production in Proton-Proton
Collisions 437
    A. Khoukaz, C. Quentmeier, H.-H. Adam, J. T. Balewski, A. Budzanowski,
    D. Grzonka, L. Jarczyk, K. Kilian, P. Kowina, N. Lang, T. Lister,
    P. Moskal, W. Oelert, R. Santo, G. Schepers, T. Sefzick, S. Sewerin,
    M. Siemaszko, J. Smyrski, A. Strzałkowski, M. Wolke, P. Wüstner,
    and W. Zipper

Composition and Properties of Mutli-Strange Hypernuclear Systems 441
    D. E. Lanskoy

KN Phase Shifts in a Model with a Spin-Orbit Interaction 445
    S. Lemaire, J. Labarsouque, and B. Silvestre-Brac

Progress of the Nuclotron Accelerator and the Hypernuclear Program 449
    A. I. Golokhvastov, S. A. Khorozov, J. Lukstins, A. Parfenov,
    and N. E. Vasyukhin

$^{10}_\Lambda$Be and $^{10}_\Lambda$B Hypernuclei: A Clue to Some Puzzles in Nonmesonic
Weak Decay 453
    L. Majling, A. Parreño, A. Margaryan, and L. Tang

Italicized name indicates the author who presented the paper.

$^{10}_\Lambda$Be and $^{10}_\Lambda$B Hypernuclei on the Nuclotron 457
    J. Lukstins, V. Nikitin, A. Parfenov, *L. Majling,* D. Chren, B. Sopko,
    and J. Bartke
Electromagnetic K^+ Production on the Deuteron and
Hyperon-Nucleon Interactions 459
    K. Miyagawa, T. Mart, C. Bennhold, and W. Glöckle
$\Lambda\Lambda$ Interaction Indicated by "Lambpha" ($^{6}_{\Lambda\Lambda}$He Double
Hypernucleus) 463
    K. Nakazawa (for the KEK-POS E373 Collaboration)
K^+-Meson Production in Subthreshold pA Collisions with ANKE 467
    M. Nekipelov (for the ANKE Collaboration)
Three-, Four-, and Five-Body Calculations of s-Shell Hypernuclei
with Realistic Interactions 471
    H. Nemura, Y. Suzuki, and Y. Akaishi
Non-Mesonic Decay of Λ-Hypernuclei and the Γ_n/Γ_p Ratio 475
    J. E. Palomar
Pion and Kaon Vector Form Factors and Some Applications 479
    J. E. Palomar
Recent Results on the Nonmesonic Weak Decay of Hypernuclei within
a One-Meson-Exchange Model 481
    A. Parreño, A. Ramos, and C. Bennhold
Finite Temperature Effects on the $\bar{K}$ Optical Potential 485
    L. Tolós, A. Polls, and A. Ramos
Phenomenological Value of the $\pi\Lambda\Sigma$ Coupling 489
    S. Wycech and B. Loiseau

MESON DYNAMICS

Microscopic Description of η-Photoproduction on Light Nuclei 495
    V. B. Belyaev, N. V. Shevchenko, S. A. Rakityansky, W. Sandhas, and
    S. Sofianos
Meson Photoproduction on the Proton at GRAAL 499
    J.-P. Bocquet, O. Bartalini, V. Bellini, M. Castoldi, P. Corvisiero,
    A. D'Angelo, J.-P. Didelez, R. Di Salvo, G. Gervino, F. Ghio, B. Girolami,
    M. Guidal, E. Hourany, V. Kouznetsov, R. Kunne, A. Lapik, P. Levi Sandri,
    A. Lleres, D. Moricciani, V. Nedorezov, L. Nicoletti, D. Rebreyend,
    F. Renard, N. V. Rudnev, M. Sanzone, C. Schaerf, M. L. Sperduto,
    C. M. Sutera, A. Turinge, and A. Zucchiatti
Exclusive Measurement of the $pp \to pp\pi^-\pi^+$ Reaction Close to
Threshold 503
    W. Brodowski, R. Bilger, H. Calén, *H. Clement,* C. Ekström, K. Fransson,
    S. Häggström, B. Höistad, J. Johanson, A. Johansson, T. Johansson,
    K. Kilian, S. Kullander, A. Kupść, P. Marciniewski, A. Mörtsell,
    B. Morosov, W. Oelert, J. Pätzold, R. J. M. Y. Ruber, M. Schepkin,
    J. Stepaniak, A. Sukhanov, P. Sundberg, A. Turowiecki, G. J. Wagner,
    Z. Wilhelmi, J. Zabierowski, A. Zernov, and J. Złomańczuk

Italicized name indicates the author who presented the paper.

Determination of πN Scattering Lengths from Pionic Hydrogen and
Pionic Deuterium X-Ray Data 507
    A. Deloff

η-Photoproduction with SAPHIR at ELSA 511
    J. Ernst (for the SAPHIR Collaboration)

The ITEP Study of Inclusive Pion Double Charge Exchange:
Experiment and Interpretation.. 515
    B. M. Abramov, L. Alvarez-Ruso, Y. A. Borodin, S. A. Bulychjov,
    I. A. Dukhovskoi, A. B. Kaidalov, A. I. Khanov, *A. P. Krutenkova,*
    V. V. Kulikov, M. A. Matsuk, I. A. Radkevich, A. I. Sutormin,
    E. N. Turdakina, and M. J. Vicente Vacas

Search for Medium Effects in Backward Pion Deuteron Quasielastic
Scattering on ^{6}Li... 519
    B. M. Abramov, Y. A. Borodin, S. A. Bulychjov, I. A. Dukhovskoi,
    A. I. Khanov, *A. P. Krutenkova,* V. V. Kulikov, M. A. Matsuk,
    I. A. Radkevich, and E. N. Turdakina

Vector Meson Production with Linearly Polarised Photons at
Jefferson Lab... 521
    K. Livingston

Nucleon and Meson Effective Masses in the Relativistic
Mean Field Theory... 525
    R. Mańka and I. Bednarek

Compton Scattering at GRAAL 527
    D. Moricciani, O. Bartalini, V. Bellini, J.-P. Bocquet, M. Capogni,
    M. Castoldi, A. D'Angelo, A. d'Angelo J.-P. Didelez, R. Di Salvo,
    A. Fantini, G. Gervino, F. Ghio, B. Girolami, A. Giusa, M. Guidal,
    E. Hourany, V. Kouznetsov, A. Lapik, P. Levi Sandri, A. Lleres,
    V. Nedorezov, L. Nicoletti, C. Randieri, D. Rebreyend, F. Renard,
    N. V. Rudnev, C. Schaerf, M. L. Sperduto, C. M. Sutera, A. Turinge,
    A. Zabrodin, and A. Zucchiatti

Some Features of the $pd \to ppn\pi^0$ Reaction at 1.037 GeV 531
    J. Stepaniak, H. Calén, L. Gustafsson, B. Höistad, M. Jacewicz,
    A. Johansson, T. Johansson, A. Kupść, S. Kullander, R. J. M. Y. Ruber,
    C. Ekström, K. Fransson, J. Złomańczuk, A. Turowiecki, Z. Wilhelmi,
    J. Zabierowski, J. Greiff, I. Koch, B. Morosov, R. Bilger, W. Brodowski,
    H. Clement, and B. Shwartz

On Radiative Muon Capture in Hydrogen 535
    E. Truhlík and F. C. Khanna

Pion Electroproduction Amplitude at Threshold and Nucleon Weak
Axial Form Factors 539
    E. Truhlík

Measurement of the $p + {}^6Li \to {}^7Be + \eta$ Reaction Near Threshold
Energies 543
    M. Uličný (for the GEM Collaboration)

The ηN Scattering Length and ηd Final States 547
    S. Wycech and A. M. Green

Italicized name indicates the author who presented the paper.

APPENDIX

Conference Program..551
List of Participants..561
Author Index..573

PREFACE

These proceedings contain invited talks, contributed papers, and poster contributions presented at the International Conference on "Mesons and Light Nuclei", held in Prague, July 2-6, 2001. This meeting was the 8th in a series organized by the Nuclear Physics Institute in Řež since 1974 (Liblice 1974 and 1978, Bechyně 1985 and 1988, Prague 1991, Stráž pod Ralskem 1995, and Pruhonice 1998).

Participants from 20 countries discussed topics that included electroweak and hadronic probes, few body systems, hadron dynamics, hypernuclear physics, meson production, and relativity in light nuclei. A balanced combination of both theoretical and experimental results was presented. It was gratifying to see such high interest in current problems of intermediate energy physics, as reflected in highly attended plenary and contributed talks, in numerous discussions, as well as in the diversity of the topics presented during the meeting.

In addition to the Nuclear Physics Institute, Řež, and the Czech Technical University, Prague, the conference was supported by the Czech Power Company ČEZ, the Thomas Jefferson National Accelerator Facility, and the Czech Committee for Cooperation with JINR Dubna. We thank each of these institutions for their generous support which enabled us to offer scholarships to 24 students and young scientists.

Many people helped make this conference successful. We would like to express our special thanks to the International Advisory Committee for valuable advice and assistance. We are particularly grateful to our secretary, Ms. Alena Wagnerova. The smooth running of the meeting was in large part due to her efforts. Of course, the ultimate success of the meeting depends on its participants. We are grateful to all the speakers for their well prepared, interesting presentations and their contributions to discussions.

We look forward to meeting you at the 9th International Conference "Mesons and Light Nuclei 2004".

Řež, September, 2001   Editors

OPENING ADDRESS

Ladies and gentlemen, dear colleagues, dear friends,

It is a great pleasure for me to open the conference "Mesons and Light Nuclei 2001" and to welcome you on behalf of the Nuclear Physics Institute of the Academy of Sciences of the Czech Republic.

This year's conference is the 8th meeting in the series. The first one was held in 1974, then meetings have followed at various places in Bohemia, mostly with three- or four-year frequency. This year, for the first time, we co-organize the meeting together with and on the grounds of the Czech Technical University.

I have to say that such a long series of international meetings is unique in the context of our institute and perhaps even in the context of the Czech physical community. Of course, I must also add that such a long series under the same and traditional name "Mesons and Light Nuclei" does not mean that conference topics have been petrified. In fact, looking at the contents of the conference proceedings one can clearly trace a movement in the field of the intermediate-energy nuclear physics. One can see that nuclear physics has entered areas which were usually considered domains of subnuclear physics. During this week, meson production, hadron dynamics, or relativistic effects will be discussed together with traditional topics like electroweak and hadronic probes of nuclei, few body systems, or hypernuclear physics.

I am sure that an important signature of the topicality of the physics to be discussed is the number of participants attending the meeting. I am particularly pleased by the presence of many young colleagues in the lecture hall. Here, I would like to gratefully acknowledge the generous support offered by the Czech Power Company CEZ, by the Thomas Jefferson National Accelerator Facility, and by the Czech Committee for Cooperation with JINR Dubna. Thanks to this support we could provide scholarships to 24 students and young scientists.

Let me remind you that our institute organizes a satellite school on a particular subject of the present meeting and under the title "Understanding the Structure of Hadrons" in this place next week.

I am confident that interesting physics will be discussed this week during the conference, that we will learn the latest results, that fresh ideas will be exchanged and new impetuses will be given to all of us. I hope as well that participants from abroad will find time to visit some of the numerous Prague monuments, to taste Czech beer, and to enjoy summertime in Bohemia. I wish our conference the best success!

J. Dobeš
Director NPI, Rez

INVITED TALKS

Nucleon structure in the resonance region

Volker D. Burkert

*Thomas Jefferson National Accelerator Facility,
12000 Jefferson Avenue, Newport News, VA23606, USA*

Abstract. I discuss recent results of inclusive and exclusive electroproduction experiments at Jefferson Lab. They include measurements of the spin response for protons and neutrons in the resonance region, exclusive single pion and multiple pion production to measure resonance transition multipoles, and searches for missing quark model states. A brief outlook to the new domain of Generalized Parton Distributions is given as well.

INTRODUCTION

Studies of the nucleon structure for over 30 years have focused on the deep inelastic regime to determine the quark momentum and spin distributions, and to test fundamental sum rules. One of the surprising findings was that less than 25% of the nucleon spin is accounted for by the spin of quarks [1]. This result is in strong contradiction to expectations, which shows that we are far from having a realistic picture of the intrinsic structure of the nucleon. Moreover, the nucleon structure has hardly been explored in the regime of confinement, which is the true domain of strong QCD. Our understanding of nucleon structure is not complete if the nucleon is not also probed and fundamentally described at large or medium distances. This is the domain where current experiments at JLab have their biggest impact. It is only through a concerted effort of precise experiments and new approaches in theory that we will be able to understand nucleon structure from the smallest to the largest distances within a consistent framework. Experiments at JLab aim at providing precise data as the basis for such an endeavor.

SPIN RESPONSE OF THE PROTON AND NEUTRON

The inclusive doubly polarized electron-nucleon cross section can be written as:

$$\frac{1}{\Gamma_T} \frac{d\sigma}{d\Omega dE'} = \sigma_T + \varepsilon\sigma_L + P_e P_t [\sqrt{1-\varepsilon^2} A_1 \sigma_T \cos\psi + \sqrt{2\varepsilon(1+\varepsilon)} A_2 \sigma_T \sin\psi] \quad (1)$$

where A_1 and A_2 are the spin-dependent asymmetries, ψ is the angle between the nucleon polarization vector and the $\vec{q}$ vector, ε the polarization parameter of the virtual photon, and σ_T and σ_L are the total absorption cross sections for transverse and longitudinal virtual photons. Experiments usually measure the asymmetry

$$A_{\exp} = P_e P_t D \frac{A_1 + \eta A_2}{1 + \varepsilon R} \quad (2)$$

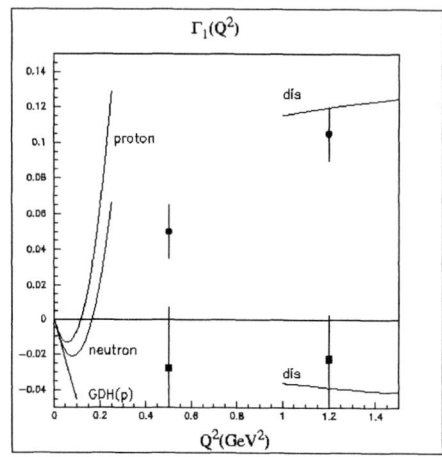 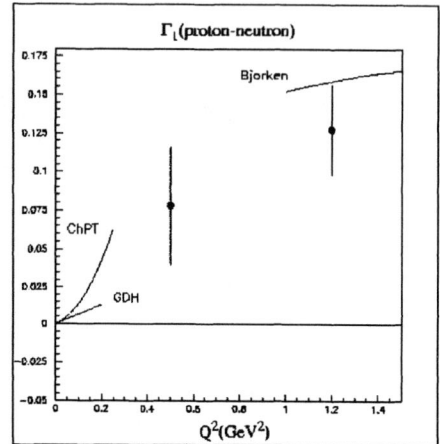

FIGURE 1. First moments of the spin structure function $g_1(x,Q^2)$ for the proton and neutron (left), and for the proton-neutron difference (right). The curves above $Q^2 = 1\text{GeV}^2$ are pQCD evolutions of the measured Γ_1 for proton and neutron, and the pQCD evolution for the Bjorken sum rule, respectively. The straight lines near $Q^2 = 0$ indicate the slopes given by the GDH sum rule. The curves at small Q^2 represent the NLO HBChPT results.

where D is a kinematical factor describing the polarization transfer from the electron to the photon. A_1 and A_2 are related to the spin structure function g_1 by

$$g_1(x,Q^2) = \frac{\tau}{1+\tau}[A_1 + \frac{1}{\sqrt{\tau}}A_2]F_1(x,Q^2) \quad (3)$$

where F_1 is the usual unpolarized structure function, and $\tau \equiv \frac{v^2}{Q^2}$. An important quantity is the first moment $\Gamma_1(Q^2) = \int g_1(x,Q^2)dx$. The Gerasimov-Drell-Hearn (GDH) sum rule [2, 3], and Bjorken sum rule $\Gamma_1^p - \Gamma_1^n = \frac{1}{6}g_A$ for the proton-neutron difference, provide constraints for Γ_1 at the kinematical endpoints $Q^2 \to 0$, and $Q^2 \to \infty$. The evolution of the Bjorken sum rule to finite values of Q^2 using pQCD and the Operator-Product-Expansion (OPE) connects experimental values measured at finite Q^2 to the endpoint. At the opposite end, the GDH sum rule defines the slope of Γ_1:

$$2M^2 \frac{d\Gamma_1}{dQ^2}(Q^2 \to 0) = -\frac{1}{4}\kappa^2 \quad (4)$$

where κ is the anomalous magnetic moment of the target nucleon. Heavy Baryon Chiral Perturbation Theory (HBChPT) may be used to evolve the GDH sum rule to $Q^2 \neq 0$ [10]. The challenge of nucleon structure physics is to test the validity of these evolutions, and to bridge the remaining gap. Lattice QCD may play an important role in describing resonance contributions to the moments of spin structure functions. Using just the constraints given by the two endpoint sum rules we may already get a qualitative picture of $\Gamma_1^p(Q^2)$ and $\Gamma_1^n(Q^2)$. There is no sum rule for the proton and neutron separately that

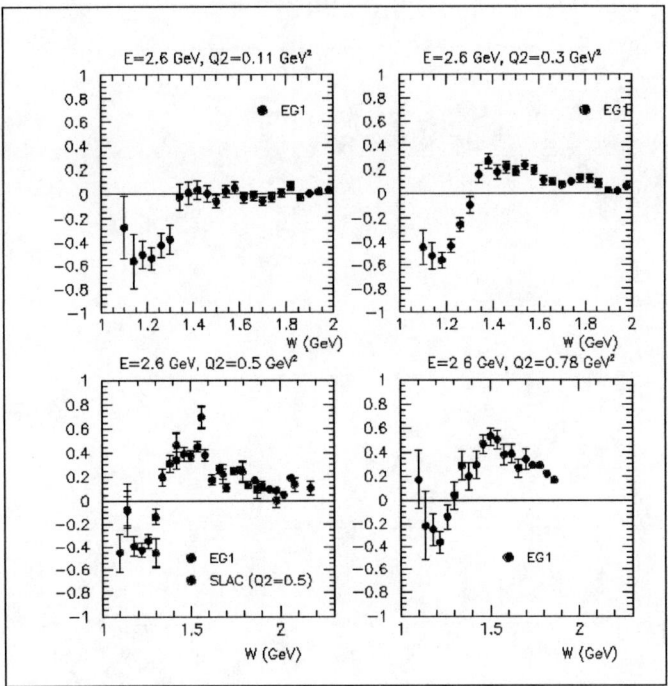

FIGURE 2. Asymmetry $A_1 + \eta A_2$ for protons. The panels show preliminary results from CLAS at a beam energy of 2.6 GeV and for different Q^2 values.

has been verified, however, experiments have determined the asymptotic limits with sufficient confidence for the proton and the neutron. At large Q^2, Γ_1 is expected to approach this limit following the pQCD evolution from finite values of Q^2. At small Q^2, Γ_1 must approach zero with a slope given by the GDH sum rule (assuming the sum rule will be verified). The situation is depicted in Figure 1, where also the next-to-leading HBChPT evolution at small Q^2 and the pQCD evolution to order α_s^3 at high Q^2 are shown. As the slope at $Q^2 = 0$ is < 0, and the asymptotic value is > 0, Γ_1^p must change sign at some value $Q^2 < 1$ GeV^2. We note that the HBChPT evolution [10] cannot give a good description of the trend shown by the existing data, for $Q^2 > 0.1$ GeV^2. However, for the proton-neutron difference the situation is quite different [11]; the HBChPT curve describes the general trend of the data quite well, and over a significantly larger range in Q^2 than for proton and neutron separately.

The first moment $\Gamma_1(Q^2)$ for the proton.

Inclusive double polarization experiments have been carried out on polarized hydrogen [13] using $\vec{NH_3}$ as polarized target material. In Figure 2 the asymmetry is shown for

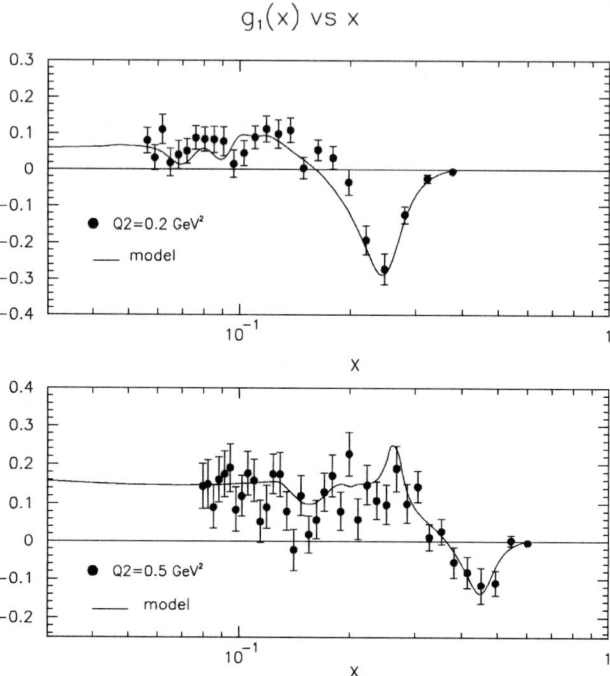

FIGURE 3. Preliminary CLAS results on the spin structure function $g_1(x,Q^2)$ for the proton. The curve labeled "model" is used for radiative corrections, and to extrapolate to $x = 0$ for the evaluation of Γ_1.

various bins in Q^2. For the lowest Q^2 bin the asymmetry is dominated by the excitation of the $\Delta(1232)$, resulting in a strong negative asymmetry. At higher Q^2 the asymmetry in the $\Delta(1232)$ region remains negative, but quickly becomes positive and large at higher W, reaching peak values of about 0.6 at $Q^2 = 0.8$ GeV^2 and W=1.5 GeV. Evaluations of resonance contributions show that this is largely driven by the $S_{11}(1535)$ $A_{1/2}$ amplitude, and by the rapidly changing helicity structure of the strong $D_{13}(1520)$ state. The latter resonance is known to have a dominant $A_{3/2}$ amplitude at the photon point, but is rapidly changing to $A_{1/2}$ dominance for $Q^2 > 0.5$ GeV^2 [12].

Using a parametrization of world data on $F_1(x,Q^2)$ and $A_2(x,Q^2)$ we can extract $g_1(x,Q^2)$ from (5). Examples of $g_1(x,Q^2)$ are shown in Figure 3. The main feature at low Q^2 is due to the negative contribution of the $\Delta(1232)$ resonance. The graphs also show a model parametrization of $g_1(x,Q^2)$ which was used to extrapolate to $x \to 0$. The extrapolation is needed to evaluate the first moment $\Gamma_1(Q^2)$ which is shown in Fig. 4. The characteristic feature is the strong Q^2 dependence for $Q^2 < 1$ GeV^2, with a zero crossing near $Q^2 = 0.3$ GeV^2. Although this result is still preliminary, the qualitative features of the data will not change. Measurements on ND$_3$ have also been carried out with CLAS [14], and on ^{3}He in JLab Hall A [15], to measure the corresponding integrals for the neutron.

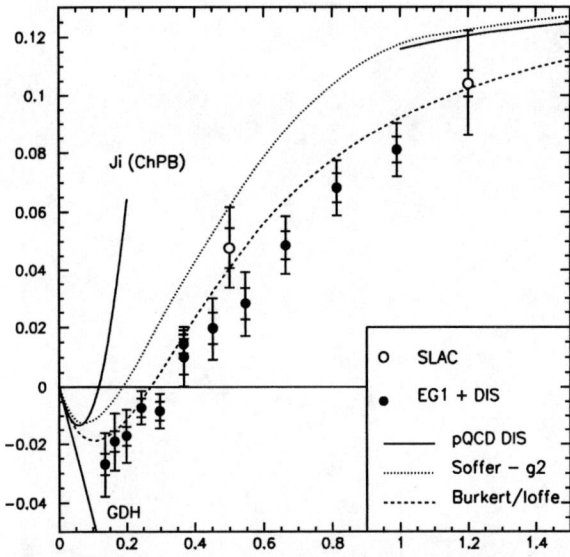

FIGURE 4. The first moment $\Gamma_1(Q^2)$ for the proton. The full symbols are preliminary results from CLAS. Data from SLAC are shown for comparison. The curves are from ref. [8], [9]

Generalized Gerasimov-Drell-Hearn sum rule for neutrons

Data were taken with the JLab Hall A spectrometers using a polarized ^{3}He target. Since the data were taken at fixed scattering angle, Q^2 and ν are correlated. Cross sections at fixed Q^2 are determined by an interpolation between measurements at different beam energies. Both longitudinal and transverse settings of the target polarization were used. Therefore, no assumptions about A_2 are necessary in this case. The GDH integrand for ^{3}He is shown in Figure 5 for various Q^2. The most remarkable feature of these data is the strong negative contribution from the $\Delta(1232)$. In contrast to the proton case, the integrand above the $\Delta(1232)$ region remains negative and small for all Q^2. The GDH integral for ^{3}He was corrected for nuclear effects to extract the integral for neutrons using the prescription by Ciofi degli Atti [16]. Preliminary results are shown in Figure 6. The integral is evaluated over the region from pion threshold (on a free neutron) to $W = 2$ GeV, to cover the resonance region only. The approach to the GDH sum rule value is slower, and the Q^2 dependence less steep than in the proton case. Part of this behavior is due to differences in the helicity structure of the dominant neutron and proton resonance excitations.

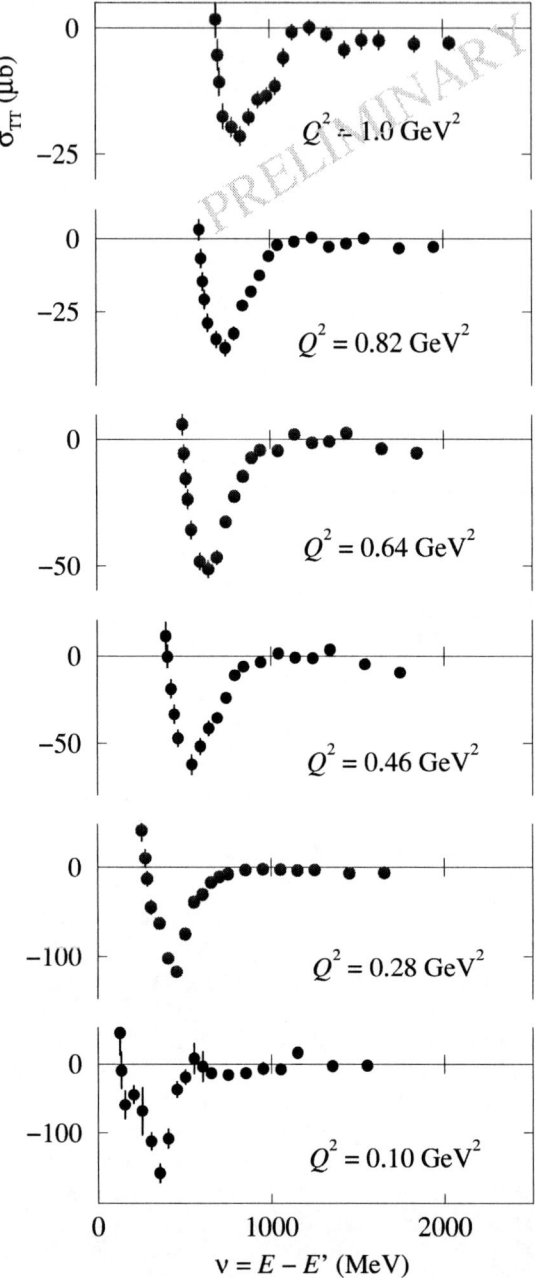

FIGURE 5. Preliminary results from experiment E94-010 on the integrand σ_{TT} for the generalized GDH integral on ^{3}He. The large negative asymmetry is due to the $\Delta(1232)$.

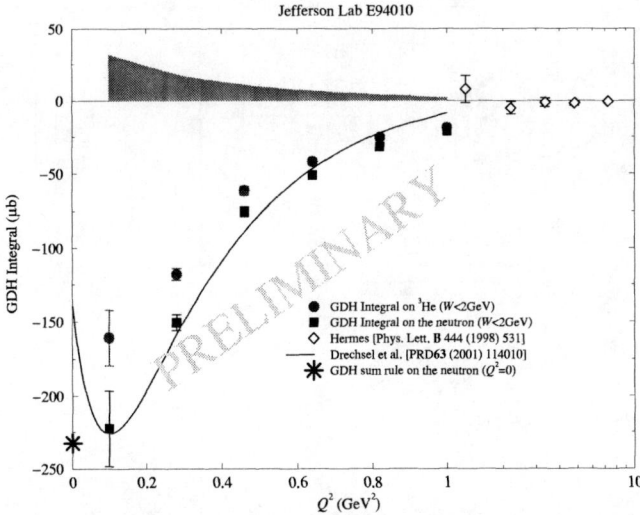

FIGURE 6. Preliminary results on the generalized GDH integral for ^{3}He and neutrons. The shaded area represents the systematic error estimate.

ELECTROPRODUCTION OF MESONS IN THE NUCLEON RESONANCE REGION

A detailed study of nucleon resonance transitions requires measurement of exclusive final states. Current CLAS results in the region of the $\Delta(1232)$ and the $N^*(1535)S_{11}$ are from single π^0 and η production, respectively. The neutral meson is inferred from the missing mass determined due to the overconstrained kinematics of the reaction. The search for "missing" resonances is systematically conducted in $N\pi\pi$ and $N\omega$ channels.

The $\gamma N\Delta(1232)$ transition multipole ratios R_{EM} and R_{SM}

The $\gamma N\Delta(1232)$ transition has been the subject of research for many years. The dominance of the magnetic dipole transition M_{1+} has been known for three decades. The magnitudes of the quadrupole transitions, however, remained poorly determined until recently. The ratio $R_{EM} = E_{1+}/M_{1+}$ was found to have a larger magnitude at the real photon point [17, 18] than constituent quark models predicted. New model developments that take into account explicit pion contributions also predict larger values, and a strong Q^2 dependence for the scalar quadrupole ratio $R_{SM} = S_{1+}/M_{1+}$, while the R_{EM} was predicted to remain nearly constant. This made a study of the Q^2-dependence of the quadrupole transition contributions very interesting.

Single pion production is most sensitive to the $\gamma N\Delta(1232)$ transition. The CLAS detector is well suited for this as it covers a large Q^2 and W range as well as the full

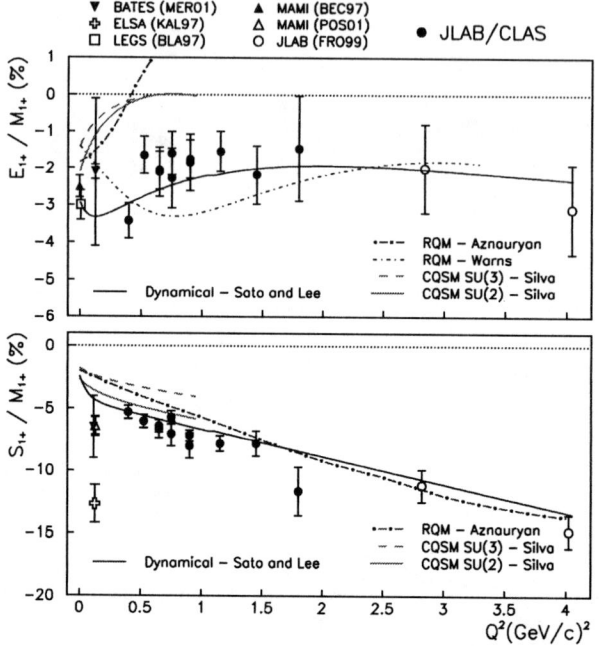

FIGURE 7. Results from CLAS on the R_{EM} and R_{SM} multipole ratios for the $\gamma N\Delta(1232)$ transition.

azimuthal and polar angle distributions of the $N\pi$ system. The azimuthal distribution is fitted to determine the response functions $\sigma_T + \varepsilon\sigma_L$, σ_{TT}, and σ_{TL}, which are then analyzed in terms of multipoles. The results are presented in Figure 7. Included are various relativized quark models and dynamical models with pionic degrees of freedom. Only models that include pions explicitly seem to be able to describe the Q^2 dependence for both the R_{EM} and R_{SM} simultaneously, while constituent quark models may describe one or the other but not both within the same model. It should be noted that dynamical models have been fitted to the photon point and to the two highest Q^2 data points. Also, chiral quark soliton models, while describing roughly the trend of R_{SM}, predict a fast falloff of R_{EM} with Q^2 which is not seen in the CLAS data. We also do not see any trend towards significant leading order contributions from pQCD which require $R_{EM} \to 1$. What is lacking are precise first principle QCD calculations of the $\gamma N\Delta(1232)$ transition multipoles.

The second resonance region

A natural candidate for detailed studies beyond the $\Delta(1232)$ is the Roper resonance $N(1440)P_{11}$. However, more than 35 years after its discovery the structure of this state is still unknown. The non-relativistic constituent quark model (nrCQM) puts its mass

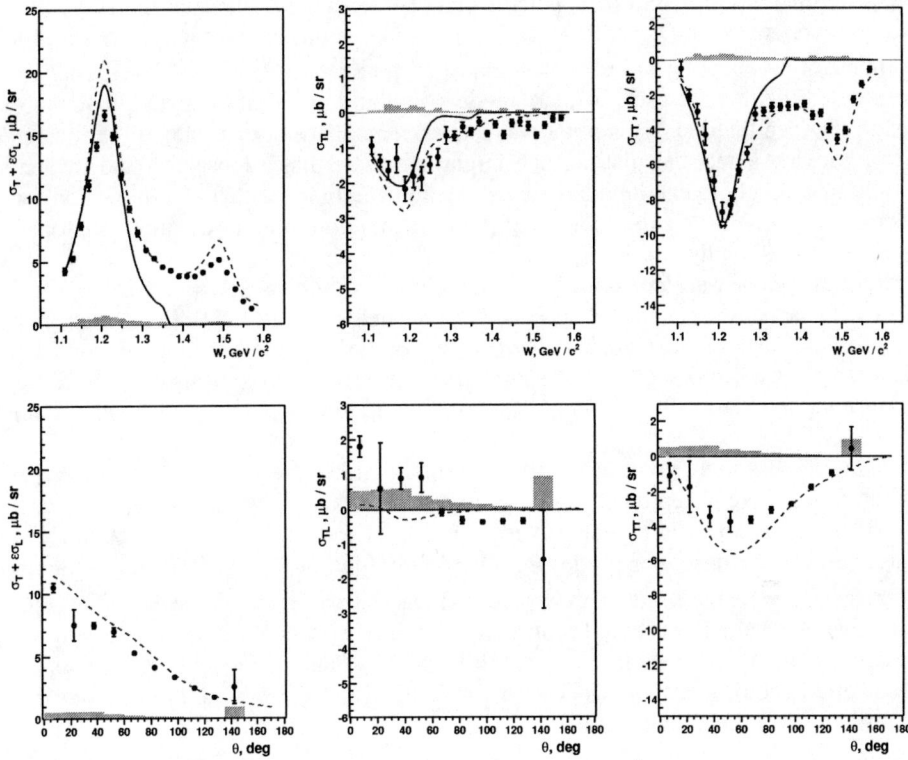

FIGURE 8. Response functions for $n\pi^+$ measured with CLAS at $Q^2 = 0.3$ GeV^2 and $\theta^* = 82.5^o$ as a function of W (top), and as a function of θ^* at W = 1.45 GeV (bottom). The dashed line represents the MAID2000 calculation [19], the solid line is from the dynamical model of Sato and Lee [20], which includes only the $\Delta(1232)$ as a resonant state.

above 1600 MeV, the photocoupling amplitudes are not described well, and the transition form factors, although poorly determined, are far off. Relativized variations of the nrCQM improved the situation only modestly. To obtain a better description of the data a number of alternative models have been proposed. Does the Roper have a large gluonic component [21]? Does it have a small quark core with a large pion cloud [22]? Or is it a nucleon-sigma molecule [23]? It is crucial to get more precise electroproduction data, as it is the Q^2 dependence where the models differ strongly. The study of the "Roper" in the $p\pi^0$ channel is hampered by the presence of the dominant $\Delta(1232)$. Better sensitivity, due to the isospin $\frac{1}{2}$ nature of the state, should be obtained if the $n\pi^+$ channel is included in the analysis. The first $n\pi^+$ data with nearly complete kinematic coverage are becoming available from CLAS. Figure 8 shows the response functions in that channel measured throughout the first and second resonance regions. The combined analysis of these data with the $p\pi^0$ channel is currently underway.

Another topic in the second resonance region has been properties of the S_{11} resonance. Analysis of single pion data gave results for the $A_{1/2}(0)$ photocoupling amplitude which were significantly different from what is obtained from the analysis of the eta channel. More importantly, the Q^2 dependence of $A_{1/2}(Q^2)$ exhibits an unusually hard transition form factor dropping by a factor of less than 2.5 over a range $Q^2 < 3$ GeV^2. This behavior has been difficult to describe in quark models. In addition, the unusual πN phase motion led to the idea that the S_{11} is not a real 3-quark resonance but possibly a $\bar{K}\Sigma$ molecule [24]. Lacking a real calculation, one might speculate that a loosely bound molecule would be unlikely to exhibit a large cross section combined with a hard transition form factor. Revisiting the Q^2 dependence of $A_{1/2}$ has therefore become an important topic of nucleon structure physics.

Measurements were performed with CLAS covering a range from $Q^2 = 0.3 - 4.0$ GeV^2. Data below 1.5 GeV^2 have been published recently [25]. They confirm the trend of the earlier data, showing a hard transition form factor. Preliminary new CLAS data covering the range $Q^2 = 0.2 - 3.0$ GeV^2 give also a very consistent picture [28], confirming the slow fall-off with Q^2, and linking up the photon data [26] with the high Q^2 data [27].

There is also some good news from the theory side. The calculation by the Genoa group [29] is able to reproduce the slow form factor fall-off within a constituent quark model, using a Coulomb-type hypercentral potential and linear confining potential. The same model also describes the leading $A_{1/2}$ amplitude of the $N^*(1520)D_{13}$ in a large Q^2 range. However, the model underpredicts the sub-leading amplitude $A_{3/2}$. This raises the question whether pion cloud contributions are more prevalent in the sub-leading $A_{3/2}$ amplitudes than in the leading $A_{1/2}$ amplitudes. A dynamical model that includes pion cloud effects could answer this question. Lattice QCD may also be able to estimate these contributions at the photon point.

Missing resonances

The so-called missing resonances [30] have been a focus of nucleon structure studies at intermediate energies for a number of years. It is only now that the first experimental results have become available, and serious studies are being undertaken to address the issue. The importance of the topic is due to the fact that these states are predicted within any model having (broken) $SU(6) \times O(3)$ symmetry, reflecting a symmetric arrangement of the 3-quark system. Other symmetry schemes [31] predict a smaller number of states, as for example a quark-diquark configuration. Search for at least some of the states predicted in one but not the other scheme is important, as it will test fundamental symmetry properties which are at the foundation of baryon structure in the domain of confinement and strong QCD. Two final states, $N\omega$ and $p\pi^+\pi^-$, show promise in the study of higher mass nucleon resonances, and the search for missing states. These are currently under intense study with the CLAS detector.

Figure 9 shows total cross section data for the $\gamma^* p \to p\pi^+\pi^-$, showing for the first time resonance structure in this channel for masses greater than 1.6 GeV. The comparison with the model [33] containing the most advanced resonance parametrization for this mass range [12] shows large missing strength in the mass range near 1.71 GeV. While there is no missing state predicted in this mass range, it nevertheless shows the

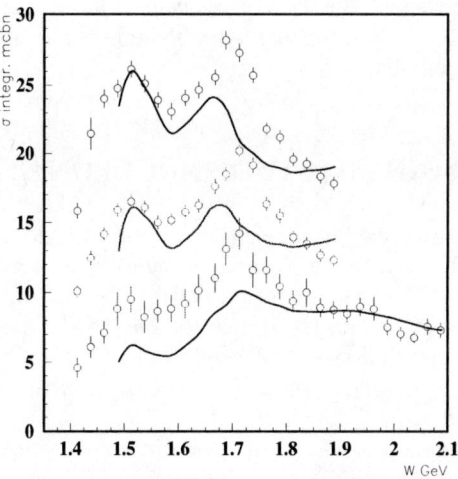

FIGURE 9. Total cross section for $\gamma^* p \to p\pi^+\pi^-$ for different Q^2. The curves represent predictions based on an isobar model containing resonance parametrizations from the analysis of single pion and eta experiments. The various data sets from top to bottom, correspond to $Q^2 = 0.65$ GeV^2, 0.95 GeV^2, and 1.25 GeV^2, respectively.

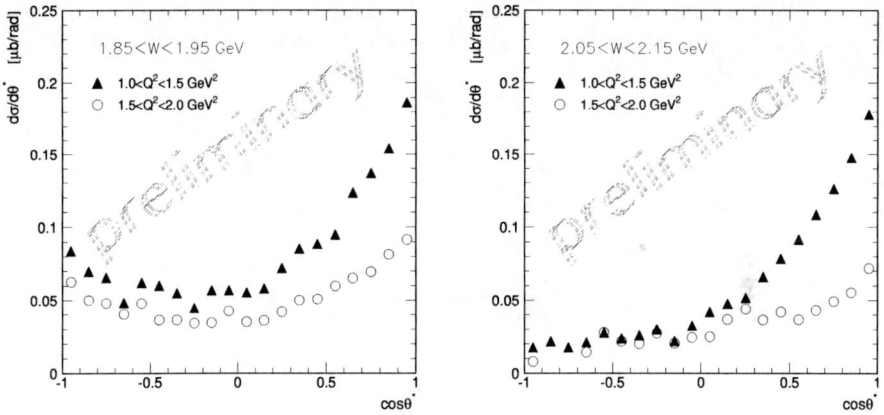

FIGURE 10. Angular distributions for $\gamma^* p \to p\omega$ at different values of the hadronic system W.

sensitivity of this channel to resonance excitations. The data above 1.9 GeV are currently limited to low statistics, high Q^2 data, and do not allow conclusions regarding resonance production in the 1.9-2.1 GeV mass region where most of the missing states are predicted.

Figure 10 shows angular distributions for the $p\omega$ final state at different hadronic masses. This channel is expected to be dominated by t-channel processes at forward angles and nucleon pole and resonance contributions at large angles. The data at high W show mostly t-channel behavior, while in the mass range below 2 GeV significant other contributions are visible. The detailed analysis of these data is currently underway. Any resonant state found in this channel would be interesting as no nucleon resonance is currently known to couple to $p\omega$.

DVCS - A TOOL TO STUDY NUCLEON STRUCTURE

A major goal of measuring exclusive reactions in the resonance region is to study the nucleon wave function which requires measurements at different distance scales. The interpretation of these reactions is complicated by the fact that the virtuality of the photon probe and the momentum transfer to the nucleon or excited state are strongly coupled leading to a correlation of the resolution of the probe and the momentum transfer to the recoiling baryon system. The recently established framework of hard exclusive reactions and generalized parton distributions (GPDs) offers the possiblity of studying resonance excitations where the virtuality of the photon probe is decoupled from the momentum transfer to the baryonic system. For certain kinematics exclusive processes have been shown to factorize into a hard scattering process governed by QED and pQCD vertices, and the soft nucleon structure described by GPDs [34, 35]. In the simplest reaction, the Deeply Virtual Compton Scattering (DVCS) $\gamma^* p \to \gamma p(\Delta, N^*)$ the virtual photon (γ^*) has to have a sufficiently high virtuality (Q^2) for the process to scale. Under these conditions the transition from the proton to the recoil baryon is probed at the parton level, controlled by the momentum transfer t, which can be varied independently of Q^2. Calculations within the GPD formalism for processes such as $\gamma^* p \to \gamma(\Delta(1232), N^*(1520), N^*(1535))$ will be needed to enter this new area of baryon spectroscopy.

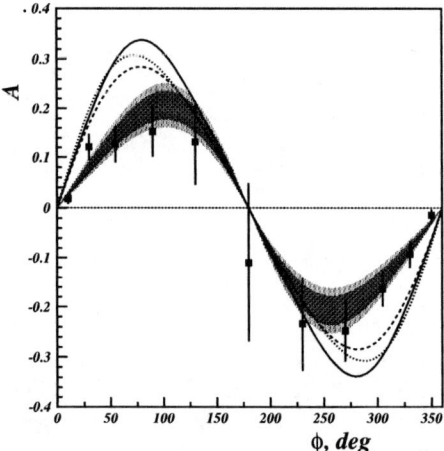

FIGURE 11. Beam spin asymmetry for the DVCS process measured with CLAS at 4.3 GeV beam energy, $Q^2 = 1.3$ GeV^2, $<t> = 0.11$ GeV^2, $x = 0.21$. The curves are predictions of twist-2 and twist-3 calculations and for different parametrizations of GPDs.

A fully exclusive measurement of the DVCS elastic process ($\gamma^* p \to p\gamma$) was recently completed at CLAS [36], using a 4.3 GeV incident polarized electron beam. The polarized beam was used to exploit the interference between the DVCS and the Bethe-Heitler (BH) processes which results in a strong beam spin asymmetry proportional to the imaginary (absorptive) part of the DVCS amplitude. The results are shown in Figure 11 in comparison with theoretical curves describing the reaction based on the hard scattering formalism and models for the GPDs.

This result marks a successful foray into the uncharted territory of GPDs. Measurements at higher energies and with much higher statistics [37] are planned for the near future. Use of the inelastic DVCS process may lead to a promising new avenue of *hard baryon spectroscopy* at the parton level.

CONCLUSIONS AND OUTLOOK

Hadron physics at JLab addresses the transition from the domain of hadronic degrees of freedom and constituent quarks to the single parton regime. The first measurements of double polarization asymmetries have been carried out in a range of Q^2 not covered in previous experiments. The results show large contributions from resonance excitations with rapidly changing helicity structure. The first moment $\Gamma_1^p(Q^2)$ of the spin structure function $g_1(x, Q^2)$ shows a dramatic change with Q^2, including a sign change near $Q^2 = 0.3$ GeV. This marks the dominance of resonance excitations and hadronic degrees of freedom over the single parton domain. The Q^2 dependence of the generalized GDH integral for the neutron shows dominant contributions from the $\Delta(1232)$. In this case no sign change is expected as the asymptotic value $\Gamma_1(Q^2 \to \infty) < 0$ for the neutron.

New data have been taken both on hydrogen and deuterium with nearly 10 times more statistics, higher target polarization, and over a larger range of energies from 1.6 GeV to 5.75 GeV. These data will cover a Q^2 range from 0.05 to 2.5 GeV^2, and a larger portion of the deep inelastic regime. This will greatly reduce systematic uncertainties related to the extrapolation to $x = 0$. The greatly increased precision, and measurements at different energies, will give information on both A_1 and A_2.

There is also a program underway in JLab Hall A to measure the GDH sum rule for neutrons down to Q^2 values near the real photon point, and to measure neutron asymmetries at high x.

Measurements of various exclusive processes in CLAS allows detailed studies of resonance excitations. Precise measurements of the transition multipoles in the $\Delta(1232)$ region show the importance of explicit pion contributions in the transition. New measurements of the $S_{11}(1535)$ transition form factors show a consistent behavior over the entire Q^2 range from 0 to 3.5 GeV^2. The highly topical question of missing resonances is being addressed in the study of multipion and vector meson channels. Both channels show great sensitivity to resonance production, and structures in the data strongly suggest s-channel resonance contributions.

The framework of GPDs and hard scattering phenomenology has opened up a new avenue for the study of the nucleon wave function at the parton level.

The Southeastern University Research Association (SURA) operates JLab for the U.S. Department of Energy under Contract No. DE-AC05-84ER40150.

REFERENCES

1. For a recent review see: Filippone, B.W., and Ji, X., hep-ph/0101224.
2. Gerasimov, S.B., *Sov. J. Nucl. Phys.* **2**, 430 (1966).
3. Drell, S.D., and Hearn, A.C., *Phys. Rev. Lett.* **16**, 908 (1966).
4. Bjorken, J.D., *Phys. Rev.* **179**, 1547 (1969).
5. Ji, X., and Osborne, J., *J. Phys. G* **27**, 127 (2001).
6. Abe, K., et al., *Phys. Rev. D* **58**, 2003 (1998).
7. Burkert, V., and Li, Zh., *Phys. Rev. D* **47**, 46 (1993).
8. Soffer, J., and Teryaev, O.V., *Phys. Rev. Lett.* **70**, 3371 (1993).
9. Burkert, V., and Ioffe, B., *Phys. Lett. B* **296**, 223 (1992); *J. Exp. Theo. Phys.* **78**, 619 (1994).
10. Ji, X., Kao, C.W., and Osborne, J., *Phys. Lett. B* **472**, 1 (2000).
11. Burkert, V., *Phys. Rev. D* **63**, 97904 (2001).
12. Burkert, V., *Czech. J. Phys.* **46**, 627 (1996).
13. Burkert, V., Crabb, D., and Minehart, R., et al., JLab experiment 91-023.
14. Kuhn, S., Dodge, G., and Taiuti, M., et al., JLab experiment 93-009.
15. Meziani, Z.E., et al., JLab experiment E94-010.
16. Ciofi degli Atti, C., and Scopetta, S., *Phys. Lett. B* **404**, 223 (1997).
17. Beck, R., et al., *Phys. Rev. Lett.* **78**, 606 (1997), *Phys. Rev. C* **61**, 035204 (2000).
18. Blanpied, G., et al., *Phys. Rev. C* **64**, 025203 (2001).
19. Drechsel, D., Hanstein, O., Kamalov, S., and Tiator, L., *Nucl. Phys.* **A645**, 145 (1999).
20. Sato, T., and Lee, T.-S.H., *Phys. Rev. C* **63**, 055201 (2001).
21. Li, Zp., Burkert, V., and Li, Zh., *Phys. Rev. D* **47**, 46 (1993).
22. Cano, F., and Gonzalez, P., *Phys. Lett. B* **431**, 270 (1998).
23. Krehl, O., Hanhart, C., Krewald, S., and Speth, J., *Phys. Rev. C* **62**, 025207 (2000).
24. Kaiser, N., Waas, T., and Weise, W., *Nucl. Phys.* **A612**, 297 (1997).
25. Thompson, R., et al. (CLAS coll.), *Phys. Rev. Lett.* **86**, 1702 (2001).
26. Krusche, B., et al., *Phys. Rev. Lett.* **74**, 3736 (1995).
27. Armstrong, C.S., et al., *Phys. Rev. D* **60**, 052004 (1999).
28. Dytman, S., private communications.
29. Giannini, M.M., Santopinto, E., *Few Body Syst. Suppl.* **11**, 37 (1999).
30. Isgur, N., and Koniuk, R., *Phys. Rev. Lett.* **44**, 845 (1980).
31. Kirchbach, M., *Mod. Phys. Lett. A* **12**, 3177 (1997).
32. Burkert, V., and Li, Zh., *Phys. Rev. D* **46**, 47 (1993); DeVita, R., private communication.
33. Ripani, M., et al., *Nucl. Phys.* **A672**, 220 (2000).
34. Ji, X., *Phys. Rev. Lett.* **78**, 610 (1997), *Phys. Rev. D* **55**, 7114 (1997).
35. Radyushkin, A., *Phys. Lett. B* **380**, 417 (1996), Phys. Rev. D **56**, 5524 (1997)
36. Stepanyan, S., Burkert, V., Elouadrhiri, L., et al., hep-ex/0107043.
37. Burkert, V., Elouadrhiri, L., Garcon, M., Stepanyan, S., et al., JLab experiment E01-113.

Chiral dynamics in few−nucleon systems

Evgeny Epelbaum*, Ulf-G. Meißner, Walter Glöckle,
C. Elster, H. Kamada, A. Nogga, H. Witała

*Forschungszentrum Jülich, Institut für Kernphysik (Th), D-52425 Jülich, Germany

Abstract. We report on recent progress achieved in calculating various few−nucleon low−energy observables from effective field theory. Our discussion includes scattering and bound states in the 2N, 3N and 4N systems and isospin violating effects in the 2N system. We also establish a link between the nucleon−nucleon potential derived in chiral effective field theory and various modern high−precision potentials.

INTRODUCTION

Chiral effective field theory (EFT) offers a systematic and controlled method to study the dynamics of few−nucleon systems. In the approach proposed by Weinberg [1, 2], one starts from an effective Lagrangian for nucleon and pion fields as well as external sources, in harmony with chiral and gauge invariance. The effective nucleon−nucleon (NN) potential is constructed using a unitary transformation to avoid any energy-dependence and applying systematic power counting, as described in ref. [3]. To leading order, one has the one−pion exchange (OPE) together with two contact terms accompanied by low−energy constants (LECs). At next−to−leading order (NLO), renormalizations of the OPE, the leading two−pion exchange (TPE) diagrams and seven more contact operators appear. At NNLO, one has to include the subleading TPE with one insertion of dimension two pion−nucleon vertices (the corresponding coupling constants are denoted by $c_{1,3,4}$). The potential is then used in a properly regularized Lippmann−Schwinger equation (e.g. by a sharp or exponential momentum cut-off) to generate the bound and scattering states, as detailed in ref. [4]. The iteration of the potential leads to a non−perturbative treatment of the pion exchange which is of major importance to properly describe the NN tensor force. The nine LECs are to be determined from a fit to the low np partial waves. Here, we discuss implications of the uncertainties in the values of the c_i's to various properties of chiral forces. Apart from the already published NNLO potential with numerically large values of the c_i's, taken from ref. [5], we construct the NNLO* version with (numerically) reduced and fine tuned values of c_3, c_4. We also discuss physical mechanisms, which can explain such smaller values of these low−energy constants. In contrast to the NNLO version, the NNLO* chiral potential does not lead to deeply lying bound states. This makes it more suitable for many−body applications. Both versions reproduce np phase shifts in most partial waves fairly well up to $E_{\text{lab}} \sim 200$ MeV. In the same framework, one can also include charge symmetry breaking and charge dependence of the nuclear force by including the light quark mass difference and elecromagnetic corrections [6]. This allows to predict differences between some phase shifts in pp and nn systems. The extension to three– and four–nucleon systems has also been started using the NLO potential [7]. At this order, no 3N forces (3NF) appear and one can make parameter–free predictions. We also present some first NNLO* results including only two–nucleon forces. We further show that the numerical values of the LECs can be understood on the basis of phenomenological one–boson–exchange (OBE) models [8]. We also extract these values from various modern high accuracy NN potentials and demonstrate their consistency and remarkable agreement with the values in the chiral effective field theory approach. This paves a way for estimating the low–energy constants of operators with more nucleon fields and/or external probes.

FEW–NUCLEON FORCES IN CHIRAL EFT

Chiral EFT is a powerful tool, which allows to calculate low–energy observables performing an expansion in powers of the low–energy scale Q associated with momenta of external particles.[1] The pion mass M_π is treated on the same footing as Q. To select the relevant diagrams contributing to the S–matrix at a certain order, one makes use of power counting. This consists of a set of rules to calculate the power ν of the low–energy scale Q for any given diagram. The precise meaning of the power counting scheme depends on the system one is investigating. In what follows, we will concentrate on various low–energy processes between few (2, 3 and 4) nucleons. The essential complication in that case compared to pion–pion or pion–nucleon scattering is given by the fact, that the nucleon–nucleon interaction is nonperturbative even at very low energies. To deal with this problem Weinberg proposed to use time–dependent ("old–fashioned") perturbation theory instead of the covariant one. The expansion of the S–matrix, obtained using this formalism, has the form of a Lippmann–Schwinger equation with an effective potential, defined as the sum of all diagrams without pure nucleonic intermediate states. Such states would lead in "old–fashioned" perturbation theory to energy denominators, which are by a factor of Q/m smaller than those from the states with pions and which destroy the power counting. Here and below, m denotes the nucleon mass. The effective potential is free from such small energy denominators and can, in principle, be calculated perturbatively to any given precision. This strategy has been followed in the pioneering work by Ordóñez et al. [9]. The effective NN Hamiltonian derived in this formalism possesses some unpleasant properties: it depends, in general, explicitly on the energy of incoming nucleons and is not Hermitean. This complicates its application to few–nucleon systems. In order to avoid these problems we construct an effective NN Hamiltonian using the method of unitary transformation (projection formalism), as described in ref. [3]. For that we modify Weinberg's power counting in an appropriate way. All details are given in [3]. In what follows, we will show the qualitative and the quantitative results for the effective potential in the first few orders of the chiral expansion.

---

[1] To be more precise, Q is associated with the 4–momenta of external pions and 3–momenta of external nucleons.

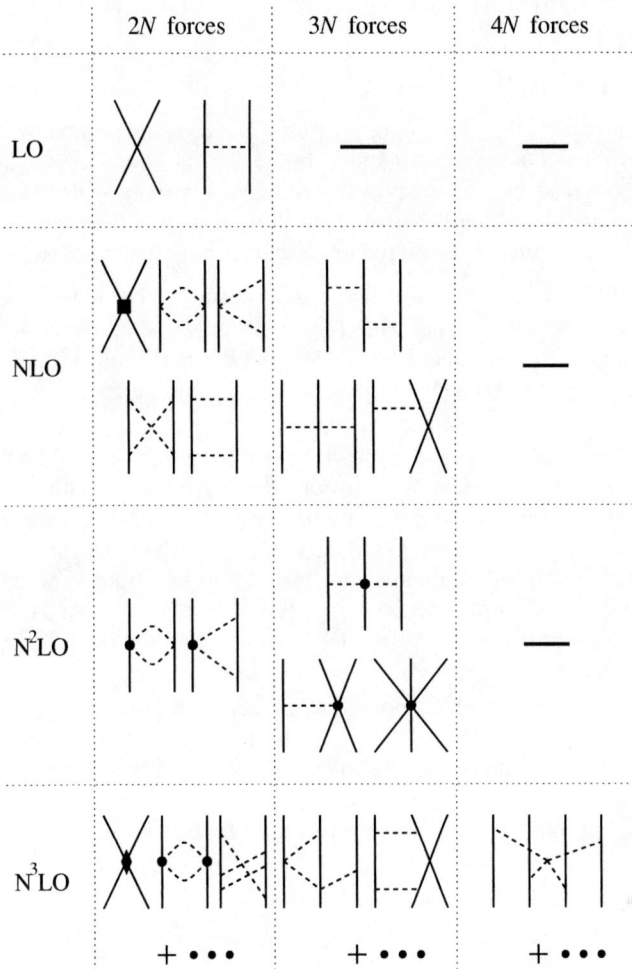

FIGURE 1. First orders in the chiral expansion for few–nucleon forces, as explained in the text. Solid (dashed) lines denote nucleons (pions). The heavy dots, filled square and filled diamond denote the $\Delta_i = 1$, $\Delta_i = 2$ and $\Delta_i = 4$ vertices, respectively.

Let us begin with the power counting. As pointed out above, any (time–ordered) diagram T contributing to the few–nucleon scattering process scales as:

$$T \sim (Q/\Lambda_\chi)^\nu , \qquad (1)$$

where $\Lambda_\chi \sim 1$ GeV is the typical scale of chiral symmetry breaking. For any diagram with E_n nucleons, L loops and V_i vertices of type i one has (we only consider connected diagrams):

$$\nu = -4 + 2E_n + 2L + \sum_i V_i \Delta_i , \qquad (2)$$

where each vertex carries the index Δ_i (also called chiral dimension) given by

$$\Delta_i = d_i + \frac{1}{2}n_i - 2. \tag{3}$$

Here, n_i is the number of nucleon field operators and d_i is the number of derivatives (or pion mass insertions). Due to spontaneously broken chiral symmetry, the index Δ_i cannot be negative. As a consequence, the power ν of the low–energy scale Q is bounded from below and a systematic and perturbative expansion for an effective Hamiltonian becomes possible. Let us now concentrate on the first few orders in the chiral expansion.

- **LO** ($\nu = 0$) The power ν of the low–energy scale Q takes its minimal value $\nu = 0$ for tree diagrams with two nucleons ($E_n = 2$) and with all interactions of dimension $\Delta_i = 0$. Thus, the potential at LO is given by OPE and contact interactions without derivatives, see Fig. 1. There are no 3N and 4N forces at LO.
- **NLO** ($\nu = 2$) The first corrections to the 2N potential are given by tree diagrams with one insertion of $\Delta_i = 2$ interaction (seven independent contact interactions with two derivatives) as well as by one–loop graphs with all leading vertices[2] (leading chiral TPE). Nominally, one has also three–nucleon forces given by tree diagrams with $\Delta_i = 0$ vertices. In the projection formalism it turnes out, that the total contribution from those diagrams vanishes. If "old–fashioned" perturbation theory is used to derive the effective potential, the contribution of the corresponding 3N force cancels against energy dependent part of once iterated LO potential [1, 10]. Thus, there are no 3N and 4N forces at this order.
- **N²LO** ($\nu = 3$) The N²LO 2N potential is given by the subleading chiral TPE with one $\pi\pi NN$ vertex of dimension $\Delta_i = 1$. There are three independent structures in the Lagrangian, which contribute to NN scattering. The corresponding LECs are denoted by $c_{1,3,4}$ [11]. Note that no additional contact interactions appear at that order. There are first nonvanishing three–nucleon forces.
- **N³LO** ($\nu = 4$) The chiral potential at this order has not yet been worked out completely. Some calculations were performed by Kaiser [12]. At this order one has first 4N forces as well as a large number of various 3N and 2N interactions (including two– and three–pion exchange graphs).

Note that one can easily write down the contributions to the effective potential within "old–fashioned" perturbation theory, which correspond to the diagrams shown in Fig. 1. The contributions in the projection formalism are, however, not easily recoverable without knowledge of the explicit operator expressions. The diagrams of Fig. 1 should therefore serve only as a guidance. For the precise numerical prefactors as well as the energy denominators corresponding to the graphs shown the reader is referred to ref. [3]. We would like to point out one of the most important qualitative findings of chiral EFT applied to few–nucleon systems [2, 10]. As shown in Fig. 1, the chiral power counting eq. (2) suggests, that 3N forces are weaker than 2N ones, 4N forces are weaker than 3N ones etc..

---

[2] Only irreducible topologies have to be taken into account in "old–fashioned" perturbation theory. On the contrary, in the projection formalism one has also to include reducible topologies, as explained below.

APPLICATIONS

In what follows, we will discuss application of the described formalism to 2N, 3N and 4N systems, consider some isospin violating effects and establish connection between chiral NN forces and OBE models.

Two nucleons

Let us now concentrate on the parameters entering the NN potential. The two unknown LECs at LO associated with the contact interactions without derivatives have to be fixed by a fit to S–wave phase shifts at low energies. The leading chiral TPE at NLO is parameter–free. At this order one has in addition seven unknown LECs, which corresponds to contact terms with two derivatives. Those constants as well as the two LO LECs are fixed from fit to phase shifts in S– and P–waves and to the mixing angle ε_1. Thus, one has nine unknown constants at this order, see ref. [4] for more details. As already stressed before, the subleading TPE at NNLO depends on the LECs $c_{1,3,4}$, which corresponds to $\pi\pi NN$ vertices of dimension $\Delta_i = 1$. Precise numerical values of these constants are crucial for some properties of the effective potential. Clearly, the subleading $\pi\pi NN$ verices represent an important link between NN scattering and other processes, such as πN scattering. Ideally, one would like, therefore, to take their values from the analysis of the πN system. Various calculations for πN scattering have been performed and are published. From the Q^2 analyses [13] one gets: $c_1 = -0.64, c_3 = -3.90, c_4 = 2.25$. Here all values are given in GeV^{-1}. From various Q^3 calculations [13, 14, 15, 16, 5] one gets the following bands for the c_i's:

$$c_1 = -0.81\ldots-1.53, \quad c_3 = -4.70\ldots-6.19, \quad c_4 = 3.25\ldots4.12. \quad (4)$$

Recently, the results from the Q^4 analysis have become available [17]. At this order the S–matrix is sensitive to 14 LECs (including $c_{1,3,4}$), which have been fixed from a fit to πN phase shifts. It turns out that different available phase shift analyses (PSA) from refs. [18, 19, 20] lead to sizable variation in the actual values of the LECs. In particular, the dimension two LECs acquire a quark mass renormalization. The corresponding shifts are proportional to M_π^2. The renormalized c_i's are denoted by $\tilde{c}_i$. A typical fit based on the phases of ref. [19] leads to:

$$\tilde{c}_1 = -0.27 \pm 0.01, \quad \tilde{c}_3 = -1.44 \pm 0.03, \quad \tilde{c}_4 = 3.53 \pm 0.08. \quad (5)$$

However, using the older Karlsruhe or the VPI phases as input, one finds sizeable variation in the $\tilde{c}_i$. Alternatively, one can also keep the c_i at their third order values and fit the fourth order corrections separately, see [17]. Due to the fitting range chosen and uncertainties in the isoscalar amplitudes, these pieces are not very well determined. In principle, these fits could be improved by including the scattering lengths determined from pionic hydrogen/deuterium. It should also be stressed that numbers consistent with the band given in eq.(4) have been obtained in [23] using IR regularized baryon chiral perturbation theory and dispersion relations. Comparable values of $c_{1,3,4}$ have also been obtained by Rentmeester et al. from analysis of the pp data [21].[3] A few comments are in order. First of all, the numerical values of some of the c_i's appear to be quite large. Indeed, from the naive dimensional analysis one would expect, for example, c_3 to scale like:

$$c_3 \sim \ell/2m \sim \ell/2\Lambda_\chi, \tag{6}$$

where ℓ is some number of order one. Taking the value $c_3 = -4.70$ from ref. [5] and $\Lambda = M_\rho = 770$ MeV we end up with $\ell \sim -7.5$. Such a large value can be partially explained by the fact that the $c_{3,4}$ are to a large extent saturated by the Δ–excitation. This implies that the different and smaller scale, namely $m_\Delta - m \sim 293$ MeV, enters the values of these constants, see [14]. More work on pion-nucleon scattering (dispersive versus chiral representation), new dispersive analyses and more precise low-energy data are needed to pin down these LECs to the precision required here. Applied to NN scattering, the large numerical values of the c_i's might lead to a slow convergence of the low–momentum expansion. Another consequence of the large c_i's for the NN system is the appearance of spurious deeply bound states in low partial waves, which can be traced back to a very strong attractive central potential related to the subleading TPE. The spurious states do not influence low–energy observables, as explained in [4]. The important consequence is, however, that strong 3NF are needed[4]. Note that despite this very different scenario from what is expected in conventional nuclear physics (small 3N forces), the individual contributions of the 2N and 3N forces to observables can, in principle, not be observed experimentally. Since the actual values of the c_i's may possibly change in future analyses of the πN system (higher orders, more precise PSA, etc.), we constructed the NNLO* version of the NN potential, in which we basically subtracted the Δ–contributions to these LECs and allowed for some fine tuning. This results in numerically reduced values of the $c_{3,4}$: $c_3 = -1.15 \,\text{GeV}^{-1}$, $c_4 = 1.20 \,\text{GeV}^{-1}$. As a consequence, no unphysical deeply bound states appear This is partially motivated by the fact that the Δ is not included as an explicit degree of freedom in existing OBE models leading to very good quantitative description of observables. Some steps along a deeper understanding of the surprisingly modest role of the Δ in the NN system within boson exchange models have been undertaken in the framework of the Bonn potential [24, 25]. It was pointed out that there are strong cancellations between the TPE, whose dominant part is given by diagrams with intermediate Δ–excitations, with the $\pi\rho$–exchange. Such a cancellation should ultimately also be observed in EFT studies, but that would require a consistent power counting scheme including vector mesons. Such a scheme has not yet been constructed. Let us finally point out that we do not know at the moment, whether the NNLO or NNLO* versions are closer to reality. Further studies of different processes as well as going to higher orders in the NN system may shed some light on to this question. In what follows, we will use the NNLO* potential in our 3N and 4N calculations as well as for comparison with conventional models.[5] We are now in the position to give quantitative results. In Fig. 2 we show the two np S-wave phase shift

---

[3] Note, however, that it is not possible to fix all three constants in this process. For that reason the constant c_1 has been fixed at the value $c_1 = -0.77 \,\text{GeV}^{-1}$.
[4] Calculated with only 2N forces, the triton is underbound by about 4 MeV [22].
[5] The application of the NNLO potential to few–nucleon systems is technically more complicated.

$^1S_0$ and $^3S_1$ and the $^3S_1 - ^3D_1$ mixing parameter at NLO (left panel) and NNLO* (right panel) in comparison to the Nijmegen PSA. To regularize the LS equation, we have used an exponential regulator $f_R(\vec{p}) = \exp(-p^4/\Lambda^4)$ (for details, see [4]). The two lines correspond to cut-offs $\Lambda = 500$ and 600 MeV. We note that the description of the phases improves when going from NLO to NNLO* and that also the cut-off dependence gets weaker (especially at low energies). This is to be expected from a converging EFT and we emphasize again that this is not the result of an increasing number of free parameters. For the other phase shifts and a more detailed discussion, see [22].

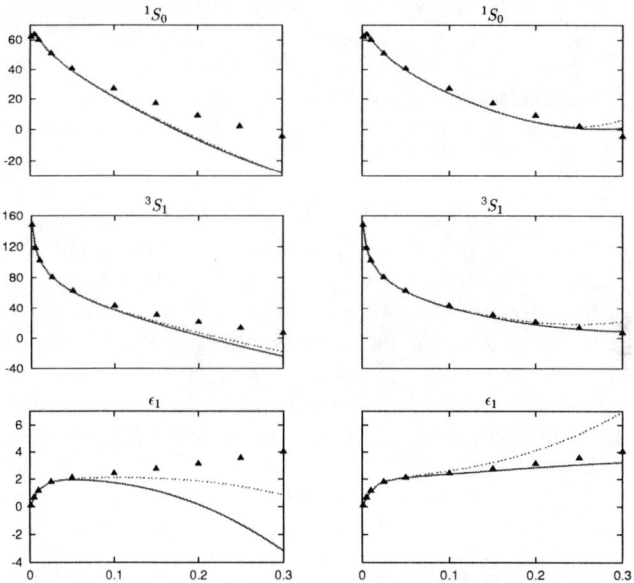

FIGURE 2. Phase shifts at NLO (left panel) and NNLO* (right panel) versus the lab energy (in GeV) in comparison to the Nijmegen PSA. The solid (dashed) line corresponds to $\Lambda = 500$ ($\Lambda = 600$) MeV.

Isospin violation

It is well known that charge symmetry (CS) and charge independence (CI) of the nuclear force are violated. In the Standard Model, these are manifestations of isospin violation (IV). There are two distinct sources of IV. In pure QCD, the light quark mass difference $m_u - m_d$ is the only source of IV. This can easily be incorporated in the EFT by means of an external scalar source. Since the charges of the quarks are also different, there is an additional electromagnetic contribution to IV. The electromagnetic effects due to hard photons are represented by local contact interactions. Soft photons appear in loop diagrams and also generate the long–range Coulomb potential. In [6], these effects have been studied in some detail, extending earlier work of van Kolck and collaborators [26, 27]. The power counting is extended to include the electromagnetic interactions with the fine structure constant α serving as the additional small parameter. With this, one can construct the various contributions to the NN potential. It consists of

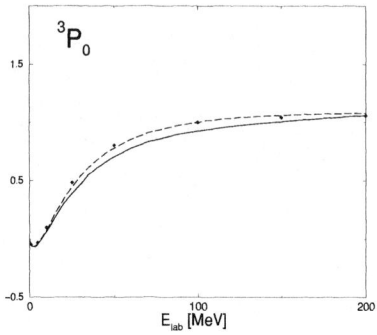 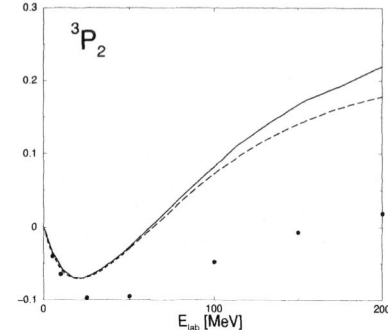

FIGURE 3. Phase shift difference $\delta_{pp} - \delta_{np}$ for the $^3P_0$ and $^3P_2$ partial waves in comparison to the Nijmegen PSA (dashed lines) and results from the CD-Bonn potential (filled circles).

two distinct pieces, the strong (nuclear) potential including isospin violating effects and the Coulomb potential. The nuclear potential consists of one– and two–pion exchange graphs (with different pion and nucleon masses), $\pi\gamma$ exchange diagrams and a set of local four–nucleon operators (some of which are isospin symmetric, some depend on the quark charges and some on the quark mass difference). Since the nuclear effective potential is naturally formulated in momentum space, the matching procedure developed in ref. [28] to incorporate the correct asymptotical Coulomb states was employed in [6]. More precisely, the classification of the IV contributions to the NN potential is as follows: To leading order one has to consider the pion mass difference in the OPE and the Coulomb potential. NLO IV corrections stem from the pion mass difference in the TPE, from $\pi\gamma$ exchange and from two four-nucleon contact interactions with no derivatives which have the generic structure:

$$O_{\text{CSB}} \sim (N^\dagger \tau_3 N)(N^\dagger N), \quad O_{\text{CIB}} \sim (N^\dagger \tau_3 N)^2 \,. \tag{7}$$

These operators parameterize non-pionic CS breaking and CI breaking effects. The low–energy constants accompanying the contact interactions and the cut–off Λ were determined in [6] by a *simultaneous* best fit to the S– and P–waves of the Nijmegen phase shift analysis in the np and the pp systems for laboratory energies below 50 MeV. This allows to predict these partial waves at higher energies and all higher partial waves. Most physical observables come out independent of the sharp cut–off for Λ between 300 and 500 MeV. The upper limit on this range is determined by the pp interactions.

In Fig. 3, we show the predictions for two P-waves in comparison to the Nijmegen PSA and the CD-Bonn potential. The range expansion for the np and the pp system was also studied in [6]. For the range of cut–offs, the pp scattering length varies modestly with Λ due to the scheme–dependent separation of the nuclear and the Coulomb part, $|\delta a_{pp}/a_{pp}| \simeq 0.6$ fm/17 fm $\simeq 3\%$. This shows that the arbitrariness in separating the Coulomb and the nuclear part to a_{pp} is strongly reduced within EFT because only a certain range of cut-offs allows to simultaneously describe np and pp scattering. Finally, it is worth to stress that CSB is largely driven by a contact interaction. The corresponding LEC can be understood to some extent in terms of $\rho - \omega$ mixing, see [27].

Three and four nucleons

The NLO np potential was applied to systems of three and four nucleons in [7]. At this order, one has no 3NF and thus obtains parameter–free predictions. One also gets a first indication about the theoretical accuracy of this approach. Consider first the binding energy. For changing the cut–off between 540 and 600 MeV, the ^{3}H and ^{4}He binding energies vary between $-7.55\ldots-8.28$ and $-24.0\ldots-28.1$ MeV, respectively, showing the level of accuracy of the NLO approximation for these observables. As discussed in detail in [7], the chiral predictions for most observables like the differential cross section or the tensor analyzing powers T_{ij} for elastic nd scattering as well as break-up observables are in agreement with what is found for high precision potentials like CD-Bonn. One also observes a clear improvement in the analyzing power A_y for low and moderate energies, compare the left and the right panels in Fig. 4. The cut-off dependence for the canonical range of Λ is moderate, as indicated by the band in Fig. 4. These calculations have now been extended to NNLO, more precisely, using the NNLO* potential and neglecting all 3NFs, which appear at that order. For a detailed discussion of the results we refer to [22]. Here, we only remark that the cut-off sensitivity of the 3N and 4N binding energies is sizeably reduced at NNLO*, one finds e.g. $-29.96\ldots-27.87$ MeV for $E_B(^4\text{He})$ for even extended cut–off range between 500 and 600 MeV (the overbinding is partly due to the fact that only an np force is used). Also, the description of A_y at 3 and 10 MeV is worse than at NLO, but it is still somewhat better than for the high-precision potentials and, more importantly, the cut-off dependence is much weaker as compared to NLO. Another crucial observation is that at NNLO* we are able to go to much higher energies than at NLO: our NNLO* prediction for A_y at 65 MeV is in excellent agreemnet with the data. Of course, final conclusions can only be drawn when the 3NFs have been included. Work along these lines is underway. For some first steps in this direction see ref. [29].

Connection to "realistic" potentials

Finally, we wish to provide a bridge between the EFT and traditional nuclear physics approaches. In the latter case, one constructs (semi)phenomenological potentials such that one can describe the NN data very precisely. One particular class are the boson exchange potentials, which besides OPE have heavy meson exchanges ($\sigma, \rho, \omega, \ldots$) to generate the intermediate range attraction, short range repulsion and so on. In most modern high-precision potentials (which lead to fits with a χ^2/datum ~ 1) the short-distance physics is parametized in different ways, say by boundary conditions, r-space fit functions or partial wave dependent boson exchanges. In [8] it was demonstrated that existing one–boson–exchange (or more phenomenologically constructed) models of NN force allow to explain the LECs in the chiral EFT potential in terms of resonance parameters. To be specific, consider a heavy meson exchange graph in a generic OBE potential. In the limit of large meson masses M_R, keeping the ratio of coupling constant g to mass fixed, one can interpret such exchange diagrams as a sum of local operators with increasing number of derivatives (momentum insertions). In a highly symbolic

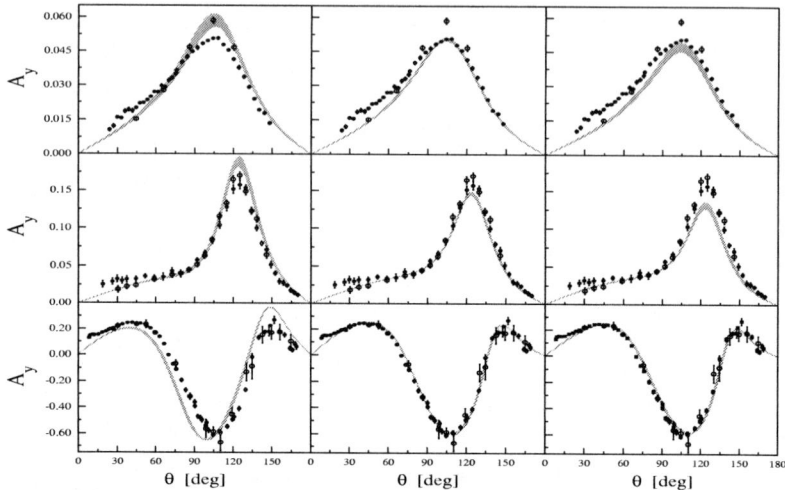

FIGURE 4. Analyzing power A_y for elastic nd scattering, for $E_{\text{lab}} = 3, 10, 65$ MeV (top to bottom). Results at NLO (left panel) and NNLO* (middle panel). The band corresponds to the range $\Lambda = 500$ to 600 MeV. Results based on the high-precision potentials (CD-Bonn, AV-18, Nijm-93, Nijm-I,II) are shown in the right panel. Here the band refers to the uncertainty using the various potentials.

relativistic notation, this reads,

$$(\bar{N}P_iN)\left(\frac{g^2\delta^{ij}}{M_R^2-t}\right)(\bar{N}P_jN) = \left(\frac{g^2}{M_R^2}\right)(\bar{N}P_iN)(\bar{N}P^iN) + \left(\frac{g^2 t}{M_R^4}\right)(\bar{N}P_iN)(\bar{N}P^iN) + \dots, \quad (8)$$

where P_i are projectors on the appropriate quantum numbers for a given meson exchange (including also Dirac matrices if needed). Clearly, it is easy to express the dimension zero and two LECs, corresponding to the two terms of the r.h.s. of eq. (8), in terms of meson masses, coupling constants (and form factor scales, if required by the model). In the chiral EFT, one has to expand the TPE in a similar way and adjust its contribution to the various LECs accordingly. One can also repeat this procedure for the high-precision potentials, this has to be done numerically for the various partial waves. In Fig. 5 we show a comparison between the 4 S- and the 5 P-wave LECs obtained at NLO (leftmost bands) and NNLO* (central bands) and extracted from various OBE and high-precision potentials (see the inset in the figure). The agreement is rather stunning since one would have expected higher dimension operators to play a more prominent role in the phenomenological potentials (related e.g. to the cut-off sensitivity of the πN form factor like in the Bonn potential). Note also that the theoretical uncertainties determined in EFT are a) small compared to their average values and b) are smaller than the band spanned by the potential models (even if one only includes the high–precision ones). Note that the comparison has been performed for the partial wave projected set of coupling constants. Let us now discuss the important issue of naturalness of the various LECs related to contact interactions. For that it turns out to be more appropriate to work directly with the coupling constants $C_S, C_T, C_{1...7}$, which enter the chiral Lagrangian, see [8]. Here, the

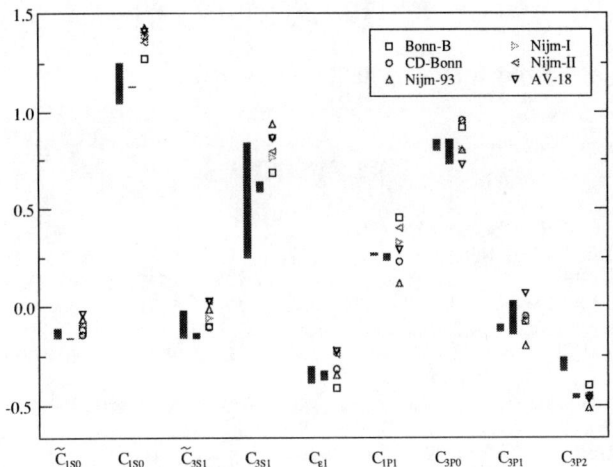

FIGURE 5. LECs from phenomenological models and chiral EFT. The leftmost band refers to NLO (the length reflects the variation with the cut-off), the middle bar is NNLO*, and the symbols correspond to the indicated potentials (see inset).

constants C_S and C_T correspond to operators without derivatives, whereas the remaining ones to contact terms with two derivatives. Dimensional arguments suggest the following scaling properties of these LECs: $C_{S,T} \sim l_{S,T}/(f_\pi^2)$, $C_{1...7} \sim l_{1...7}/(f_\pi^2 \Lambda_\chi^2)$, where the l's are some numbers of order one. These scaling properties of the LECs are crucial for the convergence of the low–momentum expansion. Note that one has to take into account numerical prefactors which accompany the various terms of the Lagrangian, see ref. [8] for more details. Taking $\Lambda_\chi = 1$ GeV it turns out that the values of the l's fluctuate between 0.3 and 3.5, i.e. the values found for these LECs are indeed natural, with the notable exception of $f_\pi^2 C_T$, which is much smaller than one: $f_\pi^2 C_T = -0.002...0.147$ at NLO and $f_\pi^2 C_T = 0.002...0.040$ at NNLO*. These unnaturally small numbers can be traced back to the SU(4) symmetry of the NN interaction, proposed about 65 years ago by Wigner [30]. For the recent discussion on that subject within the EFT approach see ref. [31].

SUMMARY AND OUTLOOK

Despite the original scepticism by its creator [2], chiral effective field theory not only offers qualitative but also *quantitative* insight into the dynamics of few-nucleon systems, as should have become clear from the discussed topics. In addition, it is the only framework at present in which one can address the question of the size of three (four) nucleon forces in a truly systematic manner. It is therefore evident that a vast effort has to be undertaken to pin down the chiral 3NF and apply it not only to few- but also many-body systems. Exciting times are ahead of us.

ACKNOWLEDGMENTS

E.E. thanks the organizers for the invitation and support.

REFERENCES

1. Weinberg, S., *Phys. Lett. B* **251**, 288 (1990).
2. Weinberg, S., *Nucl. Phys.* **B363**, 3 (1991).
3. Epelbaum, E., Glöckle, W., and Meißner, U.-G., *Nucl. Phys.* **A637**, 107 (1998).
4. Epelbaum, E., Glöckle, W., and Meißner, U.-G., *Nucl. Phys.* **A671**, 295 (2000).
5. Büttiker, P., and Meißner, U.-G., *Nucl. Phys.* **A668**, 97 (2000).
6. Walzl, M., Meißner, U.-G., and Epelbaum, E., *Nucl. Phys.* **A693**, 663 (2001).
7. Epelbaum, E., et al., *Phys. Rev. Lett.* **86**, 4787 (2001).
8. Epelbaum, E., Glöckle, W., , Meißner, U.-G., and Elster, Ch., nucl-th/0106007.
9. Ordóñez, C., Ray, L., and van Kolck, U., *Phys. Rev. C* **53**, 2086 (1996).
10. van Kolck, U., *Phys. Rev. C* **49**, 2932 (1994).
11. Bernard, V., Kaiser, N., and Meißner, U.-G., *Int. J. Mod. Phys. E* **4**, 193 (1995).
12. Kaiser, N., *Phys. Rev. C* **61**, 014003 (2000); *C* **62**, 024001 (2000); *C* **63**, 004010 (2001); nucl-th/0107064.
13. Bernard, V., Kaiser, N., and Meißner, U.-G., *Nucl. Phys.* **B457**, 147 (1995).
14. Bernard, V., Kaiser, N., and Meißner, U.-G., *Nucl. Phys.* **A615**, 483 (1997).
15. Mojžiš, M., *Eur. Phys. J. C* **2**, 181 (1998).
16. Fettes, N., Meißner, U.-G., and Steininger, S., *Phys. Lett. B* **640**, 199 (1998).
17. Fettes, N., and Meißner, U.-G., *Nucl. Phys.* **A676**, 311 (2000).
18. Koch, R., *Nucl. Phys.* **A448**, 707 (1986).
19. Matsinos, E., *Phys. Rev. C* **56**, 3014 (1997).
20. SAID on–line program, Arndt, R.A., et al., see website http://gwdac.phys.gwu.edu/.
21. Rentmeester, M.C.M., Timmermans, R.G.E., Friar, J.L., and de Swart, J.J., Phys. Rev. Lett. **82**, 4992 (1999).
22. Epelbaum, E., et al., in preparation.
23. Becher, T., and Leutwyler, H., *JHEP* **0106**, 017 (2001).
24. Machleidt, R., Holinde, K., and Elster, Ch., *Phys. Rep.* **149**, 1 (1987).
25. Janssen, G., Holinde, K., and Speth, J., *Phys. Rev. C* **54**, 2218 (1996).
26. van Kolck, U., *Few-Body Systems Suppl.* **9**, 444 1995.
27. van Kolck, U., Friar, J.L., and Goldman, T., *Phys. Lett. B* **371**, 169 (1996).
28. Vincent, C.M., and Phatak, S.C., *Phys. Rev. C* **10**, 391 (1974).
29. Hüber, D., et al., *Few–Body Syst.* **30**, 95 (2001).
30. Wigner, E., *Phys. Rev.* **51**, 106 (1937), **51**, 947 (1937), **56**, 519 (1939).
31. Mehen, T., Stewart, I.W., and Wise, M.B., *Phys. Rev. Lett.* **83**, 931 (1999).

Pion pairs in nuclear matter

E. Fragiacomo (for the CHAOS collaboration)

University and Infn Trieste, 34127 Trieste, Italy

Abstract. The pion-production reaction $\pi^+ A \to \pi^+\pi^\pm A'$ as a function of A and T has been studied with the CHAOS spectrometer at the TRIUMF meson facility. Charged pion pairs have been detected in coincidence to ensure a unique identification of $\pi^+\pi^\pm$ events.
The $\pi^+\pi^-$ invariant mass $M^A_{\pi^+\pi^-}$ has been found to peak increasingly toward the $2m_\pi$ threshold as the nucleus mass number increases, while the $M^A_{\pi^+\pi^+}$ yield shows a negligible A-dependence near threshold. The $\pi^+\pi^-$ pairs have been found to predominantly couple to S-wave, i.e. $J = 0$. Theoretical works on the $(\pi\pi)_{I=J=0}$ interaction in nuclear matter, have shown that such enhancement is the combined effect of standard many-body correlations and the restoration of chiral symmetry in nuclear matter, i.e. the appearance of the σ-meson.
New data for the pion-production reaction on ^{45}Sc at three incident kinetic energies (240, 280, 300 MeV) have been analyzed and preliminary results are presented. The results for ^{45}Sc at 280 MeV are compared with the ^{40}Ca results, previously published by the CHAOS collaboration. The ^{45}Sc results at different energies highlight the T-dependence of the $\pi\pi$ interaction in nuclear matter.

INTRODUCTION

The $\pi\pi$ interaction in nuclei was studied at TRIUMF by means of the $\pi^+ A \to \pi^+\pi^\pm A'$ ($\pi 2\pi$) reactions, which were measured simultaneously. In order to exploit the A- and T-dependence of the $\pi\pi$ interaction, the $(\pi, 2\pi)$ reaction was studied on four different nuclei (^{2}H, ^{12}C, ^{40}Ca, ^{208}Pb) at an incident kinetic energy $T_\pi = 280$ MeV and on ^{45}Sc at three different incident kinetic energies $T_\pi = 240, 280,$ and 300 MeV.

In early experimental works on $\pi 2\pi$ in nuclei, it was observed that (i) pions preferentially populate the low-energy part of the kinetic energy spectra [1, 2, 3], and (ii) the $\pi^+\pi^-$ invariant mass $M^A_{\pi^+\pi^-}$ peaks increasingly toward the $2m_\pi$ threshold as the nucleus mass number increases [3]. The property (i) could not be explained in terms of the $\pi^+ A \to \pi^+\pi^- p[A-1]$ phase space [1]. Nor was a detailed model of the $A(\pi^+, \pi^+\pi^-)$ reaction [4] able to reproduce the low-energy pion yield, although it embodied the kinematical limits of the experimental apparatus. The same model, however, could correctly predict the total cross section of the $\pi 2\pi$ reaction in nuclei [5], as well as many-fold differential cross sections [2]. A novel approach [6] was then employed to explain the near threshold behaviour of $M^A_{\pi^+\pi^-}$, i.e. the property (ii). In this approach a $\pi\pi$ pair is considered as a strongly interacting system when the two pions couple to the $I = J = 0$ quantum numbers. The $(\pi\pi)_{I=J=0}$ properties in nuclear matter are studied by dressing the single-pion propagator to account for the P-wave coupling of pions to *particle – hole* and Δ – *hole* configurations. The model is able to explain the general features of the $M^A_{\pi^+\pi^-}$ distributions, which are predicted to increasingly accumulate strength near the $2m_\pi$ threshold for ρ, the nuclear medium density, approaching $ρ_0$, the saturation density.

However, within the same theoretical framework, the near-threshold strength of the $\pi\pi$ T–matrix is considerably reduced when the $\pi\pi$ interaction is constrained to be chiral symmetric [7]. This may indicate that effects other than the in-medium $(\pi\pi)_{I=J=0}$ interaction contribute to the observed strength. Conversely, the absence of any in-medium modification of the $(\pi\pi)_{I=J=0}$ interaction leads to $M^A_{\pi^+\pi^-}$ distributions which lack of strength at threshold. This is shown in the work of [3], which compares the experimental results with the model predictions of [4].

The results for the ^{45}Sc show that the $\pi^+\pi^-$ invariant mass $M^A_{\pi^+\pi^-}(T)$ peaks toward the $2m_\pi$ threshold for all the considered incident kinetic energies, confirming the behaviour seen for the heavier nuclei at $T_\pi = 280$ MeV.

THE CHAOS SPECTROMETER AND TRIGGERING SYSTEM

CHAOS is a magnetic spectrometer which was designed for the detection of multi-particle events in the medium-energy range [8]. The magnetic field is generated by a dipole whose pole tip is 66 cm in diameter. The magnet is capable of producing a field intensity up to 1.6 T with an uniformity of about 1%. There is a 12 cm bore at the centre for the insertion of targets. Fig. 1 illustrates a reconstructed $\pi_i^+ \to \pi^+\pi^- p$ event

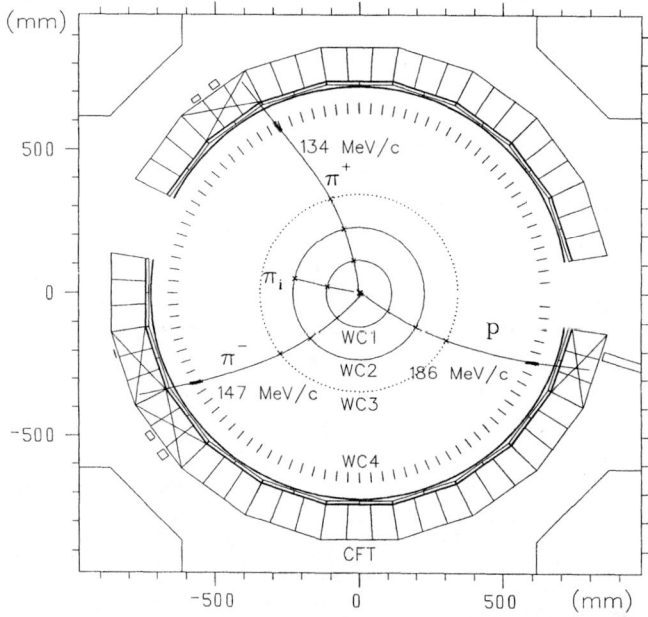

FIGURE 1. Reconstructed particle trajectories in CHAOS for the $\pi_i^+ \to \pi^+\pi^- p$ reaction on ^{12}C, the geometrical disposition of the wire chambers (WC), the first level trigger hardware (CFT) and the magnet return yokes in the corners. Two CFT segments are removed to permit the particle beam (π_i) to traverse the spectrometer. The CFT segments which are hit by particles are marked with crosses, and the energy deposited in the first two layers ($\Delta E1$ and $\Delta E2$) is indicated by boxes. The proton has a momentum slightly above the CHAOS threshold (185 MeV/c), and its energy is fully deposited in $\Delta E1$.

on ^{12}C, and shows the geometrical disposition of the wire chambers (WC), the CHAOS first level trigger hardware (CFT), and the magnet return yokes in the corners. WC1 and WC2 are multiwire proportional chambers, cylindrical in shape, with diameters of 22.8 cm and 45.8 cm, respectively. WC3 is a cylindrical drift chamber designed to operate in a magnetic field [9]. Its diameter is 68.6 cm. The outermost chamber, WC4, is a vector drift chamber 122.6 cm in diameter, which operates in the tail of the magnetic field of CHAOS. The CFT hardware consists of three adjacent cylindrical layers of fast-counting detectors [10]. The first two layers ($\Delta E1$ and $\Delta E2$) are NE110 plastic scintillators 0.3 cm and 1.2 cm thick, respectively. $\Delta E1$ is 72 cm from the magnet centre and spans a zenithal angle of $\pm 7°$; thus, it defines the geometrical solid angle of CHAOS $\Omega=1.5$ sr. The third layer is a SF5 lead-glass 12.5 cm thick, about 5 radiation lengths, which is used as a Cerenkov counter. The three layers were segmented in order to provide an efficient triggering system for multi-particle events. Each segment covered an azimuthal angle (i.e., in-the-reaction plane) of $18°$. During data taking two segments were removed in order to allow the passage of the particle beam (see Fig. 1).

The experiment relied on two on-line triggers: the first ($1^{st}LT$) [10] and second ($2^{nd}LT$) [11] level trigger. The logical signals which were issued by the CFT following $\pi^+A \rightarrow \pi^+\pi^\pm A'$ events were initially pipe-lined and then filtered by requiring a coincidence $M_b \times (\Delta E1 \times \Delta E2)_i \times (\Delta E1 \times \Delta E2)_j$ with $i,j = 1,..,18$ and $i \neq j$. Such a logic analysis was performed at a rate of about 30 MHz. Despite of the fast filtering, the $1^{st}LT$ trigger was overwhelmed (>99.99%) by multi-particle events from competing reactions: quasi-elastic scattering (qes) $\pi^+ \rightarrow \pi^+p(d)$, and pion absorption (abs) $\pi^+ \rightarrow pp$. This called for a second and more selective triggering system, which was based on the information delivered by the innermost wire chambers. For each event passed by the $1^{st}LT$, the $2^{nd}LT$ trigger calculated the momentum of each track, its polarity and the reaction vertex topology. It then compared a combination of these parameters with predefined criteria. When the criteria were fulfilled the event was accepted and recorded on tape, otherwise a fast clear signal was sent to the readout electronics, and the two triggers were reset. An entire $2^{nd}LT$ cycle was accomplished in $< 10\mu s$, depending on the number of reconstructed tracks. The event rate recorded on tape ranged from 30 to 100 Hz depending on the nucleus studied. Roughly 0.1% of the recorded events were $\pi \rightarrow \pi\pi$ as determined by detailed off-line analysis.

THE CHAOS ACCEPTANCE AND PERFORMANCE

In the present measurement CHAOS was operated with a magnetic field of 0.5 T. The field and the energy deposited by the particles emerging from the target in the WC's and $\Delta E1$ media determined the CHAOS threshold, which was 57 MeV/c for pions.

Moreover, for both positive and negative pions, the removal of the CFT segment which allows the particle beam to enter CHAOS (strip at around $180°$) negligibly affects the reaction phase space. The missing CFT segment at the exit (strip at around $0°$) somewhat restricts the available $\pi \rightarrow \pi\pi$ phase space. However, the limitation is only for an angular interval of $14°$.

The acceptance of CHAOS is irregular in the (p_π, Θ_π) plane due to the finite segmentation of the CFT, the two missing CFT segments, and the pion decays inside CHAOS. All these sources of non-uniformity were accounted for by assigning a *weight* to each $\pi \to \pi\pi$ event. The weights were determined using GEANT Monte Carlo simulations, as described in more detail in Ref. [12].

The angular and momentum resolutions were measured during the CHAOS commissioning, which just preceded the present measurement. The angular resolution was $< 0.5°$, and the momentum resolution was 1% (σ) for a 1 T magnetic field and 225 MeV/c pions [8]. However, for the pion-production experiment, the pion mean momentum was ≈ 130 MeV/c, and the CHAOS field was set at 0.5 T. Simulations show that under these conditions, the angular resolution, which was dominated by the multiple scattering, is below $2°$, and the momentum resolution is $\sim 4\%$ (σ) for 130 MeV/c pions.

RESULTS OF THE $\pi \to \pi\pi$ REACTION IN NUCLEI

A general property of the $\pi 2\pi$ process on nuclei in the low-energy $M_{\pi\pi}$ regime was outlined by previous experimental works: it is a quasi-free process both when it occurs on deuterium [13] and on complex nuclei [14]. Furthermore, a common reaction mechanism underlies the process whether it occurs on a nucleon or a nucleus [15]. Thus the study of the $\pi^+\ {}^2H \to \pi^+\pi^\pm NN$ reaction is dynamically equivalent to studying the elementary $\pi^+ n \to \pi^+\pi^- p$ and $\pi^+ p \to \pi^+\pi^+ n$ reactions separately.

In the present measurement, the $\pi^+ A \to \pi^+\pi^\pm A'$ reactions were studied under the same experimental conditions. Thus for a given observable the distributions are directly comparable. The error bars explicitly shown on the spectra represent the statistical uncertainties, which must be added in quadrature to the systematic one $\sim 11\%$ (σ) to obtain the overall uncertainty associated with the data points. The only exception is for the $\cos\Theta_{\pi\pi}$ distributions in Fig. 5, where the overall uncertainties are shown. The distributions for the ^{45}Sc are given in arbitrary units. The error bars represent the statistical uncertainties.

THE $\pi\pi$ INVARIANT MASS

Figures 2, 3, and 4 show the single differential cross sections (diamonds) as a function of the $\pi\pi$ invariant mass ($M_{\pi\pi}$, MeV) for the two reaction channels $\pi^+ A \to \pi^+\pi^- A'$ and $\pi^+ A \to \pi^+\pi^+ A'$. In particular, Fig. 2 refers to the reaction on ^{2}H, ^{12}C, ^{40}Ca, and ^{208}Pb at $T_\pi = 280$ MeV. Figure 3 refers to the reaction on ^{45}Sc at $T_\pi = 280$ MeV, for which the distribution is arbitrarly normalized to the distribution on ^{40}Ca. Figure 4 refers to the reaction on ^{45}Sc at the other two incident kinetic energies, $T_\pi = 240$ and 280 MeV. Here the distributions are given in number of events per bin. In Figures 2 and 4 the $\pi A \to \pi\pi N[A-1]$ phase space simulations are also shown, and are normalized to the same area as the experimental distributions.

Regardless of the nuclear mass number and the incident kinetic energy, the invariant mass distributions for the $\pi^+ A \to \pi^+\pi^+ A'$ channel closely follow phase space, and the

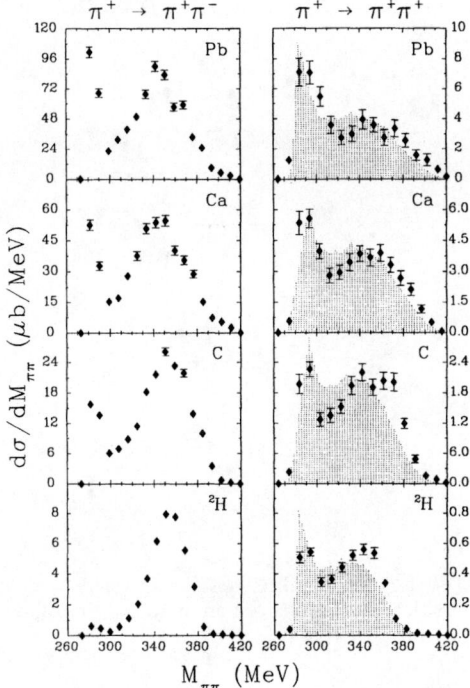

FIGURE 2. Invariant mass distributions (diamonds) for the $\pi^+A \to \pi^+\pi^- A'$ and $\pi^+A \to \pi^+\pi^+ A'$ reactions on ^{2}H, ^{12}C, ^{40}Ca, and ^{208}Pb. The shaded regions represent the results of phase-space simulations for the pion-production reaction $\pi A \to \pi\pi N[A-1]$.

energy maximum increases with both increasing A and T, that is, with increasing nuclear Fermi momentum and available energy. The $\pi^+A \to \pi^+\pi^- A'$ channel displays a different behaviour. Compared to phase space, the ^{2}H invariant mass displays little strength from $2m_\pi$ to 310 MeV, while, in the same energy interval, the ^{12}C, ^{40}Ca, and ^{208}Pb $\pi^+\pi^-$ invariant mass distributions increasingly peak as A increases. A similar behaviour is also seen for ^{45}Sc at all energies.

In order to explain the nature of the reaction mechanism contributing to the peak structure, it is useful to examine the $\cos\Theta^{CM}_{\pi\pi}$ distributions for those events with invariant masses in the region of the peak. $\Theta^{CM}_{\pi\pi}$ is the angle between the direction of a final pion and the direction of the incoming pion beam in the $\pi^+\pi^-$ rest frame. Figure 5 shows the $\cos\Theta^{CM}_{\pi\pi}$ distributions for $2m_\pi \leq M_{\pi^+\pi^-} \leq 310$ MeV and $310 < M_{\pi^+\pi^-} \leq 420$ MeV, the latter being shown for comparison. The vertical error bars represent the overall uncertainties, which are the systematic and statistical uncertainties summed in quadrature. The solid lines in Fig. 5 represent the best-fit to the differential distributions, which was obtained with a partial wave expansion limited to the three lowest waves, i.e. S, P, and D. The best-fit results along with χ^2_ν, which was evaluated using the overall uncertainties, is reported in [16]. For all the nuclei studied $\chi^2_\nu \leq 1$ which indicates that a proper number of waves was used in the expansion. In the case of heavier nuclei,

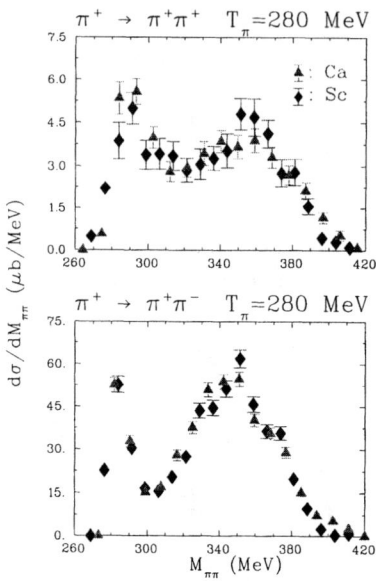

FIGURE 3. Invariant mass distributions (diamonds) for the $\pi^+A \to \pi^+\pi^-A'$ and $\pi^+A \to \pi^+\pi^+A'$ reactions on ^{45}Sc at $T_\pi = 280$ MeV. The distributions are arbitrarly normalized to ^{40}Ca (triangles).

the $\pi^+\pi^-$ system predominantly couples in S−wave $\sim 95\%$ (or $\ell=0$ relative angular momentum) and a remaining 5% is spent in a D−wave ($\ell=2$) state. Furthermore, within the sensitivity of the χ^2_ν−method, any P−wave ($\ell=1$) coupling of the two pions is excluded.

MODELS OF THE $\pi 2\pi$ REACTION AND THE $\pi\pi$ INTERACTION IN THE NUCLEAR MEDIUM

This experiment employed the $\pi^+A \to \pi^+\pi^\pm A'$ reaction to study the near-threshold $\pi\pi$ interaction as a function of the (average) nuclear density. Other reactions can, in principle, be used for similar studies, e.g. the $p \to \pi\pi$ and $\gamma \to \pi\pi$ reactions. For any reaction and for a reliable interpretation of the $\pi\pi$ dynamics in nuclear matter, theoretical models should account for both the reaction mechanism, and nuclear distortions (i.e. pion absorption and others). In the case of $\pi 2\pi$, this approach was followed in the work of Refs. [17, 18], which also included the kinematical limits of the experimental apparatus. For those theoretical works which addressed only the dynamics of a $(\pi\pi)_{I=J=0}$ interacting system in nuclear matter, it was useful to define the observable $C^A_{\pi\pi}$ since it is only weakly dependent on the reaction mechanism and nuclear distortions [15].

Before discussing the data and the comparison with the available theoretical calculations for both the $\pi 2\pi$ reaction and the behaviour of a $\pi\pi$ interacting system in nuclear matter, some details of the models are presented below.

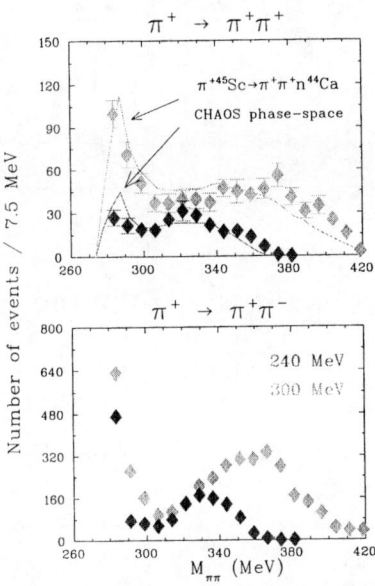

FIGURE 4. Invariant mass distributions for the $\pi^+ A \to \pi^+\pi^- A'$ and $\pi^+ A \to \pi^+\pi^+ A'$ reactions on ^{45}Sc at $T_\pi = 240$ and 300 MeV.

$\mathcal{M}1$: In order to understand the main features of medium modification on the $\pi\pi$ interacting system [18], the pion-production reaction on nuclei is modelled by means of the two leading reactions: 1) the one-pion exchange (OPE) reaction, $\pi^+ N \to \pi^+\pi^\pm N$, which contributes to both the isoscalar and isotensor channels, and 2) the $N^*(1440)$ resonance excitation which is restricted to decay only to isoscalar $\pi\pi$ states, $\pi^+ N \to N^* \to \varepsilon N \to N(\pi\pi)_{I=J=0}$. The in-vacuum $\pi\pi$ interaction is accounted for by the chirally-improved Jülich model [19], which in the GeV region is able to describe the $\pi\pi \to \pi\pi$ scattering and to reproduce the $I = J = 0$ $\pi\pi$ phase shifts. The S-wave $\pi\pi$ (final state) interaction is modified in the nuclear medium through standard renormalization of the pion propagator (see Fig. 6) which determines the following: in the $\pi^+\pi^-$ channel, the isoscalar $\pi\pi$ amplitude is strongly reshaped which finally provides the near-threshold $M_{\pi^+\pi^-}$ enhancement observed in the CHAOS data; in the $\pi^+\pi^+$ channel, the nuclear density has little effect on the pure isotensor $\pi\pi$ amplitude. In order to have a realistic comparison with the CHAOS results, the model deals with common nuclear effects like Fermi motion and pion absorption although using approximate approaches, and calculates the many-fold differential cross sections by accounting for the CHAOS acceptance.

$\mathcal{M}2$: The effort in Ref. [17] is to model the $\pi^+ A \to \pi^+\pi^\pm A'$ reaction through a microscopic description of the elementary $\pi N \to \pi\pi N$ reaction and a detailed study of the production reaction in nuclei. The elementary pion-production reaction relies on five Feynmann diagrams, which have N's, Δ's, and N^*'s as intermediate isobars. The relative scattering amplitudes are derived from chiral lagrangian. The purely mesonic $\pi\pi$ amplitude is calculated by means of the coupled-channel Bethe-

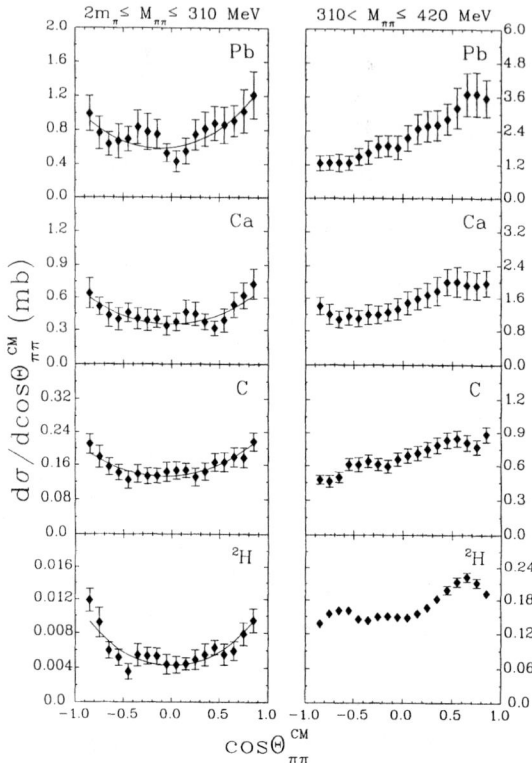

FIGURE 5. Distribution of $\cos\Theta$ in the $\pi^+\pi^-$ centre-of-mass frame for the $\pi^+ A \to \pi^+\pi^- A'$ reaction for $2m_\pi \leq M_{\pi\pi} \leq 310$ MeV (left frame), and for $310 < M_{\pi\pi} \leq 420$ MeV (right frame). The solid lines are best-fits to the data including S, P, and D waves.

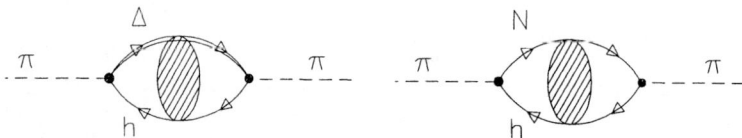

FIGURE 6. Diagrams depicting the P-wave coupling of pions to $p-h$ and $\Delta-h$ correlated states.

Salpeter equation, and the resulting amplitude is capable of predicting the experimental phase shifts of the scalar-isoscalar channel up to 1.2 GeV. With this detailed approach to the $\pi N \to \pi\pi N$ reaction, the model is able to predict the total cross sections for the $\pi^+\pi^-$, $\pi^0\pi^0$ and $\pi^+\pi^+$ channels from threshold up to 300 MeV. In the nuclear medium, the $(\pi\pi)_{I=J=0}$ interaction is strongly modified due to the coupling of pions to $p-h$ and $\Delta-h$ excitations, which are represented by the diagrams in Fig. 6. This results in a displacement of the Im $T_{\pi\pi}$ strength toward the $2m_\pi$ threshold as the nuclear density increases. Conversely, the nuclear medium weakly affects the $(\pi\pi)_{I=2,J=0}$ interaction.

The model includes several nuclear effects: Fermi motion, Pauli blocking, pion absorption and quasi-elastic scattering. The last two interactions are found to largely remove $\pi 2\pi$ pions from the incident flux, which ultimately confines the pion-production process on the nuclear skin [20].

$\mathcal{M}3$: The work of Ref. [21] aims at studing the possibility of the σ-meson identification in nuclear matter, where the elusive particle might be detected as a consequence of the partial restoration of chiral symmetry. Due to the strong interaction of σ with the nuclear medium, the description of the σ properties in terms of parameters like m_σ and Γ_σ appears inadequate, and a proper observable becomes the σ spectral function ρ_σ. By using a simple but general approach, [21] proves that in the proximity of the $2m_\pi$ threshold the partial restoration of the chiral symmetry implies $\rho_\sigma(\omega \sim 2m_\pi) = -\pi^{-1}/\mathrm{Im}\Sigma_\sigma(\omega, \rho)$, where Σ_σ is the σ self-energy, ω is the total energy and ρ the nuclear density. Near $2m_\pi$ threshold, the $\mathrm{Im}\Sigma_\sigma$ is proportional to the phase space factor $(1 - \frac{2m_\pi^2}{\omega^2})^{1/2}$, thus $\rho_\sigma(\omega \sim 2m_\pi) \propto \theta(\omega - 2m_\pi)/(1 - \frac{2m_\pi^2}{\omega^2})^{1/2}$. A direct consequence is that the σ spectral function strength enhances as the σ total energy approaches $2m_\pi$. In order to render this general finding more quantitative, the theory uses the SU(2) linear sigma model. The mean field correction in nuclear matter of $\Sigma_\sigma(\rho)$ is accounted for by the leading diagram sketched in Fig. 7. Furthermore, the model does not include collective pionic modes to derive a possible source of near-threshold strength from the pion coupling to $p-h$ and $\Delta-h$ correlated states.

$\mathcal{M}4$: This work [22] studies the $(\pi\pi)_{I=J=0}$ interaction in nuclear matter by constructing a microscopic theory for both the σ-meson propagator (D_σ) and the $\pi\pi$ T-matrices in the framework of the linear sigma model. In this approach the basic chiral symmetry constraints, i.e. the vanishing of the scattering length in the chiral limit and the Ward's identities are satisfied, and in addition, the theory is capable of reproducing the experimental phase shift in the scalar-isoscalar channel. In-medium pions are renormalized through their P-wave coupling to correlated $p-h$ and $\Delta-h$ states, as depicted in Fig. 6. Bare sigmas, of the linear sigma model, are coupled to the nuclear medium via the tad-pole diagram, which is sketched in Fig. 7. The P-wave renormalization of sigmas leads to a downward shifts of the σ mass, which ultimately produces a strong enhancement of the σ-meson mass distribution (i.e. $\mathrm{Im}D_\sigma$) around the $2m_\pi$ threshold. This enhancement is further increased by a factor of 2 to 3 by the S-wave renormalization of the bare σ-meson mass (i.e. by the partial restoration of chiral symmetry), which presents a similar behavior near the $2m_\pi$ threshold.

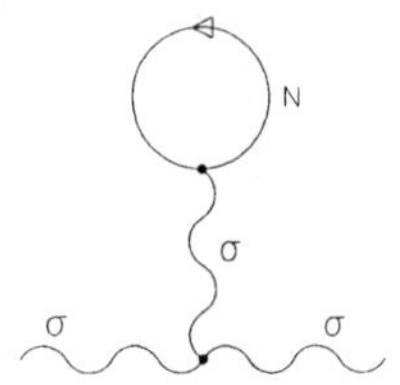

FIGURE 7. Diagram (tad-pole) describing the S-wave coupling of σ to the nuclear medium.

THE $C^A_{\pi\pi}$ AND $C_{\pi\pi}(T)$ RATIOS

$C^A_{\pi\pi}$ is defined as the composite ratio $\frac{M^A_{\pi\pi}}{\sigma^A_T}/\frac{M^N_{\pi\pi}}{\sigma^N_T}$, where σ^A_T (σ^N_T) is the measured total cross section of the $\pi 2\pi$ process in nuclei (nucleon). This observable has the property of yielding the net effect of nuclear matter on the $(\pi\pi)_{I=J=0}$ interacting system regardless of the $\pi 2\pi$ reaction mechanism used to produce the pion pair [15]. Therefore, $C^A_{\pi\pi}$ can be compared not only with the $\mathcal{M}1$ and $\mathcal{M}2$ predictions which explicitly calculate both $M^{Ca}_{\pi\pi}$ and $M^{2H}_{\pi\pi}$, but also with the theories described in $\mathcal{M}3$ and $\mathcal{M}4$ because they calculate the mass distribution of an interacting $(\pi\pi)_{I=J=0}$ system (i.e. ImD_σ) both in vacuum and in nuclear matter. Since the calculations are reported either in arbitrary units [17, 18] or in units which are complex to scale [21, 22], theoretical predictions are normalized to the experimental distributions at $M_{\pi\pi}=350\pm10$ MeV, where the $C^A_{\pi\pi}$ distribution is flat.

For both reaction channels, the full [17] and dotted [18] curves in Fig. 8 are obtained by simply dividing $M^{Ca}_{\pi\pi}/M^{2H}_{\pi\pi}$. It is worthwhile recalling that for [18] the option $\rho=0.5\rho_n$ is used while for [17] the mean density is $\rho = 0.24\rho_n$. Furthermore, for both approaches the underlying medium effect is the P-wave coupling of π's to $p-h$ and $\Delta-h$ configurations, which accounts for the near-threshold enhancement. When applied to the $C^{Ca}_{\pi\pi}$, both $\mathcal{M}1$ and $\mathcal{M}2$ predict the same result, which describes the behaviour of C^{Ca}_{++} fairly well throughout most of the $M_{\pi\pi}$ energy range, but only reproduces a small part of the near-threshold strength for C^{Ca}_{+-}.

$\mathcal{M}3$ [21] and $\mathcal{M}4$ [22] examine the medium modifications on the scalar-isoscalar meson, the σ-meson. Nuclear matter is assumed to partially restore chiral symmetry, and consequently m_σ is assumed to vary with ρ. The variation is parametrized as $1-p\frac{\rho}{\rho_n}$, where p is $0.1\leq p \leq 0.3$ for $\mathcal{M}3$, and $0.2\leq p \leq 0.3$ for $\mathcal{M}4$, and for both is $\rho=\rho_n$. Both models are capable of yielding large strength near the $2m_\pi$ threshold, therefore the results for C^A_{+-} are compared for a common minimum value of the parameter $p=0.2$. In addition, the choice $\rho=\rho_n$ may result appropriate for ^{208}Pb but it is not for medium (i.e. ^{40}Ca) and light (i.e. ^{12}C) nuclei. In Figure 8 the predictions of $\mathcal{M}3$ and $\mathcal{M}4$ are shown as the dash-dotted curves and dashed curves, respectively. $\mathcal{M}4$ predicts a larger near-threshold strength, which is due to the combined contributions of the in-medium P-wave coupling of pions to $p-h$ and $\Delta-h$ configurations, and to the partial restoration of chiral symmetry in nuclear matter. These two models, however, are still too schematic for a conclusive comparison to the present data, and full theoretical calculations are called for.

$C_{\pi\pi}(T)$ is the $C^A_{\pi\pi}$ ratio for the ^{45}Sc at the three incident kinetic energies $T_\pi = 240$, 280, and 300 MeV. The values in the denominators are taken from [23]. The open diamonds in Fig. 9 refer to the ratio to ^{2}H as in Fig. 8 and are given for comparison. The flat region at higher invariant masses is taken for normalizing the distributions. Following this criterium, the distributions are normalized to the one at $T_\pi = 280$ MeV, which, in turn, is normalized to the ^{40}Ca at $T_\pi = 280$ MeV. A clear enhancement at the $2m_\pi$ threshold is seen at all three energies in the $\pi^+\pi^-$ channel. In order to study the T-dependence of this enhancement, theoretical calculations are called for.

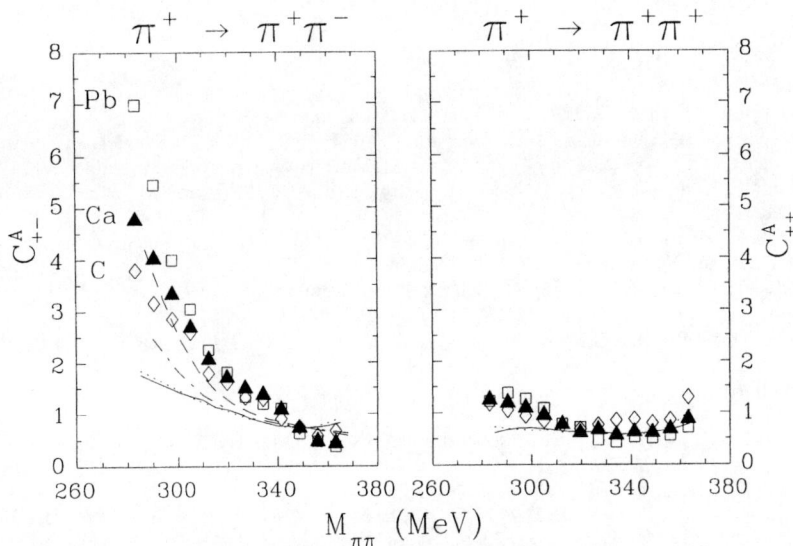

FIGURE 8. The composite ratios $C^A_{\pi\pi}$ for ^{12}C, ^{40}Ca, and ^{208}Pb. The curves are taken from [17] (full) and the model is described in $\mathcal{M}2$, [21] (dash-dotted) $\mathcal{M}3$, [18] (dotted) $\mathcal{M}1$ and [22] (dashed) $\mathcal{M}4$. Further details are reported in the text.

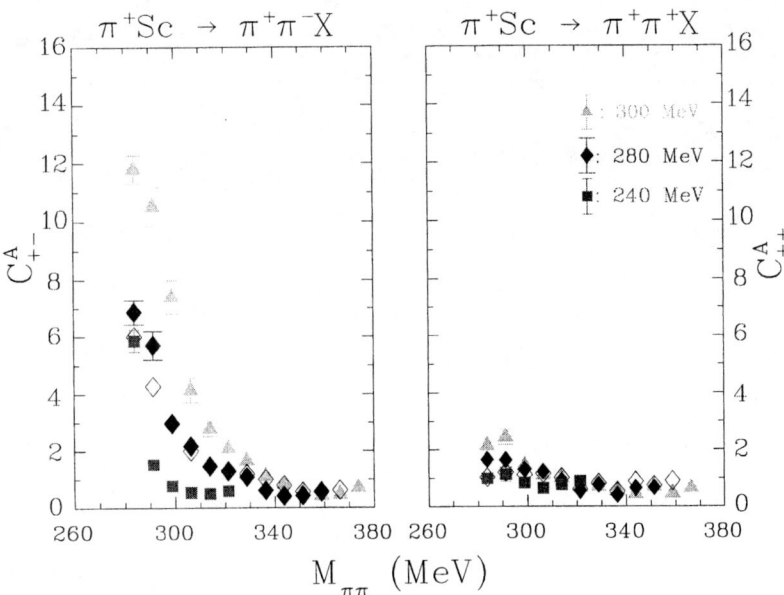

FIGURE 9. The composite ratios $C^A_{\pi\pi}$ for ^{45}Sc at $T_\pi = 240$, 280, and 300 MeV.

REFERENCES

1. Grion, N., et al., *Phys. Rev. Lett.* **59**, 1080 (1987).
2. Grion, N., et al., *Nucl. Phys.* **A492**, 509 (1989).
3. Camerini, P., Grion, N., Rui, R., and Vetterli, D., *Nucl. Phys.* **A552**, 451 (1993).
4. Oset, E., and Vicente-Vacas, M.J., *Nucl. Phys.* **A454**, 637 (1986). The model was improved to calculate total cross-sections for $N \neq Z$ nuclei (M. Vicente-Vacas, private communication, 1990).
5. Bonutti, F., Camerini, P., Grion, N., Rui, R., and Vetterli, D., and Rozon, F.M., *Phys. Rev. C* **47**, 863 (1993).
6. Schuck, P., Nörenberg, W., and Chanfray, G., *Z. Phys. A* **330**, 119 (1988); Chanfray, G., Aouissat, Z., Schuck, P., and Nörenberg, W., *Phys. Lett. B* **256**, 325 (1991).
7. Aouissat, Z., Rapp, R., Chanfray, G., Schuck, P., and Wambach, J., *Nucl. Phys.* **A581**, 471 (1995).
8. Smith, G.R., et al., *Nucl. Instr. and Meth. in Phys. Res. A* **362**, 349 (1995).
9. Hofman, G.J., Brack, J.T., Amaudruz, P.A., and Smith, G.R., *Nucl. Instr. and Meth. in Phys. Res. A* **325**, 384 (1993).
10. Bonutti, F., Buttazzoni, S., Camerini, P., Grion, N., and Rui, R., *Nucl. Instr. and Meth. in Phys. Res. A* **350**, 136 (1994).
11. Raywood, K.J., McFarland, S.J., Amaudruz, P.A., Smith, G.R., and Sevior, M.E., *Nucl. Instr. and Meth. in Phys. Res. A* **357**, 296 (1995).
12. Bonutti, F., et al., *Nucl. Phys.* **A638**, 729 (1998).
13. Rui, R., et al., *Nucl. Phys.* **A517**, 445 (1990), Bjork, C.W., et al., *Phys. Rev. Lett.* **44**, 62 (1980), Lichtenstadt, J., et al., *Phys. Rev. C* **33**, 655 (1986), Sossi, V., et al., *Nucl. Phys.* **A548**, 562 (1992).
14. Bonutti, F., et al., *Phys. Rev. C* **55**, 2998 (1997).
15. Bonutti, F., et al., *Phys. Rev. C* **60**, 018201 (1999).
16. Bonutti, F., et al., *Nucl. Phys.* **A677**, 213 (2000).
17. Vicente-Vacas, M.J., and Oset, E., *Phys. Rev. C* **60**, 064621 (1999).
18. Rapp, R., Durso, J.W., Aouissat, Z., Chanfray, G., Krehl, O., Schuck, P., Speth, J., and Wambach, J., *Phys. Rev. C* **59**, R1237 (1999).
19. Rapp, R., Durso, J.W., and Wambach, J., *Nucl. Phys.* **A596**, 436 (1996).
20. Vicente-Vacas, M.J., private communication, Jan 2000; and Vicente-Vacas, M.J. and Oset, E., nucl-th/0002010 Feb. 2000.
21. Hatsuda, T., and Kunihiro, T., *Phys. Lett. B* **185** 304 (1987); Hatsuda, T., and Kunihiro, T., *Phys. Rep.* **247** 221 (1994); Kunihiro, T., *Prog. of Theor. Phys. Suppl.* **120** 75 (1995); Chiku, S., and Hatsuda, T., *Phys. Rev. D* **57**, 6 (1998); Hatsuda, T., Kunihiro, T., and Shimizu, H., *Phys. Rev. Lett.* **82** 2840 (1999).
22. Aouissat, Z., Chanfray, G., Schuck, P., and Wambach, J., Nucl-th/9908076 v2 31 Aug 1999; Davesne, D., Zhang, Y.J., and Chanfray, G., Nucl-th/9909032 15 Sept 1999.
23. Kermani, M., et al., *Phys. Rev. C* **58**, 3419 (1998).

An introduction to few nucleon systems in effective field theory

Harald W. Grießhammer

Institut für Theoretische Physik (T39), Physik-Department der Technischen Universität München, D-85748 Garching, Germany

Abstract. Progress in the Effective Field Theory of two and three nucleon systems is sketched, concentrating mainly on the low energy version in which pions are integrated out as explicit degrees of freedom. Examples given are: the extraction of nucleon polarisabilities from deuteron Compton scattering at very low energies; the energy dependence of the nucleon polarisabilities; three body forces and the triton; and *nd* partial waves at momenta below the pion cut.

FOUNDATIONS OF EFFECTIVE NUCLEAR THEORY

This presentation is a cartoon of the Effective Field Theory (EFT) of two and three nucleon systems as it emerged in the last three years, using a lot of words and figures, and a few cheats. For details, I refer to the bibliography of a recent review [1], and to papers with Th.R. Hemmert [2], G. Rupak [3], J.-W. Chen, R.P. Springer and M.J. Savage [4], P.F. Bedaque [5, 6], and F. Gabbiani [6]. I will mainly concentrate on the theory in which pions are integrated out as explicit degrees of freedom, but also comment on the extensions to include pions. Of the plenary talks covering related subjects, E. Epelbaum's contribution [7] investigates Weinberg's proposal to include pions in more detail for the few nucleon system, D.R. Phillips [8] concentrates on electro-magnetic reactions on the deuteron, R. Timmermans [9] explains how chiral symmetry is seen in a modern potential and phase shift analysis, and M. Birse finally provides background on the renormalisation group point of view [10].

For want of free neutron targets, one cannot naïvely extract fundamental iso-scalar and iso-vector properties of the nucleon separately from experiment. On the other hand, of all nuclei, the deuteron comes closest to an iso-scalar target and hence appears well suited to extract nucleon and – after removing proton effects – neutron properties. Still, the analysis is not straightforward because, however small the deuteron binding energy seems, binding effects are often not negligible at low energies where properties of the static nucleon are tested. Although the impulse approximation of treating the nucleons inside the deuteron as quasi-free is bound to become the better the higher the typical momentum scale of the process is, the improved resolution also necessitates a more detailed description of both the binding between and structure of the nucleons: Effects from meson exchanges and excited states are resolved at intermediate scales, and at even higher energies, the nucleons and mesons themselves dissolve into quarks and gluons.

Nonetheless, model-independent predictions and extractions at low energies can succeed because nuclear physics provides a separation of scales. This observation is a cornerstone of the Effective Field Theory approach.

Effective Field Theory methods are largely used in many branches of physics where a separation of scales exists. In low energy nuclear systems, the scales are, on one side, the low scales of the typical momentum of the process considered and the pion mass, and on the other side the higher scales associated with chiral symmetry and confinement. This separation of scales produces a low energy expansion, resulting in a description of strongly interacting particles which is systematic and rigorous. It is also model independent (meaning, independent of assumptions about the non-perturbative QCD dynamics): Given that QCD is the theory of strong interactions and that chiral symmetry is broken via the Goldstone mechanism, Wilson's renormalisation group arguments show that there is only one local low energy field theory which originates from it: Chiral Perturbation Theory and its extension to the many-nucleon system discussed here.

The Lagrangean

Three main ingredients enter the construction of an EFT: the Lagrangean, the power counting and a regularisation scheme. First, the relevant degrees of freedom are identified. In his original suggestion how to extend EFT methods to systems containing two or more nucleons, Weinberg [11] noticed that below the Δ production scale, only nucleons and pions need to be retained as the infrared relevant degrees of freedom of low energy QCD. Because at these scales the momenta of the nucleons are small compared to their rest mass, the theory becomes non-relativistic at leading order in the velocity expansion, with relativistic corrections systematically included at higher orders. The most general chirally invariant Lagrangean consists hence of contact interactions between non-relativistic nucleons, and between nucleons and pions, with the first terms reading

$$\begin{aligned}L_{NN} &= N^\dagger(iD_0 + \frac{\vec{D}^2}{2M})N + \frac{f_\pi^2}{8}\,\mathrm{tr}[(D_\mu\Sigma^\dagger)(D^\mu\Sigma)] + g_A N^\dagger \vec{A}\cdot\vec{\sigma}N \qquad (1)\\ &\quad - C_0(N^T P^i N)^\dagger (N^T P^i N) + \frac{C_2}{8}\left[(N^T P^i N)^\dagger (N^T P^i(\vec{D}-\overleftarrow{D})^2 N) + \mathrm{H.c.}\right] + \ldots,\end{aligned}$$

where $N = \binom{p}{n}$ is the nucleon doublet of two-component spinors and P^i is the projector onto the iso-scalar-vector channel, $P^{i,b\beta}_{a\alpha} = \frac{1}{\sqrt{8}}(\sigma_2\sigma^i)^\beta_\alpha(\tau_2)^b_a$. The iso-vector-scalar part of the NN Lagrangean introduces more constants C_i and interactions and has not been displayed for convenience. The field $\xi(x) = \sqrt{\Sigma} = e^{i\Pi/f_\pi}$ describes the pion with a decay constant $f_\pi = 130$ MeV. $D_\mu = \partial_\mu + V_\mu$ is the chirally covariant derivative, and $A_\mu = \frac{1}{2}(\xi\partial_\mu\xi^\dagger - \xi^\dagger\partial_\mu\xi)$ ($V_\mu = \frac{1}{2}(\xi\partial_\mu\xi^\dagger + \xi^\dagger\partial_\mu\xi)$) the axial (vector) pionic current. The interactions involving pions are severely restricted by chiral invariance. As such, the theory is an extension of Chiral Perturbation Theory and Heavy Baryon Chiral Perturbation Theory to the many nucleon system: The terms in the first line couple only pions amongst themselves, and to one nucleon, providing the familiar nuclear long range force. The terms of the second line couple two nucleons to each other and to pions via point-like interactions. Like in its cousins, the coefficients of the low energy Lagrangean encode all short distance physics – branes and strings, quarks and gluons, resonances like the Δ or ρ – as strengths of the point-like interactions between particles. As it is not possible yet to derive these constants by solving QCD e.g. on the lattice, the most practical way to determine them is by fitting to experiment.

The Power Counting

Because the Lagrangean (1) consists of infinitely many terms only restricted by symmetry, an EFT may at first sight suffer from lack of predictive power. Indeed, as the second part of an EFT formulation, predictive power is ensured by establishing a power counting scheme, i.e. a way to determine at which order in a momentum expansion different contributions will appear, and keeping only and all the terms up to a given order. The dimensionless, small parameter on which the expansion is based is the typical momentum Q of the process in units of the scale Λ_{NN} at which the theory is expected to break down. Values for Λ_{NN} and Q have to be determined from comparison to experiments and are a priori unknown. Assuming that all contributions are of natural size, i.e. ordered by powers of Q, the systematic power counting ensures that the sum of all terms left out when calculating to a certain order in Q is smaller than the last order retained, allowing for an error estimate of the final result.

For extremely small momenta $Q \ll m_\pi$, the pion does not enter as explicit degree of freedom describing long range forces. All its effects are absorbed into the coefficients $\pi\!\!\!/ C_i$, while formally $m_\pi \to \infty$ in (1). The only forces between nucleons are thus point-like two and more nucleon interactions with strengths $\pi\!\!\!/ C_i$. This Effective Nuclear Theory with pions integrated out (ENT($\pi\!\!\!/$), [12]) was recently pushed to very high orders in the two-nucleon sector where accuracies of the order of 1% were obtained. It can be viewed as a systematisation of Effective Range Theory with the inclusion of relativistic and short distance effects traditionally left out in that approach. Because of non-analytic contributions from the pion cut, the breakdown scale of this theory must be of the order $\pi\!\!\!/ \Lambda_{NN} \sim m_\pi$. For the ENT with explicit pions, we would suspect the breakdown scale to be of the order of $M_\Delta - M$ or m_ρ, as Δ and ρ are not explicit degrees of freedom in (1).

Even if calculations of nuclear properties were possible starting from the underlying QCD Lagrangean, EFT simplifies the problem considerably by factorising it into a long distance part which contains the infrared-relevant physics and is dealt with by EFT methods and a short distance part, subsumed into the coefficients of the Lagrangean. QCD therefore "only" has to provide these constants, avoiding full-scale calculations of e.g. bound state properties of two nuclear systems using quarks and gluons. EFT provides an answer of finite accuracy because higher order corrections are systematically calculable and suppressed in powers of Q. Hence, the power counting allows for an error estimate of the final result, with the natural size of all neglected terms known to be of higher order. Relativistic effects, chiral dynamics and external currents are included systematically, and extensions to include e.g. parity violating effects are straightforward. Gauged interactions and exchange currents are unambiguous. Results obtained with EFT are easily dissected for the relative importance of the various terms. Because only S-matrix elements between on-shell states are observables, ambiguities nesting in "off-shell effects" are absent. On the other hand, because only symmetry considerations enter

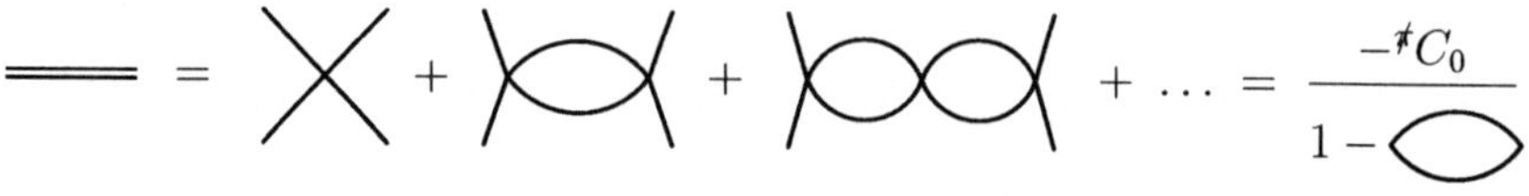

FIGURE 1. Re-summation of the contact interactions into the deuteron propagator.

the construction of the Lagrangean, EFTs are less restrictive as no assumption about the underlying QCD dynamics is incorporated. Hence the proverbial quib that "EFT parameterises our ignorance".

In systems involving two or more nucleons, establishing a power counting is complicated because unnaturally large scales have to be accommodated: Given that the typical low energy scale in the problem should be the mass of the pion as the lightest particle emerging from QCD, fine tuning is required to produce the large scattering lengths in the S wave channels ($1/a^{^1S_0} = -8.3$ MeV, $1/a^{^3S_1} = 36$ MeV). Since there is a bound state in the $^3S_1$ channel with a binding energy $B = 2.225$ MeV and hence a typical binding momentum $\gamma = \sqrt{MB} \simeq 46$ MeV well below the scale Λ_{NN} at which the theory should break down, it is also clear that at least some processes have to be treated non-perturbatively in order to accommodate the deuteron, i.e. an infinite number of diagrams has to be summed or equivalently, a Schrödinger equation needs to be solved.

For simplicity, let us first turn to ENT($\pi\!\!\!/$) in which the pion is integrated out. Here, a way to incorporate this fine tuning into the power counting was suggested by Kaplan, Savage and Wise [13], and by van Kolck [14]. At very low momenta, contact interactions with several derivatives – like $p^2 \, \pi\!\!\!/ C_2$ – should become unimportant, and we are left only with the contact interactions proportional to $\pi\!\!\!/ C_0$. The dominating contribution to nucleons scattering in an S wave comes hence from two nucleon contact interactions and is summed geometrically in Fig. 1 in order to produce the shallow real bound state.

How to justify this? Dimensional analysis allows the size of any diagram to be estimated by scaling momenta by a factor of Q and non-relativistic kinetic energies by a factor of Q^2/M. The remaining integral includes no dimensions and is taken to be of the order Q^0 and of natural size. This scaling implies the rule that nucleon propagators contribute one power of M/Q^2 and each loop a power of Q^5/M. Assuming that

$$\pi\!\!\!/ C_0 \sim \frac{1}{MQ} \quad , \quad \pi\!\!\!/ C_2 \sim \frac{1}{M \, \pi\!\!\!/ \Lambda_{NN} Q^2} \, , \tag{2}$$

the diagrams contributing at leading order to the deuteron propagator are indeed an infinite number as shown in Fig. 1, each one of order $1/(MQ)$. The deuteron propagator

$$\frac{4\pi}{M} \frac{-i}{\frac{4\pi}{M \pi\!\!\!/ C_0} + \mu - \sqrt{\frac{\vec{p}^2}{4} - M p_0 - i\varepsilon}} \tag{3}$$

has the correct pole position and cut structure when one chooses $\pi\!\!\!/ C_0(\mu) = \frac{4\pi}{M} \frac{1}{\gamma - \mu}$.

$\pi\!\!\!/ C_0$ becomes dependent on an arbitrary scale μ because of the regulator dependent, linear UV divergence in each of the bubble diagrams in Fig. 1. Indeed, when choosing

$\mu \sim Q$, the leading order contact interaction scales as in (2). As expected for a physical observable, the NN scattering amplitude becomes independent of μ, the renormalisation scale or cut-off chosen. All other coefficients $^\# C_i$ can be shown to be higher order, so that the scheme is self-consistent. Observables are independent of the cut-off chosen.

The linear divergence of each bubble in Fig. 1 does not show in dimensional regularisation as a pole in 4 dimensions, but it does appear as a pole in 3 dimensions which we subtract following the Power Divergence Subtraction scheme [13]. Dimensional regularisation is chosen to explicitly preserve the systematic power counting as well as all symmetries (esp. chiral invariance) at each order in every step of the calculation. At leading (LO), next-to-leading order (NLO) and often even N^{3}LO in the two nucleon system, it also allows for simple, closed answers whose analytic structure is readily asserted. Power Divergence Subtraction moves hence a somewhat arbitrary amount of the short distance contributions from loops to counterterms and makes precise cancellations manifest which arise from fine tuning [14].

Including Pions

The power counting of the zero and one nucleon sector of the theory are fixed by Chiral Perturbation Theory and its extension to the one baryon sector. Therefore, the question to be posed is: How does the power counting of the contact terms $^\# C_i$ change above the pion cut, i.e. when the pion must be included as explicit degree of freedom?

If pulling out pion effects does not affect the running of C_0 too much, one surprising result arises: Because chiral symmetry implies a derivative coupling of the pion to the nucleon at leading order, the instantaneous one pion exchange scales as Q^0 and is *smaller* than the contact piece $^{KSW}C_0 \sim {}^\# C_0 \sim Q^{-1}$. Pion exchange and higher derivative contact terms appear hence only as perturbations at higher orders. The LO contribution in this scheme is still given by the geometric series in Fig. 1. In contradistinction to iterative potential model approaches, each higher order contribution is inserted only once. In this scheme, the only non-perturbative physics responsible for nuclear binding is extremely simple, and the more complicated pion contributions are at each order given by a finite number of diagrams. For example, the NLO contributions to NN scattering in the $^1S_0$ channel are the one instantaneous pion exchange and the two nucleon interaction with two derivatives, Fig. 2. The constants are determined e.g. by demanding the correct scattering length and effective range. This approach is known as the "KSW" counting of ENT, ENT(KSW) [13].

If on the other hand, the power counting for $^\# C_0$ is dramatically modified when one includes pions, both the C_0 interactions and the one pion exchange might have

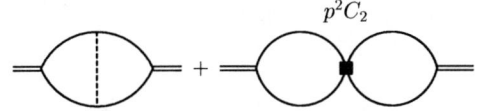

FIGURE 2. The next-to-leading order in ENT(KSW).

FIGURE 3. The leading order in ENT(Weinberg).

to be iterated. In his paper, Weinberg [11] therefore suggested to power count not the amplitude – as is usually done in EFT – but the potential, and then to solve a Schrödinger equation with a chiral potential, as pictorially represented in Fig. 3. E. Epelbaum's talk [7] gives more details on the results obtained so far in this approach.

Both power countings are presently under investigation for consistency and convergence, and each has its advantages and shortcomings [1]. Recently, Beane et al. [15] demonstrated that Weinberg's power counting is not self-consistent in the spin doublet channel of NN scattering and that pions are perturbative there, although convergence is slow. On the other hand, the strongly attractive $1/r^3$ part of the pionic tensor force in the spin triplet channel necessitates a non-perturbative renormalisation of the one pion exchange. In that channel, the KSW counting is hence inconsistent, and Weinberg's proposal is chosen by Nature. It is not yet clear how this observation is to be extended to the power counting of counter terms in which two or more nucleons couple on external currents. After all, Nuclear Physics in the two body system is more than NN scattering.

Although in general process dependent, the expansion parameter is found to be of the order of $\frac{1}{3}$ in both approaches and in most applications, so that NLO calculations can be expected to be accurate to about 10%, and N^2LO calculations to about 4%. In all cases, experimental agreement is within the estimated theoretical uncertainties, and in some cases, previously unknown counterterms could be determined.

APPLICATIONS

Polarisabilities from Low Energy Deuteron Compton Scattering

The first example we turn to demonstrates not only how simple it is to compute processes involving both external gauge and exchange currents in a self-consistent way, but also how "effective" the power counting is to estimate theoretical uncertainties. The following calculation is model-independent and performed in ENT(π) [3].

The iso-scalar, scalar electric and magnetic dipole nucleon polarisabilities parameterise the deformation of the nucleon in an external electro-magnetic dipole field:

$$L_{\text{pol}} = 2\pi \, (\alpha_0 \vec{E}^2 + \beta_0 \vec{B}^2) \, N^\dagger N \;, \qquad (4)$$

where $\alpha_0 := (\alpha^{(p)} + \alpha^{(n)})/2$ and $\beta_0 := (\beta^{(p)} + \beta^{(n)})/2$. How accurate does one have to measure elastic Compton scattering of photons on the deuteron to see them, and how severe are the binding effects which distinguish the deuteron from an iso-scalar target? Compare the graphs containing these contact interactions of *a priori* unknown strengths to the Thomson term which constitutes LO in accordance with the low energy theorem.

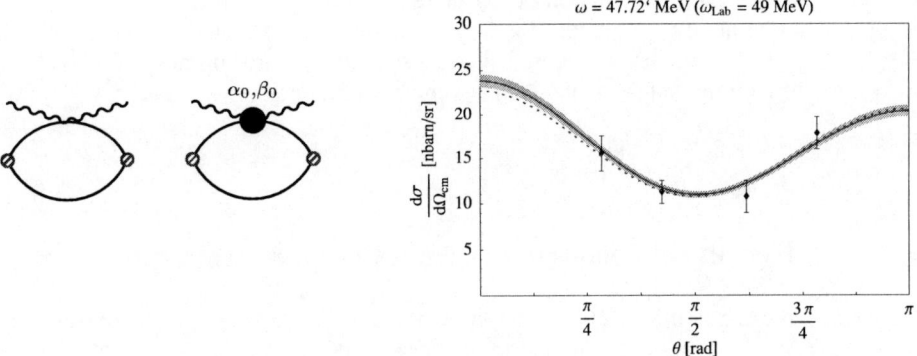

FIGURE 4. Left: Thomson term and polarisabilities contribution in deuteron Compton scattering. Right: ENT($\pi\!\!\!/$) result for the differential cross section, fitted to data. Dashed: α_0 and β_0 fitted independently; solid: α_0 and β_0 constrained by the Baldin sum rule. The estimate of the accuracy of the calculation at N²LO (±3%) is indicated by the shaded area.

A quick look on the left hand side of Fig. 4 reveals that the former is suppressed against the latter by a factor $\omega^2/{\pi\!\!\!/}\Lambda_{NN}^2$, where ω is the photon energy and the breakdown scale of ENT($\pi\!\!\!/$) enters to get a dimensionless ratio. Thus, the lower the photon energy ω in Compton scattering, the less do α and β contribute. On the other hand, at $\omega \sim \gamma$, the polarisabilities contribute about $\frac{1}{3}^2 \sim 10\%$ to the Compton cross section, while higher energies introduce large theoretical errors since corrections from non-analytic pion contributions are not suppressed sufficiently strong any more.

Results of an analytic calculation of the differential cross section to N²LO, i.e. to an accuracy of $\sim 3\%$, at photon energies ω in the window of opportunity at 15 – 50 MeV are presented in Fig. 4. The polarisabilities α_0 and β_0 enter as the only free parameters and carry a theoretical uncertainty of about 20%. As a feasibility study, we used data at $\omega_{Lab} = 49$ MeV to find $\alpha_0 = 8.4 \pm 3.0(\exp) \pm 1.7(\text{theor})$, $\beta_0 = 8.9 \pm 3.9(\exp) \pm 1.8(\text{theor})$, each in units of 10^{-4} fm³. With the experimental constraint for the iso-scalar Baldin sum rule ($\alpha_0 + \beta_0 = 14$), $\alpha_0 = 7.2 \pm 2.1(\exp) \pm 1.6(\text{theor})$, $\beta_0 = 6.9 \mp 2.1(\exp) \mp 1.6(\text{theor})$. These values differ from the proton ones, $\alpha^{(p)} = 12 \approx 10\,\beta^{(p)}$. The error bars from the experimental uncertainty in our extraction are large; furthermore, several conflicting measurements exist for the nucleon (and neutron) polarisabilities, see [3] for details. A more accurate result can be achieved (i) predominantly by better data, and (ii) by a higher order theoretical calculation including contributions from so far undetermined two-nucleon-two-photon operators. Although the scarcity of data at low energies is a big hindrance, the comparison of the calculation to experiment shows that it is quite appropriate to determine scalar polarisabilities from very low energy Compton scattering. The energy régime proposed is hence an interesting window of opportunity to determine nucleon polarisabilities in a model-independent way without having to deal with pions as explicit degrees of freedom.

Compton scattering has also been investigated in ENT(KSW) [4] and in ENT(Weinberg), see also [8]. Surprisingly, the results all agree within the theoretical uncertainty of the calculations even at relatively high momenta. This seems to suggest that non-analytic pion contributions from meson exchange diagrams are small. However, the higher the photon energy, the less seem the polarisabilities extracted to be in agreement with the numbers obtained at very low photon energy, see e.g. [8].

Dynamical Polarisabilities and Compton Scattering

One has to keep in mind [2] the well known fact that polarisabilities themselves are energy dependent, probing the temporal response of the system (so-called "dynamical polarisabilities"). One expects that the polarisabilities are enhanced around the pion production threshold $\omega \sim m_\pi$ since on-shell pions can then be produced out of the virtual pion cloud around the nucleon. In addition, a resonance is expected at $\omega \sim$ 300 MeV because of contributions from the $\Delta(1232)$ as resonance state of the nucleon. At very low energies, these effects contribute corrections of order $\omega^2/(m_\pi, M_\Delta - M_N)^2$, but as $\omega \to m_\pi$, they are expected to be large. For the proton and nucleon magnetic dipole polarisability, they will be especially pronounced because it is anomalously small compared to $\alpha^{(p)}$, and because the cancellation of dia-magnetic contributions from the pion tail with para-magnetic contributions from the Δ and pion core which is observed at zero energy for the proton does not have to hold for the neutron, nor at finite energies. As a first estimate, Fig. 5 presents the predictions of LO Heavy Baryon Chiral Perturbation Theory (no dynamical Δs) and also estimates the effect of neglecting quadrupole and octupole polarisabilities in the extraction. The effect on β is sizeable. Thus, a series of Compton scattering experiments at energies up to $\omega \sim m_\pi$ can give valuable information on the competing dia- and para-magnetic effects inside the nucleon.

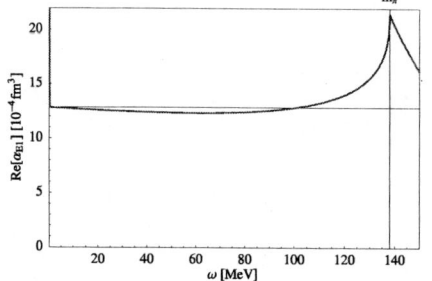
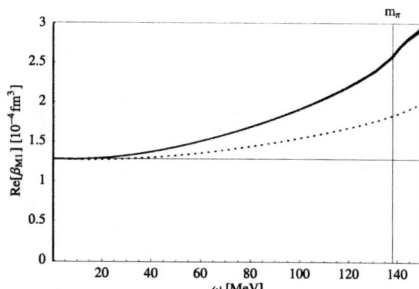

FIGURE 5. Leading order HBχPT predictions of the dependence of the dynamical electric and magnetic dipole polarisabilities on the photon energy. While there is no visible dependence of $\alpha_{E1}(\omega)$ on the number of multipoles included (left figure), the prediction for $\beta_{M1}(\omega)$ changes drastically when the extraction is truncated at the dipole polarisabilities (doted line in right figure), while including octupole polarisabilities makes no effect.

Three Body Forces and the Triton

Since all interactions permitted by the symmetries must be included into the Lagrangean, EFT dictates that in the three body sector, interactions like

$$L_{3body} = H\,(N^\dagger N)^3 \tag{5}$$

with unknown strength H are present. Turning again to ENT($\pi\!\!\!/$), we therefore have to ask at which order in the power counting they will start to contribute. Bedaque, Hammer and van Kolck [16, 17] found the surprising result that an unusual renormalisation makes the three body force of leading order in the triton channel.

In order to understand this finding, let us first consider the LO diagrams which come from two nucleon interactions. The absence of Coulomb interactions in the nd system ensures that only properties of the strong interactions are probed. All graphs involving only $\pi\!\!\!/C_0$ interactions are of the same order and form a double series which cannot be written down in closed form. Summing all "bubble-chain" sub-graphs into the deuteron propagator, one can however obtain the solution numerically from the integral equation pictorially shown in Fig. 6 within seconds on a personal computer.

Because one nucleon is exchanged in the intermediate state, each diagram is of the order of the nucleon propagator, i.e. Q^{-2}, while the first three buddy force (5) seems to be of order Q^0. We are therefore tempted to assume that three body forces are at worst N^2LO, i.e. of the order 10%. In the quartet channel, power counting suggests even N^4LO because the Pauli principle forbids three body forces without derivatives. However, such naïve counting was already fallacious in the two body sector because of the presence of a low lying bound state and of a linear divergence in each of the bubble graphs making up Fig. 1. We therefore investigate more carefully the UV behaviour of the amplitude $A(k,p)$ at half off-shell momenta $p \gg k,\gamma$. The homogeneous part of the integral equation simplifies then to

$$A_{(l,s)}(0,p) = \frac{4\,\lambda(s)}{\sqrt{3}\,\pi}\int_0^\infty \frac{dq}{p}\,A_{(l,s)}(0,q)\,Q_l\!\left[\frac{p}{q}+\frac{q}{p}\right], \tag{6}$$

where $\lambda(s) = -\frac{1}{2}$ in the spin quartet channel, and 1 in the doublet. Q_l is the Legendre Polynomial of the second kind, l the angular momentum of the partial wave investigated.

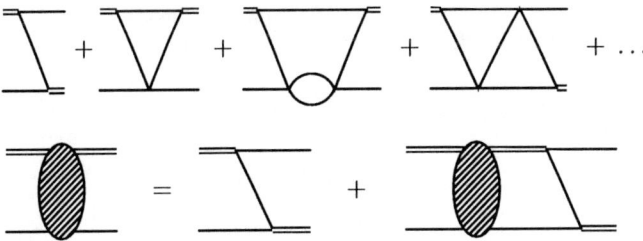

FIGURE 6. The double infinite series of LO "pinball" diagrams, some of which are shown in the first line, is equivalent to the solution of the Faddeev equation of the second line.

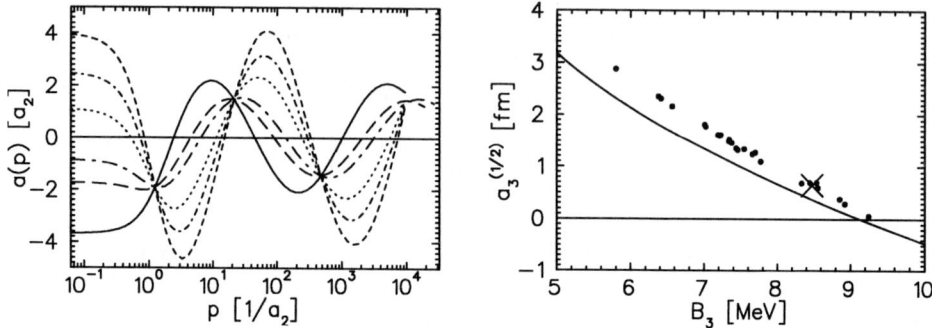

FIGURE 7. Left: Variation of the cut-off by a factor 3 in a numerical study of the naïve off-shell amplitude Fig. 6 in the triton channel changes the scattering length $a(p = 0)$ dramatically. Right: Comparing the ENT prediction for the Phillips line with results from various potential models. (Figures via private communication by the authors of Ref. [17]).

The equation is easily solved by a Mellin transformation, $A_{(l,s)}(0,p) = p^{-1+s_0(l,s)}$, and indeed in most channels s_0 is real so that only one solution exists which vanishes at infinite momentum and hence is cut-off independent. However, the fact that the kernel of (6) is not compact makes one suspicious whether this is always the case. Indeed, there exists one and only one partial wave in which two linearly independent solutions are found: the doublet S wave (triton) channel, where $s_0(l = 0, s = \frac{1}{2}) = \pm 1.0062\ldots i$. Therefore in this channel, any superposition with an arbitrary phase δ is also a solution,

$$A_{l=0,s=\frac{1}{2}}(k \to 0, p) \propto \frac{\cos[1.0062 \ln p + \delta]}{p} . \qquad (7)$$

As each value of the phase provides a different boundary condition for the solution of the full integral equation Fig. 6, the on-shell amplitude $A(k, p = k)$ depends crucially on δ. This means that the on-shell amplitude seems sensitive to off-shell physics, and what is more, to a phase which stems from arbitrarily high momenta. A numerical study of the full half off-shell amplitude confirms these findings, Fig. 7. This cannot be.

Since physics must be independent of the cut-off chosen, this sensitivity of the on-shell amplitude on UV properties of the solution to (6) must be remedied by adding a counter term. And since the power counting in the two nucleon sector is fixed, a necessary and sufficient condition to render cut-off independent results is to promote the three body force (5) to LO, $H(\delta) \sim Q^{-2}$, and absorb all phase dependence into it, see Fig. 8.

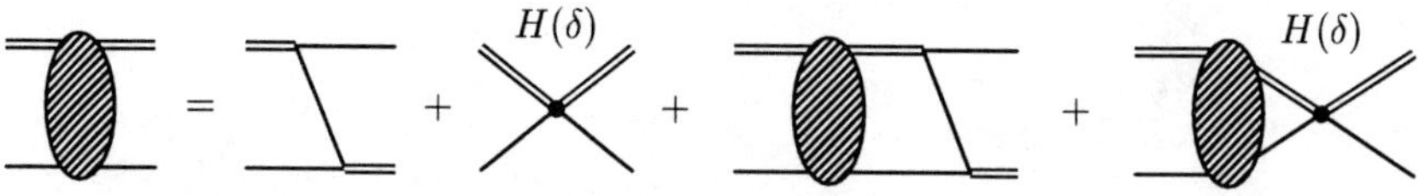

FIGURE 8. The cut-off independent Faddeev equation in the triton channel.

How $H(\delta)$ varies with the cut-off or phase is known analytically from (7), but its initial value is unknown. Therefore, one physical scale $\bar{\delta}$ must be determined experimentally. This one, new free parameter explains why potential models which provide an accurate description of NN scattering can vary significantly in their predictions of the triton binding energy B_3 and three body scattering length $a_3^{(1/2)}$ in the triton channel, although all of them lie on a curve in the $(B_3, a_3^{(1/2)})$ plane, known as the Phillips line, Fig. 7. Determining $H(\bar{\delta})$ by fixing the three body scattering length to its physical value, the triton binding energy is found in ENT($\not{\pi}$) to be 8.0 MeV at LO, 8.8 MeV at NLO, and the phase shift in the triton channel is well in agreement with experiment [18].

It must again be stressed that the three body force of strength $H(\delta)$ was added not out of phenomenological needs. It cures the arbitrariness in the off-shell and UV behaviour of the two body interactions which would otherwise contaminate the on-shell amplitude. Just as the off-shell behaviour of the two body amplitude Fig. 6, the strength $H(\delta)$ is arbitrary, and only the sum of two and three body graphs is physically meaningful.

Summing the deuteron bubbles, each graph in the upper line of Fig. 6 behaves like p^{-2} in the UV. But the solution of the Faddeev equation goes like p^{-1+s_0} with irrational (or even complex) $s_0(l,s)$ [19] and is hence more than the naïve sum of graphs.

The limit cycle thus encountered in the triton channel is a new renormalisation group phenomenon and also explains the Efimov and Thomas effects [1, 16, 17]. In all other partial waves, three body forces enter only at higher orders [19]. In contradistinction to the explanation given above, J. Gegelia's view on the triton is that a unique solution can be constructed using involved numerical methods, and that the three body force is not LO [20]. I thank him for intense and detailed discussions on that point during this conference, even though no complete agreement could be reached.

Partial Waves in Neutron–Deuteron Scattering

In the three body sector, the equations to be solved in ENT($\not{\pi}$) and ENT(KSW) are computationally trivial and can furthermore be improved systematically by higher order correction which involve only (partially analytic, partially numerical) integrations, in contradistinction to many-dimensional integral equations arising in other approaches. A comparative study between the theory with explicit, perturbative pions (ENT(KSW)) and the one with pions integrated out was performed [5] in the spin quartet S wave for momenta of up to 300 MeV in the centre-of-mass frame ($E_{cm} \approx 70$ MeV). As seen above, the two formulations are identical at LO. Because three body forces enter only at high orders, this channel is completely determined by two body properties at the first few orders and no new, free parameters enter.

The calculation with/without explicit pions to NLO/N^{2}LO shows convergence: For example, the scattering length is $a(\text{LO}) = (5.1 \pm 1.5)$ fm, $a(\text{NLO}, \not{\pi}) = (6.7 \pm 0.7)$ fm, and $a(\text{N}^2\text{LO}, \not{\pi}) = (6.33 \pm 0.1)$ fm [16, 17]. The experimental value is $a(^4\text{S}_{\frac{3}{2}}, \exp) = (6.35 \pm 0.02)$ fm. Comparing the correction of the scattering length as each order is added provides one with the familiar error estimate at N^{2}LO: $(\frac{1}{3})^3 \approx 4\%$. The N^{2}LO calculation is inside the error ascertained to the NLO calculation. The calculation of

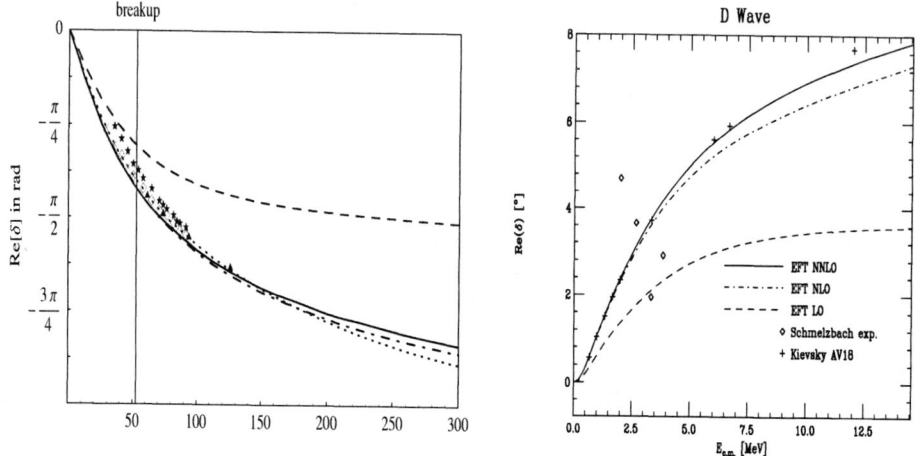

FIGURE 9. Real parts of the quartet S [5] and doublet D [6] wave phase shifts in nd scattering. Legend left: Dashed: LO; solid (dot-dashed) line: NLO with perturbative pions (pions integrated out); dotted: N^2LO without pions. Realistic potential models: squares, crosses, triangles. Stars: pd phase shift analysis.

pionic corrections in ENT(KSW) shows that they are – although formally NLO – indeed much weaker. The difference to ENT($\pi\!\!\!/$) should appear for momenta larger than m_π because of non-analytic contributions of the pion cut, but those seem to be very moderate, see Fig. 9. This and the lack of data makes it difficult to assess whether the KSW power counting scheme to include pions as perturbative increases the range of validity over the pion-less theory.

Finally, the real and imaginary parts of the higher partial waves $l = 1,\ldots,4$ in the spin quartet and doublet channel were found [6] in a parameter-free calculation in ENT($\pi\!\!\!/$), see Fig. 9. Within the range of validity, convergence is good, and the results agree with potential model calculations (as available) within the theoretical uncertainty. That makes one optimistic about carrying out higher order calculations of problematic spin observables like the nucleon-deuteron vector analysing power A_y where EFT will differ from potential model calculations due to the inclusion of three-body forces.

OUTLOOK

Many questions remain open: Which is the the power counting Nature chose in the pion-ful theory for the coupling of two and more nucleons to external currents? Does one there include pions perturbatively or non-perturbatively? At the moment, some people advocate a mixture. To investigate processes with pions in the initial or final state might be helpful [21]. How to extend the analysis of ENT($\pi\!\!\!/$) in the triton channel systematically to higher orders? Technically, how to regularise a Faddeev equation numerically with external currents coupled in a field theory? How do four body forces scale? Extend to nuclear matter!

REFERENCES

1. Beane, S.R., Bedaque, P.F., Haxton, W.C., Phillips, D.R. and Savage, M.J., nucl-th/0008064.
2. Grießhammer, H.W., and Hemmert, Th.R., in preparation.
3. G. Rupak and H.W. Grießhammer, nucl-th/0012096.
4. Chen, J.-W., Grießhammer, H.W., Savage, M.J., and Springer, R.P., *Nucl. Phys.* **A644**, 221 (1998); *Nucl. Phys.* **A644**, 245 (1998).
5. Bedaque, P.F., and Grießhammer, H.W., *Nucl. Phys.* **A671**, 357 (2000).
6. Gabbiani, F., Bedaque, P.F., and Grießhammer, H.W., *Nucl. Phys.* **A675**, 601 (2000).
7. Epelbaum, E., in these proceedings.
8. Phillips, D.R., in these proceedings.
9. Timmermans, R., in these proceedings.
10. Birse, M., in these proceedings.
11. Weinberg, S., *Nucl. Phys.* **B363**, 3 (1991).
12. Chen, J.-W., Rupak, G., and Savage, M.J., *Nucl. Phys.* **A653**, 386 (1999).
13. Kaplan, D.B., Savage, M.J., and Wise, M.B., *Phys. Lett. B* **424**, 390 (1998); *Nucl. Phys.* **B534**, 329 (1998).
14. van Kolck, U., *Nucl. Phys.* **A645**, 273 (1999).
15. Beane, S.R., Bedaque, P.F., Savage, M.J., and van Kolck, U., nucl-th/0104030.
16. Bedaque, P.F., Hammer, H.W., and van Kolck, U., *Phys. Rev. Lett.* **82**, 463 (1999); *Nucl. Phys.* **A646**, 444 (1999).
17. Bedaque, P.F., Hammer, H.W., and van Kolck, U., *Nucl. Phys.* **A676**, 357 (2000).
18. Hammer, H.W., and Mehen, T., nucl-th/0105072.
19. Grießhammer, H.W., in preparation.
20. Gegelia, J., these proceedings.
21. Borasoy, B., and Grießhammer, H.W., nucl-th/0105048.

The deuteron: a mini-review

Franz Gross[*][†] and R. Gilman[**][†]

[*]*Department of Physics, College of William and Mary, Williamsburg, VA 23187, USA*
[†]*Jefferson Laboratory, 12000 Jefferson Avenue, Newport News, VA 23606, USA*
[**]*Rutgers University, 136 Frelinghuysen Rd, Piscataway, NJ 08855, USA*

Abstract. We review[1] some recent results for elastic electron deuteron scattering (deuteron form factors) and photodisintegration of the deuteron, with emphasis on the recent high energy data from Jefferson Laboratory (JLab).

DEUTERON WAVE FUNCTIONS AND FORM FACTORS

Calculations of deuteron form factors and photo and electrodisintegration to the NN final state require a deuteron wave function. The best nonrelativistic wave functions are calculated from the Schrödinger equation using a potential adjusted to fit the NN scattering data for lab energies from 0 to 350 MeV. The quality of realistic potentials have improved steadily, and now the best potentials give fits to the NN database with a χ^2/d.o.f $\simeq 1$. The Paris potential [2] was among the first potentials to be determined from such realistic fits, and it has since been replaced by the Argonne V18 potential (denoted by AV18) [3], the Nijmegen potentials [4], and most recently by the CD Bonn potentials [5, 6]. The momentum space S and D state wave functions determined from three of these models and two relativistic models (Model IIB [7] and Model W16, one of a family of models with varying amounts of off-shell sigma coupling introduced in connection with relativistic calculations of the triton binding energy described in Ref. [8]) are shown in Fig. 1. The figure shows that the S and D-state components of all of these models are almost identical (i.e.variations of less than 10%) for momenta below about 400 MeV, and vary by less than a factor of 2 as the momenta reaches 1 GeV.

Elastic electron–deuteron scattering is described in the one-photon exchange approximation by three deuteron form factors [9, 10, 11]. In its most general form, the relativistic deuteron current can be written [9, 12]

$$-\langle d'|J^\mu|d\rangle = \left\{ G_1(Q^2)\left[\xi'^* \cdot \xi\right] - G_3(Q^2)\frac{(\xi'^* \cdot q)(\xi \cdot q)}{2m_d^2} \right\} (d^\mu + d'^\mu)$$
$$+ G_M(Q^2)\left[\xi^\mu(\xi'^* \cdot q) - \xi'^{*\mu}(\xi \cdot q)\right], \quad (1)$$

where ξ, d (ξ', d') are the polarization and momentum vectors of the incoming (outgoing) deuterons, and the form factors $G_i(Q^2)$, $i = 1 - 3$, are all functions of $Q^2 = -q^2$, the

---

[1] This talk is a shorter version of a review being prepared for *Journal of Physics G* [1].

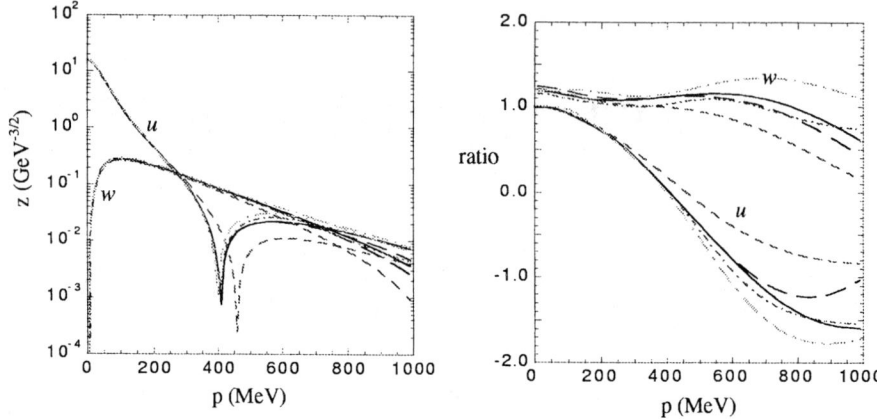

FIGURE 1. Momentum space wave functions for five models mentioned in the text: AV18 (solid), Paris (long dashed), CD Bonn (short dashed), IIB (short dot-dashed), and W16 (long dot-dashed) The wave functions in the right panel have been divided by scaling functions for easy comparison (see Ref. [1] for details).

square of the *four*-momentum transferred by the electron, with $q = d' - d$. In practice, G_1 and G_3 are replaced by a more physical choice of form factors

$$G_C = G_1 + \frac{2}{3}\eta G_Q$$
$$G_Q = G_1 - G_M + (1+\eta)G_3, \qquad (2)$$

with $\eta = Q^2/4m_d^2$. At $Q^2 = 0$, the form factors G_C, G_M, and G_Q give the charge, magnetic and quadrupole moments of the deuteron

$$\begin{aligned} G_C(0) &= 1 & \text{(in units of } e\text{)} \\ G_Q(0) &= Q_d & \text{(in units of } e/m_d^2\text{)} \\ G_M(0) &= \mu_d & \text{(in units of } e/2m_d\text{)}. \end{aligned} \qquad (3)$$

The structure functions A and B, and the polarization transfer coefficient T_{20} depend on the three electromagnetic form factors

$$A(Q^2) = G_C^2(Q^2) + \frac{8}{9}\eta^2 G_Q^2(Q^2) + \frac{2}{3}\eta G_M^2(Q^2)$$
$$B(Q^2) = \frac{4}{3}\eta(1+\eta)G_M^2(Q^2)$$
$$\tilde{T}_{20} = -\sqrt{2}\,\frac{y(2+y)}{1+2y^2}, \qquad (4)$$

where $y = 2\eta G_Q/3G_C$, and $\tilde{T}_{20}$ is T_{20} with the magnetic contributions removed.

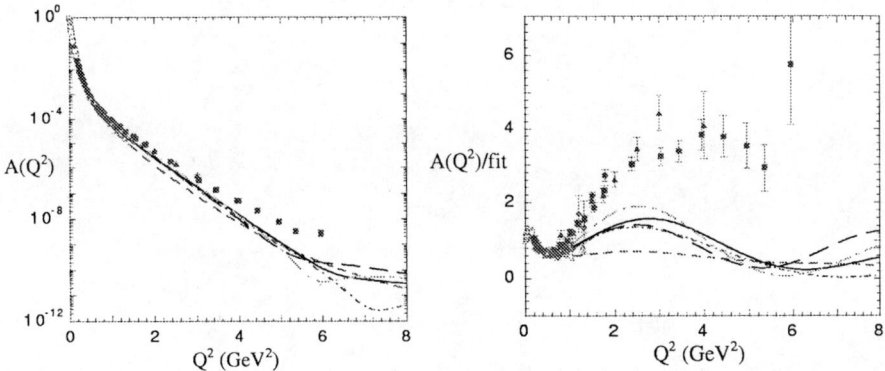

FIGURE 2. The structure function A for five nonrelativistic models using the MMD nucleon form factors. The models are labeled as in Fig. 1. The right panel shows data and models divided by a "fit" described in Ref. [1]. The data are fully referenced in [1].

Comparison of nonrelativistic theory to data

In the nonrelativistic theory, *without exchange currents or $(v/c)^2$ corrections*, the deuteron form factors are

$$G_C = G_E^s D_C$$
$$G_Q = G_E^s D_Q$$
$$G_M = \frac{m_d}{2m_p}\left[G_M^s D_M + G_E^s D_E\right], \tag{5}$$

where G_E^s and G_M^s are the *nucleon isoscalar* form factors, and the Ds are the *body* form factors. All are functions of Q^2. Hence the study of deuteron form factors is complicated by the fact that they are a *product* of the *nucleon isoscalar* form factors and *body* form factors. The dependence of the deuteron form factors on older models of the nucleon form factors is well discussed in Ref. [13]. A year ago the model of Mergell, Meissner and Drechsel [14] (referred to as MMD) gave a good fit, and could have been adopted as a standard (the new G_{Ep}/G_{Mp} data from JLab [15] may change this view). The high Q^2 data for A provide the most stringent test. In Fig. 2 we compare the data for A with nonrelativistic calculations using the five nonrelativistic wave functions shown in Fig. 1. The calculations use Eq. (5) with MMD isoscalar nucleon form factors and nonrelativistic body form factors. In the right panel the data and models have been divided by the "fit" described in Ref. [1].

It is easy to see that the nonrelativistic models *are a factor 4 to 8 smaller than the data for $Q^2 > 2$ GeV^2*. Furthermore, since the difference between different deuteron models is substantially smaller than this discrepancy, it is unlikely that any *realistic* nonrelativistic model can be found that will agree with the data. If the nucleon isoscalar charge form factor were larger than the MMD model by a factor of 2 to 3 it might

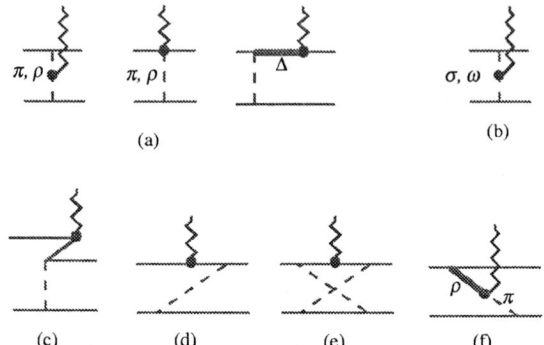

FIGURE 3. Exchange currents that might play a role in meson theories. (a) Large $I = 1$ π,ρ, and Δ currents that do not contribute to the deuteron form factors, and (b) possible $I = 0$ currents that are identically zero. The currents that do contribute to the deuteron form factors are shown in the second row: (c) "pair" currents from nucleon Z-graphs; (d) "recoil" corrections; (e) two pion exchange (TPE) currents; and (f) the famous $\rho\pi\gamma$ exchange current.

explain the data, but this is also unlikely since the variation between nucleon form factor models is substantially smaller than this. We are forced to conclude that these high Q^2 measurements *cannot be explained by nonrelativistic physics and present very strong evidence for the presence of interaction currents, relativistic effects, or possibly new physics.*

Relativistic calculations and new physics

The differences between the data and the nonrelativistic theory can only be explained by a combination of the following effects

- interaction (or meson exchange) currents;
- relativistic effects; or
- new (quark) physics.

The only possibilities excluded from this list are variations in models of the nucleon form factors, or model dependence of the deuteron wave functions. Previously we have argued that *neither* the current uncertainty in our knowledge of the nucleon form factors, *nor* the model dependence of the nonrelativistic deuteron wave functions is sufficient to provide an explanation for the discrepancies.

Possible interaction currents that might account for the discrepancy are shown in Fig. 3. Because the deuteron is an isoscalar system, the familiar large $I = 1$ exchange currents are "filtered" out and only $I = 0$ exchange currents can contribute to the form factors. The $I = 0$ currents tend to be smaller and of a more subtle origin. The nucleon

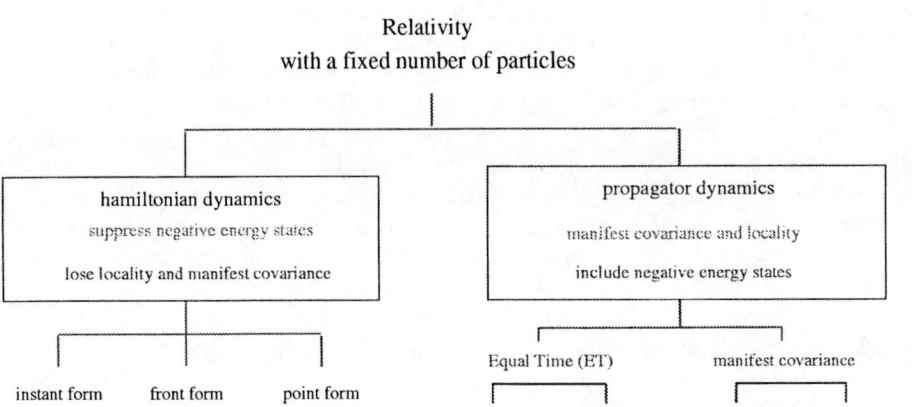

FIGURE 4. The relativistic decision tree discussed in the text.

Z-graphs, Fig. 3c, and the recoil corrections, Fig. 3d, are both of relativistic origin. (The recoil graphs will give a large, incorrect answer unless they are renormalized [16, 17, 18].) The two-meson exchange currents should be omitted unless the force also contains TPE forces. The famous $\rho\pi\gamma$ exchange current is very sensitive to the choice of $\rho\pi\gamma$ form factor, which is hard to estimate and could easily be a placeholder for new physics arising from quark degrees of freedom.

In most calculations based on meson theory, the two pion exchange (TPE) forces and currents are excluded, and, except for the $\rho\pi\gamma$ current (which we will regard as new physics), the exchange currents are of relativistic origin. Additional relativistic effects arise from boosts of the wave functions, the currents, and the potentials, which can be calculated in closed form or expanded in powers of $(v/c)^2$, depended on the method used. At low Q^2 calculations may be done using effective field theories in which a small parameter is identified, and the most general (i.e. exact) theory is expanded in a power series in this small parameter. In these calculations, an expansion of relativistic effects in a power series in $(v/c)^2$ is automatically included. *Hence*, with the exception of effective field theories, *any improvement on nonrelativistic theory using nucleon degrees of freedom leads us to relativistic theory.*

Alternatively, one may seek to explain the discrepancy using quark degrees of freedom (new physics). When two nucleons overlap, their quarks can intermingle, leading to the creation of new *NN* channels with different quantum numbers (states with nucleon isobars, or even, perhaps, so-called "hidden color" states). These models require that assumptions be made about the behavior of QCD in the nonperturbative domain, and are difficult to construct, motivate, and constrain. At very high momentum transfers it may be possible to estimate the interactions using perturbative QCD (pQCD). Very little has been done using other approaches firmly based in QCD, such as lattice gauge theory or Skrymions.

In deciding which relativistic method to use, it is first necessary to decide whether or not to allow *antiparticle, or negative energy* nucleons to propagate as part of the virtual intermediate state. Since nucleons are heavy and composite, so that their antiparticle states are very far from the region of interest, some physicists believe that intermediate states should be built only from positive energy nucleons, and that all negative energy effects (if any) should be included in the interaction. These methods are referred to collectively as *hamiltonian dynamics* and are represented by the left hand branch shown in Fig. 4. Unfortunately, it turns out that this choice precludes the possibility of retaining the properties of locality and manifest covariance enjoyed by field theory. Alternatively, in order to keep the locality and manifest covariance of the original field theory, other physicists are willing to allow negative energy states into the propagators. These methods, represented by the right-hand branch of the figure, are referred to collectively as *propagator dynamics*. However, including negative energy states tends to make calculations technically more difficult and harder to interpret physically, and those who advocate the use of hamiltonian dynamics do not believe the advantages of exact covariance justify the work it requires.

Unfortunately, these two methods are so fundamentally different that many physicists do not realize that the limitations of one may not apply to the other. For example, for some choices of propagator dynamics all 10 of the generators of the Poincaré group will depend only on the kinematics, and the Poincaré transformations of *all amplitudes can be done exactly*. With hamiltonian dynamics this is not the case; some of the 10 generators must depend on the interaction, and transformation of matrix elements under these "dynamical" transformations must be calculated. Comparison of the two methods is therefore very difficult; the language and issues of each are very different and one can be easily misled by the different appearance of the results.

Comparison of relativistic calculations with data

The high Q^2 predictions for 7 relativistic models using NN degrees of freedom and one quark cluster model are shown in Figs. 5–6. They include

- two propagator calculations: VGO [19] (using the Spectator equation), and PWM [20] (using the modified Mandelsweig-Wallace equation);
- two instant-form calculations: FSR [21] (without a v/c expansion) and ARW [22] (using a v/c expansion);
- two front-form calculations: CK [23] (with the light front retained as an unphysical degree of freedom) and LPS [24] (using a specially constructed current operator);
- a point-form calcualtion: AKP [25]; and
- a quark model calculation: DB [26].

The model dependence of the eight calculations is large. Figure 5 shows the predictions for $A(Q^2)$. In these figures we have intentionally left out the model dependent $\rho\pi\gamma$ exchange current from all of the calculations. All of the models except the AKP point-form calculation give a reasonable description of A out to $Q^2 \sim 3$ GeV^2, beyond which they begin to depart strongly from each other and the data. Taking into account that the $\rho\pi\gamma$ exchange current *could be added to any of these models, and that this contribution tends to increase A above $Q^2 \sim 3$ GeV^2*, four models seem to have the right general

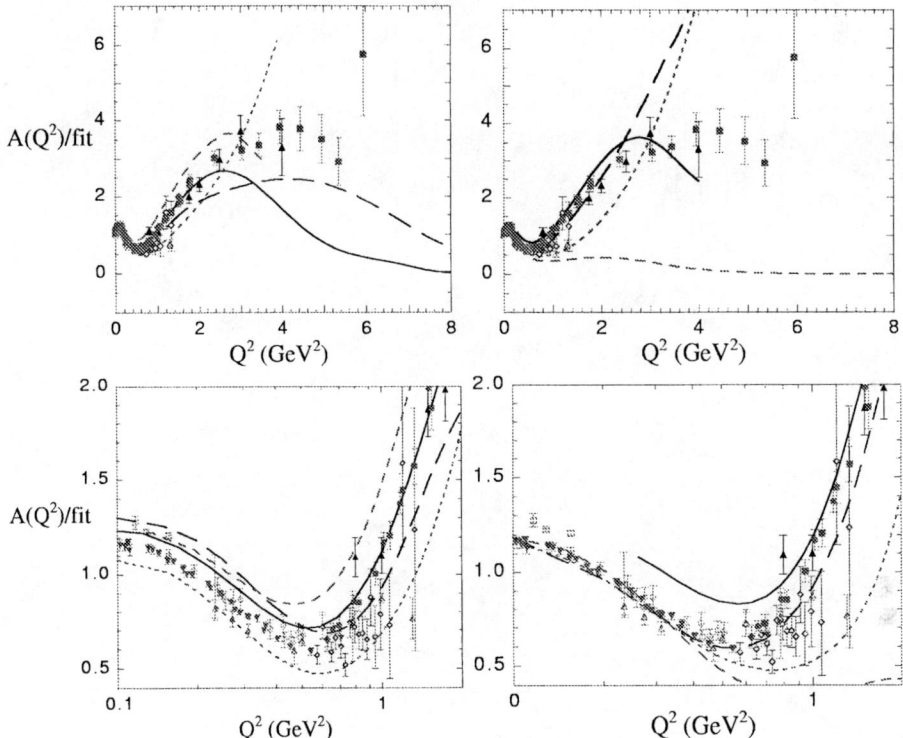

FIGURE 5. The structure function A for the eight models discussed in the text. Left panels show the propagator and instant-form results: FSR (solid line), VOG in RIA approximation (long dashed line), ARW (short dashed line), and PWM (dotted line). Right panels show the front-form CK (long dashed line) and LPS (short dashed line), the point-form AKP (medium dashed line) and the quark model calculation DB (solid line). In every case the calculations have been divided by a scaling function given in Ref. [1]. The data are labeled as in Fig 2.

behavior: the VOG, FSR, ARW and the quark model of DB (but there are no results for this model beyond $Q^2 = 4$ GeV^2). None of these models fit the data without a $\rho\pi\gamma$ exchange current, and all models would be improved by adding such a current contribution, *showing that there is some evidence for new physics at high Q^2*. Ironically, none of the models favored by the high Q^2 data does as well at low Q^2 as the three "unfavored" models shown in the right panels (unless the Platchkov [27] data are systematically too low).

Finally, Fig. 6 shows the predictions for the structure functions A, B, and T_{20} for the eight models. The LPS calculation shows a large discrepancy with the T_{20} data, but the most striking feature of these plots is the *large model dependence* of the predictions for $B(Q^2)$. The magnetic structure function provides the most stringent test of the models,

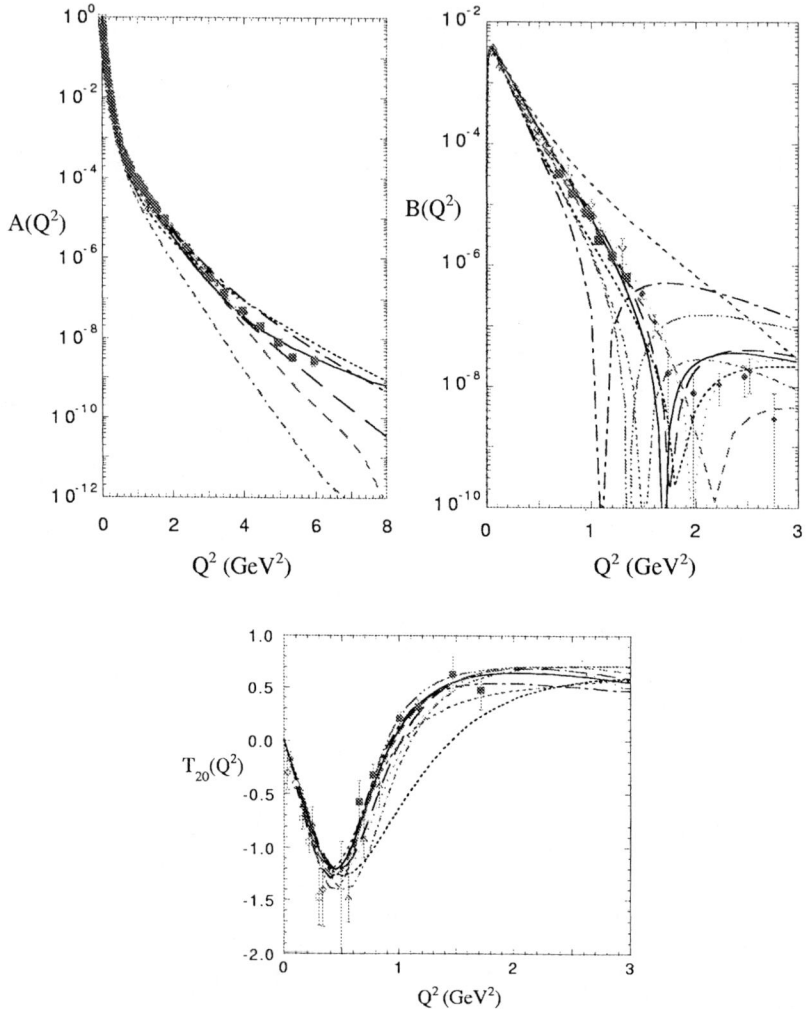

FIGURE 6. The structure functions A, B, and T_{20} for the eight models discussed in the text. VOG full calculation (CIA plus $\rho\pi\gamma$ – solid line); VOG in RIA (long dashed line); FSR (medium dashed line); ARW (short dashed line); DB (widely spaced dotted line); CK (long dot-dashed line); AKP (short dot-dashed line); PWM (dashed double-dotted line), and LPS (dotted line). The data are as labeled in Fig. 2.

and the predictions are comparatively free of the $\rho\pi\gamma$ exchange current (which gives only a small contribution to B). Examination of the figure shows that the B predictions of the PWM, ARW, AKP, CK models fare the worst. In all, taking the predictions for the three structure functions together, the best results are obtained with the FSR, VOG, and DB models.

Conclusions (deuteron form factors)

What have we learned from our measurements of the deuteron form factors? Our comparison of theory and experiment leads to the following conclusions:

- Nonrelativistic quantum mechanics (without exchange currents or relativistic effects) is ruled out by the $A(Q^2)$ data at high Q^2. Reasonable variations in nucleon form factors or uncertainties in the nonrelativistic wave functions cannot remove the discrepancies.
- In approaches using NN degrees of freedom only, relativistic effects (or $I = 0$ meson exchange currents) could be large enough to explain the data.
- Some models that include relativistic effects (or meson exchange currents) and use NN degrees of freedom with realistic forces are close to the data. None are entirely satisfactory.
- The model dependence of relativistic effects (or meson exchange currents) is larger than the errors in the data, even at low Q^2, and is not understood.
- There is evidence that new physics (either in the from of the $\rho\pi\gamma$ exchange current or something else) is beginning to show up in the A structure function above Q^2 of 2 - 3 GeV^2.
- The deuteron form factors provide no evidence for the onset of pQCD, but quark cluster models could explain the data.

Study of the experimental situation leads to the following conclusions:

- The minimum of B is very sensitive to details of the models, and improved measurements of B for Q^2 in the region 1.5 - 4 GeV^2 are particularly compelling. It is important to accurately map out the zero in the B structure function.
- Detailed disagreements between theories and different data sets suggests the need for precision studies at low Q^2.

THRESHOLD ELECTRODISINTEGRATION

Threshold deuteron electrodisintegration measures the $d(e,e')pn$ reaction in kinematics in which the proton and neutron, rather than remaining bound, are unbound with a few MeV of relative kinetic energy in their center of mass system. If the final state energy is low enough, the final state will be dominated by transitions to the $^1S_0$ final state, and will be a pure $\Delta S = 1$, $\Delta I = 1$, $M1$ transition, similar to the $N \to \Delta(1232)$ transition. This transition is a companion to the B structure function; both are magnetic transitions and both are filters for exchange currents with only one isospin ($d \to d$ is $\Delta I = 0$ and $d \to {}^1S_0$ is $\Delta I = 1$). To see the similarity, compare the top right panel of Fig. 6 with the threshold measurements shown in left panel of Fig. 7. Both have a similar shape, and in both cases the uncertainties in the theoretical predictions are large.

The similarity of these two processes (elastic and threshold inelastic) also holds for the theory. These two processes can be used to separately determine the precise details of the $I = 0$ and $I = 1$ exchange currents. Once the exchange currents are fixed, they can

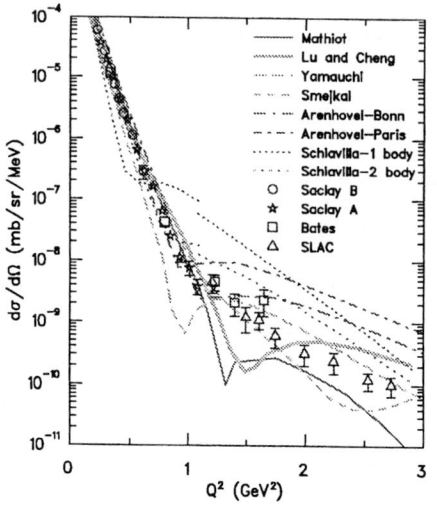 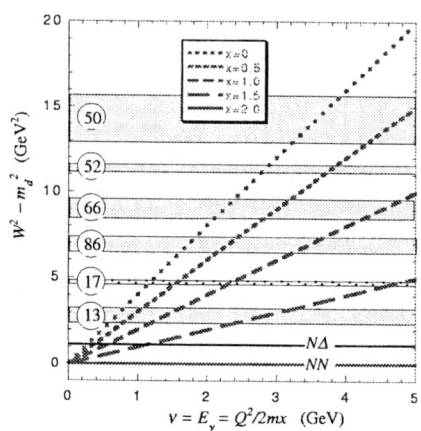

FIGURE 7. Left panel: The cross section for threshold electrodisintegration of the deuteron. (See Ref. [1] for discussion of the data and the theory.) Right panel: The variation of W^2 with the photon energy ν for various values of x. The shaded regions show the approximate thresholds for the production of bands of nucleon resonances. The numbers in the small circles are the number of distinct channels in each band.

be used to predict the results of $d(e,e'p)n$ over a wide kinematic region. Any theoretical approach that works for the form factors should also work equally well for threshold electrodisintegration, yet very few of the groups who have calculated form factors have also calculated the threshold process. These calculations, when completed, will provide a more definitive test of the various relativistic approaches discussed in the previous sections.

DEUTERON PHOTODISINTEGRATION

Since *both* the recent deuteron form factor measurements *and* the recent high energy deuteron photodisintegration measurements have been made with 4 GeV electron beams, it is sometimes assumed that the same theory should work for both. This need not be the case, because the kinematics of elastic electron-deuteron scattering and deuteron photodisintegration are very different, and the physics being explored by these two measurements is also very different. The implications of this remarkable feature of electronuclear physics is often not fully appreciated.

The kinematics of elastic scattering and photodisintegration are compared in the right panel of Fig. 7, which shows $W^2 - m_d^2$ as a function of the photon (real or virtual) energy. Here $x = Q^2/(2m\nu)$ is the familiar Bjorken scaling variable. The mass of the final excited state increases rapidly as x decreases below its maximum allowed value of $x = m_d/m \simeq 2$. For any energy ν or any Q^2, elastic ed scattering leaves the pn

system bound, with no internal excitation energy added to the two nucleons. As $x \to 0$ (the real photon limit) the maximum value of W is reached for any given beam energy. The 24 well established nucleon resonances, and the bands of thresholds at which these resonances are excited, are show in Fig. 7. At $E_\gamma = 4$ GeV, the final state mass is approximately 4.5 GeV, and at least 286 thresholds for the production of pairs of baryon resonances have been crossed (and there are probably more from unseen or weakly established resonances). A photon energy of 4 GeV corresponds to np scattering with an np laboratory kinetic energy of about 8 GeV!

It is clearly very difficult (if not impossible) to construct a theory of high energy photoproduction in which all of these resonances and their corresponding 286 production thresholds are treated microscopically. By contrast, elastic electron deuteron scattering requires a microscopic treatment of only *one channel*. All of the 286 channels also contribute to elastic scattering, of course, but in this case they are *not explicitly excited*, and can probably be well described by slowly varying short-range terms included in a meson exchange (or potential) model. In photodisintegration, *each of these channels is excited explicitly* and an alternate framework that *averages over the effects of many hadronic states* is needed. The alternatives are to use a Glauber-like approach, or to borrow from our knowledge of DIS and build models that rely on the underlying quark degrees of freedom.

High Energy Photodisintegration

Figure 8 shows the published high energy photodisintegration data, from experiments NE8 [28, 29] and NE17 [30] at SLAC, and E89-012 [31] and E96-003 [32] at CEBAF. These experiments determine cross sections for $\theta_{cm} \approx 36°, 52°, 69°,$ and $89°$ at energies from about 0.7 to 5.5 GeV; there are also some backward angle data up to 1.6 GeV from NE8.

The main feature of the data above about 1 GeV is the s^{-11} (s^{-10}) fall off (where $s = (p_1 + p_2)^2$ is the square of the cm. energy) of the cross sections $d\sigma/dt$ ($d\sigma/d\Omega$) at $\theta_{cm} = 90$ and $69°$, in agreement with perturbative QCD expectations. In contrast, the cross sections at the forward angles 36 and $52°$, fall off more slowly, with $\approx s^{-9}$ scaling at lower energies, until the onset of the s^{-11} behavior at about 4 and 3 GeV beam energy, respectively. At each angle the onset of the s^{-11} behavior corresponds to a perpendicular momentum, p_T, of approximately 1 GeV^2.

The highest energy polarization measurements are of p_y (the induced polarization of the proton), and $C_{x'}$ and $C_{z'}$ (the transfer of circular polarization from the photon to the proton) from CEBAF E89-019 [33]. The left panel of Fig. 9 shows the striking feature that the induced polarization p_y is consistent with zero (as predicted by pQCD) at energies above about 1 GeV, the same energy at which the s^{-11} cross section scaling begins. The right panel of the same figure shows that the polarization transfer observables both appear to peak near 1 GeV, and decrease at higher energies.

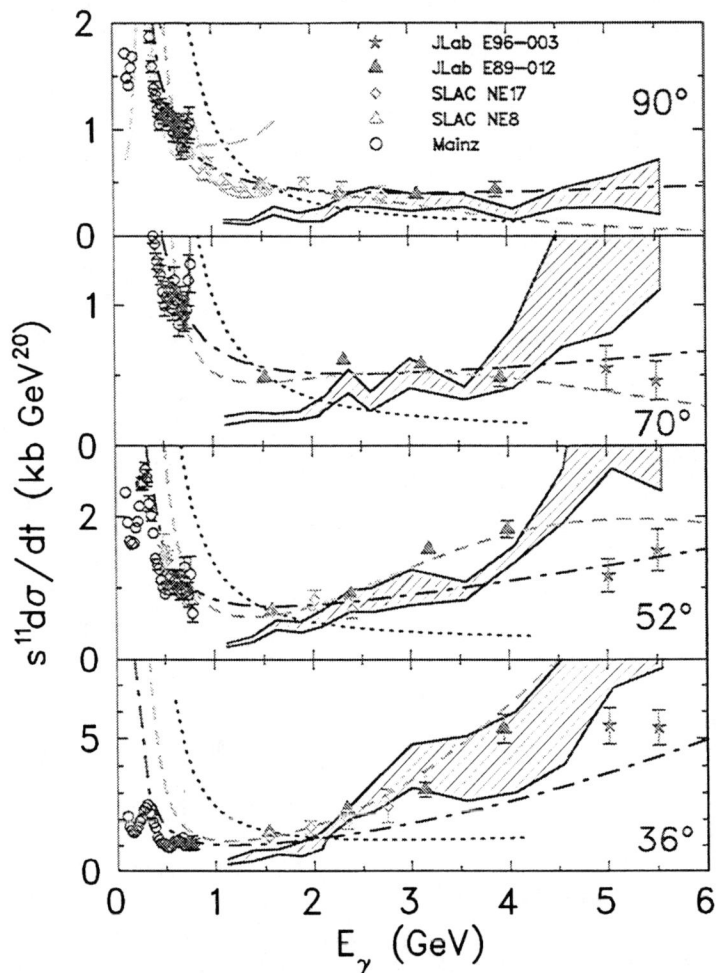

FIGURE 8. Photodisintegration cross section $s^{11}d\sigma/dt$ versus incident lab photon energy. The calculations are from Kang, Erbs, Pfeil and Rollnik (solid line), Lee (dashed line), Raydushkin quark exchange (dot-dashed line), reduced nuclear amplitudes of Brodsky Hiller (dotted line), quark gluon string model (short dashed line), and Frankfurt, Miller, Strikman and Sargsian QCD rescattering (shaded region). (See Ref. [1] for references and discussion of the data and theory.)

Conclusions (deuteron photodisintegration)

Review of deuteron photodisintegration suggests the following:

- A microscopic meson-baryon theory of deuteron photodisintegration must describe the NN interaction at high energies, including pion production and the contributions of hundreds of N^* channels. It is unlikely that such a theory will be constructed in the foreseeable future.

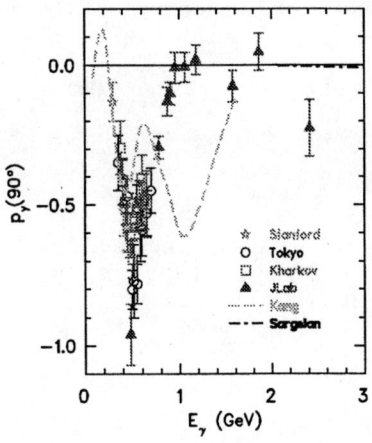
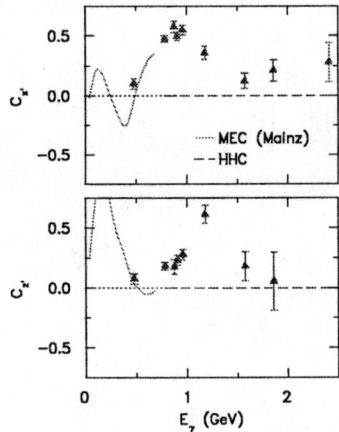

FIGURE 9. Left panel: Induced polarization for deuteron photodisintegration at $\theta_{cm} = 90°$. The calculations are from Kang, Erbs, Pfeil and Rollnik (dashed line) and from Sargsian (dot-dashed line). Right panel: Polarization transfer for deuteron photodisintegration at $\theta_{cm} = 90°$. The calculation is from Schwamb and Arenhövel.

- For $p_T^2 > 1$ GeV^2, cross sections appear to follow the constituent counting rules, but it is expected that an absolute pQCD calculation would greatly underpredict the data. Similar observations may be made for other photoreactions, and it remains to be seen how this behavior arises, and if there is a general explanation for it.
- Some nonperturbative quark models do well describing the data qualitatively. Further theoretical development and experimental tests of these models would be desirable.

OVERALL CONCLUSIONS

Our overall conclusions from the study of form factors and high energy photodisintegration can be briefly summarized as follows:

- Meson theory works well in cases where all of the *active* hadronic channels that can contribute to a process are included. This has been done for the deuteron form factors (where only the NN channel is active), but is impossible for high energy deuteron photodisintegration where 100's of N^*N^* channels are active. At high energy, any successful meson theory must include relativistic effects.
- New approaches, probably using quark degrees of freedom, are needed for high energy deuteron photodisintegration.
- Meson theory behaves as might be expected. It works for the deuteron form factors and does not work for photodisintegration (because of the number of channels).
- Perturbative QCD does not behave as might be expected. Theoretically it should work (or not work) equally well for the deuteron form factors at high Q^2 as it does for photodisintegration at high $p_\perp^2$ (both are exclusive processes). However

the scaling laws seem to work qualitatively for photodisintegration and to fail badly for elastic scattering.

Please refer to Ref. [1] for a more complete discussion of all of the points in this talk.

ACKNOWLEDGMENTS

This work was supported in part by the US Department of Energy. The Southeastern Universities Research Association (SURA) operates the Thomas Jefferson National Accelerator Facility under DOE contract DE-AC05-84ER40150.

REFERENCES

1. Gilman, R., and Gross, F., to be submitted to *Journal of Physics G*.
2. Lancombe, M., Loiseau, B., Richard, J.M., Vinh Mau, R., Côté, J., Pirès, P., and de Tourreil, R., *Phys. Rev.* C **21**, 861 (1980).
3. Wiringa, R.B., Stoks, V.G.J., and Schiavilla, R., *Phys. Rev.* C **51**, 38 (1995).
4. Stoks, V.G.J., Klomp, R.A.M., Terheggen, C.P.F., and de Swart, J.J., *Phys. Rev.* C **49**, 2950 (1994).
5. Machleidt, R., Sammarruca, F., and Song, Y., *Phys. Rev.* C **53**, R1483 (1996).
6. Machleidt, R., *Phys. Rev.* C **63**, 024001 (2001).
7. Gross, F., Van Orden, J.W., and Holinde, K., *Phys. Rev.* C **45**, 2094 (1992).
8. Stadler, A., and Gross, F., *Phys. Rev. Lett.* **78**, 26 (1997).
9. Gross, F., *Phys. Rev.* **136**, B140 (1965).
10. Arnold, R.E., Carlson, C.E., and Gross, F., *Phys. Rev.* C **21**, 1426 (1980).
11. Donnelly, T.W., and Raskin, A.S., *Ann. Phys. (N.Y.)* **169**, 247 (1986).
12. Jones, H., *Nuovo Cimento* **26**, 790 (1962).
13. Garçon, M., and Van Orden, J.W., preprint nucl-th/0102049, *Advances in Nucl. Phys.* **26**, (2001).
14. Mergell, P., Meissner, Ulf-G., and Drechsel, D., *Nucl. Phys.* A **596**, 367 (1996).
15. Jones, M., *et al.*, *Phys. Rev. Lett.* **84**, 1398 (2000).
16. Gross, F., published in *Modern Topics in Electron Scattering* (ed. Frois, B., and Sick, I., 1991), p. 219
17. Jackson, A.J., Lande, A., and Riska, D.O., *Phys. Lett.* B **55**, 23 (1975).
18. Thompson, R.H., and Heller, H., *Phys. Rev.* C **7**, 2355 (1973).
19. Van Orden, J.W., Devine, N., and Gross, F., *Phys. Rev. Lett.* **75**, 4369 (1995).
20. Phillips, D.R., Wallace, S.J., and Devine, N.K., *Phys. Rev.* C **58**, 2261 (1998).
21. Forest, J., and Schiavilla, R., (private communication – to be published)
22. Arenhövel, H., Ritz, F., and Wilbois, T., *Phys. Rev.* C **61**, 034002 (2000).
23. Carbonell, J., and Karmanov, V.A., *Eur. Phys. J* A **6**, 9 (1999).
24. Lev, F.M., Pace, E., and Salmé, G., *Phys. Rev.* C **62**, 064004 (2000).
25. Allen, T.W., Klink, W.H., and Polyzou, W.N., *Phys. Rev.* C **63**, 034002 (2001).
26. Dijk, H., and Bakker, B.L.G., *Nucl. Phys.* A **494**, 438 (1988).
27. Platchkov, S., *et al.*, *Nucl. Phys.* A **510**, 740 (1990).
28. Napolitano, J., *et al.*, *Phys. Rev. Lett.* **61**, 2530 (1988).
29. Freedman, S.J., *et al.*, *Phys. Rev.* C **48**, 1864 (1993).
30. Belz, J.E., *et al.*, *Phys. Rev. Lett.* **74**, 646 (1995).
31. Bochna, C., *et al.*, *Phys. Rev. Lett.* **81**, 4576 (1998).
32. Schulte, E., *et al.*, *Phys. Rev. Lett.* **87**, 102302 (2001).
33. Wijesooriya, K., *et al.*, *Phys. Rev. Lett.* **86**, 2975 (2001).

Recent progress of spectroscopy of light hypernuclei

Ken'ichi Imai

Department of Physics, Kyoto University, Kyoto 606 Japan

Abstract. High resolution spectroscopy of light hypernuclei has been carried out with a newly constructed Ge detector called *Hyperball*. The fine structure of hypernuclei due to the Λ-N spin-spin and spin-orbit interaction were clearly observed for $^{7}_{\Lambda}$Li and $^{9}_{\Lambda}$Be. We will discuss these results and current status of the on-going experiment at BNL. Recent observation of a double hypernucleus, Lambpha ($^{6}_{\Lambda\Lambda}$He), at KEK has a great impact on the study of S=−2 nuclei. The result is presented and further important questions about S=−2 nuclei are discussed. We will also briefly describe the present status of the high intensity proton accelerator (JHF) being constructed by KEK-JAERI in Tokai.

INTRODUCTION

The spectroscopy of hypernuclei has been made at first by emulsion in 1960's and then by using (K^-,π^-) and (π^+,K^+) reactions with magnetic spectrometers. The SKS spectrometer which has a large acceptance and high resolution has been quite successful for the study of the structure and weak decay of hypernuclei. By the (π^+,K^+) reaction from various nuclei, the Λ shell structure was observed, from s to h shell in Pb for example. With the 1.5 MeV resolution, the core excited states were also observed [1]. The structure of hypernuclei provide the unique opportunity to investigate the Λ-N interaction. The effective Λ-N interaction includes spin-spin, spin-orbit and tensor force as well as the central spin-independent force. The strength of the central force is relatively well known from the energy levels of the Λ shell orbits. However, the spin-dependent parts are not known experimentally. It is, therefore, important to study the fine structures due to the weak spin-dependence of the Λ-N interaction. The sensitivity of each spin-dependent force depends on the structure of a state. The energy resolution of MeV is not good enough to observe such fine structures. For the ordinary nuclei, γ spectroscopy with keV resolution has been a central tool for the study of nuclear structure. The γ spectroscopy can also give electromagnetic property of hypernuclei such as charge radius and magnetic moments. The keV resolution γ spectroscopy has, therefore, been long awaited for the study of hypernuclei. The recent success of the Ge detector called Hyperball opened a new window in the hypernuclear physics.

S=−2 nuclei have been studied in relation with the H-dibaryon. In spite of extensive search experiments, no convincing evidence for the H has been reported [2]. The existence of deeply bound H is very unlikely also due to the fact that the weak decay of S=−2 hypernuclei was confirmed [3]. Theoretical studies have been made for Ξ-hypernuclei and double Λ-hypernuclei. From the spectroscopy of Ξ-hypernuclei and double Λ-hypernuclei, one can study Ξ-N and Λ-Λ interactions. However, there is only a few observation of S=−2 hypernuclei until now and we do not have a consistent picture of the ground state of S=−2 hypernuclei or the binding energy of double Λ-hypernuclei [3, 4]. In order to observe 10 times more S=−2 hypernuclei and determine the binding energies at least for several nuclides, a new hybrid-emulsion experiment (E373) was carried out at KEK. A recently discovered event which is assigned as $^{6}_{\Lambda\Lambda}$He is discussed.

The improvement of these experiments and physics are possible with higher intensity kaon beam of better qualities. The high intensity proton accelerator, especially, 50 GeV PS of which construction has just started will open a new era in this field.

γ SPECTROSCOPY OF LIGHT HYPERNUCLEI

We have constructed a high resolution and large acceptance γ detector system called Hyperball, specifically for the γ spectroscopy of hypernuclei with keV resolution. It consists of 14 high purity Ge detectors with BGO Compton suppressors. Each Ge has a relative efficiency of 60% and the photo-peak efficiency of the total detector system is about 3% for 1 MeV γ rays. The energy resolution of a typical Ge was about 3 keV (FWHM) for 1.33 MeV peak of ^{60}Co. The signal is processed with a fast-reset preamplifier and a fast shaping amplifier in order to obtain high efficiency at high rate environment.

Recently, we have carried out two experiments with the Hyperball, one at KEK (E419) and the other at BNL (E930) [5]. We will describe the results and present status.

γ spectroscopy of $^{7}_{\Lambda}$Li

In E419, we have studied $^{7}_{\Lambda}$Li. The Hyperball detects γ rays in coincidence with the ^{7}Li$(\pi^{+}, K^{+})^{7}_{\Lambda}$Li* reaction which is identified by the SKS spectrometer. The observed γ ray spectrum for the bound region of $^{7}_{\Lambda}$Li is shown in Fig. 1 The γ peaks seen only in the spectrum for the bound region are considered γ rays originated from $^{7}_{\Lambda}$Li. The two peaks at 692 keV and 2050 keV are clearly seen and can be assigned as M1($\frac{3}{2}^{+} \to \frac{1}{2}^{+}$) and E2($\frac{5}{2}^{+} \to \frac{1}{2}^{+}$) transitions of $^{7}_{\Lambda}$Li, respectively [6]. We also observed two peaks at 3877 keV and at 3186 keV with 3σ significance which are not shown in the figure. They are assigned to the M1 transitions ($\frac{1}{2}^{+}(T=1) \to \frac{3}{2}^{+}$ and $\frac{1}{2}^{+}$) (see Fig. 2).

The M1 peak is broad due to the Doppler shift, but it becomes sharp peak after the event-by-event correction of the Doppler shift. The peak at 430 keV is interpreted as γ transition of the daughter nuclei ^{7}Be(430) of the $^{7}_{\Lambda}$Li weak π^{-} mesic decay. The well identified γ rays from hypernuclei were, thus, observed for the first time with Ge detector. The level scheme of $^{7}_{\Lambda}$Li as well as ^{6}Li is shown in Fig. 2. The observed peak yields are consistent with the expected yields by the calculation [7]. It confirms the present assignment of the transitions.

The E2 transition of $^{7}_{\Lambda}$Li is basically that of the core nucleus ^{6}Li($3^{+} \to 1^{+}$). The E2 peak in the spectrum has sharp and broad components. The broad part is due to the

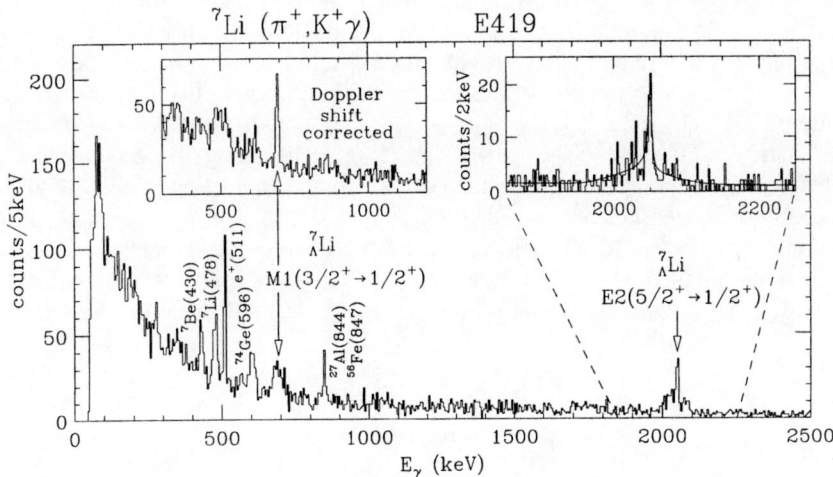

FIGURE 1. γ ray spectrum for the bound region of $^{7}_{\Lambda}$Li. Two peaks due to M1 and E2 transitions are clearly observed (see text).

Doppler shift. From the ratio of two components, we determined the lifetime of the parent state $\frac{5}{2}^+$ (Doppler shift attenuation method). Then the result was converted into the reduced transition probability B(E2) [8]. This is also the first measurement of the γ transition probability of hypernuclei. The value of B(E2) is $3.6 \pm 0.5 \pm 0.5$ e^{2}fm^4, which is much smaller than that of corresponding E2 transition for ^{6}Li nucleus, 8.6 ± 0.7 e^{2}fm^4.

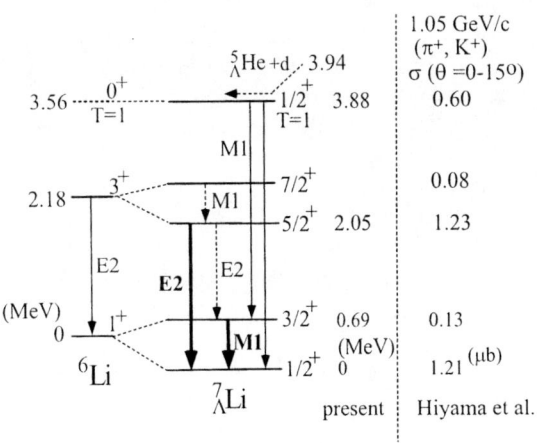

FIGURE 2. Level scheme and transitions of $^{7}_{\Lambda}$Li. The calculated production cross sections are also shown.

The transition probability B(E2) is approximately proportional to fourth power of the nuclear size. This difference can, therefore, be attributed to the shrinkage of the size of the core nucleus in $^{7}_{\Lambda}$Li by 19%, caused by the existence of a Λ particle. The shrinkage of loosely-bound nuclei by Λ was predicted by T. Motoba et al., [9]. We have confirmed this impurity effect for the first time.

The energy of the M1 peak provides the level splitting of the ground state spin doublet which is mainly caused by the Λ-N spin-spin interaction, since the ^{6}Li core has ^{3}S configuration. The energy value is very important to determine the Λ-N spin-spin interaction. The present result ($691.7 \pm 0.6 \pm 1.0$ keV) provides crucial information on the Λ-N spin-spin interaction. The theoretical calculations of this energy were made with the 4-body cluster model by Hiyama et al. [10] and the shell model by Millener et al. [11].

γ spectroscopy of $^{9}_{\Lambda}$Be

At BNL-E930, we have measured γ rays by the Hyperball in coincidence with the ^{9}Be$(K^{-}, \pi^{-})^{9}_{\Lambda}$Be* reaction which was identified with the BNL D-line spectrometer. The core nucleus ^{8}Be is a well known α cluster and unstable. However, due to the glue-like role of Λ, $^{9}_{\Lambda}$Be is stable against strong decay even for the excited state, where the core ^{8}Be is in the first 2^{+} state. In $^{9}_{\Lambda}$Be, this 2^{+} state splits into the doublet states, $\frac{3}{2}^{+}$ and $\frac{5}{2}^{+}$, due to the Λ-N spin-orbit interaction, because the 2^{+} state has almost pure $L = 2$ structure and a Λ is in s-orbit. The other spin dependent forces give negligible contribution to this level splitting.

We have observed the E2 transitions ($\frac{3}{2}^{+}, \frac{5}{2}^{+} \to \frac{1}{2}$) at around 3.05 MeV, when the bound region of $^{9}_{\Lambda}$Be is selected by using the spectrometer data. The energy spectra obtained in this experiment are shown in Fig. 3.

It is clear there are two peaks around 3.05 MeV. The energies of the two peaks were determined by fitting with simulated peak shapes, taking into account the energy resolution of Hyperball and Doppler broadening effect calculated for various lifetimes of the parent states. Their energies were derived as $3029 \pm 2 \pm 1$ keV and $3060 \pm 2 \pm 1$ keV and the energy splitting as 31.4 keV. The lifetime was also obtained as $0.51 + 0.28 - 0.14$ psec [12].

This result gives crucial information on the Λ-N spin-orbit interaction. This energy splitting is very small and suggests very small Λ-N spin-orbit interaction, almost two order of magnitude smaller than the N-N spin-orbit interaction. A cluster-model calculation by Hiyama et al. [13], predicted 80–200 keV energy splitting using the Λ-N interaction of Nijmegen OBE models. On the other hand they predicted 35–40 keV by using that of a quark model [14].

This result of very small spin-orbit interaction is consistent with the recent observation of the energy splitting in $^{13}_{\Lambda}$C made also at BNL [15]. They measured γ transitions from $p_{\Lambda}(1/2)$ and $p_{\Lambda}(3/2)$ states of $^{13}_{\Lambda}$C with NaI arrays. Their observed splitting is $152 \pm 54 \pm 36$ keV which is also smaller than the calculated value using the Nijmegen OBE models (0.36–0.96 MeV).

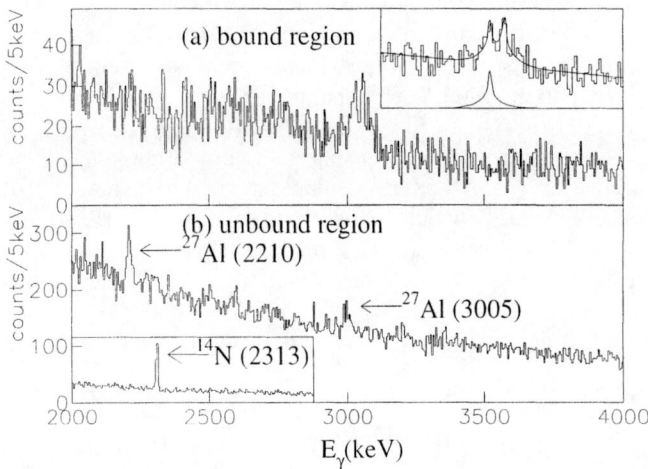

FIGURE 3. Energy spectra of γ rays. (a) is for the bound region of $^9_\Lambda$Be and (b) is for the unbound region. The double peak structure corresponding to the excited states of $^9_\Lambda$Be is observed at around 3.05 MeV only in (a).

Future prospects; γ − γ correlation

The splitting of $^9_\Lambda$Be is sensitive to the Λ-spin-dependent LS force. There is the other term, N-spin-dependent LS force. One can also decompose into symmetric and antisymmetric LS forces. The tensor force is not well determined yet. Moreover, in order to determine all the Λ-N spin dependent interactions from the structure of hypernuclei with confidence, it is very important to study at least several nuclei systematically and confirm the consistency, since the calculation depends on the structure of hypernuclei. Therefore, we plan to do the γ spectroscopy of most of p-shell hypernuclei such as $^7_\Lambda$Li, $^{10}_\Lambda$B, $^{12}_\Lambda$C, $^{15}_\Lambda$N and $^{16}_\Lambda$O with either (π^+, K^+) or (K^-, π^-) reaction [16].

Impurity effect such as the shrinkage of nuclei by Λ is one of the motive of hypernuclear physics. Another one is to study how the hadron property changes in nuclear matter, since Λ hyperon stays stable in hypernuclei. The magnetic moments of hypernuclei are very interesting parameter to be measured. But it is very difficult with present technology. The M1 transition probability B(M1) is related to the magnetic moment of Λ through the M1 operator. It is, therefore, very interesting to measure B(M1). We have to find out appropriate reaction (momentum transfer) and transition (lifetime) where the Doppler attenuation method is applicable. This is another motive of the E930.

By using a coincidence with a magnetic spectrometer, one can do a clean experiment but the species of nuclide is limited and the statistics is limited in the present technology. It is well known that the stopping K^- is the most efficient way to produce hypernuclei, almost 10% per stopping K^-. The γ spectroscopy was tried in early time of hypernuclear physics of 1960's and failed. The problem of this γ spectroscopy is that it is difficult to identify the transition even if one finds a peak, because the nuclide can not be specified without identification of the reaction by a magnetic spectrometer.

Based on the simulation, we have proposed that with a large acceptance Ge detector one can identify a transition of unknown hypernuclei by $\gamma-\gamma$ correlation in the case of stopping K^- method [17]. The $\gamma-\gamma$ correlation with a larger acceptance Ge detector provides better S/N ration for γ peaks. By using $\gamma-\gamma$ correlation to a known γ peak, one can identify the transition one by one. It was demonstrated that some transitions of $^7_\Lambda$Li which were not observed in E419 could be observed in the $\gamma-\gamma$ correlation to the γ from the daughter nucleus ^{7}Be by the stopping K^- method. This method makes it possible to extend the γ spectroscopy of hypernuclei fragmented from the K^- induced reactions. The study of neutron rich hypernuclei such as $^8_\Lambda$Li, $^9_\Lambda$Li, $^{10}_\Lambda$Be, and $^{11}_\Lambda$Be becomes possible. The Hyperball is to be upgraded to have twice efficiency for the future experiments.

DOUBLE HYPERNUCLEUS

It is naive to consider that the ground states of S=−2 hypernuclei are double Λ-hypernuclei. The binding energy of two Λ hyperons, $B_{\Lambda\Lambda}$, provides important information on the Λ-Λ interaction. From the value of $B_{\Lambda\Lambda}$, one can deduce the Λ-Λ interaction energy $\Delta B_{\Lambda\Lambda}$ as follows, $\Delta B_{\Lambda\Lambda}(^A_{\Lambda\Lambda}Z) = B_{\Lambda\Lambda}(^A_{\Lambda\Lambda}Z) - 2B_\Lambda(^{A-1}_\Lambda Z)$, where the B_Λ is the binding energy of a single Λ hypernucleus. There are three reports of the observation of double Λ-hypernuclei all in the emulsion. The latest one (KEK-E176) employed the hybrid emulsion method and found one event which decays mesonically among 80 stopping Ξ^- events. However the interpretation for the species was not unique. The Λ-Λ interaction energy can be positive or negative depending on the interpretation [3]. The determination of $\Delta B_{\Lambda\Lambda}$ has, therefore, been long demanded.

A new hybrid-emulsion experiment was recently carried out at KEK. It employed fiber-bundle detectors to measure the track of the Ξ^- hyperons produced in the (K^-, K^+) reaction [18]. The Ξ^- hyperons are traced to their stopping points in the emulsion and double hypernuclei are searched. To scan many tracks in the emulsion in a reasonable time period, automatic scanning system was developed. These technology developments made a discovery of $^6_{\Lambda\Lambda}$He called Nagara event possible. A photograph and schematic drawing of the event are shown in Fig. 4. This is a cleanest event we have ever observed. The details of the experiment and the analysis of the event are discussed by K. Nakazawa [19, 20].

After examining all possible assignments to the observed particles by using momentum and energy conservation, taking neutron(s) emission into account, we found only one interpretation was possible. That is a following. The Ξ^- hyperon is captured by ^{12}C at point A, and $^6_{\Lambda\Lambda}$He (track 1), ^{4}He (track 3) and ^{3}H (track 4) are emitted. The $^6_{\Lambda\Lambda}$He then decays into $^5_\Lambda$He (track 2), proton (track 5) and π^- (track 6). The $^5_\Lambda$He decays into two hydrogen isotopes and neutron(s) by non-mesic decay. The value of $\Delta B_{\Lambda\Lambda}$ obtained from the decay kinematics was found as 0.69 ± 0.54 MeV. By using both the production and decay kinematics, it was 1.01 ± 0.20 MeV where the Ξ^- is assumed to be absorbed at an atomic 3D state (0.13 MeV) of carbon as the most probable case. This value indicates the Λ-Λ interaction is very weak attractive. This result is not consistent with the

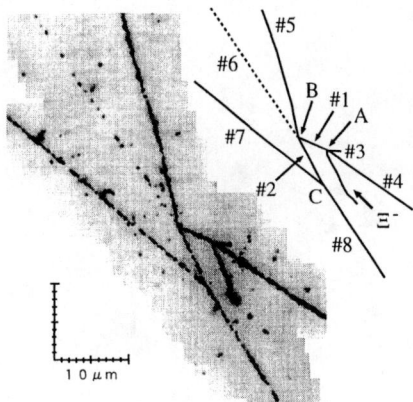

FIGURE 4. Photograph and schematic drawing of Nagara event. The track 1 is $^{6}_{\Lambda\Lambda}$He. See text for the detail.

reported values [3, 4]. However, if single hypernuclei are produced in excited states in both reported events, the present result is not inconsistent with those two events. The observed ratio of the production probability of the twin and double hypernuclei seems to favor the present result [21].

Since only 10% of the emulsion has been analyzed yet, we can expect more discoveries of double hypernuclei near future. The present result set a new lower limit of the H dibaryon mass as 2223.7 MeV/c^2. However, there still is a possibility of the existence of the H near the two Λ mass. There still is a possibility that the H-nuclei are the ground state of S=−2 nuclei rather than double Λ-hypernuclei. It may be a mixed state. To answer the question, we need more uniquely identified events. We have to have at least several nuclei of which binding energies are uniquely determined to study the A-dependence of the binding energy. The weak decay of double hypernuclei is also important, since the H-nuclei and double Λ-hypernuclei are expected to have different decay branching especially in Σ-N and Λ-n decays. Theoretically Ξ-hypernuclei are also expected to be well identified due to their narrow width.

The further experimental progress in S=−2 nuclear physics needs high intensity and high quality kaon beams as well as development of new detector technologies.

FUTURE PROSPECTS OF HYPERNUCLEAR PHYSICS; JHF

The Japan Hadron Facility (JHF) was proposed as an accelerator complex at KEK which consists of a 200 MeV LINAC, 3 GeV synchrotron and 50 GeV synchrotron. Both 3 GeV synchrotron and 50 GeV synchrotron provide high intensity proton beams of 0.6 and 0.5 MW power, respectively, for the production of various secondary particle beams. Although the essential design was not changed, after the JHF became the KEK-JAERI joint project and was decided to be built at JAERI's Tokai laboratory, the machine parameters have been modified due to the site limitation and other considerations. The

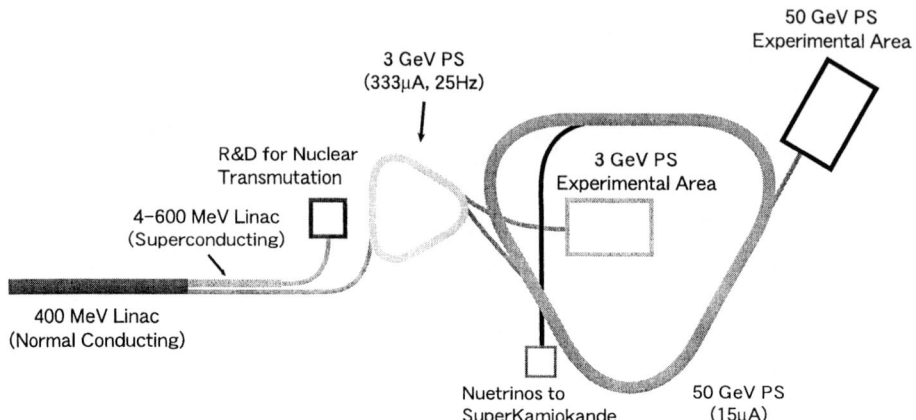

FIGURE 5. A schematic layout of the accelerator complex being constructed at Tokai Japan.

conceptual layout of the accelerator complex is shown in Fig. 5.

The 3 GeV synchrotron provides an average beam current of 333 μA with 25 Hz high repetition rate. The design goal of the total beam power is now upgraded to 1.0 MW from the original JHF design of 0.6 MW, due to the energy upgrade of the LINAC. The 50 GeV proton synchrotron provides an average beam current of 15 μA at the repetition rate of 0.3 Hz with the beam spill length of 0.7 sec. The total beam power is 0.75 MW. The circumference of the ring is about 1.4 km. The beam power of both synchrotrons are the highest among the world in GeV energy region.

Compared with other accelerators in the world, one order of improvement in the beam power above GeV region can be achieved by this project. Since the intensity of the secondary beams are roughly proportional to the beam power, we expect this facility can provide secondary beams of pion, kaon, anti-proton, neutron, muon and neutrino with intensities orders of magnitude higher than the existing accelerators.

High intensity pulsed neutrons and muons are abundantly produced with the 3 GeV synchrotron. The neutrons are used for wide range of studies on the material science such as magnetic materials, high-Tc superconductors, fractals, liquid, polymers etc. Biological science is expected to be another big user of neutrons. High intensity muons are used for μSR studies on material science, muonium physics and muon catalyzed fusion. The high intensity neutrons and muons can be used not only for material sciences but also particle physics, such as the measurements of muon rare-decays and neutron dipole-moment. In the second phase, the accelerator is planned to upgrade to 5 MW by introducing a superconducting linac.

At 50 GeV PS, various high intensity secondary beams, such as kaon, pion, anti-proton, hyperons, neutrino etc., are used for nuclear and particle physics. Major topics to be studied in nuclear and hadronic physics are: 1) Spectroscopy of hypernuclei (both single and double hypernuclei), 2) Hyperon-nucleon scattering, 3) Hadron spectroscopy, 4) Hadrons in nuclear matter. Possible programs in the field of strangeness nuclear physics as an extension of current AGS and KEK-PS programs have been proposed in various workshops [22]. This field will be greatly benefited by the much higher intensity kaon beams.

The gluonic degree of freedom should play important roles in the hadron states. The exotic hadrons such as glueballs (0^{++} and 2^{++}) and hybrids ($s\bar{s}g, c\bar{c}g$ etc.) can be studied as well as charmonium and charmed baryons. For these studies, a separated kaon beam up to 10 GeV/c and a pion beam up to 25 GeV/c are anticipated, where anti-proton beams are also useful. In future, we plan to build an anti-proton storage ring for the precision spectroscopy of charmonium and charmed hadrons, as well as various physics with cooled anti-proton beams. The hypernuclear physics is a major nuclear physics program here. High intensity kaon beams and high resolution (100 keV) pion beam will expand the field of hypernuclear physics. High resolution spectroscopy of not only Λ but Ξ^- and $\Lambda\Lambda$ nuclei becomes possible with high efficiency. We also anticipate the hyperon-nucleon scattering experiments to determine the hyperon-nucleon interaction. The first beamlines to be built are 1 GeV and 2 GeV high resolution kaon and pion beam lines for hypernuclear physics. Partial restoration of chiral symmetry in nuclear matter is a hot topic today. The production of various hadrons in nuclei and detection of their rare decays such as leptonic decays, which are important for this physics, can be made by high intensity hadron beams including protons.

Major topics in particle physics are: 1) Kaon rare decays to study CP violation, 2) Flavor mixing and topics beyond the standard model, 3) Neutrino oscillation experiment using the Super-Kamiokande. The evidence of the oscillation of the atmospheric neutrino observed at Super-Kamiokande motivated the long-baseline neutrino oscillation experiments using accelerators. The K2K experiment is now measuring neutrinos from KEK to Kamioka. It detects ν_μ disappearance around the region of $\Delta m^2 = 10^{-2} \text{eV}^2$. The 50 GeV PS provides much higher intensity neutrino beam for the high precision measurements neutrino oscillation. As possible future programs, muon factory and neutrino factory, by cooling and accelerating high intensity muon beams, are being studied. Primary beams such as polarized proton beams and heavy ion beams are also planned in the future upgrade for nuclear physics programs.

For this project, extensive studies have been made for the development of the high intensity proton accelerator. One of the key technologies of the high intensity synchrotron is the realization of the high field-gradient RF cavity for the quick acceleration. Recently, Y. Mori et al., succeeded to obtain the high field-gradient of 50 kV/m with use of a new magnetic alloy called FINEMET [23]. It provides five times greater field gradient than the currently used ferrite loaded cavities. It was a quite important progress in the accelerator technology to achieve rapid cycling synchrotrons and high intensity synchrotrons.

The proposal to build the JHF including 50 GeV PS was made in 1995 mainly by the nuclear physics community in Japan as a new facility of the Institute for Nuclear Study (INS) of University of Tokyo. After the extensive efforts, INS and KEK were unified and the new KEK were formed in 1996 in order to realize the JHF. From the new KEK, the proposal of JHF was, then, submitted to the government and the research and development for the JHF have started. In 1998 the government suggested to combine the two projects, JHF of KEK and Neutron Science Project of JAERI.

In 2000 the two institutes signed Memorandum of Understanding to make the joint project to build the high intensity proton accelerators. The site of the joint project was decided to be the JAERI's Tokai laboratory, which is located north of KEK and near the seashore of the Pacific Ocean. The government finally approved this project last year. The construction has officially started this April.

The budget of 135 billion Yen (about 120 million dollar) was approved out of the proposed budget of about 189 billion Yen. We planned to complete the whole construction in 6 years. However, because of the budget limitation, the construction of beamlines and detectors may be delayed. We are now asking various laboratories to provide unused magnets for this project. Already many magnets such as AA ring of CERN were moved to KEK. Since this facility is unique and open for the international community of science, we really hope that the new facility is supported and widely used by the international science community.

ACKNOWLEDGMENTS

The author would like to thank the collaborators of KEK-E373,419 and BNL-E930.

REFERENCES

1. Hasegawa, T., et al., *Phys. Rev. Lett.* **74**, 224 (1995); *Phys. Rev. C* **53**, 1210 (1996).
2. Imai, K., Proceedings of Physics and detectors at DAFENE, ed. Bianco et al., 1999, p.711.
3. Aoki, S., et al., *Prog. Theor. Phys.* **85**, 1287 (1991).
4. Dalitz, R.H., et al., *Proc. Roy. Soc. Lond. A* **426**, 1 (1989).
5. Tamura, H., *Nucl. Phys.* **A639**, 83c (1998).
6. Tamura, H., et al., *Phys. Rev. Lett.* **84**, 5963 (2000).
7. Hiyama, E., et al., *Phys. Rev. C* **53**, 2351 (1999).
8. Tanida, K., PhD thesis (Univ. of Tokyo) unpublished, Tanida, K., et al., *Phys. Rev. Lett.* **86**, 1982 (2001).
9. Motoba, T., Bando, H., and Ikeda, K., *Prog. Theor. Phys.* **80**, 189 (1983).
10. Hiyama, E., et al., *Nucl. Phys.* **A639**, 173c (1998).
11. Millener, D.J., et al., *Phys. Rev. C* **31**, 499 (1985).
12. Akikawa, H., et al., to be submitted.
13. Hiyama, E., et al., *Phys. Rev. Lett* **85**, 270 (2000).
14. Fujiwara, Y., Nakamoto, C., and Suzuki, Y., *Phys. Rev. Lett* **76**, 2242 (1996).
15. Ajimura, S., et al., *Phys. Rev. Lett.* **86**, 4255 (2001).
16. Tamura, H., BNL-E930 proposal.
17. Imai, K., and Zhu, L.H., Proceedings of HYP00, *Nucl. Phys.* in press.
18. Ichikawa, A., et al., *Nucl. Instr. Meth. A* **417**, 220 (1998).
19. Nakazawa, K., E373 proposal, in this Proceedings.
20. Takahashi, H., et al., submitted to *Phys. Rev. Lett.*.
21. Ichikawa, A., et al., *Phys. Lett. B* bf 500, 37 (2001).
22. Proceedings of the international workshop on physics at 50 GeV PS, *JHF Suppl.* **18** (1996).
23. Mori, Y., Proc. of Euro. Part. Accel. conference Stockholm, 1998.

Point form analysis of deuteron and nucleon form factors

W.H. Klink

Department of Physics and Astronomy University of Iowa, Iowa City, Iowa, USA

Abstract. The framework of point form relativistic quantum mechanics is used to calculate elastic deuteron and nucleon form factors. Deuteron wave functions from Argonne v_{18} and Reid'93 potentials are used in conjunction with form factors for the proton and neutron in the point form spectator approximation. For the nucleon form factors wave functions from the Goldstone Boson Exchange model with pointlike constituent quarks and no anomalous magnetic moment give very good fits to the data. Point form electrodynamics is used to show how to construct few-body current operators, needed to improve the fit with data for the deuteron form factors.

INTRODUCTION: POINT FORM RELATIVISTIC QUANTUM MECHANICS

Electron scattering experiments provide one of the best tools for investigating the structure of hadrons, both at the nuclear and quark levels. In order to interpret electron scattering data, it is important to have good models for computing form factors, models which incorporate the general principles of relativity and quantum mechanics, as well as incorporating known physical insight into the nature of the constituents making up the objects being probed. This in effect means trying to find models with good mass and current operators, since matrix elements of current operators acting on mass operator eigenstates are the form factors that provide the link with experimental data.

In this paper the framework of point form relativistic quantum mechanics will be used to construct mass and current operators. Since the point form is not as well known as the other forms of relativistic quantum mechanics listed by Dirac[1], this first section deals with some of the main features of the point form, including how to construct Bakamjian-Thomas mass operators[2] and a one-body approximation to the current operator called the point form spectator approximation[3].

Two calculations of form factors using the point form are given in sections 2 and 3, the first elastic deuteron form factors in which the mass operator arises from well-known nonrelativistic potentials, the second elastic nucleon form factors in which the mass operator comes from the Goldstone Boson Exchange model[4]. Both of these calculations use one-body current operators, and the comparison of the calculations with experimental data leads to the issue of constructing few-body current operators in the context of the point form, a subject taken up in section 4.

In the point form all interactions are put into the four-momentum operator, while the Lorentz generators are kinematic (contain no interactions). It is then convenient to write the Poincaré commutation relations not in terms of the ten generators, but rather in terms of the four-momentum operator that contains the interactions, and global Lorentz transformations:

$$[P_\mu, P_\nu] = 0,\qquad(1)$$

$$U_\Lambda P_\mu U_\Lambda^{-1} = (\Lambda^{-1})^\nu_\mu P_\nu,\qquad(2)$$

where U_Λ is a unitary operator representing the Lorentz transformation Λ. These rewritten Poincaré relations will be called the point form equations, and are the fundamental equations that have to be satisfied for the system of interest. The mass operator is given by $M = \sqrt{P \cdot P}$ and must have a spectrum that is bounded from below.

The simplest example of a system satisfying the point form equations is a one-particle system with mass m and spin j. If $|p,\sigma>$ is an eigenstate of four-momentum p (with $p \cdot p = m^2$) and spin projection σ, then

$$P_\mu |p,\sigma> = p_\mu |p,\sigma>,\qquad(3)$$

$$U_\Lambda |p,\sigma> = \sum_{\sigma'} |\Lambda p,\sigma'> D^j_{\sigma'\sigma}(R_W),\qquad(4)$$

with R_W a Wigner rotation defined by $R_W = B^{-1}(\Lambda v)\Lambda B(v)$, and $B(v)$ a canonical spin (rotationless) boost (see reference [5]) with argument $v = p/m$. Many particle operators with the same transformation properties as the single particle ones are conveniently obtained by introducing creation and annihilation operators. Let $a^\dagger(p,\sigma)$ be the operator that creates the state $|p,\sigma>$ from the vacuum. If $a(p,\sigma)$ is its adjoint, these operators must satisfy the following relations:

$$[a(p,\sigma), a^\dagger(p',\sigma')]_\pm = E\delta^3(p-p')\delta_{\sigma,\sigma'},\qquad(5)$$

$$U_a a^\dagger(p,\sigma) U_a^{-1} = e^{ip\cdot a} a^\dagger(p,\sigma),\qquad(6)$$

$$P_\mu(fr) = \sum \int \frac{d^3p}{E} p_\mu a^\dagger(p,\sigma) a(p,\sigma),\qquad(7)$$

$$U_\Lambda a^\dagger(p,\sigma) U_\Lambda^{-1} = \sum a^\dagger(\Lambda p,\sigma') D^j_{\sigma'\sigma}(R_W).\qquad(8)$$

Here $P_\mu(fr)$ is the free four-momentum operator and plays a role analogous to the free Hamiltonian in nonrelativistic quantum mechanics. Again it is straightforward to show that P_μ satisfies the point form equations. U_a in Eq.6 is the unitary operator representing the four-translation a.

To prepare for the construction of interacting four-momentum operators, built from interacting mass operators, it is convenient to introduce velocity states, states with simple Lorentz transformation properties. If a Lorentz transformation is applied to a many-particle state,

$$|p_1,\sigma_1...p_n,\sigma_n> = a^\dagger(p_1,\sigma_1)...a^\dagger(p_n,\sigma_n)|0>,$$

then it is not possible to couple all the momenta and spins together to form spin or orbital angular momentum states, because the Wigner rotations for each momentum state are different. However, velocity states, defined as n-particle states in their overall rest frame boosted to a four-velocity v, will have the desired Lorentz transformation properties:

$$|v,\vec{k}_i,\mu_i> : \ = \ U_{B(v)}|k_1,\mu_1...k_n,\mu_n> \quad (9)$$

$$= \ \sum |p_1,\sigma_1...p_n,\sigma_n> \prod D^{j_i}_{\sigma_i\mu_i}(R_{W_i}) \ . \quad (10)$$

$$U_\Lambda |v,\vec{k}_i,\mu_i> \ = \ U_\Lambda U_{B(v)}|k_1,\mu_1...k_n,\mu_n>$$

$$= \ U_{B(\Lambda v)} U_{R_W}|k_1,\mu_1...k_n,\mu_n>$$

$$= \ \sum |\Lambda v, R_W \vec{k}_i, \mu_i'> \prod D^{j_i}_{\mu_i'\mu_i}(R_W). \quad (11)$$

Unlike the Lorentz transformation of an n-particle state, where all the Wigner rotations of the D functions are different, in Eq.11 it is seen that the Wigner rotations in the D functions are all the same and given by Eq.4. Moreover the same Wigner rotation also multiplies the internal momentum vectors, which means that for velocity states, spin and orbital angular momentum can be coupled together exactly as is done nonrelativistically (for more details see reference [5]). The relationship between single particle and internal momenta is given by $p_i = B(v)k_i$, $\sum \vec{k}_i = 0$ and R_{W_i} in Eq.10 by replacing p by k_i and Λ by $B(v)$ in Eq.4. From the definition of velocity states it then follows that

$$V_\mu|v,\vec{k}_i,\mu_i> \ = \ v_\mu|v,\vec{k}_i,\mu_i> \ , \quad (12)$$

$$M_{fr}|v,\vec{k}_i,\mu_i> \ = \ m_n|v,\vec{k}_i,\mu_i> \ , \quad (13)$$

$$P_\mu(fr)|v,\vec{k}_i,\mu_i> \ = \ m_n v_\mu|v,\vec{k}_i,\mu_i> \ , \quad (14)$$

with $m_n = \sum \sqrt{m_i^2 + \vec{k}_i^2}$ the mass of the n-particle velocity state and $P_\mu(fr) = M_{fr}V_\mu$. On velocity states the free four-momentum operator has been written as the product of the four-velocity operator times the free mass operator, which is the Bakamjian-Thomas construction in the point form.

To introduce interactions, the four-momentum operator is written as $P_\mu = MV_\mu$, with the mass operator the sum of free and interacting mass operators, $M = M_{fr} + M_I$. Such a four-momentum operator will satisfy the point form equations if the velocity state kernel, $<v', \vec{k}_i', \mu_i'|M_I|v,\vec{k}_i,\mu_i>$ is independent of v and rotationally invariant (which is the same as the nonrelativistic condition on potentials). With such a four-momentum operator, the point form equations become a mass eigenvalue equation:

$$M\Psi \ = \ m\Psi \ , \quad (15)$$

which gives the bound and scattering wavefunctions.

Besides the mass operator, the other operator needed to compute form factors is the current operator. Current operators must satisfy general properties such as Poincaré covariance and current conservation. In the point form the current operator at the space-time point 0 plays a special role in that it determines the Poincaré covariance and conservation properties at an arbitrary space-time point. In fact it is easy to see that if $J_\mu(0)$ satisfies

$$U_\Lambda J_\mu(0) U_\Lambda^{-1} \ = \ (\Lambda^\nu_\mu)^{-1} J_\nu(0) \ ,$$

$$[P^\mu, J_\mu(0)] \ = \ 0 \ ,$$

then $J_\mu(x) := e^{iP\cdot x} J_\mu(0) e^{-iP\cdot x}$ is Poincaré covariant and is conserved.

Form factors are current operator matrix elements. If the states are chosen to be eigenstates of the four-momentum operator, then the covariance properties of the states and current operators make it possible to greatly simplify the structure of the form factors. As shown in reference [6] current operators are irreducible tensor operators of the Poincaré group, so that a generalized Wigner-Eckart theorem can be used to decompose current matrix elements into Clebsch-Gordan coefficients times reduced matrix elements, which are the invariant form factors. There is a natural frame in which the Clebsch-Gordan coefficients are one, namely the Breit frame, indicated by $p(st)$ (st=standard=Breit) below:

$$\begin{aligned}
<p'j'\sigma'I'|J^\mu(0)|pj\sigma I> &= \sum \Lambda_\nu^\mu(p',p) D^j_{\sigma'\,r'}(R_W'), \\
&= F^\nu_{r'r}(Q^2) D^j_{r\sigma}(R_W^{-1}), \quad (16)\\
<p'(st)j'r'I'|J^\mu(0)|p(st)jrI> &= F^\mu_{r'r}(Q^2), \quad (17)\\
p'(st) &= m'(ch\Delta,0,0,sh\Delta), \\
p(st) &= m(ch\Delta,0,0,-sh\Delta), \\
Q^2 &= (p'(st)-p(st))^2 \\
&= (m'-m)^2 - 4m'm\,sh\Delta^2 \\
p' &= \Lambda(p',p)p'(st) \\
p &= \Lambda(p',p)p(st).
\end{aligned}$$

$\Lambda(p',p)$ is a Lorentz transformation that carries the two standard four-momenta to arbitrary four-momenta, while the Wigner rotations in Eq.16 are formed from these four-momenta with $\Lambda(p',p)$.

It can then be shown that the invariant form factors in Eq.17, indexed by the spin projection labels r' and r, always give the correct number of independent form factors[6]. In fact $F^{\mu=0}_{r'r}(Q^2)$ is a diagonal matrix giving the electric form factors, $F^{\mu=1,2}_{r'r}(Q^2)$ is an off-diagonal matrix giving the magnetic form factors, and $F^{\mu=3}_{r'r}(Q^2)=0$ is an expression of current conservation in the Breit frame.

To actually compute an invariant form factor using Eq.17 a choice for the current operator must be made; usually one begins with a one-body current operator, resulting in what is called the point form spectator approximation (PFSA) [3]. This means that the four-momenta of the unstruck constituents do not change, which has the consequence that the momentum transfer to the struck constituent is greater than the momentum transfer to the object as a whole. More precisely, for the deuteron with nucleon constituents,

$$|(p'_1-p_1)^2| > Q^2 \frac{4m_N^2}{m_D^2}(1+\frac{Q^2}{4m_D^2}) > Q^2.$$

As will be seen in the next section, this has important consequences for the behavior of the form factors as a function of the momentum transfer Q^2.

With the assumption of a one-body current operator, Eq.17 can be written more explicitly as

$$F^\mu_{r'r}(Q^2) = \sum \int \mathcal{J} \, d^3\vec{k}'_i \, \mathcal{J}' \, d^3\vec{k}'_i \, \Psi^*_{m'j'r'}(\vec{k}'_i, \mu'_i)$$
$$u(p'_1\sigma'_1)\gamma^\mu u(p_1\sigma_1) \, F((p'_1 - p_1)^2)$$
$$E_{i\neq 1} \, \delta^3(p'_i - p_i)\delta_{\sigma'\sigma}\Psi_{mjr}(\vec{k}_i\mu_i) \,, \qquad (18)$$

where Ψ is an eigenfunction of the mass operator, $\mathcal{J}$ and $\mathcal{J}'$ are Jacobian factors, and the delta functions express the fact that the momenta of the unstruck constituents do not change. The one-body current matrix element in Eq.18 has been chosen for a spin 1/2 particle with form factor F.

ELASTIC DEUTERON FORM FACTORS

To compute elastic deuteron form factors using the point form it is necessary to have a mass operator that will generate the deuteron wave functions. To make use of the many nonrelativistic potentials that generate good deuteron wave functions, the mass operator, a sum of relativistic kinetic energy and interaction, is squared and then rewritten in the form of a nonrelativistic Schrödinger equation [7]:

$$M = 2\sqrt{m_N^2 + \vec{k}^2} + M_{int} \,, \qquad (19)$$
$$M^2 = 4(m_N^2 + \vec{k}^2) + 4m_N V_{N-N} \,,$$
$$M^2\Psi = (4m_N^2 + 4\vec{k}^2 + 4m_N V_{N-N})\Psi$$
$$= m_D^2 \Psi \,,$$
$$\left(\frac{\vec{k}^2}{m_N} + V_{N-N}\right)\Psi = \left(\frac{m_D^2}{4m_n} - m_N\right)\Psi \,; \qquad (20)$$

in this work the Argonne v_{18} and Reid'93 potentials were used to obtain the deuteron wave functions.

Since the nucleons that make up the deuteron themselves have internal structure, it is necessary to choose form factors for them. In this calculation the one-body current operators were determined by form factors given by Gari, Krümpelmann [8] and Mergell, Meissner, and Drechsel [9].

The results of these calculations have been published in reference [3]. Collaborators are T. Allen and W. Polyzou, with much help from F. Coester and G. Payne. A comparison of our results with those of other calculations is given by F. Gross in these proceedings [10], where it is seen that the A structure function falls off too fast in comparison with experimental data, while the results for the tensor polarization agree reasonably well with data. These results show the need for including two-body currents, as will be discussed in section 4.

NUCLEON FORM FACTORS

To compute nucleon form factors the mass operator is obtained in a rather different way as compared with the deuteron mass operator. In this case the three quark mass operator comes from a "semi-relativistic" Hamiltonian, the sum of relativistic kinetic energy, linear confinement potential and hyperfine interaction (Goldstone Boson Exchange model [4]):

$$H \longrightarrow M = \sum \sqrt{m^2 + \vec{k}_i^2} + \sum V(conf) + \sum V(HF).$$

That is, the "semi-relativistic" Hamiltonian can be reinterpreted as a point form mass operator and the eigenfunctions previously calculated can be used to compute form factors. Thus, the bound state problem for three quarks, $M\Psi = m\Psi$, gives the wave functions and a good spectroscopic fit [11]. Finally the current operator is a point-like Dirac current with no anomalous magnetic moment.

When these eigenfunctions and current operators are put into Eq.18, excellent agreement with data is obtained. The graphs and static properties will not be given here, since they already appear in the article by Wagenbrunn in these proceedings [12]. Published results are given in reference [13]. Collaborators in this project include S. Boffi, L. Glozmann, W. Plessas, M. Radici, and R. Wagenbrunn. It should also be noted that form factors for the weak interactions have been calculated and give excellent agreement with experiment [14].

POINT FORM ELECTRODYNAMICS AND THE CONSTRUCTION OF CONSERVED CURRENT OPERATORS

The form factors computed in the previous sections all made use of one-body current operators. It is important to learn of the effects of few-body operators, for deuteron form factors because they are needed to get better agreement with experiment, and for nucleon form factors, because they should only have a small effect on the static properties. Aside from their Poincaré and current conservation properties, few-body current operators should not renormalize the charge, and go to zero when the interactions are turned off (cluster property).

To see how current operators with these properties might be constructed, the current operator at the space-time point x is related to the current operator at the space-time point 0 by the four-momentum operator, which carries all the interactions in the point form:

$$J_\mu(x) := e^{iP \cdot x} J_\mu(0) e^{-iP \cdot x}.$$

If the current operator at the space-time point 0 transforms as a four-vector,

$$U_\Lambda J_\mu(0) U_\Lambda^{-1} = (\Lambda_\mu^\nu)^{-1} J_\nu(0),$$

then the current operator at the space-time point x will transform as a four-vector density:

$$U_\Lambda J_\mu(x) U_\Lambda^{-1} = (\Lambda_\mu^\nu)^{-1} J_\nu(\Lambda x).$$

The goal now is to show that a few-body current operator $J_\mu(0)$ can always be related to the electromagnetic field tensor at the space-time point 0. To that end set

$$iF_{\mu\nu}(0) := [A_\mu(0), P_\nu] - [A_\nu(0), P_\mu],$$

then

$$F_{\mu\nu}(x) = \frac{\partial A_\mu}{\partial x^\nu} - \frac{\partial A_\nu}{\partial x^\mu}$$

where these field quantities are all related to their values at the space-time point 0 by translations with the four-momentum operator. Further

$$iJ_\mu(0) := [F_{\mu\nu}(0), P^\nu] \quad \text{implies} \quad J_\mu(x) = \frac{\partial F_{\mu\nu}}{\partial x_\nu} \tag{21}$$

and

$$[P^\mu, J_\mu(0)] = 0 \quad \text{implies} \quad \frac{\partial J^\mu}{\partial x^\mu} = 0, \tag{22}$$

that is current conservation. With these definitions it is now easy to show that if $F_{\mu\nu}(0)$ is antisymmetric, and the components of the four-momentum operator commute among themselves (the point form Eq.1) then the current, Eq.21, is always conserved. The proof, given in reference [15], is straightforward and makes use of the Jacobi identity for three operators. Other properties of the field tensor and vector potential, such as gauge invariance, are also given in reference [15].

The current operator is now written as the sum of one-body and few-body current operators. One-body currents $J_\mu^{(1)}(0)$ need not be constructed from Eq.21; they can be generated, for example, as Noether currents of free fields and conserved with respect to the free four-momentum operator:

$$[P^\mu(free), J_\mu^{(1)}(0)] = 0 ;$$

for example:

$$< p'\sigma'|J_\mu^{(1)}(0)|p\sigma > = \bar{u}(p'\sigma')\gamma^\mu u(p\sigma).$$

For the full hadronic four-momentum operator,

$$P^\mu(h) = P^\mu(free) + P^\mu(I),$$

the one-body current in general is not conserved, $[P^\mu(h), J_\mu^{(1)}(0)] \neq 0$. But it can be modified so that it is conserved with respect to the full hadronic four-momentum operator. Let $J_\mu^{(1)}(Q)$ be the four dimensional Fourier transform of the one-body current operator. Since the current is not conserved, $Q^\mu J_\mu^{(1)}(Q) \neq 0$. To make it zero, modify the current operator to be

$$\tilde{J}_\mu^{(1)}(Q) = J_\mu^{(1)}(Q) - Q_\mu \frac{Q \cdot J^{(1)}(Q)}{Q^2}, \tag{23}$$

so that it is conserved. When matrix elements of such an operator are computed on eigenstates of the four-momentum operator in the Breit frame, the form factors will not change and the third component will be zero. Moreover there is no pole at $Q^2 = 0$ as is often claimed. In the Breit frame Q_μ has only a third component (call it q) and it is clear from Eq.23 that the dependence of q in the numerator and denominator cancels out for any value of q, including q=0.

The few-body current operator is defined using Eq.21:

$$iJ_\mu^{(2)}(0) = [F_{\mu\nu}(0), P^\nu(h)], \qquad (24)$$

and is conserved. Moreover, it carries no charge, as can be seen by computing matrix elements of the charge operator:

$$\begin{aligned}
<p'j\sigma'|Q|pj\sigma> &= \int d^3x <p'j\sigma'|J_{\mu=0}^{(2)}(\vec{x})|pj\sigma> \qquad (25)\\
&= \int d^3x <p'j\sigma'|e^{i\vec{P}\cdot\vec{x}} J_{\mu=0}^{(2)}(0) e^{-i\vec{P}\cdot\vec{x}}|pj\sigma>\\
&= \delta^3(\vec{p}'-\vec{p}) <p'j\sigma'|[F_{\mu=0,\nu}(0), P^\nu(h)]|pj\sigma>\\
&= 0. \qquad (26)
\end{aligned}$$

The only remaining property to be satisfied is the cluster property, whereby the current operator should go to zero when the interaction is turned off. The easiest way to satisfy this property is to ignore the definition of $F_{\mu\nu}(0)$ as the commutator of the four-vector potential with the four-momentum operator(see reference[15]) and simply choose it to be

$$iF_{\mu\nu}(0) = [I_\mu, P_\nu(I)] - [I_\nu, P_\mu(I)], \qquad (27)$$

for then, when $P_\mu(I) = 0$, $F_{\mu\nu}(0)$ will also be zero. I_μ is any current operator that transforms as a four-vector under Lorentz transformations. It need not even be conserved. The simplest choice is the original one-body current, which then is used to generate a few-body current. But any Feynman diagram with an external photon line in which the particles are on-mass-shell can serve as an I_μ kernel.

CONCLUSION

There are a variety of different frameworks for calculating form factors of hadrons. In this paper the point form of relativistic quantum mechanics has been used to calculate elastic deuteron and nucleon form factors.

The point form has several notable features that set it apart from other frameworks. First, all of the interactions are in the four-momentum operator, the components of which commute among themselves, and whose eigenvalues are commonly used observables. That means the other Poincaré generators, namely the Lorentz generators, are kinematic; as a consequence it is easy to Lorentz transform between different reference frames, and the theory is manifestly Lorentz covariant.

In the Bakamjian-Thomas construction [2] the four-momentum is written as a mass operator times a four-velocity operator, where the four-velocity operator is purely kinematic and all the dynamics resides in the mass operator. This is similar to nonrelativistic quantum mechanics, where phenomenological potentials are used to get bound and scattering wavefunctions. Sections 2 and 3 have outlined how mass operators are constructed for the deuteron and nucleon systems. The methods used there indicate the need for a more systematic procedure for calculating mass operators; reference [16] shows how to construct mass operators from field theory interaction Lagrangians.

Second, multiparticle states have the undesirable property that angular momentum cannot be coupled together, because under Lorentz transformations, the Wigner rotations of individual particles are all different. Velocity states, states boosted to an overall four-velocity from the center of momentum frame have the desirable property that under Lorentz transformations, all of the spin and internal momentum variables undergo the same Wigner rotation, and hence the angular momentum of such states can be coupled in the same way as is done nonrelativistically. Moreover, kernels of interacting mass operators in velocity state variables satisfy the point form Eqs.1,2 if they are independent of the four-velocity and rotationally invariant, the same conditions required of nonrelativistic Hamiltonians (in order that they be Galilei invariant). As a consequence nonrelativistic Hamiltonians with "semi relativistic" kinetic energies and Galilei invariant potentials can be reinterpreted as relativistic mass operators, which is the way the Goldstone Boson Exchange model was made relativistic in section 3.

Though the angular momentum operators are kinematic in the point form (also in the instant form) the operator that determines the relativistic spin structure is the Pauli-Lubanski operator,

$$W_\mu = \varepsilon_{\alpha\beta\mu\nu} J^{\alpha\beta} P^\nu.$$

Since this operator involves all ten generators of the Poincaré algebra, it might seem that it is always dynamical. However, in the point form Bakamjian-Thomas construction, the mass operator can always be factored out, leaving a modified Pauli-Lubanski operator that is kinematic.

Besides the four-momentum operator, the other operator needed to compute form factors is the current operator. In the point form the current operator at the space-time point 0, $J_\mu(0)$, takes on a special role in that the current operator at any other space-time point x is defined by translating with the four-momentum operator from the space-time point 0. This has the consequence that not only is translational covariance guaranteed, but all properties of the current operator, such as charge conservation, are formulated as commutator conditions on $J_\mu(0)$. Further, since $J_\mu(0)$ must transform as a four-vector under Lorentz transformations, form factors, matrix elements of the current operator, can be factored into a product of covariants times invariants, which is the Wigner-Eckart theorem for tensor operators under the Poincaré group. The covariants, Clebsch-Gordan coefficients of the Poincaré group [6], take on their simplest form in the Breit frame. Hence the calculation of deuteron and nucleon form factors are all done in the Breit frame. Further, since the invariant form factors are reduced matrix elements, one always obtains the correct number of independent form factors, with the $\mu = 0$ component giving the electric form factors, the $\mu = 1,2$ components giving the magnetic form factors, and the $\mu = 3$ component expressing charge conservation.

To calculate deuteron and nucleon form factors, the current operator is chosen as a one-body operator, which leads to the point form spectator approximation, wherein the momenta of unstruck constituents remain unchanged. Since the four-velocity is conserved, the impulse given to the struck constituent is different from the impulse given to the object as a whole, which has the consequence that point form form factors tend to lie below their nonrelativistic counterparts, leading to the A structure function for deuterons lying below the experimental data, while for nucleon form factors the fits with experimental data are very good.

It then becomes important to be able to construct point form few-body current operators. Section 5 outlines a procedure by which the few-body currents are given in terms of the electromagnetic field tensor at the space-time point 0. By writing Maxwell's equations as commutator equations for operators at the space-time point 0, a procedure has been given in which few-body current operators satisfying charge conservation, no charge renormalization, and cluster properties can be given in terms of another current, called I_μ. It remains to investigate different choices of I_μ and see if the resulting few-body currents are able to improve the deuteron form factors, yet leave the nucleon form factors relatively unchanged.

REFERENCES

1. Dirac, P.A.M., *Rev. Mod. Phys.* **21**, (1949) 392.
2. Bakamjian, B., Thomas, L.H., *Phys. Rev.* **92**, (1953) 1300;
   Keister, B.D., Polyzou, W.N., *Adv. Nucl. Phys.* **20**, (1991) 225.
3. Allen, T.W., Klink, W.H., Polyzou, W.N., *Phys. Rev. C* **63**, (2001) 034002.
4. Glozman, L.Ya., Riska, D.O., *Phys. Rep.* **268**, (1996) 263.
5. Klink, W.H., *Phys. Rev. C* **58**, (1998) 3617.
6. Klink, W.H., *Phys. Rev. C* **58**, (1998) 3587.
7. Coester, F., Pieper, S.C., Serduke, F.J.D., *Phys. Rev. C* **11**, (1975) 1.
8. Gari, M., Krümpelmann, W., *Phys. Rev. B* **45**, (1992) 1817.
9. Mergell, P., Meissner, U.-G., Drechsel, D., *Nucl. Phys.* **A596**, (1996) 367.
10. Gross, F.L., in these proceedings.
11. Glozman, L.Ya., Plessas, W., Varga, K., Wagenbrunn, R.F., *Phys. Rev. D* **58**, (1998) 094030.
12. Wagenbrunn, R.F., in these proceedings.
13. Wagenbrunn, R.F., Boffi, S., Klink, W., Plessas, W., Radici, M., *Phys. Lett. B* **511**, (2001) 33.
14. Glozman, L.Ya., Radici, M., Wagenbrunn, R.F., Boffi, S., Klink, W., Plessas, W., *Phys. Lett.* (to be published); nucl-th/0105028.
15. Klink, W.H., "Point Form Electrodynamics and the Construction of Conserved Current Operators", nucl-th/0012033.
16. Klink, W.H., "Constructing Mass Operators from Interaction Lagrangians", nucl-th/0012031.

Electroweak structure of few-body nuclei

L. E. Marcucci

Department of Physics, University of Pisa, Via Buonarroti 2, I-56100 Pisa, Italy
INFN- Sezione di Pisa, I-56100 Pisa, Italy

Abstract. The present status of our understanding of the electromagnetic structure of three-body nuclei and of proton radiative and weak capture on deuteron and ^{3}He is here reviewed, in the framework of the standard nuclear physics approach.

INTRODUCTION

In the present contribution, I will report about the progress made in the last few years in our understanding of nuclear few-body systems, in particular with $A=3$ and 4, and their interactions with electroweak probes. The non-relativistic framework used in these studies, which I will refer to as the "Standard Nuclear Physics Approach" (SNPA), has two main ingredients: (i) high quality Hamiltonian models and accurate initial and final state wave functions; (ii) realistic models for the nuclear electroweak currents.

Among the many studies which have been performed within the SNPA, I will review here the ones of Refs. [1, 2, 3], concerning the trinucleon electromagnetic structure, the reactions ^{2}H$(p,\gamma)^3$He and ^{3}He$(p,e^+\nu_e)^4$He in the low-energy regime, respectively. For this last process, I will finally present some recent results [4] obtained in a different framework, based on the SNPA for the description of the initial and final nuclear states, but on effective field theory for the construction of the nuclear transition operators.

THE WAVE FUNCTIONS

The nuclear Hamiltonian used in the calculations reported here, is written as sum of the non-relativistic kinetic energy operator and of two- and three-nucleon interactions. The two- and three-nucleon interactions here used are the Argonne v_{18} [5] (AV18) and the Urbana IX [6] (UIX), respectively. The AV18 two-nucleon interaction model is able to reproduce the NN low-energy database available from the early 1990s with a χ^2 per datum close to 1. Moreover, the full AV18/UIX Hamiltonian model reproduces the experimental binding energies and charge radii of the trinucleons and ^{4}He in exact Green's function Monte Carlo calculations [7, 8]. However, this Hamiltonian model presents some failures: (i) some of the elastic scattering polarization observables in three- and four-nucleon systems are not well reproduced (see, for instance, the well-known "A_y-puzzle"); (ii) the experimental values of the binding energies and spin-orbit splittings in the $A = 6 - 8$ spectrum are underpredicted by a few percent. Due to this second observation, in fact, a new three-nucleon interaction model, which improves the

TABLE 1. Three- and four-body binding energies in MeV, calculated with the CHH method using AV18 and AV18/UIX Hamiltonian models. Also listed are the corresponding experimental values.

Model	B(^{3}H)	B(^{3}He)	B(^{4}He)
AV18	7.64	6.93	24.01
AV18/UIX	8.49	7.75	28.1
expt.	8.48	7.73	28.3

TABLE 2. S-wave scattering lengths in fm for three- and four-nucleon systems, calculated with the CHH method using the AV18/UIX Hamiltonian model. Also listed are the corresponding experimental values from Ref. [12] for the nd system and from Ref. [13] for the p^3He system.

	spin-channel	CHH	expt.
nd	1/2	0.63	0.65±0.04
	3/2	6.33	6.35±0.02
pd	1/2	-0.02	
	3/2	13.7	
p^3He	0	11.5	10.8±2.6
	1	9.13	8.1±0.5
			10.2±1.5

existing UIX, has been recently developed [9]. For the present calculations, however, the AV18/UIX is to be considered a rather accurate model for the three- and four-body systems.

Although the nuclear interaction mentioned above is very simple to write down, the solution of the bound- and scattering-state problem for light nuclei has turned out to be rather difficult. In the present studies, the initial and final state wave functions have been obtained with the correlated-hyperspherical-harmonics (CHH) method, as developed both for the three- and four-body problem by the Pisa group [10, 11]. The CHH method essentially consists in expanding the wave function on a suitable basis, and in determining variationally the expansion coefficients, applying the Rayleigh-Ritz and Kohn variational principles for the bound-state and scattering-state problems, respectively. A collection of results for the three- and four-body systems are given in Tables 1 and 2 and are compared with the available experimental data [12, 13], showing an overall good agreement between the CHH calculation and the experiment. Although not shown in the tables, the results obtained using other techniques also agree within a few percent with the CHH results.

THE NUCLEAR TRANSITION OPERATORS

To describe the interaction of nuclei with electroweak probes, it is necessary to construct a realistic model for the nuclear transition operators. Since both electromagnetic and weak processes are here reviewed, the operators of interest are: (i) electromagnetic current and charge operators; (ii) weak vector and axial-vector current and charge operators. All these operators are written as sum of one- and many-body components that operate on the nucleon degrees of freedom. All the one-body terms can be obtained in a standard way from a non-relativistic reduction of the covariant single-nucleon current, including terms proportional to $1/m^2$. The most important many-body contributions arise from two-body operators and will be reviewed for the electromagnetic and weak case.

The electromagnetic two-body contributions

In the electromagnetic case, the longitudinal component of the two-body current operators can be linked to the underlying two-nucleon interaction by current conservation. Therefore these operators have to be seen as "model-independent" (MI), since, given the nuclear interaction, there is "no freedom" in their determination. Instead the transverse components of the two-body current operators cannot be constrained by current conservation, and they have to be seen as "model-dependent" (MD). Among the MD currents, those associated with excitation of Δ isobars are the most important ones in the momentum-transfer regime discussed here. These currents have been treated using the transition-correlation-operator (TCO) scheme, originally developed in Ref. [14] and further extended in Ref. [1]. In the TCO scheme–essentially, a scaled-down approach to a full $N+\Delta$ coupled-channel treatment–the Δ degrees of freedom are explicitly included in the nuclear wave functions. The Δ-currents, although being the most important MD ones, are still much smaller than the leading MI terms.

While the main two-body contributions to the electromagnetic current are linked to the form of the nucleon-nucleon interaction through the continuity equation, the most important two-body electromagnetic charge operators are model dependent, and should be viewed as relativistic corrections. The model commonly used [15] for the two-body charge operators includes the π-, ρ-, and ω-meson exchange terms with both isoscalar and isovector components, as well as the (isoscalar) $\rho\pi\gamma$ and (isovector) $\omega\pi\gamma$ charge transition couplings. At moderate values of momentum transfer ($q < 5$ fm^{-1}), the contribution due to the π-meson exchange charge operator has been found to be typically an order of magnitude larger than that of any of the remaining two-body mechanisms and one-body relativistic corrections [1].

The weak two-body contributions

For weak processes, we have to distinguish between the vector and the axial-vector operators. For the weak vector current and charge operators, we can use the conserved-vector-current hypothesis, which relates these operators to the isovector part of the corresponding electromagnetic ones. Therefore, the model used is the same as the one discussed in the previous subsection. However, in the weak vector charge, only the π- and ρ-meson exchange operators have been retained, since they give the most important contributions at low momentum-transfer.

The two-body axial charge operators can also be constrained by low-energy theorems and the partially-conserved-axial-current relation. The leading contribution arises from the long-range π-exchange term [16]. There are also short-range terms, the leading part of which has been constructed from the central and spin-orbit components of the nucleon-nucleon interaction, following the prescription of Ref. [17]. The Δ-excitation terms have also been included, but they have been found to be unimportant [3].

In contrast to the electromagnetic current, the axial current operator is not conserved. Thus, its two-body components cannot be linked to the nucleon-nucleon interaction and, in this sense, should be viewed as completely model dependent. Among the two-body axial current operators, those due to π- and ρ-meson exchanges and to the $\rho\pi$-transition mechanism have been included. However, the leading two-body terms in the axial current are due to Δ-isobar excitation, in particular to the $N\Delta$ transition. These contributions have been again treated non-perturbatively, using the TCO scheme. Being the $N\Delta$ transition axial coupling constant poorly known, the axial current presents a large model dependence. This model dependence can be strongly reduced, studying the process of tritium β-decay. In fact, from the experimental decay rate, it is possible to extract the value for the Gamow-Teller (GT) matrix element [18]. The theoretical prediction of GT matrix element requires the calculation of $\langle ^3\text{He}||O||^3\text{H}\rangle$, where O is the axial current operator described above. Since the one-body contribution alone underpredicts the experimental value of about 4%, this discrepancy can be used to fit the $N\Delta$ transition axial coupling constant [3, 18]. While this procedure is model dependent, its actual model dependence is in fact very weak, as it has been shown in Refs. [3, 18].

THE TRINUCLEON ELECTROMAGNETIC STRUCTURE

The electromagnetic structure of ^{3}H and ^{3}He has been studied using AV18/UIX CHH wave functions in Ref. [1]. Since the observables related to the charge operator are extremely well reproduced, I review here only the results for the magnetic moments and the magnetic form factors. The magnetic moments are given in Table 3, and the magnetic form factors in Figure 1. The experimental data are from Ref. [19].

A comparison between predicted and measured results leads to the following conclusions: (i) the predicted magnetic moments for both ^{3}H and ^{3}He are within less than 1% from the experimental data, if the model for the current operator includes the one-body and all the many-body (meson-exchange and Δ-isobar) contributions; the same accounts for the magnetic radii [1]. (ii) The predicted ^{3}H magnetic form factor is in quite good agreement with the experimental data over the whole range of momentum transfer q considered. (iii) The predicted ^{3}He magnetic form factor agrees with the experimental data for $q \leq 3$ fm^{-1}. For higher values of q, and in particular in the first diffraction region, there is strong disagreement between theory and experiment. The origin of this discrepancy is still under investigation.

TABLE 3. Trinucleon magnetic moments in nuclear magnetons, compared with the experimental results. The rows labelled "one-body", "mesonic", and "full" correspond, respectively, to the result obtained with single-nucleon current, with inclusion of meson-exchange currents, and of Δ-isobar contributions.

	$\mu(^3\text{He})$	$\mu(^3\text{H})$
one-body	-1.757	2.571
mesonic	-2.077	2.961
full	-2.112	2.994
expt.	-2.127	2.979

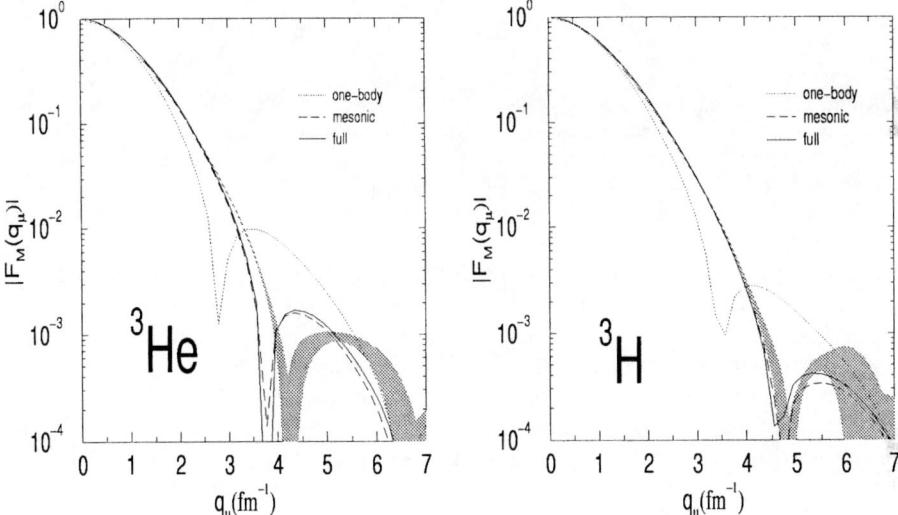

FIGURE 1. The magnetic form factors of ^{3}He and ^{3}H, compared with the experimental data. Notation as in Table 3.

THE *PD* RADIATIVE CAPTURE

The large body of available experimental data on the proton radiative capture reaction includes differential cross sections, vector and tensor analyzing powers, photon polarization coefficients, as well as the astrophysical *S*-factor up to energies of 10 MeV in the center-of-mass (c.m.) reference frame. With the approach described so far, a thorough comparison between theory and experiment has been performed in Ref. [2]. Here, I only summarized some of the results for the astrophysical *S*-factor. The *S*-factor, calculated

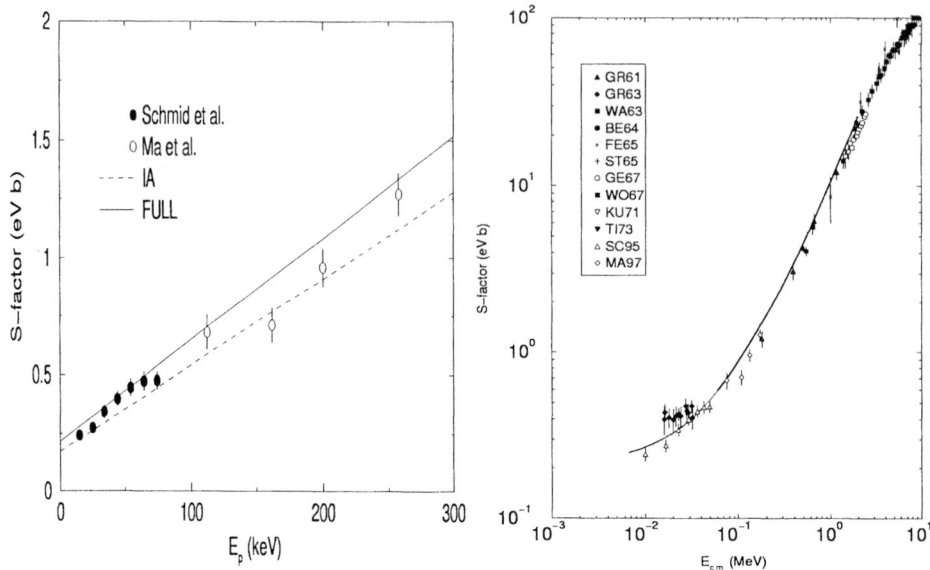

FIGURE 2. The S-factor for the ^{2}H$(p,\gamma)^3$He reaction in the c.m. range 0–2 MeV, obtained with the AV18/UIX Hamiltonian model and one-body only (dashed line) or both one- and many-body (solid line) currents. In the right panel, only the results of the full calculation are shown. The experimental data are taken from the web site http://pntpm.ulb.ac.be/nacre.htm.

using CHH wave functions with the AV18/UIX Hamiltonian model, in the c.m. energy range 0–2 MeV, is compared with the available data in Fig. 2. By inspection of the figure, it can be observed that: (i) the inclusion of many-body components in the model for the electromagnetic current operator is important to obtain a rather good agreement between theory and experiment. (ii) There are still some discrepancies at very low energies, where the full calculation is slightly higher than the data. A thorough study of the origins of these discrepancies has been performed in Ref. [2], providing useful insights in the process of refining and improving the model for the nuclear interaction and/or nuclear electromagnetic current operator. Further investigations are currently underway.

THE *HEP* REACTION

The process ^{3}He$(p,e^+\nu_e)^4$He, also known as *hep* reaction, is one of the nuclear reactions of the solar *pp*-chain which produces neutrinos. The particularity of the *hep* neutrinos is that they are the most energetic ones: their end-point energy is even larger than that of the ^{8}B neutrinos. However, the *hep* neutrino flux is much smaller than that of the ^{8}B neutrinos. At some level, therefore, there can be a significant distortion of the ^{8}B neutrino spectrum due to the *hep* neutrinos. Indeed, the first results, which the Super-Kamiokande (SK) collaboration was presenting in the late 1990's for the ^{8}B neutrino spectrum, showed an enhancement of events in the highest-energy bin [20, 21]. The

TABLE 4. The *hep* S-factor, in units of 10^{-20} keV-b, calculated with CHH wave functions corresponding to the AV18/UIX Hamiltonian model, at $p\,^3$He c.m. energies $E=0$, 5, and 10 keV. The rows labelled "one-body" and "full" list the contributions obtained by retaining the one-body only and both one- and many-body terms in the nuclear weak current respectively. The contributions due the $^3S_1$ channel only and all S- and P-wave channels are listed separately.

	$E=0$ keV		$E=5$ keV		$E=10$ keV	
	$^3S_1$	S+P	$^3S_1$	S+P	$^3S_1$	S+P
one-body	26.4	29.0	25.9	28.7	26.2	29.3
full	6.38	9.64	6.20	9.70	6.36	10.1

situation in 1999 after 825 days of data acquisition was the following [22]: (i) the ratio between the number of measured and predicted events on the base of the Standard Solar Model (SSM98) [23] was $\simeq 0.47$, as obtained from the low-energy part of the spectrum; (ii) the excess of events in the high-energy part of the spectrum could be explained by an enhancement of the *hep* SSM98 flux by a factor of $\simeq 17$.

The SSM98 estimation for the *hep* neutrino flux is based on the calculation of Ref. [14] for the astrophysical S-factor at zero energy in the c.m. frame. This calculation of 1992 is the last one of a long series of studies of the *hep* reaction [1]. Its main features are: (i) it is the first *microscopic* calculation, together with the one of Ref. [25]; (ii) within the SNPA, it uses the variational Monte Carlo (VMC) technique to calculate the initial and final state wave functions, with the older Argonne v_{14} two-nucleon [26] and Urbana VIII three-nucleon [27] interaction (AV14/UVIII); (iii) in a partial wave expansion of the initial $p\,^3$He state, it includes only the $^3S_1$ channel, and neglects any dependence on the momentum transfer q of the reaction, being considered about zero. Based on this calculation, the value for the astrophysical S-factor used in the SSM98 is 2.3×10^{-20} keV-b.

The new calculation of Ref. [3] for the *hep* S-factor, which I review here, has the following characteristics: (i) it uses CHH wave functions obtained with the AV18/UIX Hamiltonian model, to be considered a more accurate description of the initial and final states than the one obtained with AV14/UVIII VMC wave functions; (ii) it includes all S- and P-wave capture states; (iii) it retaines explicitly the q-dependence of the nuclear transition operator, going beyond the simple long-wavelength approximation.

The results are summarized in Tables 4 and 5. By inspection of Table 4, where the values for three different c.m. energies are listed, we can conclude that the energy dependence is rather weak: the value at 10 keV is only about 4 % larger than that at 0 keV. Furthermore, the P-wave capture states are found to be important, contributing about 40 % of the calculated S-factor. However, the contributions from D-wave channels are expected to be very small [3]. Finally, the many-body axial currents play a crucial role in the (dominant) $^3S_1$ capture, where they reduce the S-factor by more than a factor

---

[1] For a historical review of the different calculations for the *hep* reaction, see Refs. [3, 24].

TABLE 5. Contributions to the *hep* S-factor, in units of 10^{-20} keV-b, from individual partial waves calculated with CHH wave functions corresponding to the AV18/UIX, AV18 and the AV14/UVIII Hamiltonian models, at zero $p\,^3$He c.m. energy.

	AV18/UIX	AV18	AV14/UVIII
$^1S_0$	0.02	0.01	0.01
$^3S_1$	6.38	7.69	6.60
$^3P_0$	0.82	0.89	0.79
$^1P_1$	1.00	1.14	1.05
$^3P_1$	0.30	0.52	0.38
$^3P_2$	0.97	1.78	1.24
Total	9.64	12.1	10.1

of four.

To check the model dependence of these results, the calculation has been repeated, using different models for the nuclear interaction and refitting each time the $N\Delta$ axial coupling constant to the GT matrix element in tritium β-decay, as explained above. The contributions from the different partial waves obtained using AV18/UIX, AV18 alone, and AV14/UVIII at zero c.m. energy are listed in Table 5. The last row of the table lists the result when all the partial waves are included. It slightly differs from the sum of all the contributions, because of small interference between the different partial wave terms.

By comparing the AV18 and AV18/UIX results, we note that inclusion of three-nucleon interaction reduces the total S-factor by about 20 %. This difference emphasizes the need for performing the calculation using a Hamiltonian model that accurately reproduces the initial and final state wave functions. This is true for the AV18/UIX model, but not for the AV18 model, as can be seen, for instance, from Table 1. The S- and P-wave contributions to the S-factor are not significantly different from the AV18/UIX results, if the older AV14/UVIII Hamiltonian model is used. In particular, the most important S-wave contribution, the $^3S_1$ term, differs for only $\simeq 3$ %. Note that the AV14/UVIII Hamiltonian also reproduces the low-energy properties for the three- and four-nucleon systems.

The chief conclusion of this discussion, therefore, is that the best estimate for the S-factor at 10 keV, close to the Gamow-peak energy, is 10.1×10^{-20} keV-b. This value is $\simeq 4.5$ times larger than the SSM98 value of 2.3×10^{-20} keV-b mentioned above.

Finally, to study the implications of these results for the SK energy spectrum of electrons recoiling from scattering with solar neutrinos, a parameter α has been introduced, defined as $\alpha \equiv (S_{\text{new}}/S_{\text{SSM98}}) \times P_{\text{osc}}$, where P_{osc} is the *hep* neutrino suppression constant. Presently, $\alpha = (10.1 \times 10^{-20}\text{keV} - \text{b})/(2.3 \times 10^{-20}\text{keV} - \text{b}) = 4.4$, if *hep* neutrinos oscillations are ignored ($P_{\text{osc}}=1$). The SK data available till 1999 after 825 days of data acquisition, were presented as ratio of the measured electron spectrum to that expected in the SSM98 with no neutrino oscillations. The high energy results are shown in Fig. 3 by the filled points. The error bars denote the combined statistical and systematic error. Also shown in Fig. 3, with opaque squares, are the more recent 1117-day data.

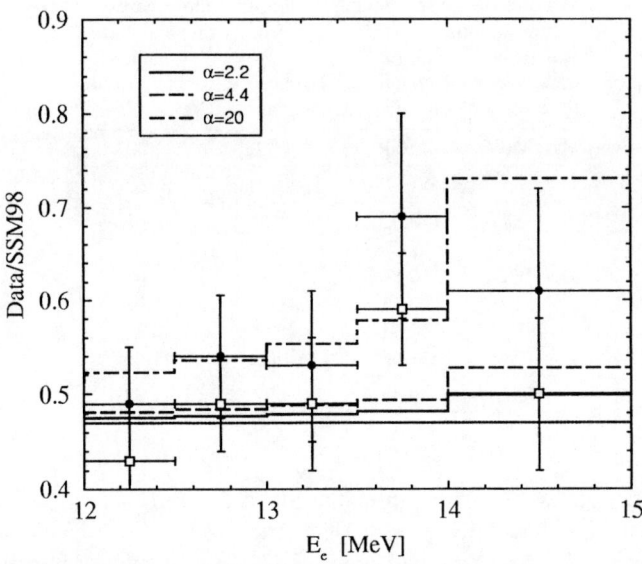

FIGURE 3. Highest energy part of the SK electron energy spectrum. The 825-day data were extracted graphically from Fig. 8 of Ref. [22], and are shown as points, while the 1117-day data are extracted from Fig. 8 of Ref. [28] and are shown by the squares. The curves correspond respectively to no *hep* contribution (dotted line), and to $\alpha = 2.2$ (solid line), 4.4 (long-dashed line), and 20 (dot-dashed line).

The effect of three different values of α is shown by the three lines. The values of α considered are 4.4 and 2.2, which correspond to $P_{osc} = 1$ and 0.5 respectively, and 20, close to the factor of 17 mentioned in Ref. [22]. From Fig. 3, it appears that the prediction of Ref. [3] are important to explain the latest SK data.

Finally, it is interesting to note that the SSM98 has been recently revisited [29] (SSM00) and the new value for the *hep* flux has been renormalized following the calculation here reviewed. Consequently, the most recent SK 1258-day data [30], are presented as ratio between the measured and the SSM00 predicted events. Considering also the effects of a new measurement of the ^{8}B spectrum [31], no high energy enhancement is anymore noticeable.

A new calculation of the *hep* reaction

Recently, a new calculation of the *hep* reaction has been performed in a different approach, which tries to combine the strengths of the SNPA and of effective field theory (EFT) [4]. This new approach (SNPA+EFT) has scored quite good success in predicting the solar *S*-factor for *pp* fusion [32], as well as the polarization observables in *np* thermal capture [33]. The starting point for SNPA+EFT is the observation that, to high accuracy, the leading-order single-particle matrix elements in SNPA and EFT are identical, and can be reliably estimated with the use of realistic SNPA wave functions for the initial

TABLE 6. Contributions to the *hep* zero energy S-factor, in units of 10^{-20} keV-b, from individual partial waves as function of Λ. The corresponding values for $\hat{d}_R$ as obtained by fit to the GT matrix element in tritium β-decay are also listed.

Λ (MeV)	500	600	800
$\hat{d}_R$	1.00 ± 0.07	1.78 ± 0.08	3.90 ± 0.10
$^1S_0$	0.02	0.02	0.02
$^3S_1$	7.00	6.37	4.30
$^3P_0$	0.67	0.66	0.66
$^1P_1$	0.85	0.88	0.91
$^3P_1$	0.34	0.34	0.34
$^3P_2$	1.06	1.06	1.06
Total	9.95	9.37	7.32

and final nuclear states. The strength of the SNPA is used therefore to describe very accurately the nuclear few-body systems. The next observation, on which this approach is based, is that in EFT the many-body contributions to the nuclear transition operators can be controlled by systematic chiral expansion in heavy-baryon chiral perturbation theory [34].

The EFT weak transition operators used in this new calculation have been obtained up to next-to-next-to-next-to-leading order (N^{3}LO), adopting a cutoff regularization. The explicit degrees of freedom in this scheme are the nucleon and the pion, with all other degrees of freedom integrated out. Therefore, a reasonable range of the cutoff Λ would be somewhere between 500 and 800 MeV. The explicit expressions for the two-body contributions to the vector and axial-vector current and charge operators can be found in Refs. [4, 32]. Here, I only review the main aspects of them: (i) the two-body vector current and axial charge are *chiral protected*, i.e. the dominant contribution is given by one-soft-pion exchange mechanism, which is uniquely dictated by chiral symmetry and therefore theoretically controlled. Furthermore, higher chiral-order terms are suppressed. (ii) At N^{3}LO, in this low energy and low momentum transfer regime, two-body vector charge contributions are zero. (iii) The expression for the two-body axial current contains a parameter ($\hat{d}_R$) to be fit from experiment. This parameter represents the strength of a contact term, and it can be adjusted by fit to the experimental value for the GT matrix element for tritium β-decay, in a procedure similar to the one used in the SNPA calculation of Ref. [3].

The results of this calculation are summarized in Table 6, where the contributions to the zero c.m. energy S-factor from individual initial partial waves are listed for three values of the cutoff Λ. Also listed in the table are the values of the parameter $\hat{d}_R$.

From inspection of the table, we can conclude that all the channels other than $^3S_1$ present a very small Λ-dependence. The Λ-dependence in the $^3S_1$ channel is the result of strong cancellations between the different contributions. This can be seen in Table 7, where the individual contributions from one- and two-body operators to the reduced matrix element $\bar{L}_1(q;A)$ of the longitudinal component of the axial current at momentum transfer q of 19.2 MeV are listed.

TABLE 7. Values of $\overline{L}_1(q;A)$ (in fm$^{3/2}$) calculated as function of the cutoff Λ. The individual contributions from the one-body (1b) and two-body (2b) operators are also listed.

Λ (MeV)	500	600	800
$\overline{L}_1(q;A)$ (fm$^{3/2}$)	−0.032	−0.029	−0.022
1b	−0.081	−0.081	−0.081
2b-tot	0.049	0.052	0.059
2b (without $\hat{d}_R$)	0.093	0.122	0.166
2b ($\propto \hat{d}_R$)	−0.044	−0.070	−0.107

Summarizing the results of Ref. [4], the new quantitative prediction for the *hep* S-factor is (8.6 ± 1.3) in units of 10^{-20} keV-b, where the "error" spans the range of the Λ-dependence for $\Lambda = 500 - 800$ MeV. This result corroborates the one of Ref. [3], which remains within the estimated error. It would be however quite desirable to decrease the uncertainty of about 20 % of this new estimate. According to a general *tenet* of EFT, the Λ-dependence should diminish including higher order terms. A preliminary study [35] indicates that it is indeed possible to reduce the Λ-dependence significantly by including N^{4}LO corrections. Further investigations are currently underway.

ACKNOWLEDGMENTS

I wish to thank R. Schiavilla, M. Viviani, A. Kievsky, S. Rosati, T.-S. Park, K. Kubodera, D.-P. Min, M. Rho, and J.F. Beacon for their many important contributions to the work reported here.

REFERENCES

1. Marcucci, L.E., Riska, D.O., and Schiavilla, R., *Phys. Rev. C* **58**, 3069 (1998).
2. Viviani, M., *et al.*, *Phys. Rev. C* **61**, 064001 (2000).
3. Marcucci, L.E., *et al.*, *Phys. Rev. Lett.* **84**, 5959 (2000); Marcucci, L.E., *et al.*, *Phys. Rev. C* **63**, 015801 (2001).
4. Park, T.-S., *et al.*, nucl-th/0107012, submitted to *Phys. Rev. Lett.*
5. Wiringa, R.B., Stoks, V.G.J., and Schiavilla, R., *Phys. Rev. C* **51**, 38 (1995).
6. Pudliner, B.S., *et al.*, *Phys. Rev. Lett.* **74**, 4396 (1995).
7. Pudliner, B.S., *et al.*, *Phys. Rev. C* **56**, 1720 (1997).
8. Carlson, J., private communication.
9. Pieper, S.C., Pandharipande, V.R., Wiringa, R.B., and Carlson, J., *Phys. Rev. C* **64**, 014001 (2001).
10. Kievsky, A., *et al.*, *Nucl. Phys.* **A551**, 241 (1993); *Nucl. Phys.* **A577**, 511 (1994); *Phys. Rev. C* **58**, 3085 (1998).
11. Viviani, M., *et al.*, *Few-Body Syst.* **18**, 25 (1995); *Phys. Rev. Lett.* **81**, 1580 (1998).
12. Dilg, W., *et al.*, *Phys. Lett.* **36B**, 208 (1971).
13. Tegnér, P.E., and Bargholtz, C., *Astrophys. J.* **272**, 311 (1983); Alley, M.T., and Knutson, L.D., *Phys. Rev. C* **48**, 1901 (1993).
14. Schiavilla, R., *et al.*, *Phys. Rev. C* **45**, 2628 (1992).

15. Schiavilla, R., Riska, D.O., and Pandharipande, V.R., *Phys. Rev. C* **41**, 309 (1990).
16. Kubodera, K., Delorme, J., and Rho, M., *Phys. Rev. Lett.* **40**, 755 (1978).
17. Kirchbach, M., Riska, D.O., and Tsushima, K., *Nucl. Phys.* **A542**, 616 (1992).
18. Schiavilla, R., *et al.*, *Phys. Rev. C* **58**, 1263 (1998).
19. Amroun, A., *et al.*, *Nucl. Phys.* **A579**, 596 (1994).
20. Fukuda, Y., *et al.*, *Phys. Rev. Lett.* **82**, 2430 (1999).
21. Smy, M., hep-ex/9903034.
22. Suzuki, Y., contribution to Lepton-Photon Symposium 99 (1999), `http://www-sk.icrr.u-tokyo.ac.jp/doc/sk/pub/index.html`.
23. Bahcall, J.N., Basu, S., and Pinsonneault, M.H., *Phys. Lett. B* **433**, 1 (1998).
24. Bahcall, J.N., and Krastev, P.I., *Phys. Lett. B* **436**, 243 (1998).
25. Carlson, J., *et al.*, *Phys. Rev. C* **44**, 619 (1991).
26. Wiringa, R.B., Smith, R.A., and Ainsworth, T.L., *Phys. Rev. C* **29**, 1207 (1984).
27. Wiringa, R.B., *Phys. Rev. C* **43**, 1585 (1991).
28. Svoboda, R., contribution to Tau 2000 (2000), `http://www-sk.icrr.u-tokyo.ac.jp/doc/sk/pub/index.html`.
29. Bahcall, J.N., *et al.*, atro-ph/0010346.
30. Fukuda, S., *et al.*, *Phys. Rev. Lett.* **86**, 5651 (2001).
31. Ortiz, C.E., *et al.*, *Phys. Rev. Lett.* **85**, 2909 (2000).
32. Park, T.-S., *et al.*, nucl-th/0106025, submitted to *Phys. Rev. Lett.*
33. Park, T.-S., *et al.*, *Phys. Lett. B* **472**, 232 (2000).
34. Park, T.-S., *et al.*, *Nucl. Phys.* **A684**, 101 (2001); nucl-th/9904053.
35. Park, T.-S., *et al.*, private communication.

Recent progress in light front dynamics

J.-F. Mathiot

Laboratoire de Physique Corpusculaire, Université Blaise-Pascal, CNRS/IN2P3
24 avenue des Landais, F-63177 Aubière Cedex, France

Abstract. We detail the various steps needed to calculate physical observables involving relativistic bound states, as formulated in Light-Front Dynamics (LFD). We emphasize in this context the advantages of a covariant formulation of LFD in order, first, to define a bound state of given angular momentum J on the light-front, and, second, to calculate its properties as revealed for instance by electron scattering.

INTRODUCTION

The relevance of a coherent relativistic description of few-body systems, both for bound and scattering states, is now well recognized in nuclear as well as in particle physics. This is already clear in particle physics for the understanding of the wave functions of the valence quarks in the nucleon or in the pion, as revealed for instance in exclusive reactions at very high momentum transfer. The need for a coherent relativistic description of few-body systems has also become clear in nuclear physics, for instance in order to check the validity of the standard description of the microscopic structure of nuclei in terms of mesons exchanged between nucleons. In this case, electromagnetic interactions play also a central role in "seeing" meson exchanges in nuclei. In both domains, it is obligatory to have a relativistic description of the bound and scattering states. It is also necessary to have a consistent description of the electromagnetic current operator needed to probe the system. This is mandatory in order to have meaningful predictions for the various physical observables.

A few relativistic approaches have been developed in the past ten years in order to meet these goals. Among them, two have received particular attention in the last few years. The first one is based on the Bethe-Salpeter formalism or its three-dimensional reductions. The Bethe-Salpeter formalism is four-dimensional and explicitly Lorentz covariant. The calculational technique to evaluate electromagnetic amplitudes is based on Feynman diagrams and associated rules. Three-dimensional reductions result in equations of the quasi-potential type.

The second one is Light-Front Dynamics (LFD) [1, 2]. In this case, the state vector describing the system is expanded in Fock components with increasing number of particles. The state vector is defined on a surface in four-dimensional space-time which should be indicated explicitly. The Fock components – the relativistic wave functions in this formalism – are the direct generalization of the non-relativistic wave functions. This is the aim of the present short review to detail recent progress made on this formulation.

THE INTEREST OF LIGHT-FRONT DYNAMICS FOR BOUND STATES

The description of nonrelativistic two-body bound states by the Schrödinger equation is already well known. For time-independent interactions, one can find stationary state solutions of the form:

$$\Phi(r,t) = \phi(r)e^{-iEt}. \tag{1}$$

The wave function of the system, Φ, is thus defined on time slices $t = const$. According to the Dirac classification [1], any operator changing the plane $t = const$ is dynamical. One needs to know the energy of the system in order to calculate its action on the wavefunction. This is of course true for translations in time, as given by (1). This is also true for boost operators. Rotations in three dimensional space on the other hand are kinematical operators. One can therefore define the total angular momentum J and the projection λ of a state without knowing the complete dynamics.

In LFD we consider the evolution of the system along the Light-Front (LF) time $\sigma = t + z$. Any boost along z leaves therefore the LF plane invariant. It is a kinematical operator. In momentum space, the light front momentum $P^+ = E + p_z$ is a kinematical operator, while the light front "energy" $P^- = E - p_z$ is a dynamical operator.

However, the LF plane is not invariant under any rotations in ordinary three-dimensional space : the angular momentum operator becomes now dynamical. One needs therefore a way to extract this dynamical part in order to define the total angular momentum of a bound state. This is one of the aim of the explicit Covariant formulation of Light-Front Dynamics (CLFD) [3, 4], as we shall detail below.

According to LF kinematics, $P^+ > 0$ and $P^- > 0$. The ground state is thus the trivial vacuum. Therefore one can legitimate a Fock state decomposition of any bound state [2]. However one should be very careful for theories with spontaneous symmetry breaking [5]. The appearance of zero modes in that case should be treated with great care. This subject will not be considered in the present review.

HOW TO DEFINE A BOUND STATE ON THE LIGHT-FRONT

Structure of a bound state of angular momentum J

We are interested here in the state vector of a bound system. It corresponds to a definite mass M, four-momentum p, total angular momentum J with projection λ onto the z axis in the rest frame, i.e., the state vector forms a representation of the Poincaré group. This means that it satisfies the following eigenvalue equations:

$$\hat{P}_\mu \phi^{J\lambda}(p) = p_\mu \phi^{J\lambda}(p), \tag{2}$$
$$\hat{P}^2 \phi^{J\lambda}(p) = M^2 \phi^{J\lambda}(p), \tag{3}$$
$$\hat{S}^2 \phi^{J\lambda}(p) = -M^2 J(J+1) \phi^{J\lambda}(p), \tag{4}$$
$$\hat{S}_3 \phi^{J\lambda}(p) = M\lambda \phi^{J\lambda}(p), \tag{5}$$

where $\hat{S}_\mu$ is the Pauli-Lubanski vector:

$$\hat{S}_\mu = \frac{1}{2}\varepsilon_{\mu\nu\rho\gamma}\hat{P}^\nu \hat{J}^{\rho\gamma}. \tag{6}$$

The angular momentum operator is now decomposed according to:

$$\hat{J}_{\mu\nu} = \hat{J}^0_{\mu\nu} + \hat{J}^{int}_{\mu\nu}, \tag{7}$$

where the 0 and int superscripts indicate the free and interacting parts of the operators respectively. We shall now see how CLFD can provide a well defined framework in order to calculate $\hat{J}^{int}_{\mu\nu}$.

Explicit covariant formulation of Light-Front Dynamics

The general idea beyond CLFD is very simple. One should keep track of the position of the LF plane at any step of the calculation. This can be achieved by the following definition of the LF time

$$\sigma = \omega \cdot x, \tag{8}$$

with $\omega^2 = 0$. In a given representation (see below), one can write $\omega = (1, \vec{n})$, with $\vec{n}$ being a unit vector in ordinary space. It characterizes the position of the LF plane tangent to the light cone. The four vector ω is fixed, but arbitrary, so that the state vector depends now on ω : $\Phi^{J\lambda}(p,\omega)$.

As a consequence, the electromagnetic current in particular, and all physical operators in general, do depend also on ω in such a way that any physical amplitude should not depend on the particular position of the light-front, i.e. on ω, unless approximations have been made. In that case, which is almost always true in practice, the explicit covariance of the approach enables us to exhibit the ω dependence of the amplitude and to extract the physical part from the non-physical ω-dependent one, as we shall explain below for simple examples.

Angular condition

Within CLFD, the operators associated to the four-momentum and four-dimensional angular momentum are expressed in terms of integrals of the energy-momentum $T_{\mu\nu}$ and the angular momentum $M^\rho_{\mu\nu}$ tensors over the LF plane $\omega \cdot x$, according to:

$$\hat{P}_\mu = \int T_{\mu\nu}\omega^\nu \delta(\omega \cdot x - \sigma)d^4x, \tag{9}$$

$$\hat{J}_{\mu\nu} = \int M^\rho_{\mu\nu}\omega_\rho \delta(\omega \cdot x - \sigma)d^4x. \tag{10}$$

By looking at the transformation properties of the state vector, ϕ, one can show that, under a rotation, one has:

$$\hat{J}^{int}_{\mu\nu} \phi_\omega(\sigma) = \hat{L}_{\mu\nu}(\omega)\phi_\omega(\sigma), \tag{11}$$

where:
$$\hat{L}_{\mu N}(\omega) = i\left(\omega_\mu \frac{\partial}{\partial \omega^\nu} - \omega_\nu \frac{\partial}{\partial \omega^\mu}\right). \quad (12)$$

Equation (11) is called the *angular condition*. We can use this angular condition and replace the operator $\hat{J}^{int}_{\mu N}$, which is contained in $\hat{J}_{\mu N}$ in (6), by $\hat{L}_{\mu N}(\omega)$. Introducing the notations:

$$\hat{M}_{\mu N} = \hat{J}^0_{\mu N} + \hat{L}_{\mu N}(\omega), \quad (13)$$

$$\hat{W}_\mu = \frac{1}{2}\varepsilon_{\mu\nu\rho\gamma}\hat{P}^\nu \hat{M}^{\rho\gamma}, \quad (14)$$

we obtain instead of eqs.(4) and (5):

$$\hat{W}^2 \phi^{J\lambda}(p) = -M^2 J(J+1) \phi^{J\lambda}(p), \quad (15)$$

$$\hat{W}_3 \phi^{J\lambda}(p) = M\lambda \phi^{J\lambda}(p). \quad (16)$$

These equations do not contain the interaction Hamiltonian, once ϕ satisfies (2) and (3). The construction of states with definite angular momentum becomes therefore a purely kinematical problem.

Two body wave function

The state vector $\phi(p)$ describing any bound state system is decomposed in Fock components according to (with $\sigma = \omega.x = 0$ for simplicity):

$$\begin{aligned}
\phi(p) &= (2\pi)^{3/2}\int \phi_1(k_1,p,\omega) a^\dagger(\vec{k}_1)|0\rangle \delta^{(4)}(k_1-p-\omega\tau)2(\omega\cdot p)d\tau \frac{d^3k_1}{(2\pi)^{3/2}\sqrt{2\varepsilon_{k_1}}} \\
&+ (2\pi)^{3/2}\int \phi_2(k_1,k_2,p,\omega)a^\dagger(\vec{k}_1)a^\dagger(\vec{k}_2)|0\rangle \\
&\times \delta^{(4)}(k_1+k_2-p-\omega\tau)2(\omega\cdot p)d\tau \frac{d^3k_1}{(2\pi)^{3/2}\sqrt{2\varepsilon_{k_1}}}\frac{d^3k_2}{(2\pi)^{3/2}\sqrt{2\varepsilon_{k_2}}} \\
&+ (2\pi)^{3/2}\int \phi_3(k_1,k_2,k_3,p,\omega)a^\dagger(\vec{k}_1)a^\dagger(\vec{k}_2)a^\dagger(\vec{k}_3)|0\rangle \\
&\times \delta^{(4)}(k_1+k_2+k_2-p-\omega\tau)2(\omega\cdot p)d\tau \\
&\times \frac{d^3k_1}{(2\pi)^{3/2}\sqrt{2\varepsilon_{k_1}}}\frac{d^3k_2}{(2\pi)^{3/2}\sqrt{2\varepsilon_{k_2}}}\frac{d^3k_3}{(2\pi)^{3/2}\sqrt{2\varepsilon_{k_3}}} + \cdots, \quad (17)
\end{aligned}$$

where $\varepsilon_{k_i} = \sqrt{\vec{k}_i^2 + m_i^2}$ and m_i is the mass of the particle i of momentum k_i. Note that in this decomposition the momentum conservation law involves the position ω of the LF plane. For the two-body component for instance:

$$p = k_1 + k_2 - \omega\tau. \quad (18)$$

The parameter τ, conjugate to an energy, is uniquely determined from the on-mass-shell conditions for the elementary particles. It measures how far the particles are off-energy shell.

As an example, let us first write down the general structure of a scalar bound state made of two scalar particles. Its wave function should depend on two invariant variables. These can be of course $s = (k_1 + k_2)^2$ and $t = (k_1 - k_2)^2$. More conveniently, it is appropriate [4] to use the variables $\vec{k}$ and $\vec{n}$ defined as:

$$\begin{aligned} \vec{k} &= \mathcal{L}^{-1}(\vec{\mathcal{P}})\vec{k}_1 \,, \\ \vec{n} &= \mathcal{L}^{-1}(\vec{\mathcal{P}})\vec{\omega} \,. \end{aligned} \qquad (19)$$

where $\mathcal{L}$ is the Lorentz boost and $\vec{\mathcal{P}} = \vec{k}_1 + \vec{k}_2$. It is interesting to note that $\vec{k}$ coincides, in the CM system, with the relative momentum between particles 1 and 2. It can be shown [4] that through a Lorentz boost, or a rotation, the vectors $\vec{k}$ and $\vec{n}$ are rotated only. This implies that one can choose as the two invariants, $\vec{k}^2$ and $\vec{k}\cdot\vec{n}$. Note that the usual light-cone coordinates can easily be defined in terms of these variables:

$$x = \frac{\omega \cdot k_1}{\omega \cdot p} = \frac{1}{2}\left(1 - \frac{\vec{n}\cdot\vec{k}}{\varepsilon_k}\right) \,, \quad \vec{k}_T^{\,2} = \vec{k}^2 - (\vec{k}\cdot\vec{n})^2 \,. \qquad (20)$$

We can then parametrize the two-body scalar wave function by :

$$\varphi(\vec{k}^2, \vec{k}\cdot\vec{n}) \,. \qquad (21)$$

In the non relativistic limit, defined by letting c go to infinity, we have $\vec{n} \equiv \vec{n}/c \to 0$ when $c \to \infty$, so that the relativistic wave function (21) tends to the Schrödinger one $\varphi(\vec{k}^2)$. This two-body wave function is solution of the following LF integral equation [4] (written here for equal constituent masses m):

$$\left[4(\vec{k}^2 + m^2) - M^2\right] \varphi(\vec{k}^2, \vec{k}\cdot\vec{n}) = -\frac{m^2}{2\pi^3} \int \varphi(\vec{k'}^2, \vec{k'}\cdot\vec{n}) K(\vec{k'}, \vec{k}, \vec{n}, M^2) \frac{d^3 k'}{\varepsilon_{k'}} \,, \qquad (22)$$

where the kernel of the interaction has been denoted by K. This is an explicit three-dimensional formalism, although no particular reduction has been made, but with an extra degree of freedom given by the position, ω, of the LF plane.

The pion wave function

The structure of the pion wave function in the simple constituent quark model, i.e. a scalar bound system made of two spin 1/2 particles, can now easily be found. The general decomposition of the wave function, compatible with the quantum numbers of the pion is:

$$\psi = \frac{1}{\sqrt{2}} \bar{u}(k_2) \left[A_1 + A_2 \frac{\slashed{\omega}}{\omega \cdot p}\right] \gamma_5 \, v(k_1) \,, \qquad (23)$$

where u and v are the usual Dirac spinors for particle and antiparticle respectively. For simplicity, we do not take into account isospin. The pion wave function has thus two independent components on the light-front. All other structures involving for instance the individual momenta k_1 or k_2 can be expressed in terms of the above ones using the Dirac equation for the on-mass shell constituent quarks. The representation of this wave function in terms of the variables $\vec{k}$ and $\vec{n}$ is thus:

$$\psi = \frac{1}{\sqrt{2}} w_2^t \left(g_1 + \frac{i\vec{\sigma}\cdot[\vec{n}\times\vec{k}]}{k} g_2 \right) w_1 , \qquad (24)$$

where w is the two-component nucleon spinor normalized to $w^\dagger w = 1$. The component proportional to g_1 is the only one which survives in the non-relativistic limit, while the component g_2 is purely of relativistic origin.

HOW TO DETERMINE BOUND STATE PROPERTIES ON THE LIGHT-FRONT

Electromagnetic observables

The question of extracting the physical form factors from the elementary electromagnetic amplitude in LFD is rather general and appears for a system with any spin. The matrix element of the exact current, $J_\rho = \langle p'|J_\rho(0)|p\rangle$, can be represented as the sum of all possible contributions:

$$J_\rho = J_\rho^{(1)} + J_\rho^{(2)} + \cdots + J_\rho^{(n)} + \cdots , \qquad (25)$$

where the superscripts is a short notation for indicating the various levels of approximation. In the case of the deuteron form factors for instance, the terms in (25) correspond to the impulse approximation, meson exchange current contribution, or the contribution of an increasing number of Fock components. Any contribution in (25) consists of the sum of two parts: a ω-independent one denoted by $F_\rho^{(n)}$, and a ω-dependent one denoted by $B_\rho^{(n)}(\omega)$:

$$J_\rho^{(n)} = F_\rho^{(n)} + B_\rho^{(n)}(\omega) . \qquad (26)$$

Since the matrix element of the exact current, which is a physical observables, does not depend on ω, the ω-dependent parts should cancel each other:

$$B_\rho^{(1)}(\omega) + B_\rho^{(2)}(\omega) + \cdots + B_\rho^{(n)}(\omega) + \cdots = 0 , \qquad (27)$$

and, hence, the matrix element is determined by the sum of the ω-independent parts only:

$$J_\rho = F_\rho^{(1)} + F_\rho^{(2)} + \cdots + F_\rho^{(n)} + \cdots . \qquad (28)$$

It follows that *the problem of restoring the independence of any physical amplitude on ω is reduced to the separation, and omission, of the "pure" ω-dependent terms.* We emphasize that this definition of the form factors, after separating out ω-dependent contributions, still leads to approximate form factors. However, our procedure guarantees that the form factors do not receive any spurious, non-physical contributions. Since these spurious, ω-dependent, contributions can be as large as the physical ones, in particular at high momentum transfer, our procedure is the most adequate when one wants to draw any conclusion on a particular model, or to compare various relativistic approaches among themselves. Moreover, it leads to an unambiguous determination of the form factors, independently of the orientation of the light front or of the choice of the component of the current which is used to calculate the amplitude.

The case of scalar bound states is trivial. The general decomposition of the current matrix element is given here by:

$$J_\rho = (p+p')_\rho F(Q^2) + \omega_\rho B_1(Q^2) . \tag{29}$$

The invariant functions F and B_1 depend on Q^2 only [4]. It is then obvious to extract the physical form factor $F(Q^2)$. One has just to contract J_ρ with ω^ρ:

$$F(Q^2) = \frac{J \cdot \omega}{2\omega \cdot p} . \tag{30}$$

This is the procedure used in standard LFD where $J \cdot \omega \to J^+$.

Elastic form factors of the deuteron

The electromagnetic form factors of the deuteron are defined by the following decomposition of the electromagnetic vertex:

$$\langle \lambda' | J_\rho | \lambda \rangle = e_\mu^{*\lambda'}(p') T_\rho^{\mu\nu} e_\nu^\lambda(p) . \tag{31}$$

Here $e_\mu^\lambda(p)$ is the deuteron polarization vector, p and p' are the initial and final deuteron momenta, λ and λ' are the corresponding helicities. The physical tensor $T_\rho^{\mu\nu}$ is now decomposed into a physical part $J_\rho^{\mu\nu}$ which gives the physical form factors of the deuteron (charge, magnetic and quadrupole), and the spurious contributions $B_\rho^{\mu\nu}$. These latter have the general form:

$$\begin{aligned} B_\rho^{\mu\nu} &= \frac{M^2}{2(\omega \cdot p)} \omega_\rho \left[B_1 g^{\mu\nu} + B_2 \frac{q^\mu q^\nu}{M^2} + B_3 M^2 \frac{\omega^\mu \omega^\nu}{(\omega \cdot p)^2} + B_4 \frac{q^\mu \omega^\nu - q^\nu \omega^\mu}{2\omega \cdot p} \right] \\ &+ B_5 P_\rho M^2 \frac{\omega^\mu \omega^\nu}{(\omega \cdot p)^2} + B_6 P_\rho \frac{q^\mu \omega^\nu - q^\nu \omega^\mu}{2\omega \cdot p} + B_7 M^2 \frac{g_\rho^\mu \omega^\nu + g_\rho^\nu \omega^\mu}{\omega \cdot p} \\ &+ B_8 q_\rho \frac{q^\mu \omega^\nu + q^\nu \omega^\mu}{2\omega \cdot p} . \end{aligned} \tag{32}$$

One can thus calculate the tensor $J_\rho^{\mu\nu}$ from the complete current $T_\rho^{\mu\nu}$, estimated in any model, by standard algebraic manipulations [6]. Note that contrarily to the pion form factor, one needs to extract the B's in order to get unambiguous determination of the physical form factors. The standard procedure which consists in calculating $\langle\lambda'|J^+|\lambda\rangle$ for various helicity states, in terms of the physical form factors only, assuming all B's to be zero, is not well defined in the sense that the physical form factors thus depend on the particular choice made for the four helicity states in order to calculate the three form factors [6]. Numerical results for the charge, magnetic and quadrupole form factors can be found in refs.[7, 4]. The link between meson-exchange currents associated with Z diagrams in non-relativistic approaches and CLFD has been done in ref.[8].

The nucleon form factors

Like in the case of spin 0, the dependence of the corresponding amplitude on the extra four-vector ω, for any approximate calculation, increases the number of independent terms in the decomposition of the amplitude [9]. We can write, for spin $1/2$ systems:

$$J_\rho = \bar{u}' \Gamma_\rho u,$$
$$\Gamma_\rho = F_1 \gamma_\rho + \frac{iF_2}{2M}\sigma_{\rho\nu}q^\nu$$
$$+ B_1\left(\frac{\not{\omega}}{\omega \cdot p} - \frac{1}{(1+\eta)M}\right)P_\rho + B_2\frac{M}{\omega \cdot p}\omega_\rho + B_3\frac{M^2}{(\omega \cdot p)^2}\not{\omega}\,\omega_\rho. \quad (33)$$

In the usual formulation of LFD on the plane $t+z=0$, the form factors of spin $1/2$ systems are found from the plus-component of the current [2], i.e., with our notation, from the contraction of J_ρ in eq.(33), with ω_ρ. *This contraction eliminates the contributions of $B_{2,3}$, but does not eliminate the term with B_1.* The form factors F_1' and F_2' deduced in this way are thus given by:

$$J \cdot \omega = \bar{u}'[F_1\not{\omega} + \frac{iF_2}{2M}\sigma_{\rho\nu}\omega^\rho q^\nu + 2B_1(\not{\omega} - \frac{\omega \cdot p}{(1+\eta)M})]u$$
$$\equiv \bar{u}'[F_1'\gamma_\rho + \frac{iF_2'}{2M}\sigma_{\rho\nu}q^\nu]u\,\omega^\rho. \quad (34)$$

where

$$F_1' = F_1 + \frac{2\eta B_1}{1+\eta}, \quad F_2' = F_2 + \frac{2}{1+\eta}B_1. \quad (35)$$

Equation (34) shows that the structure of $J \cdot \omega$ (or J^+) indeed coincides with the standard representation of the nucleon electromagnetic vertex. However, F_1' and F_2' in (34) are not the physical form factors, but their superposition (35) with the non-physical contribution B_1. It is therefore necessary, in order to draw any physical conclusion from model calculations, to explicitly extract the non-physical form factors. This procedure is detailed in ref.[9].

TOWARDS EXACT NON-PERTURBATIVE CALCULATIONS

Given the nice properties of CLFD we developed in the previous sections, one can now think of investigating non-perturbative calculations. To see how things go, one should however first check that standard perturbative observables are well reproduced within CLFD.

Perturbative calculations

Perturbative renormalization on LF

The first, text book, example of perturbative renormalization is the calculation of the electron self energy in QED. The general spin structure of the self energy in CLFD is very simple. It is given by:

$$\Sigma(p,\omega) = A_1(p^2) + B_1(p^2)\frac{\slashed{p}}{m} + C_1(p^2)\slashed{\omega}, \tag{36}$$

where m is the electron mass. The ω-dependent structures should not contribute to observables, like for instance the renormalized mass. We introduce therefore the amputated self energy $\tilde{\Sigma}(p)$ defined by:

$$\tilde{\Sigma}(p) = \Sigma(p,\omega) - C_1(p^2)\slashed{\omega} = A_1(p^2) + B_1(p^2)\frac{\slashed{p}}{m}. \tag{37}$$

The standard procedure of renormalization of Feynman diagrams relies on two counterterms: the mass counterterm δm^2 and the wave function renormalization proportional to Z_2. Alternatively we can define the renormalized self energy $\Sigma_R(p)$ as the part of $\tilde{\Sigma}(p)$ which is of second order in the variable $(\slashed{p}-m)$. Without loss of generality, we can rewrite $\tilde{\Sigma}(p)$ in the form:

$$\tilde{\Sigma}(p) = A_0 + (\slashed{p}-m)B_0 + \Sigma_R(p). \tag{38}$$

Here A_0, B_0 are constants (they do not depend on p^2), and $\Sigma_R(p)$ is the renormalized self-energy written as:

$$\Sigma_R(p) = (\slashed{p}-m)^2 \mathcal{M}(p), \tag{39}$$

where the matrix $\mathcal{M}(p)$ can be represented as:

$$\mathcal{M}(p) = a + (\slashed{p}+m)b. \tag{40}$$

The calculation of $\mathcal{M}(p)$ has been done in ref.[10]. It leads very easily to the same expression as the one obtained with the standard Feynman rules.

Leptonic decay width of J/Ψ

Kinematical relativistic corrections to the leptonic decay width of the J/Ψ can be calculated without any knowledge of the dynamical origin of the two-body wave function.

Dynamical corrections, on the other hand, correspond to relativistic corrections to the wave-function itself. These corrections can be written in the form:

$$\frac{\Gamma_1}{\Gamma_0} = R_K \left(1 - \frac{16\alpha_S}{3\pi} R_D \right) . \qquad (41)$$

where Γ_1 is the relativistic decay width to first order in α_S while Γ_0 is the non-relativistic limit in leading order in α_S given by the Van Royen-Weisskopf formula. The ratio R_K is simply the kinematical relativistic corrections. The dynamical corrections have been written in the form (41) in order to explicitly exhibit the non-relativistic limit to the radiative correction when the two-body bound state wave function is restricted to very small momenta. In that case, R_D should go to unity.

As shown in ref.[11], relativistic corrections are very large in the charmonium sector, and not negligible in the bottomonium sector. Kinematical relativistic corrections lead to a large reduction of the leptonic decay width. On the other hand, dynamical relativistic corrections, which correspond to relativistic corrections to the two-body bound state itself, lead to a sizeable reduction of the standard correction ($-\frac{16\alpha_s}{3\pi}$) found in the non-relativistic limit. This result is particularly important since it shows that an expansion in α_s becomes now meaningful. Indeed, the first order correction in α_s now amounts to about 10-15 % correction in case of the J/Ψ, as compared to 50% in the non relativistic limit.

Non-perturbative calculations in particular kinematical conditions

B semi-leptonic decays

The leptonic decay and semi-leptonic transition form factors of heavy mesons are a subject of particular interest for several reasons. In the framework of the standard model, they are directly proportional to matrix elements of the CKM matrix, and therefore can serve as constraints on their determination. This however implies that the uncertainties coming from the calculation of the hadronic transition amplitudes can be minimized as much as possible. This is indeed the case if one considers heavy mesons containing c or b quarks or antiquarks. In that case, and in leading order in $1/m$, where m is the mass of the heavy quark, several properties arise from the so-called heavy quark symmetry.

Such a symmetry, which implies various sum rules, can be used for instance to constraint the slope of the Isgur-Wise function $\xi(w)$ as $w \to 1$. It is now well known that such constraints can only be satisfied if the relevant matrix elements are described covariantly. In other words, one needs a coherent relativistic framework to describe the initial and final state, as well as the electroweak operator.

Several attempts have been made in the past to satisfy these constraints at least in $1/m$ order. Among them, the construction of covariant amplitudes using the Bakamjian-Thomas (BT) construction is the most widely used. We have shown in ref.[12] that in leading $1/m$ order, both LFD and BT approaches are identical.

Pion form factor

It is well known that QCD provides a $1/Q^2$ asymptotical behavior for the pion form factor. As it was shown above, the pion wave function is expressed in CLFD by two independent components. The first one leads to the usual non-relativistic wave function, while the second one is of purely relativistic origin and ω-dependent. We have shown that the asymptotical $1/Q^2$ behavior of the pion form factor is entirely determined by this last ω-dependent component of the pion wave function [4]. This implies in particular that one should not fit with the same component both the low-energy observables and the pion form factor at high Q^2 for instance.

Exact non-perturbative calculations

Non-perturbative calculations can be carried out in CLFD with a simple model involving two coupled scalar particles: a scalar "nucleon" N radiates scalar "pions" π [13]. In this simple example, the Fock space is restricted to N, $N\pi$ and $N\pi\pi$ states. Represented as a series of graphs in perturbation theory, it contains an infinite number of irreducible contributions to the self energy. These contributions diverge and require renormalization. At large value of the coupling constant this system cannot be solved perturbatively.

TABLE 1. Normalization of each Fock state component (one, two and three particles) as a function of the coupling constant α

α	1	3	10	30	100	1000
N_1	0.661	0.325	0.081	0.016	2. E-3	3. E-5
N_2	0.292	0.457	0.366	0.186	0.068	8. E-3
N_3	0.047	0.218	0.553	0.798	0.930	0.992

We have derived in ref.[13] the general equation to be solved in CLFD. We indicate in Table 1 the contribution of each Fock sector to the normalization of the "scalar" nucleon state. The extension of these studies to the implementation of higher Fock components is under study.

CONCLUSIONS

As it is already well known, LFD has many nice properties. This should not however hintered some of its limitations, and in particular the question of rotational invariance, and its consequences for the definition of the total angular momentum of a bound state system. We have briefly reviewed in this article how one can deal with these questions in an explicit covariant formulation of LFD.

The most interesting applications of CLFD are indeed for bound state systems: decomposition of the bound state system in Fock state components, and calculation of electromagnetic observables in a well defined and controlled manner. In the near future, new developments are to be expected in non-perturbative calculations, and among them the question of renormalization.

ACKNOWLEDGMENTS

I would like to thank my collaborators, F. Bissey, J. Carbonell, Th. Cousin, B. Desplanques, J.-J. Dugne and V. Karmanov for numerous discussions, comments and advises on the subject presented in this talk.

REFERENCES

1. Dirac, P.A.M., *Rev. Mod. Phys.* **21**, 392 (1949).
2. Brodsky, S.J., Pauli, H.C., and Pinsky, S.S., *Phys. Reports* **301**, 299 (1998).
3. Karmanov, V.A., *Zh. Eksp. Teor. Fiz.* **71**, 399 (1976) [transl. *JETP* **44**, 210 (1976)].
4. Carbonell, J., Desplanques, B., Karmanov, V.A., and Mathiot, J.-F., *Phys. Reports* **300**, 215 (1998).
5. Heinzl, Th., Krusche, St., and Werner, E., *Phys. Lett. B* **256**, 55 (1991).
6. Karmanov, V.A., and Smirnov, A.V., *Nucl. Phys.* **A546**, 691 (1992);
   Karmanov, V.A., and Smirnov, A.V., *Nucl. Phys.* **A575**, 520 (1994).
7. Carbonell, J., and Karmanov, V.A., *Nucl. Phys.* **A581**, 625 (1995).
8. Desplanques, B., Karmanov, V.A., and Mathiot, J.-F., *Nucl. Phys.* **A589**, 697 (1995).
9. Karmanov, V.A., and Mathiot, J.-F., *Nucl. Phys.* **A602**, 388 (1996).
10. Dugne, J.-J., Karmanov, V.A., and Mathiot, J.-F., *The charge and mass perturbative renormalization in explicitly covariant LFD*, hep-ph/0101156
11. Bissey, F., Dugne, J.-J., and Mathiot, J.-F., *Dynamical relativistic corrections to the leptonic decay width of heavy quarkonia*, to be published,
    Louise, S., Bissey, F., Dugne, J.-J., Mathiot, J.-F., *Phys. Lett. B* **472**, 357 (2000).
12. Bissey, F., and Mathiot, J.-F., *Eur. Phys. J.* **16**, 131 (2000).
13. Bernard, D., Cousin, Th., Karmanov, V.A., Mathiot, J.-F., *Nonperturbative renormalization in a scalar model within Light-Front Dynamics*, to be published

Search for a three-nucleon force

H.O. Meyer

Indiana University, Bloomington, IN 47405

Abstract: We discuss some basic aspects of searching for a three-nucleon force, and describe an experimental program at the Indiana Cooler to measure analyzing powers and spin correlation coefficients in *pd* scattering at 135 and 200 MeV proton energy. The experiment makes use of the PINTEX internal target facility, which was recently upgraded to provide a vector- or tensor-polarized deuterium target. Preliminary results for a number of polarization observables are in qualitative agreement with Faddeev calculations. Including a three-nucleon force in the calculation tends to alleviate the remaining discrepancies. Future plans for *pd* break-up measurements are discussed.

ABOUT THE THREE-NUCLEON FORCE

Basic Questions

Suppose you are given the task to study 'boobies'. Since such an assignment needs clarification, immediately the following questions come to mind: What do we mean by 'booby'? Do boobies exist (at least in principle)? What do we know about them? Has anybody ever seen one? How large are boobies? Where can they be found? How does one best study them?

Those of us who instead have taken on the job to study the 'three-nucleon force' do well by asking the same questions with respect to their subject of research. Some of these questions have an answer, some don't, but all are useful when trying to organize the research.

What do we mean by Three-Nucleon Force?

Let us define the three-nucleon force (3NF) that acts in a system of nucleons as that *part of the interaction that cannot be described by pair-wise interactions* (i.e., by the sum of the NN interaction for all pairs of nucleons in the system). This requires a potential that depends on three nucleons simultaneously. In principle, the definition includes higher order forces as well, but we simply ignore these. It is important to note that our definition refers to a 'description' and therefore has a meaning only in the framework of a particular model: what we mean by 3NF depends on the model.

What do we know about the Three-Nucleon Force?

Any theory of the strong interaction that is based on the exchange of virtual quanta leads to many-body forces, and thus, a 3NF is of fundamental significance and it *must* be present at least on *some* level.

The traditional approach to nuclear physics is based on interactions between *pairs* of nucleons (2NF). This assumption greatly helps to reduce the complexity of the theoretical task, but there is actually little justification for it, except for the fact that 2NF models are surprisingly successful in explaining (at least qualitatively) a huge body of data on nuclear structure, scattering and reactions. Because of that we know that *3N forces are relatively weak*.

The triton binding energy (8.482 MeV) is underestimated by 2N calculations by about 800 keV [1], while different NN potentials yield a spread of calculated binding energies of ±190 keV. These potentials differ in the treatment of locality, the tensor force, and charge independence, but they are all adjusted to fit the body of NN observables and the properties of the deuteron. Inclusion of a 3NF makes it possible to reproduce the measured binding energy. Usually, this fact is accepted as *experimental evidence for a 3NF*. On the other hand, the question whether there is no reasonable *other* way to explain the under-binding is still being investigated [2].

What would we accept as Evidence for a Three-Nucleon Force?

By now it should be clear that the 3NF is *not* an experimental observable! We can only learn about it by comparing an experiment with a 2N model that tells us how nature would behave (in the framework of that model) *without* a 3NF. Obviously, the uncertainties of such a calculation must be less than the effect caused by the presence of a 3NF. Calculations that most probably satisfy this requirement are now possible for *nd* scattering and reactions (see below: Faddeev calculations).

Obviously, discrepancies between an experiment and the corresponding 2N calculation *could* be due to the missing 3N potential. However, to be sure we would have to demonstrate that addition of a 3NF reduces or removes these discrepancies. We would expect that such a reduction is observed for all available observables: the larger their number, the stronger the argument.

Where to look for 3NF Evidence?

Because of the availability of exact calculations it seems clear that the most promising hunting ground for 3NF evidence are reactions in the three-nucleon system.

At energies below 50-100 MeV Faddeev calculations reproduce the existing data rather well (see, e.g., ref. [3]), even though there are some persistent, unexplained discrepancies, such as the 'A_y puzzle' [4]. At these energies, the calculated 3NF effects scale with the triton binding energy and become very small when the latter is fixed to the experimental value.

At energies beyond 100 MeV the calculated 3NF effects are growing due to the increased role of nucleon excited states. In addition, the sensitivity to the Coulomb force is much reduced, making *pd* data admissible for comparison with *nd* calculations that ignore the Coulomb interaction. For the *pd* elastic scattering experiment reported here, we have chosen the bombarding energies 135 and 200 MeV. Measurements are carried out at *two* energies in order to explore the energy dependence.

The formulation of 3N potentials is so complex that it is impossible to develop an intuitive understanding of what observables would be most sensitive to it. The best approach then is to investigate as many polarization observables as feasible. This will test many different combinations of amplitudes, and offers the chance that small 3NF-sensitive amplitudes might interfere with larger amplitudes, thus amplifying the 3NF effects.

Faddeev Calculations

Status and Limitations

Already 40 years ago, Faddeev showed a way to *exactly* solve the Schrödinger equation for three nucleons with a general underlying NN potential [5]. However, it was only recently that the necessary computing power became available to solve the problem numerically. The main contribution to this effort is the work of the Bochum-Cracow group, which is summarized in a review article [4]. Faddeev calculations of *nd* scattering and break-up ($nd \rightarrow nnp$) observables are now possible for neutron bombarding energies up to 200 MeV. In order to extend the calculations beyond 200 MeV, accurate NN forces would have to be developed that allow for pion production and absorption.

It has been shown that Faddeev calculations depend only weakly on which of the commonly used NN potential is employed. However, one has to keep in mind that the NN interaction has a component, the so-called off-shell part that cannot be tested in experiments with two nucleons, but contributes in many-body systems. This off-shell part is linked to 3N forces and would disappear when NN and 3N forces were derived consistently [6].

Faddeev calculations neglect the Coulomb force, but are often compared with *pd* scattering and break-up experiments. This is usually justified at proton energies beyond 100 MeV where the Coulomb force manifests itself only at extreme forward angles in elastic scattering, or in reactions with two protons at low relative energy in the exit channel.

Three-Nucleon Potentials

Three-nucleon potentials have been constructed by taking into account multiple exchanges of mesons between all three nucleons. One of these potentials is the so-called Tucson-Melbourne three-nucleon force (TM-3NF) [7]. Until recently, this 3NF has been the standard tool to assess 3NF effects in Faddeev calculations. It is now known that the original TM-3NF is in conflict with chiral symmetry [8], and a new TM-3NF that is free of this problem has been tested in Faddeev calculations [9]. An alternative 3N force has been developed by the Urbana group [10]. It features a phenomenological short-range part that is adjusted to the binding energies of light nuclei This 'Urbana-3NF' has recently been incorporated into the code of the Bochum-Cracow group, making it possible to compare different approaches at constructing a 3NF [9].

EXPERIMENTS WITH POLARIZED BEAM AND TARGET

Observables

Description of Proton and Deuteron Polarization

The polarization of a (spin ½) proton can be described by a vector $\vec{Q}$. The magnitude Q of the vector - the beam polarization - is the difference of the sub-state populations (normalized to 1) along or opposite the direction of the vector.

The situation is more complicated in the case of deuterons (spin 1) that have three magnetic substates $m_d = -1, 0, +1$ with respect to some quantization axis. If N_{md} are the normalized populations of the three states ($N_+ + N_0 + N_- = 1$), we define $P_\xi = N_+ - N_-$ as the *vector* polarization of the target, while $P_{\xi\xi} = 1 - 3N_0$ is called the *tensor* polarization. This description of the tensor polarization assumes that the ensemble is rotationally symmetric around the quantization axis (which is then called the 'spin alignment axis'). In general, however, three independent 'tensor moments' are needed and their value depends on the orientation of the spin alignment axis relative to the reaction plane. Additional complications arise when there is no reaction plane (such as in break-up reactions). More on this topic can be found, e.g., in refs. [11,12].

Cross Section with Polarized Beam and Target

The polarizations of the proton beam and the deuteron target are described in a right-handed Cartesian coordinate system, which is fixed in space and centered at the target (we take the x-axis to the left, the y-axis up, and the z-axis along the beam). The scattering plane contains the z-axis and is rotated by an angle φ relative to the positive x-axis.

The cross section of a reaction with *polarized* collision partners acquires a dependence on the azimuth φ of the reaction plane. As an example, we inspect the special case of the cross section that is observed when the proton, as well as the deuteron vector polarization is in the direction of the y-axis (up), and the (positive) tensor polarization is given by a vertical spin alignment axis. In this case,

$$\sigma = \sigma_0 \left(1 + Q A_y^p \cos\varphi + P_\xi A_y^d \cos\varphi - \tfrac{1}{4} P_{\xi\xi} [\{A_{xx} - A_{yy}\} \cos 2\varphi + A_{zz}]\right.$$

$$+ \tfrac{3}{4} P_\xi Q [C_{x,x} + C_{y,y} - \{C_{x,x} - C_{y,y}\} \cos 2\varphi]$$

$$- \tfrac{1}{4} P_{\xi\xi} Q [C_{zz,y} + C_{xx,y} - C_{yy,y}] \cos\varphi$$

$$\left. + \tfrac{1}{2} P_{\xi\xi} Q [C_{xy,x} + \tfrac{1}{2} \{C_{xx,y} - C_{yy,y}\}] \sin\varphi \sin 2\varphi\right).$$

In this expression, σ_0 is the unpolarized differential cross section and the A's and C's are the polarization observables to be measured. They depend on the scattering angle θ. Even though all terms in the equation contribute simultaneously, it is still possible to determine them individually by taking data with different signs of Q, P_ξ and $P_{\xi\xi}$, with different orientations of the target guide field and the beam polarization, and by making use of the known φ dependence of a given term. This process is fairly complicated but straightforward. More about the analysis of experiments with polarized collision partners can be found in ref. [12].

List of Polarization Observables in pd Scattering

In the following, we list all 17 analyzing powers and spin correlation coefficients that can be measured in *pd* elastic scattering. All of these observables are a function of bombarding energy and scattering angle θ.

- Proton analyzing power $\qquad\qquad A_y^p$
- Deuteron vector analyzing power $\quad A_y^d$
- Deuteron tensor analyzing powers $\;\; A_{xz}\;\;(A_{xx}-A_{yy})\;\;A_{zz}$
- Vector correlation coefficients $\qquad (C_{x,x}+C_{y,y})\;\;(C_{x,x}-C_{y,y})\;\;C_{z,x}\;\;C_{x,z}\;\;C_{z,z}$
- Tensor correlation coefficients $\qquad C_{xy,x}\;\;C_{yz,x}\;\;C_{xz,y}\;\;(C_{xx,y}-C_{yy,y})\;\;C_{zz,y}\;\;C_{xy,z}\;\;C_{yz,z}$

In the experiment described in the following, all of these observables have been measured except for the last two tensor correlation coefficients. The first subscript of the spin correlation coefficients $C_{m,n}$ refers to the deuteron polarization and the second to that of the proton. We list the observables in the so-called Cartesian notation but often the tensor analyzing powers are quoted in the spherical tensor notation:

$$iT_{11} = \frac{\sqrt{3}}{2} A_y^d \;;\quad T_{20} = \frac{1}{\sqrt{2}} A_{zz} \;;\quad T_{21} = -\frac{1}{\sqrt{3}} A_{xz} \;;\quad T_{22} = \frac{1}{2\sqrt{3}}(A_{xx} - A_{yy}).$$

Note that if the final momenta and the incident beam direction are not in one plane (as is possible for *pd* break-up), some additional observables that are normally forbidden by parity conservation can be non-zero.

Experiments at the Indiana Cooler

The Beam

A layout of the Indiana University Cooler facility (IUCF) is shown in Fig. 1. The beam from a pulsed source for negative, polarized H⁻ or D⁻ ions (CIPIOS) is accelerated to 7 MeV in a radio-frequency quadrupole cavity (RFQ) and a drift-tube linac (DTL), and strip-injected into the injector synchrotron (CIS). After accelerating the beam stored in CIS to 100-200 MeV, it is transferred to the Cooler in a single bunch. Repeating this process, in excess of 1 mA of stored beam is accumulated in the Cooler in a few seconds. Subsequently, the Cooler beam may be accelerated up to 500 MeV for protons, 270 MeV for deuterons. It is also possible to flip the polarization direction using a resonance transition method. The direction of the polarization at the target is determined by the spin-closed orbit of the storage ring, which can be made to point in a direction other than vertical, using precession solenoids (so-called Siberian snakes).

Both, CIS and CIPIOS are recent additions to the Cooler facility. With these new devices the stored beam intensity has been improved by about a factor of five, and beam accumulation in the ring has become much more efficient than previously, when the Cyclotron served as injector.

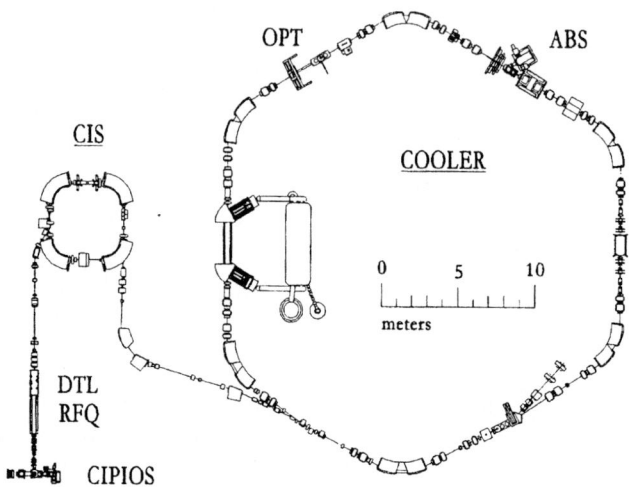

FIGURE 1. Layout of the IUCF Cooler facility. The components are explained in the text. The polarized internal target is located in straight section A.

The Target

The PINTEX [13] internal target is produced by a source of polarized atoms (ABS in Fig. 1). In this device first a beam of D atoms is formed, which is then manipulated as follows.

Deuterium atoms consist of an electron (spin ½) and a deuteron (spin 1). With respect to a quantization axis (e.g., the atomic beam axis) there are six combinations of magnetic quantum numbers ('spin states'). The atomic beam becomes polarized by eliminating the populations of selected spin states or by moving them from one state to another. Sextupole magnets are used to eliminate atoms as follows: the radial field gradient in such a magnet causes a radial force on the magnetic moment of the atoms; the atoms with their moment (and thus the electron spin) along the beam are focused while those with the moment opposite are defocused and rejected. In a magnetic field the six spin states have different energies and one can use an RF field to induce transitions of the atoms from one state to another. By applying several such a transitions in conjunction with passing the atomic beam through sextupole magnets, it is possible to obtain a beam of atoms that populate only certain desired spin states in such a way that either pure vector or pure tensor polarization results. The ABS in use at the Indiana Cooler has been constructed at the University of Wisconsin [14]. More information on the preparation of polarized atoms can be found in ref. [11].

To prepare an internal polarized target for use in a storage ring, the atomic beam is aimed into a 'fill' tube (b in Fig. 2), which is part of a 'storage cell'. The main part of the storage cell [15] is a thin-walled aluminum tube through which the stored beam passes (a in Fig. 2). The purpose of the cell is to enhance the target density.

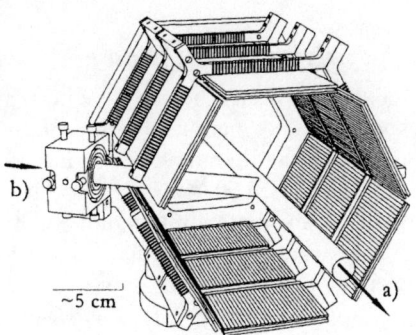

The spin alignment axis of the atomic beam (originally the ABS axis) adiabatically adjusts to the local magnetic field direction as the atoms travel into the storage cell. Helmholtz coils mounted on the outside of the target box produce the 'guide field' (about 1mT) at the location of the storage cell. In the course of an experiment, this direction is changed typically every two seconds, pointing up, down, left, right, and along or opposite to the beam direction.

FIGURE 2. The storage cell target and the surrounding array of silicon micro-strip detectors for detecting the low-energy recoils. Also indicated are the stored polarized proton beam (a), and the polarized atomic deuteron beam that generates the target (b).

Note, that changing the sign of the guide field changes the sign of the vector polarization but not of the tensor polarization. In order to change the sign of the tensor polarization, a different mix of spin states has to be prepared by the ABS. This is done during the experiment by remotely changing the operating parameters of the RF transitions, typically every 30 s. For certain observables it is necessary to incline the guide field 45° towards the proton beam axis in order to generate the appropriate tensor moments.

Detector setup

The detector setup that is used in experiments at the polarized target station is quite simple and consists of two main parts. In the forward direction, covering a cone of about 45° opening angle, a stack of scintillators and wire chambers measures the energy and the direction of outgoing charged particles. Protons of up to 200 MeV are stopped in this detector. On the other hand, recoiling low-energy charged particles emerging sideways through the thin walls of the target cell are registered by an array of silicon micro-strip detectors. The recoil detector assembly is shown in Fig. 2.

Despite the simplicity of the principle of detection, one faces the difficulty that in the backward direction the *pd* scattering cross section is as low as 100 μb/sr while the total reaction cross section (mainly break-up) is about 80 mb. In order to cleanly extract the events of interest one has to pull many tricks, making use of the coplanarity of elastic scattering, the energy deposited in all detectors and the time-of-flight between scintillators in the forward stack.

P+D ELASTIC SCATTERING: 100 – 200 MEV

What do we already know?

Polarization Data between 100 and 200 MeV Proton Energy

Polarized proton beams have been available for some 40 years and a fair number of experiments to measure the proton analyzing power A_y^p in pd elastic scattering are reported in the literature. These include partial angular distributions at 120 MeV [16], 146 MeV [17], 150 MeV [18], 155 MeV [19], 190 MeV [18] and 200 MeV [16,20,21] as well as an excitation function at a fixed deuteron lab angle of 42.6° [22].

The deuteron vector and tensor analyzing powers have been measured with a polarized deuteron beam on a hydrogen target at corresponding proton energy of 198 MeV (deuteron energy T_d = 395 MeV) at Saclay [23] with rather large errors. More recently, a similar experiment at Riken [24,25] at a proton energy of 135 MeV (T_d = 270 MeV) has yielded complete and accurate angular distributions of the vector and tensor analyzing powers A_y^d, A_{xz}, A_{xx} and A_{yy}. Also from Saclay, the 180° tensor analyzing power T_{20} is reported at 150 and 200 MeV (T_d = 300, 400 MeV) [26]. Finally, a limited angular distribution of the deuteron vector analyzing power has been measured at 200 MeV proton beam in the Indiana Cooler using an optically pumped polarized target [21]. In the same experiment the vector spin correlation coefficient $C_{y,y}$ was measured at 200 MeV. Except for an exploratory measurements by the PINTEX group and a measurement at lower energy [27], these are so far the only pd scattering spin correlation data in existence below 200 MeV.

Confrontation with Faddeev calculations

Faddeev calculations consistently underestimate the nd (or, pd) scattering cross section in the minimum of the angular distribution (around θ_{cms}=100°), and overestimate the proton analyzing power A_y^p and the deuteron vector analyzing power A_y^d in that same angular range. Inclusion of a three-nucleon force in the calculation alleviates this discrepancy but does not reproduce the data quantitatively.

What are we going to learn from the Present Experiment?

The data taking for the pd scattering experiment described here has been concluded. Eventually, we will present reasonably complete angular distributions of the 5 analyzing powers, the 5 vector spin correlation coefficients and 5 of the tensor spin correlation coefficients at 135 and 200 MeV proton bombarding energy. At this time, preliminary data for 10 observables are available (some of those have been shown at the meeting in Prague).

Qualitatively we can summarize a comparison of our preliminary pd elastic scattering results with the most recent Faddeev calculations [9] as follows. The general features of all observables are represented rather well by the 2N calculations. Overall, the discrepancies between experiment and 2N calculations are most pronounced for center-of-mass angles beyond 70° and forward of 30° (the latter presumably because of Coulomb effects). The observables with the largest discrepancies are the tensor analyzing powers. Inclusion of either the Tucson-Melbourne 3NF or the Argonne IX 3NF often (but not always) reduces the discrepancy.

P+D BREAK-UP

Break-up versus Elastic Scattering

Let us now turn to the pd break-up process ($pd \rightarrow ppn$). In the search for 3NF effects, there is a clear advantage of the break-up reaction over elastic scattering, since it allows one to select final-state configurations, which differ in their sensitivity to 3N forces. For instance, early 3NF searches compared the so-called star configuration (the three nucleons emerge from the reaction volume at 120^0 relative to each other) with the collinear configuration (one nucleon is at rest, shielding the other two from each other). While these studies failed to reveal the simple picture one hoped for, it is still true that dynamic constraints are an important tool. More recently for instance, another final-state configuration, called 'FSI', has attracted attention in which two of the outgoing nucleons are at relative rest. It turns out that for this configuration the TM-3NF has a very large effect, in particular for spin correlation observables.

3NF-Sensitive Observables?

It has been argued that three-body potentials involve spin operators of a type not allowed for ordinary two-body NN interactions [28]. Based on this, a class of polarization observables were identified that may have enhanced sensitivity to a three-body force. One of these observables is the longitudinal analyzing power A_z. In reactions with two outgoing particles this observable must vanish by parity conservation, but it is unconstrained when there are three or more outgoing particles with lab momenta that are *not* in a common plane. That fact that A_z may well be large in reactions with a three-body final state has recently been established in a measurement of the longitudinal analyzing power in the reaction $pp \rightarrow pp\pi^0$ [29].

An attempt to measure A_z in $pd \rightarrow ppn$ with a longitudinally polarized 9 MeV proton beam [30] resulted in an upper limit of 0.003. This is in accordance with a Faddeev calculation that yields values of 0.002 or less. Somewhat surprisingly, Faddeev predictions of A_z at a higher energy (T_p=135 MeV) [31] predict values that are a factor of 10^3 larger than at 9 MeV. This opens the exciting possibility of an experiment that addresses the 3NF and that is motivated by an inspired theoretical argument.

Future Plans

When using the PINTEX detector system to study the break-up reaction, one makes use of the fact that the scintillators in the forward detector stack are segmented arrays. It is thus possible to trigger on the two coincident, outgoing protons from the $pd \rightarrow ppn$ reaction, provided they fall within the $45°$ acceptance of the forward detector. The acceptance of the detector can be improved significantly when one employs reverse kinematics, bombarding a proton *target* with a deuteron *beam*. In this case, the present detector system covers about 66% of the total *pd* break-up phase space. Thus, future break-up experiments at the Cooler will make use of the polarized deuteron beam that has recently been developed. A project to supplement the forward detector with a BGO-micro-strip barrel is under way. This detector would cover angles backward of 45° and would give practically full coverage of the three-body phase space.

In the time until the end of 2002, a measurement of A_z will take place. This experiment will be carried out with a $T_d = 270$ MeV deuteron beam (corresponding to 135 MeV equivalent proton energy) on a longitudinally polarized proton target. When a polarized deuteron beam becomes routinely available, we will make an attempt to measure spin correlation coefficients for break-up in selected regions of phase space.

At the end of 2002 the Indiana Cooler will be shut down.

CONCLUSIONS

The PINTEX facility during the final 1½ years of operation of the Indiana Cooler will be devoted to providing detailed and fairly complete empirical information on *pd* elastic scattering and break-up and the energy dependence of these processes between 100 and 200 MeV. This effort will produce a broad and diverse database documenting the systematics of the difference between nature and 2NF Faddeev calculations.

We already know that the inclusion of any of the available 3NF representations tends to move calculation in the right direction for most of the currently available observables. This might be taken as an indication that we are on the right track, but much more theoretical work is needed to *quantitatively* explain the disparity between experiment and 2NF calculations. In the search for possible remedies, theorists will have to study refinements to the 3N force (in particular its spin dependence), the importance of relativity, the role of the off-shell part of the NN interaction, etc. Here, one can expect that chiral perturbation theory will serve as a guide in formulating the 2N and 3N interactions consistently [32]. It may also be worthwhile to search for a *phenomenological* three-nucleon potential that would explain the data, much along the lines of the phenomenological potentials that we employ to describe the NN interaction.

This brings us back to the boobies. It turns out that ornithologists know a great deal about them [33]. Fig. 3 shows a picture of one of several species of boobies. Let us hope that before long nuclear physicists will have an equally clear picture of the three-nucleon force.

FIGURE 3. Sula sula rubripes (red-footed booby)

The experimental work described here has been made possible by the hard work of many members of the PINTEX group at IUCF [13]. The spokes-person of the *pd* elastic scattering experiment is B. v.Przewoski (przewoski@iucf.indiana.edu). The work has been supported by the National Science Foundation and by the Department of Energy. We are also grateful to our colleagues W. Glőckle (Bochum) and H. Witała (Cracow) for their theoretical input, and their support of three-nucleon experiments at the Indiana Cooler.

REFERENCES

1. Friar, J.L., Payne, G.L. , Stoks, V.G.J., and de Swart, J.J., *Phys. Lett. B* **311**, 4 (1993)
2. Doleschall, P., and Borbely, I., *Phys. Rev. C* **62**, 054004 (2000)
3. Knutson, L.D., *Nucl. Phys.* **A631**, 9c (1998)
4. Glőckle, W., Witała, H., Hűber, D., Kamada, H., and Golak, J., *Phys. Rep.* **274**, 107 (1996)
5. Faddeev, L.D., *Sov. Phys. JETP* **12**, 1014 (1961)
6. Polyzou, W.N., and Glőckle, W., *Few Body Systems* **9**, 97 (1997)
7. Coon, S.A., Scadron, M.D., McNamee, P.C., Barrett, B.R., Blatt, D.W.E., and McKellar, B.H.J., *Nucl. Phys.* **A317**, 242 (1979)
8. Friar, J.L., Hűber, D., and van Kolck, U., *Phys. Rev. C* **59**, 53 (1999)
9. Witala, H. *et al.*, *Phys. Rev. C* **63**, 024007 (2001)
10. Carlson, J., *et al.*, *Nucl. Phys.* **A401**, 59 (1983)
11. Haeberli, W., *Ann. Rev. Nucl. Sci.* **17**, 373 (1967)
12. Ohlsen, G.G., *Rep. Prog. Phys.* **35**, 717 (1972)
13. PINTEX: Polarized INternal Target EXperiments http://www.iucf.indiana.edu/experiments/PINTEX/pintex.html
14. Wise, T., Roberts, A.D., and Haeberli, W., *Nucl. Instr. Meth.* **A336**, 410 (1993)
15. Haeberli, W., in *Proc. Workshop on Nuclear Physics with Stored Cooled Beams*, edited by P. Schwandt and H.O. Meyer, AIP Conference Proceedings **128**, New York, 1985, pp.251-267
16. Wells, S.P. *et al.*, *Nucl. Instr. Meth. A* **325**, 205 (1993)
17. Postma, H., and Wilson, R., *Phys. Rev.* **121**, 1129 (1961)
18. Bieber, R., *et al.*, *Phys. Rev. Lett.* **84**, 606 (2000)
19. Kuroda, K., Michalowicz, A., and Poulet, M., *Nucl. Phys.* **88**, 33 (1966)
20. Adelberger, R.E., and Brown, C.N., *Phys. Rev.* **D5**, 2139 (1972)
21. Cadman, R.V., *et al.*, *Phys. Rev. Lett.* **86**, 967 (2001)
22. Stephenson, E.J., *et al.*, *Phys. Rev. C* **60**, 061001 (1999)
23. Garçon, M., *et al.*, *Nucl. Phys.* **A458**, 287 (1986)
24. Sakamoto, N., *et al.*, *Phys. Lett. B* **367**, 60 (1996)
25. Sakai, H., *et al.*, *Phys. Rev. Lett.* **84**, 5288 (2000)
26. Arvieux, J., *et al.*, *Phys. Rev. Lett.* **50**, 19 (1983)
27. Chauvin, J., Garetta, D., and Fruneau, M., *Nucl. Phys.* **A247**, 335 (1975)
28. Knutson, L.D., *Phys. Rev. Lett.* **23**, 3062 (1994)
29. Meyer, H.O., *et al.*, *Phys. Lett. B* **480**, 7 (2000)
30. George, E.A., *et al.*, *Phys. Rev. C* **54**, 1523 (1996)
31. Witała, H., private communication
32. van Kolck, U., *Phys. Rev. C* **49**, 2923 (1994)
33. Alsop, F.J., *Birds of North America*, DK Publishing Inc., New York, 2001, pp. 69-73; http://midway.fws.gov/wildlife/rfbo.html

Recent topics in hypernuclear spectroscopy

Toshio Motoba

Laboratory of Physics, Osaka Electro-Communication University, Nayagawa 572-8530, Japan[1]
Institute for Nuclear Theory, University of Washington, Seattle, WA 98195-1550, USA

Abstract. Recent progress of hypernuclear spectroscopy is discussed in connection with the properties of ΛN spin-dependent interactions. The results of structure calculations using ΛN meson theoretical potentials are presented. Novel characteristics of the hypernuclear photoproduction are also discussed by showing typical predictions.

INTRODUCTION

One of the goals of hypernuclear physics is to study the properties of baryon-baryon interactions within the octet family in a wider and unified perspective. In contrast to the nucleon-nucleon scattering, however, the elementary hyperon-nucleon (Y-N) and hyperon-hyperon (Y-Y) scattering data are very scarce and, in addition, it is also hard to expect such experiments in near future due to the practical difficulty in making use of hyperon beams strong enough for its purpose. Thus hypernuclei and exotic atom phenomena play important role in providing us with strong and weak interactions of baryons (hadrons) having strangeness. On the basis of the knowledge on ordinary nuclear structures and reactions, hypernuclear structural analyses are quite important to get an insight into the Y-N and Y-Y interaction properties: There might be several bridges with both directions between hypernuclear phenomena and basic properties of the meson theoretical potentials.

Furthermore, in view of the baryon-many body problems, hypernuclear systems have several novel features: Participation of hyperon(s) gives rise to more important role than what is meant by the term "impurity": As a hyperon is free from the nucleon Pauli principle, the coupling to the nuclear motion is rather simplified and it can probe deeply inside the nucleus which is hardly known only through the nucleon itself. One of the examples is the appearance of the "supersymmetric states" predicted many years ago typically in $^{9}_{\Lambda}$Be [1, 2]. These states have been also called "genuine hypernuclear states" in the microscopic cluster model study [3], since such states cannot be realized in the ordinary nuclei. In fact the existence has been confirmed recently in the (π^+, K^+) reaction experiment (E336) done at KEK by using much better energy resolution [4, 5, 6]. As there are several hyperons with different properties, we have many kinds of exotic systems which should provide us with opportunities for the study of many-body dynamics.

---

[1] Permanent address. E-mail: motoba@isc.osakac.ac.jp

Following the Brookhaven (π^+, K^+) experiment [7], an important progress has been achieved with the superconducting Kaon spectrometer [4, 5, 6]. In addition to the expected large peaks, they disclosed interesting side peaks which are quite meaningful in view of the core-excitation. On the other hand, higher-resolution experimental data through the γ-ray measurements appeared quite recently, giving high precision level energies of some hypernuclei [8, 9, 10]. Furthermore people are planning to perform another experiments to measure more γ-rays systematically. We also expect new data will come soon from another spectroscopy using $(e, e'K^+)$ reaction at JLab.

Thus hypernuclear study is now entering into a new stage as far as the Λ-hypernuclei are concerned. It is therefore timely that Rijken *et al.* [11] proposed new meson theoretical potentials in which the unknown parameters are allowed to change systematically so as to see the effect on hypernuclear structure outputs. Thus detailed comparison of hypernuclear structures and basic properties of realistic ΛN potentials becomes more and more meaningful in understanding strangeness many-body systems. In this report we will focus our attention to spin-dependent interactions and we will also discuss typical topics in Λ-hypernuclear study.

NUCLEAR SHRINKAGE DUE TO Λ PARTICIPATION

For light *p*-shell hypernuclei, the microscopic $\alpha + x + \Lambda$ cluster model ($x = N$, d, ^{3}H, ^{3}He, α) has been extensively applied to predict various observables [3]. One of the interesting predictions was that the shrinkage of nuclear core ($\alpha + x$) should be sizable: The Λ-addition gives rise to 10–18% contraction of the $\alpha - x$ distance, so that hypernuclear B(E2) values are reduced to be about half of the corresponding nuclear B(E2) values. This remarkable effect was called the *glue-like role of* Λ. The clearest way to confirm this effect would be the measurement of γ-transition *rates* in $^7_\Lambda$Li, since the core-transition rate in ^{6}Li has been known experimentally from the Coulomb excitation ^{6}Li(E2; $1^+ \to 3^+$). This was the original motivation of the experiment which has been successfully carried out at KEK recently by H. Tamura *et al.* through the ^{7}Li$(\pi^+, K^+\gamma)^7_\Lambda$Li coincidence experiment [8].

Just before the experimental result came, the theoretical predictions for $^7_\Lambda$Li had been updated by applying the $^5_\Lambda$He+p+n model [12] in which proton and neutron are allowed to move freely rather than the "deuteron" model assumed previously [3]. All the rearrangement channels are taken into account together with the Gaussian basis, making the results more reliable in view of the numerical accuracy. The calculation shows such essentially new results that only the relative density distribution $\rho(R_{core} - (np)_{CM})$ between the core (α) and the center-of-mass of the $n - p$ pair is changed remarkably when the Λ particle participates, while the $n - p$ relative density distribution $\rho(r_{n-p})$, i.e. the internal motion of the $n - p$ pair, remains almost unchanged. Thus the angle part of the E2 operator is not affected by the contraction of core-(np) distance, so that the B(E2) value depends on the distance only.

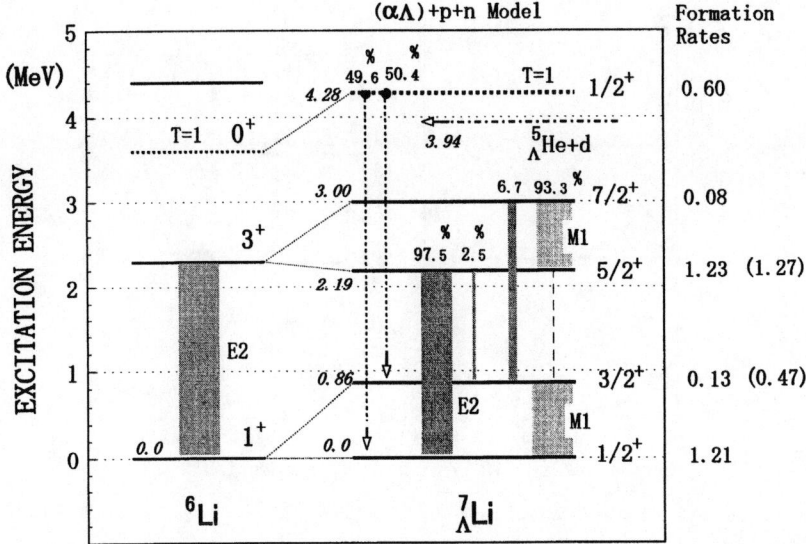

FIGURE 1. Theoretical γ-decay scheme of $^{7}_{\Lambda}$Li. On the extreme right the population rates are shown of the states calculated in combination with the calculated (π^+, K^+) cross sections.

In order to see the amount of shrinkage due to the glue-like role of Λ, we propose to check the ratio defined as

$$\Gamma_B \equiv \frac{B(E2; {}^{7}_{\Lambda}\text{Li}, 5/2^+ \to 1/2^+)}{(7/9)B(E2; {}^{6}\text{Li}, 3^+ \to 1^+)} = \frac{B(E2; 3_c^+ \to 1_c^+)_H}{B(E2; {}^{6}\text{Li}, 3^+ \to 1^+)} \quad (1)$$

Here the factor of 7/9 in the denominator takes account of the branching relation between the hypernuclear $B(E2)$ and the "core transition", and the subscript c in the right hand part denotes the transition occurred between core states in $^{7}_{\Lambda}$Li as deduced from the result mentioned above.

Before deducing the size contraction, we show the theoretical γ decay scheme of $^{7}_{\Lambda}$Li in Fig. 1. The level energies and branching ratios were all calculated before the experiment. The most important results concerned here are the theoretical branching ratios: the $5/2^+$ decays 97.5% to $1/2_{gs}$(E2) and 2.5% to $3/2^+$(M1), while the $7/2^+$ state decays 93.3% to $5/2^+$(M1) and 6.7% to $3/2^+$, as clearly shown in Fig. 1. Thus the lifetime of the $5/2^+$ state is determined by the E2 transition to the ground state, and that of the $3/2^+$ state by the M1 transition. In fact the experimental γ-ray yields are related to these transitions.

In order to make a realistic comparison with the γ-ray yields and lifetime data of the $^{7}\text{Li}(\pi^+, K^+\gamma)^{7}_{\Lambda}\text{Li}$ experiment, we combine γ-ray transition probabilities with the (π^+, K^+) cross sections (cf. Fig. 2(left)). The cross sections integrated over $\theta_K = 0° - 15°$ give: $\bar{\sigma}(J) = 1.21\mu\text{b}(1/2_{gs}^+), 0.13\mu\text{b}(3/2^+), 1.23\mu\text{b}(5/2^+)$, and $0.60\mu\text{b}(1/2_{T=1}^+)$. It is quite interesting that we can reproduce the experimental yield ratio between the E2($5/2^+ \to 1/2_{gs}^+$) and M1($3/2^+ \to 1/2_{gs}^+$) transitions (161±23 : 148±31) [8] only when we take γ-cascades from the $1/2_{T=1}^+$ state down to $3/2^+$ into account. The theo-

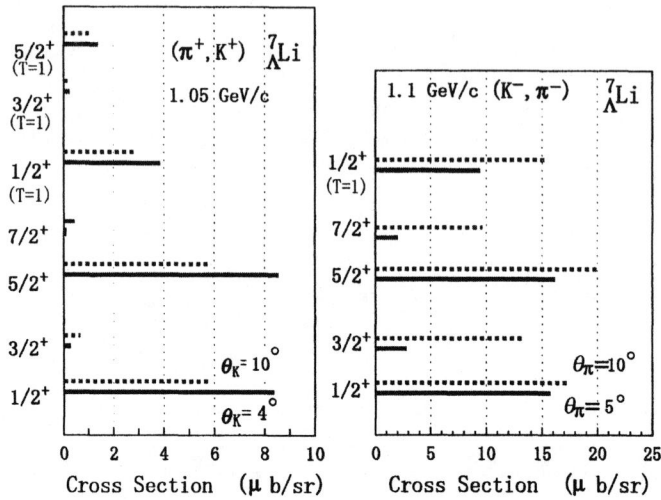

FIGURE 2. Calculated cross sections for the (π^+, K^+) and (K^-, π^-) reactions leading to $^7_\Lambda$Li. The former values (*left*) were used to estimate the level populations. The latter (*right*) demonstrates the ability of spin-flip excitation of $3/2^+$ and $7/2^+$.

TABLE 1. Predictions of electromagnetic properties of $^7_\Lambda$Li and the experiment.

	Dalitz-Gal (1978) [13] Shell model	Motoba et al. (1983) [3] ^{4}He+d+Λ	Hiyama et al. (1998) [12] $^5_\Lambda$He+p+n	Tamura et al. (1998) [8] Exp. E419
$B(M1; 3/2^+ \to 1/2^+)$ [μ_N^2]	0.364	0.352	0.322	–
$B(E2; 5/2^+ \to 1/2^+)$ [e^2fm^4]	8.6	2.46	2.42	4.1 ±1.1
$B(E2; 5/2^+ \to 3/2^+)$ [e^2fm^4]	3.1	0.40	0.74	–
$B(E2; 5/2^+ \to \text{sum})/B_c{}^*$	1†	0.44	0.33	–
Γ_B as of Eq.(1)	–	0.49	0.32	–
$R_{c-d}(^7_\Lambda\text{Li}) / R_{\alpha-d}(^6\text{Li})$	–	0.83	0.75	0.87
Lifetime $\tau(5/2^+)$ [ps]	–	6.56	6.67	5.2±1.4

* B_c denotes $B(E2; ^6\text{Li}, 3^+ \to 1^+)$ in ^{6}Li
† In Ref. [13] this ratio is assumed to be 1

retical γ-ray yields for the interesting three transitions are obtained as (in arbitrary units) $Y(E2; 5/2^+ \to 1/2^+_{gs}) : Y(M1; 3/2^+ \to 1/2^+_{gs}) : Y(M1; 7/2^+ \to 5/2^+) = 1.27 : 0.47 : 0.07$.

The major results are summarized in Table 1. The new cluster model results [12] are not much different from the original estimates [3]. In the E419 experiment [8], the observed lifetime of the $5/2^+$ state was deduced to be $\tau(5/2^+) = 5.2 \pm 1.4$ ps, which corresponds to $B(E2; 5/2^+ \to 1/2^+) = 4.1 \pm 1.1$ e^2fm^4 [8]. This experimental $B(E2)$ value is substantially smaller than the shell-model prediction (8.6 e^2fm^4) [13]. Therefore the dynamical shrinkage effect embodied in the cluster models is essential to explain the transition rate observed for the first time in the hypernuclear system. Thus prediction of the glue-like role of the Λ hyperon has been verified in this E419 experiment, although the cluster models seem to give a slight overestimation for the shrinkage effect (see Table 1). This is related with possible fine tuning of the ΛN interaction to be more appropriate to reproduce the known level energies.

DIFFERENCE RETAINED IN MESON THEORETICAL ΛN POTENTIALS

The recent high-resolution γ-ray measurements [8, 9] in $^9_\Lambda$Be and $^{13}_\Lambda$C suggest very small ΛN spin-orbit interaction. In order to get firm information on other components of ΛN potential, however, more data are necessary, because still only little is known for the ground-state doublets splittings in which both spin-spin and spin-orbit interactions are effective. On the other hand, the existing meson theoretical ΛN potentials, which all satisfy the elementary data, have different nature in their interaction components. This fact suggests that they lead to different hypernuclear outputs which should be tested in the actual experiments.

Here we adopt the meson theoretical hyperon-nucleon potentials supplied by the effort of the Nijmegen group as model-D, model-F and the soft-core models [14, 15] and also by the Jülich group as the version-A and -B [16, 17]. Here we denote these potentials as ND, NF, NSC89, JA and JB, respectively. Starting with these potentials, the G-matrices have been derived in nuclear matter, and then they are expressed by Yamamoto [18] in terms of three-range Gaussians. The strength parameters have density dependence through the nuclear Fermi momentum k_F:

$$\text{YNG}(\Lambda N): v_\alpha(r) = \sum_{i=1}^{3}(a_i + b_i k_F + c_i k_F^2)\exp[-(r/\beta_i)^2].$$

Here α distinguishes the central, tensor, LS, and antisymmetric LS interactions. Their basic properties have been discussed in Refs. [19, 18]. As shown in the upper half of Table 2 where the divided contributions to the Λ one-body potential are listed, these typical potential models give very different nature: The spin-triplet attraction is very strong in JA and JB, while in NSC89 the spin-singlet interaction becomes stronger. One also notices the difference in the even/odd shares. These differences should affect hypernuclear structures, since such character persists especially in light systems.

As often emphasized by Rijken [11], there are only about 35 data points for the

TABLE 2. ΛN-state contributions to the Λ one-body potential U_Λ.

	$^1S_0$	$^3S_1$	P	SUM(U_Λ)	$|^1E|:|^3E|$
JA	-3.6	-27.2	1.8	-29.0	1 : 7.6
JB	-0.5	-34.4	3.6	-31.3	1 : 68.8
ND	-7.4	-25.2	-8.0	-40.5	1 : 3.4
NF	-10.0	-20.7	-0.9	-31.6	1 : 2.0
NSC89	-15.3	-8.7	0.2	-23.8	1 : 0.6
NSC97a	-3.8	-30.7	0.6	-33.9	1 : 8.1
b	-5.5	-30.0	1.4	-34.1	1 : 5.5
c	-7.8	-29.7	2.1	-35.3	1 : 3.8
d	-11.0	-27.7	3.5	-35.1	1 : 2.5
e	-12.8	-26.0	4.5	-34.3	1 : 2.0
f	-14.4	-22.9	6.2	-31.1	1 : 1.6

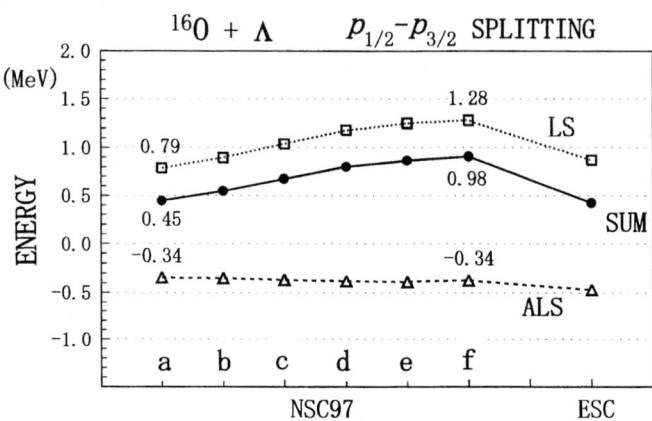

FIGURE 3. Spin-orbit splittings $\Delta E = E(p^\Lambda_{1/2}) - E(p^\Lambda_{3/2})$ calculated for $^{17}_\Lambda$O and their LS and ALS contributions.

YN scattering with mostly low energies, while for NN there are thousands scattering data points. Therefore it is important to feed back the results of hypernuclear structure analyses with the meson-theoretical YN potentials. In order to make such comparison useful, the potential models should have also some allowance of changing unknown parameters. To meet this requirement, Rijken et al. [11] presented the new soft-core model which consists of 6 versions (NSC97a, b, c, d, e, f) depending on the different $F/(F+D)$ ratios and meson mixing angles. It is note that all of these 6 versions can reproduce the YN scattering data equally well. Miyagawa et al. [20] have tested these potentials in the 3-body calculation of the simplest hypernucleus $^3_\Lambda$H and shown that only the NSC97f is acceptable in getting the Λ binding energy. It should be noted, however, that the $^3_\Lambda$H ground state provides us with only $S = 1/2$ information, so that the calculation of $^4_\Lambda$H is quite important [21].

The Λ one-body potential has been also calculated with these potentials. From the lower half of Table 2, the singlet-even contribution increases gradually as going from NSC97a to f, while the triplet-even one decreases and the repulsive p-state contribution increases. These changes should be reflected in the change of the spin-spin and spin-orbit components and hence the change of hypernuclear energy levels.

RESULTS FOR SOME P-SHELL HYPERNUCLEI

In Table 3, as the first example, we list the calculated energy splittings [22] for the $3/2^+$ and $5/2^+$ states in $^9_\Lambda$Be which have the dominant structure of $[^8\text{Be}(2^+) \times s^\Lambda_{1/2}]$. The results amount to $\Delta E = 80 - 200$ keV depending sensitively on the adopted potential models. It should be remarked that only ΛN spin-orbit interaction between Λ and the p-state nucleons play a decisive role for the splitting and the spin-spin one is negligible because of the very good $\alpha + \alpha$ description for the spinless ^{8}Be core. One sees that the antisymmetric LS interaction (ALS) reduces $20 - 40\%$ of the splittings with LS only.

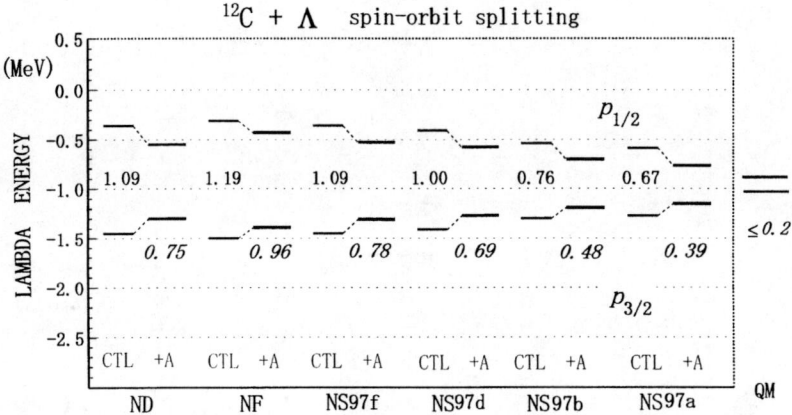

FIGURE 4. Spin-orbit splittings calculated for the Λ p-state in $^{13}_\Lambda$C. CTL denotes the central+tensor+LS interaction and "+A" means the addition of the ALS interaction. On the extreme right the result to simulate the quark model interaction [24] is inserted for reference.

However, the recent data for the splitting is much smaller than these predictions: $\Delta E^{\text{exp}} = 30.1^{+3.4}_{-3.1}$ keV [10]. This fact suggests that the theoretical LS interaction in these potential models should be much smaller or the ALS one should be larger so as to "cancel" the LS contribution.

Here, by employing the sample hypernucleus consisting of ^{16}O+Λ, we show in Fig. 3 the theoretical splittings for Λ in p-state due to LS and ALS components of 6 versions of NSC97 [11]. In each version of YNG interactions, the k_F parameter is chosen to reproduce the p-orbit binding energy in $^{16}_\Lambda$O. The LS contribution increases gradually as going from NSC97a to f versions, while the ALS one remains almost constant with negative sign and the absolute values of ALS amounts to 43%−27% of the LS strength. As a result the total spin-orbit splitting increases from NSC97a to reach 0.98 MeV at NSC97f. We have mentioned that only the NSC97f version is acceptable in view of the test of $^3_\Lambda$H binding energy in which, however, the spin-orbit interactions are not relevant. In the same figure we insert a preliminary values calculated from the extended soft-core model (ESC) [23].

As Ajimura et al. [9] succeeded recently in measuring 'two' γ rays from the Λ single-particle states ($p_{3/2}$ and $p_{1/2}$) in the $^{13}_\Lambda$C hypernucleus, it seems more realistic to compare the theoretical results which were calculated before the experiment. In the analysis of the ^{13}C($K^-,\pi^-\gamma$)$^{13}_\Lambda$C reaction data, they made use of the characteristic

TABLE 3. Energy level splitting ΔE (MeV) of the doublet $3/2^+ - 5/2^+$ in $^9_\Lambda$Be calculated with LS+ALS (LS only in parentheses) [22]. $E(J^\pi)$ is measured with respect to the $^5_\Lambda$He+α threshold. The spins are not specified in the experiment.

$E(J^\pi)$	ND	NF	NSC97a	NSC97f	EXP[10]
$E(3/2^+)$	−0.33(−0.29)	−0.31(−0.28)	−0.37(−0.34)	−0.33(−0.29)	[−0.44]
$E(5/2^+)$	−0.48(−0.52)	−0.51(−0.53)	−0.45(−0.48)	−0.49(−0.52)	[−0.47]
ΔE	0.15(0.23)	0.20(0.25)	0.08(0.14)	0.16(0.23)	0.03

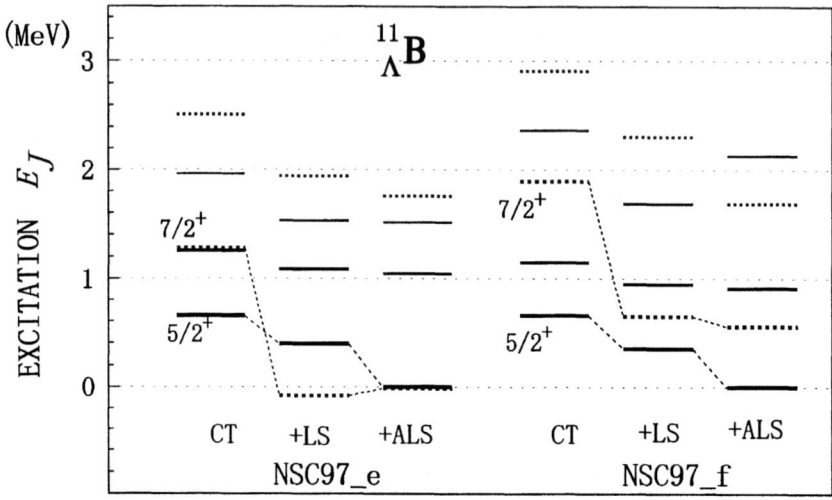

FIGURE 5. Change of energy levels of $^{11}_{\Lambda}$B when the LS and ALS interactions are added to the central+tensor result (CT).

angular distribution which was calculated in DWIA [25]. In the forward angle the $1/2^-$ states is predominantly excited due to the recoilless condition, while at larger angle the share of the $3/2^-$ states increases. They deduced the Λ p-state spin-orbit splitting of $\Delta E = 0.154 \pm 0.054$ MeV [9]. How about the theoretical values when we adopt the NSC97 potential models. As summarized in Fig. 4 the general trend of the theoretical splittings is similar to the case of $^{17}_{\Lambda}$O shown above: The splittings due to LS+ALS results in the range 0.39 MeV (NSC97a) through 0.78 MeV (NSC97f). The comparison again suggests to adopt a factor of about 3 smaller LS strength or a facotr of 3 larger ALS strength in these meson theoretical models.

Next, let us forcus our attention to the ground-state doublets and low-lying levels of other typical p-shell hypernuclei in which the s-state Λ couples to core nuclear states. In general the energy levels are quite sensitive not only to the spin-spin interaction but also to the spin-orbit interaction. In order to demonstrate the interplay, here we adopt the $^{11}_{\Lambda}$B case. Because the ground state of the core nucleus ^{10}B has $J_c = 3^+$ consisting of 'large' orbital angular momentum ($L = 2$) and 'large' spin ($S = 1$), we can expect clear effect of difference in the spin-dependent interaction when we see the behavior of $J_H = J_c \pm 1/2 = 5/2^+$ and $7/2^+$ states. In fact we have shown in Ref. [25] that the relative positions of the doublet change quite sensitively as going from NSC97a to NSC97f. In NSC97$a - c$ versions, the $7/2^+$ state comes below the $5/2^+$ partner, because, as seen from Table 2, the $^3S_1$ interaction which favors $7/2^+$ is the stronger on the a side. On the other hand, the $^1S_0$ interaction becomes larger in negative sign as going from a to f so that this favors the energy gain of $J = 5/2^+$ and disfavor $7/2^+$. From the pionic decay data, the ground state of $^{11}_{\Lambda}$B is known to have $J = 5/2^+$, although we have no information on the position of $J = 7/2^+$. Thus we adopt the e and f case and show in Fig.5 the individual roles of the spin-spin (central), LS, and ALS interactions, respectively. It is remarkable to see in both cases that the LS interaction affects the

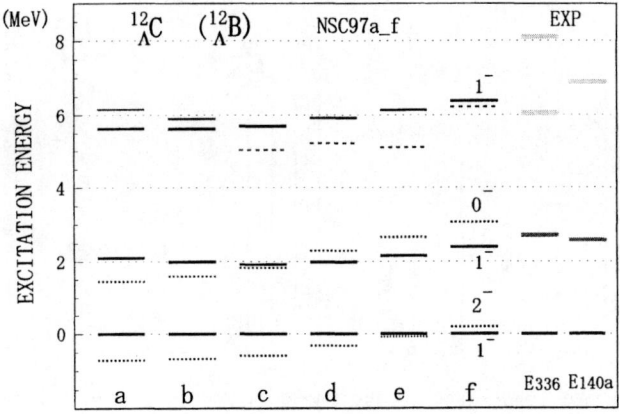

FIGURE 6. Change of low-lying energy levels of $^{12}_{\Lambda}$C calculated with 6 versions of the NSC97 potentials. Relative excitations are shown with respect to the 1^-_1 state.

relative position of $7/2^+$ quite sensitively. The present results suggest the importance of finding the energy position of $7/2^+$.

In Figure 6 we display the behavior of the low-lying states in $^{12}_{\Lambda}$C (or $^{12}_{\Lambda}$B) depending on the 6 versions of NSC97. Again we obtain a systematic change of energy positions of the ground-state doublet (1^-_1 and 2^-_1), although the splittings are not so remarkable as in the $^{11}_{\Lambda}$B case [25]. On the other hand, it is quite interesting to see the change of relative excitation energies of the 1^-_2, 1^-_3, 2^-_2, etc., as they are actually excited in the (π^+, K^+) and/or $(e, e'K^+)$ reactions.

NOVELTY OF HYPERNUCLEAR PHOTOPRODUCTION

The (K^-, π^-) and (π^+, K^+) reactions have been often used so far to produce Λ-hypernuclei. Recently, the photo-/electro-production of hypernuclei attract much attention, because high-intensity electron beams are now available at JLab, and more importantly, this process has unique characters: First, as the momentum transfer amounts to 350-400 MeV/c if we choose $E_\gamma = 1.3$ GeV/c, hypernuclear highest-spin states should be preferentially excited similarly in the (π^+, K^+) reaction. Secondly, the unique character of the process comes from its spin-flip dominance in the transition interactions. By choosing typical models [26, 27, 28] to describe the elementary amplitudes, we list the numerical values in Table 4 for the four Lab amplitudes: $\mathcal{M} \equiv <\mathbf{k} - \mathbf{p}, \mathbf{p}|t|\mathbf{k}, 0>_L = \varepsilon_0(f_0 + g_0\sigma_0) + \varepsilon_x(g_{+1}\sigma_1 + g_{-1}\sigma_{-1})$. One clearly notice that spin-dependent terms (g_0, g_{+1}, g_{-1}) dominate over the spin-independent term (f_0). These two characters give rise to the unique excitation of hypernuclear high-spin states with unnatural parity.

In order to demonstrate the unique charcteristics, the calculated results of the ^{28}Si$(\gamma, K^+)^{28}_{\Lambda}$Al reaction at $E_\gamma = 1.3$ GeV and $\theta_L^K = 3$ [29] are displayed on Fig. 7, where the strengths are divided into the Λ particle-proton-hole angular momenta of the hypernuclear configuration $[(lj)_p^{-1}(lj)^{\Lambda}]_J$. The most notable fact to be emphasized is the selective excitation of the highest-spin state within each p-h multiplet

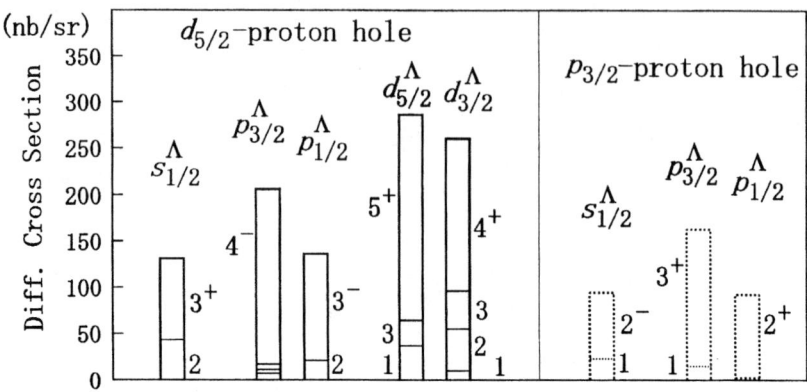

FIGURE 7. Divided contributions to the particle-hole state $[j_p^{-1}j^\Lambda]_J$ as calculated for the ^{28}Si$(\gamma,K^+)^{28}_\Lambda$Al reaction at $E_\gamma = 1.3$ GeV and $\theta_K^{Lab} = 3$ deg.

as clearly seen in Fig. 7. Among others the most strongly excited state within the same $\hbar\omega$ group is the highest-spin state with unnatural parity such as $[d_{5/2}^{-1}s_{1/2}^\Lambda]_{3^+}$, $[d_{5/2}^{-1}p_{3/2}^\Lambda]_{4^-}$, and $[d_{5/2}^{-1}d_{5/2}^\Lambda]_{5^+}$. These p-h combinations are characterized with $[j_>^{-1}j_>^\Lambda]_J$ having $J = J_{max} = j_> + j_>^\Lambda = l_N + l_\Lambda + 1 = L_{max} + 1$. The preferential excitation of the unnatural parity highest-spin states is a novel characteristic feature that is not seen in the other hypernuclear production processes such as (π^+,K^+) and (K^-,π^-) reactions. This is attributed to the spin-flip transition interactions dominated in the hyperon photoproduction reaction. It should be also noted that, in combination of the $[j_>^{-1}j_<^\Lambda]_J$ type or the $[j_<^{-1}j_>^\Lambda]_J$ type, the highest spin is limited to $J'_{max} = l_N + l_\Lambda$ and therefore it has a natural parity $(-)^{l_N+l_\Lambda}$.

Corresponding to the actual $(e,e'K^+)$ experiment to be done first at the Jefferson Lab [30], we show the calculated spectrum for the ^{12}C$(\gamma,K^+)^{12}_\Lambda$B at $E_\gamma = 1.3$ GeV in Fig. 8. This is the update of the former prediction [31] by using the NSC97f ΛN interaction in the mixed configuration and the smaller smearing width. It will be quite interesting if the positions of side peaks, as well as the unnatural parity 2^- and 3^+ states, are clearly determined with higher energy resolution than in the (π^+,K^+) reaction. Figure 9 shows another example of producing a proton-defficient hypernucleus, which can be used for weak decay study and compared with the $^{10}_\Lambda$B case.

TABLE 4. Elementary amplitudes for $\gamma p \to \Lambda K^+$ at $E_\gamma = 1.3$ GeV.

θ_L^K	Model [Ref]	$|f_0|^2$	$|g_0|^2$	$|g_{+1}|^2$	$|g_{-1}|^2$	Re($f_0 g_0^*$)	$d\sigma/d\Omega$	Pol(Λ)
3 deg.	S6B [26]	0.004	0.381	0.170	0.217	-0.006	1.96	-0.081
	AS1 [27]	0.006	0.399	0.189	0.207	-0.016	2.03	-0.061
	C4 [28]	0.0002	0.572	0.272	0.281	-0.006	2.85	-0.019
10 deg.	S6B [26]	0.035	0.325	0.127	0.263	-0.043	1.82	-0.296
	AS1 [27]	0.048	0.374	0.166	0.204	-0.045	1.93	-0.161
	C4 [28]	0.001	0.534	0.179	0.204	$-0,018$	2.23	-0.066

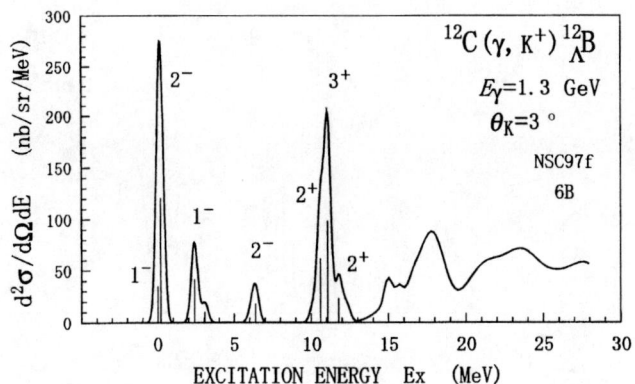

FIGURE 8. Calculated spectrum for the $^{12}C(\gamma, K^+)^{12}_\Lambda B$ reaction at $E_\gamma = 1.3$ GeV and $\theta_K^{Lab} = 3$ deg.

CONCLUDING REMARKS

Corresponding to the high-resolution measurements achieved in the recent experiments, first we discussed on the confirmation of nuclear shrinkage effect due to Λ participation as an example of new aspects of strangeness many-body syatems. Secondly we emphasized the importance of knowing detailed level structures in hypernuclei so as to get firm information on the ΛN interactions. In this relation we discussed the basic properties of several meson theoretical potentials and demonstrated some hypernuclear outputs with an emphasis on the role of spin-spin, LS, and ALS interactions. Finally, we pointed out useful characteristics of photoproduction of hypernuclei, showing theoretical predictions to be compared with the expected outcome at JLab.

ACKNOWLEDGEMENTS

The author acknowledges P. Bydžovský, O. Hashimoto, E. Hiyama, K. Itonaga, M. Kamimura, T. Rijken, M. Sotona, T. Yamada, and Y. Yamamoto for the several kinds of collaborations underlying the present article. He thanks W. Haxton and E. Henley and the Institute for Nuclear Theory, University of Washington, for the hospitality and for partial support. He is also grateful to J. Dobeš and the Nuclear Physics Institute of Czech Academy of Sciences for the hospitality with partial support. He is indebted to Richard J. Seymour of Nuclear Physics Laboratory for assistance in the computer graphics file treatment. The work has been done as a part of the JSPS binational collaboration project.

REFERENCES

1. Dalitz, R.H., and Gal, A., *Phys. Rev. Lett.* **36**, 362 (1976).
2. Zhang, Z.Y., Li,G.L., and Yu, Y.W. *Phys. Lett. B* **108**, 261 (1982).
3. Motoba, T., Bando, H., and Ikeda, K. *Prog. Theor. Phys.* **70**, 189 (1983); Motoba, T. *et al.*, ibid. Suppl. **81**, 42 (1985).

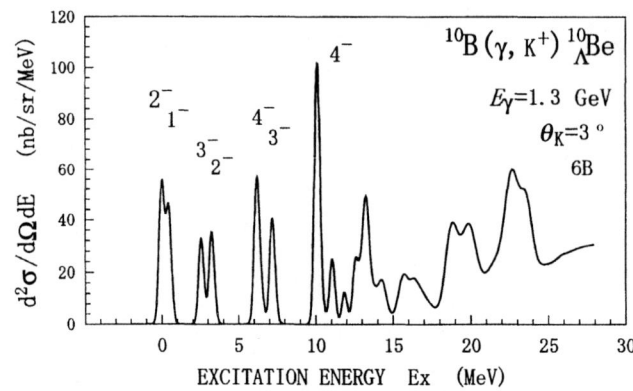

FIGURE 9. Calculated spectrum for the $^{10}B(\gamma,K^+)^{10}_\Lambda Be$ reaction at $E_\gamma = 1.3$ GeV and $\theta_K^{Lab} = 3$ deg.

4. Hasegawa,T. et al., *Phys. Rev. Lett.* **74**, 224 (1995); *Phys. Rev. C* **53**,1210 (1996).
5. Hashimoto, O., *Proc. Strangeness Nuclear Physics*, ed. by Il-T. Cheon, S.W. Hong and T. Motoba. World Sci., Singapore, 1999, pp. 116-125 and references therein.
6. Nagae, T., ibid, pp. 110-115.
7. Pile, P.H. et al., *Phys. Rev. Lett.* **66**, 2585 (1991).
8. Tamura, H., *Nucl. Phys.* **A639**, 83c (1998); Tamura, H. et al. *Phys. Rev. Lett.* **84**, 5963 (2000); Tanida, K. et al., *Phys. Rev. Lett.* **86**, 1982 (2001).
9. Ajimura, S. et al., *Phys. Rev. Lett.* **86**, 4255 (2001).
10. Tamura, H., *Proc. Int. Conf. Hypernuclear and Strange Particle Physics*, Torio (2000), in press; Akikawa, H., et al.,, to be published.
11. Rijken, Th.A., Yamamoto,Y. and Stoks, V.G.J., *Phys. Rev. C* **59**, 21 (1999); Rijken, Th.A. et al., *Strangeness Nuclear Physics*. World Sci., Singapore, 1999, pp.5-12.
12. Hiyama, E., Kamimura, M., Miyazaki, K., and Motoba, T. *Phys. Rev. C* **59**, 2351 (1999).
13. Dalitz, R.H., and Gal, A., *Ann. Phys.(N.Y.)* **116**, 167 (1978); *J. Phys. G* **4**, 889 (1978).
14. Nagels, M.M., et al., *Phys. Rev. D* **12**, 744 (1975); **15**, 2547 (1977); **D20**, 1633 (1979).
15. Maessen, P.M.M., et al., *Phys. Rev. C* **40**, 2226 (1989).
16. Holzenkamp, B., et al., *Nucl. Phys.* **A500**, 485 (1989).
17. Reuber, A., et al.,*Czech. J. Phys.* **42**, 1115 (1992) and references therein.
18. Yamamoto, Y., Motoba, T., Himeno, H., Ikeda, K., and Nagata, S., *Prog. Theor. Phys. Suppl.* **117** (1994) 361, and references therein.
19. Yamamoto, Y. et al., *Czech J. Phys.* **42**, 1249 (1992); Motoba, T. and Yamamoto, Y. *Nucl. Phys.* **A585**, 29c (1995).
20. Miyagawa, K., et al., *Phys. Rev. C* **51**, 2905 (1995) and private communication.
21. Hiyama, E., *Proc. Int. Conf. Hypernuclear and Strange Particle Physics*, Torio (2000), in press.
22. Hiyama, E., Kamimura, M., Motoba, T., Yamada, T., Yamamoto, Y. *Phys. Rev. Lett.* **85**, 270 (2000).
23. Rijken, Th.A. and Yamamoto, Y., private communication (2000).
24. Fujiwara, Y., Nakamoto, C., Suzuki, Y. *Phys. Rev. Lett.* **76**, 2242 (1996).
25. Motoba, T., *Proc. Intern. Symposium on Hadrons and Nuclei*, Seoul (Feb.,2001), in press.
26. Schorsch, W., Tietge, J. and Weilenböck, W., *Nucl. Phys.* **B25**, 179 (1970).
27. Adelseck, R.A., Saghai, B. *Phys. Rev. C* **42**, 108 (1990).
28. Williams, R.A. et al., *Phys. Rev. C* **46**, 1617 (1992).
29. Bydžovský, P., Sotona, M., Motoba, T., Itonaga, K., and Hashimoto, O., in preparation.
30. Hashimoto, O., Hungerford, Ed., and Tang, L., private communication.
31. Motoba, T., Sotona, M., and Itonaga, K., *Prog. Theor. Phys. Suppl.* **117**, 123 (1994).

Probing the ηN interaction with η production reactions

M.T. Peña

Departamento de Física and Centro de Física das Interacções Fundamentais, IST
Av. Rovisco Pais, P-1049-001, Lisboa, Portugal

Abstract. I address the general features of the ηN interaction, which can be obtained from the theoretical study and experimental scrutiny of the $pn \to \eta d$ reaction. I discuss in particular the origin of the enhancement of the $np \to \eta d$ cross section observed at threshold. I show how the interplay of the ηN and (realistic) NN interactions invalidates the interpretation of the marked threshold effect as a signal of the existence of an ηNN quasi-bound state. Furthermore, I provide other links between the η production reaction and the short-range behavior of the nuclear interaction, when parameterized in terms of heavy-meson exchanges.

INTRODUCTION

In this talk I present recent results on meson production. The study of meson production is interesting not only for the knowledge on the production mechanisms, but, above all, because meson production processes directly relate to the short-range behavior of the nucleon-nucleon and three-nucleon interactions, as already illustrated at this Conference, for example, in the talks of E. Epelbaum and R. Timmermans. I will specifically focus on the $pn \to \eta d$ reaction, which has been measured recently with great accuracy in the region near threshold [1]. The interest in this study is triggered by the enhancement of the cross-section observed at threshold, as compared with the predictions based on a two-body phase space, for the ηd system. It has been speculated that this pronounced threshold effect could be a signal for an ηNN quasi-bound state, predicted in the beginning of the 90's by Wilkin [2]. This conclusion is of interest for astrophysics, with consequences namely for the models underlying the equation of state and the structure of a neutron star. Indeed, it could allow for the possibility of an exotic ground state of matter, in density regimes where the meson fields become true dynamical degrees of freedom, and turn into more than the simple carriers of the interactions between baryons at low energies. In addition, an ab-initio calculation of the $pn \to \eta d$ reaction provides some information on the spread of the theoretical values of ηN scattering length, for which the several data analysis do not coincide, and also of the ηNN coupling constant, which is not sufficiently constrained by the nucleon-nucleon scattering data [3]. From other calculations [4], where approximation schemes had to be used, such information implies larger uncertainties.

TABLE 1. A sample of values obtained by several authors for the ηN scattering length $a_{\eta N}$. The table illustrates the wide spread of the results.

Bhalerau and Liu	Krusche	Batinic et al.	Green and Wycech
0.27+i0.22	0.579+i0.394	0.75+i0.271	0.87+i0.27

THE IMPORTANCE OF THE FINAL STATE INTERACTION

We tested, in a three-body calculation for the ηNN system, different ηN dynamical models constructed upon recent data analysis of the coupled reactions $\pi N \to \eta N$, $\eta N \to \eta N$ and $\gamma N \to \eta N$. Table I shows some of the values for the ηN scattering length, obtained by different groups and from different data analysis, and clearly evidences the lack of unanimity of the different determinations.

The bottom line of this talk is that, in spite of this poor knowledge on the ηN interaction, part of the dynamical input of the three-body equations, an exact three-body calculation allows to draw some universal conclusions, which are even qualitatively in agreement with the conclusions of other works [4, 5, 6].

Even if the smallest of the values presented on Table I is the one which corresponds to reality, the scattering length for ηN scattering is about seven times larger than the one for πN scattering. In the η production case, this requires to go beyond the DWBA description, which is commonly used in the description of pion production. In this latter approximation scheme, one assumes that the meson production process factorizes into a pure nucleonic initial and final state distortion and an intermediate production amplitude. Given the strength of the ηN interaction, for the reaction $np \to \eta d$ reaction near threshold, it turns out that it is important to describe well the $\eta d \to \eta d$ three-body transition matrix, describing the multiple scattering series for the ηd final-state interaction.

The theoretical description of the reaction is then achieved by considering an impulse term consisting of the direct production mechanism, together with a "box"-type diagram where the η is produced after excitation of a nucleonic resonance, mediated by the exchange of the η and other mesons between the two nucleons (as represented Figure 1 (a)). After the emission of the η meson, the nuclear state is distorted by the final-state ηd interaction. Theoretically, this one may be obtained by solving the standard nonrelativistic Faddeev equations with separable potentials [7, 8]. These Faddeev equations for the ηNN system are represented diagrammatically in Fig. 1(b). As for the meson-nucleon amplitude depicted by the heavy line in the diagrams of Figure 1, the small deviation of the $S11(1535)$ resonance mass from the sum of the η and nucleon masses, on top of the almost equal decay probabilities to the ηN and πN channels, give this resonance the dominant role in the production process, at energies near the η production threshold. The resonance may be phenomenologically described by a separable model [7, 8], which is generated dynamically from solving a coupled-channel Lippmann-Schwinger equation with separable potentials.

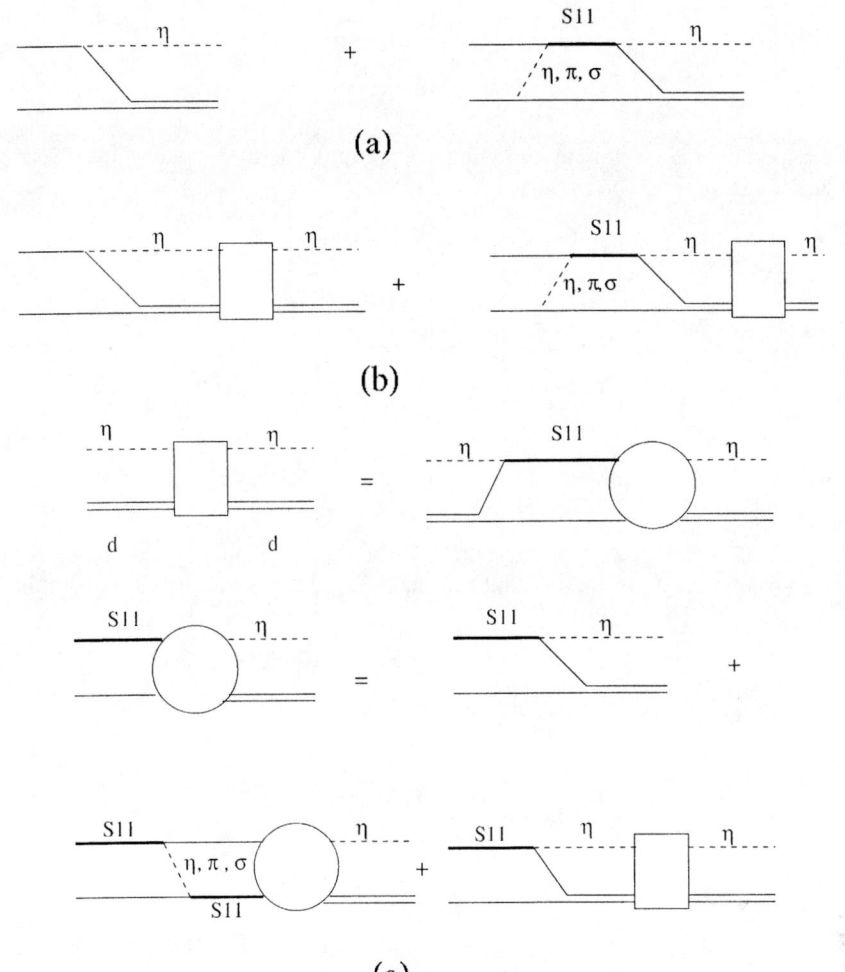

FIGURE 1. The theoretical model to describe the $np \to \eta d$ reaction

The potentials describing the meson-nucleon transitions were taken in Refs. [7, 8] to be of the form

$$<p|V_{ii}|p'> = -g_i(p)g_i(p'), \quad (i = \eta, \pi, \sigma) \quad (1)$$
$$<p|V_{ij}|p'> = \pm g_i(p)g_j(p'), \quad (i,j = \eta, \pi, \sigma) \quad (2)$$

with

$$g_\eta(p) = \sqrt{\lambda_\eta} \frac{A + p^2}{(\alpha_\eta^2 + p^2)^2}, \quad (3)$$

$$g_\pi(p) = \sqrt{\lambda_\pi}\,\frac{1}{\alpha_\pi^2+p^2}, \tag{4}$$

$$g_\sigma(p) = \sqrt{\lambda_\sigma}\,\frac{p}{\alpha_\sigma^2+p^2}. \tag{5}$$

The coupled-channel Lippmann-Schwinger equations for the meson-nucleon system generate the meson-nucleon t matrices from these potentials (1)-(2), as

$$<p|t_{\eta\eta}(E)|p'> = g_\eta(p)\,\tau_2(E)\,g_\eta(p'), \tag{6}$$

$$<p|t_{\pi\pi}(E)|p'> = \frac{\lambda_\pi}{\lambda_\eta}\,g_\pi(p)\,\tau_2(E)\,g_\pi(p'), \tag{7}$$

$$<p|t_{\eta\pi}(E)|p'> = \pm\sqrt{\frac{\lambda_\pi}{\lambda_\eta}}\,g_\eta(p)\,\tau_2(E)\,g_\pi(p'), \tag{8}$$

$$<p|t_{\eta\sigma}(E)|p'> = \pm\sqrt{\frac{\lambda_\sigma}{\lambda_\eta}}\,g_\eta(p)\,\tau_2(E)\,g_\sigma(p'). \tag{9}$$

The inverse of $\tau_2(E)$ gives the dressed S_{11} resonance propagator, corresponding to a sum of a Dyson series, for the ηN and πN channels, plus the unitarity cut from the σN channel.

$$\frac{1}{\tau_2(E)} = \frac{1}{\lambda_\eta} - G_2(E) - \frac{\lambda_\pi}{\lambda_\eta}G_\pi(E) + i\pi\mu_\eta\,p_2\,g_\sigma^2(p_2), \tag{10}$$

with p_2 denoting the σN relative on-shell-momentum, and

$$G_2(E) = \int_0^\infty p^2 dp\,\frac{g_\eta^2(p)}{E - p^2/2\mu_\eta + i\varepsilon}, \tag{11}$$

$$G_\pi(E) = \int_0^\infty p^2 dp\,\frac{g_\pi^2(p)}{E + p_0^2/2\mu_\pi - p^2/2\mu_\pi + i\varepsilon}. \tag{12}$$

In the equations above, μ_η and μ_π are the ηN and πN reduced masses respectively, while p_0 is the πN relative momentum at the ηN threshold, i.e.,

$$p_0^2 = \frac{[s_0 - (m_\pi + m_N)^2][s_0 - (m_\pi - m_N)^2]}{4s_0},\quad s_0 = (m_\eta + m_N)^2. \tag{13}$$

For the σ meson we took the range $\alpha_\sigma = 2.5$ fm^{-1}, and the strength was determined by requiring that 9% of the decay width of the S_{11} isobar is due to this meson, i.e.,

$$\lambda_\sigma = 0.1\,\frac{\lambda_2\,p_2\,\mu_2\,g_\eta^2(p_2) + \lambda_\pi\,p_\pi\,\mu_\pi\,g_\pi^2(p_\pi)}{p_2\,\mu_2\,g_\sigma^2(p_2)}. \tag{14}$$

The relative sign of the ηN-ηN and πN-ηN amplitudes cannot be fixed by unitarity. This ambiguity is explicit in equations (8) and (9). It happens to have little consequences for the three-body final state ηd T-matrix [7] but it is crucial for $np \to \eta d$ process, due to the "box" driving terms in Fig. 1a.

TABLE 2. Three-body ηd scattering length, in fermi, as a function of the two-body ηN scattering length. $Y_{\eta N}$ and $R_{\eta N}$ denote two different descriptions of the phenomenological ηN amplitudes, from reference [8]. The $R_{\eta N}$ describes the whole "resonant" width of the $S11$. The nucleon-nucleon models are labeled Y_{NN} and P_{NN}, an Yamaguchi-separable model and the more realistic Paris potential respectively. The signature of the three-body quasi-bound state is the negative sign for the ηd scattering length. The short-range nucleon-nucleon correlations induced by the Paris potential wipes out the quasi-bound state, when the ηN interaction would allow it.

Model*	$a_{\eta N}[fm]$	$Y_{\eta N}Y_{NN}$	$Y_{\eta N}P_{NN}$	$R_{\eta N}Y_{NN}$	$R_{\eta N}P_{NN}$
1	0.72+i0.26	2.38+i3.04	2.59+i2.37	2.43+i1.60	2.46+i1.62
2	0.75+i0.27	2.38+i3.41	2.70+i2.64	2.56+i1.65	2.61*1.72
3	0.83+i0.27	2.57+i4.51	3.21+i3.34	3.00+i1.87	3.10+i2.03
4	0.87+i0.27	2.54+i5.20	3.47+i3.78	3.20+i1.94	3.36+12.03
5	1.05+i0.27	0.06+i8.66	4.27+i7.05	4.20+i2.24	4.81+i3.19
6	1.07+i0.26	-0.42+i9.21	4.53+i7.60	4.21+i2.04	5.02+13.14

* from reference [8]

For the construction of the "box"-type diagram in Fig. 1.a, we described the η-nucleon-nucleon vertex by

$$V_{\eta NN} = \frac{f_\eta}{m_\eta}\left(1+\frac{m_\eta}{m_N}\right)\vec{\sigma}\cdot\vec{p}, \tag{15}$$

and the π-nucleon-nucleon and σ-nucleon-nucleon vertices by

$$V_{\pi NN} = \frac{f_\pi}{m_\pi}\left(1+\frac{m_\pi}{m_N}\right)\vec{\sigma}\cdot\vec{p}\,\vec{\tau}\cdot\vec{\phi}_\pi, \tag{16}$$

$$V_{\sigma NN} = f_\sigma, \tag{17}$$

where $\vec{p}$ is the meson-nucleon relative momentum. The πNN coupling constant is $f_\pi = 0.079$, while the ηNN coupling constant f_η will be extracted from the calculation. We took $m_\sigma = m_\eta$ and $f_\sigma = 0.798$, as given in reference [9]. We point out that the results are sensitive to the effect of the $\pi\pi N$ inelastic channel (mocked by an effective sigma exchange) on the S_{11} propagator, but almost insensitive to the value of f_σ. In Ref. [8] we constructed 6 different phenomenological models of the coupled ηN-πN subsystem, which were fitted to the amplitudes of recent data analysis [10, 11, 12, 13]. Here we extended the calculation to the third coupled-channel $(\pi\pi)\sigma N$ of the decay of the S_{11}.

The parameters of the ηN and πN form-factors given by equations 3 and 4 are the same as in Ref.[8]. Following references [7, 8], we probe several ηN interaction models based on the amplitude analysis of [10, 11, 12, 13], which we will keep labeling from 1 to 6, as in [7, 8].

Figure 2 represents the results obtained in this way in Ref. [7] for the experimentally inaccessible elastic $\eta d \to \eta d$ reaction. At variance with the impulse approximation curves, all the curves corresponding to calculations which include the three-body final state interaction show a strong and rapid energy dependence at threshold. This happens

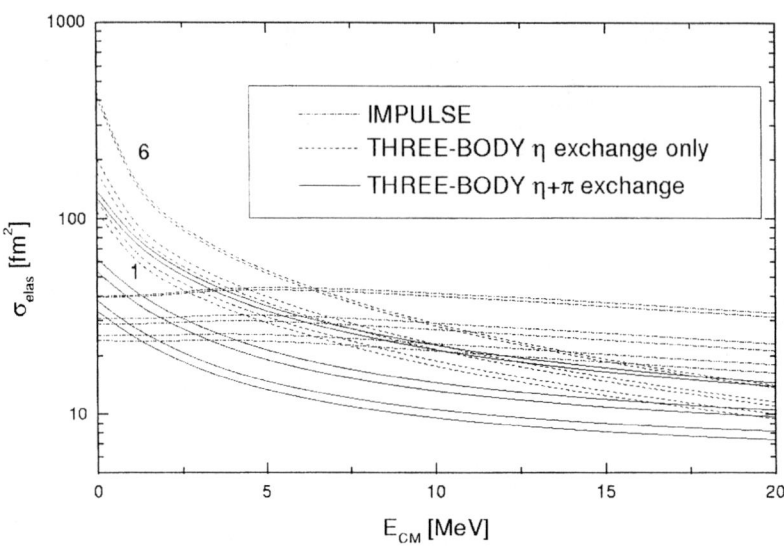

FIGURE 2. Results for the $\eta d \to \eta d$ elastic cross-section obtained with the 6 different combinations of ηN and NN dynamical models from the last column of Table II. The models correspond to values of the real part of the ηN scattering length, which run from 0.72 fm (model 1) to 1.07 fm (model 6). None of the combinations imply a three-body quasi-bound state. There are 3 groups with 6 curves each. The group of full lines show the results of the complete calculation ($\eta + \pi$ exchange). The group of dashed lines show the results of a calculation with η exchange only. The group of dashed-dotted lines show the results of the impulse approximation. This figure is taken from Ref. [7].

for ηN and NN two-body interactions which do not support a three-body ηNN quasi-bound state, as exhibited by the positive sign of the produced three-body ηd scattering length (Table II, last column). The results of Fig. 2 together with the ones displayed on Table II lead to the conclusion that the observed threshold effect does not have an exotic origin, but is rather explained by the ηd final-state interaction. The pion contribution for $\eta d \to \eta d$ is repulsive in our calculation, and counter-acts the effect of the η-exchange contribution. It agrees in sign with the one found by Wycech and Green [5], although its size is larger.

THE IMPORTANCE OF THE HEAVY-MESON EXCHANGE

In Refs. [7, 8], $\eta d \to \eta d$ elastic rescattering was discussed and the framework enabling the calculation of the $np \to \eta d$ process was established. For that purpose, however, the ηN-ηN, πN-ηN coupled-channel equations which describe the S_{11} had to be extended to accommodate also the inelastic coupled $(\pi\pi)\sigma N$ channel. In effect, although the decay width of the S_{11} resonance to this last channel is only around 10%, the full calculation

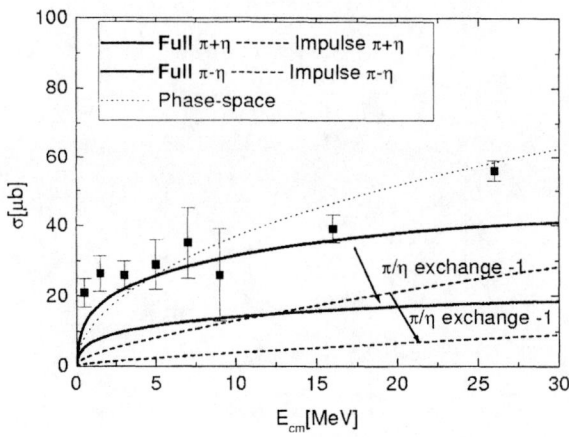

FIGURE 3. Results for the $np \to \eta d$ cross-section obtained with model 3 of Table II. The coupling to the inelastic $\pi\pi N$ channel is not included. The impulse approximation calculation is compared with the full amplitude. Two sets of curves are displayed corresponding to the two possible cases for the relative sign between the $\eta N \to \eta N$ and $\eta N \to \pi N$ amplitudes. The curve of a two-body phase-space calculation is also shown.

showed that the $pn \to \eta d$ reaction is rather sensitive to it. This is what we will see next. In Fig. 3, the full line, which represents a calculation excluding the $(\pi\pi)\sigma N$ decay channel of the resonance, fails to describe the strength and even the shape of the experimental results. The same figure also illustrates how the impulse approximation deviates from the full calculation.

The failure in reproducing the experimental strength of the cross-section at threshold energies exhibited in Fig. 3, motivates the inclusion of the inelastic $(\pi\pi)\sigma N$ channel. This inclusion is moreover justified on grounds of unitarity.

With the new extended meson-nucleon model, which includes the coupling to the $(\pi\pi)\sigma N$ inelastic channel, the energy dependence of the enhancement factor relatively to a two-body phase-space calculation is extracted and compared with data [1] in Fig. 4. It turns out that the results are in very good agreement with the data in the region very near threshold. This checks that the $(\pi\pi)\sigma N$ inelastic channel is essential for a description of the imaginary part of the ηN-ηN transition with the accuracy necessary to allow the description of the inelastic reaction $np \to \eta d$. It is worth to mention that although the $(\pi\pi)\sigma N$ channel is very important to model the resonance in the "box" diagram and its iterations correctly, the σ-exchange mechanism in the iterated series has a small effect on the final result.

As for the ambiguity in the relative sign of the direct ηN amplitude relatively to the transition amplitude πN-ηN, Fig. 5 illustrates that the data on $np \to \eta d$ clearly selects the positive sign. Our results also indicate that this relative positive sign is favored, if one wants to keep the ηNN coupling constant $g_{\eta NN}$ such that $g_{\eta NN}/4\pi$ has a value below 10, in agreement with experimental evidence from very different origins.

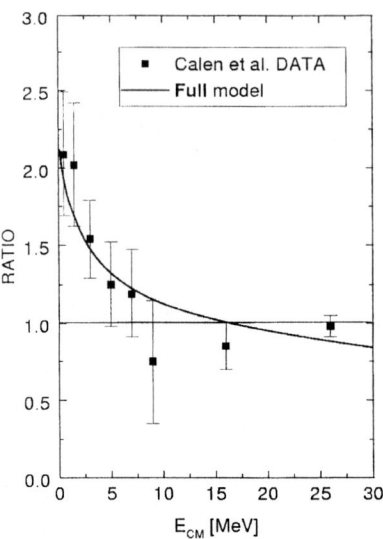

FIGURE 4. Calculated enhancement factor at threshold for the reaction $np \to \eta d$. The inclusion of the $\pi\pi N$ channel was crucial for the quality of the results obtained.

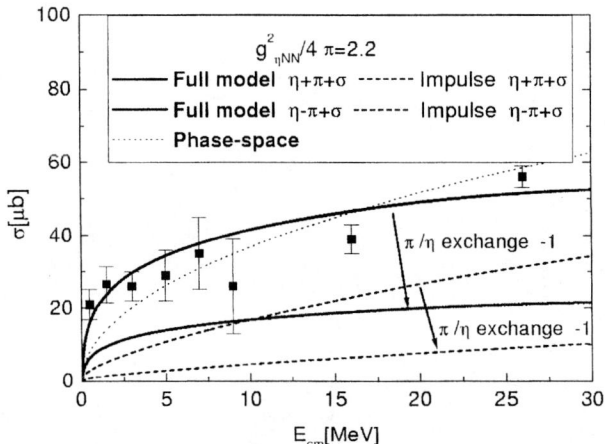

FIGURE 5. Results for the $np \to \eta d$ cross-section obtained with model 3 of Table II. The coupling to the inelastic $\pi\pi N$ channel is included. The curves have the same meaning as in Fig. 3. The value for $g^2_{\eta NN}/4\pi$ is 2.2.

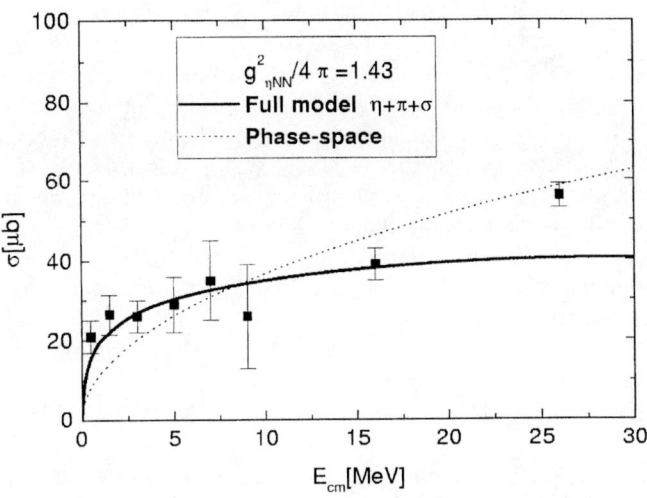

FIGURE 6. Results for the $np \to \eta d$ cross-section obtained with model 4 of Table II. The coupling to the inelastic $\pi\pi N$ channel is included. The full calculation for the case of positive relative sign between the $\eta N \to \eta N$ and $\eta N \to \pi N$ amplitudes and the two-body phase-space calculation are shown. The value for $g^2_{\eta NN}/4\pi$ is 1.43.

THE ηNN COUPLING CONSTANT

A very important quantity related to η physics is the ηNN coupling constant. This quantity is not very well known. For example, it is badly constrained by the nucleon-nucleon scattering data: in the Bonn family of NN potentials $g^2_{\eta NN}/4\pi$ lies between 2 and 7 [3]. The calculation on Fig. 5 was obtained with a coupling constant $g^2_{\eta NN}/4\pi = 2.2$, which is consistent with the meson exchange models of the nucleon-nucleon interaction, in particular the Bonn model. Another determination of the ηNN coupling constant was carried out by Benmerrouche and Mukhopadhyay [14], from the analysis of η photoproduction data near threshold, within an effective Lagrangian approach. They obtained $g^2_{\eta NN}/4\pi = 1.4$ which is smaller than our extracted value. Finally, Bernard, Kaiser, and Meißner [15] analysed the reaction $pp \to \eta pp$ near threshold within an approach based on chiral perturbation theory and meson-exchange diagrams. They obtained $g^2_{\eta NN}/4\pi = 2.24$, which is close to the findings reported in Fig. 5. We need to mention at this point, however, that the initial-state distortion was not yet considered, neither in our work, at this stage, nor in the calculation of [15]. The investigation, still under away, on the effect of the initial-state interaction seems to indicate the need for a stronger coupling. Some preliminary results which will be reported later, show that the initial state interaction suppresses the cross-section. Indeed, this effect comes from the feature of the nucleon interaction which prevents two nucleons from overlapping at short-distances (or with large relative momentum). Finally, in the other direction, an ηN model with a scattering length larger than the one we considered (model 3), would mean that the $pn \to \eta d$ data is evidence for an ηNN coupling weaker than the ones of the OBE models. To quantify the sensitivity of the results to this last aspect, we show in Fig. 6 the curves with the same meaning as in Fig. 5, but calculated with a model where the ηN scattering length is somewhat larger (model 4). The description of the $np \to \eta d$ data near threshold, in this other case, demands for $g^2_{\eta NN}/4\pi = 1.43$. We stress also that the ηNN coupling was adjusted in both cases (Figs. 5 and 6) without considering short-range sigma- and omega- exchange mechanisms, from perturbative "Z-diagrams", as the ones introduced in previous works on the $pp \to pp\eta$ and $pp \to pp\pi^0$ reactions [16]. One can expect that the effects of these diagrams would impose an extra decrease on the extracted value of the ηNN coupling strength.

OFF-SHELL BEHAVIOR: ARBITRARINESS WITHIN LIMITS

In Ref. [7], we calculated the elastic ηd cross-section which involves the three-body ηd on-energy-shell scattering T matrix. In the present work, after improving the description of the meson-nucleon sector, i.e., of the $S11$ resonance, the ηd half-off-energy-shell scattering T-matrix was calculated. The description of the cross-section for the $np \to \eta d$ reaction was then possible, by integrating the product of the reaction mechanism with the half-off-energy T-matrix.

I want to comment here on some arbitrariness of the half-off-shell-behavior of the T-matrix, which is dependent on the choice of the dynamical equation. The question arises then on how "meaningful" or how "under-control" this behavior may be, since it is not directly accessible from experiment. We show in Fig. 7 the ratio at different energies of the half-energy-shell T-matrix to the on-shell T matrix, as a function of the off-shell-momentum over which an integration has to be performed. The curves almost coincide for energies close to threshold. This feature means that the energy dependence can be factorized in good approximation, and is about the same in both the half-off-shell and on-energy-shell situations. Thus we confirm that the half-off-shell-behavior generated by the dynamical equations is well determined by the measurable on-shell behavior, at least for energies close to the threshold. In this sense, the half-off-shell properties of the T-matrix are not ad-hoc and seem to be "under control". This simple check makes our description of the $np \to \eta d$ cross-section more robust, because it shows that the energy dependence obtained, which is rooted in the energy dependence of the half-off- shell three-body T-matrix, although model dependent, is not largely arbitrary.

CONCLUSIONS

The ηN interaction is not yet known with the quantitative precision which we encounter in other more "mature" sectors, such as the description of the nucleon-nucleon interaction. The three-body calculation of the ηNN system reported in this talk, allows however to draw some general conclusions. I want to emphasize in particular those conclusions which are independent of the dynamical two-body model for the ηN interaction. The

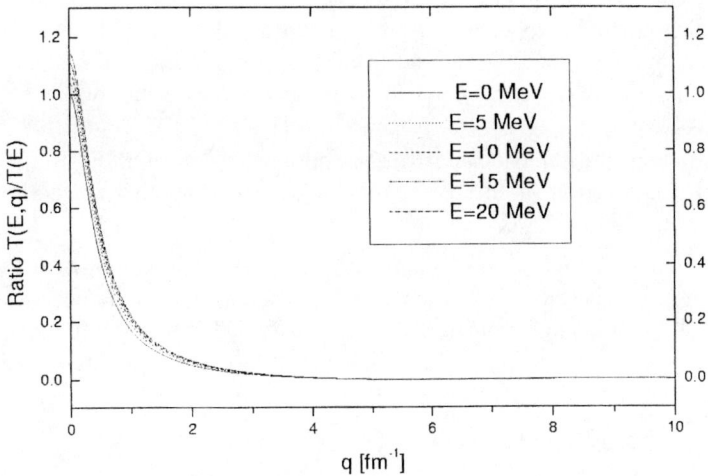

FIGURE 7. Ratio between the half-off-energy-shell and the on-shell behavior ηd T-matrix as a function of the off-shell momentum.

central conclusion is that the three-body calculation does not support the conjecture of an exotic ηNN mesic-bound nucleus:

> The ηd final state interaction determines the energy behavior of the $np \to \eta d$ reaction at threshold. The three-body dynamics by itself explains the observed threshold effect, also for underlying two-body interactions which do not generate an ηNN quasi-bound state.

Other groups [4, 5, 6], although not in precise quantitative agreement (due to differences in dynamical inputs), reach the same conclusion, based on calculational methods different from ours. Beyond this general feature, but also independently of the quantitative details of the ηN dynamics, we may also add that

> The ηNN quasi-bound state, even in the presence of a sufficiently attractive ηN interaction, is prevented by realistic short-range nucleon-nucleon interactions, which for the range of the distances involved in η production, provide enough repulsion to overcome the η-mediated attraction.

> The coupling to the two-pion inelastic channel affects the $np \to \eta d$ reaction, meaning that a fine-tuning of the ηN dynamics should not neglect it.

> A large real part for the ηN scattering length is compatible with the data for $np \to \eta d$:
> – when one does not consider the effects of sigma-omega short-range perturbative "Z-diagrams", this compatibility holds, provided that the ηNN coupling constant $g_{\eta NN}$ is close to the values obtained from modeling the nucleon-

nucleon interaction by OBE models.

- the extracted ηNN coupling strengths extracted then are not in agreement with the smaller values, suggested from η photo- and electro-production, but the extracted values decrease with decreasing values of the real parts of the ηN scattering.

At variance with an widespread assumption, the contribution of η-rescattering to the cross-section of the $np \rightarrow \eta d$ reaction is sizable, when compared to π-rescattering contribution.

Our results start to deviate from the data around 40 MeV above threshold, which may signal that at those energies heavier mesons, in particular the ρ, may become important, as well as other nucleonic resonances not yet considered at this stage. Work to include the initial state nucleon-nucleon interaction in the study of the $np \rightarrow \eta d$ reaction is under way.

ACKNOWLEDGMENTS

I want to thank H. Garcilazo for his interest in the development of the collaboration on the study of η production reactions. I also want to thank J. Haidenbauer and C. Hanhart for the discussions during my visit to Julich in May 2000. This work was supported by the Portuguese Foundation for Scientific Research under the grants PRAXIS/P/FIS/10031/1998, SAPIENS/36291/99, CERN/P/FIS/40135/2000.

REFERENCES

1. Calén, H., et al., *Phys. Rev. Lett.* **79**, 2642 (1997), Calén, H., et al., *Phys. Rev. Lett.* **80**, 2069 (1998).
2. Plouin, F., Fleury, P., Wilkin, C., *Phys. Rev. Lett.* **65**, 690 (1990);
   Ueda, T., *Phys. Rev. Lett.* **66**, 297 (1991).
3. Machleidt, R., *Adv. Nucl. Part. Phys.* **19**, 189 (1989).
4. Grishina, V.Yu., Kondratyuk, L.A., Buscher, M., Haidenbauer, J., Hanhart, C., Speth, J., *Phys. Lett.* B **475**, 9 (2000).
5. Wycech, S., and Green, A.M., nucl-th/0104053;
   Green, A.M., Niskanen, J., Wycech, S., *Phys. Rev. C* **54** page! (1996);
6. Deloff, A., *Phys. Rev. C* **61**, 024004 (2000).
7. Garcilazo, H., andPeña, M.T., *Phys. Rev. C* **63**, 021001 (2001).
8. Garcilazo, H., andPeña, M.T., *Phys. Rev. C* **61**, 064010 (2000).
9. Holinde, K., *Phys. Reports* **68**, 121 (1981).
10. Batinić, M., Šlaus, I., Švarc, A., and Nefkens, B.M.K., *Phys. Rev. C* **51**, 2310 (1995).
11. Batinić, M., Dadić, I.,Šlaus, I., Švarc, A., Nefkens, B.M.K., and Lee, T.-S.H., *Physica Scripta* **58**, 15 (1998).
12. Green, A.M., and Wycech, S., *Phys. Rev. C* **55**, R2167 (1997).
13. Green, A.M., and Wycech, S., *Phys. Rev. C* **60**, 035208 (1999).
14. Benmerrouche, M., and Mukhopadhyay, N.C., *Phys. Rev. Lett.* **67**, 1070 (1991).
15. Bernard, V., Kaiser, N., and Meißner, U.-G., *Eur. J. Phys. A* **4**, 259 (1999).
16. Peña, M.T., Garcilazo, H., Riska, D.O., *Nucl. Phys.* **A683**, 322 (2001);
    Adam, J., Stadler, A., Peña, M. T., Gross, F.L., *Phys. Lett.* B **407**, 97 (1997).

Probing the effectiveness: Chiral perturbation theory calculations of low-energy electromagnetic reactions on deuterium

Daniel R. Phillips

Department of Physics and Astronomy, Ohio University, Athens, OH 45701, U. S. A.;
Email: phillips@phy.ohiou.edu.

Abstract. I summarize three recent calculations of electromagnetic reactions on deuterium in chiral perturbation theory. All of these calculations were carried out to $O(Q^4)$, i.e. next-to-next-to-leading order. The reactions discussed here are: elastic electron-deuteron scattering, Compton scattering on deuterium, and the photoproduction of neutral pions from deuterium at threshold.

INTRODUCTION

Effective field theory (EFT) is a technique commonly used in particle physics to deal with problems involving widely-separated energy scales. It facilitates the systematic separation of the effects of high-energy physics from those of low-energy physics. In strong-interaction physics the low-energy effective theory is chiral perturbation theory (χPT) [1]. Here the low-energy physics is that of nucleons and pions interacting with each other in a way that respects the spontaneously-broken approximate chiral symmetry of QCD. Higher-energy effects of QCD appear in χPT as non-renormalizable contact operators. The EFT yields amplitudes which can be thought of as expansions in the ratio of nucleon or probe momenta (denoted here by p and q) and the pion mass to the scale of chiral-symmetry breaking, $\Lambda_{\chi SB}$, which is of order the mass of the ρ meson. The existence of a small parameter $Q = p/\Lambda_{\chi SB}$, $q/\Lambda_{\chi SB}$, $m_\pi/\Lambda_{\chi SB}$, means that hadronic processes can be computed in a controlled way.

The momentum scale of binding in light nuclei is of order m_π and so we should be able to calculate the response of such nuclei to low-energy probes using χPT. The result is a systematically-improvable, model-independent description. Here I will describe a few recent calculations using this approach. Section 2 outlines the χPT expansion, and sketches the implications of χPT for calculations of the NN interaction. Section 3 then looks at electron-deuteron scattering in χPT as a probe of deuteron structre. Section 4 examines the use of Compton scattering on the deuteron as a way to extract neutron polaraizabilities. I conclude in Section 5 with a discussion of neutral pion photoproduction on deuterium at threshold. In these last two reactions I will argue that χPT's successful reproduction of the experimental data is in no small part due to its consistent treatment of the chiral structure of the nucleon and the deuteron.

POWER COUNTING AND DEUTERON WAVE FUNCTIONS

Power counting

Consider an elastic scattering process on the deuteron whose amplitude we wish to compute. If $\hat{O}$ is the transition operator for this process then the amplitude in question is simply $\langle \psi | \hat{O} | \psi \rangle$, with $|\psi\rangle$ the deuteron wave function. In this section, we follow Weinberg [2, 3, 4], and divide the formulation of a systematic expansion for this amplitude into two parts: the expansion for $\hat{O}$, and the construction of $|\psi\rangle$.

Chiral perturbation theory gives a systematic expansion for $\hat{O}$ of the form

$$\hat{O} = \sum_{n=0}^{\infty} \hat{O}^{(n)}, \qquad (1)$$

where we have labeled the contributions to $\hat{O}$ by their order n in the small parameter Q defined above. Equation (1) is an operator statement, and the nucleon momentum operator $\hat{p}$ appears on the right-hand side. However, the only quantities which ultimately affect observables are expectation values such as $\langle \psi | \hat{p} | \psi \rangle$. For light nuclei this number is generically small compared to $\Lambda_{\chi SB}$.

To construct $\hat{O}^{(n)}$ one first writes down the vertices appearing in the chiral Lagrangian up to order n. One then draws all of the two-body, two-nucleon-irreducible, Feynman graphs for the process of interest which are of chiral order Q^n. The rules for calculating the chiral order of a particular graph are:

Each nucleon propagator scales like $1/Q$;
Each loop contributes Q^4;
Graphs in which both particles participate in the reaction acquire a factor of Q^3;
Each pion propagator scales like $1/Q^2$;
Each vertex from the nth-order piece of the chiral Lagrangian contributes Q^n.

In this way we see that more complicated graphs, involving two-body mechanisms, and/or higher-order vertices, and/or more loops, are suppressed by powers of Q.

Deuteron wave functions

There remains the problem of constructing a deuteron wave function which is consistent with the operator $\hat{O}$. Weinberg's proposal was to construct a χPT expansion in Eq. (1) for the NN potential V, and then solve the Schrödinger equation to find the deuteron (or other nuclear) wave function [2, 3, 4]. Recent calculations have shown that the NN phase shifts can be understood, and deuteron bound-state static properties reliably computed, with wave functions derived from χPT in this way [5, 6, 7, 8, 9].

Now for χPT in the Goldstone-boson and single-nucleon sector loop effects are generically suppressed by powers of the small parameter Q. In zero and one-nucleon reactions the power counting in Q applies to the amplitude, and not to the two-particle potential. However, the existence of the deuteron tells us immediately that a power counting in which loop effects are suppressed cannot be correct for the two-nucleon case, since if it were there would be no NN bound state. Weinberg's proposal to instead power-count the potential is one response to this dilemma. However, its consistency has been vigorously debated in the literature (see [10, 11] for reviews). Recently Beane *et al.* [12] have resolved this discussion, by showing that Weinberg's proposal is consistent in the $^3S_1 - {}^3D_1$ channel.

One way to understand the χPT power-counting for deuteron wave functions is to examine the deuteron wave function in three different regions. Firstly, in the region $R \gg 1/m_\pi$ the deuteron wave function is described solely by the asymptotic normalizations A_S, A_D, and the binding energy B. These quantities are observables, in the sense that they can be extracted from phase shifts by an analytic continuation to the deuteron pole.

The second region corresponds to $R \sim 1/m_\pi$. Here pion exchanges play a role in determining the NN potential V, and, associatedly, the deuteron wave functions u and w. The leading effect comes from iterated one-pion exchange—as has been known for at least fifty years. Calculations with one-pion exchange (OPE) defining the potential in this regime will be referred to below as "leading-order" (LO) calculations for the deuteron wave function. Corrections at these distances come from two-pion exchange, and these corrections can be consistently calculated in χPT. They are suppressed by powers of the small parameter Q, and in fact the "leading" two-pion exchange is suppressed by Q^2 relative to OPE. This two-pion exchange can be calculated from vertices in $\mathcal{L}_{\pi N}^{(1)}$ and its inclusion in the NN potential results in the so-called "NLO" calculation described in detail in Ref. [7]. Corrections to this two-pion-exchange result from replacing one of the NLO two-pion-exchange vertices by a vertex from $\mathcal{L}_{\pi N}^{(2)}$. This results in an additional suppression factor of Q, or an overall suppression of Q^3 relative to OPE, and an "NNLO" chiral potential [5, 6, 7, 8]. More details on this can be found in the contributions of Epelbaum and Timmermans to these proceedings.

Finally, at short distances, $R \ll 1/m_\pi$ we cutoff the chiral one and two-pion-exchange potentials and put in some short-distance potential whose parameters are arranged so as to give the correct deuteron asymptotic properties.

ELASTIC ELECTRON SCATTERING ON DEUTERIUM

One quantitative test of this picture of deuteron structure is provided by elastic electron-deuteron scattering. We thus turn our attention to the deuteron electromagnetic form factors G_C, G_Q, and G_M. These are matrix elements of the deuteron current J_μ, with:

$$G_C = \frac{1}{3e(1+\eta)} \left(\langle 1|J^0|1\rangle + \langle 0|J^0|0\rangle + \langle -1|J^0|-1\rangle \right), \quad (2)$$

$$G_Q = \frac{1}{M_d^2} \frac{1}{2e\eta(1+\eta)} \left(\langle 0|J^0|0\rangle - \langle 1|J^0|1\rangle \right) \quad (3)$$

$$G_M = \frac{1}{\sqrt{2\eta}(1+\eta)} \langle 1|J^+|0\rangle \quad (4)$$

where we have labeled these (non-relativistic) deuteron states by the projection of the deuteron spin along the direction of the momentum transfer $\mathbf{q}$ and $\eta \equiv |\mathbf{q}|^2/(4M_d^2)$. G_C, G_Q, and G_M are related to the experimentally-measured A, B, and T_{20} in the usual way, with T_{20} being primarily sensitive to G_Q/G_C and B depending only to G_M. Here we will compare calculations of the charge and quadrupole form factor with the recent extractions of G_C and G_Q from data [13].

Both of these form factors involve the zeroth-component of the deuteron four-current J^0. Here we split J^0 into two pieces: a one-body part, and a two-body part. The one-body part of J^0 begins at order Q (where we are counting the proton charge $e \sim Q$) with the impulse approximation diagram calculated with the non-relativistic single-nucleon charge operator for strutcutreless nucleons. Corrections to the single-nucleon charge operator from relativistic effects and nucleon structure are suppressed by two powers of Q, and thus arise at $O(Q^3)$, which is the next-to-leading order (NLO) for G_C and G_Q. At this order one might also expect meson-exchange current (MEC) contributions, such as those shown in Fig. 1. However, all MECs constructed with vertices from $\mathcal{L}_{\pi N}^{(1)}$ are isovector, and so the first effect does not occur until N²LO, or $O(Q^4)$, where an $NN\pi\gamma$ vertex from $\mathcal{L}_{\pi N}^{(2)}$ replaces the upper vertex in the middle graph of Fig. 1, and produces an isoscalar contribution to the deuteron charge operator. (This exchange-charge contribution was first derived by Riska [14].)

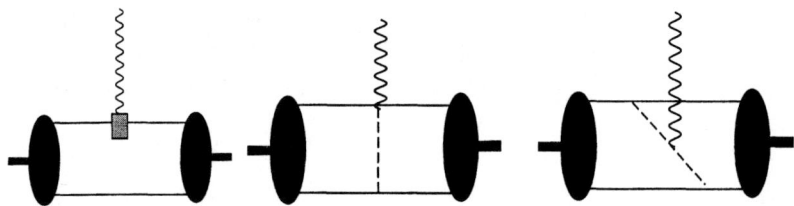

FIGURE 1. The impulse-approximation contribution to G_C and G_Q is shown on the left, while two meson-exchange current mechanisms which would contribute were the deuteron not an isoscalar target are depicted in the middle and on the right.

The most important correction that arises at NLO is the inclusion of *nucleon* structure in χPT. At $O(Q^3)$ the isoscalar form factors are dominated by short-distance physics, and so the only correction to the point-like leading-order result comes from the inclusion of the nucleon's electric radius, i.e.

$$G_E^{(s)} \text{ χPT NLO} = 1 - \tfrac{1}{6}\langle r_E^{(s)2}\rangle q^2. \tag{5}$$

This description of nucleon structure breaks down at momentum transfers q of order 300 MeV. There is a concomitant failure in the description of eD scattering data [15, 16]. Consequently, in order to focus on *deuteron* structure, in the results presented below I have chosen to circumvent this issue by using a "factorized" inclusion of nucleon structure [16]. This facilitates the inclusion of experimentally-measured single-nucleon form factors in the calculation, thereby allowing us to test how far the theory is able to describe the *two-body* dynamics that takes place in eD scattering.

The results for G_C and G_Q are shown in Fig. 2. The figure demonstrates that convergence is quite good below $q \sim 700$ MeV–especially for G_C. The results shown are for the NLO chiral wave function, but the use of the NNLO chiral wave function, or indeed of simple wave functions which include only one-pion exchange, do not modify the picture greatly below $q = 700$ MeV [16]. It is also clear that–provided information from eN

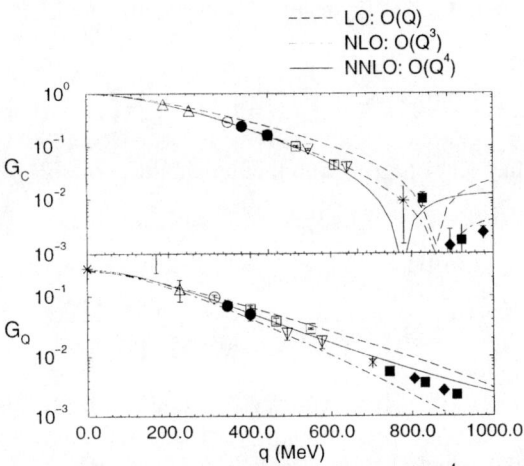

FIGURE 2. The deuteron charge and quadrupole form factors to order Q^4 in chiral perturbation theory. The experimental data is taken from the compilation of Ref. [13]. G_Q is in units of fm^2.

scattering is taken into account–χPT is perfectly capable of describing the charge and quadrupole form factors of deuterium at least as far as the minimum in G_C. This result is extremely encouraging for the application of EFT to light nuclei.

G_M can be obtained in a similar way, but, importantly, the LO contribution to G_M is $O(Q^2)$. Furthermore, no two-body mechanism enters until $O(Q^5)$, when an undetermined two-body counterterm appears [15, 17]. Results for F_M at $O(Q^4)$ turn out to be of similar quality to those for F_C [15, 16], but are somewhat more sensitive to short-distance physics, as was expected given the presence of the $O(Q^5)$ counterterm in this observable.

The static properties of the deuteron obtained in this expansion are also generically quite reasonable, and have good convergence properties. The one exception to this is the deuteron quadrupole moment, Q_d, which, is underpredicted by about 4%–as is also true in all modern potential-model calculations [18]. However, such an underprediction is not unexpected in χPT since simple estimates of the effect on Q_d of higher-order terms in the chiral Lagrangian suggest that a discrepancy of order 5% is to be expected at $O(Q^4)$. Q_d is rather sensitive to short-distance physics and it transpires that higher-order counterterms have a larger effect on it than on G_C [19, 20]. This way of understanding the "Q_d puzzle" is one example of the way in which χPT can assist in the analysis of electromagnetic currents for few-body systems. For an analogous application of χPT/EFT ideas to the important solar reaction $pp \rightarrow de^+\nu_e$ see Ref. [21] and the contribution of Marcucci to these proceedings.

COMPTON SCATTERING ON DEUTERIUM

Compton scattering on the nucleon at low energies is a fundamental probe of the long-distance structure of these hadrons. This process has been studied in χPT in

Refs. [22, 23], where the following results for the proton polarizabilities were obtained at LO:

$$\alpha_p = \frac{5e^2 g_A^2}{384\pi^2 f_\pi^2 m_\pi} = 12.2 \times 10^{-4} \text{fm}^3; \qquad \beta_p = \frac{e^2 g_A^2}{768\pi^2 f_\pi^2 m_\pi} = 1.2 \times 10^{-4} \text{fm}^3. \qquad (6)$$

Recent experimental values for the proton polarizabilities are [24]

$$\alpha_p + \beta_p = 13.23 \pm 0.86^{+0.20}_{-0.49} \times 10^{-4} \text{fm}^3,$$
$$\alpha_p - \beta_p = 10.11 \pm 1.74^{+1.22}_{-0.86} \times 10^{-4} \text{fm}^3, \qquad (7)$$

where the first error is a combined statistical and systematic error, and the second set of errors comes from the theoretical model employed. These values are in good agreement with the χPT predictions.

Chiral perturbation theory also predicts $\alpha_n = \alpha_p$, $\beta_n = \beta_p$ at this order. The neutron polarizabilities α_n and β_n are difficult to measure, due to the absence of suitable neutron targets, and so this prediction is not well tested. One way to extract α_n and β_n is to perform Compton scattering on nuclear targets. Coherent Compton scattering on a deuteron target has been measured at $E_\gamma = 49$ and 69 MeV by the Illinois group [25] and $E_\gamma = 84.2 - 104.5$ MeV at Saskatoon [26]. The amplitude for Compton scattering on the deuteron clearly involves mechanisms other than Compton scattering on the individual constituent nucleons. Hence, the desire to extract neutron polarizabilities argues for a theoretical calculation of Compton scattering on the deuteron that is under control in the sense that it accounts for *all* mechanisms to a given order in χPT.

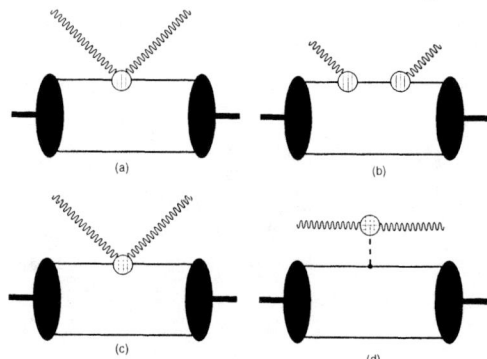

FIGURE 3. Graphs which contribute to Compton scattering on the deuteron at $O(Q^2)$ (a) and $O(Q^3)$ (b-d). The sliced and diced blobs are from $\mathcal{L}^{(3)}_{\pi N}$ (c) and $\mathcal{L}^{(4)}_{\pi \gamma}$ (d). Crossed graphs are not shown.

The Compton amplitude we wish to evaluate is (in the γd center-of-mass frame):

$$\begin{aligned} T^{\gamma d}_{M'\lambda'M\lambda}(\vec{k}',\vec{k}) &= \int \frac{d^3p}{(2\pi)^3} \, \psi_{M'}\left(\vec{p} + \frac{\vec{k}-\vec{k}'}{2}\right) T_{\gamma N_{\lambda'\lambda}}(\vec{k}',\vec{k}) \, \psi_M(\vec{p}) \\ &+ \int \frac{d^3p\, d^3p'}{(2\pi)^6} \, \psi_{M'}(\vec{p}') \, T^{2N}_{\gamma N N_{\lambda'\lambda}}(\vec{k}',\vec{k}) \, \psi_M(\vec{p}) \end{aligned} \qquad (8)$$

where M (M') is the initial (final) deuteron spin state, and λ (λ') is the initial (final) photon polarization state, and $\vec{k}$ ($\vec{k}'$) the initial (final) photon three-momentum, which are constrained to $|\vec{k}| = |\vec{k}'| = \omega$. The amplitude $T_{\gamma N}$ represents the graphs of Fig. 3 and Fig. 4b where the photon interacts with only one nucleon. The amplitude $T_{\gamma NN}^{2N}$ represents the graphs of Fig. 4a where there is an exchanged pion between the two nucleons.

The LO contribution to Compton scattering on the deuteron is shown in Fig. 3(a). This graph involves a vertex from $\mathcal{L}_{\pi N}^{(2)}$ and so is $O(Q^2)$. This contribution is simply the Thomson term for scattering on the proton. There is thus no sensitivity to either two-body contributions *or* nucleon polarizabilities at this order. At $O(Q^3)$ there are several more graphs with a spectator nucleon (Figs. 3(b),(c),(d)), as well as graphs involving an exchanged pion with leading order vertices (Fig. 4(a)) and one loop graphs with a spectator nucleon (Fig. 4(b)) [27]. Graphs such as Fig. 4(b) contain the physics of the proton and neutron polarizabilities at $O(Q^3)$ in χPT.

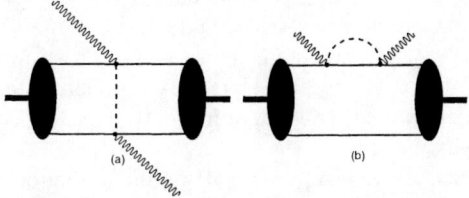

FIGURE 4. Graphs which contribute to Compton scattering on the deuteron at $O(Q^3)$. Crossed graphs are not shown.

We employed a variety of wave functions ψ, and found only moderate wave-function sensitivity. Results shown here are generated with the NLO chiral wave function of Ref. [7]. Figure 5 shows the results at $E_\gamma = 49, 69,$ and 95 MeV. For comparison we have included the calculation at $O(Q^2)$ in the kernel, where the second contribution in Eq. (8) is zero, and the single-scattering contribution is given solely by Fig. 3(a). At $O(Q^3)$ all contributions to the kernel are fixed in terms of known pion and nucleon parameters, so to this order χPT makes *predictions* for deuteron Compton scattering. We also show the $O(Q^4)$ result which will be discussed below. The curves indicate that higher-order corrections get larger as ω is increased–as expected.

We have also shown the six Illinois data points at 49 and 69 MeV [25] and the Saskatoon data at 95 MeV [26]. Statistical and systematic errors have been added in quadrature. It is quite remarkable how well the $O(Q^2)$ calculation reproduces the

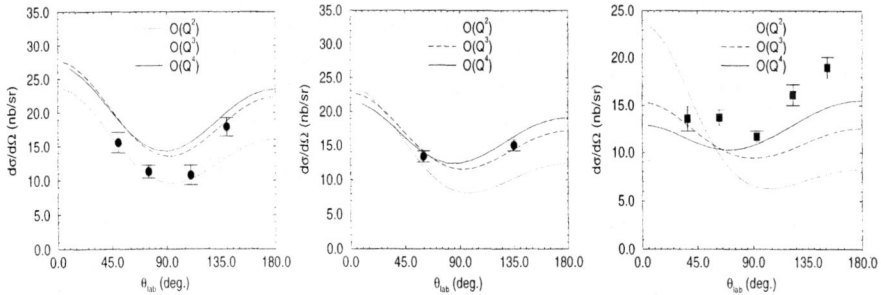

FIGURE 5. Results of the $O(Q^2)$ (dotted line), $O(Q^3)$ (dashed line), and $O(Q^4)$ (solid line) calculations for $E_\gamma = 49$ MeV, 69 MeV and 95 MeV respectively from left to right.

49 MeV data. However, the agreement is somewhat fortuitous: there are significant $O(Q^3)$ corrections. Note that at these lower photon energies Weinberg power counting begins to break down, since it is designed for $\omega \sim m_\pi$, and does *not* recover the deuteron Thomson amplitude as $\omega \to 0$. Correcting the power counting to remedy this difficulty appears to improve the description of the 49 MeV data, without significantly modifying the higher-energy results [27]. Meanwhile, the agreement of the $O(Q^3)$ calculation with the 69 MeV data is very good, although only limited conclusions can be drawn. These results are not very different from other, potential-model, calculations [28, 29, 30]. They are also quite similar to those obtained in NN EFTs without explicit pions (see Ref. [31], and the contribution of Grießhammer to these proceedings). However our calculation is the only one that does not employ the polarizability approximation for the γN amplitude.

At $O(Q^4)$ single-nucleon counterterms which shift the polarizabilities enter the calculation. However, there are still no two-body counterterms contributing to $\gamma d \to \gamma d$ at this order. In this sense Compton scattering on deuterium at $O(Q^4)$ is analogous to the reaction $\gamma d \to \pi^0 d$ discussed below: an $O(Q^4)$ calculation allows us to test the single-nucleon physics which is used to predict the results of coherent scattering on deuterium, since there are no undetermined parameters in the two-body mechanisms that enter to this order in the chiral expansion.

The $O(Q^4)$ plot shown above is a partial calculation at that order. It includes all two-body mechanisms at $O(Q^4)$, and some one-body mechanisms, the latter being calculated in ways motivated by dispersion-relation analyses. The values of the proton polarizabilities used in Fig. 5 were taken from Eq. (7). Meanwhile the (disputed) neutron-atom scattering value of Ref. [32] was employed for α_n and the Baldin sum rule used to fix β_n. To demonstrate the sensitivity to α_n, we can also use the freedom in the single-nucleon amplitude at $O(Q^4)$ to fit the SAL data using the incomplete $O(Q^4)$ calculation. A reasonable fit at backward angles can be achieved with $\alpha_n = 4.4$ and $\beta_n = 10$. These numbers are in startling disagreement with the $O(Q^3)$ χPT expectations. We have also plotted the cross-section at 69 MeV with $\alpha_n = 4.4$ and $\beta_n = 10$: this curve misses the Illinois data. (Similar results were found in a potential model in Ref. [29].) This situation poses an interesting theoretical puzzle. A full $O(Q^4)$ calculation in χPT using the recently derived single-nucleon Compton amplitude [33] is necessary before firm conclusions can be drawn. Such a calculation is in progress [34].

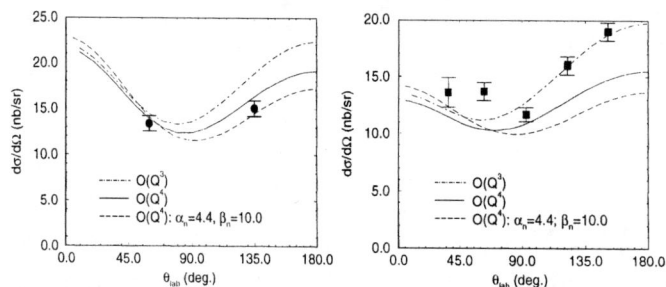

FIGURE 6. Results of $O(Q^3)$ (dashed line), partial $O(Q^4)$ (solid line), and partial $O(Q^4)$ with modified α_n (dot-dashed line), χPT calculations for $E_\gamma = 69$ MeV (left panel) and 95 MeV (right panel).

NEUTRAL PION PHOTOPRODUCTION ON DEUTERIUM

Pion photoproduction on the nucleon near threshold has been studied up to $O(Q^4)$ in χPT with the delta integrated out in Ref. [35]. The differential cross-section at threshold is given solely by E_{0+}, the electric dipole amplitude. In χPT neutral pion photoproduction on the nucleon does not begin until $O(Q^2)$, where there is a tree-level contribution from a $\gamma\pi NN$ vertex. Then at $O(Q^3)$ there are tree-level contributions involving the magnetic moment of the nucleon. The sum of these $O(Q^2)$ and $O(Q^3)$ effects reproduces an old "low-energy theorem". Also at $O(Q^3)$ there occur finite loop corrections where the photon interacts with a virtual pion which then rescatters on the nucleon. This large quantum effect was missed in the old "low-energy theorem" and is absent in most models. At $O(Q^4)$ there are loops with relativistic corrections, together with a counterterm. The proton amplitude that results from fitting this counterterm to data produces energy-dependence in relatively good agreement with the recent results from Mainz and Saskatoon. At the same order there is also a prediction for the near-threshold behavior of the $\gamma n \to \pi^0 n$ reaction; the cross-section is considerably larger than that obtained in models that omit the important pion-cloud $O(Q^3)$ diagrams. In fact, the $O(Q^4)$ prediction is that the cross section for $\gamma n \to \pi^0 n$ is about four times larger than that for neutral pion photoproduction on the proton. A good way to test this prediction is to study neutral pion photoproduction on the deuteron.

If deuterium is to be used as a target in this particular process then two-body mechanisms that contribute to the reaction must be calculated. χPT provides an ideal way to do this, as was demonstrated by Beane and collaborators [36].

Consider the reaction $\gamma(k) + d(p_1) \to \pi^0(q) + d(p_2)$ in the threshold region, $\vec{q} \simeq 0$, where the pion is in an S wave with respect to the center-of-mass (cm) frame. For real photons the threshold differential cross section is:

$$\frac{|\vec{k}|}{|\vec{q}|} \frac{d\sigma}{d\Omega}\bigg|_{|\vec{q}|=0} = \tfrac{8}{3} E_d^2. \tag{9}$$

We now present the results of the chiral expansion of the dipole amplitude E_d to $O(Q^4)$.

Two-body contributions to E_d do not begin until $O(Q^3)$. Thus, to $O(Q^4)$ we have

$$E_d = \langle\psi|\hat{O}_{ob}|\psi\rangle + \langle\psi|\hat{O}_{tb}^{(3)}|\psi\rangle + +\langle\psi|\hat{O}_{tb}^{(4)}|\psi\rangle \tag{10}$$

$$\equiv E_d^{ss} + E_d^{tb,3} + E_d^{tb,4}, \tag{11}$$

where we have explicitly isolated the $O(Q^3)$ and $O(Q^4)$ two-body contributions to the operator $\hat{O}$, and $|\psi\rangle$ is a deuteron wave function.

The single-scattering contribution to E_d, E_d^{ss}, is given by all diagrams where the photon is absorbed and the pion emitted from one nucleon with the second nucleon acting as a spectator, i. e. the impulse approximation result. Since the χPT results for the elementary S-wave pion production amplitudes to $O(Q^4)$ [35] are known there is an $O(Q^4)$ χPT *prediction* for E_d^{ss}. This is the prediction we want to test.

But, before comparing this prediction with the experimental data we must compute the two-body contributions to E_d. Those of $O(Q^3)$ which survive at threshold, in the Coulomb gauge, are shown in Fig. 7 [37].

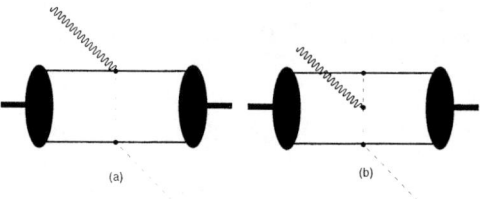

FIGURE 7. Two-nucleon graphs which contribute to neutral pion photoproduction at threshold to $O(Q^3)$ (in the Coulomb gauge). All vertices come from $\mathcal{L}_{\pi N}^{(1)}$.

At $O(Q^4)$, we have to consider the two-nucleon diagrams—some of which are shown in Fig. 8, with the blob characterizing an insertion from $\mathcal{L}_{\pi N}^{(2)}$. There are also relativistic corrections to the graphs in Fig. 7.

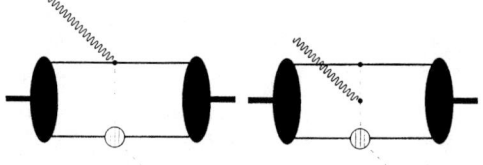

FIGURE 8. Characteristic two-nucleon graphs contributing at $O(Q^4)$ to neutral pion photoproduction. The hatched circles denote an insertion from $\mathcal{L}_{\pi N}^{(2)}$.

One can show that the only terms that survive at threshold result in insertions $\sim 1/2M$, $\sim g_A/2M$ and $\sim \kappa_{0,1}$. To $O(Q^4)$ there are no four-nucleon operators contributing to the deuteron electric dipole amplitude, and so *no new, undetermined parameters appear*. The only free parameter is fixed in $\gamma p \to \pi^0 p$, and so a genuine *prediction* can be, and was, made for the reaction $\gamma d \to \pi^0 d$ at threshold.

We now present results. Using a variety of deuteron wave functions the single-scattering contribution is found to be [36]:

$$E_d^{ss} = (0.36 \pm 0.05) \times 10^{-3}/m_{\pi^+}. \tag{12}$$

The sensitivity of the single-scattering contribution E_d^{ss} to the elementary neutron amplitude may be parameterized by:

$$E_d^{ss} = \left[0.36 - 0.38 \cdot (2.13 - E_{0+}^{\pi^0 n})\right] \times 10^{-3}/m_{\pi^+}. \tag{13}$$

The two-nucleon contribution is evaluated at $O(Q^3)$ using different deuteron wave functions [36]. The results prove to be largely insensitive to the choice of wave function, as do the two-body contributions at $O(Q^4)$. Choosing the AV18 wave function for definiteness the results are summarized in Table 1[1]. The $O(Q^4)$ contributions give

TABLE 1. Values for E_d in units of $10^{-3}/m_{\pi^+}$ from one-nucleon contributions (1N) up to $O(Q^4)$, two-nucleon kernel (2N) at $O(Q^3)$ and at $O(Q^3)$, and their sum (1N + 2N).

1N	2N		1N + 2N
$Q + Q^2 + Q^3 + Q^4$	Q^3	Q^4	$Q + Q^2 + Q^3 + Q^4$
0.36	−1.90	−0.25	−1.79

corrections of order 15% to the $O(Q^3)$ two-nucleon terms. We also observe that $E_d^{tb,4}$ is of the same size as E_d^{ss}, clearly demonstrating the need to go to this order in the expansion. This gives us confidence that the χPT expansion is controlled, and that we may compare experimental data with the $O(Q^4)$ prediction [36]:

$$E_d = (-1.8 \pm 0.2) \times 10^{-3}/m_{\pi^+}. \quad (14)$$

To see the sensitivity to the elementary neutron amplitude, we set the latter to zero and find $E_d = -2.6 \times 10^{-3}/m_{\pi^+}$ (for the AV18 potential). Thus χPT makes a prediction that differs markedly from conventional models, with that difference arising predominantly from a different result for $E_{0+}^{\pi^0 n}$. An experimental test of this prediction was carried out recently at Saskatoon [38]. The results for the pion photoproduction cross-section near threshold are shown in Fig. 9, together with the χPT prediction at threshold. The agreement with χPT to order $O(Q^4)$ is not better than a reasonable estimate of higher-order terms, but χPT is clearly superior to models. This is compelling evidence of the importance of chiral loops, and also testimony to the consistency and usefulness of χPT in analyzing low-energy reactions on deuterium.

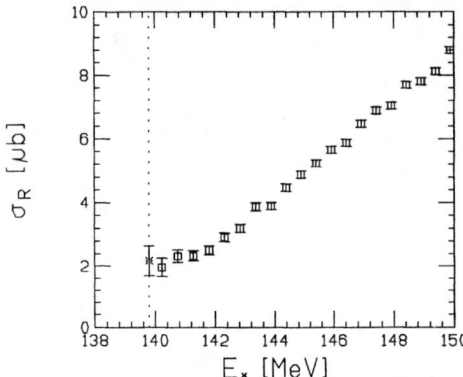

FIGURE 9. Reduced cross-section $\sigma_R = (k/q)\sigma$ in μb for neutral pion photoproduction as function of the photon energy in MeV. Threshold is marked by a dotted line. Squares are data points from Ref. [38] and the star is the χPT prediction of Ref. [36]. Figure courtesy of Ulf Meißner.

---

[1] Similar answers were obtained with the wave functions of Ref. [5], which should be equivalent to the NNLO wave function discussed above, up to higher-order terms.

ACKNOWLEDGMENTS

I thank the organizers of "Mesons and Light Nuclei" for an enjoyable meeting. I am also grateful to Silas Beane, Manuel Malheiro, and Bira van Kolck for a profitable and enjoyable collaboration on $\gamma d \to \gamma d$. Thanks are due to Silas for discussions on $\gamma d \to \pi^0 d$ and numerous other subjects in effective field theory. Finally, I am grateful to the U. S. Insitute for Nuclear Theory for its hospitality during the writing of this paper.

REFERENCES

1. Bernard, V., Kaiser, N., and Meißner, U.-G., *Int. Jour. of Mod. Phys. E* **4**, 193 (1995).
2. Weinberg, S., *Phys. Lett. B* **251**, 288 (1990).
3. Weinberg, S., *Nucl. Phys.* **B363**, 3 (1991).
4. Weinberg, S., *Phys. Lett. B* **295**, 114 (1992).
5. Ordonéz, C., Ray, L., and van Kolck, U., *Phys. Rev. C* **53**, 2086 (1996).
6. Kaiser, N., Brockmann, R., and Weise, W., *Nucl. Phys.* **A625**, 758 (1997).
7. Epelbaum, E., Glockle, W., and Meißner, U.-G., *Nucl. Phys.* **A671**, 295 (1999).
8. Rentmeester, M. C. M., Timmermans, R. G. E., Friar, J. L., and de Swart, J. J., *Phys. Rev. Lett.* **82**, 4992 (1999).
9. Entem, D. R., and Machleidt, R., nucl-th/0107057.
10. van Kolck, U., *Prog. Part. Nucl. Phys.* **43**, 409 (1999).
11. Beane, S. R., Bedaque, P. F., Haxton, W., Phillips, D. R., and Savage, M. J., in *At the frontier of particle physics–Handbook of QCD*, M. Shifman, ed. (World Scientific, Singapore, 2000).
12. Beane, S. R., Bedaque, P. F., Savage, M. J., and van Kolck, U., nucl-th/0104030.
13. Abbott, D., *et al.* [JLAB t20 Collaboration] *Eur. Phys. J.* **A7**, 421 (2000).
14. Riska, D. O., *Prog. Part. Nucl. Phys.* **11**, 199 (1984).
15. Meißner, U.-G., and Walzl, M., *Phys. Lett. B* **513**, 37 (2001).
16. Phillips, D. R., in preparation.
17. Park, T.-S., Kubodera, K., Min, D.-P., and Rho, M., *Phys. Lett. B* **472**, 232 (2000).
18. Carlson, J., and Schiavilla, R., *Rev. Mod. Phys.* **70**, 743 (1998).
19. Chen, J.-W., Rupak, G., and Savage, M., *Nucl. Phys.* **A653**, 386 (1999).
20. Phillips, D. R., and Cohen, T. D., *Nucl. Phys.* **A668**, 45 (2000).
21. Park, T. S., Marcucci, L. E., Viviani, M., Kievsky, A., Rosati, S., Kubodera, K., Min, D.-P., and Rho, M., nucl-th/0106025.
22. Bernard, V., Kaiser, N., and Meißner, U. G., *Nucl. Phys.* **B383**, 442 (1992).
23. Bernard, V., Kaiser, N., Kambor, J., and Meißner, U. G., *Nucl. Phys.* **B388**, 315 (1992).
24. Tonnison, J., Sandorfi, A. M., Hoblit, S., and Nathan, A. M., *Phys. Rev. Lett.* **80**, 4382 (1998).
25. Lucas, M. A., *Compton scattering from the deuteron at intermediate energies*, Ph.D. thesis, University of Illinois (1994), unpublished.
26. Hornidge, D. L., *et al.*, *Phys. Rev. Lett.* **84**, 2334 (2000).
27. Beane, S. R., Malheiro, M., Phillips, D. R., and van Kolck, U., *Nucl. Phys.* **A656**, 367 (1999).
28. Wilbois, T., Wilhelm, P., and Arenhovel, H., *Few Bod. Sys.* **9**, 263 (1995).
29. Levchuk, M. I., and L'vov, A. I., *Nucl. Phys.* **A674**, 449 (2000).
30. Karakowski, J. J., and Miller, G. A., *Phys. Rev. C* **60**, 014001 (1999).
31. Grießhammer, H. W., and Rupak, G., nucl-th/0012096.
32. Schmiedmayer, J., Riehs, P., Harvey, J. A., and Hill, N. W., *Phys. Rev. Lett.* **66**, 1015 (1991).
33. McGovern, J., *Phys. Rev. C* **63**, 064608 (2001), and in these proceedings.
34. Beane, S. R., Malheiro, M., McGovern, J., Phillips, D. R., and van Kolck, U., in preparation.
35. Bernard, V., Kaiser, N., and Meißner, U.-G., *Z. Phys. C* **70**, 483 (1996).
36. Beane, S. R., Bernard, V., Lee, H., Meißner, U.-G., and van Kolck, U., *Nucl. Phys.* **A618**, 381 (1997).
37. Beane, S. R., Lee, C.-Y., van Kolck, U., *Phys. Rev. C* **52**, 2914 (1995).
38. Bergstrom, J. C., Igarashi, R., Vogt, J. M., Kolb, N., Pywell, R. E., Skopik, D. M., and Korkmaz, E., *Phys. Rev. C* **57**, 3203 (1998).

Hard exclusive processes at HERMES

G. van der Steenhoven (for the HERMES Collaboration)

NIKHEF, P.O. Box 41882, 1009 DB Amsterdam, The Netherlands

Abstract. The production of mesons (or photons) in deep-inelastic lepton scattering gives access to information on the structure of hadrons that is otherwise hard to obtain. When the production process involves at least one hard scale and is exclusive, the data can be interpreted in terms of the recently introduced generalized parton distributions (GPDs). These GPDs provide a unified description of hadronic structure, which can be applied to many different reactions. It will be shown how new data collected by the HERMES experiment at DESY are used to obtain first information on GPDs. Examples of hard exclusive processes observed at HERMES include the production of pseudoscalar and vector mesons, and deeply-virtual Compton scattering.

INTRODUCTION

Recently, a renewed interest in hard exclusive processes emerged when it was realized that such processes can be interpreted in a QCD framework. Collins, Frankfurt and Strikman [1] proved that the amplitude for two-body exclusive deep-inelastic processes such as[1]

$$\gamma_L^*(q) + p \to M(q-\Delta) + B(p+\Delta), \qquad (1)$$

can be factorized in three terms: (i) a hard scattering coefficient, which is calculable in QCD; (ii) the minimal Fock component of the wave function of the produced meson M (which can also be a photon); and (iii) a generalized parton distribution (GPD) describing the interference or correlation between two partons with momentum fractions $x+\xi$ and $x-\xi$, respectively. Following their introduction [2, 3, 4] the generalized parton distributions attracted a lot of attention, as it was realized that they provide a unified formalism for the description of inclusive deep-inelastic scattering, exclusive meson production, electromagnetic form factor measurements and (deeply)-virtual Compton scattering [5, 6]. Moreover, Ji [3] has shown that the first moment of certain GPDs can be related to the total angular momentum of the quarks and the gluons in the nucleon.

Experimental information on GPDs can be obtained by measuring cross sections or asymmetries of exclusive reactions of the type shown above. By studying different final states M one is sensitive to different combinations of GPDs. Exclusive vector-meson production, for instance, is only sensitive to the unpolarized GPDs, $H(x,\xi,\Delta^2)$ and $E(x,\xi,\Delta^2)$ with $\Delta^2 = t$ representing the momentum transfer to the target. Deeply-virtual Compton scattering (DVCS) on the other hand is also sensitive to the polarized GPDs, $\tilde{H}(x,\xi,\Delta^2)$ and $\tilde{E}(x,\xi,\Delta^2)$.

---

[1] In the two-body reaction displayed in Eq. (1) the final state consists of meson (or photon) M carrying momentum $q - \Delta$ and a nucleon (or nucleon resonance) B carrying momentum $p + \Delta$.

So far, very few experimental data exist on exclusive processes. The reasons are the small cross sections involved, the high energy resolution required, and the need to extract the longitudinal cross section from the total cross section.[2]

In this contribution new experimental data are presented on some of the exclusive processes mentioned above. These data have been collected by the HERMES experiment at DESY in Hamburg. Where possible the data will be interpreted in the GPD framework. The aim of these first analyses is to demonstrate that exclusive processes can indeed be observed – despite the small cross sections involved, thus paving the way for future experiments at higher luminosity aimed at the actual extraction of GPDs (as a function of x, ξ and Δ^2.)

The paper is organized as follows. In the next section a brief description of the HERMES experiment is presented. Experimental results on vector-meson production, deeply-virtual Compton scattering and pseudoscalar-meson production are discussed in three subsequent sections. The paper is concluded by a short section in which the results are summarized and some future perspectives are given.

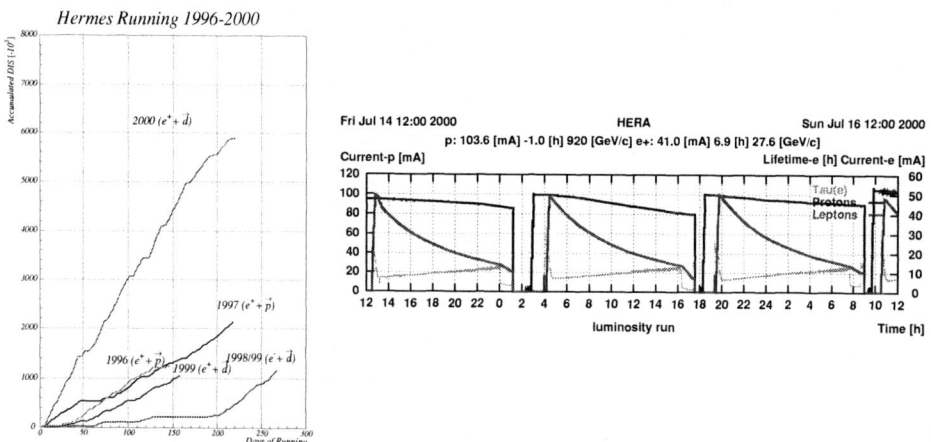

FIGURE 1. Cumulative number of polarized DIS events collected by the HERMES experiment in the period 1996 – 2000 as a function of the number of days since the beginning of the run in that year (left). Example of the intensity profile (upper curves) and lifetime (lowest curve) of the HERA beams during 48 hours in July 2000 (right). Note the sharp decrease of the positron life time at the end of each fill due to the injection of high-density unpolarized gas in the HERMES target cell.

THE HERMES EXPERIMENT

A detailed description of the HERMES experiment can be found elsewhere [7]. For the present purposes it is sufficient to remark that the experiment consists of a forward angle spectrometer with an angular acceptance ranging from 40 to 140 mrad in the vertical direction both below and above the accelerator plane, and from -170 to + 170 mrad in

---

[2] The factorization proof of Ref. [1] only applies to exclusive reactions induced by longitudinal photons.

the horizontal direction. A dipole magnet, various sets of tracking chambers, and various particle identification detectors provide information on the direction, momentum, charge and type of the particle detected. Behind these detectors a large segmented electromagnetic calorimeter is located, which is used for electron-hadron discrimination and for photon identification.

Many deep-inelastic scattering (DIS) data have been collected since the commissioning of the experiment in 1995. In figure 1 (left) the accumulated polarized DIS events are shown as a function of the number of days since the start of the run for each year. The year 2000 was particularly successful, due to the fact that a smaller target cell was used as compared to previous years. As a result the luminosity was increased by more than a factor of two. Moreover, at the end of each HERA fill, high-density unpolarized gas was injected into the target cell. This made it possible to accumulate over 14 Million unpolarized DIS events on various target nuclei. Using this running mode it has been demonstrated that the HERMES spectrometer is able to operate smoothly up to luminosities of about 4×10^{33} N/cm^2/s.

VECTOR-MESON PRODUCTION

In this section data are presented for (diffractive) production of vector mesons on hydrogen. As the HERMES collaboration has already published several papers on this subject [8, 9, 10], the emphasis is on data that have recently become availabe. The various vector mesons are discussed in separate subsections. Where possible the longitudinal cross section is extracted, such that comparisons with available GPD calculations can be made. Moreover, the relative yield of ρ, ϕ and ω vector mesons is discussed as well.

Diffractive ρ^0 production

The reaction mechanism for ρ^0 electroproduction on hydrogen changes significantly at $W \approx 4$ GeV [8]. Well above this value the cross section rises slowy with W and can be interpreted in terms of (multiple) gluon exchange. In the transition region between roughly 4 and 10 GeV one of the first calculations based on the GPD-framework predicted that the mechanism should be dominated by quark exchange instead [11]. In order to investigate this prediction the longitudinal component of the cross section needs to be determined. This can be done by extracting the spin-density matrix elements from the (decay) angular distributions for ρ^0 production [9]. The results of such an analysis (on hydrogen) are displayed in figure 2. The extracted spin-density matrix elements (denoted by r_{ij}^{kl}) are plotted as a function of the four-momentum transfer squared Q^2. In most cases the data are compared to a solid (horizontal) line, which represents the value of that matrix element if s-channel helicity conservation (SCHC) is assumed. The SCHC assumption implies that the helicity of the virtual photon is preserved by the produced vector meson, leading to constraints on the values of certain matrix elements. As can be seen from figure 2, all data are in agreement with SCHC except for r_{00}^5. The role of this matrix element is analagous to that of the so-called out-of-plane structure functions

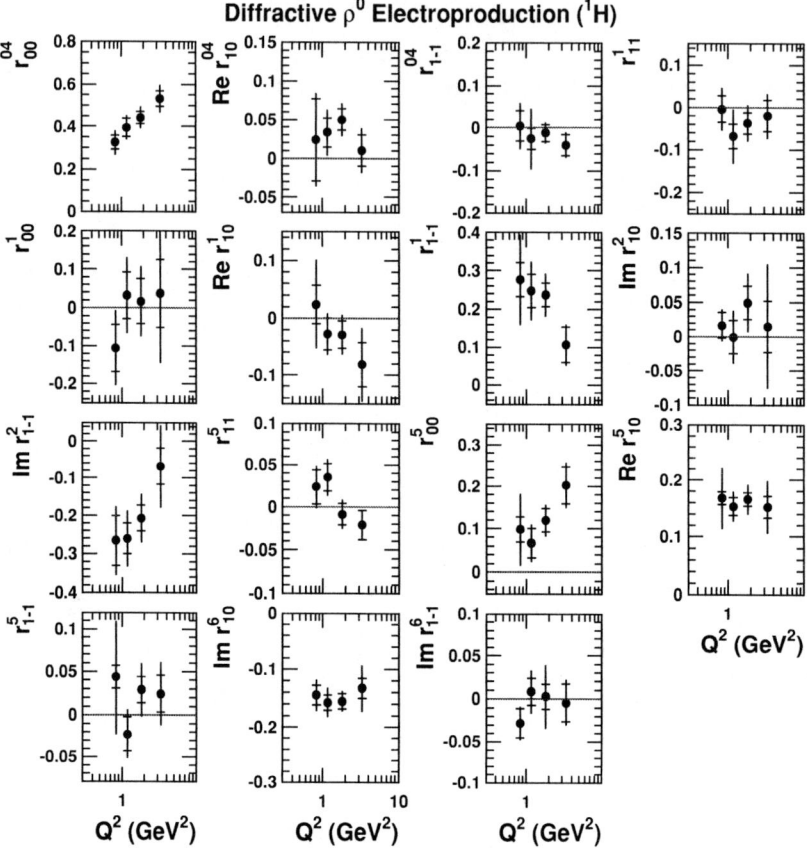

FIGURE 2. Spin-density matrix elements for ρ^0 production on hydrogen. The horizontal lines represent the expected values if s-channel helicity conservation (SCHC) is assumed.

in quasi-elastic electron scattering. This analogy suggests that the deviation from zero of r_{00}^5 might be related by some reinteraction effect of the ρ^0 vector meson in the final state. A similar deviation of r_{00}^5 from zero was reported by ZEUS [12] and H1 [13].

In order to extract information on the longitudinal ρ^0 production cross section the data on r_{00}^{04} are used. Assuming SCHS this spin-density matrix element is related to the ratio R of longitudinal over transverse production cross sections:

$$R = \frac{\sigma_L}{\sigma_T} = \frac{r_{00}^{04}}{\varepsilon(1 - r_{00}^{04})}, \quad (2)$$

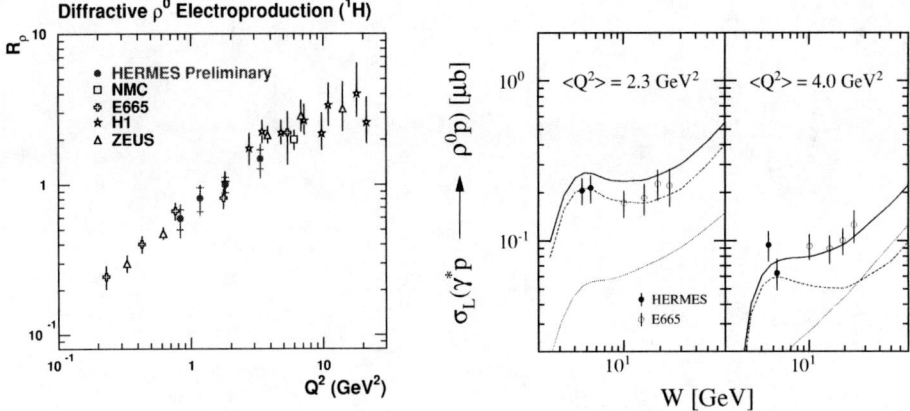

FIGURE 3. The ratio (R) of the longitudinal over transverse ρ^0 production cross sections on hydrogen (left). The HERMES data (solid circles) are compared to several existing data sets. On the right-hand side the derived longitudinal ρ^0 production cross section is compared to the results of a GPD-based calculation [11]. From bottom to top, the dotted curves represent the gluon-exchange contribution, the dashed curves the quark-exchange contribution and the solid curves their sum.

where ε represents the virtual-photon polarization parameter. Using the data of figure 2, the value of R has been determined. The results are compared to existing data[3] in figure 3 (left), showing that there is good agreement. At high values of Q^2 the longitudinal cross section is seen to dominate over the transverse one.

Using a parameterization of the data of figure 3 (left), the longitudinal ρ^0 production cross section σ_L has been determined:

$$\sigma_L = \frac{R}{1+\varepsilon R}\sigma_{total}, \qquad (3)$$

where σ_{total} represents the total measured cross section. The resulting values are shown in figure 3 (right), where they are compared to the calculations of Vanderhaeghen et al. [11]. The calculations are only in agreement with the data if both the quark-exchange and the gluon-exchange contributions are included. This result [8] confirms the predicted importance of quark-exchange in exclusive ρ^0 production at intermediate values of W.

Further evidence in support of the dominance of quark-exchange in ρ^0 production at $W \approx 5$ GeV came from a measurement of the target spin-asymmetry at HERMES. Using a polarized beam and a polarized target, the double spin asymmetry for ρ^0 production has been determined [10]. The data - although statistically limited - yield a positive

---

[3] The vector-meson production data from other experiments that are displayed in the figures of this section have almost exclusively been obtained by the H1 and ZEUS experiments at DESY. An overview of these data including all original references can be found in Ref. [14].

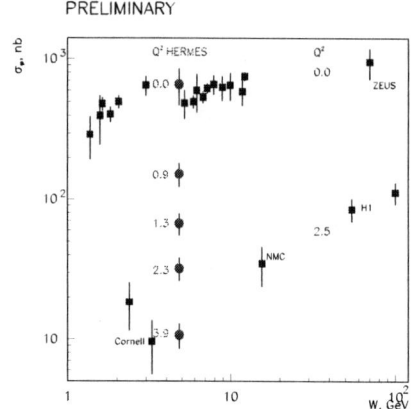 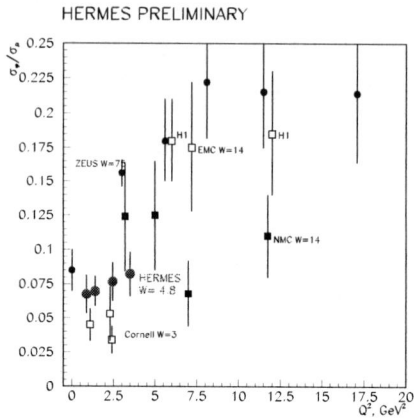

FIGURE 4. The cross section for φ production on hydrogen as a function of the invariant mass W (left). The HERMES data are represented by the solid circles. Existing data (solid squares) are also shown. On the right the ratio of the φ to ρ production cross sections is shown (large solid circles). The SU(4) prediction for this ratio is $\frac{2}{9}$.

asymmetry, which can be attributed to unnatural parity contributions such as quark-exchange.

Diffractive φ production

It is of interest to extend the studies described in the previous section to φ production, as its different mass and quark-content may give rise to a different production mechanism. The φ mesons could easily be identified in a h^+h^- invariant mass spectrum, where the hadrons were assumed to be kaons. Such spectra have been used to determine the φ-production cross section and the ratio of φ to ρ leptoproduction cross sections (see figure 4.) From the figure (left) it can be seen that the cross sections cover a largely unexplored W and Q^2 domain. After correcting the data for small Q^2 differences, it was found to be possible to describe all data at $Q^2 \approx 2.5$ GeV^2 by fitting the exponent α in $\sigma_\phi \propto W^\alpha$ yielding $\alpha = 0.53(9)$.

The ratio of φ to ρ production cross sections at $W \approx 5$ GeV is well below the SU(4) prediction of $\frac{2}{9}$ (see the right panel of figure 4). This might be caused by the aforementioned quark-exchange contributions to the ρ cross sections, assuming that quark exchange does not contribute to the φ-production cross section. This is also in accordance with the observation (from figure 4) that the ratio increases for larger values of W – and is thus getting closer to the SU(4) limit – where quark exchange is known to be negligible.

In order to investigate whether the assumed absence of quark-exchange contributions in φ production is also borne out by other data, the longitudinal φ-production cross

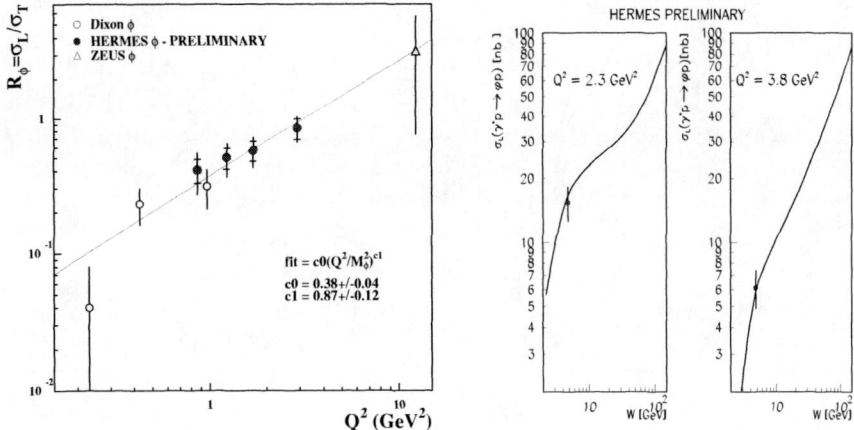

FIGURE 5. The ratio (R) of the longitudinal to transverse ϕ production cross sections on hydrogen (left). The ϕ data are represented by the solid circles. Some existing data (open circles) are also shown for comparison. The curve represents a fit to the data. In the right panel the derived longitudinal ϕ-production cross section is compared to the results of a GPD-based calculation [15].

section has been determined. To this end, the spin-density matrix element r_{00}^{04} has been determined using the ϕ-decay angular distributions. The resulting values have been converted to $R = \sigma_L/\sigma_T$ and are displayed in figure 5 (left panel). The data are in good agreement with existing data, which are also shown. Using the same expressions as were shown before for ρ production, the measured ϕ-production cross sections and R values were combined to extract the longitudinal ϕ-production cross sections. In figure 5 (right panel) the obtained values are compared to gluon exchange calculations that are similar to those displayed in figure 3, but without a quark-exchange contribution [15]. The data are well described by the calculations, thus confirming our interpretation of the relatively low ϕ-ρ ratios displayed in figure 4.

Diffractive ω production

In order to further study the flavour dependence of vector-meson production, the cross section for ω production has also been determined. The ω mesons have been identified through their main decay channel, $\omega \to \pi^+\pi^-\pi^0$, which has a branching ratio of almost 90%. The cross sections are displayed in figure 6 (left), as a function of W for various Q^2 values. Also in this case the data cover an almost completely unexplored (W,Q^2)-domain. It is noted that the photoproduction cross sections (at $Q^2 = 0$) have been determined using two independent methods. The solid circle (which is hard to see amidst the other $Q^2=0$ data points) is the result of an analysis requiring no scattered positron, while the solid triangle has been obtained by extrapolating the Q^2-dependence of the

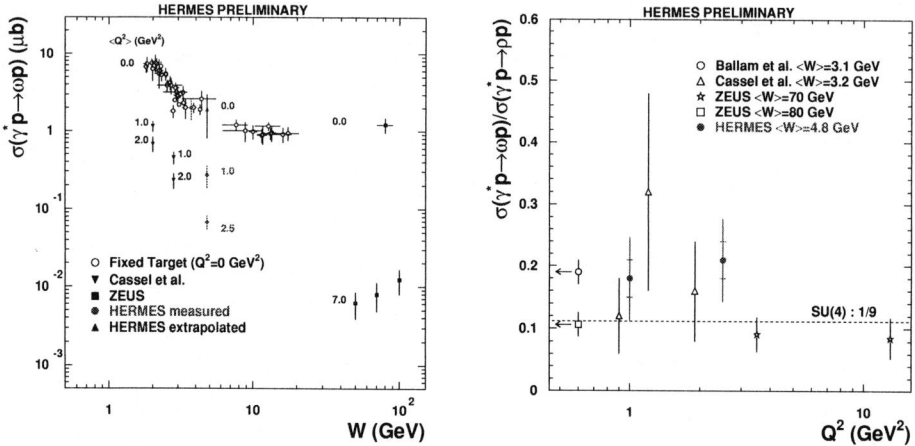

FIGURE 6. The cross section for ω production on hydrogen as a function of the invariant mass W (left). The solid data points at $W \approx 5$ GeV represent the HERMES measurements. Existing data are also shown. On the right the ratio of the ω to ρ-production cross sections is shown. The HERMES data are represented by the solid circles. The SU(4) prediction for the ω to ρ ratio is $\frac{1}{9}$.

data to $Q^2 = 0$. The two results are consistent.

In the right panel of figure 6 the ratio of ω to ρ-production cross sections is shown. The data are relatively close to the SU(4) limit of $\frac{1}{9}$, a value which was also reported by ZEUS and H1 [14]. At low Q^2 the ω-ρ ratios might even be somewhat higher than expected on the basis of SU(4). Since we know that there are substantial quark-exchange contributions to the ρ cross section, it is not unlikely that similar processes also contribute to the ω-production cross section; otherwise, the ω-ρ ratio would not be so close to $\frac{1}{9}$. In fact, preliminary calculations [15] confirm this idea. It would be desirable to further investigate this issue by also carrying out a separation of the longitudinal and transverse cross-section components for ω production.

PSEUDOSCALAR-MESON PRODUCTION

While the unpolarized GPDs, $H(x,\xi,\Delta^2)$ and $E(x,\xi,\Delta^2)$, are accessible through exclusive vector-meson production, the polarized GPDs, $\tilde{H}(x,\xi,\Delta^2)$ and $\tilde{E}(x,\xi,\Delta^2)$, can be studied through exclusive pseudoscalar-meson production. It is remarkable that such measurements do not require the use of a polarized beam or target. Unfortunately, the identification of exclusive π^+ production in an experiment such as HERMES is difficult, since there is no clear separation between exclusive and semi-inclusive π^+ production due to the limited missing mass resolution of the spectrometer. In an effort to circumvent this problem the semi-inclusive π^--production spectrum has been subtracted from the π^+ spectrum. As exclusive π^- production with a nucleon in the final state is forbidden on hydrogen by charge conservation, the difference between the two spectra should enable us to identify exclusive π^+ production on hydrogen. The result is shown in the left

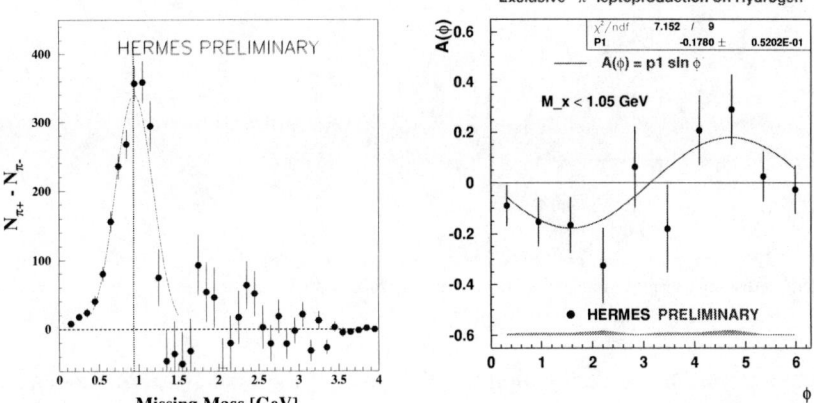

FIGURE 7. Missing mass spectrum for exclusive production of π^+ mesons on hydrogen (left). The target-spin azimuthal asymmetry for exclusive electroproduction of π^+ mesons (right). The curve represents a $\sin\phi$ fit to the data.

panel of figure 7. A clear exclusive π^+-production peak is observed centered at the mass of the proton.

It has been shown [16] that for longitudinal electroproduction of exclusive π^+ mesons from a transversely polarized target, the interference between the pseudoscalar ($\tilde{E}$) and pseudovector ($\tilde{H}$) amplitudes leads to a target-spin asymmetry of the produced π^+ mesons which depends on the azimuthal angle ϕ. Although no data have (yet) been collected with transversely polarized targets, the asymmetry $A(\phi)$ has been evaluated for data collected with a longitudinally polarized hydrogen target. The results are shown in the right panel of figure 7. The measured asymmetry is seen to be ϕ dependent, and can be described using the expression

$$A(\phi) = A_{UL}^{\sin\phi} \cdot \sin\phi \tag{4}$$

yielding $A_{UL}^{\sin\phi} = -0.18 \pm 0.05$ (stat) ± 0.01 (syst). The fitted value of the $\sin\phi$ moment $A_{UL}^{\sin\phi}$ cannot be directly compared to the prediction of Ref. [16], which is large and positive, since the present result has been obtained with a longitudinally (instead of a transversely) polarized target and a longitudinally-transversely mixed (instead of a purely longitudinal) virtual photon beam. Further measurements and calculations are needed to determine whether the possible disagreement between the data and the calculation of Ref. [16] can be attributed to these differences.

DEEPLY-VIRTUAL COMPTON SCATTERING

When studying Generalized Parton Distributions in exclusive processes it is very desirable to have a photon in the final state. The reason is that there is no remaining

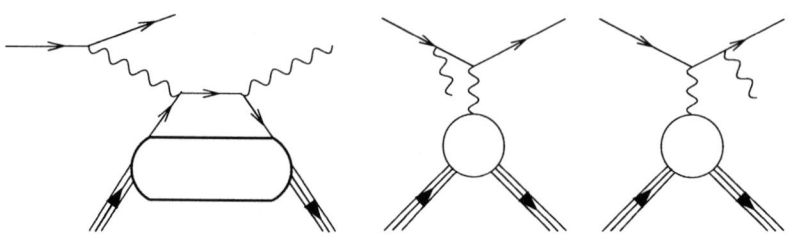

FIGURE 8. Feynman diagram for deeply-virtual Compton scattering (left), and photon radiation from the incident and scattered lepton in the Bethe-Heitler process (right).

uncertainty due to the wave function of the meson in the final state. Moreover, while vector-meson production and pseudoscalar-meson production are each sensitive to two different GPDs, deeply-virtual Compton scattering (DVCS) is sensitive to all four known GPDs, $H(x,\xi,\Delta^2)$, $E(x,\xi,\Delta^2)$, $\tilde{H}(x,\xi,\Delta^2)$ and $\tilde{E}(x,\xi,\Delta^2)$.

It is extremely difficult to observe photons that can be associated with deeply-virtual Compton scattering (DVCS) in a electron scattering environment, because of the dominance of the Bethe-Heitler (BH) process (see figure 8). However, by exploiting the interference between the DVCS and BH processes, one may obtain access to the DVCS amplitude [17]. As the interference term in the cross section depends on the beam polarization and the azimuthal angle ϕ, measurements of the ϕ-dependence of the beam-spin asymmetry are needed.

The beam-spin asymmetry in hard electroproduction of photons has been measured for the first time by the HERMES experiment at DESY using the HERA 27.6 GeV longitudinally polarized positron beam and an unpolarized hydrogen gas target. The asymmetry with respect to the helicity state of the incoming positron beam is plotted as a function of ϕ in figure 9 (left) for data with a missing energy close to the proton mass. The data show a clear $\sin\phi$ dependence, an effect that can only be caused by the DVCS-BH interference.

The missing mass (M_x) dependence of the effect has been studied by evaluating the beam-spin analyzing power

$$A_{LU}^{\sin\phi} = \frac{2}{N} \sum_{i=1}^{N} \frac{\sin\phi_i}{(P_l)_i}, \qquad (5)$$

where $(P_l)_i$ represents the beam polarization measured during the beam burst of the ith event. The results are shown in the right-hand panel of figure 9. As the missing mass resolution of the HERMES spectrometer for DVCS-like events is about 0.8 GeV, the M_x-bins left and right of $M_x = m_{proton}$ also show non-zero value of $A_{LU}^{\sin\phi}$. In the missing-mass range below 1.7 GeV the analyzing power in the $\sin\phi$ moment was measured to be -0.23 ± 0.04(stat) ± 0.03(syst). The average values of the kinematic variables corresponding to this measurement are: $\langle x \rangle = 0.11$, $\langle Q^2 \rangle = 2.6$ GeV^2 and $\langle -t \rangle = 0.27$ GeV^2. The observed analyzing power is somewhat smaller than the value of -0.37 quoted in Ref. [18] for kinematics close to those of the present experiment.

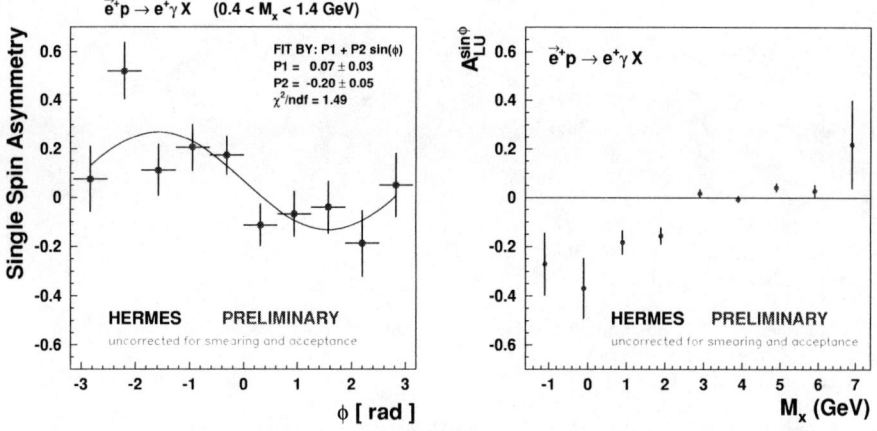

FIGURE 9. The beam-spin azimuthal asymmetry for exclusive electroproduction of real photons (left). The curve represents a $\sin\phi$ fit of the data. On the right hand side the $\sin\phi$ moment of the data is shown as a function of the missing mass.

OUTLOOK

Hard exclusive processes give access to a new type of information on nucleon structure as contained in the so-called generalized parton distributions (GPDs). The GPDs contain information on partonic correlations in the nucleon, and can be used to describe a whole range of different reactions. In this contribution first results are presented of analyses aimed at extracting data on hard exclusive processes from measurements carried out by the HERMES collaboration at DESY. It has been shown to be possible to determine cross sections and/or analyzing powers for various exclusive processes, involving vector mesons, pseudoscalar mesons and real photons. These results demonstrate that it is possible to study exclusive processes, even at the modest luminosities offered by the HERMES experiment. Future experiments with considerably higher luminosities (up to 10^{34-35} N/cm^2/s) are needed to enable a full mapping of the GPDs in various kinematical variables such as x, Q^2, ξ and Δ. Several future projects that are presently being discussed both in the U.S. and Europe are aimed at studying hard exclusive processes at high luminosities.

ACKNOWLEDGMENTS

I would like to thank my colleagues of the HERMES collaboration for making it possible to perform an experiment producing data of the quality shown in this contribution. In particular I am grateful to Moskov Amarian, Sasha Borrisov, James Ely, Eric Thomas and Michael Tytgat, who were responsible for most of the data that were discussed in this paper.

REFERENCES

1. Collins, J., Frankfurt, L., and Strikman, M., *Phys. Rev. D* **56**, 2982 (1997).
2. Müller, D., et al., *Fortsch. Phys.* **42**, 101 (1994).
3. Ji, X., *Phys. Rev. D* **55**, 7114 (1997).
4. Radyushkin, A., *Phys. Lett. B* **385**, 333 (1996).
5. Filippone, B., and Ji, X., *Adv. in Nucl. Phys.*, **in press**, hep–ph/0101224 (2001).
6. Goeke, K., Polyakov, M., and Vanderhaeghen, M., *Prog. Part. Nucl. Phys.*, **submitted** (hep-ph/0106012).
7. HERMES, *Nucl. Instr. Meth. A* **417**, 230 (1998).
8. HERMES, *Eur. Phys. J. C* **17**, 389 (2000).
9. HERMES, *Eur. Phys. J. C* **18**, 303 (2000).
10. HERMES, *Phys. Lett. B* **513**, 301 (2001).
11. Vanderhaeghen, M., Guichon, P., and Guidal, M., *Phys. Rev. Lett.* **80**, 5064 (1998).
12. ZEUS, *Eur. Phys. J. C* **12**, 393 (2000).
13. H1, *Eur. Phys. J. C* **13**, 371 (2000).
14. Abramowicz, H., *Int. J. Mod. Phys. A* **15 (Suppl.1)**, 495 (2000).
15. Vanderhaeghen, M., Guichon, P., and Guidal, M., *priv. communication* (2001).
16. Frankfurt, L., Polyakov, M., Strikman, M., and Vanderhaeghen, M., *Phys. Rev. Lett.* **56**, 2589 (2000).
17. Diehl, M., et al., *Phys. Lett. B* **411**, 193 (1997).
18. Kivel, N., et al., *Phys. Rev. D* **63**, 114014 (2001).

First experiment on spectroscopy of Λ-hypernuclei by electroproduction at JLab

L. Tang[1,10], T. Miyoshi[2], M. Sarsour[3], L. Yuan[1], X. Zhu[1], A. Ahmidouch[4], P. Ambrozewicz[5], D. Androic[6], T. Angelescu[7], R. Asaturyan[8], S. Avery[1], O.K. Baker[1], I. Bertovic[6], H. Breuer[9], R. Carlini[10], J. Cha[1], R. Chrien[11], M. Christy[1], L. Cole[1], S. Danagoulian[4], D. Dehnhard[12], M. Elaasar[13], A. Empl[14], R. Ent[10], H. Fenker[10], Y. Fujii[2], M. Furic[6], L. Gan[1], K. Garrow[10], A. Gasparian[1], P. Gueye[1], M. Harvey[1], O. Hashimoto[2], W. Hinton[1], B. Hu[1], E. Hungerford[3], C. Jackson[1], K. Johnston[15], H. Juengst[12], C. Keppel[1], K. Lan[3], Y. Liang[1], V.P. Likhachev[16], J.H. Liu[12], D. Mack[10], K. Maeda[2], A. Margaryan[8], P. Markowitz[17], J. Martoff[5], H. Mkrtchyan[8], T. Petkovic[6], J. Reinhold[17], J. Roche[18], Y. Sato[1,2], R. Sawafta[4], N. Simicevic[15], G. Smith[10], S. Stepanyan[8], V. Tadevosyan[8], T. Takahashi[2], H. Tamura[2], K. Tanida[19], M. Ukai[2], A. Uzzle[1], W. Vulcan[10], S. Wells[15], S. Wood[10], G. Xu[3], Y. Yamaguchi[2], and C. Yan[10]

1) Department of Physics, Hampton University, Hampton, VA 23668, USA
2) Department of Physics, Tohoku University, Sendai 980-8578, Japan
3) Department of Physics, University of Houston, Houston, TX 77204, USA
4) Department of Physics, North Carolina A&T State University, Greensboro, NC 27411, USA
5) Department of Physics, Temple University, Philadelphia, PA 19122, USA
6) University of Zagreb, Zagreb, Croatia
7) University of Bucharest, Bucharest, Romania
8) Yerevan Physics Institute, Yerevan, Armenia
9) Department of Physics, University of Maryland, College Park, MD 20742, USA
10) Thomas Jefferson National Accelerator Facility, Newport News, VA 23606, USA
11) Brookhaven National Laboratory, Upton, NY 11973, USA
12) Department of Physics, University of Minnesota, Minneapolis, MN 55455, USA
13) Department of Physics, Southern University at New Orleans, New Orleans, LA 70126, USA
14) Department of Physics, Rensselaer Polytechnic Institute, Troy, NY 12180, USA
15) Department of Physics, Louisiana Tech University, Ruston, LA 71272, USA
16) Institute of Physics, University of Sao Paulo, Sao Paulo, Brazil
17) Department of Physics, Florida International University, Miami, FL 27411 USA
18) Department of Physics, College of Williams and Mary, Williamsburg, VA 23187, USA
19) Department of Physics, University of Tokyo, Tokyo 113-0033, Japan

Abstract. The first experiment in Λ-hypernuclear spectroscopy using the high-precision electron beam at Jefferson laboratory (JLab) has been carried out. The hypernuclear spectrometer system (HNSS) was used to measure spectra from the $^{12}C(e, e'K^+)^{12}_\Lambda B$ reaction with sub-1-MeV resolution, the best energy resolution obtained thus far in hypernuclear spectroscopy with magnetic spectrometers. This paper describes the HNSS and the preliminary results for the $^{12}_\Lambda B$ system. The experimental spectrum is consistent with the expected strong spin-flip excitations of unnatural parity states. A program of hypernuclear physics experiments is planned for the future with much higher yield and even better energy resolution.

1. INTRODUCTION

The introduction of a new degree of freedom, strangeness, into the nuclear medium challenges our conventional models of the nuclear many-body system to their limits. Many new features of the strong interaction between hyperons and the nuclear medium and between hyperons and nucleons can be explored [1]. Analyses of experiments on light and heavy systems [2-6] have shown that the Λ particle, because it is distinguishable from nucleons, can indeed occupy any of the nuclear shells, even those filled with nucleons. Thus, the Λ can be used as an effective probe of the nuclear interior. In addition, the knowledge of an effective hyperon-nucleus interaction deduced from such studies will enable the extraction of an effective ΛN interaction that is difficult if not impossible to obtain by others means.

Traditionally, hypernuclear studies have been carried out using secondary hadronic beams (K or π mesons), producing the Λ in the nucleus by a strong interaction with a nucleon. Figure 1(a) is a simplified illustration of these processes. For the (K^-, π^-) reaction the momentum transfer is small and the cross section is relatively large. The spectroscopy is characterized predominantly by the excitation of low-spin substitutional states [7,8] where the Λ replaces the nucleon in the same shell model orbit. For the (π^+, K^+) reaction the momentum transfer is large and the cross section is relatively small [2]. This reaction preferentially populates high-spin stretched states [2,3] where a nucleon hole is coupled to a Λ. Because the Λ can be in any shell this reaction also produces deeply bound states with the Λ in the s-shell. At forward angles, neither of the two reactions has significant spin-flip amplitude so that the spectra are dominated by the transitions to the states of natural parity.

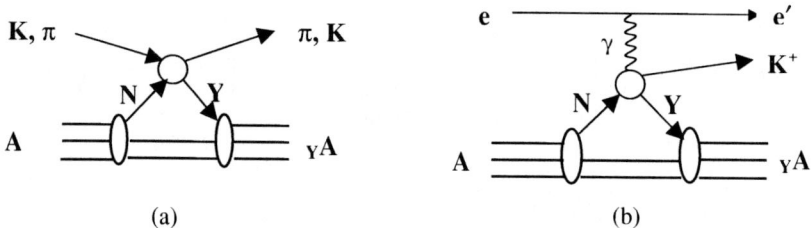

FIGURE 1. Comparison of the production processes: (a) hadronic and (b) electromagnetic.

The investigations using hadronic production have been hampered by poor energy resolution. Thus far, the best resolution (better than 2 MeV (FWHM)) with magnetic spectrometers was obtained in a study of light Λ-hypernuclei at KEK using the (π^+, K^+) reaction [9]. This work demonstrated the importance of good resolution in gaining significantly new information. Much better resolution than in experiments with magnetic spectrometers alone has been achieved in experiments detecting the ejected K^+ or π^- in a magnetic spectrometer in coincidence with γ's from hypernuclear decay using high-resolution Germanium crystals [10,11]. However, such studies are limited to particle stable states. Thus there is continued interest to obtain better and better resolution in experiments that measure singles spectra of the residual nucleus excited to either particle bound or unbound states. High resolution is of special importance for extracting parameters of the spin-dependent Λ-nucleus interaction and for studying the single particle motion of the Λ in the strongly interacting nuclear medium [12].

Electroproduction of hypernuclei using a high-duty factor (100%) and high-intensity electron beam, such as the CEBAF beam at JLab, has long been known for its many unique features [13] that deserve to be exploited. In Fig. 1 we compare the diagram (1b) for the electroproduction process, $A(e, e'K^+)_Y A$, with that (1a) for the hadronic processes. Here the subscript Y indicates a hyperon in the nucleus. The electro-magnetic process with an exchange of a colorless virtual photon is much better understood theoretically than the strong process. Even at far forward angles the spin-flip amplitude is large in contrast to the hadronic process where the spin-flip amplitude is small. The momentum transfer is high ($q \geq 300$ MeV/c), similar to that of the (π^+, K^+) reaction. Therefore, the spectra are expected to show strong spin-flip transitions to high-spin stretched states of unnatural parity as well as transitions to natural parity states and to deeply bound states.

The very different features of the different reactions make it possible to study a large variety of hypernuclear states [14]. In addition, the electromagnetic process changes a proton in the nucleus into a Λ, creating a proton-hole—Λ-particle state whereas the frequently studied hadronic reactions change a neutron into a Λ resulting in a neutron-hole—Λ-particle state. For targets of equal number of neutrons and protons, the reactions induced by the electromagnetic and hadronic processes lead to mirror Λ hypernuclei thus allowing studies of charge symmetry breaking in the effective Λ-nucleon interaction.

Cross sections for electromagnetic production are at least two orders of magnitude smaller than that for hadronic processes, but this can be compensated by much higher beam intensity. Due to the high quality of the primary electron beam and the ability to transport it to the target without losing its high precision, there is no need for tracking the incident electrons. Energy and angle straggling in the target are minimized by use of a very thin target. Estimates show that it should be possible to reach an energy resolution in the sub-1-MeV range. (See below). With such precision, the $(e, e'K^+)$ reaction is a powerful probe for a systematic study of hypernuclei.

2. HNSS EXPERIMENT

JLab experiment E89-009, employing the HNSS, is the first high-resolution hyper-nuclear spectroscopy experiment using electromagnetic production of strangeness. This experiment probes certain expectations for electroproduction of hypernuclei and tests our experimental techniques.

2.1 Experimental considerations

In electroproduction, the Λ and K^+ particles are created associatively via an interaction between a virtual photon and a proton in the nucleus. The hypernucleus $_Y A$

is formed by coupling the Λ to the residual nuclear core, as shown in Fig. 1(b). The energy and momentum of the virtual photon are defined as $\omega = E\text{-}E'$ and $\mathbf{k} = \mathbf{p}\text{-}\mathbf{p}'$, respectively. The four-momentum transfer of the electron is then given by $Q^2 = k^2 - \omega^2$. Since the elementary cross section for $p(e, e'K^+)\Lambda$ falls off fast with increasing Q^2, the measurements should be done at Q^2 close to zero. This requires that the electron scattering angle should be as small as possible.

To a good approximation, the electroproduction cross section can be expressed[15] by

$$\frac{d^5\sigma}{dE'd\Omega'd\Omega_K} = \Gamma \frac{d^2\sigma}{d\Omega_K}, \quad (1)$$

where Γ is the integrated virtual photon flux produced by (e, e') scattering and $d^2\sigma/d\Omega_K$ is the photoproduction cross section. As $Q^2 \to 0$, the cross section is completely dominated by the transverse component.

For the current experiment ω was chosen to be about 1.5 GeV, at which the elementary photoproduction cross section has its maximum. In order to keep the rate of events from background K^+ production channels small, the energy (E) of the incident electrons was set to about 1.8 GeV. Thus, the scattered electron energy (E') was about 0.3 GeV. Figure 2 shows the calculated virtual photon flux factor in units of photons per electron per MeV per sr for the chosen kinematics. This factor is peaked near zero degrees and falls off rapidly as the scattering angle increases. With the electrons detected at zero degrees, even a relatively small solid angle will accept a large percentage of the scattered electrons. Thus, the chosen experimental parameters simplify the electron detection and maximize the virtual photon flux. However, near zero degrees the electron background rate from bremsstrahlung increases even faster with decreasing angle than the virtual photon flux so that the electron single-arm rate is dominated by electrons from bremsstrahlung thus limiting the usable luminosity of the beam.

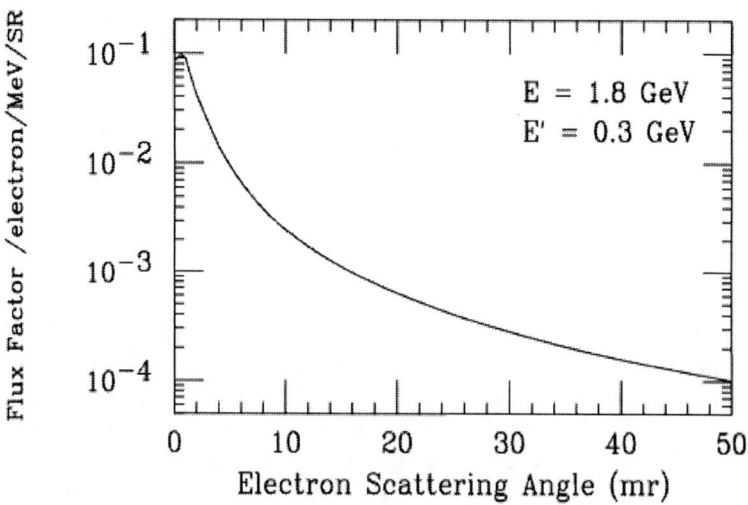

FIGURE 2. Virtual photon flux factor as a function of electron scattering angle.

In the $(e, e'K^+)$ reaction, both the scattered electron and the K^+ have to be detected in coincidence. With the kinematics chosen for the electron scattering, the K^+ momentum is about 1.2 GeV/c. The 3-momentum transfer to the associatively produced Λ is about $q \approx 300$ MeV/c when the K^+ is detected at zero degrees. The production cross sections for the hypernuclear ground state and core-excited states decrease strongly as the 3-momentum increases with the K^+ scattering angle. One advantage of using low incident electron and scattered electron energies (about 1.8 and 0.3 GeV, respectively,) is that the cross sections drop more slowly in the small forward angle region since the 3-momentum increases more slowly than in the case of higher electron energies. Thus, detecting the K^+ at angles near zero degrees ensures maximization of hypernuclear production. The relatively large momentum transfer to the Λ, similar to that of the (π^+, K^+) reaction, provides access to deeply bound and high-spin states.

2.2 Experimental setup

Figure 3 shows a schematic top view of the HNSS. In order to be able to detect both scattered electrons and positively charged kaons near zero degrees a "C" dipole, placed right behind the experimental target, served as a beam splitter. It bent the scattered electrons (centered at zero degrees) and the kaons (centered at about 2 degrees) in opposite directions by 33 and 16 degrees, respectively. The target was located at the effective field boundary of the splitter magnet.

The scattered electrons were detected by a split-pole magnetic spectrometer [16]. The central momentum of the spit-pole spectrometer was chosen to be 300 MeV/c where its momentum acceptance is about 120 MeV/c. The solid angle acceptance of the combined splitter and split-pole system was about 9 msr, which effectively tagged about 35% of the virtual photon flux within the momentum acceptance. This was possible because of the far forward peaking of the scattered electrons mentioned above.

In order to be able to handle the very high rates of scattered electrons and to keep the means of detecting the electrons simple, the focal plane detector for the split pole was made of 10 one-dimensional silicon strip detectors (SSD) with 144 strips each and a pitch width of 0.5 mm. The 10 SSD cover the full length of the 72-cm long focal plane. The position measurement provided directly the momentum of the scattered electron. Eight scintillation strip counters were placed behind each of the 10 SSD, a total of 80 strips. They were used to provide the timing for the coincidence with the kaons.

The kaons were detected by an existing short orbit spectrometer (SOS) placed so that the kaon scattering angle centered at 2 degrees. The angular acceptance of the SOS was about 6 msr covering a range of scattering angles from 0 to 4 degrees. The central momentum was set at 1.2 GeV/c and the acceptance was about 46%. Only the central ±15% of the acceptance was used where it is flat within the range of missing mass of interest. The total path length of the kaons from the target to the end of the SOS detectors was 10 m. Thus, on the average the survival rate from target to focal area was 35%.

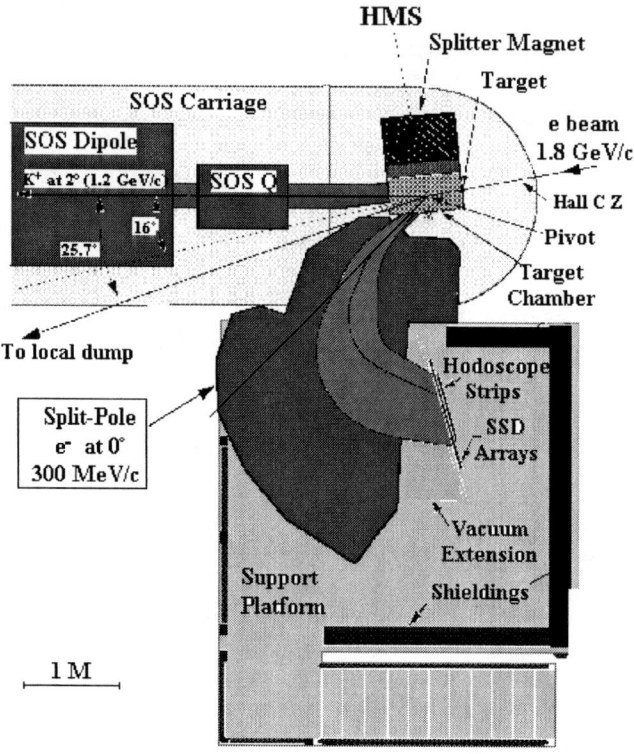

FIGURE 3. Top view of the layout of the HNSS.

The standard SOS detector system was used for kaon detection. It consists of: (1) two sets of tracking chambers that measure positions and angles at the focal plane for momentum reconstruction, (2) four scintillation hodoscope planes to provide coincidence timing to the electron arm and also to measure the time-of-flight (TOF) for separation of kaons and background particles (p, π^+, and e^+), (3) one aerogel Čerenkov (AČ) veto counter to reduce π^+ and e^+ background triggers, (4) one Lucite Čerenkov (LČ) counter to reduce the proton triggers, and (5) one gas Čerenkov (GČ) detector and 3 layers of shower counters to reject the e^+ triggers.

The JLab beam has a bunch width of 1.67 ps and a separation between bunches of 2 ns. The coincident time resolution for the two arms was $\sigma \approx 400$ ps after path length and signal size corrections for both arms. This resolution was sufficient to identify the real and accidental coincidence peaks individually in the coincidence time spectrum. The final e'/K^+ coincidence events were selected by a two-dimensional cut on the real coincidence window of 2 ns and the velocity measurement (from TOF in the SOS). The events selected from eight nearby accidental coincidence windows were used to obtain the shape and magnitude of the accidental background spectra to high accuracy (see below).

Three different thin target foils were employed, CH_2, ^{12}C, and ^7Li. The CH_2 foil (100 μm thick) was used for calibration and optical tuning using events from the $p(e, e' K^+)\Lambda$ and $p(e, e' K^+)\Sigma^0$ reactions. $^{12}_\Lambda B$ and $^7_\Lambda He$ hypernuclear spectra were obtained from the ^{12}C (22 mg/cm^2) and ^7Li (19 mg/cm^2) targets. In the HNSS, a complete vacuum system coupled the beam line, target chamber, and spectrometers. Thus multiple scattering in vacuum windows happened only in the exit windows of the spectrometers where multiple scattering effects were minor since these foils were located right in front of the first tracking detector. Table 1 lists the sources of contributions to the energy resolution and the expected overall energy resolution for this experiment. The SOS contribution was expected to dominate.

TABLE 1. Sources of contribution and expected overall energy resolution for HNSS.

Source	Contribution	Resolution (keV) (FWHM)
Beam Energy Uncertainties	$\leq 10^{-4}$	≤ 180
SOS momentum Uncertainty	$\leq 5 \times 10^{-4}$	≤ 600
e' Arm Momentum Uncertainty	$\leq 5 \times 10^{-4}$	≤ 150
K^+ Scattering Angle Uncertainty	10 mr	≤ 200 (^{12}C)
Target Energy Loss Uncertainty	1.7 keV/mg/cm^2	38 (^{12}C)
Total		≤ 678

2.3 Rates, background, and calibrations

The singles rate in the electron arm reached about 2×10^8/sec. It was primarily due to background of bremsstrahlung electrons that cannot be distinguished from the coincident electrons. Therefore, the experiment used the much less frequent kaon arm events in the trigger. Coincidence spectra were obtained later in off-line analysis. The high electron rate caused large accidental background in the spectra. The SSD and scintillation hodoscopes worked well under the high rates.

The positrons from e^+/e^- pair production that were emitted near zero degrees dominated the kaon arm singles rate. Since the SOS covered an angular range from 0 to 4 degrees, these positrons were accepted by the SOS but the combined use of vetoes from AČ, GČ, and shower counters eliminated this background. The rate from background protons and π^+'s were low (~1 kHz) after on-line rejection by the AČ and LČ detectors. The remaining background protons and π^+'s were eliminated in off-line analysis using TOF information.

Due to the high rate in the electron arm, about 95% of the background in the spectra was from accidental coincidences. A precise measurement of this background was obtained as follows. The analysis of the raw data generates a spectrum of the time difference between the emission of the K^+ and the electron from the target. In addition to a peak corresponding to the true coincidences there are many peaks containing only accidental coincidences, 2 ns apart according to the time structure of the electron beam. The analysis of the events from the accidental coincidence peaks in the time spectrum under the same condition as those from the real coincidence peak provided a high-statistics background spectrum that could be subtracted from the missing-mass spectrum (containing the real and the accidental coincidences) after proper normalization. The remainder of the background (5%) was from the real coincidences with π^+'s. The full path length TOF separation between π^+'s and K^+'s was about 2 ns. Therefore the real coincidental π^+'s should be contained in a coincident peak next to the real K^+ coincidences according to the beam bunch structure. The timing resolution of $\sigma \approx 400$ ps allowed an overlap of the tails from the real K^+ and π^+ coincidence peaks. The magnitude of this background was determined by analyzing the overlap. The shape of the background in the missing-mass spectra was then obtained by an analysis of the coincident π^+ assuming $(e, e'K^+)$ kinematics. The absolute magnitude of the background was obtained by normalizing the spectrum to the number of background events in the spectrum.

An upper limit for the HNSS resolution was obtained by investigating the coincident events of the $A(e, e'e^+e^-)A$ reaction. The beam energy was reconstructed simply by a summation of the energies of the two electrons (the scattered electron e' and the electron from pair production) in the electron arm and a positron (from the pair production) in the SOS. The result is shown in Fig. 4. This method includes the contributions to the resolution from the extra electron in the split pole that does not exist in the $(e'K^+)$ final state. The 820-keV (FWHM) resolution in Fig. 4 is therefore an upper limit of what we expect for the missing mass spectra from the $(e, e'K^+)$ reaction.

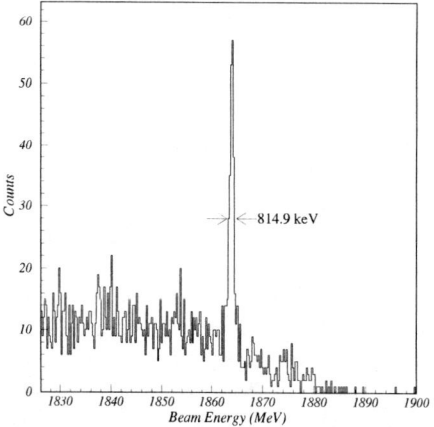

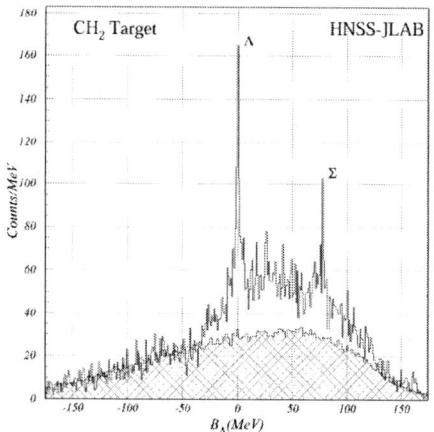

FIGURE 4. Reconstructed beam energy, which provided the upper limit of the HNSS resolution.

FIGURE 5. Missing mass spectrum with Λ and Σ^0 peaks from the $p(e, e'K^+)$ reaction using the CH_2 target.

Figure 5 shows a missing mass spectrum obtained with the CH_2 target. The $(e,e'K^+)$ reaction on the protons in the target produced both Λ and Σ^0. On the missing mass scale the Λ mass was subtracted which places the "missing mass" of the Λ at 0 MeV and that of the Σ^0 at 77 MeV. The shaded area is the background from the accidentals. The broad distribution above the accidental background other than the two peaks are events from the carbon in the CH_2 target. The beam was defocused to 4x4 mm^2 by the fast raster for the beam position on target and was kept below 1.5 µA to avoid target melting. The peak width (FWHM) of the hyperon peaks is about 3 MeV. This large width is due to the kinematic broadening within the angular resolution (10 mr) of the SOS for the scattered kaons and the large beam size. This broadening effect is much smaller (<200 keV) for the carbon target.

The missing mass scale, which depends on the beam energy and the central momenta of both electron and kaon arms, was calibrated using the measured positions of the Λ and Σ^0 peaks and their known masses. The systematic error from this calibration in the determination of the binding energy of the hypernuclear system is about 125 keV. This error is mainly due to the low statistics in the Λ and Σ^0 peaks. The width of the peaks has no effect on the calibration. The spectrum of Fig. 5 also provides a first measurement of the cross section of the electroproduction of hyperons on protons near $Q^2 = 0$.

3. EXPERIMENTAL RESULTS

The HNSS experiment obtained data for both $^{12}_\Lambda$B and $^{7}_\Lambda$He hypernuclei using ^{12}C and ^{7}Li targets. Our analysis is currently focusing on $^{12}_\Lambda$B. Fig. 6 shows a preliminary missing mass spectrum of the $^{12}_\Lambda$B hypernucleus, plotted in terms of the Λ binding energy with the background (shaded area) included. The background spectrum of accidentals (≈95% of all background) was obtained with very good statistics using events from eight accidental coincidence peaks. The remaining background (about 5%) was from a contamination by real coincident π^+'s, as mentioned previously. The high accidental rate was from the bremsstrahlung electrons that are peaked strongly at zero degrees. This high rate limited the maximum usable luminosity and thus the good event rate. Fig. 7 shows the spectrum with the background subtracted and with half the bin size of the spectrum in Fig. 6. The error bars are the statistical errors with the contribution from the background subtraction included.

Only specific hypernuclear states are expected [14,17] to have significant strength in the $(e, e'K^+)$ reaction, whereas others may be strongly excited in the (π^+, K^+) or (K^-, π^-) reactions. In our preliminary spectrum the most prominent peak, located at $B_\Lambda \approx 11.5$ MeV, is from the transition to the ground state doublet of $^{12}_\Lambda$B. This unresolved $(1^-_1/2^-_1)$ doublet is made primarily by coupling a Λ in the s shell to the ground state of ^{11}B (primarily a proton hole in the $p_{3/2}$ shell). The 2^- state can be reached only by spin-flip of Λ and is thus expected to be more strongly excited by the $(e, e'K^+)$ reaction than the 1^- state. Theory predicts a spacing between the two states (resulting from spin-dependent parts of the interaction) of roughly 100 keV, too close to be resolved in our experiment. We note that in the study [9] of the ^{12}C(π^+, K^+) reaction the 1^- state of the g.s. doublet in the mirror hypernucleus $^{12}_\Lambda$C is the most strongly excited state. The second clearly visible peak, at $B_\Lambda \approx -5.9$ MeV, we tentatively interpret to be the predicted 2^-_3 state arising from coupling the Λ in the s shell to the first excited $3/2^-$ state of the ^{11}B core. This 2^-_3 state is also a member of a $(1^-/2^-)$ doublet with its 1^- member observed in the (π^+, K^+) reaction.

There is an indication of a peak at $B_\Lambda \approx -8.9$ MeV, between the other two s- shell Λ states mentioned above, where the 1^-_2 state of a $(0^-/1^-)$ doublet, made by $(p_{1/2}^{-1}, s^\Lambda)$ coupling, is expected but the statistics is not sufficient to make a solid claim.

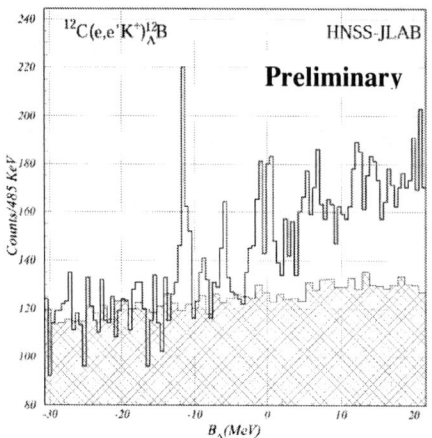 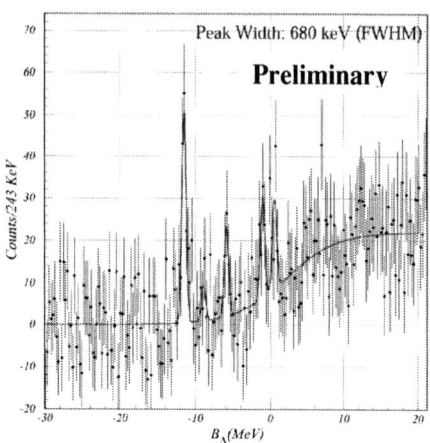

FIGURE 6. Preliminary $^{12}_\Lambda$B missing mass spectrum including background.

FIGURE 7. Background-subtracted $^{12}_\Lambda$B spectrum. Five peaks are fitted with a common peak width.

Near breakup threshold, between $B_\Lambda = -2$ and $+1$ MeV, several states arising from coupling a Λ in the $p_{3/2}$ shell to the $3/2^-$ g.s. of ^{11}B are expected. These are positive parity states of spin/parity 0^+, 1^+, 2^+, and 3^+, of which the 3^+ unnatural parity state is predicted to be most strongly excited and the 2^+ to contain about half the strength of that of the 3^+ state. It is not clear whether we can speak of two resolved peaks here. But the yield between $B_\Lambda = -2$ and $+1$ MeV must contain the predicted strongly excited 3^+ and 2^+ states.

Configuration mixing of the ^{11}B core states of the p shell hypernuclear states is expected to be significant [17]. This leads to the splitting of the total strength among many states. Nevertheless, theory predicts that a large fraction of the strength resides in the 3^+_1 state. The complex of states near breakup threshold is of great interest and will be the subject of more detailed investigation.

The experimental spectrum in Fig. 7 is the result of subtracting the background, indicated in Fig. 6 by the shaded area, from the total spectrum. Note that the bin size in Fig. 7 (243 keV) is half that of Fig. 6. The region from $B_\Lambda = -30$ to $+20$ MeV was fitted assuming five states and a background primarily from quasifree Λ production. The extracted peak positions are listed in Table 2 for a comparison with the most recent structure prediction provided by Millener [17]. The prediction lists all possible states from the coupling of a Λ to ^{11}B core states. The energies of the levels for the states with the Λ in the p shell had been given [16] relative to the lowest 2^+_1 state. In order to facilitate the comparison between experiment and theory, an energy separation of 10.69 MeV between the ground state and the 2^+_1 state was added to the theoretical values. 10.69 MeV had been reported as the excitation energy of the 2^+_1 state in $^{12}_\Lambda$C observed in the study [9] of the (π^+, K^+) reaction. We use this value for the 2^+_1 mirror state in $^{12}_\Lambda$B since the difference between the ^{11}C and ^{11}B core states is small. The extracted experimental values are listed in the table next to the closest theoretical values for comparison only. It does not mean an actual claim of experimental observation of a predicted state. Full interpretation of the experimental spectrum will have to rely on detailed and complete theoretical studies. The uncertainty of the binding energy is dominated by the uncertainty in the missing mass scale calibration using the positions of the Λ and Σ^0 peaks in the relatively low-statistics spectrum from the CH_2 target.

TABLE 2. Comparison of theoretically predicted [17] and preliminary experimental values.

State	Core State in ^{11}B	E_x (MeV) Theory	E_x (MeV) Experiment		B_Λ (MeV) Experiment
1^-_1 (Λ_s)	$3/2^-_1$	0.000			
2^-_1 (Λ_s)	$3/2^-_1$	0.165	#1	0.00	-11.53 ± 0.13
1^-_2 (Λ_s)	$1/2^-_1$	2.728	(#1'	3.06)	(-8.47 ± 0.13)
0^-_1 (Λ_s)	$1/2^-_1$	2.752			
2^-_2 (Λ_s)	$5/2^-_1$	4.553			
2^-_3 (Λ_s)	$3/2^-_2$	5.829	#2	5.62	-5.91 ± 0.13
1^-_3 (Λ_s)	$3/2^-_2$	5.894			
2^+_1 (Λ_p)		10.69	#3	10.43	-1.10 ± 0.13
1^+_1 (Λ_p)		10.72			
2^+_2 (Λ_p)		11.15			
3^+_1 (Λ_p)		11.23	#4	11.85	0.32 ± 0.13
0^+_1 (Λ_p)		11.31			
1^+_2 (Λ_p)		11.75			
2^+_3 (Λ_p)		13.00			
1^+_3 (Λ_p)		13.09			
1^+_4 (Λ_p)		13.37			

4. A NEW HNSS

From the current experiment we attained valuable information for planning future spectroscopic studies using the $(e, e'K^+)$ reaction with a much-improved experimental setup. Fig. 8 shows the layout of a new generation experiment that has been proposed to and approved by the JLab program advisory committee [18]. The electron arm of the HNSS will be placed at an angle with respect to the floor plane so that the tagging angle of the scattered electrons is about 2.5 degrees. This will reduce the rate of forward electrons from bremsstrahlung by almost four orders of magnitude whereas the virtual photon flux will be reduced only by a factor of about 10. Thus, the new geometry allows a luminosity increase of more than a factor of 200. A new high-resolution and short-path-length kaon spectrometer (HKS) will be built. It is dedicated to the $(e,e'K^+)$ hypernuclear spectroscopy program at JLab under the construction fund by Monkasho [19] of Japan. It will improve the kaon arm momentum resolution by a factor of two and its solid angle acceptance by a factor of about 3. Overall, the yield is expected to increase by a factor of about 50 and the energy resolution may reach 350 keV (FWHM). Finally, the background will be reduced by a factor of 10. The goal of the new experiment is to carry out high precision and high statistics studies on medium mass hypernuclei, e.g. $^{28}_\Lambda Al$. $^{12}_\Lambda B$ and other p-shell hypernuclei will be measured again with better resolution and higher statistics. The $^{12}_\Lambda B$ system will serve also as a monitor and for calibrating the new HNSS system.

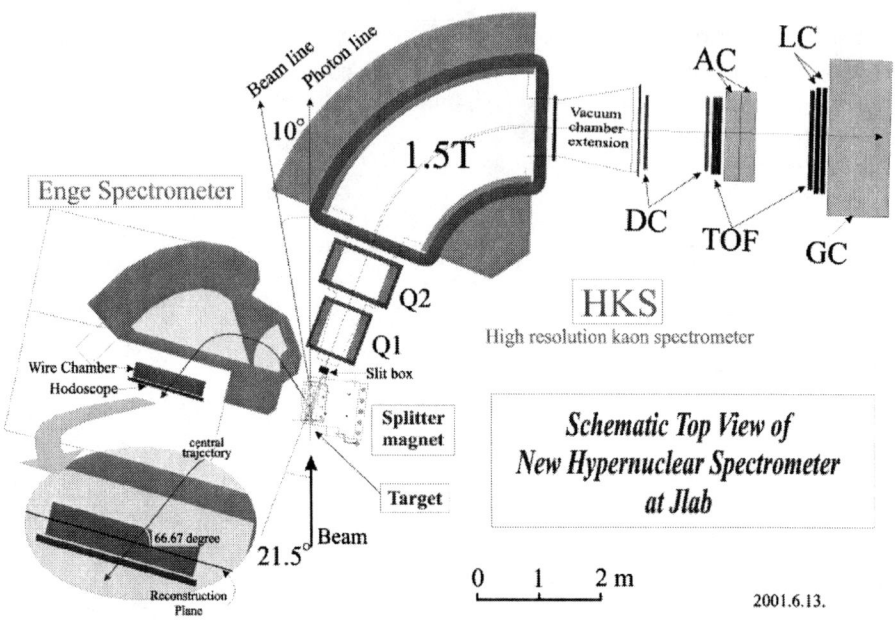

FIGURE 8. Experimental layout of the new HNSS system. The HKS replaces the SOS used in the current experiment.

5. SUMMARY

In the first experiment using the HNSS at JLab, sub-1-MeV energy resolution has been obtained in the spectrum from the $^{12}C(e, e'K^+)^{12}_\Lambda B$ reaction. The experiment succeeded in spite of an extremely high rate of electrons from bremsstrahlung and demonstrated that electroproduction can be used effectively for hypernuclear spectroscopic studies. Our preliminary spectrum shows strong peaks where spin-flip excitations are expected. Systematic studies of such transitions, produced by the large spin-flip amplitude for the electroproduction of hypernuclei, will complement hypernuclear studies by hadronic probes. The high-quality electron beam at JLab provides new opportunities for future hypernuclear studies. In addition, the new HKS which is currently being constructed and a new experimental geometry, will provide a 200-fold increase in good event rates and more than a factor of two improvement in energy resolution.

ACKNOWLEDGMENTS

We acknowledge support by the staff at both the accelerator and physics divisions at the Thomas Jefferson National Accelerator Facility (JLab). L. Tang wishes to thank Dr. D. J. Millener for many useful discussions and for making his most recent structure predictions available prior to publication. The Southeastern University

Research Association (SURA) operates JLab for the U.S. Department of Energy under Contract No. DE-AC05-84ER40150. This work is supported in part by research grants from the U.S. Department of Energy and the National Science Foundation. We would acknowledge the support by the U.S. DOE grants DE-FG02-97ER41047, DE-FG02-87ER40362, and the NSF grant HRD-9633750.

REFERENCES

1. Gibson, B.F., and Hungerford III, E.V., *Physics Reports* **257**, 349-388 (1995).
2. Milner, C., et al., *Phys. Rev. Lett.* **54**, 1237 (1985).
3. Pile, P.H., et al., *Phys. Rev. Lett.* **66**, 2585 (1991).
4. Dover, C.B., Proc. Int. Symp. on Medium Energy Physics, Beijing, World Scientific, 1987, p.257.
5. Bandō, H., Motoba, T., and Yamamoto, Y., *Phys. Rev. C* **31**, 265 (1985).
6. Likar, A., Rosina, M., and Povh, B., *Z. Phys. A* **324**, 35 (1986).
7. Gal, A., *Adv. Nucl. Sci.* **8**, 1 (1977).
8. Povh, B., *Ann. Rev. Nucl. Part. Sci.* **28**, 1 (1978).
9. Hashimoto, O., *Proc.* Int. Workshop on Strangeness Nuclear Physics, Seoul, World Scientific, 1999, p. 116;
    Hasegawa, T., et al., *Phys. Rev. Lett.* **74**, 224 (1995);
    Hasegawa, T., et al., *Phys. Rev. C* **53**, 1210 (1996).
10. Tamura, H., et al., *Phys. Rev. Lett.* **84**, 5963 (2000).
11. Tanida, K., et al., *Phys. Rev. Lett.* **86**, 1982 (2001).
12. Chrien, R.E., and Dover, C.B., *Ann. Rev. Nucl. Part. Sci.* **39**, 113 (1989).
13. Donnelly, T.W., and Cotanch, S.R., Proceedings of the 1985 CEBAF Summer Study, 1985;
    Hyde-Wright, C., et al., *Proceedings of the 1985 CEBAF Workshop*, 1985;
    Hsiao, S.S., and Cotanch, S.R., *Phys Rev. C* **28**, 1668 (1983);
    Cotanch, S.R., and Hsiao, S.S., *Nucl. Phys.* **A450**, 419 (1986).
14. Motoba, T., Stone, M., and Itonaga, K., *Progr. Theor. Phys. Suppl.* **117**, 123 (1994);
    Sotona, M., and Frullani, S., *ibid.,* 151.
15. Nozawa, S., and Lee, T.S., Nucl. Phys. **A513**, 511 (1990).
16. Spencer, J.E., and Enge, H.A., *Nucl. Instr. Meth.* **49**, 177 (1967).
17. Millener, D.J., *private communication*.
18. Hashimoto, O., Tang, L., and Reinhold, J., E01-011 experimental proposal to JLab PAC 19, 2001.
19. Hashimoto, O., *"Specially Promoted Research"* of Grant-in-Aid for Scientific Research (2000-2004), Monkasho, Japan.

Two-nucleon interaction and chiral symmetry

Rob G.E. Timmermans

*Theory Group, Kernfysisch Versneller Instituut, University of Groningen,
Zernikelaan 25, 9747 AA Groningen, The Netherlands*

Abstract. The method of energy-dependent partial-wave analysis of the low-energy two-nucleon scattering data is reviewed. Some recent additions to the database are discussed. In particular, the Uppsala np differential cross section at 162 MeV is analyzed. The first steps to improve the partial-wave analysis by implementing constraints from chiral perturbation theory are reviewed. Specifically, the long-range chiral two-pion exchange interaction has been included, and its importance has been demonstrated.

INTRODUCTION

In the energy-dependent Nijmegen PWA's of the NN and $\overline{N}N$ scattering data [1-6], the long-range forces are taken into account exactly and the short-range forces are parametrized analytically. The partial-wave scattering amplitudes are analytic functions of the energy. The nearby left-hand singularities in the complex-energy plane are due to the long-range forces; these cause the rapid energy dependence of the physical NN scattering amplitudes. The shorter-range forces are responsible for the far-away singularities, which give, in the physical region, only slow energy variations of the amplitudes. Energy-dependent PWA first of all serves as a high-precision tool for data analysis. But this specific method of PWA can also serve as a sensitive tool to investigate precisely these long-range interactions. It has been used extensively and successfully in studies of the NN electromagnetic interactions [7, 8], of the one-pion exchange (OPE) potential [9-12], and of the chiral two-pion exchange (TPE) potential [13, 14].

The great advantages of energy-dependent PWA's, as compared to single-energy PWA's or "amplitude analyses" have been explained by us extensively in the past, and do not have to be repeated here. See, in particular, Ref. [4] for specific examples that demonstrate the superiority of energy-dependent analyses over single-energy ones; and Ref. [6] for a similar discussion with respect to amplitude analyses.

The specific methods of the Nijmegen PWA's are described in detail in Refs. [2-5]. Originally started as a modified effective-range analysis [1, 2], it developed into what is essentially a boundary condition (BC) model. The long-range potentials, including the full electromagnetic interaction (relativistic Coulomb, magnetic-moment interaction, and vacuum polarization) and the longest-range strong interactions are used in the (relativistic) Schrödinger equation which is solved with a boundary condition at some $r = b$,

$$\text{BC} = b \frac{d\psi}{dr} \psi^{-1} \bigg|_{r=b}, \tag{1}$$

where $\psi(r)$ is the wave function. This BC is parametrized as an analytic function of energy for the various partial waves; in fact, the parametrization used for the BC corresponds essentially to a short-range interaction of the general form

$$V_S = C_0 + C_2 p^2 + C_4 p^4 + \ldots \qquad (2)$$

In case of NN scattering above pion-production threshold, or in the $\overline{N}N$ case [5], inelasticities are easily taken into account by using a complex BC.

The BC parameters, representing unknown short-range physics, and free parameters in the long-range forces (*e.g.* the pion-nucleon coupling constant, or the chiral constants c_i in TPE) are determined from a fit to the data. In the "standard" Nijmegen PWA's of Refs. [3, 4] the boundary is put at $b = 1.4$ fm, and the long-range strong potential outside of 1.4 fm is taken as the OPE potential supplemented by the non-OPE forces of the Nijmegen soft-core potential Nijm78 [15], which provides a good fit to the pp data. The heavy-boson exchanges were included because OPE alone did not allow for an optimal description of the data. In the most recent PWA's, these heavy-boson exchanges were removed in favor of the theoretically well-motivated chiral TPE interaction [14].

DATABASE

The pp and np database can be found on NN-OnLine [16] and on SAID [17]. The pp database below 500 MeV is of high quality, the np database is of good quality but not as varied and precise as the pp database.

During the last few years, measurements at IUCF of the pp analyzing power and spin correlations between 197 and 449 MeV (A_y, A_{xx}, A_{yy}, A_{xz}, and A_{zz}) have resulted in a significant addition to the pp database [18-21]. While, in general, the importance of "spin-data" is overestimated (see the discussion in Ref. [6]), these accurate IUCF data, especially the first sets that became available, have clearly improved the PWA. In the PWA below 500 MeV, the about 1400 data are described very well. To compare, the pp database below 350 MeV used in PWA93 contained 1787 data. Details about the description of the IUCF data in the PWA can be found on NN-OnLine [16].

The neutron-proton (np) differential cross section at neutron laboratory energies below 350 MeV has been a topic of frequent investigations. One reason for the special interest in this cross section has been the suggestion made by G.F. Chew [22] in 1958 that the pion-nucleon coupling constant could be determined from the backward np data. In Fig.1 we show the np differential cross section at $T_{\text{lab}} = 212$ MeV as predicted by the energy-dependent Nijmegen partial-wave analysis PWA93 [4], together with absolutely normalized TRIUMF data [23]. The most distinctive features of this cross section are the forward peak, due to the destructive interference between neutral-pion exchange and the rest of the amplitude, and the backward peak, similarly due to the destructive interference between charged-pion exchange and the rest of the amplitude.

The measurement of an np differential cross section is notoriously difficult. Especially the determination of the correct normalization often poses problems. These cross sections are most easily measured in the backward direction, because the forward-going recoil protons have in that case a relatively high energy. In order to cover a larger an-

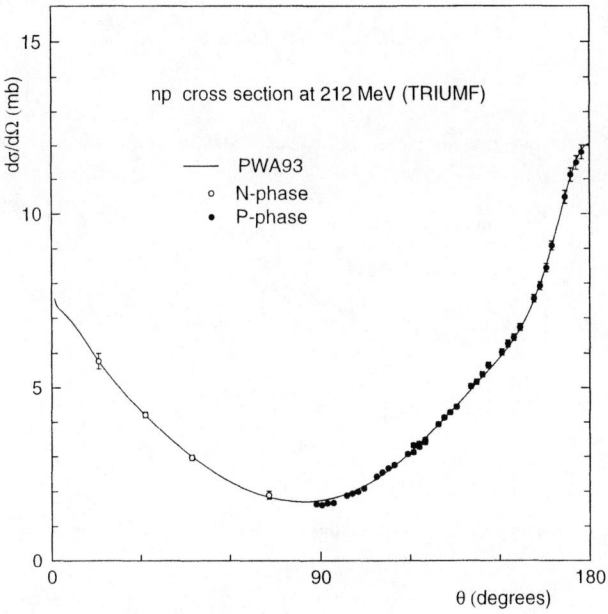

FIGURE 1. The np differential cross section at 212 MeV.

gular region, often several different settings of the detection apparatus are used. The differential cross section is then measured in different, sometimes overlapping, angular regions. The np data base [16, 17] with $T_{lab} < 350$ MeV contains a number of such data sets [23-26]. In a recent paper we analyzed these np backward cross section data, using the np PWA93 as a tool [27].

For the older LAMPF [24] and TRIUMF [23] data, we found that we could improve the description of these data in the PWA. In particular, for the LAMPF data we proposed a novel way of relative normalization [27]. We also studied carefully the more recent Uppsala data at 162 MeV [25, 26]. These data are not in agreement [28] with the PWA's of the Nijmegen and Virginia Tech groups, and strong claims have been made about this disagreement by the Uppsala group [25, 26]. In this Uppsala experiment, five different settings of the spectrometer position were chosen to cover the angular range of $\theta_{lab} = 0°$ to 54° for the recoil proton. This guaranteed a large angular overlap between the different settings. After the sets were relatively normalized, the data in the overlap regions were averaged point by point. In this way the relative differential cross section between 72° and 180° was obtained. Finally, this relative cross section was absolutely normalized with the help of PWA's.

In Fig. 2 we show the difference between the differential cross section, normalized by us, and PWA93. Comparison with PWA93 gives for the 54 data $\chi^2 = 393$, while for 54 data one would expect $\langle \chi^2 \rangle = 53 \pm 10$. Our result is 33 s.d. away from this expectation value, which is an extremely large difference. The Uppsala group has always pointed

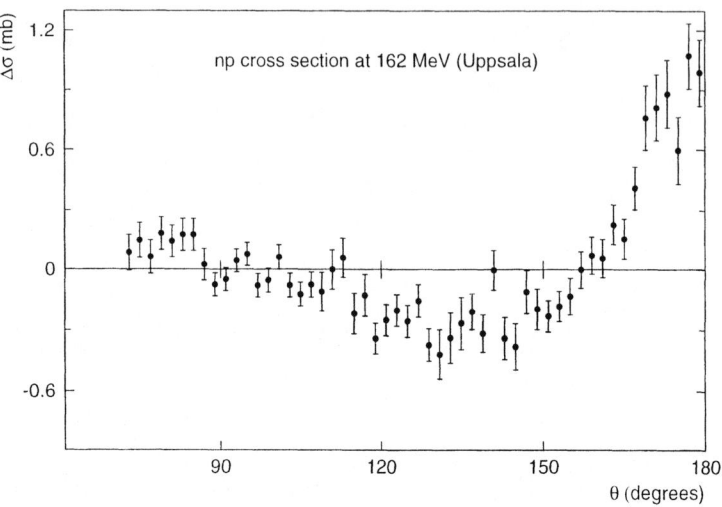

FIGURE 2. The difference $\Delta\sigma(\theta)$ between the absolutely normalized Uppsala data and PWA93.

to the LAMPF data as the culprit and claims [26] that these data dominate our solution and also the Virginia Tech solution. To study this claim we omitted all the LAMPF data [24] from our data base and included the 96 MeV and the 162 MeV Uppsala data. For the original 31 data at 162 MeV [25], we get then $\chi^2 = 246$, which is not much of an improvement! This shows that the Uppsala differential cross section is in conflict not only with the LAMPF cross sections, but also with the rest of the np data base, including asymmetries, spin correlations, etc.

A first hint why the Uppsala data produces such a large value of χ^2 can be obtained when we look at the data in the overlap region between sets 4 and 5, where set 5 contains the most backward angles. This overlap region runs from 151° to 167° and contains from each set 9 points. For each data set i the function $\Delta\sigma$ is fitted by a straight line $\Delta\sigma = a_i + b_i \theta$. These fits are statistically quite acceptable. For set 4 $\chi^2 = 10.3$ and for set 5 $\chi^2 = 4.2$. In both cases 9 points were used in the fit. For the slopes b_i of the sets 4 and 5 we find $b_4 = 0.019(9)$ and $b_5 = 0.053(9)$. We see that both slopes are different from 0, therefore the slopes do not agree with PWA93. Especially the slope of set 5 in this overlap region is almost 6 s.d. away from the slope predicted by PWA93. The Uppsala group took the relative normalization between the sets such, that in the overlap region the differential cross section is as much as possible continuous. They could not require that the slope was also continuous. The slopes b_4 and b_5 of set 4 and 5 in the overlap region turn out to be rather different. We find $b_5 - b_4 = 0.034(13)$. Therefore, in the overlap region the sets 4 and 5 appear to be in disagreement with each other.

In our PWA, we would determine for each individual data set separably an absolute normalization. We have then 5 separate calculated absolute normalizations and no relative normalizations. For the complete data set we have 88 points with $\chi^2 = 257$ with respect to PWA93. For the expectation value of χ^2 we find $\langle\chi^2\rangle = 83 \pm 13$. We can now

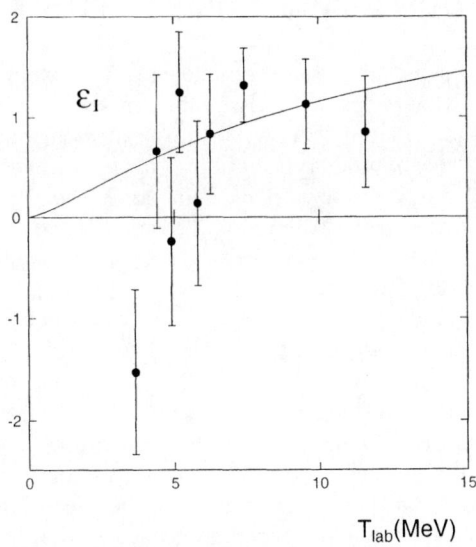

FIGURE 3. The np ε_1 mixing parameter at low energies [31]. The drawn curve is the prediction from the energy-dependent analysis PWA93. The individual points are the values resulting from the single-point analyses of the TUNL data.

look at the way the relative normalizations were determined by the Uppsala group. We demonstrate this with the straight-line approximations to the individual data sets. We require these straight lines to intersect in the middle of each overlap region. In this way a more-or-less continuous differential cross section is obtained. For the original Uppsala set of 88 points we find that $\chi^2 = 470$; see Figs. 12 and 13 of Ref. [27].

We conclude that the Uppsala data contains unexplained large systematic deviations from PWA93. The manner in which the Uppsala group performed the relative normalization of their data enhanced the negative effects of these systematic flaws.

At TUNL the spin-dependent np total cross sections were measured between 5 and 20 MeV neutron energy [29, 30]. Single-parameter phase-shift analyses were performed to extract the $^3S_1$-$^3D_1$ mixing parameter ε_1. It is claimed [29, 30] that these value support a stronger tensor force than predicted by meson-exchange NN potential models and PWA's. Such a strong claim deserve closer scrutiny. When these TUNL data are included in the PWA, properly normalized, we find that the *prediction* of PWA93 is already very good, with $\chi^2_{min}/N \simeq 0.6$. Refitting with the TUNL data included then obviously makes extremely little difference, both for the χ^2-value and for the resulting values of ε_1. For instance, before inclusion of the TUNL data ε_1 at 5 MeV was $0.6717°(42)$, and after refitting it becomes $0.6715°(42)$. Thus, these TUNL data are described (predicted) well by PWA93, and they have no effect on ε_1. In Fig. 3 is shown our result for ε_1 at low energies [31].

This discussion about ε_1 has a *déjà vu* aspect: see the claims made by the Basel group in Ref. [32] based on the analysis of their 67.5 MeV A_{zz} measurements, and the refutation of these claims in Ref. [33].

CHIRAL TWO-PION EXCHANGE

The longest-range part of the strong nucleon-nucleon (NN) interaction is the one-pion exchange (OPE) force, which became firmly established around 1960. The formulation of the two-pion exchange (TPE) force, however, has been a long-standing problem, both in field theory and in dispersion theory, and little progress was achieved between 1950 and 1990. However, starting with the work of Weinberg in 1990 [34], it has been argued in recent years that the key to the solution is the chiral symmetry of QCD, and that the long-range parts of the TPE potential can be derived model-independently by a systematic expansion of the effective chiral Lagrangian [35-38]. Following up on these developments, we have studied the long-range chiral two-pion exchange (χTPE) force in the proton-proton (pp) interaction in Ref. [13].

This is possible because, as mentioned, the specific method of PWA allows us to study specific parts of the long-range NN interaction. This can be demonstrated with some well-known parts of the long-range electromagnetic interaction: the magnetic-moment interaction [8], which contains a long-range spin-orbit and tensor force and which is needed to achieve an accurate description of the pp analyzing-power data, and the vacuum polarization potential [2], which is needed to describe properly the low-energy pp cross sections. When one omits in the standard pp PWA (with 1951 data) the magnetic-moment interaction, both from the potential and in constructing the scattering amplitude, the $\chi^2_{\min}$ increases by 390 from $\chi^2_{\min} = 1968$ to 2358. This is therefore a 20 standard deviation (s.d.) effect. Omitting vacuum polarization leads to $\chi^2_{\min} = 2181$, i.e., a rise in $\chi^2_{\min}$ of 213, which corresponds to 15 s.d. These numbers demonstrate that one can use this method of energy-dependent PWA to show the presence and the importance of these specific well-known parts of the long-range pp interaction.

A very important part of the energy dependence of the NN phase shifts comes from OPE. In the Nijmegen energy-dependent PWA's the different pion-nucleon coupling constants could be determined accurately and reliably [9-11]. In Ref. [11], we recommended for the charge-independent coupling constant the value $f^2_{NN\pi} = 0.0750(9)$, where the error includes statistical as well as systematic effects. As a systematic check, the masses of the exchanged pions were determined, with excellent results: $m_{\pi^0} = 135.6(1.0)$ MeV and $m_{\pi^+} = 139.6(1.3)$ MeV. In this way, the presence of OPE in the NN force was shown with an enormous statistical significance. In Ref. [12], also the electromagnetic corrections to the OPE potential in np scattering were investigated. These have only little effect on the quality of the fit.

The starting point to derive the OPE and TPE potentials is the effective chiral Lagrangian, the leading-order of which is the nonlinear Weinberg model,

$$\mathcal{L}^{(0)} = -\overline{N}\left[\gamma_\mu \mathcal{D}^\mu + M + g_A i \gamma_5 \gamma_\mu \vec{\tau} \cdot \vec{D}^\mu\right] N , \qquad (3)$$

with the chiral-covariant derivatives

$$\begin{aligned}\vec{D}^\mu &= D^{-1}\partial^\mu \vec{\pi}/F_\pi , \\ \mathcal{D}^\mu N &= \left(\partial^\mu + \frac{i}{F_\pi}\vec{\tau}\cdot\vec{\pi}\times\vec{D}^\mu\right) N .\end{aligned} \qquad (4)$$

Here, $D = 1 + \vec{\pi}^2/F_\pi^2$, $g_A \simeq 1.26$ is the Gamow-Teller coupling, and $F_\pi = 185$ MeV is the pion decay constant. In addition to the standard ("pseudovector") $NN\pi$ interaction, Eq. (3) contains the Weinberg-Tomozawa (WT) $NN2\pi$ seagull interaction, which gives rise to the well-known result for the isovector πN scattering lengths. In the leading-order TPE potential, the WT interaction leads to triangle and bubble ("football") TPE diagrams.

In order to derive the TPE potential in subleading order, three more $NN2\pi$ interactions are required [35], viz.

$$\mathcal{L}^{(1)} = -\overline{N} \left[8c_1 D^{-1} m_\pi^2 \vec{\pi}^2 / F_\pi^2 + 4c_3 \vec{D}_\mu \cdot \vec{D}^\mu \right.$$
$$\left. + 2c_4 \sigma_{\mu\nu} \vec{\tau} \cdot \vec{D}^\mu \times \vec{D}^\nu \right] N, \quad (5)$$

leading to additional triangle diagrams. The values of the chiral parameters ("low-energy constants") c_i ($i = 1, 3, 4$), of order $(1/M)$, are not fixed by chiral symmetry and have to be determined by a fit to experimental data. Loosely speaking, these c_i's represent "integrated-out" hadrons, such as the heavier mesons like the ε and ρ, and the N- and Δ-isobars. For instance, the (isoscalar) c_3-term can receive sizable contributions for ε-exchange and the Δ, while the (isovector) c_4-term gets contributions from ρ-exchange and the Δ. The definition Eq. (5) of the c_i's has been chosen to agree with the convention used in heavy-baryon χPT [39]. In πN scattering, an additional c_2-term appears, but this term does not contribute to the NN force in this order. The c_1-term violates chiral symmetry explicitly and is related to the famous pion-nucleon "sigma term" (see below). A systematic expansion of Eqs. (3) and (5) to order $(1/M)$ gives then the relevant part of the chiral Lagrangian [40].

The full χTPE potential has a complicated spin and isospin structure. Several features due to the chiral constants c_i are reminiscent of the well-known one-boson exchange (OBE) model of the NN interaction. The c_3-term gives a strong attractive isoscalar central interaction, in a naive OBE model parametrized by the fictitious-σ(500) exchange; the c_4-term provides an isovector tensor force opposite in sign to that of OPE, a feature also given by ρ-exchange. It is interesting to study this χTPE potential in the chiral limit $m_\pi \to 0$. If we define $\zeta = g_A^2/4\pi F_\pi^2$, then OPE reduces to an isovector tensor force of the form $W_T = \zeta/r^3$, and the leading-order TPE potentials behave like $V \sim \zeta^2/r^5$. In the subleading TPE potentials the c_1 terms disappear; in the static limit, only the terms that go like $\sim 1/r^6$ survive. Specifically, one gets

$$V_C = \zeta \frac{144 c_3}{4\pi F_\pi^2} \frac{1}{r^6},$$
$$W_{SS} = \zeta \frac{16 c_4}{4\pi F_\pi^2} \frac{1}{r^6}, \quad (6)$$
$$W_T = -\zeta \frac{16 c_4}{4\pi F_\pi^2} \frac{1}{r^6}.$$

One observes that the central attraction is especially strong, both because of the large value of c_3 and because of the large factor 144 that enters the potential. The expressions Eq. (6) illustrate the analogy of these TPE forces to the electromagnetic van der Waals forces, which correspond to two-photon exchange bubble ("football") diagrams between two neutral but polarizable systems. In fact, the triangle diagram with one c_3-vertex is partly a "chiral van der Waals force" due to the *axial* polarizability of the nucleon [41].

As demonstrated above for other well-known parts of the long-range NN interaction, we can use the PWA to study the presence and importance of the χTPE interaction. In Ref. [13], we did this for the pp PWA below 350 MeV. The main results of the various PWA's that were done are summarized in Table 1. In Ref. [14], the χTPE interaction was included in the pp and np PWA's below 500 MeV.

We started conservatively with the boundary at $b = 1.8$ fm, since beyond 1.8 fm only OPE and TPE are expected to contribute significantly. When only OPE was included as strong force, $\chi^2_{min} = 1957$ was reached at the cost of 29 BC parameters. The question then is whether the fit can be even further improved when TPE is added. When only the TPE(l.o.) potential was used, we obtained $\chi^2_{min} = 1966$ with 26 BC parameters. The next step is to include the complete TPE potential, χTPE=TPE(l.o.)+TPE(s.o.). This potential contains three *a priori* unknown constants: the chiral parameters c_i ($i = 1, 3, 4$) from Eq. (5). In the fits we obtained $c_1 = -4.4(3.4)/$GeV. The values of c_1 and c_3, appearing both only in the isoscalar central potential, are strongly correlated; c_1 is also, but less, correlated with c_4. These correlations between the chiral parameters can be summarized concisely by the equations:

$$c_3 = \left[-5.08 - 0.62(c_1 + 0.76) + 40(f_p^2 - 0.0755) \right]/\text{GeV},$$
$$c_4 = \left[+4.70 + 0.01(c_1 + 0.76) + 250(f_p^2 - 0.0755) \right]/\text{GeV}. \quad (7)$$

In order to determine reliable values for c_3 and c_4, one can use the theoretical estimate [39] for c_1 obtained from the scalar form factor $\sigma(t)$ of the proton [42, 43] at $t = 0$, viz.

$$c_1 = -\left[\sigma(0)/4m_\pi^2 + 9f^2\xi^2/16m_\pi \right] ; \quad (8)$$

$\sigma(0)$ is the famous pion-nucleon sigma term, the value of which is uncertain. In Ref. [13], we took the plausible "low" value $\sigma(0) = 35(5)$ MeV [44], which is supported by the πN PWA of Ref. [45], and which is consistent with zero strangeness content of the proton. This gave then

$$c_1 = -[0.46(7) + 0.30]/\text{GeV} = -0.76(7)/\text{GeV} ; \quad (9)$$

the error here is theoretical. Our determination of c_1 above is within errors consistent with this value. Fixing $c_1 = -0.76/$GeV, we found, with 22 BC parameters, $\chi^2_{min} = 1938$ and $f_p^2 = 0.0755(7)$. The resulting values for c_3 and c_4 were

$$c_3 = -5.08(28)/\text{GeV}, \quad c_4 = +4.70(70)/\text{GeV}, \quad (10)$$

where the errors are statistical. The improvement over only OPE is reflected, even beyond 1.8 fm, in the lower χ^2_{min} and in the 7 fewer BC parameters required.

The result found for f_p^2 is in very good agreement with the value 0.0756(4) determined in the standard 1998 pp PWA. Our values for the c_i's can be compared to the determination from the πN scattering amplitudes in Ref. [46]. Here, $c_1 = -0.93(9)/$GeV was obtained using Eq. (8), but with $\sigma(0) = 45(8)$ MeV, along

TABLE 1. PWA's with different long-range interactions. #BC is the number of BC parameters.

	$b = 1.4$ fm		$b = 1.8$ fm	
	#BC	χ^2_{min}	#BC	χ^2_{min}
Nijm78	19	1969	–	–
OPE	31	2026	29	1957
OPE + TPE(l.o.)	28	1985	26	1966
OPE + χTPE	23	1935	22	1938

with $c_3 = -5.29(25)$/GeV and $c_4 = +3.63(10)$/GeV. More recently, the values $c_1 = -0.80(7)$/GeV, $c_3 = -4.70(95)$/GeV, and $c_4 = +3.40(04)$/GeV were reported [47]. In view of the uncertainties that exist in the πN amplitudes [45], the good agreement is a significant success. It underlines, for the first time quantitatively, that the long-range NN and the low-energy πN interactions are governed by the same chiral Lagrangian.

In previous studies of the OPE potential, it has always been a useful systematic check to determine not only the $NN\pi$ coupling constant, but also the masses of the exchanged pions. In order to check explicitly that one is now actually looking at the TPE interaction, we determined the range. This was done by adding the pion mass m_π in the potential χTPE as another free parameter. We first fixed the pion coupling in OPE at $f_p^2 = 0.0755$ and the c_i's to their central values given in Eqs. (9) and (10). Then we fitted an overall scale factor λ for the potential χTPE, the pion mass m_π, and the BC parameters. The results were: $\lambda = 0.82(16)$ and $m_\pi = 125(10)$ MeV. Alternatively, we fixed c_1 and fit m_π together with f_p^2, c_3, c_4, and the BC parameters. This resulted in $m_\pi = 128(9)$ MeV, again in good agreement with the average pion mass $m_\pi = 138.04$ MeV. The very good χ^2_{min} obtained, the good values for the c_i's, and this correct pion mass constitute convincing proof for the presence of chiral TPE loops in the long-range pp interaction.

In order to investigate further the importance of χTPE, we moved the boundary inwards to $b = 1.4$ fm. When only OPE was used as the long-range force, it was possible to achieve a reasonable, but not a very good, fit: at the cost of 31 BC parameters $\chi^2_{min} = 2026$ was reached. We then added to OPE the potential TPE(l.o.). With 28 BC parameters, $\chi^2_{min} = 1985$ was obtained. Compared to only OPE, this corresponds to a drop in χ^2_{min} of 42 with 3 fewer parameters, a significant improvement. However, the fit was still not optimal. We next added also the potential TPE(s.o.). With fixed $c_1 = -0.76$/GeV, this gave with 23 BC parameters $\chi^2_{min} = 1935$, $c_3 = -4.99(21)$/GeV, and $c_4 = +5.62(59)$/GeV. These results show that OPE together with χTPE gives a very good NN force at least as far inwards as 1.4 fm.

CONCLUSIONS AND SUMMARY

In conclusion, the Nijmegen energy-dependent partial-wave analysis is a *high-precision tool* for data analysis and for studying the long-range two-nucleon interaction. We have demonstrated that the Uppsala np differential cross section data suffer from serious systematic flaws. The TUNL np spin-dependent total cross sections are in very good agreement with PWA93.

With OPE and χTPE, an excellent fit to the NN database becomes possible, a fit that is even better than that of PWA93. The chiral parameters agree with the ones found in πN scattering. Especially important to obtain the very good fit was the strong isoscalar central attraction from the c_3-term. A novel class of PWA has been established, with such a theoretically well-founded and model-independent χTPE potential included in all partial waves. Further constraints from chiral perturbation theory are being incorporated in the PWA.

In Ref. [14], this χPWA was extended to 500 MeV, for both pp and np. In that case, an independent np PWA is possible, with no input from pp for the isovector waves [48]. This implies that we can now study charge-independence breaking in all partial waves.

ACKNOWLEDGMENTS

This work reported on here was done in collaboration with M.C.M. Rentmeester and J.J. de Swart. I thank the organizers for the invitation to give this talk at Mesons and Light Nuclei '01 and for putting together such a stimulating and well-organized conference in beautiful Prague. This work was made possible by a Fellowship of the Royal Netherlands Academy of Arts and Sciences (KNAW).

REFERENCES

1. van der Sanden, W.A., Emmen, A.H., and de Swart, J.J., Nijmegen Report THEF-NYM-83-11.
2. Bergervoet, J.R., van Campen, P.C., van der Sanden, W.A., and de Swart, J.J., *Phys. Rev. C* **38**, 15 (1988).
3. Bergervoet, J.R., van Campen, P.C., Klomp, R.A.M., de Kok, J.-L., Rijken, Th.A., Stoks, V.G.J., and de Swart, J.J., *Phys. Rev. C* **41**, 1435 (1990).
4. Stoks, V.G.J., Klomp, R.A.M., Rentmeester, M.C.M., and de Swart, J.J., *Phys. Rev. C* **48**, 792 (1993).
5. Timmermans, R., Rijken, Th.A., and de Swart, J.J., *Phys. Rev. C* **50**, 48 (1994).
6. Timmermans, R., Rijken, Th.A., and de Swart, J.J., *Phys. Rev. C* **52**, 1145 (1995).
7. Austen, G.J.M., and de Swart, J.J., *Phys. Rev. Lett.* **50**, 2039 (1983).
8. Stoks, V.G.J., and de Swart, J.J., *Phys. Rev. C* **42**, 1235 (1990).
9. Stoks, V.G.J., Timmermans, R., and de Swart, J.J., *Phys. Rev. C* **47**, 512 (1993).
10. Bergervoet, J.R., van Campen, P.C., Rijken, Th.A., and de Swart, J.J., *Phys. Rev. Lett.* **59**, 2255 (1987); Klomp, R.A.M., Stoks, V.G.J., and de Swart, J.J., *Phys. Rev. C* **44**, R1258 (1991); Timmermans, R., Rijken, Th.A., and de Swart, J.J., *Phys. Rev. Lett.* **67**, 1074 (1991).
11. de Swart, J.J., Rentmeester, M.C.M., and Timmermans, R.G.E., nucl-th/9802084.
12. van Kolck, U., Rentmeester, M.C.M., Friar, J.L., Goldman, T., and de Swart, J.J., *Phys. Rev. Lett.* **80**, 4386 (1998).
13. Rentmeester, M.C.M., Timmermans, R.G.E., Friar, J.L., and de Swart, J.J., *Phys. Rev. Lett.* **82**, 4992 (1999).
14. Rentmeester, M.C.M., Timmermans, R.G.E., and de Swart, J.J., in preparation.
15. Nagels, M.M., Rijken, Th.A., and de Swart, J.J., *Phys. Rev. D* **17**, 768 (1978).
16. Nijmegen NN-Online Facility, see http://NN-OnLine.sci.kun.nl.
17. VPI/GWU SAID Facility, see http://gwdac.phys.gwu.edu.
18. Haeberli, W., *et al.*, *Phys. Rev. C* **55**, 597 (1997).
19. Rathmann, F., *et al.*, *Phys. Rev. C* **58**, 658 (1998).
20. von Przewoski, B., *et al.*, *Phys. Rev. C* **58**, 1897 (1998).
21. Lorentz, B., *et al.*, *Phys. Rev. C* **61**, 054002 (2000).
22. Chew, G.F., *Phys. Rev.* **112**, 1380 (1958).

23. Keeler, R.K., *et al.*, *Nucl. Phys.* **A377**, 529 (1982).
24. Bonner, B.E., *et al*, *Phys. Rev. Lett.* **41**, 1200 (1978).
25. Ericson, T.E.O., *et al.*, *Phys. Rev. Lett.* **75**, 1046 (1995).
26. Rahm, J., *et al.*, *Phys. Rev. C* **57**, 1077 (1998).
27. Rentmeester, M.C.M., Timmermans, R.G.E., and de Swart, J.J., *Phys. Rev. C* **64**, 034004 (2001).
28. Rentmeester, M.C.M., Klomp, R.A.M., and de Swart, J.J., *Phys. Rev. Lett.* **81**, 5253 (1998).
29. Raichle, B.W., *et al.*, *Phys. Rev. Lett.* **83**, 2711 (1999).
30. Walston, J.R., *et al.*, *Phys. Rev. C* **63**, 014004 (2001).
31. Rentmeester, M.C.M., Ph.D. thesis, University of Nijmegen, 2001.
32. Hammans, M., *et al.*, *Phys. Rev. Lett.* **66**, 2293 (1991).
33. Klomp, R.A.M., Stoks, V.G.J., and de Swart, J.J., *Phys. Rev. C* **45**, 2023 (1992).
34. Weinberg, S., *Phys. Lett. B* **251**, 288 (1990); *Nucl. Phys.* **B363**, 3 (1991).
35. Ordóñez, C., and van Kolck, U., *Phys. Lett. B* **291**, 459 (1992); Ordóñez, C., Ray, L., and van Kolck, U., *Phys. Rev. Lett.* **72**, 1982 (1994); *Phys. Rev. C* **53**, 2086 (1996).
36. Friar, J.L., and Coon, S.A., *Phys. Rev. C* **49**, 1272 (1994); Friar, J.L., *Phys. Rev. C* **60**, 034002 (1999).
37. Kaiser, N., Brockmann, R., and Weise, W., *Nucl. Phys.* **A625**, 758 (1997).
38. Epelbaum, E., Glöckle, W., and Meissner, U.-G., *Nucl. Phys.* **A637**, 107 (1998); *Nucl. Phys.* **A671**, 295 (2000).
39. Bernard, V., Kaiser, N., Kambor, J., and Meissner, U.-G., *Nucl. Phys.* **B388**, 315 (1992).
40. See Eqs. (1) and (2) of: Friar, J.L., Hüber, D., and van Kolck, U., *Phys. Rev. C* **59**, 53 (1999).
41. Tarrach, R., and Ericson, M., *Nucl. Phys.* **A294**, 417 (1978).
42. Gasser, J., Sainio, M.E., and Švarc, A., *Nucl. Phys.* **B307**, 779 (1988).
43. Gasser, J., Leutwyler, H., and Sainio, M.E., *Phys. Lett. B* **253**, 252, 260 (1991).
44. Gasser, J., *Ann. Phys. (N.Y.)* **136**, 62 (1981); Gasser, J., and Leutwyler, H., *Phys. Rep.* **87**, 77 (1982).
45. Timmermans, R.G.E., πN *Newsletter* **13**, 80 (1997).
46. Bernard, V., Kaiser, N., and Meissner, U.-G., *Nucl. Phys.* **A615**, 483 (1997).
47. Buttiker, P., and Meissner, U.-G., *Nucl. Phys.* **A668**, 97 (2000).
48. Klomp, R.A.M.M., de Kok, J.-L., Rentmeester, M.C.M., Rijken, Th.A., and de Swart, J.J., in: *Few-Body Problems in Physics, Williamsburg 1994*, AIP Conference Proceedings 334, edited by F. Gross, (AIP Press, 1995), p. 367.

Quark degrees of freedom in hadronic systems

V. Vento

Departamento de Física Teórica, Universidad de Valencia, 46100 Burjassot (Valencia), Spain

Abstract. The role of models in Quantum Chromodynamics is to produce simple physical pictures that connect the phenomenological regularities with the underlying structure. The static properties of hadrons have provided experimental input to define a variety of very successful Quark Models. We discuss applications of some of the most widely used of these models to the high energy regime, a scenario for which they were not proposed. The initial assumption underlying our presentation will be that gluon and sea bremsstrahlung connect the constituent quark momentum distributions with the partonic structure functions. The results obtained are encouraging but lead to the necessity of more complex structures at the hadronic scale. This initial hypothesis may be relaxed by introducing some non perturbative model for the constituent quarks. Within this scheme we will discuss some relevant problems in nucleon structure as seen in high energy experiments.

INTRODUCTION

The constituent quark, one of the most fruitful concepts in 20th century physics, was proposed to explain the structure of the large number of hadrons being discovered in the sixties [1]. Soon thereafter deep inelastic scattering of leptons off protons was explained in terms of pointlike constituents named partons [2]. Thus already at a very early stage of the study of hadron structure the need to connect the laboratory description, based on constituent quarks, and the light cone description, based on partons, arose as the way to understand phenomena at different scales. Sum rules and current algebra were very powerful tools to establish a conceptual link between the two descriptions.

The birth of Quantum Chromodynamics (QCD) and the proof that it is asymptotically free set the framework for an understanding of deep inelastic phenomena beyond the parton model [3]. However, the perturbative approach to QCD does not provide absolute values for the observables, it just gives their variation with momentum in terms of some unknown non perturbative matrix element. On order for the description based on the Operator Product Expansion (OPE) and QCD evolution to be predictive, these matrix elements have to be eliminated by comparing several processes or by the input of experimental data. Therefore the perturbative scheme is used, most of the time, to relate experiments at different momentum scales.

The phenomenological analysis proceeds by finding a parametrization which is appropriate at a sufficiently large momentum Q_0^2, where one expects perturbation theory to be fully applicable, and then QCD evolution techniques determine the parton distributions at a higher Q^2. As an example we show the parametrization due to Martin Sterling and Roberts ($Q_0^2 = 4\ GeV^2$) [4]:

$$\begin{aligned}
xu_v &= 2.26x^{0.559}(1-0.54\sqrt{x}+4.65x)(1-x)^{3.96} \\
xd_v &= 0.279x^{0.335}(1+6.80\sqrt{x}+1.93x)(1-x)^{4.46} \\
xS &= 0.956x^{-0.17}(1-2.55\sqrt{x}+11.2x)(1-x)^{9.63} \\
xg &= 1.94x^{-0.17}(1-1.90\sqrt{x}+4.07x)(1-x)^{5.33}
\end{aligned} \quad (1)$$

This parametrization incorporates the flavor and momentum sum rules. The distributions are defined in the $\overline{MS}$ renormalization and factorization schemes and the QCD scale paramenter Λ is found to be $0.231\ GeV$. With this fit a large body of data is reasonably described. However this parametrization is purely phenomenological with little theoretical input.

The work of Glück, Reya and Vogt [5] has shown that the high energy parton distributions when evolved to a low scale appear to indicate that a valence picture of hadron structure arises. This idea was suggested a long time ago by Parisi and Petronzio [6], who assumed that there exists a low energy scale μ_0^2 such that the glue and sea are absent, i.e., the long range part of the interaction (confinement) is composed in the P_∞ frame of only three quarks. If one turns on the short range part of the interaction (perturbative QCD), using the renormalization group one introduces gluons and the sea.

The constituent quark concept embedded in a QCD framework, leads to models that are able to reproduce in an extraordinary way the low energy properties with very few parameters [7]. The goal was to use them as substitutes for QCD at low energies. The needed ingredient was provided by Jaffe and Ross [8]. According to these authors the quark model calculation of matrix elements give their values at a hadronic scale μ_0^2 and for all larger Q^2 their coefficient functions evolve as in perturbative QCD.

We have developed a formalism, for potential quark models, based on these ideas which connects the parton distributions with the momentum distributions of the model [9]. A analogous procedure may be derived for bag models by using the bag model limit of light cone matrix elements [10]. The low energy scale μ_0^2 is determined by evolving downward from the high energy data the second moment of the valence quark distribution until it reaches the value given by the quark model describing the hadronic behavior. The model provides the matrix elements of the needed twist operators characterizing observables at the high energy scale and their values are ascribed to this hadronic scale. Then, they are evolved to high momentum transfers, where comparison with experiments takes place, using perturbative QCD.

This approach describes successfully the gross features of the DIS results [9]. In order to produce more quantitative fits different mechanisms have been proposed: *valence* gluons, sea quarks and antiquarks, relativistic kinematics, etc... We will show that some of these mechanisms appear naturally if we endow the constituent quarks with structure following the work of Altarelli et al.[11]. In our scheme *constituent* quarks are complex objects, made up of point-like partons (*current* quarks (antiquarks) and gluons), interacting by a residual interaction described by a quark model [12]. The hadron structure functions are obtained as convolutions of the constituent quark wave function with the constituent quark structure functions.

Our aim here is neither technical nor bibliographical. We will simply guide the reader to the literature by discussing the physics behind the various formalisms. In the referred literature he will find a complete account of the needed references and technicalities, so that he may be able to reconstruct the calculations presented in detail. We will elaborate on the theoretical framework, discuss some of the main results and explore future directions.

CONSTITUENT QUARKS AND PARTONS

Constituent quark models have been designed to describe the static properties of hadrons and therefore aimed at modeling the non-perturbative aspects of *QCD*. They are in general very successful in their performance. We discuss a formalism which uses them to describe high energy data, whose basis lies on the following reasoning. *QCD* perturbation theory is non predictive. The renormalization group relates different momentum scales. Experimental input is required to avoid the unknown non-perturbative properties of the theory. Our formalism substitutes the experimental input by model physics. In this way we define a predictive scheme, whose appeal lies in the relation it establishes between physics at very different scales and whose weakness is its model dependence.

Parton distributions from quark models

The basic idea in our approach arises from rephrasing the OPE which states that,

$$F_i^n(Q^2) = M_{ij}^n F_j^n(Q_0^2), \qquad (2)$$

i.e., the nth moment of structure functions at one scale are related by means of perturbatively calculable transformation matrices to the same moments at another scale [3]. If Q_0^2 is taken to be a low scale, which we have named hadronic scale, the F functions become highly non perturbative matrix elements in general. We substitute the matrix elements at the hadronic scale by the matrix elements calculated in the chosen model. In particular we are able to relate the valence quark distribution functions with the appropriate momentum distributions in the corresponding baryonic state n_q^a, i.e. with the hadronic wave functions in the model,

$$x q_V^a(x) = \frac{1}{(1-x)^2} \int d^3 p \, n_q^a(\vec{p}) \, \delta\left(\frac{x}{1-x} - \frac{p_+}{M}\right) \qquad (3)$$

where a represents the diverse degrees of freedom (unpolarized, $\uparrow, \downarrow, \ldots$), $\vec{p}$ the momentum of the constituent, $p_+ = p_0 - p_z$, x is the Bjorken variable and M the mass of the baryonic state.

In this way we have studied polarized and unpolarized structure function, transversity distributions and angular momentum distributions with various models [9, 13, 14]. The results of our calculations show that these models, with the parameters fixed by low energy properties are able to provide a qualitative description of the data and therefore

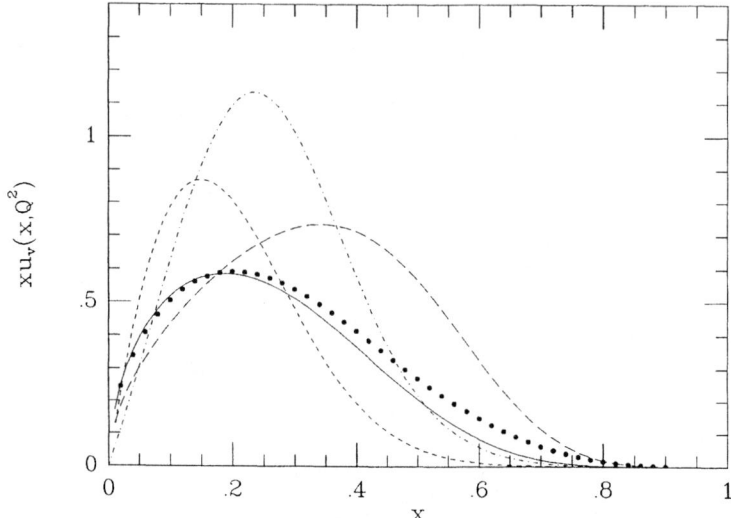

FIGURE 1. We show the unpolarized parton distribution xu_v: i) for a quark model [15] at the hadronic scale (dot dashed); ii) for the same model within the convolution approach at the hadronic scale (long dashed); iii) evolved (NLO) to the scale of the data at $10 GeV^2$ for the model in i) (dashed); iv) evolved for the convolution approach of ii) (full curve) to the scale of the data; v) as guide line through the data (dotted) [22].

the scheme becomes predictive. They are however too naive and new ingredients, not seen by low energy probes, have to be incorporated.

Applications

We comment on some of the calculations performed by stressing only the main results. We refer the reader to the figures and discussions in the given references for a complete account.

1) Parton distributions [9]

We have analyzed in this formalism the polarized and unpolarized experimental results and have shown that well-known Quark Models lead to a qualitative description of the data. The relevant features are: the original model distributions, which are vastly different from the data, evolve, via the Renormalization Group, towards them; sea quarks and gluons, initially absent, are generated by bremsstrahlung. In Fig. 1 one can see how the initial large quark model distribution at the hadronic scale approaches the data by evolution. The momentum lost by the valence quarks goes into the other components. In Fig. 2 we show the gluonic component obtained by evolution in a scheme to be specified later.

In the polarized case, the spin distribution function for the proton, which are too large

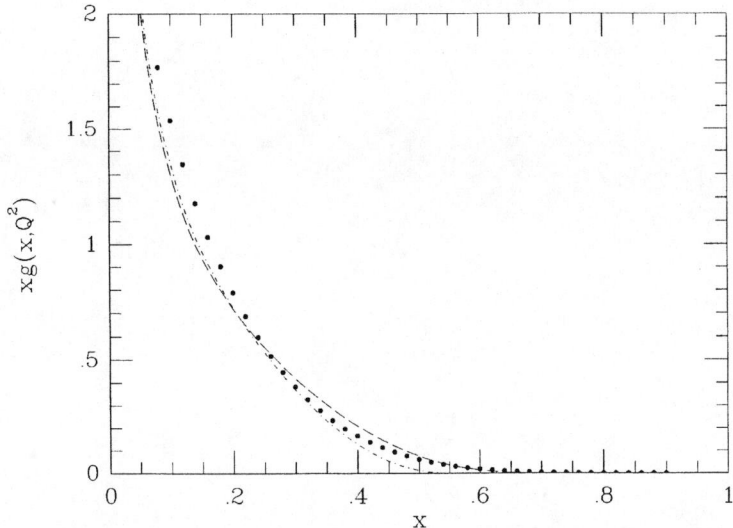

FIGURE 2. We show the gluon distribution $xg(x,Q^2)$ at $Q^2 = 10 GeV$ obtained with the ACMP approach for two different models of hadron structure [15, 23]. The data are those of ref. [22].

for the model calculation, the famous proton spin problem, decreases and approaches the data via the same RG mechanism. In this way the spin is transferred to the new components and the problem greatly disappears. In Fig. 3 the remaining discrepancy between the model calculation and the data after evolution can be seen.

If one aims at a quantitative agreement with the data, the conventional low energy models have to be modified to include, higher momentum and higher angular momentum components for the quarks, and sea components at the hadronic scale. Moreover the experimental gluon distributions, at present extracted in a very indirect way, if taken at face value, imply the need for soft gluons at the hadronic scale. Moreover in the case of the spin parton distribution, the anomaly contribution helps in the explanation of the data.

2) Transversity distribution [14]

The feasibility of measuring chiral-odd parton distribution functions in polarized Drell-Yan and semi-inclusive experiments has renewed theoretical interest in their study. Models of hadron structure have proven successful in describing the gross features of the chiral-even structure functions. Similar expectations motivated our study of the, experimentally unknown, transversity parton distributions with these models. We confirmed, by performing a NLO calculation, the diverse low x behaviors of the transversity and spin structure functions at the experimental scale and showed that it is fundamentally a consequence of the different behaviors under evolution of these functions. The inequalities of Soffer establish constraints between data and model calculations of the chiral-odd transversity function. The approximate compatibility of our model calculations with

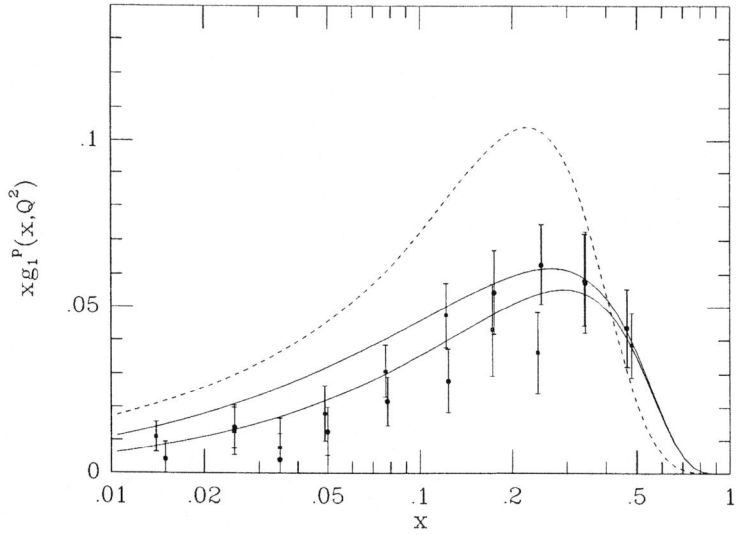

FIGURE 3. We show the spin structure function g_1 for the proton. The dashed curve represents the results of a Quark Model calculation evolved at NLO to the scale of the data ($10 GeV^2$). The full two curves represent the calculation in the *ACMP* scenario, within the same quark model, for two parametrizations of the quark structure functions. The data have been taken from [24].

these constraints confers credibility to our estimates.

3) Skewed parton distributions

A new type of observable, the so called skewed parton distributions (SPD), have been intensively studied in the last years [16]. The SPDs generalize and interpolate between the ordinary parton distributions and the elastic form factors and therefore contain rich structural information. They have been instrumental in understanding the Orbital Angular Momentum (OAM) and furthermore the Deeply Virtual Compton Scattering (DVCS) process has been proposed as a practical way to measure them. From the point of view of parton physics they appear, similarly to the conventional distributions, as light cone matrix elements of the quark-gluon operators, however here the initial and final states have different momenta, and in this way there is an additional t-dependence besides the conventional x dependence. A model calculation within the MIT bag framework has provided estimates about their magnitude which serve as guidance for future experiments [17].

i) Orbital angular momentum [13]

We have studied OAM twist-two parton distributions, for the relativistic MIT bag model and for non-relativistic potential quark models. The contribution of quarks OAM to the nucleon spin evolves at high Q^2 to a vanishingly small value, while that of the gluons increases dramatically. As expected by general arguments, the large gluon OAM contribution is almost canceled by the gluon spin contribution. At large Q^2 the gluons contribute 50% to the angular momentum and the quarks carry only spin.

ii) Twist three contributions to DVCS [18]

The study of the gauge invariance of the DVCS amplitude leads to the inclusion of higher twist components [19]. We have performed an extensive study of the DVCS amplitude within a bag model framework of the single spin asymmetry in the case of spin 0 systems. Our results imply that the choice of kinematics is crucial in order to observe certain amplitudes and therefore unravel the structure of the system.

TOWARDS AN UNIFIED PICTURE OF CONSTITUENT AND CURRENT QUARKS

Our basic assumption has been that gluon and sea bremsstrahlung are the source of difference between the constituents and the current quarks. However the data seem to indicate that the hadronic structure is more complex, with primordial sea quarks (antiquarks) and gluons. Thus the analysis thus far implies that constituent models have to be of greater complexity in order to describe, simultaneously, low and high energy data. Next, we analyze one way to generate this complex structure from a quark model, by assuming that constituent quarks are non elementary and therefore have partonic structure. These ideas where investigated a long time ago by two groups, Altarelli et al. [11], starting from a quark model scenario, and Kuti and Weisskopf [20], who defined a more complex scenario which contained sea and gluons at the hadronic scale. We have studied the consequences of the former approach.

Current structure from the Constituent Quarks

We have gone beyond the bremsstrahlung formalism by incorporating structure to the constituent quarks following the procedure we have called ACMP [11]. Within this approach constituent quarks are effective particles made up of point-like partons (current quarks, antiquarks and gluons), interacting by a residual interaction described by a quark model [12]. The structure of the hadron is obtained by a convolution of the constituent quark model wave function with the constituent quark structure function. For a proton made up of u and d quarks,

$$f(x,\mu_0^2) = \int_x^1 \frac{dz}{z}[u_0(z,\mu_0^2)\Phi_{uf}(\frac{x}{z},\mu_0^2) + d_0(z,\mu_0^2)\Phi_{df}(\frac{x}{z},\mu_0^2)], \qquad (4)$$

where μ_0^2 is the hadronic scale, $f = q_v, q_s, g$ (valence quarks, sea quarks and gluons respectively) and Φ represents the constituents probability in each quark and has been parametrized following general arguments of *QCD* as

$$\Phi_{qf}(x,\mu_0^2) = C_f x^{a_f}(1-x)^{A_f-1}. \qquad (5)$$

The constants have been fixed by Regge phenomenology and the choice of the hadronic scale ($\mu_0 = 0.34 \text{ GeV}^2$).

The discussion can be generalized to the polarized structure functions [21]. The procedure is able to reproduce the data extremely well and in this framework the so called spin problem does not arise.

Applications

1) Unpolarized parton distributions [12]

Using that the constituent quark is a composite system of point-like partons, we construct the parton distributions by a convolution between constituent quark momentum distributions and constituent quark structure functions, Eq.(4).

The different types and functional forms of the structure functions of the constituent quarks, Φ, are derived from three very natural assumptions:

i) The point-like partons are determined by *QCD*, therefore, they are quarks, antiquarks and gluons;
ii) Regge behavior for $x \to 0$ and duality ideas;
iii) invariance under charge conjugation and isospin.

These considerations define in the case of the valence quarks the following structure function,

$$\phi_{qq_v}(x, Q^2) = \frac{\Gamma(A + \frac{1}{2})}{\Gamma(\frac{1}{2})\Gamma(A)} \frac{(1-x)^{A-1}}{\sqrt{x}}. \qquad (6)$$

For the sea quarks the corresponding structure function becomes,

$$\phi_{qq_s}(x, Q^2) = \frac{C}{x}(1-x)^{D-1}, \qquad (7)$$

and in the case of the gluons we take

$$\phi_{qg}(x, Q^2) = \frac{G}{x}(1-x)^{B-1}. \qquad (8)$$

The other ingredients of the formalism, i.e., the probability distributions for each constituent quark, are defined according to the procedure of Traini et al. [9], that is, a constituent quark has a probability distribution determined by Eq.(3).

Our last assumption relates to the scale at which the constituent quark structure is defined. We choose for it the hadronic scale μ_0^2. This hypothesis fixes *all* the parameters of the approach except one, which is fixed by looking at the low x behavior of the F_2 structure function at the hadronic scale, where the sea in known to be dominant.

The resulting parton distributions and structure functions are evolved to the experimental scale and good agreement with the available DIS data is achieved (See Fig. 1). In Fig. 2 we show the gluonic components generated in the ACMP scheme for two models. The primordial sea and gluon components at the hadronic scale are instrumental in achieving a good agreement with the experimental observation.

When compared with a similar calculation using non-composite constituent quarks, the accord with experiment of the present calculation becomes impressive. We therefore conclude that DIS data are consistent with a low energy scenario dominated by composite, mainly non-relativistic constituents of the nucleon.

2) Polarized parton distributions [21]

The previous discussion can be generalized to the polarized case. The functions Φ_{ab} now specify spin and flavor.

Let

$$\Delta q(x,\mu_0^2) = q_+(x,\mu_0^2) - q_-(x,\mu_0^2) \qquad (9)$$

where $\pm$ label the quark spin projections and q represents any flavor. The generalized ACMP approach implies

$$q_i(x,\mu_0^2) = \int_x^1 \frac{dz}{z} \sum_j (u_{0j}(z,\mu_0^2)\Phi_{u_j q_i}(\frac{x}{z},\mu_0^2) + d_{0j}(z,\mu_0^2)\Phi_{d_j q_j}(\frac{x}{z},\mu_0^2)) \qquad (10)$$

where $i = \pm$ labels the partonic spin projections and $j = \pm$ the constituent quark spins. Using spin symmetry we arrive at [1]

$$\Delta q(x) = \int_x^1 \frac{dz}{z}(\Delta u_0(z)\Delta\Phi_{uq}(\frac{x}{z}) + \Delta d_0(z)\Delta\Phi_{dq}(\frac{x}{z})) \qquad (11)$$

where $\Delta q_0 = q_{0+} - q_{0-}$, and

$$\Delta\Phi_{uq} = \Phi_{u+q+} - \Phi_{u+q-} \qquad (12)$$

$$\Delta\Phi_{dq} = \Phi_{d+q+} - \Phi_{d+q-} \qquad (13)$$

We next reformulate the description in term of the conventional valence and sea quark separation, i.e.,

$$\Delta q(x) = \Delta q_v(x) + \Delta q_s(x)$$

After a series of simplifying assumptions we obtain for the various polarized parton distributions the following expressions:

$$\Delta q_v(x) = \int_x^1 \frac{dz}{z} \Delta q_0(z) \Delta\Phi_{qq_v}(\frac{x}{z}), \qquad (14)$$

for the valence quarks,

$$\Delta q_s(x) = \int_x^1 \frac{dz}{z}(\Delta u_0(z) + \Delta d_0(z))\Delta\Phi_{qq_s}(\frac{x}{z}), \qquad (15)$$

for the sea quarks, and

$$\Delta g(x) = \int_x^1 \frac{dz}{z}(\Delta u_0(z) + \Delta d_0(z))\Delta\Phi_{qg}(\frac{x}{z}) \qquad (16)$$

---

[1] We omit writing explicitly the hadronic scale dependence from now on.

for the gluons

Thus the *ACMP* procedure can be extended to the polarized case just by introducing three additional structure functions for the constituent quarks: $\Delta\Phi_{qq_v}$, $\Delta\Phi_{qq_s}$ and $\Delta\Phi_{qg}$.

In order to determine the polarized constituent structure functions we add some assumptions which will tie up the constituent structure functions for the polarized and unpolarized cases completely, reducing dramatically the number of parameters. They are:

iv) factorization assumption: $\Delta\Phi$ cannot depend upon the quark model used, i.e, cannot depend upon the particular Δq_0;

v) positivity assumption: the positivity constraint $\Delta\Phi \leq \Phi$ is saturated for $x = 1$.

These additional assumptions determine completely the parameters of the polarized constituent structure functions.

Using unpolarized data to fix the parameters we achieve good agreement with the polarization experiments for the proton (see Fig. 3), while not so for the neutron. By relaxing our assumptions for the sea distributions, we define new quark functions for the polarized case, which reproduce well the proton data and are in better agreement with the neutron data (see discussion in ref. [21]).

When our results are compared with similar calculations using non-composite constituent quarks the accord with the experiments of the present scheme is impressive. We conclude that, also in the polarized case, DIS data are consistent with a low energy scenario dominated by composite constituents of the nucleon.

CONCLUDING REMARKS

The high energy parton distributions when evolved to a low energy scale appear to indicate that a valence picture of hadron structure arises. This valence picture is well represented theoretically by Quark Models which are very successful in explaining the low energy properties of hadrons. We have developed a formalism based on a laboratory partonic description which connects the parton distributions with the momentum distributions of the constituents giving us a description of partons in terms of Quark Model wave functions. Our basic assumption is that gluon and sea bremsstrahlung are the source of difference between the various momentum scales. We have implemented the Renormalization Group program by defining a hadronic scale and using as initial, non perturbative, conditions those obtained from the parton distributions of the low energy model.

Our analysis, based on a NLO formalism of evolution, has shown that the perturbative scheme is applicable to the low energy scales of interest. The formalism used has the correct support for the parton distributions and allows the discussion of a large class of Quark Models.

The results of our calculations show that low energy models, with their parameters fixed by low energy properties, tend to give a qualitative description of the data. Fig. 1 is very clarifying in this respect. This feature allows one to use them in order to be predictive in new observations.

The next step, which our formalism allows, is to proceed to define models which describe quantitatively the data at all energy scales. Our analysis has shown that present models are too naive. The new models seem to require: primordial gluons and sea.

The limitations associated with naive Quark Models of DIS data can be overcome by a very appealing scheme where the constituent quarks are not elementary. Partons (the quarks, antiquarks and gluons of QCD) at the hadronic scale are generated by unveiling the structure of the constituent quarks. We have seen that incorporating this structure in a very physical way improves notably the agreement with the DIS data (See Fig. 1). From the point of view of the calculation, we must stress, that no parameters of the model have been changed with respect to the original fit to the low energy properties. The new parameters arising from the description of the constituent quark structure functions have been adjusted to describe the input scenario according to the hadronic scale philosophy. In this way the sea and gluon distributions are generated in a consistent way (see Fig. 2).

The same analysis can be easily performed for the polarized case. Using a physically motivated minimal prescription for the polarized case, with no additional parameters, we are able to obtain a good prediction of the the proton data (see Fig. 3). The minimal procedure fails, however, to reproduce the accurate neutron data. Relaxing the minimal procedure, with the addition of only one new parameter to define the polarized sea, we obtain a significantly improved description also for the neutron data [21]. The calculation has also clarified the role of the gluons and the valence quarks. It is clear that the gluons become important through the evolution process, i.e., it is the soft bremsstrahlung gluons which acquire a large portion of the partonic spin.

We would like to stress that within our procedure the *spin problem*, as initially presented, does not arise. The constituent quarks carry all of the polarization. When their structure is unveiled this polarization is split among their different partonic contributions in the manner we have described and which is consistent with the data. The quality of both unpolarized and polarized data thus far analyzed confirm the validity of the approach. We have showed also, that with very reasonable assumptions, the scheme becomes highly predictive, a feature which is necessary for the planning of future experiments.

We feel safe to conclude that, the current quarks seen at the parton level seem to be embedded in the composite constituent quarks seen at lower Q^2. An unified picture of current quarks, successfully describing DIS, and constituent quarks, successfully describing static properties is possible.

ACKNOWLEDGMENTS

Sergio Scopetta has been a main contributor to much of the work presented here. It has been a great pleasure to work with him. My collaboration with Marco Traini comprises a long period of time. During all these years it has been very inspiring to interact with him. I acknowledge conversations with I.V. Anikin, D. Binosi, R. Medrano and S. Noguera on Skewed Parton Distributions and gauge invariance. I thank the organizers for the invitation to present my work in this most beautiful city, and in particular Jiri Adam, for his continuous help and assistance before and during the conference. This work was supported in part by DGICYT grant PB97-1227 and La Generalitat Valenciana with a travel grant.

REFERENCES

1. Kokkedee, J.J.J., *The Quark Model*, W. A. Benjamin, New York, 1969;
   Lichtenberg, D.B., *Unitary Symmetry and Elementary Particles*, Academic Press, Inc. New York, 1978.
2. Feynman, R.P., *Photon-Hadron Interaction*, W.A. Benjamin, New York, 1972.
3. Muta, T., *Foundations of Quantum Chromodynamics*, World Scientific, Singapore 1987 ;
   Field, R.D., *Applications of perturbative QCD*, Addison Wesley Pub. Co., New York, 1989;
   Yndurain, F.J., *The Theory of the Quark and Gluon Interactions*, Springer Verlag, Heidelberg, 1999.
4. Martin, A.D., Stirling W.J., and Roberts, R.G., Ral Report 94-055; ibid Ral Report 95-021.
5. Glück, M., and Reya, E., *Phys. Rev. D* **14**, 3024 (1976); Reya, E., *Phys. Rep.* **69**, 195 (1981); Glück, M., Reya, E., and Vogt, A., *Z. Phys. C* **48**, 471 (1990); Glück, M., Reya, E., and Vogt, A., *Z. Phys. C* **53**, 127 (1992); *Z. Phys. C* **67**, 433 (1995).
6. Parisi, G., and Petronzio, R., *Phys. Lett. B* **62**, 331 (1976).
7. Alvarez-Estrada, R.F., Fernández, F., Sánchez Gómez, J.L., and Vento, V., "Model of Hadron Structure based on QCD", *Lecture Notes in Physics* **259**, Springer Verlag, Heidelberg, 1986.
8. Jaffe, R.L., and Ross, G.C., *Phys. Lett. B* **93**, 313 (1980).
9. Traini, M., Zambarda, A., and Vento, V., *Mod. Phys. Lett.* **10**, 1235 (1995); Ropele, M., Traini, M., and Vento, V., *Nucl. Phys.* **A584**, 634 (1995); Traini, M., Vento, V., Mair, A., and Zambarda, A., *Nucl. Phys.* **A614**, 472 (1997).
10. Jaffe, R.L., *Phys. Rev. D* **11**, 1953 (1975); Betz, M., and Goldflam, R., *Phys. Rev. D* **28**, 2848 (1983).
11. Altarelli, G., Cabibbo, N., Maiani, L., and Petronzio, R., *Nucl. Phys.* **B69**, 531 (1974).
12. Scopetta, S., Vento, V., and Traini, M., *Phys. Lett. B* **412**, 64 (1998).
13. Scopetta, S., and Vento, V., *Phys. Lett. B* **460**, 8 (1999), *Phys. Lett. B* **474**, 235 (2000).
14. Scopetta, S., and Vento, V., *Phys. Lett. B* **424**, 25 (1998).
15. Bijker, R., Iachello, F., and Leviatan, A., *Ann. Phys.* **236**, 69 (1994) ; *Phys. Rev. C* **54**, 1935 (1996); *Phys. Rev. D* **55**, 2862 (1997).
16. Radyuskin, A.V., *Phys. Lett. B* **380**, 417 (1996); *Phys. Lett. B* **385**, 333 (1996) ; Ji, X., *Phys. Rev. Lett.* **78**, 610 (1997); *Phys. Rev. D* **55**, 7114 (1997).
17. Ji, X., Melnitchouk, W., and Song, X., *Phys. Rev. D* **56**, 5511 (1997).
18. Anikin, I.V., Binosi, D., Medrano, R., Noguera, S., and Vento, V., work in preparation.
19. Anikin, I.V., Pire, B., and Teryaev, O.V., *Phys. Rev. D* **62**, 071501 (2000).
20. Kuti, J., and Weiskopf, V.F., *Phys. Rev. D* **11**, 3418 (1974).
21. Scopetta, S., Vento, V., and Traini, M., *Phys. Lett.* **442**, 28 (1998).
22. Lai, H.L. *et al.*, *Phys. Rev. D* **51**, 4763 (1995).
23. Isgur, N., and Karl, G., *Phys. Rev. D* **18**, 4187 (1978); **19**, 2653 (1979); **23**, 817 (1981) (E).
24. EMC Collaboration, Ashman, J., *et al.*, *Nucl. Phys.* **B328**, 1 (1989);
    SMC Collaboration, Adams, D., *et al.*, *Phys. Rev. D* **56**, 5330 (1997).

Pion production in pN collisions near threshold

J. Złomańczuk[a1]

for the WASA/PROMICE collaboration:

R. Bilger[b], W. Brodowski[b], H. Calén[c], H. Clement[b], C. Ekström[c],
K. Fransson[a], G. Fäldt[a], L. Gustafsson[a], B. Höistad[a], J. Johanson[a],
A. Johansson[a], T. Johansson[a], K. Kilian[d], S. Kullander[a], A. Kupść[c],
B. Morosov[e], W. Oelert[d], R.J.M.Y. Ruber[c], J. Stepaniak[f], A. Sukhanov[e],
P. Sundberg[a], A. Turowiecki[g], G.J. Wagner[b], Z. Wilhelmi[g], C. Wilkin[h],
J. Zabierowski[f], A. Zernov[e]

[a] *Department of Radiation Sciences, Uppsala University, S-751 21 Uppsala, Sweden*
[b] *Physikalisches Institut, Tübingen University, D-72076 Tübingen, Germany*
[c] *The Svedberg Laboratory, S-751 21 Uppsala, Sweden*
[d] *IKP - Forschungszentrum Jülich GmbH, D-5245 Jülich, Germany*
[e] *Joint Institute for Nuclear Research Dubna, 101000 Moscow, Russia*
[f] *Institute for Nuclear Studies, PL-00681 Warsaw, Poland*
[g] *Institute of experimental Physics, Warsaw University, PL-00681 Warsaw, Poland*
[h] *Physics & Astronomy Dept., University College London, London WCE 6BT, U.K.*

Abstract. Measurements of the $pp \to pp\pi^0$ reaction at 310, 320, 340, 360, 400 and 425 MeV and quasi-free $pn \to pp\pi^-$ production in pd collisions at 320 MeV have been carried out at the PROMICE/WASA facility at CELSIUS. The $pp \to pp\pi^0$ cross sections have been parameterized and used to extract the poorly known σ_{01} cross section through the relation: $\sigma_{01} = 2\sigma(pn \to pp\pi^-) - \sigma(pp \to pp\pi^0)$. The results obtained show the σ_{01} and σ_{11} to be of comparable size for CM energies between 8 and 48 MeV.

INTRODUCTION

An important and unresolved issue in intermediate energy physics is the understanding of the mechanism behind pion production and absorption in nuclear reactions. The problem was dramatized around 1991, when the $pp \to pp\pi^0$ reaction was measured with high accuracy in the threshold region [1], and the magnitude of the cross section was found to be about five times larger than the theoretical predictions available at that time [1,2]. The large discrepancy triggered much theoretical activity and the widely accepted Koltun and Reitan model [3] was modified to account for the

---

[1] jozef.zlomanczuk@tsl.uu.se

new experimental data. One explanation suggested that the effect was due to heavy meson exchange [4]. Somewhat later, an alternative solution was advanced where the large cross section was attributed to the off-shell variation of pion rescattering [5]. Which, if either, of the explanations is right has not yet been settled.

Recent theoretical models attempt unified calculations of all near threshold $NN \rightarrow NN\pi$ reactions and consider contributions from partial waves higher than Ss [6,7]. We use here the notation Ll, with L representing the relative angular momentum of the final NN pair and l the pion angular momentum with respect to this pair. In addition to total cross sections, differential cross section and polarization variables are also calculated. These theoretical extensions call for new and precise data on pion production in NN collisions in the energy range from threshold to several tens of MeV above.

Under the assumption of isospin invariance all $NN \rightarrow NN\pi$ reactions may be described in terms of the three independent cross sections, σ_{if}, where indices i and f represent the isospin, I, of the initial and final NN pairs [8] (see Table, 1). While

TABLE 1. Decomposition of pion production cross sections into elementary cross sections σ_{if}. ' denotes transition to the bound final state.

Reaction cross sections	Isospin cross sections
$\sigma(nn \rightarrow nn\pi^0) = \sigma(pp \rightarrow pp\pi^0)$	σ_{11}
$\sigma(nn \rightarrow np\pi^-) = \sigma(pp \rightarrow np\pi^+)$	σ_{11}, σ_{10}
$\sigma(np \rightarrow nn\pi^+) = \sigma(np \rightarrow pp\pi^-)$	$(\sigma_{11}, \sigma_{01})/2$
$\sigma(np \rightarrow np\pi^0)$	$(\sigma_{10}, \sigma_{01})/2$
$\sigma(np \rightarrow d\pi^0) = \sigma(pp \rightarrow d\pi^+)$	σ'_{10}

σ_{11} and σ_{10} are well measured in the threshold region, σ_{01} is poorly known. A phase shift analysis of all $NN \rightarrow NN\pi$ data below 1 GeV gave σ_{01} compatible with zero [9], whereas analysis of the NN inelastic scattering data below 600 MeV resulted in σ_{01} small but non-negligible [10]. Since the $N\Delta$ intermediate state is not allowed for the $I=0$ initial state, such small values of σ_{01} would indicate dominance of the pion production via the Δ-resonance. However, measurements of the reaction carried out for excitation energies of the pp pair below 1.5 MeV [11] and 3 MeV [12], revealed large forward-backward asymmetry of the center-of-momentum (CM) pion angular distributions, which is a signature of a large contribution from the σ_{01} cross section. Similar asymmetry has been found in the $pn \rightarrow pp\pi^-$ reaction for unconstrained pp pairs [13]. Close to threshold measurements of the $pn \rightarrow pp\pi^-$ reaction as a quasi-free process in pd collisions at 320 MeV resulted in σ_{01} values comparable to those of the σ_{11} [14].

In this contribution we first summarize recent experimental results on the $pp \rightarrow pp\pi^0$ reaction and present a parameterization of the matrix element derived in Ref. [15]. The parameterization is then used to find the σ_{01} contribution to the total and differential cross sections of the quasi-free $np \rightarrow pp\pi^-$ reaction measured in pd collisions at the CELSIUS storage ring.

ANALYSIS OF THE $pp \to pp\pi^0$ REACTION NEAR THRESOLD

In the analysis of experimental data on the $pp \to pp\pi^0$ reaction near threshold it is customary to use a phenomenological theory of pion production in NN collisions formulated nearly 30 years ago by Gell-Mann and Watson [16] and, independently, by Rosenfeld [8]. The theory is based on the assumption that the production region is small and the final state particles do not interact. As a consequence the transition amplitudes may be obtained from the properties of the spherical Bessel functions for small arguments,

$$M_{l,l} \propto q^l k^l, \qquad (1)$$

where q and k represent the two-proton relative momentum and the π^0 momentum respectively. Note, that q and k are linked by the energy conservation. The total cross section may then be written as

$$\sigma_{tot} \propto \sum_{l,l} \int_0^{k_{max}} |M_{l,l}|^2 d\rho \qquad (2)$$

where k_{max} is the maximum CM pion momentum and $d\rho$ stands for the phase space element. In non-relativistic kinematics

$$k_{max} = \sqrt{\frac{4m_p m_\pi}{(m_\pi + 2m_p)} Q}, \quad d\rho \propto k^2 q dk, \qquad (3)$$

with m_p, m_π, Q representing proton and π^0 masses, and available CM kinetic energy. Eq. (2) can be evaluated analytically, see for example Ref. [17], and gives the energy dependence of the partial cross sections near threshold

$$\sigma_{l,l} \propto \eta^{4+2L+2l}, \qquad (4)$$

with η representing the maximum pion momentum (at a given beam energy) in units of the pion mass. It should be noted, that Eq. (4) does not hold for the Ss final state due to a strong interaction between protons in the $^1S_0$ state. Following Watson [18] one may factor out this interaction and get

$$|M_{Ss}|^2 = \frac{|M_{Ss}^0|^2}{q^2(1+\cot^2\delta_0)}, \qquad (5)$$

where $|M_{Ss}^0|^2$ is the production part of the amplitude and δ_0 is the S-wave phase shift.

The above equation may be further simplified using the effective range expansion

$$q\cot\delta_0 = -\frac{1}{a} + \frac{1}{2}bq^2 + \cdots, \qquad (6)$$

and, if the second term is omitted, would lead to σ_{Ss} roughly proportional to η^2. The constants a and b above represent the pp scattering length (-7.82 fm) and the effective range (2.7 fm).

A small interaction region implies that near threshold only few partial waves are present. The number of the transitions is further reduced by requirements of the parity, angular momentum and isospin conservation. Possible transitions for σ_{11} and σ_{01} cross sections and $L, l \leq 2$ are listed in Table 2.

TABLE 2. List of allowed transitions for the $pn \to pp\pi^-$ reaction, $(\sigma_{11}+\sigma_{01})/2$.

Initial state $^{2S+1}L_J$	Final state $^{2S+1}L_jl_J$	Isospin cross section	Initial state $^{2S+1}L_J$	Final state $^{2S+1}L_jl_J$	Isospin cross section
$^3P_0$	$^1S_0s_0$	σ_{11}	$^3P_2$	$^3P_1p_2$	σ_{11}
$^3S_1$	$^1S_0p_1$	σ_{01}	$^3P_2$	$^3P_2p_2$	σ_{11}
$^3D_1$	$^1S_0p_1$	σ_{01}	$^3F_2$	$^3P_1p_2$	σ_{11}
$^1S_0$	$^3P_0s_0$	σ_{11}	$^3F_2$	$^3P_2p_2$	σ_{11}
$^1D_2$	$^3P_2s_2$	σ_{11}	$^3F_3$	$^3P_2p_3$	σ_{11}
$^3S_1$	$^3P_1s_1$	σ_{01}	$^1P_1$	$^3P_0p_1$	σ_{01}
$^3D_1$	$^3P_1s_1$	σ_{01}	$^1P_1$	$^3P_1p_1$	σ_{01}
$^3D_2$	$^3P_2s_2$	σ_{01}	$^1P_1$	$^3P_2p_1$	σ_{01}
$^3P_0$	$^3P_1p_0$	σ_{11}	$^3P_2$	$^1S_0d_2$	σ_{11}
$^3P_1$	$^3P_0p_1$	σ_{11}	$^3F_2$	$^1S_0d_2$	σ_{11}
$^3P_1$	$^3P_1p_1$	σ_{11}	$^3P_2$	$^1D_2s_2$	σ_{11}
$^3P_1$	$^3P_2p_1$	σ_{11}	$^3F_2$	$^1D_2s_2$	σ_{11}

The theory was experimentally tested using $np \to pp\pi^-$ reaction in the energy range from threshold to 440 MeV and a fair agreement was found [19].

The energy dependence of Eq. (4) has been used to decompose the total cross sections into contributions from partial waves up to $L, l=1$ at the beam energy of 325 MeV [1] and below 450 MeV [20]. While the σ_{Ss} contributions obtained in both experiments are similar, the σ_{Ps} and σ_{Pp} are not. At 325 MeV Meyer et al. finds the σ_{Ps} and σ_{Pp} to be strongly correlated [1]. However, with help of the proton angular distribution, they resolve the ambiguity and as a result they get σ_{Ps} to be close to 40% of the total cross section and σ_{Pp} small. Conversely, Stanislaus et al. at all energies finds σ_{Ps} smaller than σ_{Pp} [20].

It should be stressed that Eq. (4) is valid only in non-relativistic limit. In order to see how strong the relativistic effects are, the σ_{Ps} cross section has been calculated numerically from threshold to 425 MeV using relativistic kinematics. The k_{max} and $d\rho$ from Eq. (3) have been replaced with their relativistic counterparts:

$$k_{max} = \frac{\left[\left(s-(2m_p+m_\pi)^2\right)\left(s-(2m_p-m_\pi)^2\right)\right]^{1/2}}{2s}, \quad (7)$$

$$d\rho \propto \frac{k^2 q\, dk}{p_L \sqrt{m_p^2+q^2}\sqrt{m_\pi^2+k^2}}, \quad (8)$$

where $\sqrt{s}$ and p_L represent the total CM energy and the initial proton momentum in the laboratory system. The energy conservation provides the link between k and q needed to carry out the integration

$$k = \frac{\left[\left(s-(M_{pp}+m_\pi)^2\right)\left(s-(M_{pp}-m_\pi)^2\right)\right]^{1/2}}{2s}, \quad (9)$$

with $M_{pp} = 2\sqrt{m_p^2+q^2}$ being the invariant mass of the pp pair. The resulting dependence is compared in Fig. 1 to the non-relativistic η^6 curve expected for the Ps final state from Eq. (4). The curves, arbitrarily normalized to one at $\eta = 0.95$ (corresponding

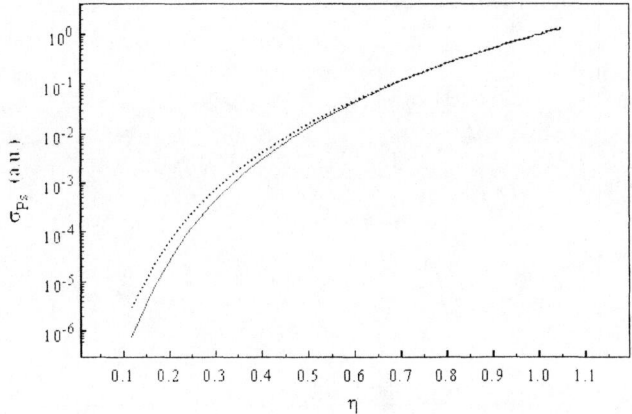

FIGURE 1. σ_{Ps} partial cross section calculated using relativistic kinematics (solid line) compared to the η^6 energy dependence (dotted line). The solid line may roughly be approximated with $\eta^{6.53}$. The curves have been normalized to one at $\eta = 0.95$, which corresponds to the beam energy of 400 MeV.

to the beam energy of 400 MeV), differ by more than a factor of four at threshold. The σ_{Ps} cross section calculated using relativistic kinematics has the energy dependence, which may be approximated with $\eta^{6.53}$, hence stronger than η^6 expected in the non-relativistic treatment.

Systematic measurements of the $pp \to pp\pi^0$ reaction carried out at the CELSIUS have shown that amplitudes given by Eq. (1) cannot describe the experimental data even if relativistic kinematics is used [15]. This suggests that a more complicated dependence on q is needed. To avoid introducing further parameters, the remedy was sought and found in the DWBA motivated treatment assuming interaction ranges corresponding to π exchange

$$M_{l,l} \propto \int_0^\infty \Psi_{\tilde{q}}^{(-)*}(r) j_l(\tfrac{1}{2}kr) \overline{Y}_1(m_\pi^* r) \Psi_{\tilde{q}}^{(+)}(r) r^2 dr . \qquad (10)$$

The $j_l(\tfrac{1}{2}kr)$ represents the l-wave pion, $\Psi_{\tilde{q}}^{(+)}$ is the wave function of the initial pp-pair, and $\Psi_{\tilde{q}}^{(-)}$ that of the final pp-pair, both calculated from the Paris potential [21]. Note, that $\Psi_{\tilde{q}}^{(+)}$ and $\Psi_{\tilde{q}}^{(-)}$ depend on the spin, orbital momentum and total angular momentum of the pp pairs so they had to be calculated separately for each of the $\sigma_{l,l}$ transitions listed in Table 2. It was found that transition probabilities leading to the same final state, e. g. those corresponding to the nine transitions to the Pp state (Table 2), have similar q dependence so they were added together and the summed probability was used in the matrix element squared [15]. The $\overline{Y}_1$ operator has the form:

$$\overline{Y}_1(m_\pi^* r) = \left(1 + \frac{1}{m_\pi^* r}\right) \frac{e^{-m_\pi^* r}}{m_\pi^* r} , \qquad (11)$$

where $m_\pi^* = (m_\pi^2 - \omega^2)^{1/2}$ and ω is the total pion energy.

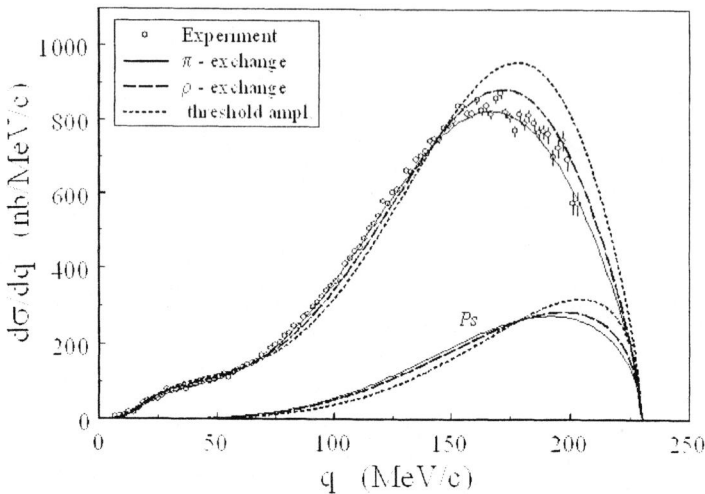

FIGURE 2. Acceptance-corrected distribution in q at 400 MeV compared to distributions obtained with amplitudes calculated using Eq. (10) with different assumptions for the Ps and Pp amplitudes and σ_{Ps} put equal to 26 μb (obtained at this energy in a model independent way in Ref. [22]). The solid and long-dashed curves have been obtained with the Ps and Pp amplitudes calculated for π and ρ exchange, respectively. The short-dashed curve corresponds to the Ps and Pp amplitudes being approximated by their threshold values q and qk. The lower set of curves represents the σ_{Ps} contributions using the same line style convention as before.

Taking as the matrix element squared the sum of transition probabilities corresponding to the Ss, Ps, Pp, Sd, Ds final states and the Ss^*Sd, Ss^*Ds interference terms, it was possible to describe the experiment very well as it is shown in Fig. 2. In addition, the resulting σ_{Ps} contributions were found to be in a reasonable agreement with those obtained in a model independent way in Ref. [22].

ANALYSIS OF THE QUASI-FREE $pp \to pp\pi^0$ AND $pn \to pp\pi^-$ REACTIONS

The analysis of the quasi-free reactions is based on the spectator model, in which it is assumed that [11]:
1. the spectator nucleon influences the interaction only in terms of the associated Fermi motion,
2. the matrix element for quasi-free pion production from a bound nucleon is identical to that for free pion production from an unbound nucleon at the same energy.

Using the spectator model it is then straightforward to extend the phenomenological model of the $pp \to pp\pi^0$ reaction developed in Ref. [15] to the quasi-free case if the energy dependence of the matrix element is known. Assuming that this dependence is well approximated by an interpolation between the energy points measured in Ref. [15], dσ/dq and the angular distributions as well may be simulated by folding in the deuteron wave function. In this work we have used the deuteron wave function taken from Ref. [23].

In the case of the $pn \to pp\pi^-$ reaction the matrix element has to be extended to include the σ_{01} transitions listed in Table 2. Taking, as in the σ_{11} case, the incoherent sums of transition probabilities leading to the Ps and Pp states one arrives at the following form of the extension of the matrix element squared:

$$f_{Sp} = C_{Sp}^{'2} |M_{Ps}|^2 \left[1 + C_{Sp}^{'k2}(\mu_k^2 - \tfrac{1}{3})\right],$$
$$f_{Ps0} = C_{Ps0}^{'2} |M_{Ps0}|^2 \left[1 + C_{Ps}^{q2}(\mu_q^2 - \tfrac{1}{3})\right],$$
$$f_{Pp0} = C_{Pp0}^{'2} |M_{Pp0}|^2 \left[1 + C_{Pp0}^{k2}(\mu_k^2 - \tfrac{1}{3}) + C_{Pp0}^{q2}(\mu_q^2 - \tfrac{1}{3})\right], \qquad (12)$$
$$f_{SsSp} = C_{SsSp} C_{Ss} C_{Sp} |M_{Ss}||M_{Sp}|\mu_k,$$
$$f_{PsPp0} = C_{PsPp0} C_{Ps} C_{Pp0} |M_{Ps}||M_{Pp0}|\mu_k,$$
$$f_{Ps0Pp} = C_{Ps0Pp} C_{Ps0} C_{Pp} |M_{Ps0}||M_{Pp}|\mu_k,$$

where μ_k, μ_q and μ are the cosines of the angles between the projectile and $\vec{k}$, projectile and $\vec{q}$, and $\vec{k}$ and $\vec{q}$, respectively (see Ref. [15]). The interference terms between the Ps and Pp states, f_{PsPp0} and f_{Ps0Pp}, are simplified with respect to the one used by Handler [19] since we restrict ourselves to studying only the one-dimensional distribution of the pion angle and μ_k is the only term which is left after integration over μ and μ_q. Since in our experiment we cannot distinguish between the f_{PsPp0} and f_{Ps0Pp}, we make a further simplification and replace the two interference terms with one:

$$f_{PsPp} = C_{PsPp}\left[C_{Ps0}C_{Pp}|M_{Ps0}||M_{Pp}| + C_{Ps0}C_{Pp}|M_{Ps0}||M_{Pp}|\right]\mu_k.$$

For the $pn \to pp\pi^-$ reaction the formula (10) has to be modified in order to account for the change of the initial state from pp to pn. This has been accomplished by replacing the wave function of the initial pp pair, $\Psi_{\vec{q}}^{(+)}(r)$, with the spherical Bessel function $j_L(qr)$, where L represents the orbital momentum of the initial pn pair. In addition, the C parameters in Eqs. (11-12) of Ref. [15] have been renormalized to give the same strength of the partial waves at a given CM kinetic energy Q for the $pn \to pp\pi^-$ reaction as one would get for the $pp \to pp\pi^0$ reaction at the same Q. The C parameters defined in Eq. (12) are to be found by fitting the experimental data.

EXPERIMENT

The experiment was carried out using an electron-cooled proton beam and the PROMICE-WASA experimental set-up at the CELSIUS storage ring of the Svedberg Laboratory. The detector system is described elsewhere [15,24] and only the main components are mentioned here. For the $pp \to pp\pi^0$ reaction, both protons were measured by the forward scintillator hodoscope and the tracker (FD), except for a small hole subtending an angle of $\approx 4.5^0$ to accommodate the beam. This is the principal source of detection inefficiency for the two protons, giving a geometrical acceptance of about 70% at 310 MeV. At higher energies the proton maximum angle is greater than 20^0, the largest angle measured in the FD, and the acceptance is consequently reduced. The overall efficiency is diminished additionally by ~20%, mainly through the interaction of protons in the scintillator. Since protons from the $pp \to pp\pi^0$ reaction for energies not exceeding 425 MeV stop in the hodoscope, their energies and angles can be measured with accuracies of 4.5% and 0.5^0 (r.m.s.) respectively. The π^0 is then reconstructed through the missing mass. Careful calibration of the FD elements leads to narrow, nearly background-free, missing mass peaks, with a FWHM ranging from 2.2 MeV/c^2 at 310 MeV to about 7 MeV/c^2 at 425 MeV.

For the quasi-free $pp \to pp\pi^0$ reaction, in addition to the two protons detected in the FD, also two photons from the π^0 decay were measured in the CsI arrays covering angles from 30^0 to 90^0 on both sides of the target. In the case of the $pn \to pp\pi^-$ reaction two protons were detected in the FD and an extra charged particle, either in the FD or in the CsI arrays.

The energy resolution for pions was not sufficient for a precise kinematical reconstruction of the missing mass. However, once a $pp\pi^0$ or $pp\pi^-$ event had been selected, it was possible to calculate the 4-momenta of the spectator nucleon and the pion using the momenta of the fast protons and the pion angle. This allowed us to calculate the invariant mass of the $pp\pi$ system, needed to find the excitation function, and reject events with large energy for the undetected proton. The latter reduced the contribution from a $pd \to ppp\pi^-$ and a $pd \to ppn\pi^0$ reactions where all three nucleons are involved.

The sample of the quasi-free $pp \to pp\pi^0$ events turned out to be nearly background-free. However, due to conversion in the exit windows of the scattering chamber, photons from this reaction, build up a few percent of background for the quasi-free $pn \to pp\pi^-$ reaction. This background was reproduced in Monte Carlo simulation and subtracted from the final distributions.

The normalization of the experimental distributions was provided by the quasi-elastic pp scattering, quasi-free $pp \to d\pi^+$ reaction and pd elastic scattering measured simultaneously with the quasi-free $pp \to pp\pi^0$ and $pn \to pp\pi^-$ reactions. The integrated luminosities obtained from these three processes agreed to within 15 %.

The limited acceptance of the experimental set-up and losses due to nuclear interactions of the measured particles with the detector material were taken care of with a careful Monte Carlo simulation using the CERN software package GEANT [25].

RESULTS AND DISCUSSION

The excellent resolution of the FD allows a reliable selection of the $pp\pi$ events by imposing conditions on the missing mass, MM_{pp}, calculated for two detected protons. As is seen in Fig. 3, presenting the distribution in MM_{pp}, a pronounced bump above $\pi+N$ mass contains nearly no background.

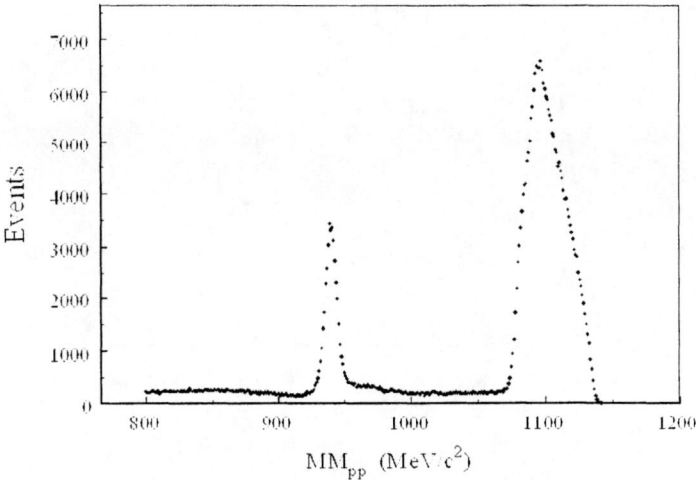

FIGURE 3. Missing mass distribution obtained for two detected protons in *pd* collisions at 320 MeV.

Quasi-free $pp \to pp\pi^0$ reaction

The detection of two protons in the FD and two photons in the CsI arrays provides a nearly background-free sample of $pd \to pp n \pi^0$ events. In order to show what fraction of events represents the quasi-free $pp \to pp\pi^0$ reaction, the experimental missing momentum distribution is presented in Fig. 4 together with the results of the Monte Carlo simulation. For comparison, the Paris-potential deuteron wave function [23], used in the simulation, is also shown. The distributions have been normalized to have the same area below 42 MeV/c. As one can see they match each other rather well in that region but the Monte Carlo underestimates the experiment at higher momenta. This may indicate a contribution from the $pn \to pn\pi^0$ reaction for which the acceptance is small but the cross section is higher than that of the $pp \to pp\pi^0$ reaction. In order to improve the selection of the quasi-free $pp \to pp\pi^0$ events a limit of 137 MeV/c (10 MeV in energy) has been set on the maximum value of the spectator momentum.

The acceptance-corrected distribution in q is shown in Fig. 5. The experimental points have been normalized using the quasi-free $pp \to d\pi^+$ reaction. The solid line represents the result of the Monte Carlo simulation based on the spectator model with the matrix element parameterized using data on the free $pp \to pp\pi^0$ reaction as discussed above. As one can see the model underestimates the experimental data and, in addition, the shape of the distribution is not reproduced as well as in the case of the free $pp \to pp\pi^0$ reaction (see Fig. 2). The excess of experimental events over the Monte Carlo ones again supports the hypothesis that the selected $pp\pi^0$ data contain some amount of the $pn \to pn\pi^0$ events.

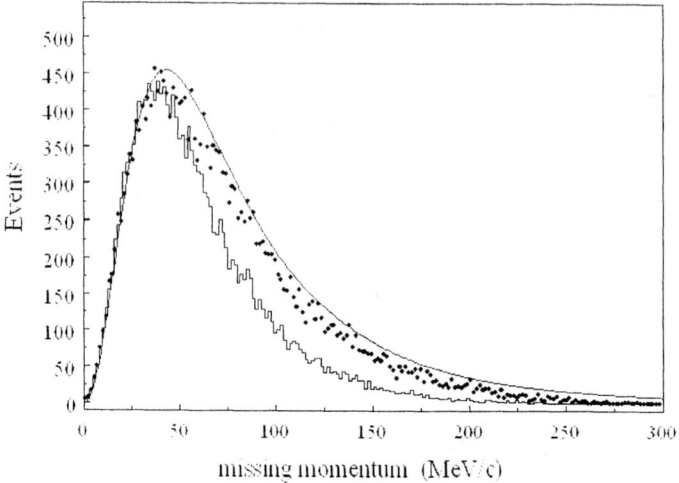

FIGURE 4. Missing momentum distribution obtained for the detected $pp\pi^0$ in pd collisions at 320 MeV (points) compared to the Monte Carlo simulation (histogram). To illustrate the influence of the detector acceptance also the deuteron wave function taken from Ref. [23] is shown (solid line).

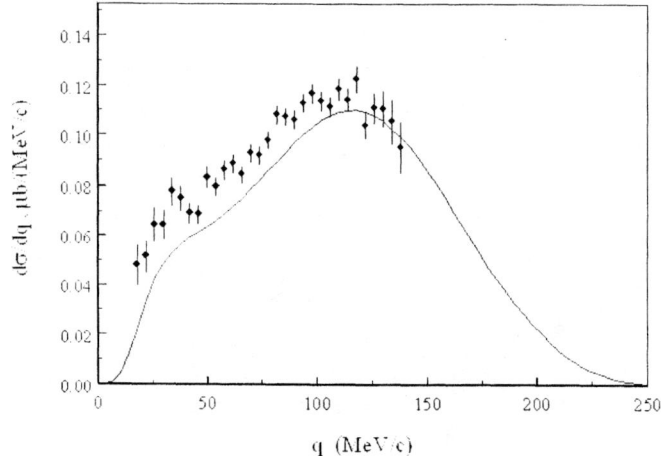

FIGURE 5. Acceptance-corrected two-proton relative momentum distribution obtained for the quasi-free $pp \rightarrow pp\pi^0$ reaction measured in pd collisions at 320 MeV compared to the Monte Carlo predictions. The experimental points have been normalized using the quasi-free $pp \rightarrow d\pi^+$ reaction measured simultaneously.

Quasi-free $pp \rightarrow pp\pi^-$ reaction

The $pd \rightarrow ppp\pi^-$ reaction was selected with a requirement of detecting two protons in the FD and an extra charged particle either in the FD or in the CsI arrays. In addition, the missing mass calculated for the two detected protons had to be larger than 1060 MeV ($\sim m_p + m_\pi$). With this selection it was possible to reject nearly all back-

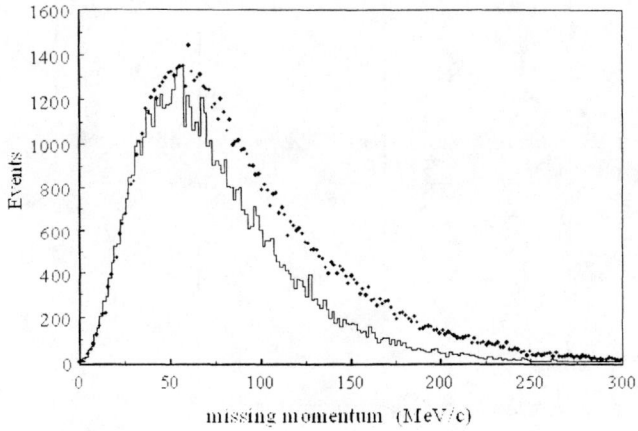

FIGURE 6. Missing momentum distribution obtained for the detected $pp\pi^-$ in pd collisions at 320 MeV (points) compared to the deuteron wave function (histogram) taken from Ref. [23].

ground events except those from the $pd \rightarrow ppn\pi^0$ reaction with one of the photons converting in the metal structure of the scattering chamber and faking a charged particle. This background, which was on the level of a few percent of the total statistics, was simulated and subtracted from the experimental distributions.

The test of the spectator model for this reaction is shown in Fig. 6, presenting a comparison between the missing momentum distribution obtained for the $pd \rightarrow pp\pi^-$ +X events and the deuteron wave function. The distributions have been normalized to the same height as in the $pd \rightarrow pp\pi^0$ +X case. The surplus of the high q events suggests that in addition to the quasi-free events there is some contribution from the mechanism when all three nucleons are involved. In order to diminish this contribution events with q greater than 137 MeV/c have been discarded.

The acceptance-corrected distributions in q are compared to the Monte Carlo predictions in Fig. 7. In order to see the dependence on the CM kinetic energy, Q, the whole accessible range has been divided into six intervals eight MeV wide. The experimental points have been normalized using the quasi-elastic pp scattering. The solid line represents $(\sigma_{11}+\sigma_{01})/2$ with σ_{11} fixed and σ_{01} fitted to reproduce the experimental points. As one can see σ_{01} (dashed line) is comparable to σ_{11} in the whole energy range.

Acceptance-corrected CM distributions of $\cos^2\theta_k$ obtained for the quasi-free $pp \rightarrow pp\pi^-$ reaction are shown in Fig. 8. As before the whole range of Q has been split into six intervals. Solid lines represent the Monte Carlo predictions. At all energies a large forward-backward anisotropy is seen, which is compatible with results obtained in Ref. [13,19]. It should be noted however, that this anisotropy is not only caused only by the $Ss*Sp$ interference as concluded in Ref. [19] but requires also a $Ps*Pp$ interference with the same negative sign as the $Ss*Sp$.

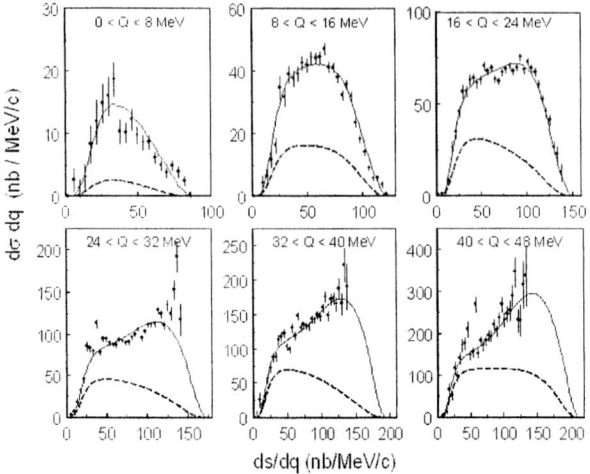

FIGURE 7. Acceptance-corrected two-proton relative momentum distributions obtained for the quasi-free $pn \to pp\pi^-$ reaction in six ranges of the CM kinetic energy Q. The solid and dashed lines represent $(\sigma_{11}+\sigma_{01})/2$ and $\sigma_{01}/2$, respectively.

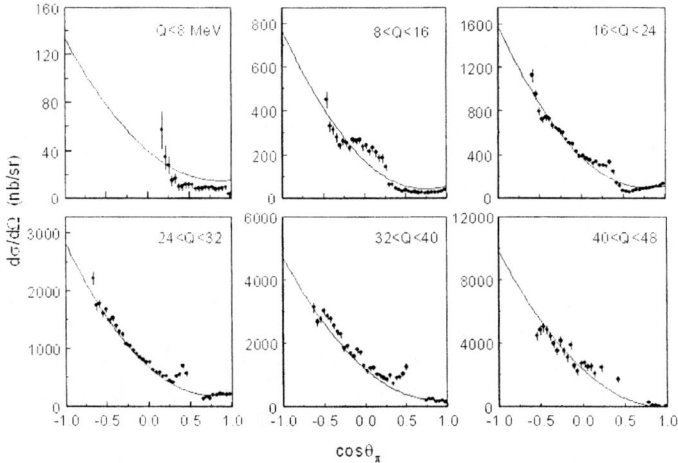

FIGURE 8. Acceptance-corrected CM distributions of $\cos^2\theta_k$ obtained for the quasi-free $pp \to pp\pi$ reaction in six ranges of the CM kinetic energy Q. Solid lines represent the Monte Carlo predictions.

The excitation function of the $pn \to pp\pi^-$ reaction extracted from our data is compared in Fig. 9 to the predictions of the Jülich model [6]. The theoretical curve matches the experimental points reasonably well so we do not have the same conflict with the theory as was in the case of the $pp \to pp\pi^0$ reaction 10 years ago.

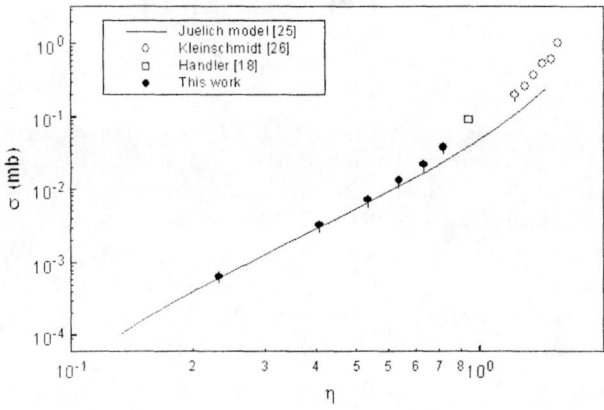

FIGURE 9. Excitation function of the $pn \to pp\pi^-$ reaction measured as a quasi-free process in pd collisions at 320 MeV. The solid line represents prediction of the Jülich model [6].

CONCLUSIONS

To summarize we would like to state that the phenomenological theory of pion production in NN collisions [16] should be limited to lower energies than commonly accepted. In the case of the $pp \to pp\pi^0$ reaction already at 400 MeV, the η^6 prediction for the σ_{Ps} is wrong by a factor of four. In addition, the assumption that the interaction radius is so small that one can use the properties of the spherical Bessel functions for small arguments to derive the amplitudes for the Ps and Pp final states does not hold. A much better description of the differential cross section is obtained if the amplitudes are calculated within the DWBA treatment assuming pion exchange.

A model based on the parameterization of the $pp \to pp\pi^0$ reaction and the spectator model does not describe the experimental data very well. The reason seems to be related to a fraction of the $pn \to pn\pi^0$ events being accepted together with the quasi-free $pp \to pp\pi^0$ events.

The CM pion angular distributions obtained for the quasi-free $pn \to pp\pi^-$ reaction show a large forward-backward asymmetry as observed before [6,19]. The distributions provide evidence that in addition to the $Ss*Sp$ interference also the $Ps*Pp$ interference is important. The excitation function extracted from the experiment agrees reasonably well with the predictions of the Jülich model [6]. The magnitude of the σ_{01} cross section is comparable to that of σ_{11}.

The analysis is still in progress and we hope to improve the normalization and also get a better estimation of background in the $pn \to pp\pi^-$ reaction, especially for small CM pion angles.

ACKNOWLEDGMENTS

We are very grateful to the TSL/ISV personnel for their continued help during the course of this work. Discussions with K. Tamura about the theoretical calculations of the $NN \rightarrow NN\pi$ reactions were much appreciated. Financial support for this experiment and its analysis was provided by the Swedish Natural Science Research Council, the Swedish Royal Academy of Science, the Swedish Institute, the Japanese Ministry of Education, Deutsche Forschung Gesellschaft (Mu 705/3 Graduiertenkolleg), the Polish Scientific Research Committee (grant 5P03B09420), the Russian Academy of Science, the German Bundesministerium für Bildung und Forschung [06TU886 and DAAD], and the European Science Exchange Programme. Part of the analysis has been done at RCNP of Osaka University and one of the authors (JZł) would like to express his gratitude for the hospitality and financial support he received during his stay at this Institute.

REFERENCES

1. Meyer, H.O., et al., Nucl. Phys. **A539**, 633 (1992).
2. Miller, G.A., and Sauer, P., Phys. Rev. C **44**, R1725 (1992).
3. Koltun, D.S., and Reitan, A., Phys. Rev. **141**, 1413 (1966).
4. Lee, T.-S., and Riska, D.O., Phys. Rev. Lett. **70**, 2237 (1993); Horowitz, C.J., Meyer, H.O., and Griegel, D.K., Phys. Rev. C **49**, 1337 (1994).
5. Hernández, E., and Oset, E., Phys. Lett. B **350**, 158 (1995).
6. Hanhart, C., Haidenbauer, J., Krehl, O., and Speth, J., Phys. Lett. B **444**, 25 (1998).
7. Tamura, K., Maeda, Y., and Matsuoka, N., Nucl. Phys. **A663**, 457c (2000).
8. Rosenfeld, A.H., Phys. Rev. **96**, 139 (1954).
9. VerWest, B.J., and Arndt, R.A., Phys. Rev. C **25**, 1979 (1982).
10. Bystricky, J., et al., J. Phys. (Paris) **48**, 1901 (1987).
11. Duncan, F., et al., Phys. Rev. Lett. **80**, 4390 (1998).
12. Maeda, Y., et al., Nucl. Phys. A **663**, 457c (2000).
13. Lacker, H., in Nuclear Physics at Storage Rings, AIP conference proceedings **512**, 34 (1999).
14. Złomańczuk, J., in Nuclear Physics at Storage Rings, AIP conference proceedings **512**, 43 (1999).
15. Bilger, R., et al., Nucl. Phys. A in print.
16. Gell-Mann, M., and Watson, K.M., Ann. Rev. Nucl. Sci. **4**, 219 (1954).
17. Meyer, H.O., in Particle Production Near Threshold, AIP conference proceedings **221**, 129 (1990).
18. Watson, K.M., Phys. Rev. **88**, 1163 (1952); Migdal, A.B., Sov. Phys. JETP **1**, 2 (1955).
19. Handler, R., Phys. Rev. B **138**, 1230 (1965).
20. Stanislaus, S., et al., Phys. Rev. C **41**, R1913 (1990).
21. Loiseau, B., and Mathelitsch, L., Z. Phys. A **358**, 435 (1997).
22. Meyer, H.O., et al., Phys. Rev. Lett. **83**, 5439 (1999).
23. Lacombe, M., et al., Phys. Rev. C **21**, 861 (1980).
24. Calen, H., et al., Nucl. Instr. Meth. A **379**, 57 (1996).
25. CERN Program Library Long Write-up W5013, GEANT - Detector Description and Simulation Tool.
26. Kleinschmidt, M., et al., Z. Phys. A **298**, 253 (1980).

CONTRIBUTED PAPERS

Effective Field Theory Approach

A renormalisation-group approach to two-body scattering with long-range forces

Thomas Barford and Michael C. Birse

Department of Physics and Astronomy, University of Manchester, Manchester, M13 9PL, U.K.

Abstract. We apply the renormalisation-group to two-body scattering by a combination of known long-range and unknown short-range forces. A crucial feature is that the low-energy effective theory is regulated by applying a cut-off in the basis of distorted waves for the long range potential. We illustrate the method by applying it to scattering in the presence of a repulsive $1/r^2$ potential. We find a trivial fixed point, describing systems with weak short-range interactions, and a unstable fixed point. The expansion around the latter corresponds to a distorted-wave effective-range expansion.

INTRODUCTION

Effective field theories (EFT's) offer the promise of a systematic treatment of few-nucleon systems at low-energies. (For a review, see: [1].) They are based on the existence of a separation of scales between those of the low-energy physics: momenta, energies, m_π (generically Q), and those of the underlying physics: m_ρ, M_N, $4\pi f_\pi$ (generically Λ_0). This makes it possible to systematically expand both the theory (defined by a Lagrangian or a potential) and physical observables in powers of Q/Λ_0. Such an expansion will be useful provided it converges rapidly enough.

For weakly interacting systems (such as chiral perturbation theory in the zero- or one-nucleon sectors) the terms in the expansion can be organised according to naive (Weinberg) power counting. The order of a term $(Q/\Lambda_0)^d$ in the theory is just d [2, 3].

In contrast for strongly interacting systems (such as s-wave nucleon-nucleon scattering) there can be new low-energy scales generated by nonperturbative dynamics. In such cases we need to resum certain terms in the theory and this leads to a new power counting [4, 5, 6, 7, 8], often referred to as KSW power counting. The theoretical tool which allows us to determine the power counting in these cases is the renormalisation group (RG).

RENORMALISATION GROUP

The RG is an extension of simple dimensional analysis which has been used to study the scaling behaviour of systems in a wide range of areas of physics. In our work we use a Wilsonian version of the RG [9]. This has the following ingredients:

- Impose a momentum cut-off, $|\mathbf{k}| < \Lambda$ on the low-energy effective theory.

- Demand that observables be independent of the cut-off Λ. This corresponds to integrating out physics on momentum scales above Λ.
- Rescale the theory, expressing all dimensioned quantities in units of Λ. Before doing this, we need to have identified all the important low-energy scales in our system.

We can then follow the flow of the various coupling constants of the theory as $\Lambda \to 0$. If there is a clear separation of scales, then the theory should flow towards a fixed point. This is because, for $\Lambda \ll \Lambda_0$, the only scale left is Λ and so the rescaled theory becomes independent of scale.

Near a fixed point, perturbations scale as powers of the cut-off, Λ^ν, and these define the power counting for the corresponding terms in the theory: $d = \nu - 1$. The sign of ν can be used to classify these terms:

- $\nu > 0$: irrelevant perturbation, flows towards the fixed point as $\Lambda \to 0$,
- $\nu = 0$: marginal perturbation (or "renormalisable" in field-theory terminology), leads to logarithmic flow with Λ,
- $\nu < 0$: relevant perturbation, flows away from the fixed point as $\Lambda \to 0$, making the point an unstable one.

The application of these ideas to two-body scattering by short-range forces can be found in Ref. [8]. Two fixed points were found: a trivial one describing a system with weak scattering, and a nontrivial one describing a system with a bound state at zero energy. The expansion around the nontrivial one is organised by KSW power counting. It is in fact just the effective-range expansion [10], reinvented in modern languange as an EFT.

DISTORTED-WAVE THEORY

Scattering by short-range interactions in the presence of a known long-range potential V_L can be treated by distorted-wave theory. We write the full scattering matrix as

$$T = T_L + (1 + V_L G_L)\tilde{T}_S(1 + G_L V_L), \tag{1}$$

where T_L is the scattering matrix for V_L alone and G_L is the corresponding Green's function. The operator $\tilde{T}_S$ describes the scattering between distorted waves of V_L. In terms of a short-range potential V_S, it satisfies the Lippmann-Schwinger equation

$$\tilde{T}_S = V_S + V_S G_L \tilde{T}_S. \tag{2}$$

We regulate this equation by cutting off G_L in the basis of distorted waves [11, 12],

$$G_L = M \int^\Lambda \frac{d^3\mathbf{q}}{(2\pi)^3} \frac{|\psi_L(q)\rangle\langle\psi_L(q)|}{p^2 - q^2 + i\varepsilon} \; (+\text{bound states}). \tag{3}$$

(Here $p = \sqrt{ME}$ denotes the on-shell relative momentum for two particles of mass M.) The use of this basis is crucial for the identification of the scaling behaviour.

Since the long-range potentials of interest are generally singular as $r \to 0$, we cannot represent the short-range interaction by a simple δ-function at the origin. Instead we choose a δ-shell form,

$$V_S = V(p, \Lambda) \frac{\delta(r-R)}{4\pi R^2}. \tag{4}$$

Note that we have allowed the potential to be energy (p) dependent. However we have not included momentum dependence since, for the pure short-range-case, this was found to affect only the off-shell behaviour of scattering amplitudes [8].

We require that $V(p,\Lambda)$ vary with Λ to keep the fully off-shell DW scattering matrix $\tilde{T}_S$ independent of cut-off. Rescaling the resulting differential equation gives us a DW version of the RG equation [12].

REPULSIVE INVERSE-SQUARE POTENTIAL

To illustrate these ideas we apply them here to a specific example: a repulsive $1/r^2$ potential,

$$V_L = \frac{\beta}{r^2}. \tag{5}$$

This potential is analogous to the interaction in a three-body system such as quartet nd scattering, in the limit of an infinite two-body scattering length [13]. It is scale independent and so should be treated as part of a fixed point (and resummed to all orders). It acts in s-waves like a centrifugal barrier with "angular momentum"

$$\lambda = \sqrt{\beta M + \tfrac{1}{4}} - \tfrac{1}{2}. \tag{6}$$

The corresponding DW's are Bessel functions $j_\lambda(kr)$ of noninteger order.

The requirement that $\tilde{T}_S$ be independent of cut-off leads to a differential equation for $V(p,\Lambda)$,

$$\frac{\partial V}{\partial \Lambda} = -\frac{MV^2}{2\pi^2} \frac{\Lambda^2 j_\lambda(\Lambda r)^2}{p^2 - \Lambda^2}. \tag{7}$$

Provided $R \ll \Lambda^{-1}$ the DW factors scale as powers of Λ: $j_\lambda(\Lambda r) \propto (\Lambda R)^\lambda$. We can then rescale, defining a dimensionless potential

$$\hat{V} \propto MR^{2\lambda} \Lambda^{2\lambda+1} V, \tag{8}$$

and on-shell momentum $\hat{p} = p/\Lambda$, to rewrite Eq. (7) as an RG equation for $\hat{V}$.

We have found two-fixed points of this equation:

- The trivial one, $\hat{V} = 0$. Perturbations around this are of the form $\Lambda^{2n+2\lambda+1} \hat{p}^{2n}$ and can be assigned an order $d = 2(n+\lambda)$.
- A nontrivial one. Perturbations around this are of the form $\Lambda^{2n-2\lambda-1} \hat{p}^{2n}$ and can be assigned an order $d = 2(n-\lambda-1)$.

All perturbations around the trivial point are irrelevant. In contrast, the nontrivial fixed point is unstable; there is always at least one relevant perturbation.

DISCUSSION

The method outlined here makes it possible to determine the power counting for the short-range interactions in the presence of a known long-range potential. A crucial feature is that the cut-off is applied in the basis of DW's. One should not try to regulate the long-range as well as the short-range potential.

For the example described here, the repulsive $1/r^2$ potential, this leads (appropriately for Prague) to a baroque power counting involving non-integer orders. Nonetheless the terms in the EFT remain in one-to-one correspondance with observables. For the expansion around the nontrivial fixed point these are the terms of a DW (or "modified") effective-range expansion [10, 14]. If we write the full phase shift as $\delta = \delta_L + \tilde{\delta}_S$, where δ_L is due to V_L alone, then this expansion has the form

$$p^{2\lambda+1}\left[\cot\tilde{\delta}_S - \cot\pi\left(\lambda+\tfrac{1}{2}\right)\right] = \sum_{n=0}^{\infty} C_{2n}\hat{p}^{2n}, \qquad (9)$$

where C_{2n} is the coefficient of the perturbation of order $2(n-\lambda-1)$ in the potential.

We have also applied this method to the Coulomb potential, elucidating the power counting for the results of Kong and Ravndal [11], and the attractive $1/r^2$ potential [12]. In the latter case, one has to make the scattering for V_L alone well-defined before the method can be applied. This is done by choosing a self-adjoint extension [15] or, more physically, by including a short-range potential as part of V_L. The need for such an interaction has also been found in the corresponding three-body problems [16].

ACKNOWLEDGMENTS

This work was supported by the EPSRC. MCB is grateful for the hospitality of the Institute for Nuclear Theory, Seattle, where this work was started.

REFERENCES

1. Beane, S.R., Bedaque, P.F., Haxton, W.C., Phillips, D.R., and Savage, M.J., nucl-th/0008064.
2. Weinberg, S., *Physica* **96A**, 327 (1979).
3. Weinberg, S., *Nucl. Phys.* **B363**, 3 (1991).
4. Bedaque, P.F., and van Kolck, U., *Phys. Lett. B* **428**, 221 (1998).
5. van Kolck, U., *Nucl. Phys.* **A645**, 273 (1999).
6. Kaplan, D.B., Savage, M.J., and Wise, M.B., *Nucl. Phys.* **B534**, 329 (1998).
7. Gegelia, J., *J. Phys. G: Nucl. Part. Phys.* **25**, 1681 (1999).
8. Birse, M.C., McGovern, J.A., and Richardson, K.G., *Phys. Lett. B* **464**, 169 (1999).
9. Wilson, K.G., and Kogut, J.G., *Phys. Rep.* **12**, 75 (1974).
10. Blatt, J.M., and Jackson, J.D., *Phys. Rev.* **76**, 18 (1949); Bethe, H.A., *Phys. Rev.* **76**, 38 (1949).
11. Kong, X., and Ravndal, F., *Nucl. Phys.* **A665**, 137 (2000).
12. Barford, T., and Birse, M.C., in preparation.
13. Efimov, V.N., *Sov. J. Nucl. Phys.* **12**, 589 (1971).
14. van Haeringen, H., and Kok, L.P., *Phys. Rev. A* **26**, 1218 (1982).
15. Perelomov, A.M., and Popov, V.S., *Theor. Math. Phys.* **4**, 664 (1970).
16. Bedaque, P.F., Hammer, H.-W. and van Kolck, U., *Nucl. Phys.* **A646**, 444 (1999); **A676**, 357 (2000).

Doublet channel neutron-deuteron scattering in leading order effective field theory

B. Blankleider* and J. Gegelia[†]

SoCPES, The Flinders University of South Australia, Bedford Park, SA 5042, Australia
[†]*INFN - Sezione di Ferrara, via Paradiso 12, 44100 Ferrara, Italy
and High Energy Physics Institute of TSU, University str. 9, Tbilisi 380086, Georgia*

Abstract. The doublet channel neutron-deuteron scattering amplitude is calculated in leading order effective field theory (EFT). It is shown that this amplitude does not depend on a constant contact interaction three-body force. Satisfactory agreement with available data is obtained when only two-body forces are included.

INTRODUCTION

An intriguing difficulty arises in the application of leading order EFT to the three-body problem. One finds that the full amplitude describing three-boson scattering, or nucleon-deuteron (nd) scattering in the $J = 1/2$ channel, is sensitive to the cutoff used to solve the scattering equations - even though each perturbation diagram, with resummed two-body interactions, is individually finite. In [1] it was argued that the addition of a one-parameter three-body force counter-term is necessary and sufficient to eliminate this cutoff dependence. On the other hand, in refs. [3] and [4] we have shown, on the example of three bosons, that the cutoff dependence is just a natural consequence of the existence of infinitely many solutions to the given scattering equation; moreover, by carefully identifying the physical amplitude amongst the infinitely many non-physical solutions, we have shown that the cutoff problem can be solved without the introduction of a three-body force.

In the present contribution we show that the physical three-body scattering amplitude in fact does not depend on the constant contact interaction three-body force at all. Further, we demonstrate that for doublet channel *nd* scattering, good agreement with experiment is obtained with the inclusion of two-body forces only.

WHY THERE IS NO THREE-BODY FORCE DEPENDENCE

To show that the leading order EFT three-body amplitude is independent of constant three-body forces, it is sufficient to restrict the discussion to the case of three bosons.

Scattering amplitude without three-body forces

In the three-boson case without three-body forces, the s-wave particle-bound-state scattering amplitude $a(p,k)$ satisfies the equation [5]

$$a(p,k) = M(p,k) + \frac{2\lambda}{\pi} \int_0^\infty dq\, M(p,q) \frac{q^2}{q^2 - k^2 - i\varepsilon} a(q,k), \quad (1)$$

where

$$M(p,q) = \frac{8}{3}\left(\frac{1}{a_2} + \sqrt{\frac{3}{4}p^2 - mE}\right)\left[\frac{1}{2pq}\ln\left(\frac{q^2 + pq + p^2 - mE}{q^2 - qp + p^2 - mE}\right)\right]. \quad (2)$$

In this equation k (p) is the incoming (outgoing) momentum magnitude, $E = 3k^2/4m - 1/ma_2^2$ is the total energy, and a_2 is the two-body scattering length. Here it is assumed that the summation of perturbation theory diagrams and loop integration can be interchanged in the sense that the difference is of higher order and hence negligible in given leading order calculations. In general such assumptions have to be investigated very carefully as they may lead to fictitious fundamental problems [2].

Eq. (1) is known as the S-TM equation [6], and in the three-boson case has $\lambda = 1$. Three nucleons in the spin $J = 1/2$ channel obey a pair of integral equations with similar properties to this bosonic equation, while the $J = 3/2$ channel corresponds to $\lambda = -1/2$. For $\lambda > 0$ Danilov's work [7] shows that the homogeneous equation corresponding to Eq. (1) has a solution for arbitrary E; in particular, there exists a solution for every energy corresponding to the *scattering* of a projectile off a two-body bound state.

The existence of these solutions implies that Eq. (1) has an infinite number of solutions. In fact the homogeneous equation has more than one solution for any given E. Writing these solutions as a_h^i where $i = 1, 2, 3, \ldots$, the most general solution of Eq. (1) can be written as $a = a_p + \sum_i C_i a_h^i$ where a_p is any particular solution. It is useful to examine the asymptotic behaviour of $a(p,k)$ for large p. Because the inhomogeneous term M behaves asymptotically as $1/p$, it follows that either (i) $a \to 0$ faster than $1/p$, or (ii) the asymptotic behaviour of a is determined by the asymptotic behaviour of the homogeneous solution a_h. In the latter case the asymptotic behaviour has the form [7]

$$a(p,k) = \sum_i A_i(k) p^{s_i} + O(1/p), \quad (3)$$

where s_i are roots of the equation

$$1 - \frac{8\lambda}{\sqrt{3}} \frac{\sin \pi s/6}{s \cos \pi s/2} = 0b. \quad (4)$$

The summation in Eq. (3) goes over all solutions of Eq. (4) for which $|\mathrm{Re}\,s| < 1$. For $\lambda = 1$ Eq. (4) has two roots for which $|\mathrm{Re}\,s| < 1$: $s = \pm i s_0$, where $s_0 \approx 1.00624$, so that Eq. (3) gives the asymptotic behaviour of the amplitude as

$$a(p,k) \sim A_1(k) p^{is_0} + A_2(k) p^{-is_0}. \quad (5)$$

By contrast, for $\lambda < 0$ the homogeneous equation has no non-trivial solution and the solution of Eq. (1) is unique; in this case the physical amplitude a must vanish asymptotically faster than $1/p$.

In refs. [3, 4] we have shown that the oscillatory behaviour of the general amplitude a for $\lambda = 1$ is simply an artifact of the homogeneous equation (corresponding to Eq. (1)) having non-zero solutions, and that the *physical* amplitude does not display this spurious behaviour; instead, it behaves just like the solution for $\lambda < 0$, namely, it vanishes asymptotically faster than $1/p$. Thus amongst the solutions given by Eq. (5), the physical solution is the one with $A_1(k) = A_2(k) = 0$. More generally, for any $\lambda > 0$ the physical amplitude is the one that has no admixtures of homogeneous equation solutions, and by the above argument, it must therefore vanish asymptotically faster than $1/p$.

Considering Eq. (1) for the case where a is the physical amplitude, since the free term of Eq. (1) behaves like $1/p$ for large p, the coefficient of $1/p$ coming from this inhomogeneous term should cancel the coefficient of a similar term coming from the integral part (we note that this argument is valid for both $\lambda = 1$ and $\lambda = -1/2$). Hence

$$0 = \frac{4}{\sqrt{3}} + \frac{8\lambda}{\sqrt{3}\pi} \int_0^\infty \frac{dq\, q^2}{q^2 - k^2 - i\varepsilon} a(q,k). \tag{6}$$

Identical scattering amplitude with a constant three-body force

Multiplying Eq. (6) by $2/\sqrt{3}(1/a_2 + \sqrt{3/4p^2 - mE})H$ and adding the result to Eq. (1), we obtain the scattering equation where the constant three-body force H is included to all orders (an H is simply added to the term in the square bracket of Eq. (2)). Hence the physical non-oscillating solution of Eq. (1) with no three-body force also satisfies the modified Eq. (1) where an arbitrary H is included. Hence the inclusion of a constant three-body force has no effect on the physical scattering amplitude.

In a recent paper [8] it has been shown that doublet channel neutron-deutron scattering amplitude in EFT with effective range parameters taken into account, does not exhibit any dependence on a constant three-body force. This result is therefore in agreement with the observations of the present work.

ND SCATTERING WITH TWO-BODY FORCES ONLY

Doublet channel neutron-deuteron scattering in leading order EFT is analogous to the scalar case and does not involve any additional problems. Starting from the EFT Lagrangian, one can obtain the following system of doublet channel neutron-deuteron scattering equations [6, 9]:

$$a(p,k) = M(p,k) + \frac{2}{\pi}\int_0^\infty dq\, M(p,q)\left[G_1(q)a(q,k) + 3G_2(q)b(q,k)\right], \tag{7}$$

$$b(p,k) = 3M(p,k) + \frac{2}{\pi}\int_0^\infty dq\, M(p,q)\left[3G_1(q)a(q,k) + G_2(q)b(q,k)\right], \tag{8}$$

where a and b are the neutron-$^3S_1$ (nd) and nucleon-$^1S_0$ amplitudes, respectively,

$$G_1(q) = \frac{q^2}{q^2 - k^2 - i\varepsilon}, \quad G_2 = \frac{3/4\, q^2}{\left(\sqrt{3/4q^2 - mE} + 1/a_2^t\right)\left(\sqrt{3/4q^2 - mE} - 1/a_2^s - i\varepsilon\right)},$$

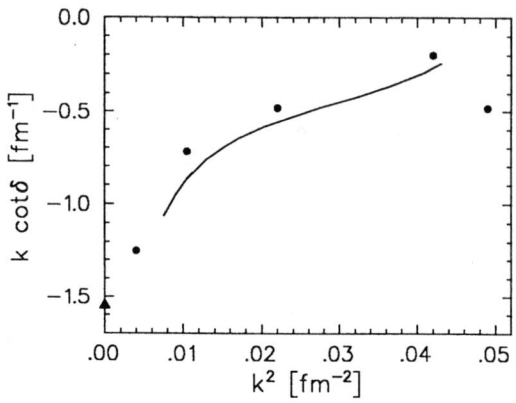

FIGURE 1. Doublet channel nd scattering phase shifts. Data are from phase shift analysis of van Oers and Seagrave (dots) and a measurement by Dilg et al. (triangle).

$a_2^{s,t}$ are the two-particle scattering lengths in singlet and triplet channels, and $E = 3k^2/4m - 1/m(a_2^t)^2$ is the total energy. Apart from a factor of 1/4, the driving term $M(p,q)$ differs from Eq. (2) only in that a_2 is replaced by a_2^t. Solving the system (7)-(8) and isolating the non-oscillating solution for $a(p,k)$ we obtain the physical amplitude for nd scattering. The results are shown in Fig. 1. In order to isolate the physical amplitude we had to solve the corresponding homogeneous system of equations. This task is especially difficult due to the existence of a continuum of solutions corresponding to the continuous spectrum of scattering energies. The method employed to achieve this numerical solution does not allow us to obtain high accuracy for low momenta [3, 4] - that is why our curve does not extend to the origin. We note that while our calculations fit the experimental data quite well, the accuracy of these data is open to question. Ref. [10] does not contain error estimates and ref. [11] claims that at least the scattering length calculated in ref. [10] may be incorrect.

This work has been supported by a grant from the Australian Research Council.

REFERENCES

1. Bedaque, P.F., Hammer, H.W., and van Kolck, U., *Phys. Rev. Lett.* **82**, 463 (1999).
2. Gegelia, J., nucl-th/9802038.
3. Gegelia, J., *Nucl. Phys.* **A680**, 303 (2000).
4. Blankleider, B., and Gegelia, J., nucl-th/0009007.
5. Bedaque, P.F., Hammer, H.W., and van Kolck, U., *Nucl. Phys.* **A646**, 444 (1999).
6. Skorniakov, G.V., and Ter-Martirosian, K.A., *Sov. Phys. JETP* **4**, 648 (1957).
7. Danilov, G.S., *Sov. Phys. JETP* **13**, 349 (1961).
8. Gabbiani, F., nucl-th/0104088.
9. Bedaque, P.F., Hammer, H.W., and van Kolck, U., *Nucl. Phys.* **A676**, 357 (2000).
10. van Oers, W.T.H., and Seagrave, J.D., *Phys. Lett. B* **24**, 562 (1967).
11. Dilg, W., Koester, L., and Nistler, W., *Phys. Lett. B* **36**, 208 (1971).

The rho meson in nuclear matter - a chiral unitary approach

D. Cabrera

Departamento de Física Teórica and IFIC, Centro Mixto Universidad de Valencia-CSIC, Institutos de Investigación de Paterna, Apdo. correos 22085, 46071, Valencia, Spain

Abstract. In this work, the properties of the ρ meson at rest in cold symmetric nuclear matter are studied. We make use of a chiral unitary approach to pion-pion scattering in the vector-isovector channel, calculated from the lowest order Chiral Perturbation Theory (χPT) lagrangian including explicit resonance fields. Low energy chiral constraints are considered by matching our expressions to those of one loop χPT. To account for the medium corrections, the ρ couples to $\pi\pi$ pairs which are properly renormalized in the nuclear medium, accounting for both $p-h$ and $\Delta-h$ excitations. The terms where the ρ couples directly to the hadrons in the $p-h$ or $\Delta-h$ excitations are also accounted for. In addition, the ρ is also allowed to couple to $N^*(1520)-h$ components.

MESON-MESON SCATTERING IN A CHIRAL UNITARY APPROACH

We study the ρ propagation properties by obtaining the $\pi\pi \to \pi\pi$ scattering amplitude in the $(I,J) = (1,1)$ channel. The model for meson-meson scattering in vacuum is fully explained in [1, 2]. We start from the $(I=1)$ $\pi\pi$, $K\bar{K}$ states in the isospin basis:

$$|\pi\pi\rangle = \frac{1}{2}|\pi^+\pi^- - \pi^-\pi^+\rangle$$
$$|K\bar{K}\rangle = \frac{1}{\sqrt{2}}|K^+K^- - K^0\bar{K}^0\rangle. \tag{1}$$

Tree level amplitudes are obtained from the lowest order χPT lagrangians [3] including explicit resonance fields [4]. We collect these amplitudes in a 2×2 symmetric K matrix and work in a coupled channel approach.

The final expression of the T matrix is obtained by unitarizing the tree level scattering amplitudes. To this end we follow the N/D method, which was adapted to the context of chiral theory in ref. [5]. We get

$$T(s) = [I + K(s) \cdot G(s)]^{-1} \cdot K(s), \tag{2}$$

where $G(s)$ is a diagonal matrix given by the loop integral of two meson propagators. In dimensional regularization its diagonal elements are given by

$$G_i^D(s) = \frac{1}{16\pi^2}\left[-2 + d_i + \sigma_i(s)\log\frac{\sigma_i(s)+1}{\sigma_i(s)-1}\right], \tag{3}$$

where the subindex i refers to the corresponding two meson state (1 for $K\bar{K}$, 2 for $\pi\pi$) and $\sigma_i(s) = \sqrt{1 - 4m_i^2/s}$ with m_i the mass of the particles in the state i. The d_i constants in eq. (3) are chosen to obey the low energy chiral constraints, and they are obtained by a matching to one loop χPT.

At this stage, the model successfully describes $\pi\pi$ P-wave phase shifts and π, K electromagnetic vector form factors up to $\sqrt{s} \lesssim 1.2$ GeV. The results for the $\pi\pi \to \pi\pi$ scattering amplitude show that the inclusion of the $K\bar{K}$ channel introduces minimum changes compared to the calculation including only pion loops. Keeping this in mind, the calculations in nuclear matter are performed ignoring the contribution of kaon loops.

By using dimensional regularization it is possible to establish connection with other approaches where tadpole terms are explicitly kept in the lagrangian [6]. One can prove that the formalism keeping tadpoles and full off shell dependence of the $\rho\pi\pi$ vertex is equivalent to the one we have followed where only the on shell part of the $\rho\pi\pi$ vertex is kept. In the medium, however, the pion propagator in the tadpole term will change. Hence, in order to stick to the gauge invariance of the vector field formalism the tadpole term is kept.

$(I,J) = (1,1)$ $\pi\pi$ SCATTERING IN THE NUCLEAR MEDIUM

The basic input for the calculation in nuclear matter is the pion self-energy. It is written as usual in terms of the Lindhard functions accounting for both $p-h$ and $\Delta-h$ excitations. Short range correlations are also accounted for with the Landau-Migdal parameter g', set to 0.7. The final expression is

$$\Pi_\pi(q,\rho) = f(\vec{q}^2)^2 \vec{q}^2 \frac{(\frac{D+F}{2f})^2 U(q,\rho)}{1 - (\frac{D+F}{2f})^2 g' U(q,\rho)}, \qquad (4)$$

where here q is the four-momentum of the pion and $U = U_N + U_\Delta$ the Lindhard function for $p-h$, $\Delta-h$ excitations [7, 8]. We use a monopole form factor $f(\vec{q}^2) = \frac{\Lambda^2}{\Lambda^2+\vec{q}^2}$ for the πNN and $\pi N\Delta$ vertices with the cut-off parameter set to $\Lambda = 1$ GeV.

In accordance to the gauge invariance of the theory, a ρ-meson-baryon contact term must be considered [9, 10, 11]. Its contribution can be derived from the set of gauge invariant diagrams in Fig. 1 (left), and together with the single insertion of a pion self-energy, it provides the set of medium correction graphs shown in Fig. 1 (right) [12]. Their contribution is readily incorporated in the T matrix by a proper substitution and redefinition of the G two meson loop function of eq. (2), and using fully dressed pion propagators. A subtraction of the contribution in vacuum is performed to cancel quadratic divergences and convergence is achieved by means of the form factors.

In the same fashion the tadpole diagram in nuclear matter is calculated, and by subtracting the contribution in vacuum quadratic divergences are removed.

In addition to the contact interactions mentioned above there are other medium corrections that arise if one sticks to the gauge vector field formalism and generate interactions via minimal coupling scheme [6]. Because of this a set of diagrams involving ρNN and

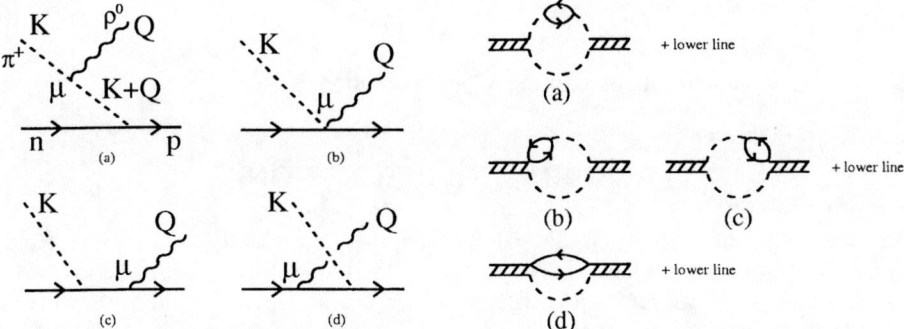

FIGURE 1. Left: Gauge invariant set of diagrams used to calculate the ρ-meson-baryon contact term. The wavy lines represent here the vector meson, single solid lines are reserved for particle-hole excitations and dashed lines are π mesons. Right: Medium correction graphs. Double solid line filled with short oblique dashes represents the K_{22} scattering amplitude, external particle lines are omitted.

ρΔΔ vertices also contribute to the ρ meson self-energy in nuclear matter. The relevant ones are shown in Fig. 2.

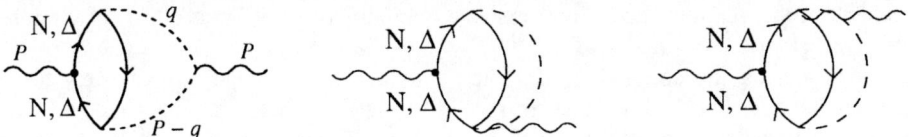

FIGURE 2. Additional medium corrections involving ρNN, ρΔΔ vertices.

A step forward is done by considering the coupling of the ρ meson to the $N^*(1520)$ resonance. A much detailed work along these lines has been done in [13], where many other resonances are included. This correction manifests as an extra self-energy term in the ρ propagator. The basic vertex involved in this effect is shown in Fig. 3a, and the lagrangian describing the interaction reads [14]

$$\mathcal{L}_{N^*N\rho} = -g_{N^*N\rho}\bar{\Psi}_N S_i \vec{\phi}_i \vec{\tau} \Psi_{N^*} + h.c., \qquad (5)$$

and the contribution of the self-energy diagram of Fig. 3b can be written as

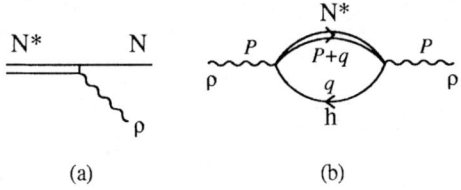

FIGURE 3. (a) ρNN* vertex. (b) $N^* - h$ bubble contributing to the selfenergy of the ρ meson.

$$\Pi_\rho^{N^*-h}(P) = \frac{2}{3}g_{N^*N\rho}^2 U_{N^*}(P), \tag{6}$$

in terms of a Lindhard function for the $N^* - h$ excitation, $U_{N^*}(P)$.

RESULTS AND DISCUSSION

We calculated in this approach the real and imaginary parts of T_{22}, the $\pi\pi \to \pi\pi$ scattering amplitude matrix element. The imaginary part shows a clear broadening of the resonance and the peak position is slightly shifted upwards by about 30 MeV, which is also observed in the position of the zero of the real part. The coupling to the $N^* - h$ components manifests as a visible bump at lower energies. As a whole much strength is spread below the resonance mass.

In order to test the model dependence on the phenomenological parameters we performed variations of Λ in the range 0.9-1.1 GeV, and of g' in the range 0.6-0.8. The results are rather stable under the first of the variations, showing uncertainties in the position of the ρ meson peak of about 10-15 MeV. Variations of g' are relevant close to the resonance maximum, and lead to uncertainties in the position of the resonance of about 20-25 MeV. This was expected since the pion self-energy, which is one of the basic ingredients of the medium corrections, is strongly modified by the short range correlations which directly depend on g'.

In considering the coupling to baryonic resonances our model does not try to be complete since many other resonances should be included, but this allows us to have an estimate of how these new channels affect the results. The calculation could be improved by following a self-consistent treatment as done in [13].

REFERENCES

1. Oller, J.A., Oset, E., and Palomar, J.E., *Phys. Rev. D* **63**, 114009 (2001) [hep-ph/0011096].
2. Palomar, J.E., in these proceedings.
3. Gasser, J., and Leutwyler, H., *Nucl. Phys.* **B250**, 465, 517, 539 (1985).
4. Ecker, G., Gasser, J., Leutwyler, H., Pich, A., and de Rafael, E., *Phys. Lett. B* **223**, 425 (1989); Ecker, G., Gasser, J., Pich, A., and de Rafael, E., *Nucl. Phys.* **B321**, 311 (1989).
5. Oller, J.A., and Oset, E., *Phys. Rev. D* **60**, 074023 (1999).
6. Urban, M., Buballa, M., Rapp, R., and Wambach, J., *Nucl. Phys.* **A641**, 433 (1998) [nucl-th/9806030].
7. Oset, E., Fernandez de Cordoba, P., Salcedo, L.L., and Brockmann, R., *Phys. Rep.* **188**, 79 (1990).
8. Chiang, H.C., Oset, E., and Vicente-Vacas, M.J., *Nucl. Phys.* **A644**, 77 (1998) [nucl-th/9712047].
9. Klingl, F., Kaiser, N., and Weise, W., *Nucl. Phys.* **A624**, 527 (1997) [hep-ph/9704398].
10. Chanfray, G., and Schuck, P., *Nucl. Phys.* **A555**, 329 (1993).
11. Herrmann, M., Friman, B.L., and Norenberg, W., *Nucl. Phys.* **A560**, 411 (1993).
12. Cabrera, D., Oset, E., and Vicente-Vacas, M.J., *Acta Phys. Polon. B* **31**, 2167 (2000) [nucl-th/0006029]; Cabrera, D., Oset, E., and Vicente Vacas, M.J., nucl-th/0011037.
13. Peters, W., Post, M., Lenske, H., Leupold, S., and Mosel, U., *Nucl. Phys.* **A632**, 109 (1998) [nucl-th/9708004].
14. Gomez Tejedor, J.A., Cano, F., and Oset, E., *Phys. Lett. B* **379**, 39 (1996).

Effective chiral theory for pseudoscalar and vector mesons

E. Gedalin*, A. Moalem* and L.Razdolskaya*

Department of Physics, Ben Gurion University, 84105, Beer Sheva, Israel

Abstract. An extended $U(3)_L \otimes U(3)_R$ chiral effective field theory which includes pseudoscalar and vector meson nonets as dynamic variables is presented. We combine the hidden symmetry approach with a general procedure of including the η' meson into chiral theory, and accounts for *direct* and *indirect* symmetry breaking effects via a mechanism based on the quark mass matrix. The theory is applied to anomalous radiative decays using particle mixing schemes, corresponding to different symmetry breaking assumptions and uniquely determined by the lagrangian presumed. Radiative decays of light flavor mesons are best explained within the framework of a one mixing angle scheme and provide evidence for $SU(3)_F$ and nonet symmetry breaking.

The decay of light flavor mesons was subject to a numerous studies. Particularly successful are effective field theories (EFT) based on the Hidden Local Symmetry (HLS) approach [1], which seem to provide an accurate and consistent framework in explaining a vast amount of data, and more importantly reflect on low energy properties of the QCD lagrangian.

Our main interest in this representation is to report on a $U(3)_L \otimes U(3)_R$ chiral theory which contains the pseudoscalar and vector nonets as dynamic variables. To this aim we combine a generalized HLS lagrangian [1] with a general procedure [2] of including the η' meson into chiral perturbation theory. The lagrangian constructed exhibits the fundamental symmetries of the QCD lagrangian and accounts for *direct* and *indirect* symmetry breaking (SB) effects through a mechanism based on the quark mass matrix. We believe that the theory extends easily to include tensor and higher spin mesons. In what follows we give a brief description of the theory and present some results from numerical analyses of anomalous radiative decays of pseudoscalar and vector mesons. Details of the calculations and the analyses are given elsewhere [3, 4].

The lagrangian. Based on the HLS approach [1], the lagrangian is written in the form,

$$L = L_A + \bar{L}_A + L_m + a(L_V + \bar{L}_V) - \frac{1}{4}\mathrm{Tr}(V_{\mu\nu}V^{\mu\nu}) + L_{WZW} + \ldots, \qquad (1)$$

where $L_A(\bar{L}_A)$ and $L_V(\bar{L}_V)$ are symmetric (direct symmetry breaking) pseudoscalar and vector mesons parts, respectively, L_m is a symmetry breaking mass term, $-\frac{1}{4}\mathrm{Tr}(V_{\mu\nu}V^{\mu\nu})$ a vector meson kinetic term, and L_{WZW} is the Wess-Zumino-Witten term. The symmetric (and likewise the asymmetric) lagrangian densities are constructed using vector and axial vector like variables,

$$\Gamma_\mu = \frac{i}{2}\left[\xi^\dagger, \partial_\mu \xi\right] + \frac{1}{2}\left(\xi^\dagger r_\mu \xi + \xi l_\mu \xi^\dagger\right), \quad \Delta_\mu = \frac{i}{2}\left\{\xi^\dagger, \partial_\mu \xi\right\} + \frac{1}{2}\left(\xi^\dagger r_\mu \xi - \xi l_\mu \xi^\dagger\right), \qquad (2)$$

where as a nonlinear representation of the fields we use, $U(x) = \xi^2 = \exp i\frac{\sqrt{2}}{F_\pi}\mathcal{P}$, with,

$$\mathcal{P} = \begin{pmatrix} \frac{\pi^0}{\sqrt{2}} + \frac{1}{\sqrt{6}}(X_\eta \eta + X_{\eta'}\eta') & \pi^+ & z_s K^+ \\ \pi^- & -\frac{\pi^0}{\sqrt{2}} + \frac{1}{\sqrt{6}}(X_\eta \eta + X_{\eta'}\eta') & z_s K^0 \\ z_s K^- & z_s \bar{K}^0 & \frac{1}{\sqrt{6}}(Y_\eta \eta + Y_{\eta'}\eta') \end{pmatrix}, \quad (3)$$

being the pseudoscalar nonet matrix. The coefficients X_i, Y_i depend on mixing angles and field rescaling parameters, all being defined through an orthogonalization procedure, transforming the intrinsic fields into the physical fields. With V_μ denoting the dynamical gauge bosons, and to lowest order, L_i and $\bar{L}_i$ involves traces of $\Delta_\mu, \Delta_\mu^2, (\Gamma_\mu - gV_\mu)$ and $(\Gamma_\mu - gV_\mu)^2$, the covariant derivative of the vacuum angle $\vartheta(x)$, and arbitrary functions of the (phase) variable $X \equiv \sqrt{6}\eta_0(x)/F_0 + \vartheta(x)$.

Our L_A, L_V (and their SB companions $\bar{L}_A, \bar{L}_V$) are different from those of Refs. [1, 5, 6]. First, they involve additional terms due to including the η' as a ninth dynamical variable. Secondly, the matrix B used to generate the SB companions is chosen to be hermitian and proportional to the quark mass matrix, allowing to maintain the same isospin to $SU(3)$ SB scale ratios as in QCD. Thirdly, the symmetry breaking scales c_i and d_i are not related in our model. In view of detailed data analyses, the assumption $d_i = c_i^2$ of Bando et al. [1] seems quite unjustified.

The kinetic and mass lagrangian densities of the Goldstone bosons determine uniquely the state (and decay constants) mixing schemes [3]. In what follows we consider two alternatives for the SB companion $\bar{L}_A$. The first, *Alternative I*, employs an octet axial vector variable Δ, allowing for $SU(3)_F$ nonet SB. The second, *Alternative II* uses the nonet axial vector, Eq. (2), allowing for $U(3)_F$ nonet SB. The kinetic and mass terms corresponding to both alternatives depart from the standard quadratic form. However, for any EFT the kinetic and mass terms can be reduced into the standard form by transforming the intrinsic fields into the physical fields through a renormalization and two unitary transformations. This orthogonalization procedure defines the state mixing in a unique way, (an observation which seems to be overlooked in previous studies). Moreover, taking the eigenvalues of the resulting mass matrix to be equal to the physical masses, much of the dependence on the model free parameters is eliminated. In the numerical analyses to be presented only two SB scales c_A and d_A and the nonet SB measure $r = F_8/F_0$ need be determined from global fit to data. Furthermore, these transformations allow expressing the pseudoscalar nonet matrix in terms of the physical states, and thus incorporating SB indirectly into the field matrix ξ. In passing by we note that the vector nonet matrix depends essentially on four parameters, all of which can be determined through a similar orthogonalization procedure. In practice, the vector mixing angle is rather close to the ideal mixing value.

Anomalous radiative decays. We now consider anomalous radiative decays. In marked difference with previous numerous analyses, we treat the pseudoscalar and vector nonets on equal footing and most importantly, we use the proper state mixing schemes corresponding to the lagrangians presumed. Following the discussion above, we factorize the anomalous lagrangian into *direct* and *indirect* SB parts,

$$L_{P\gamma\gamma} = L_{P\gamma\gamma} + c_W \bar{L}_{P\gamma\gamma}, \quad L_{VP\gamma} = L_{VP\gamma} + c_W \bar{L}_{VP\gamma}, \quad (4)$$

where $L_{P\gamma\gamma}(L_{VP\gamma})$ represents the overall contribution of the unbroken anomalous lagrangian to the $P\gamma\gamma(V\gamma\gamma)$ interactions, and $\bar{L}_{P\gamma\gamma}(\bar{L}_{VP\gamma})$ is the corresponding direct SB companion. Note that even in the limit $c_W = 0$, indirect SB due to $c_A \neq 0, d_A \neq 0, r \neq 1$ are still included in $L_{P\gamma\gamma}$. Explicitly, the corresponding symmetric and asymmetric parts are,

$$L_{VP\gamma} = g_V \frac{e}{F_\pi} \varepsilon^{\mu\nu\alpha\beta} \partial_\mu A_\nu \mathrm{Tr}\left(Q\{\partial_\alpha V_\beta, \mathcal{P}\}\right) \tag{5}$$

$$L_{P\gamma\gamma} = g_P \frac{e^2}{2F_\pi} \varepsilon^{\mu\nu\alpha\beta} \partial_\mu A_\nu \partial_\alpha A_\beta \mathrm{Tr}\left(\{Q^2, \mathcal{P}\}\right), \tag{6}$$

$$\bar{L}_{VP\gamma} = g_V \frac{e}{F_\pi} \varepsilon^{\mu\nu\alpha\beta} \partial_\mu A_\nu \mathrm{Tr}\left(Q\{B, \{\partial_\alpha V_\beta, \mathcal{P}\}\}\right), \tag{7}$$

$$\bar{L}_{P\gamma\gamma} = g_P \frac{e^2}{2F_\pi} \varepsilon^{\mu\nu\alpha\beta} \partial_\mu A_\nu \partial_\alpha A_\beta \mathrm{Tr}\left(\{Q^2, \{B, \mathcal{P}\}\}\right). \tag{8}$$

With these expressions the decay widths of the $P \to V\gamma, V \to P\gamma$ and $P \to \gamma\gamma$, with $P = \pi, K, \eta, \eta'$ and $V = \rho, K^*, \omega, \phi$, involve three SB scales (c_A, d_A, r), a direct SB scale (c_W), and two coupling constants (g_V, g_P). We may determine all of these parameters by comparing calculated decay widths of certain transitions with data. However, for better evaluation of SB effects, analyses are performed with the SB scales treated as free parameters. As a general rule the mass and decay width data are the best fit values quoted in the latest review of particle properties [7]. The pion radiative decay constant is taken to be $F_\pi = F_8 = 93$ MeV. The analyses results from using the *Alternative I* and *Alternative II* lagrangians are summarized in Tables 1 and 2. We also show the results from the so called quark flavor basis, which corresponds to our *Alternative II* scheme in the limit of exact nonet symmetry. Based on the χ^2/dof values, the one mixing angle *Alternative I* scheme provides the best explanation of the data. This observation remains valid, should we have used the older set of data [8], though the resulting fit qualities are poorer. With the parameter set of Table 1, the mixing angle $\theta_P = (-14.3 \pm 2.2)^o$.

How significant are the departures from exact symmetry ? Clearly, the exact $SU(3)_F$ symmetry values, $c_W = 0$, $c_A = d_A = 0$, $r = 1$, would be inconsistent with the measured value of the ratio $\Gamma(K^{*0}K^0\gamma)/\Gamma(K^{*+}K^+\gamma)$; with $c_W = 0$ this ratio becomes 4 as opposed to the experimental value of 2.34 ± 0.43. In addition, direct symmetry breaking alone is not sufficient; with $c_A = 0$ and $c_W = -0.2$, the calculated width $\Gamma(K^{*0}K^0\gamma)$ is $\equiv 154$ keV, some 30% higher than experimental value, (117 ± 10) keV, and well beyond the measurement accuracy. We may thus conclude that both *direct* $(c_W \neq 0)$ and *indirect* $(c_A \neq 0)$ SB are required to explain data. Moreover, the departure from nonet symmetry are small but significant. With $r = 1$, the global fit analysis gives $c_W = -0.18 \pm 0.05$, $c_A = 0.65 \pm 0.06$, $d_A = -0.32 \pm 0.04$ and $\theta_P = -(15.2 \pm 2.2)^o$. However, the fit quality deteriorates to $\chi^2/dof = 13.2/7$. Furthermore, with $z = z_K$, $d_A = -c_A/4$ the fit analysis yields, $c_W = -0.26 \pm 0.05$, $c_A = 0.56 \pm 0.06$, $d_A = -0.32 \pm 0.04$, corresponding to $\theta_P = -(8.2 \pm 0.2)^o$ and $\chi^2/dof = 8.2/6$. Again the fit quality is significantly poorer as compared with the *Alternative I* results.

Summary and Conclusions We have presented a generalized HLS based lagrangian where the pseudoscalar and vector nonets are treated on equal footing as dynamical degrees of freedom. At lowest order the lagrangian comprises two terms L_A, L_V describing

TABLE 1. Symmetry breaking scales and χ^2/dof from global fit to data. * Values kept fixed in the fit procedure.

	c_W	c_A	d_A	r	χ^2/dof
Alternative I	$-(0.20\pm 0.05)$	(0.64 ± 0.06)	-0.25 ± 0.04	0.91 ± 0.04	3.1/6
Alternative II	$-(0.27\pm 0.05)$	0.2 ± 0.05	0.1 ± 0.02	0.94 ± 0.05	18.6/6
QFB	$-(0.19\pm 0.05)$	(1.4 ± 0.1)	-1.1 ± 0.1	1	19.8/7

TABLE 2. Calculated radiative decay widths. */** Values used to fix the coupling constants g_V and g_P, respectively. $^{a)}$ Best fit values from Ref. [7].

Decay	Γ_{exp}(keV) $^{a)}$	Γ_{calc}(keV) Alternative I	Alternative II	QFB
$\rho \to \pi\gamma$	76 ± 10	76 ± 10	76 ± 10	76 ± 10
$\omega \to \pi\gamma$	716 ± 43	*716	*716	*716
$\rho \to \eta\gamma$	36 ± 12	40.0 ± 4.1	47.4 ± 4.2	52.8 ± 4.5
$\omega \to \eta\gamma$	5.5 ± 0.85	5.2 ± 0.45	6.1 ± 0.5	6.8 ± 0.6
$\phi \to \eta\gamma$	57.8 ± 1.6	60.8 ± 2.4	65.7 ± 3.5	60 ± 3
$\phi \to \eta'\gamma$	$0.30^{+0.20}_{-0.16}$	0.37 ± 0.03	0.24 ± 0.03	0.4 ± 0.05
$\eta' \to \rho\gamma$	61.2 ± 7.5	70.7 ± 5.9	70.3 ± 4.1	78.5 ± 6.6
$\eta' \to \omega\gamma$	6.12 ± 0.75	6.47 ± 0.45	6.4 ± 0.4	7.2 ± 0.5
$\pi^0 \to \gamma\gamma$	$(7.8\pm 0.55)10^{-3}$	$**7.8\cdot 10^{-3}$	$**7.8\cdot 10^{-3}$	$**7.8\cdot 10^{-3}$
$\eta \to \gamma\gamma$	0.460 ± 0.050	0.468 ± 0.03	0.58 ± 0.02	0.637 ± 0.03
$\eta' \to \gamma\gamma$	4.280 ± 0.280	4.02 ± 0.3	3.68 ± 0.24	4.52 ± 0.3
$K^{*0} \to K^0\gamma$	117 ± 10	106 ± 8.3	120.7 ± 4.6	113 ± 8
$K^{*\pm} \to K^{\pm}\gamma$	50 ± 5	48.6 ± 3.2	63.3 ± 5.2	48 ± 5

respectively, the interactions of pseudoscalar and vector meson nonets, and a "kinetic" term for the vector mesons. The direct SB companions $\bar{L}_A, \bar{L}_B$ are constructed using a hermitian matrix B taken to be proportional to the quark mass matrix. Extensive analyses of anomalous radiative decays show that the *Alternative I* lagrangian which allows for $SU(3)_F$ and nonet symmetry breaking provides the best fit to the data with a mixing angle $\theta_P = (-14.3\pm 2.2)^o$. The other alternatives of $U(3)_F$ SB with or without nonet symmetry breaking yield rather poor fits and seem unjustified.

REFERENCES

1. Bando, M., Kugo, T., and Yamawaki, K., *Nucl. Phys.* **B256**, 493 (1985); *Phys. Rep.* **164**, 215 (1988).
2. Gasser, J., and Leutwyler, H., *Nucl. Phys.* **B250**, 465 (1985).
3. Gedalin, E., Moalem, A., and Razdolskaya, L., to be published in *Phys. Rev. D*; hep-ph/0106301.
4. Gedalin, E., Moalem, A., and Razdolskaya, L., submitted for publication in *Phys. Rev. D*; hep-ph/0108124.
5. Bramon, A., Grau, A., and Pancheri, G., The second DAΦNE physics handbook, Vol.II, p.477, (Eds. L.Maiani et al.,INFN, Frascati, May 1995.).
6. Benayoun, M., et al. *Eur. Phys. J. C* **2**, 269 (1998).
7. Groom, D.E., et al. (Particle Data Group), *Eur. Phys. J. C* **15**, 1 (2000).
8. Yost, G.P., et al. (Particle Data Group), *Phys. Lett. B* **204**, 1 (1988).

Pseudoscalar meson mixing in effective field theory

E. Gedalin*, A. Moalem* and L. Razdolskaya*

Department of Physics, Ben Gurion University, 84105, Beer Sheva, Israel

Abstract. We show that for any effective field theory of colorless meson fields, the mixing schemes of particle states and decay constants are not only related but also determined exclusively by the kinetic and mass Lagrangian densities.

Mixing schemes of the pseudoscalar $\eta(547.3 \text{ MeV}) - \eta'(957 \text{ MeV})$ mesons has attracted considerable interest in recent years. The traditional approach [1] involves a single mixing angle θ_P for the octet η_8 and singlet η_0 states as well as octet (F_8) and a singlet (F_0) decay constants. Recently several authors [2, 3, 4, 5] suggested a two mixing angle scheme for the decay constants, while the particle states follow either the same transformation [5] or a single angle mixing pattern [2, 3, 4]. Feldmann et al. [4] proposed that only in the quark flavor basis (QFB), $q\bar{q} = (u\bar{u} + d\bar{d})/\sqrt{2}$ and $s\bar{s}$, the decay constants and particle states follow the same mixing pattern. However, in the octet-singlet basis, there is a need for one mixing angle for the particle states and two mixing angles for the decay constants. Here we shell show that for any effective field theory (EFT) irrespective with the field interactions presumed, the mixing schemes for the states and decay constants not only are related but also determined exclusively by the structure of the kinetic and mass Lagrangian. For any EFT the most general form of the meson kinetic and mass Lagrangian terms is,

$$L_{km} = \frac{1}{2}(\partial_\mu \Phi) K (\partial^\mu \Phi) + \frac{1}{2} \Phi M^2 \Phi , \qquad (1)$$

where Φ stands for the intrinsic field matrix, K and M^2 are the kinetic and mass matrices. Usually, K and M^2 are non-diagonal and L_{km} is bilinear rather than quadratic as invoked from the Klein-Gordon equation for the physical fields Φ_{ph}. By applying three consecutive steps : (i) the diagonalization of the kinetic matrix, (ii) rescaling of the fields, and (iii) the diagonalization of the resulting mass matrix, L_{km} can by reduced into the standard quadratic form in terms of the physical fields.

We first diagonalize K using the transformation $\Phi = \Upsilon \Phi'$. Following this step the Lagrangian becomes,

$$L_{km} = \frac{1}{2}(\partial_\mu \Phi') \hat{K} (\partial^\mu \Phi') + \frac{1}{2} \Phi' \Upsilon^{-1} \hat{M}^2 \Upsilon \Phi' , \qquad (2)$$

with $\hat{K} = diag(\kappa_1, \kappa_2, ...)$. The eigenvalues κ_i ($i = 1, 2, ...$), however, are not necessarily equal 1, and we rescale the fields in order to restore the standard normalization of

CP603, *Mesons and Light Nuclei: 8th Conference*, edited by J. Adam et al.
© 2001 American Institute of Physics 0-7354-0047-4/01/$18.00

245

the kinetic term, i.e. $\Phi' = R\Phi''$, $R = diag(1/\sqrt{\kappa_1}, 1/\sqrt{\kappa_2}, \ldots)$. The kinetic term in terms of the fields Φ'' acquires the standard quadratic form and the mass matrix is $\tilde{M}^2 = R\Upsilon^{-1}\hat{M}^2\Upsilon R$. Now we diagonalize the matrix $\tilde{M}^2$ via transformation $\Phi'' = \Omega\Phi_{ph}$. The mass matrix becomes $M_{ph}^2 = diag(m_1^2, m_2^2, \ldots)$, where the eigenvalues $m_1, m_2, \ldots$ are to be identified with the physical meson masses. Altogether the three steps transform the intrinsic fields Φ into Φ_{ph},

$$\Phi = \Theta\Phi_{ph}; \quad \Theta = \Upsilon R\Omega. \tag{3}$$

Let us consider now the meson decay constants F_{ph} transformation. The usual definition of the decay constants reads,

$$<0|J_{\mu 5}^i|\Phi_{ph}^m> \equiv iF_{ph}^{im}q_\mu, \tag{4}$$

where $J_{\mu 5}^i$ stands for the axial vector currents, and the indices i and m label the intrinsic and physical fields. From the Lagrangian Eq. (1) and by substituting the transformation (3) we may write,

$$J_{\mu 5} = FK\partial_\mu\Phi = FK\Theta\partial_\mu\Phi_{ph}, \tag{5}$$

where $F = diag(F_1, F_2, \ldots)$ is the intrinsic decay constant matrix. Thus, for the matrix of the physical decay constants we have,

$$F_{ph} = FK\Theta = \tilde{F}\Theta, \tag{6}$$

where $\tilde{F} = FK$ is the matrix of renormalized intrinsic decay constants.

Thus the transformation Θ determines the mixing schemes for the particle states and the decay constants.

In case, where the dimensions of the non-diagonal kinetic and mass sub-matrices are respectively, $k \times k$ and $n \times n$, this procedure leads to mixing schemes which involve $[k(k-1)/2] + [n(n-1)/2]$ angles and k field rescaling parameters. This observation holds true irrespective with the type of particle interactions presumed. The commonly used mixing schemes, correspond to a proper choice of the kinetic and mass matrices, and are derived as special cases. In particular, η-η' mixing, requires one angle, if and only if, the kinetic term with the intrinsic fields has a quadratic form. Obviously the parameterization of this transformation depends on the dimensions of non-diagonal sub-matrices of K and M^2.

REFERENCES

1. Groom, D.E., et al. (Particle Data Group), *Eur. Phys. J. C* **15**, 1 (2000).
2. Leutwyler, H., Proc.QCD 97, Montpellier, France, July 1997, Ed. S.Narison, *Nucl. Phys.* **B64**, 223 (1998); hep-ph/9709408.
3. Kaiser, R., and Leutwyler, H., Proc.Workshop "Nonperturbative Methods in Quantum Field Theory", NITP/CSSM, University of Adelaide, Australia, Feb.1998, Eds. A.W.Schreiber, A.G.Williams and A.W Thomas. (World Scientific, Singapore, 1998); hep-ph/9806336.
4. Feldmann, T., Kroll, P., and Stech, B., *Phys. Lett. B* **449**, 339 (1999); hep-ph/9812269; *Phys. Rev. D* **58**, 114006 (1998).
5. Escribano, R., Frère, J.-M., *Phys. Lett. B* **459**, 288 (1999).

On convergence of the χPT HFF expansion for one loop contribution to meson production in NN collisions

E. Gedalin*, A. Moalem* and L. Razdolskaya*

Department of Physics, Ben Gurion University, 84105, Beer Sheva, Israel

Abstract. It is shown that for one loop contributions the heavy fermion formalism expansion corrections for the nucleon propagator produce infinite series of correction terms which are of the same momentum power order.

It is the purpose of the present note to show that within the framework of extremely non-relativistic heavy fermion formalism (HFF χPT) and for sufficiently large momentum transfer processes, the one-to-one correspondence between the perturbative expansion and small momentum expansion is badly destroyed at tree level as well as for one loop contributions. Namely, the HFF expansion corrections for the nucleon propagator form an infinite series of terms all of which having the same momentum power.

To be specific we consider pion production via the $NN \to NN\pi$ reaction. The characteristic four momentum transferred at threshold is $q \approx (-m/2, \sqrt{Mm})$, with M and m being the masses of the nucleon and meson produced. The HFF χPT nucleon propagator (in Park et al. [1] notations)

$$S_N(l) = \frac{1}{vl + i\varepsilon} \sum \left[\frac{T}{vl + i\varepsilon}\right]^n, \qquad (1)$$

where T and v are the nucleon kinetic energy and velocity, respectively, converges only for $T/vl < 1$. For a production process $NN \to NN\pi$, the virtual nucleon a residual momentum $l' = (-m/2, \mathbf{l}')$; with $\mathbf{l}' \cdot \mathbf{l}' = Mm$, so that $T/vl' \approx -1$. Thus the corrections corresponding to each of the terms in Eq. (1) are all of the same magnitude, i.e., the power series for nucleon propagator in the Figure 1b is on the border of its convergence circle and can not be approximated by any finite sum [2, 3, 4]. Consequently, the HFF can not possibly predict the impulse term correctly.

This same conclusion holds for one loop diagrams also. As an example we consider the contribution from the loop diagram Fig. 1a, which is defined by the off mass shell amplitude corresponding to the loop diagram of Fig. 1b. To simplify calculations, we take $\vec{p}_4 = 0$, $Q^2 < 0$, $vQ = 0$, values corresponding to the reaction $NN \to NN\pi$ at threshold. After straightforward though long and tedious calculations, the n-th correction term of the amplitude is (more details may be found in [4]),

$$T_{loop}^{(n)} = \frac{g_A^2 m}{6F^4}\left[3Q^2 - q^2 - k^2 + \frac{m^2}{2}\right] N^\dagger(p_2) N(p_1) \times$$

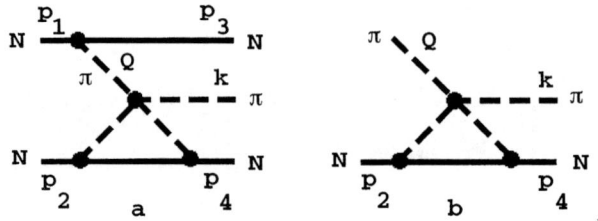

FIGURE 1.

$$\times \frac{i}{16\pi^2} \left[A_1(Q^2/m^2, n) + \frac{-Q^2}{m^2} \left(A_2(Q^2/m^2, n) + A_3(Q^2/m^2, n) \right) \right], \quad (2)$$

At threshold of the production process $-Q^2/m^2 \gg 1$. Then we obtain the estimation

$$A_1 \approx \mathcal{A}_1(n) \left(\frac{Q^2}{mM}\right)^n, \quad A_2 \approx \mathcal{A}_2(n) \left(\frac{m^2}{Q^2}\right)^2 \left(\frac{Q^2}{mM}\right)^n,$$

$$A_3 \approx \mathcal{A}_3(n) \left(\frac{m^2}{Q^2}\right) \left(\frac{Q^2}{mM}\right)^n, \quad (3)$$

where $\mathcal{A}_i(n)$ incorporate dependence on n. By substituting Eqs. (3) into Eq. (2) one finds that the n-th correction term is of order $(Q^2/mM)^n \approx 1$, i.e. at threshold of the pion production process, each correction term $T_{Loop}^{(n)}$ involves a contribution of the same low momentum power order independently of n. Thus for the one loop graph 1b (and consequently for loop graph 1a) the one to one correspondence between small momentum expansion and HFF corrections is broken, thereby violating the fundamental organizing principle of HFF χPT. This excludes the possibility that a finite chiral order of present form HFF χPT calculations can explain meson production in NN collisions.

REFERENCES

1. Park, T.S., Min, D.-P., and Rho, M., *Phys. Rep.* **233**, 341 (1993).
2. Bernard, V., Kaiser, N., and Meissner, U.f-G., *Eur. Phys. J. A* **4**, 259 (1999).
3. Gedalin, E., Moalem, A., and Razdolskaya, L., nucl-th/9812009.
4. Gedalin, E., Moalem, A., and Razdolskaya, L., nucl-th/9906025;hep-ph/0010027.

Compton scattering from the proton at NLO in the chiral expansion

Judith A. McGovern

Theoretical Physics Group, Department of Physics and Astronomy, University of Manchester, Manchester, M13 9PL, U.K.

Abstract. We present calculations of differential cross sections for Compton scattering from the proton, using amplitudes calculated to fourth order in heavy baryon chiral perturbation theory. We compare with available data up to 200 MeV. We find that the agreement for angles up to 90° is good over the whole energy range, but that at more backward angles the agreement decreases above about 100 MeV, and fails completely above the photoproduction threshold.

Heavy baryon chiral perturbation theory (HBCPT) is a systematic framework in which to describe the interactions of nucleons with pions and photons [1]. Respecting gauge and Lorentz invariance, it reproduces low-energy theorems, which have been known since the 50's and 60's, as the first terms in chiral expansions, and in a number of cases also makes further parameter-free predictions. One of the simplest and cleanest processes to which it can be applied, theoretically at least, is Compton scattering, and indeed this was first studied in the very early days of the theory [2].

In a recent paper (ref. [3]) we presented calculations of differential cross sections for Compton scattering from the proton, using amplitudes calculated to fourth order in HBCPT. At this order, the experimental spin-independent polarisabilities α and β are not predicted but the spin-dependent polarisabilities γ_i are [4]. Thus the spin-independent polarisabilities α and β are input paramenters to the calculation, as are the nucleon magnetic moments and three of the the low energy constants c_i from the second order HBCPT Lagrangian.

We compared the predictions of the theory with available data up to 200 MeV [5]. Somewhat surprisingly we found that convergence is good, in that the third and fourth order predictions are quite similar; in general the fourth-order is an improvement. (Third order predictions were given by Babusci *et al.* in ref. [6].) This convergence is something of a surprise, as the convergence for the spin polarisabilities is not good [4]. The agreement for angles up to 90° is good over the whole energy range. However at more backward angles the agreement decreases above about 100 MeV, and fails completely above the photoproduction threshold.

Of course, classic HBCPT has no explicit Δ. This is assumed to have been integrated out, and its legacy is in the LEC's of the nucleon-only theory. It is seen first at different orders in different processes: second order for pion scattering, fourth order for spin-independent and fifth order for spin dependent Compton scattering. It is expected to be important; it may improve the agreement of the spin polarisabilities, but risks spoiling the spin-independent ones. Although in the absence of an explicit Δ agreement must break down at some point, it is not obvious why that point should be so much lower for backward angles.

We explored the sensitivity of our results to the input parameters, and found that it is impossible to reproduce the steep rise above 150 MeV, and even at lower energies agreement at backward angles requires $\alpha = 8$ and $\beta = 6 \times 10^{-4}$ fm^3 (which are totally incompatible with the modern PDG value of 12.1 and 2.1 respectively). The results are fairly insensitive to the values of the spin polarisabilities, and to the second order LECs c_i (taken from fits to pion-scattering data [7]); certainly no reasonable variation of these parameters brings the predictions of HBCPT at backwards angles into agreement with the experimental data.

Since publication of ref. [3] some minor errors have been found which change a number of the graphs slightly; these will be published as errata in due course. The conclusions however are substantially unchanged.

REFERENCES

1. Bernard, V., Kaiser, N., and Meißner, U.-G., *Int. J. Mod. Phys. E* **4**, 193 (1995).
2. Bernard, V., Kaiser, N., Kambor, J., and Meißner, U.-G., *Nucl. Phys.* **B388**, 315 (1992).
3. McGovern, J.A., *Phys. Rev. C* **63**, 064608 (2001).
4. Ji, X., Kao, C-W., and Osborne, J., *Phys. Rev. D* **61**, 074003 (2000);
   Kumar, K.B.V., McGovern, J.A., and Birse, M.C., hep-ph/9909442 (1999);
   Gellas, G.C., Hemmert, T.R., and Meißner, U.-G., *Phys. Rev. Lett.* **85**, 14 (2000);
   Kumar, K.B.V., McGovern, J.A., and Birse, M.C., *Phys. Lett. B* **479**, 167 (2000).
5. Baranov, P.S., *et al.*, *Sov. J. Nucl. Phys.* **21**, 355 (1975);
   Ziegler, A., *et al.*, *Phys. Lett. B* **278**, 34 (1992);
   Federspiel, F.J., *et al.*, *Phys. Rev. Lett.* **67**, 1511 (1991);
   Hallin, E.L., *et al.*, *Phys. Rev. C* **48**, 1497 (1993);
   MacGibbon, B.E., *et al.*, *Phys. Rev. C* **52**, 2097 (1995).
6. Babusci, D., Giordano, G., and Mantone, G., *Phys. Rev. C* **55**, R1645 (1997).
7. Fettes, N., and Meißner, U.-G., *Nucl. Phys.* **A676**, 311 (2000).

DIRAC experiment: Measurement of pionium lifetime

Jan Smolík
for DIRAC collaboration

Institute of Physics ASCR, Prague, Czech Republic

Abstract. The main goal of the Dirac Experiment at CERN is a precise determination of the S-wave $\pi\pi$ scattering lengths from the measurement of the ground state lifetime of $\pi^+\pi^-$ atoms. This measurement will represent a crucial test of low-energy predictions of Chiral Perturbation Theory. The method of lifetime measurement is described and demonstrated in a preliminary analysis of the data collected in 1999.

PHYSICS MOTIVATION

The pionium is an electromagnetically bound $\pi^+\pi^-$ system ($\pi^+\pi^-$ atom) with the ground state binding energy $E_b = 1.86$ keV – corresponding to Bohr radius $a = 387$ fm or Bohr momentum $p_b = 0.51$ MeV/c. The dominant decay is into $\pi^0\pi^0$ (99.6%). It is mediated by the strong interaction process $\pi^+\pi^- \to \pi^0\pi^0$. Therefore, the inverse pionium lifetime is proportional to the square of the two–pion strong interaction amplitude.

The features of hadronic atoms were studied already in sixtees [3]. The leading order formula for the pionium lifetime reads [4]

$$\frac{1}{\tau} = \Gamma^{LO}_{A_{\pi^+\pi^-}} = \frac{2}{9}\alpha^3 p^*(a_0 - a_2)^2 ,$$

where α is fine-structure constant ($\sim 1/137$), $p^* = (M^2_{\pi^+} - M^2_{\pi^0} - \frac{1}{4}\alpha^2 M^2_{\pi^+})^{1/2}$ is the π^0 momentum in the pionium rest frame, a_0 and a_2 are the isoscalar and isotensor S-wave $\pi\pi$ scattering lengths. The recent prediction of Standard Chiral Perturbation Theory (SχPT) gives $a_0 - a_2 = 0.265 \pm 0.004$ and $\tau = (2.90 \pm 0.10) \cdot 10^{-15}s$ [5].

The expected 10% precision in lifetime measurement allows to determine the value of $|a_0 - a_2|$ with 5% accuracy. This represents improvement in the present experimental error of about a factor four [1] allowing a crucial test of SχPT.

---

[1] The account of recent data on semileptonic kaon decays from experiment E865 at BNL can improve this error by a factor of two [6].

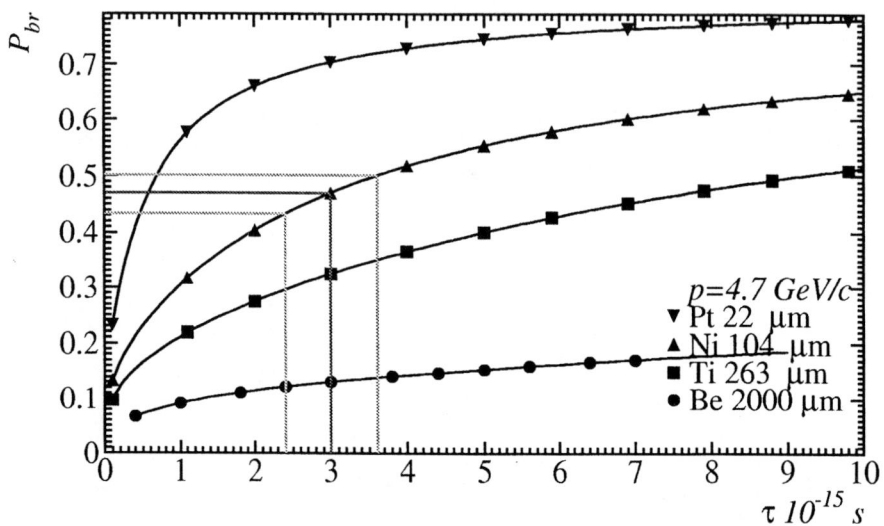

FIGURE 1. Probability of pionium breakup in the target as a function of lifetime.

THE METHOD OF MEASUREMENT

Due to the very short pionium lifetime an indirect method of measurement is used, based on the determination of the pionium breakup probability P_{br}.

There is a small but finite probability that two oppositely charged pions produced in an interaction will create the bound state - pionium. After its creation pionium flies through the target and can either decay into $\pi^0\pi^0$ or interact with atoms of the target and thus be excited or broken-up. The product of this breakup is a $\pi^+\pi^-$ pair with very small relative momentum ($Q < 3$ MeV/c).

The probability P_{br}, defined as a ration of the numbers of broken-up (n_A) and produced (N_A) pioniums, can be calculated for a given target material and thickness as a function of pionium momentum and lifetime (Fig. 1) [1]. The measurement of the probability P_{br} thus allows to determine the pionium lifetime.

Due to the very distinct features of $\pi^+\pi^-$ pairs from broken-up pioniums ($Q < 3$ MeV/c, opening angle < 3 mrad), the n_A can be measured from the excess of $\pi^+\pi^-$ pairs at very small relative momentum. As for N_A, it can be calculated using the relation between the two–pion production in continuous (free pions) and discrete (bound pions) spectrum based on the theory of final state interaction.

DETECTOR SETUP

The DIRAC detector setup represents a double arm spectrometer with magnet of 2.3 Tm bending power. It is designed to detect $\pi^+\pi^-$ pairs with very small opening angles and measure their relative momentum with very high precision (~ 1 MeV/c). It is located

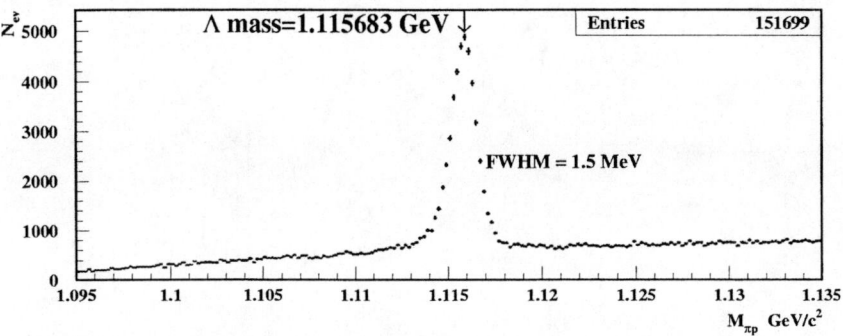

FIGURE 2. Invariant mass spectrum of $p\pi^-$ pairs.

in PS East Hall at CERN and uses high intensity ($\sim 10^{11} p/s$) PS proton beam with the energy of 24 GeV. The data collection has started in 1999.

The part upstream to the magnet consists of microstrip gas chamber, scintillating fiber detector and scintillating ionization hodoscope. In each downstream arm there are drift chambers, vertical and horizontal scintillating hodoscopes, Čerenkov counter, preshower and muon detectors.

The resolution in relative momentum was tested with the help of lambda decays. The measured invariant mass of $p\pi^-$ pairs is shown in Fig. 2. The width of the lambda peak is in agreement with the expected resolution in the relative momentum.

FIRST RESULTS

The preliminary results of the analysis of 1999 data are shown in Fig 3. The left panel shows the measured $\pi^+\pi^-$ correlation function calculated as a ratio of the relative momentum spectrum of the true pairs (correlated in time) to that of the accidental pairs (with a large time difference). The parameterization of this function in the absence of atomic pairs is known [1]. Fitting the correlation function in the region of $Q > 5$ MeV/c, free of the contribution of atomic pairs, one can estimate this contribution from the excess of the measured number of $\pi^+\pi^-$ pairs above the fitted one in the low-Q region. This excess is plotted in the right panel. It corresponds to about 234 pion pairs from the pionium breakups.

SUMMARY AND OUTLOOK

The experiment is presently running with the complete set of the detectors. The preliminary analysis of 1999 data demonstrates the applicability of the chosen method of

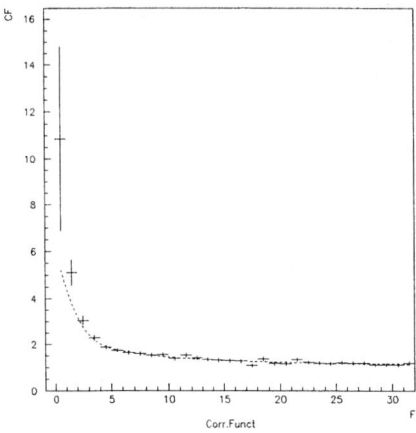 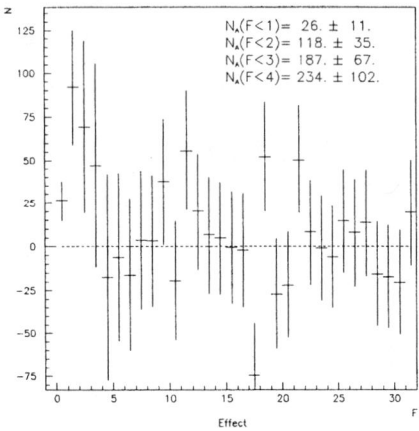

FIGURE 3. Measured (points) and fitted (curve) correlation functions (LEFT) and the excess in the number of $\pi^+\pi^-$ pairs due to pionium breakups (RIGHT) from 1999 data. The variable $F = \sqrt{(Q_x/\sigma_x)^2 + (Q_y/\sigma_y)^2 + (Q_z/\sigma_z)^2}$ is slightly higher than the relative momentum Q in MeV/c since the errors σ_i in the components of the vector **Q** are somewhat less then 1 MeV/c.

measurement. During year 2000 there were collected about 2000 atomic breakups. This year we plan to collect the statistics enough for the determination of the pionium lifetime with the precision of 15%. The goal of the experiment, the 10% precision, should be reached in 2002.

A further step in the physics of hadronic atoms can be done using a somewhat upgraded DIRAC setup to measure the lifetime of πK atoms [2].

REFERENCES

1. Adeva, B., *et al.*, Proposal to the SPSC, CERN/SPSC 95-1, SPSLC/P 284, (1994).
2. Adeva, B., *et al.*, DIRAC Addendum, CERN/SPSC 2000-032, SPSC/P284 Add. 2, (2000)
3. Deser, S., *et al.*, *Phys. Rev.* **96**, 774 (1954).
4. Uretsky, J.L., and Palfrey, T.R., Jr., *Phys. Rev.* **121**, 1798 (1961).
5. Gasser, J., *et al.*, Preprint hep-ph/0103157, (2001).
6. Pislak, S., *et al.*, Preprint hep-ex/0106071, (2001).

FEW BODY SYSTEMS AND RELATIVISTIC TECHNIQUES

Relativistic covariant model for proton-proton bremsstrahlung

M.D. Cozma*, O. Scholten*, R.G.E. Timmermans* and J.A. Tjon[†]

KVI, University of Groningen, Zernikelaan 25, 9747 AA Groningen, The Netherlands
[†]*Institute for Theoretical Physics, University of Utrecht, Utrecht, The Netherlands*

Abstract. We compare a relativistic covariant model for proton-proton bremsstrahlung with new high-quality data from KVI. The agreement in large parts of phase space is satisfactory. However, remarkably large discrepancies are observed for specific kinematic regions. These failures are shown to occur primarily when the final two-nucleon system has energies less than about 15 MeV.

Introduction

The proton-proton bremsstrahlung ($pp\gamma$) process has attracted significant attention over the years, both experimentally and theoretically. In recent years, several new experiments have been performed [1, 2, 3, 4, 5], inspiring many new theoretical investigations [6, 7, 8, 9, 10, 11, 12]. In particular, a number of microscopic models have been developed to describe the $pp\gamma$ process. Examples are the potential model of Nakayama et al. [6] and the covariant model of Martinus et al. [8]. The theoretical predictions of these models could be compared to the $pp\gamma$ cross sections and analyzing powers for the TRIUMF experiment at 280 MeV [1]. The agreement of theory with these TRIUMF data is rather good, especially for the cross sections, provided that the experimental cross sections are renormalized [1]. Some outstanding discrepancies occur, however, for certain asymmetric proton angles.

More recently, high-precision data from the KVI experiment at 190 MeV became available [5]. When comparing these data with theory, a pronounced and undisputable discrepancy between theory and experiment was observed in specific kinematic regions [5]. The size of the discrepancy between theory and experiment is disturbing, since what primarily enters are the two-nucleon (NN) interaction and the electromagnetic coupling of the photon to the NN system, both of which are believed to be known accurately at this energy. The high precision of the new KVI data allows in principle the study of smaller effects, like those arising from negative-energy states, the Δ-isobar, and meson-exchange currents. It is therefore important to identify the possible reasons for the discrepancies.

We compare here the covariant model of Ref. [8] with the KVI data available so far and analyze the discrepancies. We demonstrate that the dominant contribution to $pp\gamma$ for the specific problematic kinematic regions results when the NN interaction is evaluated at energies below about 15 MeV, and that, in fact, at least a major part of the problem resides in the low-energy behavior of the NN interaction models used.

Covariant model for bremsstrahlung

In this model the NN T matrix is obtained by solving the Bethe Salpeter equation for the two-nucleon system in the equal time approximation [18] with the one boson exchange kernel of Ref. [13]. At the time of construction the on-shell T matrix was fitted to the Virginia Tech phase shift solution [15] by adjusting the meson-nucleon coupling constants and the form factor parameters.

This covariant NN T-matrix enters the model for $pp\gamma$, in which a number of contributions can be distinguished. The most important ones in the energy regime we consider here are the "nucleonic" contributions, consisting of single-scattering terms, *i.e.* photon emission off the external proton legs, and the contribution commonly known as rescattering. The model is relativistic covariant and therefore negative-energy states are included in a natural way. The relevance of these negative-energy states is small for energies around 200 MeV. The reason [8] is that the contributions from the single-scattering diagrams where the intermediate nucleons are in a negative-energy state are canceled by similar contributions coming from the rescattering diagram. This cancellation holds for terms up to order $O(q)$, where q is the photon momentum, and is a consequence of the soft-photon theorem [19].

Contributions from the Δ-isobar and from magnetic meson-exchange currents, containing in particular the $\omega\pi\gamma$ and $\rho\pi\gamma$ decay graphs, are also taken into account. These two-body current terms are included in a perturbative way, since they are small in general. The coupling constants of the photon to the various mesons were determined from the radiative decay widths of the vector mesons. As one can observe from Fig. 1, at an energy of 190 MeV the contribution of the two-body currents is small. These terms, however, increase in size with energy and can become appreciable around the pion-production threshold at 280 MeV and above.

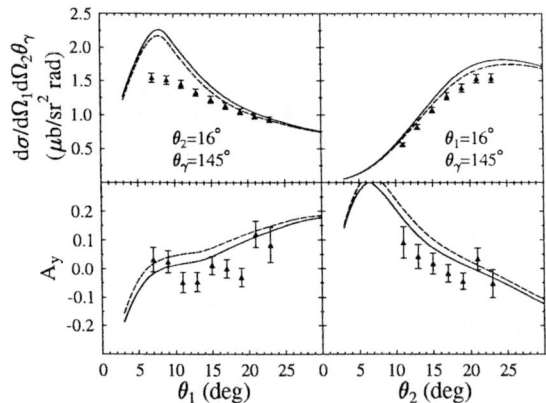

FIGURE 1. Bremsstrahlung cross sections (upper panels) and analyzing powers (lower panels) at 190 MeV incoming proton energy, and for $\theta_2=16°$, $\theta_\gamma=145°$ and $\theta_1=16°$, $\theta_\gamma=145°$ respectively. The solid curves show the results of the full model, including negative-energy states and two-body currents, while the nucleonic contribution is shown by the dashed lines. The experimental data are from the KVI experiment [5].

This covariant bremsstrahlung model [8] is theoretically well founded and many of its ingredients have been tested in other calculations such as those for electron scattering on the deuteron [14]. Therefore, one did not expect *major* discrepancies with new experimental $pp\gamma$ data at energies below 280 MeV.

Comparison with the KVI data

In the KVI experiment, $pp\gamma$ cross sections and analyzing powers were measured for 190 MeV incoming proton energy, with the scattered protons detected at small forward angles, and with the photon emitted in the backward hemisphere [5]. A typical example of the KVI data in comparison with theory is shown is Fig. 1. The data are plotted for two different kinematic situations, with θ_γ and either θ_1 or θ_2 fixed (θ_i the angle of the outgoing proton i in the laboratory frame). The theoretical predictions are from the model as published in Ref. [8].

For the angles $\theta_2 = 16°$, $\theta_\gamma = 145°$ (upper left panel) the cross section shows a large discrepancy between theory and experiment for kinematics that correspond to the backward peak in the cross section. For $\theta_1 = 16°$, $\theta_\gamma = 145°$ (upper right panel), on the other hand, the cross section shows a much better agreement between theory and data. The contribution of the two-body currents is seen to be minor, and thus the discrepancy for $\theta_2 = 16°$, $\theta_\gamma = 145°$ is unlikely to come from this source. In the following, we focus therefore on the nucleonic contribution as the cause of the problem.

To reveal the source of the problem we have studied the kinematical details of the two above mentioned cases. Namely, we have determined the energy (kinetic energy in the laboratory system) T_{lab} at which the elastic T matrix, which enters in the bremsstrahlung amplitude, is evaluated. The results are plotted in Fig. 2 (left panel). It is seen that discrepancies appear for situations in which the elastic T matrix is evaluated at very low energies, namely below 20 MeV. For such kinematical cases it is well known that the dominant contribution to the proton-proton potential is given by the $^1S_0$ partial wave.

To confirm the above result that the observed discrepancy is mainly due to an inaccurate description of the $^1S_0$ partial wave, we have determined the separate contributions of various partial waves to the differential cross sections at the position of the peak. Terms in which the elastic T matrix enters evaluated at low energies can appear in two distinct situations: in a single scattering process (when first the photon is emitted and then the strong interaction takes place) and in a rescattering diagram (where after the emission of the photon the low energy protons interact strongly). For the $\theta_2 = 16°$, $\theta_\gamma = 145°$ case the $^1S_0$ partial wave gives the dominant contribution for single scattering diagrams. P waves ($^3P_1$ and $^3P_2$) are dominant for the second kinematics, since here the energy at which the elastic T matrix never goes below 25 MeV. The good agreement between theory and experiment for the latter case, suggests that most of the observed discrepancy is due to the poor description of the $^1S_0$ wave at low energies.

At the time of its construction, the OBE model of Ref. [14] was fitted to the np phase-shift parameters of the Virginia Tech group, where the interest of the fit was at the higher energies. The description of the $pp\gamma$ reaction requires, in principle, a pp potential. The main difference between the np and pp potential $^1S_0$ phase shifts lies at low energies (10

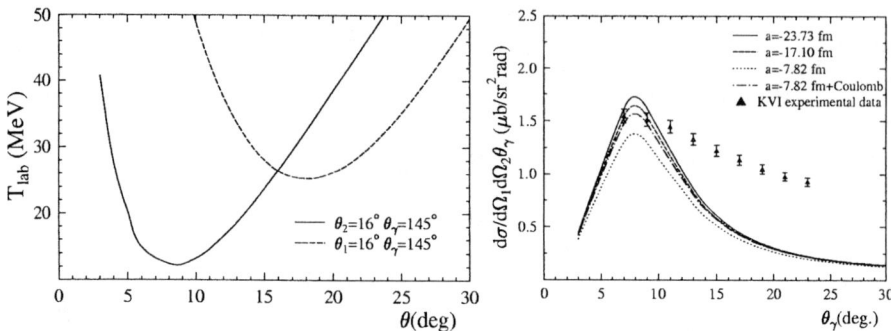

FIGURE 2. Left panel: kinetic energy of the incoming proton at which the *NN* T-matrix is evaluated for the two kinematics discussed in the text and in Fig. 2. Right panel: bremsstrahlung cross sections at 190 MeV for three different scattering lengths for $\theta_2 = 16°$, $\theta_\gamma = 145°$. Only the contribution from the $^1S_0$ partial wave is included (see text).

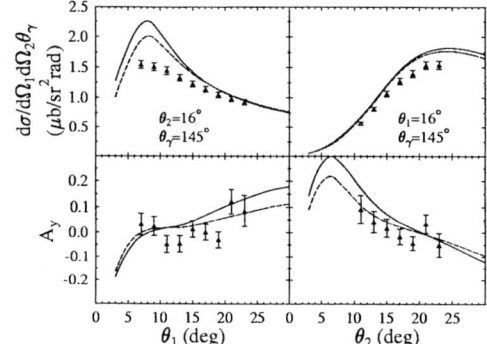

FIGURE 3. Bremsstrahlung cross sections and analyzing powers at 190 MeV, for the same kinematics as in Fig. 2. The solid (dashed) curves show the results before (after) the refit of the *NN* model.

MeV). At such low energies also the Coulomb interaction has to be taken into account. In view of our findings, one will thus expect a high sensitivity of the *pp*γ cross section on the low energy *NN* interaction.

A toy model for the *NN* interaction has been developed, by considering only contributions from the $^1S_0$ partial wave. The bremsstrahlung amplitude was also constructed (for further information see [17]). It was used to study the sensitivity of the bremsstrahlung cross section to low energy *NN* interaction for the $^1S_0$ dominated kinematics. The result for $\theta_2 = 16°$, $\theta_\gamma = 145°$ is plotted in the right panel of Figure 2. Results are plotted for a few values of the scattering length: $a = -23.73$fm (*np*), $a = -17.10$fm (*nn*) and $a = -7.82$fm (*pp*). These three calculations were performed using an elastic T matrix which would correspond to a pure strong potential. The *NN* potential was fitted to reproduce the above mentioned values for a. A pronounced sensitivity on this low energy parameter is observed. The inclusion of the Coulomb interaction (dash-dotted line) is seen to affect the *pp*γ cross section by amount less than 10 percent.

Final remarks and conclusions

In the previous sections we have identified the source of the discrepancy to reside in a poor description of the NN potential at low energies. We have also proven the sensitivity of the $pp\gamma$ cross section on the low energy NN interaction and have argued that the Coulomb interaction could affect the final results by less than 10 percent. To account for these effects we have performed a refit of the NN potential with an emphasis of obtaining better agreement with the Nijmegen pp partial-wave analysis of Ref. [16], at energies starting from 10 MeV up to 215 MeV.

In Fig. 3 we present the same calculation as in Fig. 1, but now for the refitted NN model. The description of the $pp\gamma$ cross sections in the region of the peak for $\theta_2 = 16°$, $\theta_\gamma = 145°$ has improved. This is mainly due to the improvement in the description of the $^1S_0$ wave. At the same time, the situation for $\theta_1 = 16°$, $\theta_\gamma = 145°$ has improved. This is due to the fact that also the description of the ^3P partial waves has improved. For a comparison between old and new phase shifts see [17]. In all cases, the analyzing powers are affected less by changes in the interaction.

We have also investigated other regions of the phase space which pose problems of the same type we encountered for the $\theta_2 = 16°$, $\theta_\gamma = 145°$. The descriptions in these regions is also improved once the OBE model of Ref. [14] is refitted with an emphasis on the low energy.

REFERENCES

1. Michaelian, K., *et al.*, *Phys. Rev. D* **41**, 2689 (1990).
2. Bilger, R., *et al.*, *Phys. Lett. B* **429**, 195 (1998).
3. Zlomanczuk, J., and Johansson, A., *Nucl. Phys.* **A631**, 622c (1998).
4. Yasuda, K., *et al.*, *Phys. Rev. Lett.* **82**, 4775 (1999).
5. Huisman, H., *et al.*, *Phys. Rev. Lett.* **83**, 4017 (1999); *Phys. Lett. B* **476**, 9 (2000).
6. Herrmann, V., and Nakayama, K., *Phys. Rev. C* **46**, 2199 (1992);
    de Jong, F., *et al.*, *Phys. Lett. B* **333**, 1 (1994);
    de Jong, F., Nakayama, K., and Lee, T.S.H., *Phys. Rev. C* **51**, 2334 (1995).
7. Eden, J.A., and Gari, M., *Phys. Lett. B* **347**, 187 (1995); *Phys. Rev. C* **53**, 1102 (1996).
8. Martinus, G.H., Scholten, O., and Tjon, J.A., *Phys. Lett. B* **402**, 7 (1997); *Phys. Rev. C* **56**, 2945 (1997); *ibid.* **58**, 686 (1998); *Few Body Systems* **26**, 197 (1999).
9. Liou, M.K., Timmermans. R., and Gibson, B.F., *Phys. Lett. B* **345**, 372 (1995); *ibid.* **355**, 606(E) (1995); *Phys. Rev. C* **54**, 1574 (1996).
10. Korchin, A.Yu., and Scholten, O., *Nucl. Phys.* **A581**, 493 (1995);
    Korchin, A.Yu., Scholten, O., and van Neck, D., *Nucl. Phys.* **A602**, 423 (1996).
11. Li, Yi, Liou, M.K., Timmermans, R., and Gibson, B.F., *Phys. Rev. C* **58**, R1880 (1998).
12. Kondratyuk, S., Martinus, G., and Scholten, O., *Phys. Lett. B* **418**, 20 (1998).
13. Fleischer, J., Tjon, J., *Nucl. Phys.* **B84**, 375 (1974); *Phys. Rev. D* **15**, 2537 (1977); *ibid.* **21**, 87 (1990).
14. Hummel, E. and Tjon, J.A., *Phys. Rev. C* **42**, 423 (1990); *ibid.* **49**, 21 (1994).
15. Arndt R.A., Hyslop III, J.S., and Roper, L.D., *Phys. Rev. D* **35**, 128 (1987).
16. Stoks, V.G.J., *et al.*, *Phys. Rev. C* **48**, 792 (1993);
    Rentmeester, M.C.M., *et al.*, *Phys. Rev. Lett.* **82**, 4992 (1999).
17. Cozma, M.D., Scholten, O., Timmermans, R.G.E., and Tjon, J.A., to be published.
18. Logunov, A., and Tavkhelidze, A., *Nouvo Cim.* **29**, 380 (1963);
    Blankenbecler, R., and Sugar, R., *Phys. Rev.* **142**, 1051 (1966).
19. Low, F.E., *Phys. Rev.* **110**, 974 (1958).

Quark model study of the $NN^*(1440)$ components of the deuteron

B. Juliá-Díaz*, D.R. Entem†, F. Fernández* and A. Valcarce*

*Grupo de Física Nuclear, Universidad de Salamanca, E- 37008 Salamanca, Spain.
†Department of Physics, University of Idaho, Moscow, Idaho 83844

Abstract. We present a calculation of the deuteron wave function including NN, $\Delta\Delta$, and $NN^*(1440)$ channels. All the transition potentials, as well as the direct NN potential, have been obtained from the same underlying quark model using standard techniques available in the literature. The calculated weights for the different channels are in agreement with previous theoretical estimations.

An important effort has been devoted during the last decade to the quantitative understanding of the NN interaction in constituent quark models [1, 2, 3, 4, 5, 6]. These models consider hadrons as clusters of confined non-relativistic constituent quarks interacting through forces related to QCD. The original one-gluon exchange (OGE), which provided the first explanation for the repulsive core of the NN interaction, was supplemented by Goldstone-boson exchanges at the level of quarks when it was realized that the constituent quark mass is a signal of the spontaneous chiral symmetry breaking (χSB) producing fully quark-model based nucleon-nucleon potentials.

Constituent quark models are ideally suited to study the influence of baryon resonances on NN phenomenology, because both, nucleon and baryon resonances, are described within the same scheme, the parameters involved in the calculations being fixed from the NN sector where a huge amount of data are available.

In the early days of the NN quark-model calculations it was reasonably established [2, 3] that the coupling of NN, $\Delta\Delta$ and hidden color components through the central part of the quark-quark interaction provided almost negligible contributions. However, it was soon demonstrated [5] that the tensor coupling to $N\Delta$ configurations results of tremendous importance in order to reproduce the low-partial waves experimental data. It arises therefore the question of the role of nuclear resonances in the two nucleon phenomenology.

This problem has already been undertaken by Maeda et al. [7]. They have studied S-wave scattering using the resonating group method (RGM) including not only N's and Δ's, but also their excited states, being the effects of other resonances smaller. However, due to the naive model they use, they are not able to reproduce the deuteron binding energy and thus, their conclusions should be taken with care.

BARYON-BARYON INTERACTIONS

In order to asses the effects of the resonances we require to start from a quark model that gives accurate results for the phase-shift and bound state properties of the NN system. The model of Ref. [4, 6] fulfills the above requirements and simultaneously gives a correct description of the low-lying baryon spectrum [8]. In this model the primary ingredients of the quark-quark interaction are the confining potential and the one-gluon exchange term for the long- and short-range part of the interaction.

In the intermediate region between χSB scale (~ 1 GeV) and the confinement scale (~ 0.2 GeV) QCD is formulated in terms of an effective theory of constituent quarks interacting through the Goldstone modes associated with χSB with an effective lagrangian of the form:

$$L_{ch} = g_{ch}F(q^2)\bar{\Psi}(\sigma + i\gamma_5 \vec{\tau} \cdot \vec{\pi})\Psi, \qquad (1)$$

where $F(q^2)$ is a monopole form factor

$$F(q^2) = \left[\frac{\Lambda_\chi^2}{\Lambda_\chi^2 + q^2}\right]^{\frac{1}{2}}. \qquad (2)$$

As the confining force does not contribute to the baryon-baryon interaction we do not consider it henceforth.

The form of the interaction derived from Eq. (1) after a non-relativistic reduction is [6]:

$$V_{ij}^{PS}(\vec{q}) = -\frac{g_{ch}^2}{4m_q^2}\frac{\Lambda_\chi^2}{\Lambda_\chi^2 + q^2}\frac{(\vec{\sigma}_i \cdot \vec{q})(\vec{\sigma}_j \cdot \vec{q})}{m_{PS}^2 + q^2}(\vec{\tau}_i \cdot \vec{\tau}_j), \qquad (3)$$

$$V_{ij}^{S}(\vec{q}) = -g_{ch}^2\frac{\Lambda_\chi^2}{\Lambda_\chi^2 + q^2}\frac{1}{m_S^2 + q^2}. \qquad (4)$$

and the OGE potential [9] is,

$$V_{ij}^{OGE}(\vec{q}) = \alpha_s(\vec{\lambda}_i \cdot \vec{\lambda}_j)\left\{\frac{\pi}{q^2} - \frac{\pi}{4m_q^2}\left[1 + \frac{2}{3}(\vec{\sigma}_i \cdot \vec{\sigma}_j)\right] + \frac{\pi}{4m_q^2}\frac{[\vec{q} \otimes \vec{q}]^{(2)} \cdot [\vec{\sigma}_i \otimes \vec{\sigma}_j]^{(2)}}{q^2}\right\}, \qquad (5)$$

where the λ's are the color Gell-Mann matrices and α_s is the strong coupling constant. The quark model parameters are those of Ref. [6].

From the constituent quark-model outlined above the NN as well as the $\Delta\Delta$ potentials have been calculated through the RGM using the approach explained in detail in [6]. Due to the more involved structure of the N^* wave function we use the Born-Oppenheimer (BO) approximation for the derivation of the NN^* interaction. The procedure followed for the calculation of the diagonal part of the interaction is described in [10], where the direct NN^* potential is calculated and some discussion about the properties of the partial waves is done. For the purpose of our calculation the non-diagonal terms of the NN^* potential have been obtained in the same BO approximation. The definition of the non-diagonal terms is:

$$V_{NN(LST)\to NN^*(L'S'T)}(R) = \xi_{LST}^{L'S'T}(R) - \xi_{LST}^{L'S'T}(\infty), \qquad (6)$$

where

$$\xi_{LST}^{L'S'T}(R) = \frac{\left\langle \Psi_{NN}^{L'S'T}(\vec{R}) \mid \Sigma_{i<j=1}^{6} V_{qq}(\vec{r}_{ij}) \mid \Psi_{NN^*}^{LST}(\vec{R}) \right\rangle}{\sqrt{\left\langle \Psi_{NN^*}^{L'S'T}(\vec{R}) \mid \Psi_{NN^*}^{L'S'T}(\vec{R}) \right\rangle} \sqrt{\left\langle \Psi_{NN}^{LST}(\vec{R}) \mid \Psi_{NN}^{LST}(\vec{R}) \right\rangle}} . \tag{7}$$

It is interesting to emphasize at this point that the BO approximation has been proved to give results of comparable quality to those obtained through the RGM formalism. For the sake of clarity it can be seen that the on-shell properties for the low partial waves obtained by means of the BO method are of the same quality (within 3% for energies below 300 MeV) as those obtained, using the same quark model Hamiltonian, making use of the RGM, (see Fig. 1 of Ref. [11]).

DEUTERON CALCULATION AND RESULTS

To calculate the deuteron binding energy, E_B, and the deuteron wave function we use a method based on a discretization of the integral Schrödinger equation in momentum space. The eigenvalue problem can be written, in a simplified notation:

$$\sum_j [E_i(p_i)\delta_{ij} + V_{ij} - E\delta_{ij}]\Psi_j = 0 \tag{8}$$

where i and j run for all the points of the discretized space and also for the different channels included in the calculation. E stands for the energy of the system referred to the NN threshold and $E_i(p_i) = \frac{p^2}{2\mu_i} + \Delta M_i$ where ΔM_i is the mass difference between channel i and the NN system. For the above system to have solutions different from the trivial one, the following condition has to be fulfilled,

$$|E_i(p_i)\delta_{ij} + V_{ij} - E\delta_{ij}| = 0 \tag{9}$$

the values E that satisfy Eq. (9) are the energies of the bound states of the system. Once the energies have been found the calculation of the wave function can be easily done just by solving the linear problem of Eq. (8). The whole problem has been solved on a Gauss-Legendre mesh of 48 points per channel that gives already stable results.

We include in the calculation those channels which can couple to the quantum numbers of the deuteron, $^3S_1^{NN} - {}^3D_1^{NN}, {}^3S_1^{NN*} - {}^3D_1^{NN*}, {}^3S_1^{\Delta\Delta} - {}^3D_1^{\Delta\Delta} - {}^7D_1^{\Delta\Delta} - {}^7G_1^{\Delta\Delta}$. Note that the first relevant channel, from an strictly energetic point of view, is the NN^* channel as the $N\Delta$ system cannot couple to the deuteron. The most important feature of the quark model interaction that is relevant here is the tensor part of the pseudo-scalar exchange that is responsible for all the non-diagonal parts of the baryonic interactions used in the calculation. In table 1 we show the results obtained in our calculation. The deuteron binding energy is fixed to -2.2246 MeV for all calculations by fine tunning the quark model parameters. The first thing to be outlined is that the probabilities of the NN^* channels are indeed small as compared to the $^3S_1^{\Delta\Delta}$ and $^7D_1^{\Delta\Delta}$, in agreement with the estimation of Glozman et al. [12], where they studied the spectroscopic factor for NN^* and

TABLE 1. Deuteron wave function (%)

NN		NN*		ΔΔ						
$^3S_1$	$^3D_1$	$^3S_1$	$^3D_1$	$^3S_1$	$^3D_1$	$^7D_1$	$^7G_1$	r_m(fm)	A_S(fm$^{-1/2}$)	η
95.38	4.62	-	-	-	-	-	-	1.976	0.8895	0.0251
95.20	4.56	-	-	0.11	0.0035	0.12	0.0063	1.985	0.8941	0.0250
95.17	4.53	0.027	0.024	0.13	0.0036	0.12	0.0062	1.986	0.8944	0.0250

ΔΔ channels finding an upper estimate of $<10^{-2}$ and $<10^{-3}$, respectively. The prediction we obtain is, however, a factor of 10 lower than that of Rost [13] where he found an estimate of 0.16% for the NN* channels in a meson-exchange calculation. At the same time it can be seen that the probability for the Roper channels is a factor of 10 bigger than the contributions from the $^3D_1^{\Delta\Delta}$ and $^7G_1^{\Delta\Delta}$. The inclusion of the Roper channels results on a decrease of the probabilities of the NN channels and a slight enhancement of the $^3S_1^{\Delta\Delta}$ channel. Concerning the other deuteron properties, the main changes are due to the inclusion of the Δ channels, the Roper contribution is almost negligible.

REFERENCES

1. Shimizu, K., *Rep. Prog. Phys.* **52**, 1 (1989).
2. Oka, M., Yazaki, K., *Phys. Lett. B* **90**, 41 (1980).
3. Faessler, A., Fernández, F., Lübeck, G., and Shimizu, K., *Phys. Lett. B* **112**, 201 (1982).
4. Fernández, F., Valcarce, A., Straub, U., Faessler, A., *J. Phys. G* **19**, 2013 (1993).
5. Valcarce, A., Faessler, A., Fernández, F., *Phys. Lett. B* **256**, 367 (1995).
6. Entem, D.R., Fernández, F., Valcarce, A., *Phys. Rev. C* **62**, 034002 (2000).
7. Maeda, I., Arima, M., and Masutami, K., *Phys. Lett. B* **474**, 255 (2000).
8. Garcilazo, H., Valcarce, A., and Fernández, F., *Phys. Rev. C* **63**, 035207 (2001); Garcilazo H., Valcarce, A., and Fernández, F., submitted to *Phys. Rev. C*.
9. de Rújula, A., Georgi, H., Glashow, S.L., *Phys. Rev. D* **12**, 147 (1975).
10. Juliá-Díaz, B., Valcarce, A., González, P., and Fernández, F., *Phys. Rev. C* **63**, 024006 (2001).
11. Juliá-Díaz, B., Fernández, F., Valcarce, A., and Haidenbauer, J., in these proceedings.
12. Glozman, L.Ya., and Kuchina, E.I., *Phys. Rev. C* **49**, 1149 (1994).
13. Rost, E., *Nucl. Phys.* **A249**, 510 (1975).

Quark model study of the triton bound state

B. Juliá-Díaz*, F. Fernández*, A. Valcarce* and J. Haidenbauer[†]

*Grupo de Física Nuclear, Universidad de Salamanca, E- 37008 Salamanca, Spain.
[†]Institut für Kernphysik (Theorie), Forschungszentrum Jülich, D-52425 Jülich, Germany.

Abstract. The three-nucleon bound state problem is studied employing nucleon-nucleon potentials derived from a basic quark-quark interaction. We analyze the effects of the nonlocalities generated by the quark model. The calculated triton binding energies indicate that quark-model nonlocalities can yield additional binding in the order of few hundred keV.

During the last decade the development of quark-model based interactions for the hadronic force has led to nucleon-nucleon (NN) potentials that provide a fairly reliable description of the on-shell data. As a consequence of the internal structure of the nucleon, such interaction models are characterized by the presence of nonlocalities. These nonlocalities are reflected in the off-shell properties and emerge from the underlying dynamics in a natural way.

The relevance and/or necessity of considering the nonlocal parts of nucleon-nucleon potentials in realistic interactions is still under debate. Indeed, over the past few years several studies have appeared in the literature which stress the potential importance of nonlocal effects for the quantitative understanding of few-body observables and, specifically, for the triton binding energy [1, 2, 3, 4, 5]. However, the majority of these investigations [1, 2, 3, 4] explore only nonlocalities arising from the meson-exchange picture of the NN interaction. The effects of nonlocalities resulting from the quark substructure of the nucleon have only been addressed once so far [5] and more systematic studies are lacking altogether.

In this work we study the triton binding energy by means of a local and a nonlocal potential, derived from the same constituent quark-model. We will pay special attention to the nonlocal effects originating from the quark model. The nonlocal potential is derived by means of the resonating group method (RGM) and the local one is obtained through the Born-Oppenheimer approximation. These interactions are employed in Faddeev calculations of the three-nucleon binding energy.

QUARK-MODEL BASED NN POTENTIALS

The underlying idea of the quark model we use is that the constituent quark mass is a consequence of the spontaneous chiral symmetry breaking (SCSB). Then, between the chiral symmetry breaking scale ($\Lambda_{CSB} \sim 1$ GeV) and the confinement scale, ($\Lambda_C \sim 0.2$ GeV) QCD may be simulated in terms of an effective theory of constituent quarks interacting through the Goldstone modes associated with the SCSB. Perturbative features

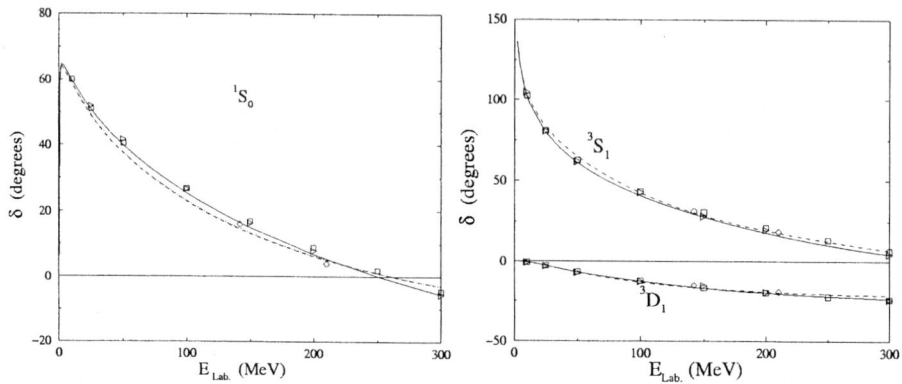

FIGURE 1. NN phase shifts. The solid line stands for the nonlocal potential, the dashed line corresponds to the local one. The squares, diamonds and triangles are the experimental data taken from Refs. [10]. The dotted line shows the result of the EST separable representation of the local model.

of QCD are incorporated through the one-gluon exchange potential. A more extensive description of the quark-model Hamiltonian can be found in the literature [7, 8, 9].

Based on the quark model Hamiltonian, two different procedures have been used in the literature to obtain baryonic interactions. The first one is the RGM. It allows to treat the inter-cluster dynamics in an exact way once the Hilbert space has been fixed. The potential derived from this method contains all the nonlocalities associated with quark antisymmetry. For the present study we will make use of the NN potential derived through a Lippmann-Schwinger formulation of the RGM equations in momentum space [9]. It reproduces the NN phase shifts up to 250 MeV lab energy with quite a good accuracy.

The second method is based on the Born-Oppenheimer approximation. It provides a clear-cut prescription for removing the nonlocalities while preserving the general properties of the interaction for lower partial waves, i.e., those coming form quark antisymmetry. This local interaction has been widely applied to a great variety of physical problems, obtaining reasonable results. In general, the phase shifts are reproduced with a comparable accuracy to the RGM results [6].

In both cases, for a correct description of the $^1S_0$ phase shift it is necessary to take into account the coupling to the $^5D_0$ $N\Delta$ channel [8], providing the required additional attraction. In order to achieve almost phase shift equivalence between the local and nonlocal interaction models, which is mandatory if one wants to reliably judge the influence of the nonlocalities, we have done a fine tuning of the potential parameters.

The results obtained for the two-body system with the local and nonlocal potentials are presented in Table 1 and Fig. 1. The $^1S_0$ and the $^3S_1 - ^3D_1$ phase shifts and the low-energy scattering parameters as well as the deuteron binding energy are practically the same for both potential models, and also in very good agreement with experimental data.

TABLE 1. NN properties

		NN Low-energy scattering parameters			Deuteron properties	
		Local	Nonlocal		Local	Nonlocal
$^1S_0$	a_s (fm)	-23.758	-23.759	ε_d (MeV)	-2.2245	-2.2242
	r_s (fm)	2.694	2.682	P_D (%)	4.79	4.85
$^3S_1$	a_t (fm)	5.464	5.461	Q_d (fm^2)	0.280	0.276
	r_t (fm)	1.779	1.820	A_S (fm$^{-1/2}$)	0.900	0.891
				A_D/A_S	0.0243	0.0257

TRITON BINDING ENERGY: RESULTS AND DISCUSSION

The three-body system is solved performing a five channel Faddeev calculation including the $^1S_0$ and $^3S_1 - ^3D_1$ NN partial waves as input. Note that since in our model there is a coupling to the $N\Delta$ system, a fully consistent calculation would require the inclusion of two more three-body channels. However, their contribution to the $3N$ binding energy is known to be rather small [11] and therefore we neglect them for simplicity reasons.

To solve the three-body Faddeev equations in momentum space we first perform a separable finite-rank expansion of the $NN(-N\Delta)$ sector utilizing the EST method [12, 13, 14]. In Ref. [14] it was shown that with a separable expansion of sufficiently high rank, reliable and accurate results on the three-body level can be achieved. In the present case it turned out that separable representations of rank 6-8 for $^1S_0 - (^5D_0)$ and rank 6 for $^3S_1 - ^3D_1$, are sufficient to get converged results.

TABLE 2. Properties of the three-nucleon bound state.

	E_B (MeV)	P_S (%)	$P_{S'}$ (%)	P_P (%)	P_D (%)
Local	-7.572	91.413	1.597	0.044	6.946
Nonlocal	-7.715	91.493	1.430	0.044	7.033

The results of the triton bound state calculation are summarized in Table 2.

Let us first emphasize that the predicted binding energies for both models are comparable to those obtained from conventional potentials of the NN interaction such as the Paris, Bonn, or Nijmegen models [13, 15]. Comparing the values for our local and nonlocal models, one observes that there is about 150 keV more binding for the nonlocal potential. Is this the enhancement we can expect from the nonlocalities due to the quark substructure of the nucleon? In order to answer this question we need to go back again to the NN results and scrutinize the on-shell properties carefully. For the $^1S_0$ partial wave the differences in the low-energy scattering parameters and in the phase shift are indeed very small, see Table 1 and Fig. 1 respectively.

Unfortunately, for the $^3S_1 - ^3D_1$ partial wave the situation is much more complicated. While the deuteron binding energy and also the $^3S_1$ and $^3D_1$ phase shifts are in excellent agreement, (see Fig. 1) this cannot be said about the mixing parameter ε_1. In this case, it is difficult to estimate reliably the effect from the obvious deviation from phase equivalence on the triton binding energy.

However, one can clearly separate the effects from the two involved partial waves. For this purpose, we carried out additional $3N$ calculations where we combined the $^1S_0$ of the local model with the $^3S_1 - {}^3D_1$ of the nonlocal model and vice versa. Corresponding results are compiled in Table 3 where we show triton binding energies (in MeV) for different combinations of the local and nonlocal models. They strongly suggest that the nonlocalities present in the $^1S_0$ alone are already responsible for the enhancement of around 150 keV in the triton binding energy. The shift in the binding energy is independent of whether we use the local or nonlocal version model for the $^3S_1 - {}^3D_1$ partial wave. On the other hand, the nonlocalities present in the $^3S_1 - {}^3D_1$ partial wave seem to even decrease the binding energy. However, we suspect that here the effect of the nonlocalities is obscured by the fact that the two models are not strictly phase equivalent.

TABLE 3.

		$^3S_1 - {}^3D_1$	
		Local	Nonlocal
$^1S_0$	Local	-7.572	-7.544
	Nonlocal	-7.745	-7.715

In summary, we have calculated the three-nucleon bound state problem utilizing *NN* potentials derived from a basic quark-quark interaction. One of these potentials was generated by means of the resonating group method so that nonlocalities resulting from the internal structure of the nucleon were preserved. The other potential is based on the Born-Oppenheimer approximation and is strictly local. These potentials are made nearly phase equivalent by fine tuning of some of the model parameters. The corresponding calculations of the triton bound state indicate that the nonlocalities resulting from the quark substructure of the nucleon yield additional attraction and, specifically, can lead to an increase of the binding energy by up to 200 keV. Thus, the effect of those nonlocalities on the three-nucleon binding energy is certainly appreciable. In particular, it is of the same magnitude as the one resulting from nonlocalities that occur in the meson-exchange picture of the *NN* interaction.

REFERENCES

1. Gibson, B.F., Kohlhoff, H., von Geramb, H.V., *Phys. Rev. C* **51**, R465 (1995).
2. Haidenbauer, J., Holinde, K., *Phys. Rev. C* **53**, R25 (1996).
3. Machleidt, R., Sammarruca, F., Song, Y., *Phys. Rev. C* **53**, R1483 (1996).
4. Elster, Ch., Evans, E.E., Kamada, H., Glöckle, W., *Few-Body Syst.* **21**, 25 (1996).
5. Takeuchi, S., Cheon, T., Redish, E.F., *Phys. Lett. B* **280**, 175 (1992).
6. Garcilazo, H., Valcarce, V., Fernández, F., *Phys. Rev. C* **60**, 044002 (1999).
7. Fernández, F., Valcarce, A., Straub, U., Faessler, A., *J. Phys. G* **19**, 2013 (1993).
8. Valcarce, A., Faessler, A., Fernández, F., *Phys. Lett. B* **345**, 367 (1995).
9. Entem, D.R., Fernández, F., Valcarce, A., *Phys. Rev. C* **62**, 034002 (2000).
10. Stoks, V.G.J., *et al.*, *Phys. Rev. C* **48**, 792 (1993); Bugg, D.V., Bryan, R.A., *Nucl. Phys.* **A540**, 449 (1992); Arndt, R.A., Hyslop, J.S., Roper, L.D., *Phys. Rev. D* **35**, 128 (1987).
11. Hajduk, Ch., Sauer, P.U., *Nucl. Phys.* **A322**, 329 (1979).
12. Ernst, D.J., Shakin, C.M., Thaler, R.M., *Phys. Rev. C* **8**, 507 (1973).
13. Schadow, W., Sandhas, W., Haidenbauer, J., Nogga, A., *Few-Body Syst.* **28**, 241 (2000).
14. Nemoto, S., Chmielewski, K., Schellingerhout, N.W., Haidenbauer, J., Oryu, S., Sauer, P.U., *Few-Body Syst.* **24**, 213 (1998).
15. Friar, J.L., Payne, G.L., Stoks, V.G.J., de Swart, J.J., *Phys. Lett. B* **311**, 4 (1993).

Stability of bound states in the light-front Yukawa model

V.A. Karmanov*, J. Carbonell† and M. Mangin-Brinet†

*Lebedev Physical Institute, Leninsky Pr. 53, 119991 Moscow, Russia
†Institut des Sciences Nucléaires, 53, Av. des Martyrs, 38026 Grenoble, France

Abstract. We show that in the system of two fermions interacting by scalar exchange, the solutions for $J^\pi=0^+$ bound states are stable without any cutoff regularization, for values of the coupling constant α below a critical value α_c. The latter one is calculated from an eigenvalue equation.

INTRODUCTION

The Yukawa model, describing a system of two fermions interacting by scalar exchange ($\mathcal{L}^{int} = g\bar{\psi}\psi\phi^{(s)}$), is instructive for studying the relativistic bound states and for developing the renormalization methods. It is also a main ingredient in building the NN interaction, which contains an important contribution of scalar meson exchanges. The bound state problem and the renormalization in the Yukawa model were studied [1] in the framework of standard light-front dynamics [2]. The relativistic two-nucleon wave functions have been also calculated perturbatively [3] in the explicitly covariant version of light-front dynamics [4], where the state vector is defined on the invariant plane $\omega \cdot x = 0$ with $\omega^2 = 0$ (see for review [5]). In this work, the Bonn NN model was used with the corresponding form factors [6] and the problem of cut-off dependence was not analyzed.

In reference [7], we investigated the stability of the bound states relative to the high momentum contributions of the kernel, when the cutoff tends to infinity. Below we present the results of our study and compare them with those obtained in [1].

THE CUTOFF DEPENDENCE OF THE BINDING ENERGY

We consider the two fermion wave function with total angular momentum $J = 0^+$. In the fermion spin indices, it is a 2×2 matrix determined, due to the parity conservation, by two independent elements – the spin components $f_{i=1,2}$. Since the wave function is defined on the light-front plane $\omega \cdot x = 0$, components $f_i(k,\theta)$ depend not only on the relative momentum k, but also on the angle θ between $\vec{k}$ and $\hat{n} = \vec{\omega}/|\vec{\omega}|$. The system of equations for f_i contains a 2×2 matrix kernel K_{ij}, calculated by using the explicitly covariant light-front graph techniques [4, 5]. These equations and the analytical expressions of kernels are given in [7]. Though their form differs from the ones used in [1], we have shown that they are strictly equivalent.

Let us consider the equations on the finite interval $0 \leq k \leq k_{max}$. The dependence of their solutions on the cutoff k_{max} in the limit $k_{max} \to \infty$ is determined by the kernels asymptotics. The kernel K_{22} is repulsive and cannot generate a collapse, whereas K_{11} is attractive. Therefore, to investigate the stability, we can consider the one channel problem for the component f_1, which satisfies the equation:

$$[M^2 - 4(k^2 + m^2)] f_1(k,\theta) = \frac{m^2}{2\pi^3} \int K_{11}(k,\theta;k',\theta') f_1(k',\theta') \frac{d^3k'}{\varepsilon_{k'}}, \tag{1}$$

where $\varepsilon_{k'} = \sqrt{m^2 + k'^2}$.

Our analysis uses the fact that at $k \to \infty$ the integral in the r.h.s. of (1) is dominated by the region $k' \propto k$, i.e. $k' \to \infty$ with a fixed ratio $k'/k = \gamma$ [8]. One can therefore replace in (1) both wave function and kernel by their asymptotics, which have the form [7]:

$$f_1(k,z) = \frac{h_1(z)}{k^{2+\beta}}, \quad K_{11} = -\frac{\pi\alpha}{m^2} \begin{cases} \sqrt{\gamma} A_{11}(z,z',\gamma), & \text{if } \gamma < 1 \\ A_{11}(z,z',1/\gamma)/\sqrt{\gamma}, & \text{if } \gamma > 1 \end{cases} \tag{2}$$

with $0 < \beta < 1$ and

$$A_{11}(z,z',\gamma) = \int_0^{2\pi} \frac{d\phi}{2\pi\sqrt{\gamma}} \frac{2\gamma(1-zz') - (1+\gamma^2)\sqrt{1-z^2}\sqrt{1-z'^2}\cos\phi}{(1+\gamma^2)(1+|z-z'|-zz') - 2\gamma\sqrt{1-z^2}\sqrt{1-z'^2}\cos\phi}, \tag{3}$$

where $\alpha = g^2/4\pi$ and $z = \cos\theta$. By setting $A_{11}(z,z',\gamma) \equiv 1$ and $\alpha = 2\pi m\alpha'$ in (2), the asymptotics of K_{11} becomes identical to asymptotics of the momentum space kernel corresponding to the non-relativistic potential $-\alpha'/r^2$. As it is well known [9], there exists for this potential a critical coupling constant $\alpha'_c = 1/(4m)$, that corresponds to $\alpha_c = \pi/2$. The inspection of (3) shows that in the Yukawa model the function A_{11} is smaller than for the $-\alpha'/r^2$ potential: $0 \leq A_{11}(z,z',\gamma) \leq 1$. Therefore in this model, one can expect a larger critical coupling constant i.e. $\alpha_c > \pi/2$, what is confirmed by numerical calculations.

Substituting (2) into equation (1), we obtain for $h_1(z)$ [8]:

$$\int_{-1}^{+1} dz' \, H_\beta(z,z') h_1(z') = \lambda h_1(z) \tag{4}$$

with $\lambda = 1/\alpha$ and

$$H_\beta(z,z') = \int_0^1 \frac{d\gamma}{2\pi\sqrt{\gamma}} A_{11}(z,z',\gamma) \cosh(\beta \log \gamma). \tag{5}$$

Equation (4) is an eigenvalue equation for λ, parametrized by β. It provides the relation between the coupling constant α and β, determining the power law (2) of the wave function asymptotics. The r.h.s. of equation (1) becomes divergent for $\beta \leq 0$. Hence, the equation $\beta(\alpha_c) = 0$ determines the critical coupling constant α_c.

RESULTS

In the numerical calculations, the constituent masses were taken equal to $m=1$ and the mass of the exchanged scalar is $\mu=0.25$. The numerical solution $\alpha(\beta)$, found from equation (4) with the function $A_{11}(\gamma,z,z')$ given by (3), is plotted in figure 1. The critical coupling constant is obtained for $\beta = 0$ for which the eigenvalue is $\lambda_c = 0.269$. It corresponds to $\alpha_c = 1/\lambda_c = 3.72$, as shown in figure 1 at $\beta = 0$.

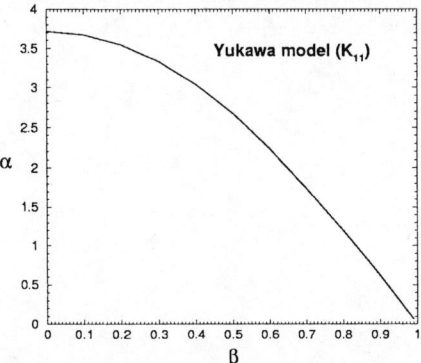

FIGURE 1. Function $\alpha(\beta)$ determined by eq. (4). The critical coupling constant is: $\alpha_c = \alpha(\beta = 0) = 3.72$. The values discussed in the text: $\alpha = 1.096, \beta = 0.819$ and $\alpha = 2.480, \beta = 0.548$ are on the curve.

In figure 2, we have plotted the mass square M^2 of the two fermion system, found from (1), as a function of the cutoff k_{max} for two fixed values of the coupling constant below and above the critical value $\alpha_c = 3.72$. One can see two dramatically different

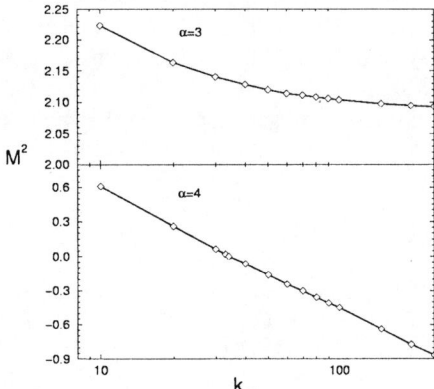

FIGURE 2. Cutoff dependence of the binding energy in the $J = 0^+$ state, in the one-channel problem (f_1), for two fixed values of the coupling constant below and above the critical value.

behaviors depending on the value of the coupling constant α. For $\alpha = 3$, i.e. $\alpha < \alpha_c$, the result is convergent. On the contrary, for $\alpha = 4$, i.e. $\alpha > \alpha_c$, the result is clearly divergent: M^2 decreases logarithmically as a function of k_{max} and becomes even negative. Though the negative values of M^2 are physically meaningless, they are formally allowed by equations. The first degree of M does not enter neither in the equation nor in the kernel, and M^2 crosses zero without any singularity.

We have examined the asymptotical behavior of the wave function $f_1(k,z)$ and found that it very accurately follows the power law (2) with the power $\beta(\alpha)$ given in figure 1. For instance for a binding energy $B = 2m - M = 0.05$ ($\alpha = 1.096$) a direct measurement in the numerical solution plotted in figure 3 gives $\beta = 0.820 \pm 0.002$ whereas the solution of equation (4) for the corresponding α gives $\beta = 0.819$. The same kind of agreement

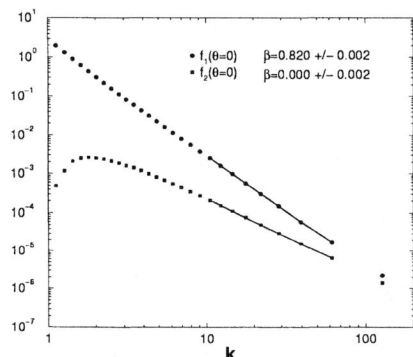

FIGURE 3. Asymptotic behavior of the $J = 0^+$ wave function components $f_{1,2}$ for B=0.05, α=1.096, μ=0.25. The slope coefficients are $\beta_1 = 0.820$ and $\beta_2 \approx 0$.

was found for $B = 0.5$ ($\alpha = 2.480$, $\beta = 0.548 \pm 0.002$).

We conclude that the $J = 0$ – or $(1+, 2-)$ state in the classification [1] – is stable (i.e. convergent relative to the cutoff $k_{max} \to \infty$) for coupling constant α below the critical value $\alpha_c = 3.72$. In this point, our conclusion differs from the one settled in [1], where it was stated that the integrals in the r.h.s. of the equations diverge logarithmically with cutoff. Above the critical value the integrals indeed diverge and the system collapses.

In the $J = 1$ state the system is found to be always unstable, as pointed out in [1].

Thus, by an analytical method, confirmed by numerical calculations, we have shown that the Yukawa model is not cutoff dependent for coupling constant below a critical value. The results obtained should be taken into account for instance in the renormalization procedures.

REFERENCES

1. Glazek, St., Harindranath, A., Pinsky, S., Shigemutsu, J., and Wilson, K., *Phys. Rev. D* **47**, 1599-1619 (1993); Glazek, St.D., and Wilson, K.G., *Phys. Rev. D* **47**, 4657-4669 (1993).
2. Brodsky, S.J., Pauli, H.-C., and Pinsky, S.S., *Phys. Rep.* **301**, 299-486 (1998).
3. Carbonell, J., and Karmanov, V.A., *Nucl. Phys.* **A581**, 625-653 (1995); *Nucl. Phys.* **A589**, 713-723 (1995).
4. Karmanov, V.A., *ZhETF* **71**, 399-416 (1976) [transl.: *JETP* **44**, 210 (1976)].
5. Carbonell, J., Desplanques, B., Karmanov, V.A., and Mathiot, J.F., *Phys. Rep.* **300**, 215-347 (1998).
6. Machleidt, R., Holinde, K., and Elster, C., *Phys. Rep.* **149**, 1 (1987).
7. Mangin-Brinet, M., Carbonell, J., and Karmanov, V.A., *Phys. Rev. D* **66**, 027701 (2001).
8. Smirnov, A.V., *private communication*.
9. Landau, L.D., and Lifshits, E.M., *Quantum mechanics*, §35, Pergamon press, 1965.

New polarization test for Delta–isobar admixture in ^{3}He

V.M. Kolybasov

Lebedev Physical Institute, Leninsky prospect 53, 119991 Moscow, Russia

Abstract. A new test is proposed for the search of states with Δ–isobar in ^{3}He and ^{4}He. The test is based on the fact that the spectator isobar generally must have tensor polarization. It is shown that for the expected dominant D-states of Δ-isobar in ^{3}He and ^{4}He its tensor polarization achieves the maximum possible value. It leads to the angular anisotropy of the form $(1+3z^2)$ where z is the cosine of the angle between the pion direction in the isobar system and the isobar direction in the laboratory system. The effects of possible contribution for S and P states is also discussed.

Isobar states in nuclei are one of the simplest manifestations for non–nucleon degrees of freedom. They were discussed during 30 years [1]. In spite of this fact, up to now we have no convincing evidence of their experimental observation. There are two reasons for that. First of all, it is difficult to discriminate between the signal (pion–nucleon pair) from "pre–existed" Δ–isobar and the signal from Δ–isobar which is produced during the interaction and, in some cases, from accidental correlation of independently flying out pion and nucleon. Secondly, the "pre–existed" isobar in nucleus is strongly out of mass shell and any dynamical estimation is unreliable.

The essential progress is possible via the application of criteria based on the conservation of discrete quantum numbers and the peculiarities of spin states. For example, in Ref. [2] it was noted that reactions on Δ–isobar configurations of ^{3}He must give large ratio of $\pi^+ p$ to $\pi^- p$ pairs. The main idea of present investigation is that the spectator isobar, generally speaking, must have nonzero tensor polarization and it leads to the anisotropy in angular distribution of isobar decay products (the deuteron case was thoroughly studied in Ref. [3]). In quasi–free approximation we have a virtual decay $^3\text{He} \rightarrow \Delta + (2N)$ with following interaction of an initial particle and (2N) system. The isobar polarization moment of the second rank can be expressed as

$$T_\Delta^{(2)M} = e Y_{2M}^*(\mathbf{n}) \tag{1}$$

where $\mathbf{n}$ is the unit vector along the relative momentum of Δ and (2N) system (that is along Δ direction in ^{3}He system) and e is some coefficient which dependes on quantum numbers of actual isobar configuration. Here the following definition is used for polarization moments of spin j particle in terms of spin density matrix ρ

$$T^{(L)M} = \rho_{\mu\mu'} C_{j\mu'LM}^{j\mu}, \tag{2}$$

$$\rho_{\mu\mu'} = \sum_{L=0}^{2j} C_{j\mu'LM}^{j\mu} \frac{2L+1}{2j+1} T^{(L)M}. \tag{3}$$

The amplitude of $\Delta \to N\pi$ decay is proportional to

$$M \sim C^{3/2\mu_\Delta}_{1/2\mu_N\,1m} Y_{1m}(\mathbf{k}), \tag{4}$$

$\mathbf{k}$ is the unit vector along the direction of pion in Δ system. It can be easily shown from (1) and (4) that the angular distribution of Δ decay products has the form

$$\sigma \sim 1 - \frac{5e}{2\sqrt{4\pi}}(3z^2 - 1) \tag{5}$$

where z is the cosine of the angle between $\mathbf{n}$ and $\mathbf{k}$. So the problem now is in the calculation of the coefficient e.

The dominant Δ–isobar configuration in ^{3}He is considered [2, 4] to be D–state with NN pair in the state $T = 1$, $L = 0$, $S = 0$, $J = 0$. (It is natural if the driving process is NN$\to \Delta$N.) So NN pair is actually equivalent to spinless particle and it must lead to the isotropy in the Treiman–Yang angle distribution. The amplitude of ^{3}He $\to \Delta$+(2N) decay has the form

$$M = \sqrt{2\pi}\, C^{1/2\mu_{He}}_{3/2\mu_\Delta\,2m} Y_{2m}(\mathbf{n}). \tag{6}$$

It can be shown that Eq. (6) gives for the coefficient e value

$$e = -\frac{\sqrt{4\pi}}{5}. \tag{7}$$

If z axis is chosen along $\mathbf{n}$ then only the term with $M = 0$ will be nonzero: $T^{(2)0} = -1/\sqrt{5}$. It is the maximum possible absolute value for tensor polarization of spin 3/2 particle as in this case $T^{(2)0}$ is expressed through the numbers of particles with different projections of spin to z axis

$$T^{(2)0} = \frac{1}{\sqrt{5}} \frac{N_{3/2} - N_{1/2} - N_{-1/2} + N_{-3/2}}{N_{3/2} + N_{1/2} + N_{-1/2} + N_{-3/2}} \tag{8}$$

and can vary from $-1/\sqrt{5}$ (when $N_{3/2} = N_{-3/2} = 0$) to $1/\sqrt{5}$ (when $N_{1/2} = N_{-1/2} = 0$). From Eqs. (5) and (7) the form of the angular distribution of isobar decay products is

$$\sigma \sim (1 + 3z^2). \tag{9}$$

One can see the characteristic sharp anisotropy of the angular distribution. Expression (9) is the main result of present investigation.

In analogy with ^{3}He case, if we take the transition NN$\to \Delta$N as the driving term, the dominant isobar configuration in ^{4}He must be D–wave with 3N system having angular momentum $L = 0$ and total spin $S = 1/2$. So 3N system is equivalent to spin 1/2 particle and it again must lead to the isotropy in Treiman–Yang angle distribution [5]. The amplitude of the decay ^{4}He $\to \Delta$+(3N) has the form

$$M \sim C^{2\mu}_{3/2\mu_\Delta\,1/2\mu_{(3N)}} C^{00}_{2\mu\,2m} Y_{2m}(\mathbf{n}). \tag{10}$$

It gives the same results (7) and (9) as in ^{3}He case.

Besides dominant D–wave isobar states, there can also be S and P configurations. S state of Δ in ^{3}He is possible with 2N system in state with total spin $J_{NN} = 2^+$ ($T = 1$, $L = 2$, $S = 0$). When calculating $T^{(2)M}$ this state does not interfere with dominant D-state due to different J_{NN} (D state has $J_{NN} = 0$). So its influence will be important only at very small values of spectator isobar momentum. Isobar S state in ^{4}He is possible if 3N system has $J_{3N} = 3/2$. Again this state does not interfere with dominant D state due to different J_{3N} (dominant D state has $J_{3N} = 1/2$).

In the case of ^{3}He it can be simply shown that "small" P and D states of Δ also cannot interfere with dominant D state in calculation of tensor polarization due to different other quantum numbers. Their influence will be explicitly considered in separate publication.

In conclusion the procedure, which can give the convincing evidence for the Δ–admixture in ^{3}He, seems in such a way:

- the application of the above test with the selection of events where the isobar goes to backward hemisphere;
- approximate isotropy of Treiman–Yang angle distribution;
- taking into account large predicted ratio of π^+ p to π^- p pairs [2];
- taking into account that the momentum distribution of spectator isobar must be rather broad and has a maximum at $350 \div 400$ MeV/c [6].

ACKNOWLEDGMENTS

This investigation was partly supported by Russian Foundation for Basic Researches (grant 99–02–17263).

REFERENCES

1. Weber, H.J., and Arenhövel, H., *Phys. Rep.* **36**, 277–348 (1978).
2. Lipkin, H.J., and Lee, T.–S.H., *Phys. Lett. B* **183**, 22–26 (1987).
3. Kolybasov, V.M., *Yad. Fiz.* **25**, 209–217 (1977).
4. Huber, G.M., Lolos, G.I., Brash, E.J., et al., *nucl–ex/9912001*.
5. Shapiro, I.S., Kolybasov, V.M., and Augst, G.R,, *Nucl. Phys.* **61**, 353–367 (1965).
6. Strueve, W., Hajduk, Ch., Sauer, P.U., and Theis, W., *Nucl. Phys.* **A465**, 651–685 (1987).

Asymptotic structure of the three-body Coulomb Green's function for the case of two charged particles

S.B. Levin[1*], S.L. Yakovlev[†] and N. Elander[*]

[*]Stockholm University, Stockholm Center for Physics, Astronomy and Biotechnology,
S-10691 Stockholm, Sweden
[†]Department of Mathematical and Computational Physics, Institute for Physics, St. Petersburg
State University, 198904 Petrodvorets, Ulyanovskaya Str. 1, St. Petersburg, Russia.

Abstract. The three-body Coulomb Green's function asymptotic structure is studied by the stationary phase method for the convolution integral for the case of two charged particles. The stationary phase points are roots of the third degree polynomial with the coefficients depending on the position in configuration space.

INTRODUCTION

The detailed description of the coordinate asymptotics of the three-body Coulomb wave function is up to now a challenging problem of few-body physics. A variety of methods were proposed as the solution of this problem. The eikonal approximation method (a variant of semiclassical approximation) seems to be an adequate tool to construct the leading terms of the three-body Coulomb wave function up to the preexponential factors in those parts of configuration space where particles are well separated [1, 2, 3]. The regions of the configuration space where the particles of a given pair are not well separated requires a special care [4, 5].

In this paper the coordinate asymptotic structure of the three-body Coulomb Green's function for the case of two charged particles is studied from the convolution integral with the stationary phase method. We consider the region of the three-body configuration space where particles of the charged pair are far from the forward scattering direction. This means that the condition

$$\sqrt{k}(|x| + |x'| - |x - x'|) \gg 1,$$

is satisfied. Here x and x' are dual relative positions and k is the relative momentum of the charged pair.

---

[1] On leave of absence from Department of Mathematical and Computational Physics, Institute for Physics, St. Petersburg State University, St. Petersburg, Russia.

ASYMPTOTICS CONSTRUCTION

The three-body Green's function for the case of two charged particles can be represented as the following convolution integral

$$R_c(X,X',E) = \frac{1}{2\pi i}\oint_{l_\sigma} r^c(x,x',\zeta)r_0(y,y',E-\zeta)d\zeta, \qquad (1)$$

where r^c is the Green's function of the two charged particles and r_0 is the free Green's function of the spectator and contour l_σ encircles in the anticlockwise direction the spectrum of r^c on the complex plane ζ. We use the usual Jacobi coordinates x, y denoting this pair as X. In order to evaluate the integral (1) we will use the following coordinate asymptotics for the Green's function r^c as $kR_- \to \infty$

$$r^c(x,x',k^2+i0) = \Omega(\gamma)\frac{\exp\left\{ik|x-x'|+i\gamma\ln\left(\frac{R_-}{R_+}\right)\right\}}{4\pi|x-x'|} + O\left(\frac{1}{(kR)^2}\right), \qquad (2)$$

where

$$\Omega(\gamma) = \frac{e^{-\pi\gamma}}{\Gamma^3(1-i\gamma)} \times$$

$$\left\{1 - i\gamma\left[i\pi + \psi(1-i\gamma) + \psi(2-i\gamma) - \psi(1+i\gamma) - \frac{1}{1-i\gamma} + \frac{e}{2\pi}\Gamma(1-i\gamma)F(\gamma)\right]\right\},$$

$$R_\pm = |x| + |x'| \pm |x-x'|,$$

$$\gamma = \frac{Z_1Z_2e^2\mu}{k}, \quad F(\gamma) = \int_{-\infty}^\infty \frac{e^{it}\ln(1+it)}{(1+it)^{1-i\gamma}}dt.$$

Here $\psi(x)$ is the Euler psi-function. Complete derivation of this formula can be found in [6]-[7].

Let us consider the phase function which determines the oscillation factor of the integral (1) if one uses the asymptotics (2)

$$S(\zeta) = \sqrt{E-\zeta} + \frac{P}{\sqrt{\zeta}} + \sqrt{\zeta}Q,$$

where

$$P = \frac{\omega}{|y-y'|}\ln\left(\frac{R_-}{R_+}\right), \quad Q = \frac{|x-x'|}{|y-y'|}, \quad \omega = Z_1Z_2e^2\mu.$$

Critical points, which define the leading order of the integral (1), obey the equation $S'(\zeta) = 0$ which can be explicitly written as a cubic equation

$$\zeta^3(1+Q^2) - \zeta^2 Q(EQ+2P) + \zeta(2EQ+P)P - EP^2 = 0. \qquad (3)$$

The Cardano formulas lead to the following values of the three roots of (3)

$$\zeta_1 = E\frac{|x-x'|^2}{|x-x'|^2 + |y-y'|^2} +$$

$$+2\left(1-\frac{1}{3}\frac{|x-x'|^2}{|x-x'|^2+|y-y'|^2}\right)\frac{\omega}{|x-x'|}\ln\left(\frac{|x|+|x'|-|x-x'|}{|x|+|x'|+|x-x'|}\right)+O(\lambda^{-3/2}),$$

$$\zeta_2 = \frac{1}{3}\frac{\omega}{|x-x'|}\frac{|x-x'|^2-3|y-y'|^2}{|x-x'|^2+|y-y'|^2}\ln\left(\frac{|x|+|x'|-|x-x'|}{|x|+|x'|+|x-x'|}\right)+O(\lambda^{-3/2}),$$

$$\zeta_3 = \frac{1}{3}\frac{\omega}{|x-x'|}\frac{|x-x'|^2-3|y-y'|^2}{|x-x'|^2+|y-y'|^2}\ln\left(\frac{|x|+|x'|-|x-x'|}{|x|+|x'|+|x-x'|}\right)-O(\lambda^{-3/2}).$$

Here we have introduced the large parameter λ as

$$\lambda = |X-X'| = \sqrt{|x-x'|^2+|y-y'|^2}.$$

As one can see the critical points ζ_i are of a different order with respect to the parameter λ, namely $\zeta_1 = O(1)$ and $\zeta_{2,3} = O(\lambda^{-1})$. Consequently, the main contribution to the integral (1) comes from ζ_1 and can be written as

$$R_c(X,X',E) = \frac{e^{-3/4i\pi}E^{3/4}}{2}\Omega\left(\frac{\omega}{\sqrt{E}}\frac{|X-X'|}{|x-x'|}\right)\times \qquad (4)$$

$$\frac{\exp\left(i\sqrt{E}|X-X'|+i\frac{\omega}{\sqrt{E}}\frac{|X-X'|}{|x-x'|}\ln\left(\frac{R_-}{R_+}\right)\right)}{(2\pi)^{5/2}|X-X'|^{5/2}}+O\left(\frac{1}{\lambda^{2+\mu}}\right).$$

Here μ ($1/2 < \mu < 1$) appears due to the contributions of stationary points $\zeta_{2,3}$. Formula (4) is the main result of this paper, it coincides with the eikonal approximation formulas of [1, 2, 3] up to the preexponential factor which can not be defined in framework of the eikonal approximation method.

CONCLUSION

The asymptotic structure of the Coulomb Green's function for the case of two charged particles was constructed from the convolution integral by the stationary phase method. The stationary phase points are the roots of the third degree polynomial with the coefficients depending on the position in configuration space. It is shown that independently on the sign of the charges the Green's function in the leading order can be defined by the eikonal approximation formula in that part of configuration space where the charged particles are far from the forward scattering direction.

ACKNOWLEDGMENTS

The authors would like to thank P. Froelich for fruitful discussions. This work was supported by the Russian Foundation for Basic Research and the Natural Swedish Research Council (NFR). S.L. acknowledges the support of the Royal Swedish Academy of Sciences and the Wenner-Gren foundations.

REFERENCES

1. Merkuriev, S.P., *Yad. Fiz.* (in Russian), **24**, 289 (1976) [*Sov. J. Nucl. Phys.* **24**, 150 (1976)].
2. Merkuriev, S.P., *Teor. Mat. Fiz.* (in Russian), **32**, 187 (1977).
3. Faddeev, L.D., and Merkuriev, S.P., *Quantum Scattering Theory for Several Particle Systems*, (Kluver, Dordrecht), 1993.
4. Alt, E.O., Mukhamedzhanov, A.M., *Phys. Rev. A* **47**, 2004 (1993).
5. Mukhamedzhanov, A.M., Lieber, M., *Phys. Rev. A* **54**, 3078 (1996).
6. Levin, S.B., Yakovlev, S.L., Elander, N., (submitted to *J. Math. Phys.*)
7. Levin, S.B., Yakovlev, S.L., Elander, N., (in preparation for *J. Math. Phys.*)
8. Sidorov, Yu.V., Fedorjuk, M.V., Shabunin, M.I., *Lectures on the complex variable functions theory*, Nauka, 1989.

Phase-shift analyses of p-^{3}He scattering at $T_L = 4.0 - 200$ MeV

M. Matsuda*, V. Limkaisang[†], J. Nagata** and H. Yoshino[‡]

*Hiroshima University, Higashi-Hiroshima, 739-8521, Japan
[†]Ratjamangala Institute of Technology, Nakornratchasima, Thailand
**Kyushu International University, Kitakyushu, 805-8512, Japan
[‡]Hiroshima International University, Kurose, 724-0695

Abstract. The result of single-energy phase-shift analyses of p-^{3}He scattering carried out at T_L=4,0, 5.5, 6.8, 8.5, 9.5, 19.5 and 200 MeV is reported. The phase shift solution at 200 MeV is given. A proposal of p-^{3}He scattering experiments at the intermediate energies is done in the relationship to RIKEN project : the study of three-body force.

In 1996, the experiment of ^{3}He$(d,p)^4$He reaction was performed where the incident kinetic energy of deuteron was 270 MeV at RIKEN[1]. The obtained data have been theoretically investigated by Oryu et al.[2] using multi-channel Faddeev equation. This project is related to revealing the three-body force. In this study, n-p-^{3}He system plays an important role. However, the nucleon-^{3}He interaction is not well known in this energy region. It is necessary to carry out a phase-shift analysis (PSA) of nucleon-^{3}He scattering in the hundreds MeV region and determine the scattering amplitudes.

Since 1950's, the experiments of p-^{3}He scattering at low energies and their PSAs have been carried out with great interest in nuclei of mass number 4. In their analyses, some assumptions or restrictions were included because of lack of experimental data. One of them was an assumption of the energy dependence in scattering amplitudes, which called as an energy dependent PSA.

We previously carried out the single-energy PSAs of p-p scattering at the incident proton energy (T_L) region 25−500 MeV for a precise determination of $g^2_{\pi pp}$[3], at T_L=500−1090 MeV for a search of dibaryons[4], at T_L=3−10 GeV to have given a suggestion of dynamical shrinkage in the spin-orbit interaction[5], and n-p scattering at T_L=500−1090 MeV for dibaryons[6]. Here, the computer program PANN[7] were used.

A single energy PSA is the method to determine the scattering amplitudes from experimental data without any assumption and restriction. Therefore, the single-energy PSA has higher quality than the energy dependent PSA. We developed the computer program PAPH[8] for the PSA of p-^{3}He scattering, and carried out the single-energy PSAs at the incident proton energy T_L= 4.0, 5.5, 6.8, 8.5, 9.5 and 19.5 MeV[9].

In our analyses, since the experimental data on inelastic cross section (σ_r) are regarded as 0 mb in the region $T_L \lesssim 10$ MeV, the reflection parameters were taken as $\eta_l = 1$. At 19.5 MeV, by use of the experimental data of $\sigma_r = 44 \pm 12$ mb, we carried out the PSA and determined the reflection parameters. We obtained the solutions which were good agreement with the experimental data at all energy points, and predicted the inelastic cross section 48 mb in comparison with the experimental data mentioned above. Our success of PSA at 19.5 MeV means a possible expansion of PSA to the higher energy region where the inelastic channel opens.

Table 1 shows the numbers of all experimental data at T_L = 85, 156 and 200 MeV. As is seen in this table, there are very few data at 85 and 156 MeV. It is too difficult to determine the scattering amplitudes at these energy points by single energy PSA. On the other hand, there are enough experimental data to perform PSA at 200 MeV. We carried out the PSA of p-^{3}He scattering only at T_L=200 MeV.

At 200 MeV, there are many experimental data on $d\sigma/d\Omega$ in the scattering-angle region in the center of mass system $\theta_c \lesssim 90°$. All of them are measured at TRIUMF. However no data are found at the angles from 90° to 180°. On the other hand, there are the experimental data in the large angle region at 156 MeV. We multiplied those data by the factor 0.4 and used as the pseudo-data on $d\sigma/d\Omega$ at 200 MeV, which are very useful to determine the scattering amplitudes.

Forward observables, σ_t, σ_r, spin-dependent total cross section differences and so on, are necessary to determine the scattering amplitude at small angles. However, there is no experimental data at 200 MeV. Especially, σ_r is essential to determine the reflection parameters. We have the data on σ_r only below 50 MeV. We extrapolated those data to estimate its value about 130 mb at 200 MeV. We added this value as a pseudo datum to the database.

After all, we carried out the single energy PSA by using the database including 151 real data, 10 pseudo data on $d\sigma/d\Omega$ and 1 pseudo datum on σ_r. Phase shifts, reflection parameters and mixing parameters with the orbital angular momentum less than 7 were used as free parameters.

Figure 1 shows the experimental data on each observable and their calculated values by the solution of PSA. The markers show the experimental data and the solid lines the evaluated values by our solution. As is seen in these figures, the evaluated values show good agreement with experimental data as a whole.

We give the obtained phase-shift solution with the χ^2 value 790 in Table 2. The χ^2 value seems to be large, which is caused by the inconsistency between experimental data. The data on A_{0n} and A_{nn} have small experimental error bars, and respectively have the differences between those measured by Brash et al.[10] and by Häusser et al.[11]. Such a discrepancy resulted in the large χ^2-value. Both of them are measured at TRIUMF, and the members are almost the same. We hope to know the reason why they are different from each other.

It should be noticed that there are no data on every observables in the large-angle region. The evaluated value of σ_r 112 mb is relatively small in comparison with 130 mb of pseudo data.

We would like to do a proposal of p-^{3}He scattering experiments at T_L= 200 MeV to measure the reaction cross section, the differential cross section at large angles and the polarized-spin experiments at large angles. Also the supplies of experimental data at 85 and 156 MeV are desired. If we determine the scattering amplitude of p-^{3}He scattering at the intermediate energies, we could take out the phenomenological p-^{3}He potential, which will be a real contribution to RIKEN project. The study of revealing the three-body force should be an exact science.

TABLE 1. The numbers of experimental data at 85, 156 and 200 MeV

T_L (MeV)	observables								total
	$d\sigma/d\Omega$	A_{n0}	A_{0n}	A_{nn}	A_{ll}	A_{mm}	A_{ml}	A_{lm}	
85	28								28
156	25	14							39
200	39	34	15	15	12	12	12	12	151

TABLE 2. Phase-shift solutions obtained by single energy PSA

Partial waves	δ	η	Partial waves	δ	η
$^1S_0$	-6.795 ± 0.379	0.6557 ± 0.0170	$^3G_3$	-4.737 ± 0.086	0.8993 ± 0.0034
$^3S_1$	-9.114 ± 0.363	0.4988 ± 0.0053	$^3G_4$	1.523 ± 0.171	0.7371 ± 0.0047
$^1P_1$	-7.048 ± 0.192	0.6809 ± 0.0050	$^3G_5$	6.603 ± 0.104	0.6439 ± 0.0024
$^3P_0$	-7.337 ± 0.275	0.8465 ± 0.0025	$^1H_5$	0.837 ± 0.188	0.9172 ± 0.0051
$^3P_1$	-16.232 ± 0.302	0.6559 ± 0.0038	$^3H_4$	-2.746 ± 0.045	0.9158 ± 0.0030
$^3P_2$	10.267 ± 0.510	0.3646 ± 0.0048	$^3H_5$	0.867 ± 0.075	0.7890 ± 0.0033
$^1D_2$	-3.917 ± 0.171	0.7184 ± 0.0045	$^3H_6$	4.243 ± 0.100	0.8087 ± 0.0026
$^3D_1$	-10.815 ± 0.345	0.8476 ± 0.0035	$^1I_6$	1.572 ± 0.149	0.9499 ± 0.0018
$^3D_2$	-5.241 ± 0.191	0.6404 ± 0.0041	$^3I_5$	-0.648 ± 0.079	0.8468 ± 0.0022
$^3D_3$	17.957 ± 0.491	0.3921 ± 0.0032	$^3I_6$	0.175 ± 0.036	0.9406 ± 0.0029
$^1F_3$	-0.022 ± 0.140	0.8165 ± 0.0081	$^3I_7$	2.373 ± 0.765	0.9069 ± 0.0027
$^3F_2$	-8.086 ± 0.195	0.8745 ± 0.0061	$^1J_7$	0.368 ± 0.085	0.9785 ± 0.0013
$^3F_3$	0.631 ± 0.169	0.6891 ± 0.0045	$^3J_6$	-1.253 ± 0.052	0.9874 ± 0.0015
$^3F_4$	12.191 ± 0.318	0.4695 ± 0.0024	$^3J_7$	0.390 ± 0.025	0.9078 ± 0.0015
$^1G_4$	1.395 ± 0.185	0.8910 ± 0.0013	$^3J_8$	0.941 ± 0.034	0.9586 ± 0.0017
ρ'_1	0.289 ± 0.002		ρ_1^{s-t}	-0.180 ± 0.005	
ρ'_2	0.175 ± 0.006		ρ_2^{s-t}	-0.216 ± 0.005	
ρ'_3	0.031 ± 0.004		ρ_3^{s-t}	-0.227 ± 0.004	
ρ'_4	0.002 ± 0.004		ρ_4^{s-t}	-0.169 ± 0.004	
ρ'_5	0.015 ± 0.001		ρ_5^{s-t}	-0.087 ± 0.004	
ρ'_6	0.002 ± 0.001		ρ_6^{s-t}	-0.043 ± 0.003	
			ρ_7^{s-t}	-0.015 ± 0.002	

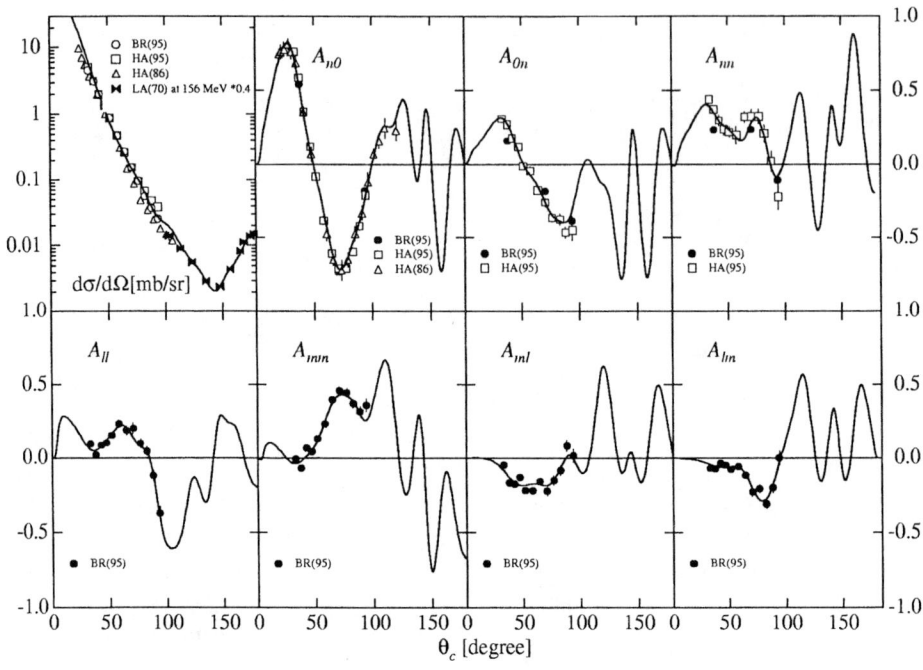

FIGURE 1. Experimental data and evaluated values by our solution on each observables.

REFERENCES

1. Uesaka, T., et al, *Nucl. Instr. and Meth. A* **402**, 212(1998).
2. Oryu, S., et al, *Nucl. Instr. and Meth. A* **402**, 402(1998).
3. Limkaisang, V., et al, *Prog. Theor. Phys.* **105**, 233 (2001).
4. Nagata, J., et al, *Prog. Theor. Phys.* **93**, 559 (1995); **95**, 691 (1996).
5. Matsuda, M., et al, *Phys. Rev. D* **33**, 2563 (1986); *Prog. Theor. Phys.* **93**, 1059 (1995).
6. Yoshino, H., et al, *Prog. Theor. Phys.* **95**, 577 (1996).
7. Matsuda, M., et al, *Comp. Phys. Comm.* **131**, 225 (2000).
8. Matsuda, M., et al, *Comp. Phys. Comm.* **131**, 264 (2000).
9. Yoshino, Y., et al, *Prog. Theor. Phys.* **103**, 107(2000).
10. Brash, E. J., et al, *Phys. Rev. C* **52**, 807(1995).
11. Häusser, O., et al, *Phys. Lett. B* **343**, 36(1995).

Meson dynamics and the resulting "3-Nucleon-Force" diagrams: Results from a simplified test case

L. Canton*, T. Melde† and J.P. Svenne**

INFN and University of Padova, Padova, Italy, I-35131
†*Institute for Theoretical Physics, University of Graz, 8010 Graz, Austria*
**WITP and University of Manitoba, Winnipeg, MB, Canada R3T 2N2*

Abstract. A simplified 1D (one-dimensional) model for a generalized 3N system is considered, as a testing ground for the explicit treatment of the meson dynamics in this system. We focus attention on the irreducible diagrams generated by the pion dynamics in the 3N system, and in particular to a new type of three-nucleon force discussed recently in the literature, and generated by the one-pion-exchange mechanism in presence of a nucleon-nucleon correlation. It is found that these new terms in the simplified model have an approximately 30% effect compared to the standard three-nucleon force terms in a 'Triton' binding energy calculation. It is suggested that this effect should also not be ignored in realistic calculations.

INTRODUCTION

An important aspect in modern three-nucleon dynamics is the explicit inclusion of mesonic degrees of freedom that can not be described by conventional two-nucleon potentials. These additional effects can be described by three-nucleon forces (3NF), which describe the addition of possible irreducible mesonic contributions. The construction and implementation of these terms is not trivial and care has been taken to avoid double-contributions. The standard 3NF approaches turned out to be able to account for the underbinding of the three-nucleon bound state. However, so far they could not fully explain the puzzle of the vector analyzing powers in nucleon-deuteron scattering at low energies.
Recently, starting from the rigorous four-body theory of Grassberger-Sandhas [1] and Yakubovsky [2], a new method to describe the coupled $NNN - \pi NNN$ system has been developed [3]. In a subsequent paper, an approximation scheme for this new method was derived by the authors [4]. It is based on reasonable physical and mathematical approximations and a procedure to freeze-out the pion channel.
After the approximations were performed and the pionic channel was projected out, it was shown that the residual pion-dynamics produced correction terms to the standard Faddeev-Alt-Grassberger-Sandhas 3N equation. These correction terms can be interpreted as three-nucleon force diagrams (3NF). The approach developed has the distinctive feature that these irreducible diagrams naturally appear in the Faddeev equation, and the corresponding 3NF is generated consistently with the 3N dynamical equation used for the actual calculations. To our knowledge, this level of consistency has here been

obtained for the first time.
A simplified one-dimensional model was developed, which describes in first approximation the dynamics of the three-nucleon system [5]. It is based on a spinless one-dimensional scattering theory with a potential which is the $1D$ analogue of the standard Malfliet-Tjon potential. The simplicity of the model allowed a straightforward investigation of the effects due to the residual pion dynamics.

THE SIMPLIFIED TEST CASE

An approximation scheme for reducing the $\pi - NNN$ system to a tractable set of equations has been recently developed [4]. At the lowest order, the residual effects of the pion dynamics result in irreducible corrections to the driving term of the Lovelace type $3N$ equation. The driving term now reads

$$Z_{ab} = Z_{ab}^{AGS} + Z_{ab}^{\pi} \qquad (1)$$

where a,b are the standard Faddeev labels. The first term has the usual structure of an AGS-type driving term, namely it is vanishing for $a = b$, while for $a \neq b$ it is

$$Z_{12}^{AGS} = \langle N_1(N_2N_3)|g_0|N_2(N_1N_3)\rangle \qquad (2)$$

and the other term corresponds to the correction that comes from the residual pion dynamics. Another type of correction term results from projecting out the pion channel. It describes the intermediate propagation of a pion under the presence of a correlated three nucleon cluster. In this study such an additional correction is not taken into account.
To the lowest order, the correction term Z_{ab}^{π} contains two topologically different contributions, for $a \neq b$,

$$Z_{12}^{\pi} = \langle N_1(N_2N_3\pi); N_1N_2(N_3\pi)|\tau_{(N_3\pi)}|N_2(N_3N_1\pi); N_1N_2(N_3\pi)\rangle \qquad (3)$$

and, for $a = b$,

$$Z_{11}^{\pi} = \langle N_1(N_2N_3\pi); N_1(N_2N_3)\pi|\tau_{(N_2N_3)}|(N_1\pi)(N_2N_3); N_1N_3(N_1\pi)\rangle \qquad (4)$$

$$+ \langle (N_1\pi)(N_2N_3); N_1(N_2N_3)\pi|\tau_{(N_2N_3)}|N_1(N_2N_3\pi); N_1N_3(N_1\pi)\rangle$$

The static approximation of the first term corresponds to a correction term usually attributed to a three-nucleon force diagram of the Fujita-Miyazawa type [6]. The second term corresponds to a topologically different type of correction term that describes the propagation of an intermediate pion under the presence of a correlated two nucleon cluster. Furthermore, it should be noted that the first correction term has contributions to the off-diagonal elements of the driving term only. On the other hand, the second type has contributions to the diagonal elements of the driving term only, which in the standard AGS driving term are always zero.
A simple test model was designed from a one-dimensional scattering theory describing symmetric spinless particles. The strength parameters of the $1D$ NN-potential were

chosen in a way to reproduce the deuteron binding energy. The potential also shows a similar spatial behaviour as observed in more realistic nuclear potentials, which makes it a good candidate for our test-calculation. The t-matrices are described by the Unitary Pole Approximation (UPA), which is expected to be a good approximation due to the simplicity of the toy-model. The 'deuteron' binding energies are found using a sturmian procedure. Subsequently we calculated the 'triton' binding energy of this system without the correction terms and found a value of $-7.28 MeV$.

In the calculation the first type of correction term was included in a static approximation which allowed the interpretation of the dynamical terms as $3NF$. This correction term represents the one-dimensional analogue of the standard 2π-exchange $3NF$ contribution. Standard 2π-$3NF$, like the Tucson-Melbourne or Brasil $3NF$ [7, 8], have at most four parameters, a, b, c, d. In the one-dimensional spinless model we only have one coefficient a_1, which depends on the πN scattering length and is defined in our simplified model by the expression

$$a_1 = -\frac{2\pi\mu}{\hbar^2} t_{\pi N}(p=0, p'=0, E=0) \qquad (5)$$

The threshold πN t-matrix is approximated by the corresponding NN t-matrix times an adjustable parameter $c_{\pi N}$. This parameter $c_{\pi N}$ can be chosen freely over a certain range to recover the triton binding energy. However, we also know that the corresponding threshold scattering lengths differ by a factor of 0.01 and we choose the parameter $c_{\pi N} = 0.01$ as a first guess. With this value we find a 'triton' binding energy of $-8.17 MeV$. We then varied the parameter $c_{\pi N}$ and found that the correction term depends strongly on this adjustable parameter. It is observed that it is in principle possible to adjust the parameter $c_{\pi N}$ in order to recover the experimental 'triton' binding energy.

When calculating the effect of the second type of correction terms a delicate cancellation has to be observed in order to avoid overcounting. This cancellation is described by Canton and Schadow [9] and is of the type $t - v$. Namely, we have to subtract the potential from the full t-matrix and it turns out that this subtraction does not result in a complete cancellation at low energies. For more details we refer to the publications [9, 10, 11]. We then calculated the 'triton' energy including both types of correction terms, where for the adjustable parameter $c_{\pi N}$ a value of 0.01 was chosen, thereby obtaining $-8.48 MeV$ for the model 'triton' binding energy. It should be noted that the effect of the 'new' correction terms in respect to the Fujita-Miyazawa type terms is approximately 30%. If we use the parameter $c_{\pi N}$ that was already adjusted to recover the 'triton' binding energy, then the new type of correction term leads to an over binding. It is therefore important to fit the adjustable parameter $c_{\pi N}$ with both types of correction terms present.

CONCLUSIONS

We developed a model that included one pion degree of freedom explicitly in an AGS type three body equation. It was shown that the effect of the pion dynamics resulted in correction terms in the driving terms of the two-cluster equations. We calculated the explicit effects of these terms, which could be interpreted as 3NF terms, on the 'triton' binding energy. In a simplified model calculation we show that the correction terms can account for an underbinding of the 'three-nucleon' bound state. However, a new type of correction term also has an approximately 30% effect on the 'triton' binding energy. A third topological type of correction term still needs to be investigated, but it is expected that the effect of this term is not as important.

The calculations we presented were done only with a rather simple model. However, one might expect that the new type of correction terms could have an effect also in more realistic calculations. This suggestion has been recently confirmed by Canton and Schadow in a more realistic study [9, 10] on the *nd* vector analyzing powers, where it was shown that a new 3NF term of tensor structure can be generated by these new correction terms, and could indeed hold the key to solve the A_y-puzzle. In conclusion, a new way of handling the pion dynamics in $3N$ calculations has been developed. Calculations suggest that a newly developed type of 3NF should also be included in realistic three-nucleon calculations.

ACKNOWLEDGMENTS

L.C. thanks for support from the Italian MURST-PRIN Project "Fisica Teorica del Nucleo e dei Sistemi a Piu Corpi". T.M acknowledges support from the University of Manitoba and the conference organizing committee for waiving the conference fee. J.P.S. acknowledges support from the Natural Science and Engineering Research Council of Canada. The authors also would like to thank the Institutions of INFN (Padova), University of Manitoba, and Universita di Padova for hospitality during reciprocal visits.

REFERENCES

1. Grassberger, P., and Sandhas, W., *Nucl. Phys.* **B2**, 181 (1967).
2. Yakubovsky, O.A., *Sov. J. Nucl. Phys.* **5**, 937 (1967).
3. Canton, L., *Phys. Rev. C* **58**, 3121 (1998).
4. Canton, L., Melde T., and Svenne, J.P., *Phys. Rev. C* **63**, 034004 (2001).
5. Melde, T., PhD Thesis, University of Manitoba (2001).
6. Fujita, J., and Miyazawa, H., *Prog. Theor. Phys.* **17**, 360 (1957).
7. Coon, S.A., Scadron, M.D., McNamee, P.C., Barrett, B.R., Blatt, D.W.E., and Kellar, B.H.J., *Nucl. Phys.* **A317**, 242 (1979).
8. Robilotta, M.R., and Coelho, H.T., *Nucl. Phys.* **A460**, 645 (1986).
9. Canton, L., Schadow, W., *Phys. Rev. C* **62**, 044005 (2000).
10. Canton, L., Schadow, W., *Phys. Rev. C* **64**, 031001 (2001).
11. Canton, L., Pisent, G., Schadow, W., Melde, T., Svenne, J.P., in *Theoretical Nuclear Physics in Italy*, World Scientific, Singapore, 249 (2001).

Retardation effects from quark confinement on low-energy hadron dynamics

Renat Kh. Gainutdinov and Aigul A. Mutygullina

Department of Physics, Kazan State University, 18 Kremlevskaya St, Kazan 420008, Russia

Abstract. We investigate effects from quark confinement on low-energy hadron dynamics. These effects are shown to give rise to the fact that nucleon dynamics is governed by a generalized dynamical equation with a nonlocal-in-time interaction operator. We show that renormalization of equations governing nucleon dynamics leads to the same dynamical situation.

Understanding how nuclear forces emerge from the fundamental theory of quantum chromodynamics (QCD) is one of the most important problem of quantum physics. To study hadron dynamics at scales where QCD is strongly coupled, it is useful to employ effective field theories (EFT's), an invaluable tool for computing physical quantities in the theories with disparate energy scales. In order to describe low-energy processes involving nucleons and pions, all possible interaction operators consistent with the symmetries of QCD are included in an effective Lagrangian of an EFT. However such a Lagrangian leads to ultraviolet (UV) divergences that must be regulated and a renormalization scheme defined. A fundamental difficulty in an EFT description of nuclear forces is that they are nonperturbative, so that an infinite series of Feynman diagrams must be summed. Summing the relevant diagrams is equivalent to solving a Schrödinger equation. However, an EFT yields graphs which are divergent, and gives rise to a singular Schrödinger potential. For this reason NN potentials are regulated and renormalized couplings are defined. Another aspect of the problem is that due to existence of the quark and gluon degrees of freedom hadron interactions should be nonlocal in time. As is well known, such a nonlocality is a consequence of the fact that some degrees of freedom are not taken into account in describing a quantum system. Usually the nonlocality in time of the interaction in a quantum system is associated with a loss of probability from the system, since there is nonzero probability to find the system in a state corresponding to the above degrees of freedom. At the same time, due to confinement the existence of the quark and gluon degrees of freedom must not lead to a loss of probabilities from a system of hadrons, despite that they play an important role in hadron interactions. However, as is well known, within the Hamiltonian formalism the interaction generating the dynamics of a quantum system may be nonlocal in time only in the case where the system is open, and, as a result, the evolution is not unitary. To solve this problem quantum dynamics need to be extended to describe the evolution of a closed quantum system the dynamics of which is generated by nonlocal-in-time interaction. This problem was solved in Ref. [1], where it was shown that the use of the Feynman approach to quantum theory in combination with the canonical approach allows such extension of quantum dynamics. A generalized quantum dynamics (GQD) developed in this way was shown [3, 4] to open new possibilities for describing hadron dynamics. The models developed in Refs. [3, 4] allow one to take into account the quark-gluon retardation effects in describing nucleon dynamics. In the present paper we investigate the effects from quark confinement on low-energy hadron dynamics.

In the GQD the evolution operator $U(t,t_0)$ in the interaction picture is represented in the form [1]

$$\langle \psi_2 | U(t,t_0) | \psi_1 \rangle = \langle \psi_2 | \psi_1 \rangle + \int_{t_0}^{t} dt_2 \int_{t_0}^{t_2} dt_1 \langle \psi_2 | \tilde{S}(t_2,t_1) | \psi_1 \rangle, \qquad (1)$$

where $\langle \psi_2 | \tilde{S}(t_2,t_1) | \psi_1 \rangle$ is the probability amplitude that if at time t_1 the system was in the state $|\psi_1\rangle$, then the interaction in the system will begin at time t_1 and will end at time t_2, and at this time the system will be in the state $|\psi_2\rangle$. This equation represents the Feynman superposition principle according to which the probability amplitude of an event which can happen in several different ways is a sum of contributions from each alternative way. Here subprocesses with definite instants of the beginning and end of the interaction in the system are used as alternative ways of realization of the corresponding evolution process, and $\langle \psi_2 | \tilde{S}(t_2,t_1) | \psi_1 \rangle$ represents the contribution to the evolution operator from the subprocess in which the interaction begins at time t_1 and ends at time t_2. The first term on the right-hand side of (1) is the contribution from the process in which the particles of the system do not interact. In the case of an isolated system the operator $\tilde{S}(t_2,t_1)$ can be represented in the form [1] $\tilde{S}(t_2,t_1) = \exp(iH_0 t_2)\tilde{T}(t_2 - t_1)\exp(-iH_0 t_1)$, H_0 being the free Hamiltonian. It was shown in Ref. [1] that for the evolution operator $U(t,t_0)$ given by (1) to be unitary for any times t_0 and t, the operator $\tilde{S}(t_2,t_1)$ must satisfy the following equation:

$$(t_2 - t_1)\tilde{S}(t_2,t_1) = \int_{t_1}^{t_2} dt_4 \int_{t_1}^{t_4} dt_3 (t_4 - t_3) \tilde{S}(t_2,t_4) \tilde{S}(t_3,t_1). \qquad (2)$$

This equation allows one to obtain the operators $\tilde{S}(t_2,t_1)$ for any t_1 and t_2, if the operators $\tilde{S}(t_2',t_1')$ corresponding to infinitesimal duration times $\tau = t_2' - t_1'$ of interaction are known. It is natural to assume that most of the contribution to the evolution operator in the limit $t_2 \to t_1$ comes from the processes associated with the fundamental interaction in the system under study. Denoting this contribution by $H_{int}(t_2,t_1)$, we can write

$$\tilde{S}(t_2,t_1) \underset{t_2 \to t_1}{\to} H_{int}(t_2,t_1) + o(\tau^\varepsilon). \qquad (3)$$

Within the GQD the operator $H_{int}(t_2,t_1)$ plays the role which the interaction Hamiltonian plays in the ordinary formulation of quantum theory: It generates dynamics of a system. Being a generalization of the interaction Hamiltonian, this operator is called the generalized interaction operator. The parameter ε is determined by demanding that $H_{int}(t_2,t_1)$ must be so close to the solution of Eq. (2) in the limit $t_2 \to t_1$ that this equation has a unique solution having the behavior (3) near the point $t_2 = t_1$.

If $H_{int}(t_2,t_1)$ is specified, Eq. (2) allows one to find the operator $\tilde{S}(t_2,t_1)$. Formula (1) can then be used to construct the evolution operator $U(t,t_0)$ and accordingly the state vector $|\psi(t)\rangle = |\psi(t_0)\rangle + \int_{t_0}^{t} dt_2 \int_{t_0}^{t_2} dt_1 \tilde{S}(t_2,t_1)|\psi(t_0)\rangle$ at any time t. Thus Eq. (2) can be regarded as an equation of motion for states of a quantum system. By using (2), the evolution operator can be expressed in the form [1]

$$\langle n|U(t,t_0)|n'\rangle = \langle n|n'\rangle + \frac{i}{2\pi} \int_{-\infty}^{\infty} dx \frac{\exp[-i(z-E_n)t]\langle n|T(z)|n'\rangle \exp[i(z-E_{n'})t_0]}{(z-E_n)(z-E_{n'})}$$

where $z = x + iy$, and $y > 0$, n stands for the entire set of discrete and continuous variables that characterize the system in full, $|n\rangle$ are the eigenvectors of the free Hamiltonian, and $\langle n|T(z)|n'\rangle$ is defined by

$$\langle n|T(z)|n'\rangle = i \int_0^{\infty} d\tau \exp(iz\tau)\langle n|\tilde{T}(\tau)|n'\rangle. \tag{4}$$

The equation of motion (2) is equivalent to the following equation for the T matrix [1]:

$$\frac{d\langle n_2|T(z)|n_1\rangle}{dz} = -\sum_n \frac{\langle n_2|T(z)|n\rangle\langle n|T(z)|n_1\rangle}{(z-E_n)^2}. \tag{5}$$

It was shown in Ref. [1] that the dynamics governed by Eq. (2) is equivalent to the Hamiltonian dynamics in the case where the interaction operator is of the form

$$H_{int}(t_2,t_1) = -2i\delta(t_2-t_1)H_I(t_1), \tag{6}$$

$H_I(t_1)$ being the interaction Hamiltonian in the interaction picture. In this case the state vector $|\psi(t)\rangle$ satisfies the Schrödinger equation $\frac{d|\psi(t)\rangle}{dt} = -iH_I(t)|\psi(t)\rangle$. The delta function $\delta(\tau)$ in (6) emphasizes the fact that in this case the fundamental interaction is instantaneous. At the same time, Eq. (2) permits the generalization to the case where the interaction generating the dynamics of a system is nonlocal in time [1,3]. This point has been illustrated on a toy model in which the NN interaction is described by the separable interaction operator $\langle \mathbf{p}_2|H_{int}^{(s)}(\tau)|\mathbf{p}_1\rangle = \psi^*(\mathbf{p}_2)\psi(\mathbf{p}_1)f(\tau)$, $H_{int}^{(s)}(\tau)$ being the interaction operator in the interaction picture $H_{int}^{(s)}(t_2-t_1) = \exp(-iH_0t_2)H_{int}(t_2,t_1)\exp(iH_0t_1)$, $\mathbf{p}$ the relative momentum of two nucleons, $f(\tau)$ some function of duration time of interaction τ. It was shown that there is one-to-one correspondence between the form of the generalized interaction operator and the UV behavior of the form factor $\psi(\mathbf{p})$. In the case where $\psi(\mathbf{p}) \sim |\mathbf{p}|^{-\alpha}$, $(|\mathbf{p}| \to \infty)$ with $\alpha > \frac{1}{2}$, the interaction operator would necessarily have the form (6). In the case $\alpha \leq \frac{1}{2}$, the interaction operator must be nonlocal in time:

$$\langle \mathbf{p}|H_{int}^{(s)}(\tau)|\mathbf{p}'\rangle = \psi^*(\mathbf{p})\psi(\mathbf{p}')\left(a_1\tau^{-\alpha-\frac{1}{2}} + a_2\tau^{-2\alpha}\right), \quad \alpha < \frac{1}{2}, \tag{7}$$

$$\langle \mathbf{p}|H_{int}^{(s)}(\tau)|\mathbf{p}'\rangle = \psi^*(\mathbf{p})\psi(\mathbf{p}') \int_{-\infty}^{\infty} dx \exp(-iz\tau)\left(\frac{b_1}{\ln(-z)} + \frac{b_2}{\ln^2(-z)}\right), \quad \alpha = \frac{1}{2}, \tag{8}$$

where $a_1 = i(2\pi^2 c_1^2)^{-1} \cos(\alpha\pi)\Gamma^{-1}(1-2\alpha)\exp[\frac{i\pi}{4}(1-2\alpha)](2\mu)^{\alpha-\frac{3}{2}}$, and $b_1 = i(8\pi^2\mu)^{-1}$, μ being the reduced mass. Thus in the case of the nonlocal-in-time NN interaction the form factor $\psi(\mathbf{p})$ would necessarily have the "bad" UV behavior leading to the UV divergences in the Born series. From this one can conclude that the existence of the quarks and gluons confined within hadrons must lead to the "bad" UV behavior of the matrix elements of the evolution operator as a function of the momenta of nucleons. Note that EFT's lead to the same conclusion: Within the EFT approach the quark and gluon degrees of freedom manifest themselves in the form of Lagrangians consistent with the symmetries of QCD which gives rise to the UV divergences.

Let us consider the case $\psi(\mathbf{p}) = (d^2 + p^2)^{-\frac{1}{4}}$. In this case the interaction operator is of the form (8) which contains only one free parameter b_2. However, if there is a bound state in the channel under study, then the parameter b_2 is completely determined by demanding that the T matrix has a pole at energy $E_B = -2.2246$ MeV. In this case $b_2 = b_1 \ln(-E_B)$, and the corresponding solution of Eq. (5) is of the form

$$\langle \mathbf{p}_2 | T(z) | \mathbf{p}_1 \rangle = \psi^*(\mathbf{p}_2)\psi(\mathbf{p}_1)t(z), \tag{9}$$

where

$$[t(z)]^{-1} = (z - E_B) \int d^3k \frac{|\psi(\mathbf{k})|^2}{(z - E_k)(E_B - E_k)}.$$

Note that the same result has been obtained in Ref. [5] by renormalizing the Lippmann-Schwinger (LS) equation where the singular potential $\langle \mathbf{p}_2 | V | \mathbf{p}_1 \rangle = \lambda \psi^*(\mathbf{p}_2)\psi(\mathbf{p}_1)$ with $\psi(\mathbf{p}) = (d^2 + p^2)^{-\frac{1}{4}}$. Thus renormalization of the LS equation with the above singular potential leads to the T matrix we have obtained by solving Eq. (2) with nonlocal-in-time interaction operator. This T matrix satisfies Eq. (5), but does not satisfy the LS equation. Correspondingly the Schrödinger equation is not valid in this case.

To conclude, we have shown that the quark-gluon retardation effects gives rise to the fact that low-energy nucleon dynamics is not governed by the Schrödinger equation, but is governed by the generalized dynamical equation (2). We have shown that the same dynamical situation arises after renormalization of equations governing nucleon dynamics. After renormalization nucleon dynamics is governed by the generalized dynamical equation (2) with a nonlocal-in-time interaction operator. This gives reason to hope that the use of the GQD and parametrization of the *NN* forces like (8) can open new possibilities for applying the EFT approach to the description of low-energy nucleon dynamics. By using an EFT, one can construct the generalized interaction operator consistent with the symmetries of QCD. This operator can then be used in Eq. (2) for describing nucleon dynamics.

REFERENCES

1. Gainutdinov, R. Kh., *J. Phys. A: Math. Gen.* **32**, 5657-5678 (1999).
2. Feynman, R. P., *Rev. Mod. Phys.* **20**, 367-387 (1948).
3. Gainutdinov, R. Kh., and Mutygullina, A. A, *Yad. Fiz.* **60**, 938-945 (1997).
4. Gainutdinov, R. Kh., and Mutygullina, A. A, *Yad. Fiz.* **62**, 2061-2070 (1999).
5. Phillips, D. R., Afnan, I. R., and Henry-Edwards, A. G., *Phys. Rev. C.* **61**, 044002-1–7 (2000).

Quark-gluon retardation effects and an anomalous off-shell behavior of the two-nucleon amplitudes

Renat Kh. Gainutdinov and Aigul A. Mutygullina

Department of Physics, Kazan State University, 18 Kremlevskaya St, Kazan 420008, Russia

Abstract. We show that nonlocality in time of the nucleon-nucleon interaction caused by quark-gluon retardation effects gives rise to an anomalous off-shell behavior of the two-nucleon amplitudes. This anomalous off-shell behavior can have significant effects on three- and many-nucleon results.

Study of the effects of quarks and gluons confined within hadrons on nucleon dynamics is of fundamental importance in understanding the strong interaction. These effects are important for describing the short-range (SR) part of the NN interactions. The effects of quark-gluon retardation on nucleon dynamics differ profoundly from the well-known meson-retardation effects that give rise to nonlocality in time of the NN interaction, and hence to an energy dependence of the effective potentials describing these interactions. Nonlocality in time of such interactions is an expression of a loss of probability from the two-nucleon subspace of the Hilbert space of hadron states. Obviously, quark-gluon retardation, that has to be taken into account in describing hadron dynamics, should also result in nonlocality in time of hadron interactions. However, due to confinement this must not lead to a loss of probability from the Hilbert space of hadron states. On the other hand, as is well known, within the Hamiltonian formalism the interaction generating the dynamics of a quantum system may be nonlocal in time only in the case when the system is not closed, and as a result the evolution is not unitary. To solve this problem quantum dynamics needs to be extended to describe the unitary evolution of a closed system in the case where the interaction generating the dynamics of the system is nonlocal in time. For the first time, this problem was solved in Ref. [1], where it was shown that the use of the Feynman approach [2] to quantum theory in combination with the canonical approach allows such extension of quantum dynamics. The generalized quantum dynamics (GQD) developed in this way was shown [3, 4] to open new possibilities for describing hadron dynamics. In the present paper the GQD is used for investigating quark-gluon retardation effects on low-energy hadron dynamics. We show that retardation from quark confinement results in an anomalous off-shell behavior of the two-nucleon amplitudes.

In the GQD the evolution operator $U(t,t_0)$ is represented in the form [1]

$$\langle \psi_2|U(t,t_0)|\psi_1\rangle = \langle \psi_2|\psi_1\rangle + \int_{t_0}^{t} dt_2 \int_{t_0}^{t_2} dt_1 \langle \psi_2|\tilde{S}(t_2,t_1)|\psi_1\rangle, \qquad (1)$$

where $<\psi_2|\tilde{S}(t_2,t_1)|\psi_1>$ is the probability amplitude that if at time t_1 the system was in the state $|\psi_1>$, then the interaction in the system will begin at time t_1 and will end at time t_2, and at this time the system will be in the state $|\psi_2>$. In the case of an isolated system the operator $\tilde{S}(t_2,t_1)$ can be represented in the form [1] $\tilde{S}(t_2,t_1) = exp(iH_0t_2)\tilde{T}(t_2-t_1)exp(-iH_0t_1)$, H_0 being the free Hamiltonian. It was shown in Ref. [1] that for the evolution operator $U(t,t_0)$ given by (1) to be unitary for any times t_0 and t, the operator $\tilde{S}(t_2,t_1)$ must satisfy the following equation:

$$(t_2-t_1)\tilde{S}(t_2,t_1) = \int_{t_1}^{t_2} dt_4 \int_{t_1}^{t_4} dt_3 (t_4-t_3)\tilde{S}(t_2,t_4)\tilde{S}(t_3,t_1). \tag{2}$$

This equation allows one to obtain the operators $\tilde{S}(t_2,t_1)$ for any t_1 and t_2, if the operators $\tilde{S}(t'_2,t'_1)$ corresponding to infinitesimal duration times $\tau = t'_2 - t'_1$ of interaction are known. It is natural to assume that most of the contribution to the evolution operator in the limit $t_2 \to t_1$ comes from the processes associated with the fundamental interaction in the system under study. Denoting this contribution by $H_{int}(t_2,t_1)$, we can write

$$\tilde{S}(t_2,t_1) \underset{t_2 \to t_1}{\to} H_{int}(t_2,t_1) + o(\tau^\varepsilon). \tag{3}$$

Being a generalization of the interaction Hamiltonian, the operator $H_{int}(t_2,t_1)$ is called the generalized interaction operator. The parameter ε is determined by demanding that $H_{int}(t_2,t_1)$ must be so close to the solution of Eq. (2) in the limit $t_2 \to t_1$ that this equation has a unique solution having the behavior (3) near the point $t_2 = t_1$. If $H_{int}(t_2,t_1)$ is specified, Eq. (2) allows one to find the operator $\tilde{S}(t_2,t_1)$. Formula (1) can then be used to construct the evolution operator $U(t,t_0)$ and accordingly the state vector

$$|\psi(t)\rangle = |\psi(t_0)\rangle + \int_{t_0}^{t} dt_2 \int_{t_0}^{t_2} dt_1 \tilde{S}(t_2,t_1)|\psi(t_0)\rangle$$

at any time t. Thus Eq. (2) can be regarded as an equation of motion for states of a quantum system. It was shown in Ref. [1] that the dynamics governed by Eq. (2) is equivalent to the Hamiltonian dynamics in the case where the interaction operator is of the form

$$H_{int}(t_2,t_1) = -2i\delta(t_2-t_1)H_I(t_1), \tag{4}$$

$H_I(t_1)$ being the interaction Hamiltonian in the interaction picture. In this case the state vector $|\psi(t)\rangle$ satisfies the Schrödinger equation. The delta function $\delta(\tau)$ in (4) emphasizes the fact that in this case the fundamental interaction is instantaneous. At the same time, Eq. (2) permits the generalization to the case where the interaction generating the dynamics of a system is nonlocal in time [3, 4]. In Ref. [4] this point was illustrated on a toy model in which the NN interaction is described by the separable interaction operator

$$\langle \mathbf{p}_2|H_{int}^{(s)}(\tau)|\mathbf{p}_1\rangle = \psi^*(\mathbf{p}_2)\psi(\mathbf{p}_1)f(\tau),$$

$H_{int}^{(s)}(\tau)$ being the interaction operator in the interaction picture $H_{int}^{(s)}(t_2-t_1) = exp(-iH_0t_2)H_{int}(t_2,t_1)exp(iH_0t_1)$, $\mathbf{p}$ the relative momentum of two nucleons, $f(\tau)$ some function of duration time of interaction τ.

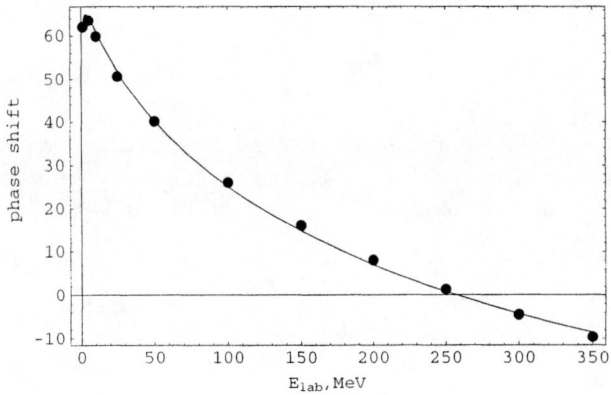

FIGURE 1. Phase shift (solid line) in the $^1S_0$ channel for np scattering, compared to the experimental data (points) [5].

In Ref. [3, 4] the generalized interaction operator

$$\langle \mathbf{p}_2|H_{int}^{(s)}(\tau)|\mathbf{p}_1\rangle = \psi^*(\mathbf{p}_2)\psi(\mathbf{p}_1)\left(a_1\tau^{-\alpha-\frac{1}{2}} + a_2\tau^{-2\alpha}\right), \quad \alpha < \frac{1}{2}, \tag{5}$$

where $a_1 = i(2\pi^2 c_1^2)^{-1}\cos(\alpha\pi)\Gamma^{-1}(1-2\alpha)\exp[\frac{i\pi}{4}(1-2\alpha)](2\mu)^{\alpha-\frac{3}{2}}$ was used for the parametrization of the NN forces. The motivation for using such parametrization is the fact that due to the quark and gluon degrees of freedom the NN interaction must be nonlocal in time. The form factor in (5) is of the form $\varphi(\mathbf{p}) = c_1\left(p^2+d^2\right)^{-\frac{\alpha}{2}} + c_2 g_Y(\mathbf{p})$, $0 < \alpha < \frac{1}{2}$ with $g_Y(\mathbf{p})$ being the Yamaguchi form factor $g_Y(\mathbf{p}) = \left(p^2+\beta^2\right)^{-1}$. Here d, c_1 c_2 and β the parameters given in [4]. Being the generalization of the Yamaguchi model to the case where the NN interaction is nonlocal in time, our model yields nucleon-nucleon phase shifts in good agreement with experiment (see Fig. 1). However, the main advantage of this model is that it allows one to investigate the effects of the retardation in the NN interaction caused by the existence of the quark and gluon degrees of freedom on nucleon dynamics. As is well known, the dynamics of many nucleon systems depends on the off-shell properties of the two-nucleon amplitudes. For this reason, let us consider the effects of nonlocality of the NN interaction on these properties. In the nonlocal case, the matrix elements of the evolution operator as functions of momenta do not go to zero as momenta tend to infinity so fast as it is required by ordinary quantum mechanics [1], and within the Hamiltonian formalism this leads to the ultraviolet divergences. Correspondingly in the nonlocal case the two-nucleon amplitudes $\langle \mathbf{p}_2|T(z)|\mathbf{p}_1\rangle$ do not go to zero as momenta tend to infinity fast enough to make, for example, the Faddeev equation well-behaved. Another consequence of nonlocality in time of the NN interaction is that for fixed momenta $\mathbf{p}_1$ and $\mathbf{p}_2$ the matrix elements $\langle \mathbf{p}_2|T(z)|\mathbf{p}_1\rangle$ tend to zero as $|z| \to \infty$, while, in the local case, they tend to $\langle \mathbf{p}_2|V|\mathbf{p}_1\rangle$ in this limit. To illustrate this point, we present in Fig. 2 the off-shell behavior $\langle \mathbf{p}_2|T(z)|\mathbf{p}_1\rangle$ in the limit $|z| \to \infty$. Thus, nonlocality in

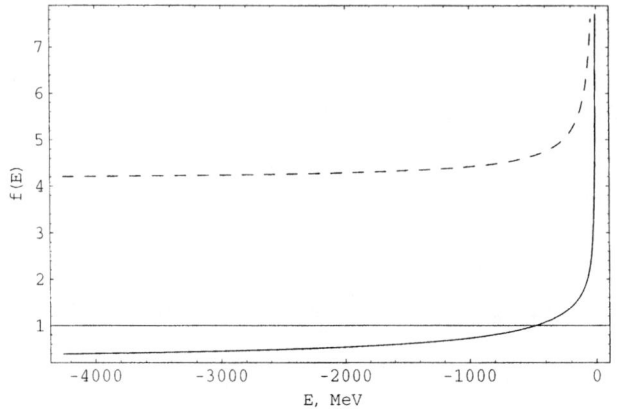

FIGURE 2. The off-shell behavior of $f(E) = \langle \mathbf{p}|T(z)|\mathbf{p}\rangle$, $|\mathbf{p}| = 500$ MeV in the $^3S_1$ channel for np scattering. The solid curves corresponds to the model with generalized interaction operator (5), compared to the model with Yamaguchi potential with parameters given in [6] (dashed line).

time of the NN interaction caused by quark-gluon retardation effects gives rise to an anomalous off-shell behavior of the two-nucleon amplitudes. The off-shell properties of the amplitudes for the ordinary interaction operator and the operator containing the nonlocal term are qualitatively different. This is true even when the two interaction operators have approximately the same phase shifts. Such a large variation in the off-shell behavior of the amplitudes, even when the interaction operators are identical on-shell, can have significant effects on three- and many-body results. This gives reason to expect that the anomalous off-shell behavior of the two-nucleon amplitudes can also have significant effects on nucleon matter properties.

To conclude, we demonstrated that the quark-gluon retardation effects give rise to an anomalous off-shell behavior of the two-nucleon amplitudes. This means that one cannot neglect these effects when describing the NN interaction. As it was shown, the natural way to take into account these effects is the use of the generalized dynamical equation (2) and parametrization of the NN forces like (5).

REFERENCES

1. Gainutdinov, R. Kh., *J. Phys. A: Math. Gen.* **32**, 5657-5678 (1999).
2. Feynman, R. P., *Rev. Mod. Phys.* **20**, 367-387 (1948).
3. Gainutdinov, R. Kh., and Mutygullina, A. A, *Yad. Fiz.* **60**, 938-945 (1997).
4. Gainutdinov, R. Kh., and Mutygullina, A. A, *Yad. Fiz.* **62**, 2061-2070 (1999).
5. Stoks, V. G. J., Klomp, R. A. M., Rentmeester, M. C. M., and de Swart, J. J., *Phys. Rev. C* **48**, 792-812 (1993).
6. Yamaguchi Y., *Phys.Rev.*, **95**, 1628-1643 (1954).

The two-nucleon system in the Δ region including full meson retardation[1]

M. Schwamb* and H. Arenhövel*

Institut für Kernphysik, Johannes Gutenberg Universität, D-55099 Mainz, Germany

Abstract. A model is developed for the hadronic and electromagnetic interaction in the two-nucleon system above pion threshold in the framework of meson, nucleon and Δ degrees of freedom. It is based on time-ordered perturbation theory and includes full meson retardation in potentials and exchange currents as well as loop contributions to the nucleonic one-body current. Results for NN scattering and deuteron photodisintegration are presented.

INTRODUCTION

At present, a very interesting topic in the field of medium energy physics is devoted to the role of effective degrees of freedom (d.o.f.) in hadronic systems in terms of nucleon, meson and isobar d.o.f. and their connection to the underlying quark-gluon dynamics of QCD. For the study of this basic question, the two-nucleon system provides an important test laboratory, because it is the simplest nuclear system for the study of the nucleon-nucleon interaction. Moreover, the role of medium effects due to two-body operators and the role of offshell effects, i.e. the change of single particle properties in the nuclear medium, can be investigated most precisely.

State-of-the-art models for describing hadronic and electromagnetic reactions on the two-nucleon system up to the Δ region should incorporate – among other things – a dynamical treatment of the Δ isobar. Moreover, gauge invariance and unitarity should be fulfilled at least approximately. Within a unitary model, the various possible reactions cannot be treated independently, because they are linked by the optical theorem. For example, for energies up to the two-pion threshold the forward Compton scattering amplitude is related via

$$\text{Im}\, T(\gamma d \to \gamma d; \theta = 0) \;\sim\; \sigma_{tot}(\gamma d \to NN, \pi d, \pi NN) \tag{1}$$

to the sum of the cross sections of photodisintegration and coherent and incoherent pion photoproduction. Therefore, all reactions on the two-nucleon system should be described within one consistent framework. In the past years, we have started to realize this ambitious project within a *retarded* coupled channel $NN/N\Delta$-approach based on three-body scattering theory with nucleon, Δ and meson degrees of freedom [1, 2, 3, 4, 5].

---

[1] Supported by the Deutsche Forschungsgemeinschaft (SFB 443).

THE MODEL

In order to motivate why retardation should be taken into account above pion threshold, let us consider an arbitrary two-body meson-exchange operator like the ordinary one-pion exchange potential. In time-ordered perturbation theory, the propagation of the intermediate πNN system is described by the retarded propagator $G_0^{ret}(E+i\varepsilon) = (E+i\varepsilon - H_0)^{-1}$, where E is the invariant energy of the system and H_0 denotes the kinetic Hamilton operator for the intermediate πNN system. Due to its nonhermiticity, nonlocality and the existence of singularities above pion threshold, an exact treatment of G_0^{ret} is quite complicated. Therefore, in most practical applications a low energy approximation, the so-called *static limit* is used by neglecting the energy transfer between the nucleons by the pion. The corresponding propagator G_0^{stat} is much easier to handle. One encounters on the other hand at least two serious problems. Above pion-threshold, unitarity is violated due to the absence of singularities in G_0^{stat}. Moreover, in the past it turned out that even the simplest photonuclear reaction, namely deuteron photodisintegration, cannot be described even qualitatively within a consistent static framework [6, 7].

For incorporating retardation, the mesons generating the nucleon-nucleon interacton and the meson-exchange currents have to be treated as explicit degrees of freedom. Therefore, the model Hilbert space consists then of three orthogonal subspaces $\mathcal{H}^{[2]} = \mathcal{H}_{\tilde{N}}^{[2]} \oplus \mathcal{H}_{\Delta}^{[2]} \oplus \mathcal{H}_{X}^{[2]}$, where $\mathcal{H}_{\tilde{N}}^{[2]}$ contains two bare nucleons, $\mathcal{H}_{\Delta}^{[2]}$ one nucleon and one Δ resonance, and $\mathcal{H}_{X}^{[2]}$ two nucleons and one meson $X \in \{\pi, \rho, \sigma, \delta, \omega, \eta\}$. Concerning the hadronic part, the basic interactions in our model are XNN and $\pi N\Delta$ vertices. Inserting these into the Lippmann-Schwinger equation, one obtains after some straightforward algebra [3] effective hadronic interactions acting in $\mathcal{H}_{\tilde{N}}^{[2]} \oplus \mathcal{H}_{\Delta}^{[2]}$ which contain the desired *retarded* one-boson exchange (OBE) mechanisms describing the transitions $NN \to NN$, $NN \to N\Delta$ and $N\Delta \leftrightarrow N\Delta$. This strategy of starting with vertices as basic interaction terms in the enlarged Hilbert space $\mathcal{H}^{[2]}$ has the advantage that no inconsistencies occur. In this context, note especially that the vertices are hermitean. On the other hand, if one started with a retarded OBE as basic interaction in $\mathcal{H}_{\tilde{N}}^{[2]} \oplus \mathcal{H}_{\Delta}^{[2]}$, one would loose hermiticity and therefore the solid grounds of quantum mechanics.

In our explicit realization, we use for the parametrization of the retarded NN interaction the Elster potential [8] which takes into account in addition one-pion loop diagrams in order to fulfill unitarity above pion threshold. Therefore, one has to distinguish between bare and physical nucleons (see [3] for details). Concerning the transitions $NN \to N\Delta$ and $N\Delta \leftrightarrow N\Delta$, we take besides retarded pion exchange static ρ exchange into account. Moreover, the interaction of two nucleons in the deuteron channel in presence of a spectator pion (the so-called πd channel) is also considered. By a suitable box renormalization [9], we are able to obtain approximate phase equivalence between the Elster potential and our coupled channel approach below pion threshold.

Similarly, the basic electromagnetic interactions consist of baryonic and mesonic one-body currents as well as vertex and Kroll-Rudermann contributions [4, 5]. These currents are, together with the πNN vertex, the basic building blocks of the corresponding effective current operators. The latter contain beside the ordinary spin-, convection-

and spin-orbit current full retarded pionic meson exchange currents and electromagnetic loop contributions, where the latter can be interpreted as off-shell contributions to the baryonic one-body current [5]. Moreover, static ρ MEC as well as Δ MEC contributions are taken into account. It can be shown [5] that concerning the pionic part gauge invariance is fulfilled in leading order of $1/M_N$.

RESULTS

The hadronic Δ parameters and the M1 $\gamma N\Delta$ coupling are simultaneously fitted to the $M_{1+}(3/2)$ multipole of pion photoproduction on the nucleon, the P_{33} channel in pion-nucleon scattering and the $^1D_2$ channel in nucleon-nucleon scattering [3, 4]. In Fig. 1, our results for the $^1D_2$ phase shift and inelasticity are depicted. We obtain a good description at least up to about $T_{lab} = 800$ MeV. Concering the other partial NN waves, the overall description is fairly well but needs some further improvement in the future [3]. Therefore, at present we are constructing from scratch an improved hadronic interaction model whose parameters are fitted to the phase shifts and inelasticities of *all* relevant NN scattering partial waves for T_{lab} energies up to about 1 GeV.

As next, we discuss very briefly deuteron photodisintegration. The starting point of our consideration is the static approach of Wilhelm *et al.* [7] which is based on the Bonn-OBEPR potential [10]. Similar to our present approach, there is no free parameter in the calculation of the photodisintegration process in [7]. As is evident form Fig. 2, Wilhelm *et al.* clearly fail in describing the data. One obtains a considerable underestimation of the cross section in the Δ peak. Moreover, a dip structure around 90° occurs at higher energies which is not present in the data. On the other hand, these problems in the differential cross section vanish almost completely in a retarded approach [4, 5]. However, some discrepancies in polarization observables like the linear photon asymmetry Σ or the polarization $P_y(p)$ of the outgoing proton are still present, and which need further consideration [4, 5].

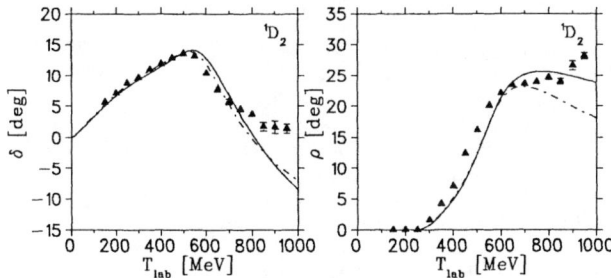

FIGURE 1. Phase shift δ and inelasticity ρ for the $^1D_2$ NN-channel in comparison with experiment (solution SM97 of Arndt *et al.* [11]) for two potential models: dash-dotted curve: static approach, based on the Bonn-OBEPR potential [10], full curve: retarded approach. See [3] for further details.

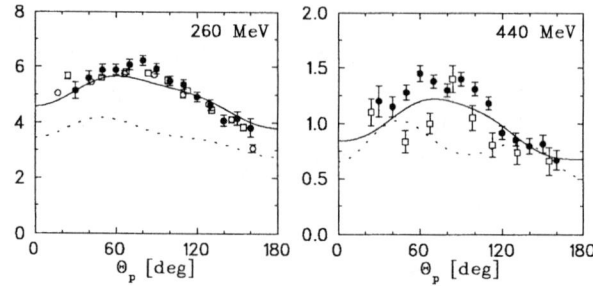

FIGURE 2. Differential cross section of deuteron photodisintegration for two photon energies k_{lab} as function of the c.m. proton angle θ_p: dotted curve: result of Wilhelm *et al.* in static approach [7], full curve: retarded approach of [4]. Offshell contributions to the nucleonic one-body current are included, too [5]. Experimental data from [12] (●), [13] (open box) and [14] (○).

OUTLOOK

A very interesting topic to be studied in the future is the exploration of the spin asymmetry of the total cross section on the nucleon which determines the GDH-sum rule [15, 16] and which is at present under investigation experimentally [17]. Due to the lack of a free neutron target, a measurement on the deuteron is of specific significance because it may serve as an effective neutron target. However, the extraction of the neutron contributions relies on the basic assumption that final state interactions and MEC can be neglected and that proton and neutron contribute incoherently. First, still preliminary results show that these assumptions are quite crude. In the future, we plan to apply the present model to other reactions, especially electrodisintegration. Conceptually, we have to improve our hadronic interaction model. Moreover, additional d.o.f. like the Roper, the D_{13} and the S_{11} resonance should be taken into account if one wants to consider higher energies.

REFERENCES

1. Schwamb, M., Arenhövel, H., Wilhelm, P., and Wilbois, Th., *Phys. Lett.* B **420**, 255 (1998).
2. Schwamb, M., Arenhövel, H., Wilhelm, P., and Wilbois, Th., *Nucl. Phys.* **A631**, 583c (1998).
3. Schwamb, M., and Arenhövel, H., *Nucl. Phys.* **A690**, 647 (2001).
4. Schwamb, M., and Arenhövel, H., *Nucl. Phys.* **A690**, 682 (2001).
5. Schwamb, M., and Arenhövel, H., *nucl-th/0105033*, to be published in Nuclear Physics
6. Tanabe, H., and Ohta, K., *Phys. Rev.* C **40**, 1905 (1989).
7. Wilhelm, P., and Arenhövel, H., *Phys. Lett.* B **318**, 410 (1993).
8. Elster, C., Ferchländer, W., Holinde, K., Schütte, D., and Machleidt, R., *Phys. Rev.* C **37**, 1647 (1988).
9. Green, A M., and Sainio, M.E., *J. Phys.* G **8**, 1337 (1982).
10. Machleidt, R., Holinde, K., and Elster, Ch., *Phys. Rep.* **149**, 1 (1987).
11. Arndt, R.A., *et al.*, program SAID
12. Crawford, R., *et al.*, *Nucl. Phys.* **A603**, 303 (1996).
13. Arends, J., *et al.*, *Nucl. Phys.* **A412**, 509 (1984).
14. Blanpied, G., *et al.*, *Phys. Rev.* C **52**, R455 (1995); *Phys. Rev.* C **61**, 024604 (2000).
15. Gerasimov, S.B., *Yad. Fiz.* **2**, 598 (1965).
16. Drell, S.D., and Hearn, A.C., *Phys. Rev. Lett.* **16**, 908 (1966).
17. Ahrens, J., *et al.*, *Phys. Rev. Lett.* **87**, 022003 (2001).

Neutron-proton and neutron-neutron quasi-free scattering in the nd breakup reaction at 26 MeV

A. Siepe*, W. Glöckle[†], V. Huhn*, L. Wätzold*, Ch. Weber*, H. Wilała**
and W. von Witsch *

*Institut für Strahlen- und Kernphysik, Universität Bonn, D-53115 Bonn, Germany
[†]Institut für Theoretische Physik II, Ruhr-Universität Bochum, D-44780 Bochum, Germany
**Institute of Physics, Jagellonian University, Reymonta 4, PL-30059 Cracow, Poland

Abstract. The regions of neutron-proton and neutron-neutron quasi-free scattering (QFS) have been investigated in the nd breakup reaction at 25.8 MeV, measuring absolute cross sections with an accuracy of a few per cent. The data were analyzed by means of detailed Monte-Carlo simulations based on rigorous three-body calculations with realistic nucleon-nucleon potentials. For np QFS, good agreement between theory and experiment was found. In contrast, for nn QFS the experimentally determined cross section is at least six standard deviations larger than the theoretical predictions.

INTRODUCTION

Quasi-free scattering (QFS) represents a prominent kinematical situation in three-body breakup reactions. Of particular interest is the neutron-induced breakup of the deuteron. There are only three nucleons in this system, two of which are neutrons. Because there are no Coulomb forces, dynamically exact Faddeev-type calculations [1] can be performed with realistic nucleon-nucleon (NN) interactions, making neutron-neutron QFS in the ^{2}H$(n,nn)p$ reaction an important tool for the investigation of the nn interaction. In this paper we report on two experiments in which nn and np QFS were investigated at 25.8 MeV, using the nd breakup reaction.

EXPERIMENTAL DETAILS

The experiments were performed at the cyclotron of the Institut für Strahlen- und Kernphysik at the University of Bonn. The neutron beam was produced *via* the ^{2}H$(d,n)^3$He reaction with 27.2 MeV deuterons incident on a liquid-nitrogen cooled gas target, operated at a pressure of 40 bar. The primary beam was stopped directly behind the gas target which served as a Faraday cup. The neutrons were collimated at 0^0 to form a well-defined circular beam with a diameter of 31 mm (FWHM) at the reaction target which was positioned 195 cm from the center of the gas target. The neutron flux on the target in the quasi-monoenergetic high-energy (HE) peak from the ^{2}H$(d,n)^3$He reaction was 3.7×10^5 /s, with an average energy $E_0 = 25.8$ MeV and an energy spread $\Delta E_0 = 4.0$ MeV. As a beam monitor, a double proton-recoil telescope (PRT) was placed in the n-

beam 148 cm from the gas target to detect protons emitted from a CH_2 target at angles of $\pm 35^0$ with respect to the n-beam. The PRT was used for the absolute normalization of the HE neutron beam as described below.

For the measurement of np QFS, the neutrons were detected at $\Theta_n = 39^0$ and protons at $\Theta_p = 42^0$, with $\Phi_{np} = 180^0$. The reaction target was a deuterated polyethylene foil with a thickness of 48 mg/cm^2. At the position of the CD_2 target, the n-beam had a plateau of constant intensity with a diameter >25 mm, and the whole target was illuminated homogeneously. At a distance of 8 cm from the target and outside of the neutron beam, a thin NE104 scintillator foil was positioned in an Al reflector with thin entrance and exit windows, viewed from above by a photo-multiplier. The charged-particle signals produced in this transmission foil detector (TFD) were used as start signals for all TOF measurements. The protons were detected with a plastic scintillator of 10 cm diameter and 5 mm thickness, positioned 70 cm from the CD_2 target and viewed by a 5 in. photomultiplier. The target and transmission detector were mounted in an evacuated pipe, called "proton arm", which was sealed at the front end with a Be window for the n-beam. The n-detector was positioned at a distance of 75 cm from the CD_2 target. It had a nominal diameter of 5 in. and a thickness of 3 in., and was equipped with $n - \gamma$ pulse-shape discrimination. All detectors were provided with LED pulsers to monitor gain shifts, dead times and pile-up. For the nn measurement, the proton arm was replaced by a second n-detector. Both n-detectors were positioned symmetrically at $\Theta_n = 42^0$. The target now consisted of a thin-walled, upright Al cylinder, 65 mm high and 44 mm in diameter, filled with C_6D_{12} liquid scintillator. It was closed at the bottom with a quartz window and viewed from below by a photomultiplier.

The neutron fluence F_n in the intensity plateau of the beam was measured by means of np scattering, for which the CD_2 target in the proton arm was replaced by a CH_2 foil of equal size. A second, independent value for F_n was obtained from the simultaneous PRT measurement. The integrated HE neutron flux for the subsequent measurements with the CD_2 target could thus be determined with an absolute accuracy of 1.1%, using the PRT as a relative monitor. The integrated beam-target luminosity BT for the nn measurement was also determined through np scattering. For this the C_6D_{12} target cylinder was replaced by an identical one filled with C_6H_6. Scattered neutrons were detected at $\Theta_n = 42^0$, and the luminosity was determined relative to the number of counts in the PRT. The luminosity was also *calculated* with a detailed Monte-Carlo (MC) simulation based on the accurately known value of F_n. Both results for BT agreed within 0.6%, and the overall error for this quantity is estimated to be 1.2%.

The *central* efficiency of the n-detectors was measured using again the setup with the CH_2 target. The n-detector was positioned at 90^0 with respect to the proton arm, close to the target to assure that all np neutrons hit the detector near its center in a narrow cone defined by the solid angle of the p-detector. Windows were set off-line in the TOF spectrum of the incoming neutron beam to select bins of energies for the scattered neutrons for which the efficiency was then determined from the number of free proton counts vs. the number of pn coincidences. The results were compared with MC calculations based on an expanded version [2] of the PTB Monte-Carlo efficiency program developed by Dietze and Klein [3], and the agreement was very good for all energies. The *average* efficiency ε was then calculated with the PTB program and normalized to the measured central one. The total error for ε was ±1.4%.

In the np measurement, triple coincidences were required, with the TFD providing the start signal for all TOF measurements. In the nn QFS experiment, only two-fold coincidences were required between the two n-detectors, and the energy of the neutrons was measured *via* their TOF with respect to the rf of the cyclotron. In addition, the time between the signals in the two n-detectors, TOF_{12}, was measured to allow for the subtraction of accidental background, while the target signals were used for diagnostic purposes only. A background run was made in both measurements with the deuterium target replaced by a hydrogen target of equal size.

For np scattering, the raw data were first reduced by selecting the HE part of the n-beam. A lower threshold of 60 keV equivalent electron energy (keVee) was set in the dynode of the n-detector, and γ-rays were removed via pulse shape analysis. Coincidences with deuterons in the p-detector were excluded by an appropriate window in the $(E_p$ vs. $TOF_p)$ - matrix. For nn scattering, thresholds of 1 MeVee were required for the dynode signals of the n-detectors in order to remove the ambiguities caused by the 28 MHz repetition rate of the rf signal from the accelerator. For the data of the nn measurement the most important correction - apart from the efficiency - was due to multiple scattering in the target which was estimated by means of Monte Carlo (MC) simulations.

Absolute theoretical spectra were produced with finite-geometry MC calculations using 3N breakup cross sections obtained from rigorous, fully charge-dependent Faddeev-type calculations in momentum space, with the CD-Bonn potential [4] as input for the NN interaction. The possible effect of a three-nucleon force was estimated by adding the Tucson-Melbourne 3NF to the NN interactions, and was found to be negligibly small.

RESULTS

For np QFS the measured cross section agrees with the theoretical prediction within $(2.0 \pm 2.5)\%$. On the other hand, for nn QFS there is a large discrepancy between the experimentally measured yield and the MC prediction. As in the np case, the shape of the QFS peak is well described, but the absolute yield is underestimated by 17.8% using CD-Bonn [4], and by 19.7% with the Argonne v_{18} potential [5]. The total experimental error is 3.0%. Considering the near-perfect agreement with theory for np QFS, the large difference found in the nn experiment is surprising. Unfortunately, there are no other high-precision data for nn QFS in the literature. Quasi-free scattering in pd breakup can not be used for comparison because of Coulomb effects which are appreciable in the region of QFS [6], and pd calculations with a realistic nuclear input are not yet available. An independent nn measurement is urgently needed to verify the present result.

ACKNOWLEDGEMENTS

This work was supported by the Deutsche Forschungsgemeinschaft under Grant No. WI 1144/5-2, and by the Polish Committee for Scientific Research under Grant No. 2P03B02818. The three-body calculations were performed on the CRAY T90 and T3E of the John von Neumann Institute for Computing in Jülich, Germany.

REFERENCES

1. Glöckle, W., Witała, H., Hüber, D., Kamada, H., and Golak, J., *Phys. Rep.* **274**, 107-317 (1996).
2. Fabry, I., Diploma thesis, University of Bonn, 1998 (unpublished).
3. Dietze, G., and Klein, H., Technical Report PTB-ND-22 (1982), Physikalisch-Technisch Bundesanstalt Braunschweig, Germany.
4. Machleidt, R., Sammarruca, F., and Song, Y., *Phys. Rev. C* **53**, 1483-1487 (1996).
5. Wiringa, R.B., Stoks, V.G.J., and Schiavilla, R., *Phys. Rev. C* **51**, 38-51 (1995).
6. Alt, E.O., and Rauh, M., *Few-Body Syst.* **17**, 121-133 (1994).

Recent measurements with the out-of-plane spectrometer system at MIT-Bates

Simon Širca, for the OOPS Collaboration

MIT, Laboratory for Nuclear Science, 77 Massachusetts Avenue, Cambridge, MA 02139, USA

Abstract. The recent experimental program with the out-of-plane spectrometer system (OOPS) at MIT-Bates encompassed an extensive set of $d(\vec{e},e'p)$ measurements, investigations of the $N \to \Delta$ transition using $p(\vec{e},e'p)\pi^0$ and $p(\vec{e},e'\pi^+)n$ reaction channels, and studies of virtual Compton scattering (VCS) $p(e,e'p)\gamma$ below the pion threshold. Preliminary results are presented.

Measurements of $d(\vec{e},e'p)$ in the dip region

Early measurements of unpolarised responses for deuteron electro-disintegration have yielded a substantial, but inconsistent body of data which could not be adequately described by theoretical models (see [1] for a review). To provide a richer input to theories, the deuteron program at Bates has been dedicated to separations of interference responses in a variety of kinematical settings. This could be achieved by simultaneous out-of-plane detection of ejected hadrons about the momentum transfer, thereby minimising systematic uncertainties and allowing for separation of both asymmetries and absolute responses [2, 3]. With this novel technique, competing effect in deuteron electro-disintegration (final-state interactions (FSI), meson-exchange currents (MEC), isobar configurations (IC), and relativistic corrections (RC)) can be probed precisely and selectively.

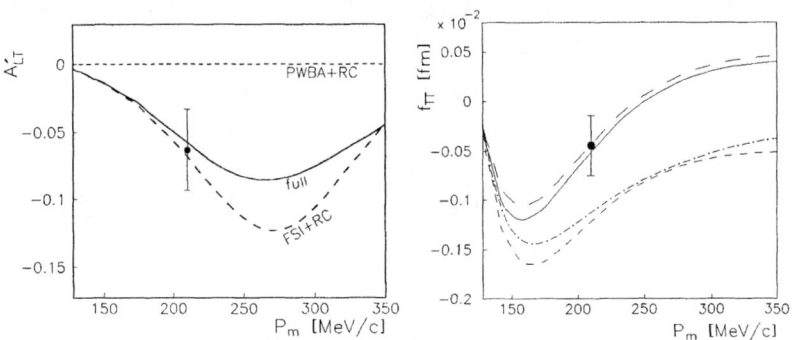

FIGURE 1. Preliminary results for the asymmetry A'_{LT} and the f_{TT} response as functions of p_m. Left panel: calculations of ref. [6] with ingredients marked in the figure. Right panel: calculations of ref. [6]: relativistic PWBA+FSI+RC (dashed), PWBA+FSI+MEC+IC (long-dashed), and full (solid), and of ref. [7]: PWBA+FSI (dashed-dotted). Note that both of these observables require out-of-plane detection.

Early deuteron electro-disintegration studies at MIT-Bates were focused on the quasi-elastic region [4, 5]. We report here on the measurements of $d(\vec{e}, e'p)$ in the dip region, where the asymmetries A_{LT}, A'_{LT}, and A_{TT}, as well as the responses f_{LT}, f_{LT}, and f_{TT} were measured at a beam energy of 800 MeV, a four-momentum transfer of $Q^2 = 0.15\,\text{GeV}^2$, and a missing momentum of 210 MeV/c. Preliminary results on the asymmetry A'_{LT} and the f_{TT} response are given in figure 1. The asymmetry A'_{LT} reflects the imaginary part of the LT interference term, and is thus highly sensitive to FSI. The f_{TT} response is very sensitive to an accurate inclusion of MEC and IC, and exhibits almost no dependence on relativity, contrary to A_{LT} and f_{LT}. These results have been submitted for publication in Phys. Rev. Lett.

These measurements were performed with pulsed ($\simeq 1\%$-duty-factor) beam, while the recent availability of the high-duty-factor extracted beam has challenged the OOPS Collaboration to continue the pursuit of the out-of-plane program, to extend it to higher missing momenta and into the Δ-region. At higher energy transfers and higher missing momenta, an enhanced sensitivity to relativistic effects in A_{LT} and f_{LT}, and to MEC and IC in A_{TT} and f_{TT} is expected. In the same kinematical region, the A'_{LT} and f'_{LT} will improve our understanding of $\Delta - N$ interactions in the final state. In addition, these studies can be complemented with an enlarged set of polarisation observables, and can be performed with a significant decrease in experimental uncertainties.

Upgrade and commissioning of the MIT-Bates South Hall and OOPS

For the two most recent experimental efforts at MIT-Bates which required a high-duty-factor beam, major instrumental developments have taken place in the South Hall of the facility. A gantry support system allowing for out-of-plane positioning of two OOPS modules has been completed, and the fourth OOPS module has been commissioned. The OHIPS spectrometer has been upgraded for a momentum bite of 14% and outfitted with a new vertical drift-chamber and additional scintillators and lead-glass detectors. Substantial modifications were made to the beam-line and readout electronics to conform to the CW beam extracted from the South-Hall storage ring [8].

Studies of the N $\to \Delta$ transition

Studies of the $\gamma^* N \to \Delta$ transition using OOPS have a long tradition [9]. They were originally aimed at precise extractions of the E2/M1 and C2/M1 multipole amplitude ratios which quantify the deviation of the nucleon or the Δ from a spherical shape such as that assumed in the naive non-relativistic quark model or in models with a spherically-symmetric pion field. The initial effort to disentangle the quadrupole amplitudes from the dominating magnetic dipole component has now been augmented by more complete investigations. The LT, LT', and TT interference responses and the corresponding asymmetries have been isolated in wide kinematical regions. The $p(e, e'p)\pi^0$ process has been studied in a range of invariant masses in the vicinity of the Δ-resonance peak to explore the W-dependence of resonance and background contributions. A handle on the isospin

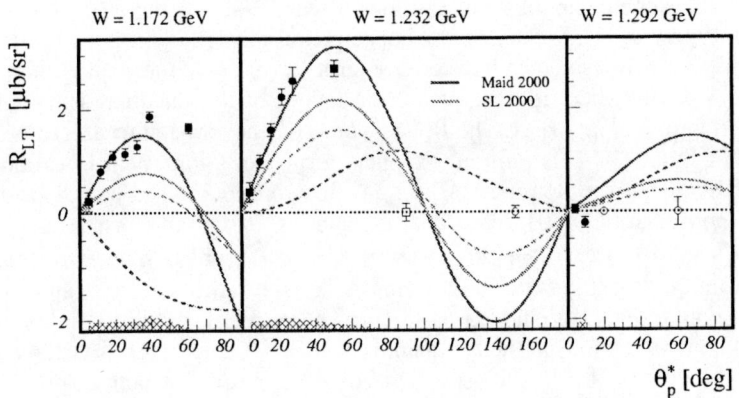

FIGURE 2. The R_{LT} response at $Q^2 = 0.126\,\text{GeV}^2$ and $W = 1172$, 1232 and 1292 MeV. The open square and the open circles denote the expected statistical uncertainties for the 2001 run. The shaded areas represent the systematic uncertainties. Full circles and squares: previously taken data [10, 11]. Model predictions are by Sato and Lee [12] and MAID [13]. The dashed curves correspond to calculations without the C2 contribution.

structure of the N → Δ transition has been obtained by measurements of the concurrent $p(e,e'n)\pi^+$ process. Angular distributions in $\theta^\star_{pq}$ (or $\theta^\star_{\pi q}$) of protons (or π^+) have been measured to allow for a multipole decomposition of the responses.

During the 2001 run, the OOPS system was operated in the full, four-module setup for the first time, with a beam energy of 950 MeV at high duty-factor currents of up to 10 μA. Two modules mounted on the gantry, in conjunction with the two in-plane modules, allowed for a decomposition of all unpolarised response functions, including R_{TT}. Running at forward angles and low proton momenta permitted a significant ex-

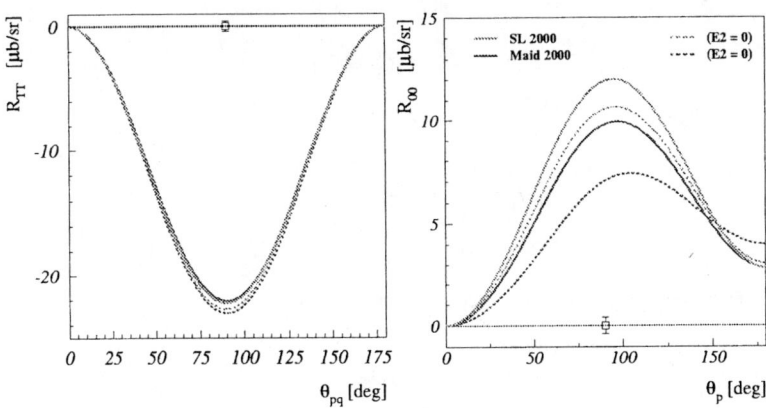

FIGURE 3. Expected statistical uncertainties for R_{TT} and R_{00} (see text) in the π^0-channel at $W = 1232\,\text{MeV}$ and $Q^2 = 0.126\,\text{GeV}^2$. The empty square in the right panel corresponds to about a third of the data taken in the run of 2001. (Data at $\theta^\star_{pq} = 151°$ was also taken.) For curves, see caption to figure 2.

tension of the range of angular distributions. Figure 2 shows the R_{LT} response which offers a most precise test of the existing phenomenological models. It is highly sensitive to the poorly-known resonance interference term $\mathrm{Re}(S^\star_{1+}M_{1+})$, except at the particular angle of $\theta_{pq} = 90°$ where it is suppressed in favour of the maximal sensitivity to the background term $\mathrm{Re}(S^\star_{0+}M_{1+})$. The R_{TT} response has never been measured before (see left panel of figure 3). This pure out-of-plane response is important in calibrating the absolute strength of the dominant M_{1+} transition multipole. The particular combination of responses at $\theta^\star_{pq} = 0°$ (in parallel kinematics) and at an emission angle of $90°$, $R_{00} \equiv (R_T + \varepsilon_L R_L)(90°) + R_{TT}(90°) - (R_T + \varepsilon_L R_L)(0°)$ will also be extracted, since it exhibits a unique sensitivity to the E2 amplitude (see right panel of figure 3). Excellent beam conditions also enabled us to carry out a measurement of interference responses and cross-sections in the π^+-channel at $W = 1232\,\mathrm{MeV}$, $Q^2 = 0.126\,\mathrm{GeV}^2$, and $\theta^\star_{\pi q} = 44.5°$, which is being analysed. The combined results of both reaction channels will yield information on the isospin structure of the $N \to \Delta$ transition.

Virtual Compton scattering

The Virtual Compton scattering experiment [14] was one of the highlights of this year's extracted-beam program. The process $p(e,e'p)\gamma$ has been measured at five beam energies, corresponding to the outgoing-photon energies ranging from 28 to 115 MeV/c, at four-momentum transfer of $Q^2 = 0.05\,\mathrm{GeV}^2$. The physical goal was to measure the generalised polarisabilities of the proton at low momentum transfers and test several competing theories with great accuracy and minimal systematical uncertainties. The OOPS setup is excellently matched to access these observables at low momentum transfers, in particular due to a great out-of-plane suppression of the Bethe-Heitler background. Contrary to in-plane experiments which are primarily sensitive to the magnetic generalised polarisabilities, out-of-plane detection uniquely allows for a clear isolation of the electric part. A detailed analysis of the data is underway.

REFERENCES

1. Gilad, S., Bertozzi, W., Zhou, Z.-L., *Nucl. Phys.* **A631**, 276c (1998).
2. Dolfini, S.M., et al., *Nucl. Instr. Meth. A* **344**, 571 (1994).
3. Mandeville, J.B., et al., *Nucl. Instr. Meth. A* **344**, 583 (1994).
4. Dolfini, S.M., et al., *Phys. Rev. C* **51**, 3479 (1995).
5. Jordan, D., McIlvain, T., et al., *Phys. Rev. Lett.* **76**, 1579 (1996).
6. Ritz, F., Göller, H., Wilbois, Th., Arenhövel, H., *Phys. Rev. C* **55**, 2214 (1997).
7. Hummel, E., Tjon, J.A., *Phys. Rev. C* **42**, 423 (1990); **49**, 21 (1994).
8. The MIT-Bates Linear Accelerator Center Annual Report, 2000.
9. Bates Proposal 87–09 (C.N. Papanicolas, spokesperson), Bates Proposal 97–04 (M.O. Distler, A.M. Bernstein, spokespersons).
10. Mertz, C., et al., *Phys. Rev. Lett.* **86**, 2963 (2001).
11. Kunz, C., Doctoral Dissertation, MIT, 2000.
12. Sato, T., Lee, T.-S.H., *Phys. Rev. C* **63**, 055201 (2001).
13. Drechsel, D., Hanstein, O., Kamalov, S.S., Tiator, L., *Nucl. Phys.* **A645**, 145 (1999).
14. Bates Proposal 97–03, (J. Shaw, R. Miskimen, spokespersons).

Electromagnetic isoscalar $\rho\pi\gamma$ exchange current and the anomalous action

E. Truhlík*, J. Smejkal* and F.C. Khanna[†]

*Institute of Nuclear Physics, Academy of Sciences of the Czech Republic, CZ-25068 Řež, Czechia
[†]Theoretical Physics Institute, Department of Physics, University of Alberta, Edmonton, Alberta, Canada, T6G 2J1 and TRIUMF, 4004 Wesbrook Mall, Vancouver, BC, Canada, V6T 2A3

Abstract. Using modern data, we first refit constants needed to complete an anomalous $\pi\rho\omega a_1$ Lagrangian obtained within the approach of hidden local symmetries. Then we derive from this Lagrangian electromagnetic isoscalar $\rho\pi\gamma$ and $\rho a_1\gamma$ exchange currents needed in calculations of the deuteron electromagnetic form factors at large momentum transfers.

INTRODUCTION

In calculations of the electromagnetic form factors of the deuteron, the $\rho\pi\gamma$ exchange current plays an important role [1, 2, 3]. According to [4, 5, 6], it is derived from a $\rho\pi\gamma$ vertex

$$< \pi^m(q_2)|J^{e.m.}_\lambda|\rho^l_\mu(q_1) > = i\frac{e\, g_{\rho\pi\gamma}K_{\rho\pi\gamma}(q^2)}{m_\rho}\varepsilon_{\lambda\nu\sigma\mu}q_{2\nu}q_\sigma\delta_{ml}, \qquad (1)$$

where $q = q_2 - q_1$ and the constant $g_{\rho\pi\gamma}$ is extracted from the data on the width of the decay $\rho \to \pi + \gamma$.

In Ref. [7], we consider the task of constructing this current by using an anomalous Lagrangian. Such a Lagrangian of the $\pi\rho\omega a_1 f_1$ system was constructed in [8] within the approach of hidden local symmetries. It includes also the external electromagnetic field. The subsystem $\pi\rho\omega a_1$ was considered later in [9], with both the external electromagnetic and weak vector and axial–vector fields included, however. In the Lagrangian, several terms are present, which yield, besides the current (1), a correction to it and also a new current $\rho a_1\gamma$. The Lagrangian is characterized by several parameters, which should be determined from the data on various reactions. Experimental data available 10 years ago allowed Kaiser and Meissner [8] to extract only some of them. The present experimental situation [10] and results of the recent work [11] allow us to improve the analysis. As a result, all 4 parameters entering the Lagrangian of the system $\pi\rho\omega a_1$ are now available.

We used [7] further the constructed anomalous Lagrangian to derive the electromagnetic isoscalar $\rho\pi\gamma$ and $\rho a_1\gamma$ exchange currents, which can be employed in calculations of the electromagnetic form factors of the deuteron. They differ from the standard approach by the presence of an additional momentum dependence in the electromagnetic form factor of the $\rho\pi\gamma$ current and by the completely new $\rho a_1\gamma$ current. The considered model fully respects chiral invariance and vector dominance and it is consistent with the present experimental knowledge of elementary processes such as radiative decays $\rho \to \pi\gamma$, $\omega \to \pi\gamma$, $f_1 \to \rho\gamma$ and $f_1 \to \rho\pi\pi$.

EXTRACTING THE CONSTANTS FROM THE DATA

The $\rho\pi\gamma$ and $\rho a_1 \gamma$ vertices and the associated exchange currents were derived [7] from an anomalous $\pi\rho\omega a_1$ Lagrangian [8, 9], which includes all contributions arising from the homogeneous terms of the anomalous action due to the presence of the external electroweak interactions changing the natural parity. The total Yukawa-type part of this Lagrangian is given by the sum [7]

$$\bar{\mathcal{L}}_{an} = \sum_{i=7,10} \bar{c}_i \bar{\mathcal{L}}_i, \qquad (2)$$

where the constants, $\bar{c}_i$, according to [9] are

$$\bar{c}_7 = \tilde{c}_7 + \frac{1}{2}\tilde{c}_8, \quad \bar{c}_8 = \tilde{c}_8, \quad \bar{c}_9 = \tilde{c}_9 + \frac{1}{2}\tilde{c}_{10}, \quad \bar{c}_{10} = \tilde{c}_{10}, \qquad (3)$$

and the constants $\tilde{c}_i$ are given in [8] [1]. However, new data on several reactions relevant to the analysis aiming to find these constants has recently been published [10] and also a new work [11], in addition to [8], has appeared, which enabled us [7] to get more reliable values of them.

The constants $\tilde{c}_7$ and $\tilde{c}_9$ are related to the effective constants $g_{\rho\omega\pi}$, $g_{\rho\pi\gamma}$ and $g_{\omega\pi\gamma}$ as

$$g_{\rho\omega\pi} = 4g^2 \tilde{c}_7, \quad g_{\rho\pi\gamma} = \frac{2gm_\rho}{3f_\pi}(\tilde{c}_7 + \tilde{c}_9), \quad g_{\omega\pi\gamma} = \frac{2gm_\omega}{f_\pi}(\tilde{c}_7 + \tilde{c}_9). \qquad (4)$$

As found in [11], the constant $g_{\rho\omega\pi} = 1.2$, whereas the other two decay constants enter the radiative decay width of a vector meson, V, for the decay into a pseudoscalar meson P [12]. Using the data [10] for the widths $\Gamma_{\rho^\pm \to \pi^\pm \gamma}$ and $\Gamma_{\omega \to \pi^0 \gamma}$ we have

$$\tilde{c}_7 = 8.64 \times 10^{-3}, \quad \tilde{c}_9 = 9.23 \times 10^{-3}. \qquad (5)$$

In comparison with $\tilde{c}_9$ [8], the new value (5) has the opposite sign and it is larger by a factor of ~ 2.

We can also use the same procedure with the neutral ρ meson decay, which provides $\tilde{c}_9 = 1.25 \times 10^{-2}$. This value of $\tilde{c}_9$ differs from the one of Eq. (5) by $\approx 10\%$. However, the difference in the effective coupling constants $g_{\rho\pi\gamma}$ and $g_{\omega\pi\gamma}$ is only 4% [7].

The constants $\tilde{c}_8$ and $\tilde{c}_{10}$ are needed to calculate the anomalous processes with the a_1 meson. However, the decays of this meson are not well measured and we use the f_1 meson decays to extract them from the data. In contrast to the experimental situation

---

[1] The constants $\tilde{c}_i$ used here differ from those of Ref. [8] by the factor g.

in late eighties, the decay $f_1 \to \rho\pi\pi$ is now well measured [10] and good data for the radiative decay $f_1 \to \rho\gamma$ has appeared [10]. Using the new data yields

$$\tilde{c}_8 = -1.02 \times 10^{-1}, \quad \tilde{c}_{10} = 1.29 \times 10^{-1}. \tag{6}$$

The new value of $\tilde{c}_8$ is lower by a factor of ~ 2 than the old one [8] and $\tilde{c}_{10}$ is in absolute value even larger than $\tilde{c}_8$.

Electromagnetic form factors

The Lagrangian of Eq. (2) yields the vertex $\rho\pi\gamma$, Eq. (1), with the form factor

$$K_{\rho\pi\gamma}(q^2) = F_\omega(q^2) + \left(\frac{\tilde{c}_9 - \tilde{c}_7}{\tilde{c}_9 + \tilde{c}_7}\right)\left[1 - F_\omega(q^2)\right], \tag{7}$$

where the form factor $F_\omega(q^2)$ is given by the vector dominance. On the other side, the terms $\tilde{L}_8$ and $\tilde{L}_{10}$ in Eq. (2) imply the existence of the vertices $\rho a_1 \gamma$, which give rise to the corresponding $\rho a_1 \gamma$ currents. In analogy with Eq. (1) we have

$$< a_{1\nu}^m(q_2)|J_\lambda^{e.m.}|\rho_\mu^l(q_1)> = eg_{\rho a_1 \gamma} \varepsilon_{\lambda\nu\sigma\mu}\left[K_1(q^2)q_\sigma + K_2(q^2)q_{1\sigma}\right]\delta_{ml}, \tag{8}$$

where the form factors $K_{1,2}(q^2)$ are defined as

$$K_1(q^2) = \frac{1}{\tilde{c}_8 + \tilde{c}_{10}}\left[\tilde{c}_8 F_\omega(q^2) + \tilde{c}_{10}\right], \quad K_2(q^2) = \frac{\tilde{c}_8}{\tilde{c}_8 + \tilde{c}_{10}}\left[F_\omega(q^2) - 1\right]. \tag{9}$$

This gives a new contribution to the current.

THE $\rho\pi\gamma$ AND $\rho A_1 \gamma$ EXCHANGE CURRENTS

The general structure of the exchange currents which follows from our Lagrangian (2) is given in Fig. 1. The current of the pion range is

$$J_{\pi\mu} = ig_{\pi NN}\frac{g}{2}\frac{g_{\rho\pi\gamma}}{m_\rho}K_{\rho\pi\gamma}(q^2)\varepsilon_{\mu\kappa\beta\nu}q_\kappa \mathcal{P}_\beta(1,2)q_{2\nu} + (1 \leftrightarrow 2), \tag{10}$$

where

$$\mathcal{P}_\beta(1,2) = \Gamma_{1,\beta}^a \Delta_F^\rho(q_1^2)\Delta_F^\pi(q_2^2)\Gamma_{2,5}^a, \tag{11}$$

and

$$\Gamma_{i,\beta}^a = \bar{u}(p_i')(\gamma_\beta - \frac{\kappa_V}{2M}\sigma_{\beta\delta}q_{i\delta})\tau^a u(p_i), \quad \Gamma_{j,5}^a = \bar{u}(p_j')\gamma_5 \tau^a u(p_j), \tag{12}$$

with the effective coupling constant $g_{\rho\pi\gamma}$ and the form factor $K_{\rho\pi\gamma}(q^2)$ being defined in Eqs. (4) and (7), respectively.

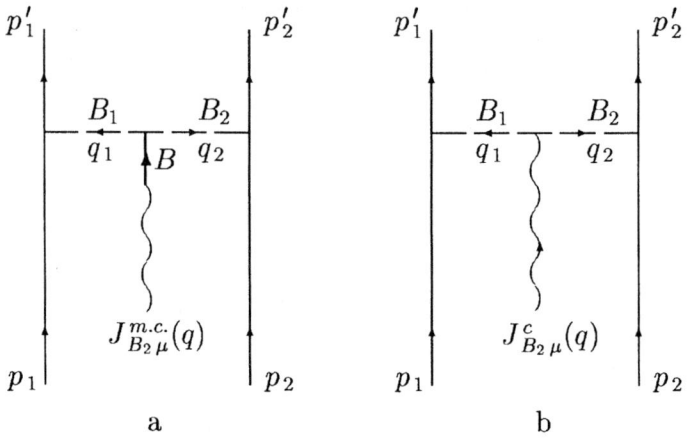

FIGURE 1. The general structure of the electromagnetic isoscalar $\rho\pi\gamma$ and $\rho a_1\gamma$ exchange currents $J_\mu(q)$ considered in this paper. The meson B_1 is the ρ meson. The range of the current is given by the meson B_2 which is either π or a_1 meson.

The currents of the a_1 meson range is

$$J_{a_1\mu} = -\frac{g_A}{2} g^2 g_{\rho a_1 \gamma} \varepsilon_{\mu\kappa\beta\nu} \left[K_1(q^2) q_\kappa + K_2(q^2) q_{1\kappa} \right] \mathcal{P}_{\beta\nu}(1,2) + (1 \leftrightarrow 2), \tag{13}$$

where the weak axial coupling constant $g_A = 1.26$, and

$$\mathcal{P}_{\beta\nu}(1,2) = \Gamma^a_{1,\beta} \Delta^\rho_F(q_1^2) \Delta^{a_1}_F(q_2^2) \Gamma^a_{2,5\nu}, \quad \Gamma^a_{j,5\nu} = \bar{u}(p'_j) \gamma_\nu \gamma_5 \tau^a u(p_j), \tag{14}$$

with the effective coupling constant $g_{\rho a_1 \gamma} = 2g^2(\tilde{c}_8 + \tilde{c}_{10})/3$ and the form factors $K_{1,2}(q^2)$ given in Eq. (9).

This work is supported in part by the grant GA ČR 202/00/1669. The research of F. C. K. is supported in part by NSERCC.

REFERENCES

1. Platchkov, S., et al., *Nucl. Phys.* **A510**, 740 (1990).
2. Mosconi, B. and Ricci, P., *Few–Body Systems* **6**, 63 (1989).
3. Sarriguren, P., Martorell, J. and Sprung, D., *Phys. Lett.* B **228**, 285 (1989).
4. Adler, R.J., *Phys. Rev.* **141**, 1499 (1966).
5. Chemtob, M., Moniz, E. and Rho, M., *Phys. Rev.* C **10**, 344 (1974).
6. Gari, M. and Hyuga, H., *Nucl. Phys.* **A264**, 409 (1976).
7. Truhlík, E., Smejkal, J. and Khanna, F.C., *Nucl. Phys.* **A689**, 741 (2001).
8. Kaiser, N. and Meißner, U.-G., *Nucl. Phys.* **A519**, 671 (1990).
9. Smejkal, J., Truhlík, E. and Khanna, F.C., Non-abelian anomaly and the Lagrangian for radiative muon capture, in: Adam, J., Jr. et al., eds., *Proceedings Sevenths International Conference 'Mesons and Light Nuclei 98', Prague–Pruhonice, 1998* (World Scientific, Singapore, 1999), p.490.
10. Caso, C. et al., (Particle Data Group), *Eur. Phys. J.* C **3**, 1 (1998).
11. Klingl, F., Kaiser, N. and Weise, W., *Z. Phys.* A **356**, 193 (1996).
12. Dumbrais, O. et al., *Nucl. Phys.* **B216**, 277 (1983).

Nuclear forces and quark degrees of freedom

M. Lacombe*, B. Loiseau*, R. Vinh Mau*, P. Demetriou[†],
J.P.B.C. de Melo** and C. Semay[‡]

*Laboratoire de Physique Nucléaire et des Hautes Energies, LPTPE, Université Pierre et Marie Curie, 4 Place Jussieu, F-75252 Paris CEDEX 05, France.
[†]Institute of Nuclear Physics, NCSR "Demokritos", GR-15310 Athens, Greece.
**Instituto de Fisica Teorica, IFT, (UNESP) rua Pamplona, 145, Bela Vista, Cep: 01405-900, Sao Paulo, Brazil.
[‡]Université de Mons-Hainaut, Faculté des Sciences, 20 Place du Parc, B-7000 Mons, Belgium.

Abstract. The description of the short-range part of the nucleon-nucleon forces in terms of quark degrees of freedom is tested against experimental observables. The model considered for this purpose consists of short range forces given by the quark cluster model and long and medium range forces by well established meson exchanges. The investigation is performed using different quark cluster models coming from different sets of quark-quark interaction. The predictions of this model are compared directly with the experimental observables. Agreement with the existing pp and np world set of data is poor. This suggests that the current description of the nucleon-nucleon interaction, at short distances, in the framework of the non-relativistic quark models, is at present only qualitative.

THE MODEL

Several attempts [1, 2, 3, 4] have been devoted to the derivation of a NN potential from the quark degrees of freedom, namely in the framework of the so called quark cluster model (QCM). One of the outcomes of these works is that the dominant potential obtained is repulsive for all internucleon distances. This property, of course, is desirable for short distances but the lack of attraction in the medium range is very troublesome. Subsequent works [5, 6] remedy this defect by adding to the QCM some meson exchanges. However, these meson exchanges are *adjusted* to give a best fit of the NN phase shifts. The result is that the obtained medium range NN forces are then too strongly attractive to be realistic as noted in Refs. [7, 8].

In contrast, the long and medium range (LR+MR) nucleon-nucleon forces provided by meson exchanges are nowadays well established and well tested against the low energy data [9]. They must be taken as fixed. In this work, we adopt this viewpoint, and investigate a model in which these LR+MR forces are supplemented with the SR forces derived form the quark cluster models. We believe that this procedure brings a better insight into the role played by quark degrees of freedom in the nucleon-nucleon interaction, and provides a more meaningful test of the QCM.

Our model amounts to the Schrödinger equation

$$-\frac{1}{m_N}\nabla^2\psi(\mathbf{r}) + V_P(\mathbf{r},E)[1-f(r)]\psi(\mathbf{r}) + \int d\mathbf{r}' V_{QCM}(\mathbf{r},\mathbf{r}')\sqrt{f(r)f(r')}\psi(\mathbf{r}') = E\,\psi(\mathbf{r}), \quad (1)$$

where $V_P(\mathbf{r},E)$ is the theoretical LR+MR part of the Paris potential, given by the one-pion-exchange, the uncorrelated and correlated two-pion exchange contributions [10]. This potential is local but energy dependent. $V_{QCM}(\mathbf{r},\mathbf{r}')$ is the QCM nucleon-nucleon potential due to the quark confinement plus the one-gluon exchange contributions. It contains a local and a non-local part but is energy independent. $f(r) = [1 + (r/R_c)^p]^{-1}$ is a cut off function designed to make a clear separation between the SR and LR+MR parts of the interaction.

RESULTS

We have built a general numerical code to generate the $V_{QCM}(r,r')$ from the different quark-quark interaction models of Refs. [1, 2, 6, 11, 12], and we have solved equation (1) using these different QCM potentials V_{QCM}, together with the LR+MR parts of Paris potential [10]. The only remaining free parameters are those of the cut off function $f(r)$, namely p and R_c. To make a clear separation between the short range potential and the meson exchange potential, we chose $p = 10$. Regarding the cut off radius R_c, we determine it, as in Ref. [7], by fitting to the $^1S_0$ phase shift at 25 MeV in the isospin $T = 1$ channel, and to the deuteron binding energy ε in the isospin $T = 0$ channel. The values of R_c obtained with the different QCM potentials lie between 0.8 and 0.9 fm (see Ref. [13]. Interestingly enough they are close to the value 0.8 fm we adopted for the separation between the theoretical and phenomenological parts of the Paris potential.

COMPARISON WITH SCATTERING OBSERVABLES

We believe that comparison with phase shifts does not provide a severe enough model testing bench. For a more meaningful test we confront the predictions directly with the data on observables. We have performed such a comparison using the presently available world set of NN scattering data up to 350 MeV. All observables have been calculated for pp as well as np scatterings and examples are shown in Figs. 1, 2. As it can be seen, the agreement with experiment is again poor. This is generally true for the other observables as well, leading to the values of the total χ^2/data listed in Table 1. The model fails to reproduce not only spin observables but also cross sections. It is worth noticing that the differences in the quark-quark interactions show up more manifestly in the spin observables. One might argue that choosing R_c to be dependent only on isospin states is a too drastic prescription. We have tried to leave it free and to carry out a best fit of each partial wave. This leads to values lying between 0.6 and 1.2 fm. It results in a better χ^2/data, shown also in Table 1, but without improving really the fit to the observables.

TABLE 1. χ^2/data for pp and np observables. The models correspond to the different QCM, supplemented with the LR+MR part of the Paris potential. The fit performed by Takeuchi *et al.* in Ref. [6], the Paris potential results, and the phase shift analyzis (PSA) are also shown

Models	pp (1353 data from 25 to 333 MeV)	np (2268 data from 25 to 325 MeV)
Oka [1]	140.28	20.34
Yamauchi [11]	230.57	36.00
Takeuchi [6]	143.20	19.00
Faessler [2]	228.06	34.00
Harvey [12]	360.6	30.36
Takeuchi [6] (with phase-dependent R_c)	13.34	13.36
Takeuchi [6] + (adjustable meson exchanges)	13.25	25.24
Paris potential [10]	1.96	2.83
PSA	1.40	1.61

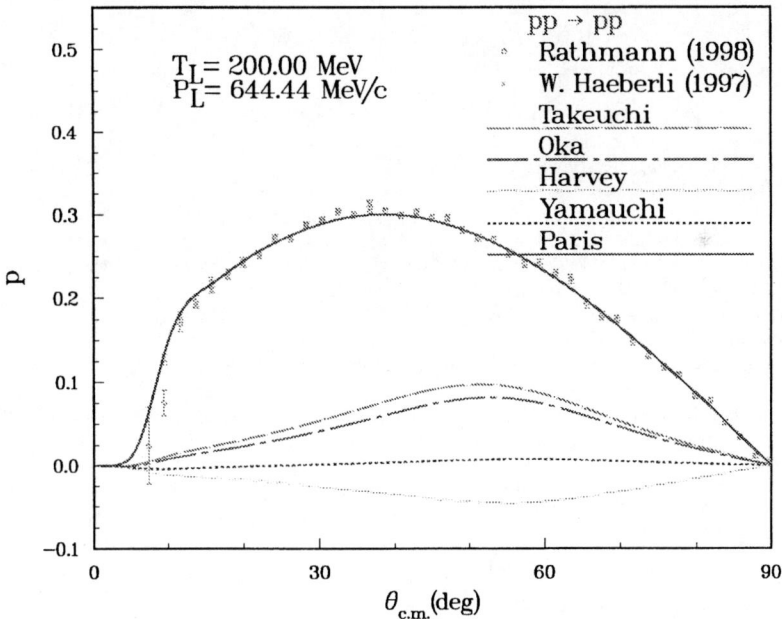

FIGURE 1. pp elastic polarization at $T_{lab} = 200$ MeV. The curves are labeled as in Table 1. The data are from Refs. [14, 15]

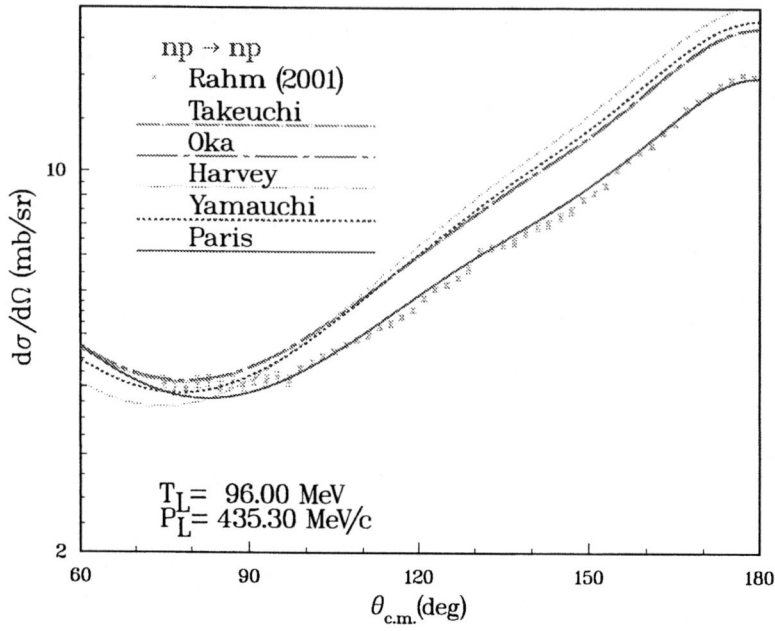

FIGURE 2. np elastic differential cross section at $T_{lab} = 96$ MeV The curves are labeled as in Table 1. The data are from Refs. [16]

REFERENCES

1. Oka, M., and Yazaki, K., *Phys. Lett. B* **90**, 41 (1980); *Prog. Theor. Phys.* **66**, 556 (1981); *Nucl. Phys.* **A402**, 477 (1983).
2. Faessler, M.A., et al., *Phys. Lett.* **112B**, 201 (1982); *Nucl. Phys.* **A402**, 555 (1983).
3. Suzuki, Y., and Hecht, K.T., *Phys. Rev. C* **28**, 1458 (1983); *Phys. Rev. C* **29**, 1586 (1984).
4. Fujiwara, Y., and Hecht, K.T., *Nucl. Phys.* **A444**, 541 (1985); *Nucl. Phys.* **A451**, 625 (1986); *Nucl. Phys.* **A456**, 699 (1986); *Phys. Lett. B* **171**, 17 (1986).
5. Yamauchi, Y., Yamamoto, R., and Wakamatsu, M., *Nucl. Phys.* **A443**, 628 (1985).
6. Takeuchi, S., Shimizu, K., and Yazaki, K., *Nucl. Phys.* **A504**, 777 (1989).
7. Vinh Mau, R., Semay, C., Loiseau, B., and Lacombe, M., *Phys. Rev. Lett.* **67**, 1392 (1991).
8. Hadjimichef, D., Haidenbauer, J., and Krein, G., *Phys. Rev. C* **63**, 035204 (2001).
9. See, for example, de Swart, J.J., et al., in *Few-Body-Systems Suppl.* **8**, 437 (1995); Vinh Mau, R., in *Recent Advances in Hadron Physics*, edited by K. Kang et al., World Scientific, p. 295 (1998); Machleidt, R., and Slaus, I., *J. Phys. G* **27**, 69 (2001).
10. Lacombe, M., et al., *Phys. Rev. D* **12**, 1495 (1975); Lacombe, M., et al., *Phys. Rev. C* **21**, 861 (1980).
11. Yamauchi, Y., Tsushima, K., and Faessler, M.A., *Few-Body Systems*, **12**, 69 (1992).
12. Harvey, M., *Nucl. Phys.* **A352**, 326 (1981).
13. Lacombe, M., et al., to be published.
14. Haeberli, W., et al., *Phys. Rev. C* **55**, 597 (1997).
15. Rathmann, F., et al., *Phys. Rev. C* **58**, 658 (1998).
16. Rahm, J., et al., *Phys. Rev. C* **63**, 044001 (2001).

Covariant electromagnetic and axial form factors of the nucleons from a chiral quark model

R.F. Wagenbrunn[*], S. Boffi[†], L. Ya. Glozman[*], W. Klink[**], W. Plessas[*] and M. Radici[†]

[*]*Institut für Theoretische Physik, Universität Graz, A-8010 Graz, Austria*
[†]*Dipartimento di Fisica Nucleare e Teorica, Università di Pavia, I-27100 Pavia, Italy
and INFN, Sezione di Pavia, I-27100 Pavia, Italy*
[**]*Department of Physics and Astronomy, University of Iowa, Iowa City, IA 52242, USA*

Abstract. We discuss new results for the nucleon elastic electromagnetic as well as axial form factors as predicted by the Goldstone-boson-exchange constituent quark model within a covariant calculation in the point-form approach to relativistic quantum mechanics.

The Goldstone-boson-exchange (GBE) constituent quark model (CQM) [1] relies on a three-quark Hamiltonian consisting of a relativistic kinetic-energy operator, a linear confinement, and a hyperfine interaction deduced from pseudoscalar boson exchange. The model respects essential properties of low-energy quantum chromodynamics, such as the spontaneous breaking of chiral symmetry, and meets the requirements of Lorentz covariance. It has been shown to provide a unified description of the excitation spectra of all light and strange baryons in good agreement with phenomenology [1, 2].

Beyond spectroscopy, however, a CQM is naturally expected to describe also the wealth of low-energy hadron reactions as far as possible. We have now investigated the performance of the GBE CQM with regard to the electromagnetic and axial structure of the nucleons. In particular, we have calculated the elastic electric and magnetic form factors of the proton and neutron [3] as well as the axial and induced pseudoscalar nucleon form factors [4]. We have produced covariant results using the point form of Poincaré-invariant quantum mechanics [5, 6]. Within this framework we are able to perform the necessary Lorentz boost transformations exactly. At this stage we have employed only one-body current operators; this corresponds to the so-called point-form spectator approximation (PFSA). The constituent quarks have been treated as pointlike and no additional parameters have been introduced. All the results have been obtained directly by employing the nucleon wave functions as produced by the GBE CQM.

In Fig. 1 we show the electromagnetic nucleon form factors and in Table 1 the charge radii and magnetic moments. In all cases the PFSA results fall remarkably close to the experimental data. Huge differences are detected in comparison to the nonrelativistic impulse approximation (NRIA). This demonstrates the important role of relativistic effects. A nonrelativistic treatment of the nucleon form factors is evidently ruled out. The influence of the chiral hyperfine interaction can be estimated by comparing to the case with the confinement interaction only. Especially when considering the neutron form factors a decisive dependence on subtle ingredients in the wave functions becomes

TABLE 1. Proton and neutron charge radii and magnetic moments as well as nucleon axial charge and axial radius as predicted by the GBE CQM in PFSA. A comparison is given also to the results in NRIA and to the case with the confinement interaction only.

	PFSA	NRIA	Conf.	Experiment
r_p^2 [fm^2]	0.81	0.1	0.37	0.774(27) [9], 0.780(25) [10]
r_n^2 [fm^2]	−0.13	−0.01	−0.01	−0.113(7) [11]
μ_p [n.m.]	2.70	2.74	1.84	2.792847337(29) [12]
μ_n [n.m.]	−1.70	−1.82	−1.20	−1.91304270(5) [12]
g_A	1.15	1.65	1.29	1.255(6) [12]
r_A [fm]	0.53	0.36	0.43	0.635(23), 0.65(7) [13]

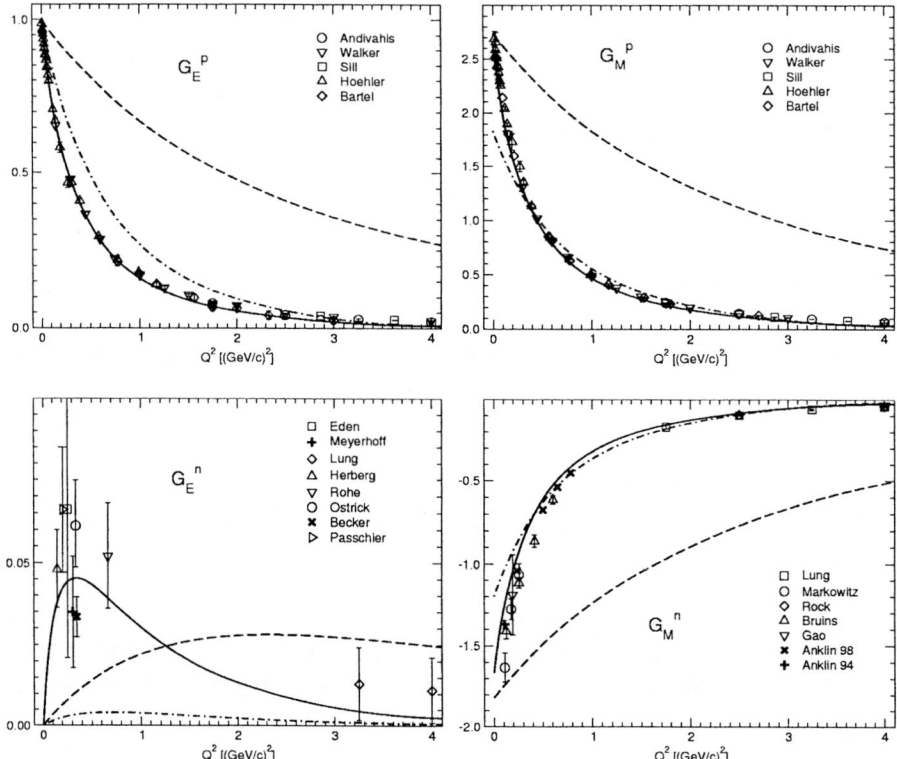

FIGURE 1. Proton (top) and neutron (bottom) electric (left) and magnetic (right) form factors as predicted by the GBE CQM [1] in PFSA (solid). A comparison is given to the results in NRIA (dashed) and the case with the confinement interaction only (dashed-dotted). The experimental data are from Refs. [7].

apparent. Specifically, small mixed-symmetry spatial configurations, which are lacking in the wave function obtained with the confinement interaction only, are necessary to achieve a reasonable agreement with the data.

For the electric form factor of the proton and the magnetic form factors of both the

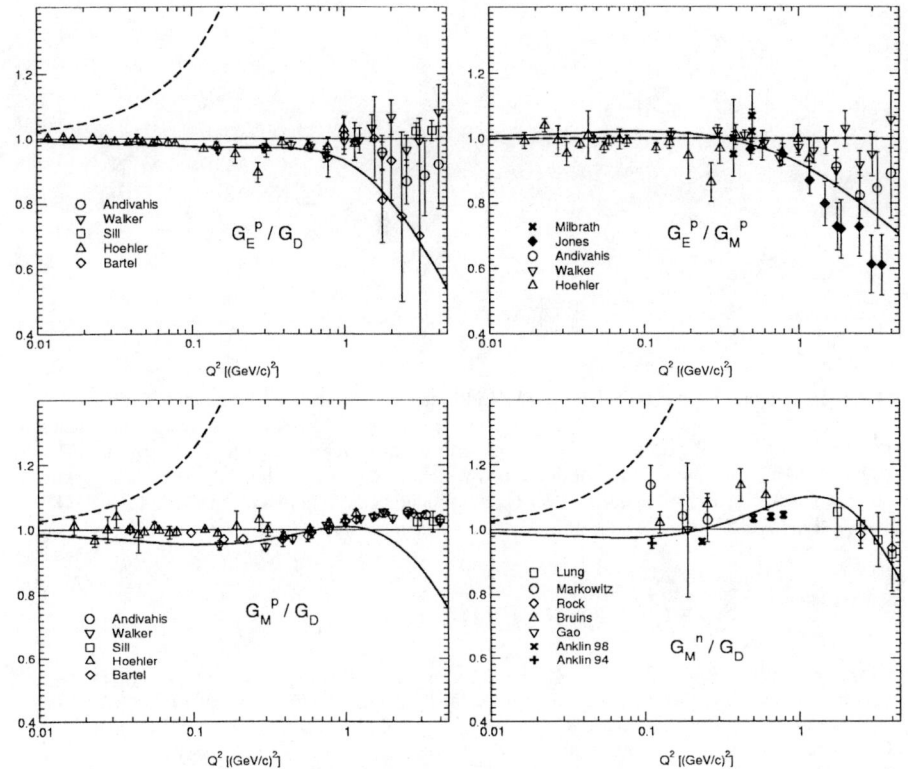

FIGURE 2. Ratios of the proton electric form factor to the dipole form, $G_D(Q^2) = (1 + Q^2/(0.71 \text{ GeV}^2))^{-2}$, (top left) and to the proton magnetic form factor (top right) as well as ratios of proton and neutron magnetic form factors to the dipole form (bottom). All ratios are normalized to 1 at $Q^2 = 0$. Solid lines: PFSA, dashed lines: NRIA. Experimental data as in Fig. 1 and from ref. [8].

proton and the neutron we also show their ratios to the dipole form factor in Fig. 2; in addition the ratio of the proton electric to the magnetic form factor is demonstrated there. In all cases the theoretical predictions agree rather well with the available experimental data, at least up to $Q^2 \sim 1 \text{ GeV}^2$. For the ratio G_E^p/G_M^p even the trend of the recent JLAB experiments [8] is followed.

In Fig. 3 we present the results for the axial and induced pseudoscalar form factors. The axial charge, $g_A = G_A(0)$, and the axial radius are given in Table 1. For the induced pseudoscalar form factor a pion-pole term was used together with the same quark-pion coupling constant as occurring in the GBE CQM. For the weak form factors in Fig. 3 essentially the same agreement with experimental data is found as before with the electromagnetic form factors.

In summary, the GBE CQM yields quite a consistent picture of the electroweak nucleon structure. The usage of a relativistic approach appears as most important. It is remarkable how the inclusion of relativistic effects in PFSA brings the theoretical predictions close to experiments.

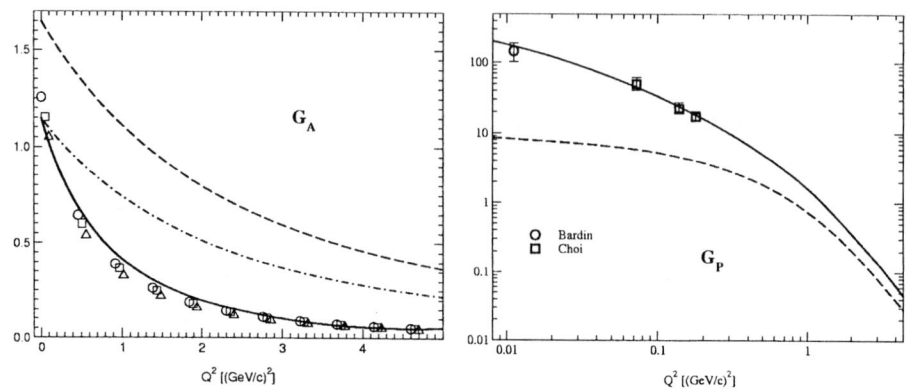

FIGURE 3. Nucleon axial form factor (left) and induced pseudoscalar form factor (right). The solid lines are the PFSA predictions of the GBE CQM. In the left plot the dashed curve is the NRIA result and the dashed-dotted curve the result without Lorentz boosts but with a relativistic axial current. In the right plot the dashed curve is the PFSA result obtained without the pion-pole term in the axial current of the constituent quark. The experimental data for G_A are plotted according to a dipole form $(1 + Q^2/M_A^2)^{-2}$ with three slightly different values for M_A [13]. The experimental data for G_P are from ref. [14].

REFERENCES

1. Glozman, L.Ya., Plessas, W., Varga, K., and Wagenbrunn, R.F., *Phys. Rev. D* **58**, 094030 (1998).
2. Glozman, L.Ya., Papp, Z., Plessas, W., Varga, K., and Wagenbrunn, R.F., *Phys. Rev. C* **57**, 3406 (1998).
3. Wagenbrunn, R.F., Boffi, S., Klink, W., Plessas, W., and Radici, M., *Phys. Lett. B* **511**, 33 (2001).
4. Glozman, L.Ya., Radici, M., Wagenbrunn, R.F., Boffi, S., Klink, W., and Plessas, W., *Phys. Lett. B* **516**, 183 (2001).
5. Dirac, P.A.M., *Rev. Mod. Phys.* **21**, 392 (1949).
6. Klink, W.H., *Phys. Rev. C* **58**, 3587 (1998); Klink, W.H., *Phys. Rev. C* **58**, 3617 (1998).
7. Andivahis, L., *et al.*, *Phys. Rev. D* **50**, 5491 (1994); Walker, R.C., *et al.*, *Phys. Rev. D* **49** 5671 (1994); Sill, A.F., *et al.*, *Phys. Rev. D* **48**, 29 (1993); Höhler, G., *et al.*, *Nucl. Phys.* **B114**, 505 (1976); Bartel, W., *et al.*, *Nucl. Phys.* **B58**, 429 (1973); Eden, T., *et al.*, *Phys. Rev. C* **50**, R1749 (1994); Meyerhoff, M., *et al.*, *Phys. Lett. B* **327**, 201 (1994); Lung, A., *et al.*, *Phys. Rev. Lett.* **70**, 718 (1993); Herberg, C., *et al.*, *Eur. Phys. J. A* **5**, 131 (1999); Passchier, I., *et al.*, *Phys. Rev. Lett.* **82**, 4988 (1999); Rohe, D., *et al.*, *Phys. Rev. Lett.* **83**, 4257 (1999); Ostrick, M., *et al.*, *Phys. Rev. Lett.* **83**, 276 (1999); Becker, J., *et al.*, *Eur. Phys. J. A* **6**, 329 (1999); Markowitz, P., *et al.*, *Phys. Rev. C* **48**, R5 (1993); Rock, S., *et al.*, *Phys. Rev. Lett.* **49**, 1139 (1982); Bruins, E.E.W., *et al.*, *Phys. Rev. Lett.* **75**, 21 (1995); Gao, H., *et al.*, *Phys. Rev. C* **50**, R546 (1994); Anklin, H., *et al.*, *Phys. Lett. B* **428**, 248 (1998); Anklin, H., *et al.*, *Phys. Lett. B* **336**, 313 (1994); Milbrath, B.D., *et al.*, *Phys. Rev. Lett.* **80**, 452 (1998) [Erratum-ibid. **82**, 2221 (1998)].
8. Jones, M.K., *et al.*, *Phys. Rev. Lett.* **84**, 1398 (2000).
9. Rosenfelder, R., *Phys. Lett. B* **479**, 381 (2000).
10. Melnikov, K., and van Ritbergen, T., *Phys. Rev. Lett.* **84**, 1673 (2000).
11. Kopecky, S., Riehs, P., Harvey, J.A., and Hill, N.W., *Phys. Rev. Lett.* **74**, 2427 (1995).
12. Groom, D.E., *et al.*, *Eur. Phys. J. C* **15**, 1 (2000).
13. Liesenfeld, A., *et al.*, *Phys. Lett. B* **468**, 20 (1999); Kitagaki, T., *et al.*, *Phys. Rev. D* **28**, 436 (1983).
14. Bardin, G., *et al.*, *Phys. Lett. B* **104**, 320 (1981); Choi, S., *et al.*, *Phys. Rev. Lett.* **71**, 3927 (1993).

Trinucleon two-body photo disintegration with Δ-isobar excitation

L.P. Yuan*, K. Chmielewski*, M. Oelsner*, P.U. Sauer*, J. Adam Jr.[†] and A.C. Fonseca**

*Institut für Theoretische Physik, Universität Hannover, Appelstraße 2, D-30167 Hannover, Germany
[†]Nuclear Physics Institute, Academy of Sciences of Czech Republic, Řež, CZ-25068, Czech Republic
**Centró Física Nuclear da Universidade de Lisboa, Av. Prof. Gama Pinto, N°2, P-1699 Lisboa, Portugal

Nucleon-deuteron radiative capture and electron scattering from the trinucleon bound state are described in Ref.[1]; the work includes final state interaction, meson exchange currents and Δ-isobar excitation; a characteristic result is given in Fig.1. In this con-

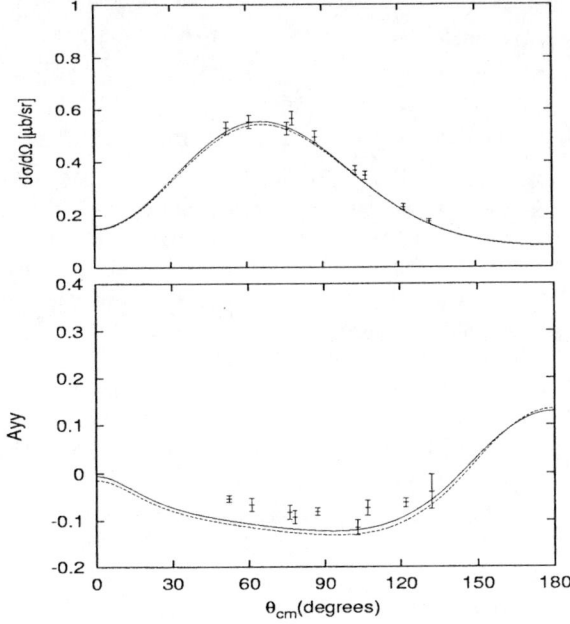

FIGURE 1. Radiative capture at 95 MeV deuteron lab energy as function of the c.m. photon angle with respect to the direction of the proton. The differential cross section is given on the top. The spin observable A_{yy} is given on the bottom; The full results for the interaction with Δ-isobar excitation are shown as solid lines, whereas the results for the purely nucleonic reference potential, the Paris potential, are shown as dashed lines. Both sets of experimental data are taken from Ref.[2].

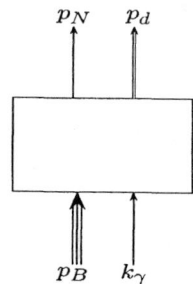

FIGURE 2. Schematic description of two-body photo disintegration of the three-nucleon bound state. The lines for the two-baryon and three-baryon particles are drawn in a special form to indicate their compositeness.

tribution the same description is applied to two-body photo disintegration of the three-nucleon bound state. The kinematics of photo disintegration is schematically shown as in Fig.2. The corresponding S-matrix element has the form:

$$\langle f|S^{dis}|i\rangle = (-2\pi i)\delta(E_N(\mathbf{p}_N) + E_d(\mathbf{p}_d) - k_{\gamma 0}c - E_B(\mathbf{p}_B))$$
$$\times \delta(\mathbf{p}_N + \mathbf{p}_d - \mathbf{k}_\gamma - \mathbf{p}_B) \frac{1}{c}\frac{\sqrt{4\pi}\hbar c}{\sqrt{2k_{\gamma 0}c}} \frac{1}{(2\pi\hbar)^{3/2}}$$
$$\times \varepsilon_\mu(\mathbf{k}_\gamma \lambda)\langle \psi_\alpha^{(-)}(\mathbf{q}_f)v_{\alpha_f}(Nd)|j^\mu(\mathbf{k}_\gamma,\mathbf{p}_B)|B\rangle\Big|_{\mathbf{q}_f = \frac{\mathbf{p}_d - 2\mathbf{p}_N}{3}} . \quad (1)$$

In Eq. (1) $\varepsilon_\mu(\mathbf{k}_\gamma\lambda)$ is the polarization vector of the photon with helicity λ; the other symbols have obvious meanings. The AGS equations are solved for the trinucleon bound state $|B\rangle$ as in Ref.[3] and for the nucleon-deuteron scattering states $|\psi_\alpha^{(+)}(\mathbf{q}_f)v_{\alpha_f}(Nd)\rangle$ as in Ref.[4]. The calculations are based on a coupled-channel two-baryon potential which allows for single Δ-isobar excitation. The employed current j^μ, required for the photo disintegration matrix element, couples nucleonic states and states with a Δ-isobar. It includes one-baryon and two-baryon contributions and is expanded into multipoles. In addition to the Siegert operator, the contributions from meson exchange are explicitly calculated. Electric and magnetic multipoles up to $E2$ and $M2$ are taken into account. The convergence of the results with respect to the used multipole expansion is checked. In Fig.3 we show the differential cross section at 90° for two-body photo disintegration of the triton as a function of the photon energy E_γ; in Fig.4 we show the complete differential cross section for ^{3}He at two distinct photon energies. The observed Δ-isobar effect is small at lower energies and becomes moderate at larger energies. The observed shift of the peak is in part an indirect Δ-isobar effect: The triton binding energy gets increased and the cross section changes due to scaling. The Δ-isobar effect seen in Figs.3 and 4 is studied in more detail in Fig.5 for triton two-body photo disintegration; the study refers to two selected photon energies, one below and one above the peak of the 90° cross section as shown in Fig.3. Fig.5 demonstrates that at the lower energy the Δ-isobar effect comes mainly from changes in the binding energy and in the nucleonic components of the bound-state wave function, at the higher energy mainly from the scattering state and the Δ-components of the e.m. current.

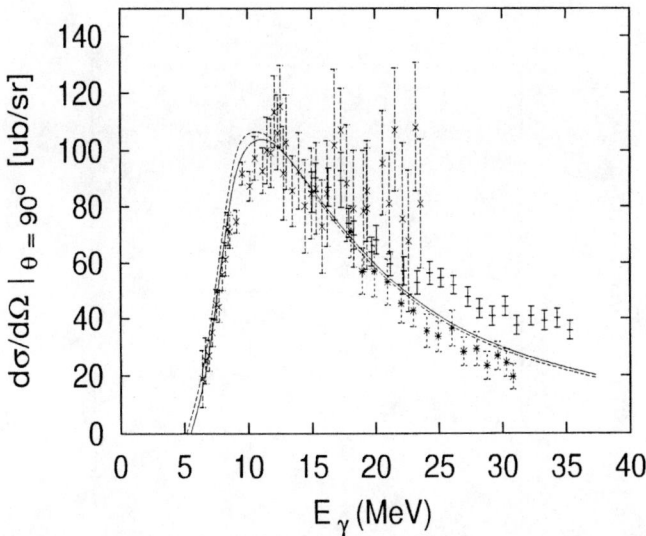

FIGURE 3. Differential cross section at 90° for two-body photo disintegration of the triton in the lab system; The full results for the interaction with Δ-isobar excitation are shown as solid lines, whereas the results for the purely nucleonic reference potential, the Paris potential, are shown as dashed lines. Experimental data are taken from Refs.[5, 6, 7].

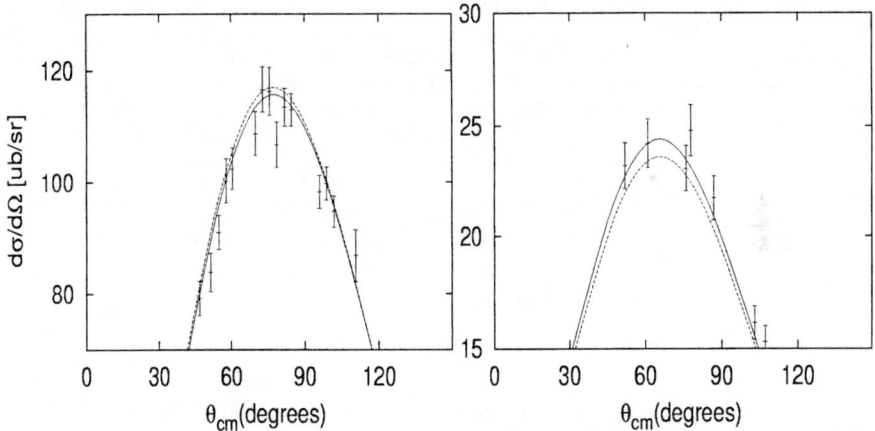

FIGURE 4. C.M. differential cross section of two-body photo disintegration of ^{3}He at 12.16 MeV (left) and 36.85 MeV (right) photon c.m. energy as function of the c.m. proton angle with respect to the direction of the photon. The full results for the interaction with Δ-isobar excitation are shown as solid lines, whereas the results for the purely nucleonic reference potential, the Paris potential, are shown as dashed lines. Experimental data are taken from Ref.[8] (left) and Ref.[2] (right).

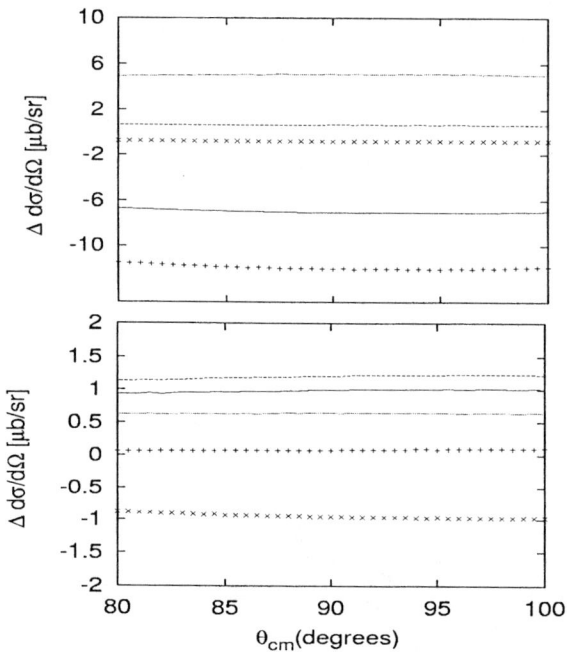

FIGURE 5. Contributions to the Δ-isobar effect for c.m. differential cross section of triton two-body photo disintegration at 8.91 MeV (top) and 36.85 MeV (bottom). The full shift in the prediction arising from the different predictions of the nucleonic reference potential and the coupled-channel potential is shown as solid lines. That full shift is split up into four parts, i.e., the contribution arising from the theoretical change of the triton binding energy due to scaling (dotted lines), the contribution arising from the change of the nucleonic components in the bound-state wave function (crosses +), the contribution arising from the change of the nucleonic components in the scattering wave function (crosses ×) and the contribution arising from the Δ-components in both hadronic wave functions and the e.m. current (dashed lines); those four contributions add up to the full shift, making up the complete Δ-isobar effect.

ACKNOWLEDGMENTS

L. P. Yuan is supported in part by the DFG grant Sa 247/19-1 and by a graduate-student grant of Lower Saxony, M. Oelsner by the DFG grant Sa 247/20-1/2. J. Adam Jr. is supported by the grant GA CR 202/00/1669. The calculations are performed at Regionales Rechenzentrum für Niedersachen.

REFERENCES

1. Yuan, L.P., Chmielewski, K., Oelsner, M., Sauer, P.U., Fonseca, A.C., and Adam, J., Jr., to appear in *Nucl. Phys.* **A**.
2. Pitts, W.K., *et al.*, *Phys. Rev. C* **37**, 1 (1988).
3. Nemoto, S., Chmielewski, K., Haidenbauer, J., Oryu, S., Sauer, P.U., and Schellingerhout, N.W., *Few-Body Systems* **24**, 213 (1998).
4. Chmielewski, K., Fonseca, A.C., Nemoto, S., Sauer, P.U., *Few-Body Systems, Suppl.* **10**, 335 (1998).
5. Kosiek, R.D.M., and Pfeiffer, R., *Phys. Lett.* **21**, 199 (1966).
6. Faul, D.D., *et al*, *Phys. Rev. Lett.* **44**, 129 (1980).
7. Skopik, D.M., *et al*, *Phys. Rev. C* **24**, 1791 (1981).
8. Belt, B.D., *et al*, *Phys. Rev. Lett.* **24**, 1120 (1970).

Trinucleon magnetic form factors

L.P. Yuan*, J. Adam Jr.†, K. Chmielewski*, H. Henning*,
S. Nemoto*, M. Oelsner* and P. U. Sauer*

*Institut für Theoretische Physik, Universität Hannover, Appelstraße 2, D-30167 Hannover,
Germany
†Nuclear Physics Institute, Academy of Sciences of Czech Republic, CZ-25068, Řež,
Czech Republic

Abstract. The trinucleon bound state is obtained from a coupled-channel two-baryon potential which allows for single Δ-isobar excitation [2]. The e.m. current consistent with that potential also couples nucleonic states and states with a Δ-isobar. One-baryon and two-baryon contributions to the e.m. current are taken into account.

The trinucleon magnetic form factors, given previously [1] by some of the present authors, are recalculated. The reason for the recalculation is the fact that the underlying current is presently used for describing inelastic e.m. processes in the three-nucleon system; thus the e.m. current operator needs a final check.

The trinucleon bound state is obtained from a coupled-channel two-baryon potential which allows for single Δ-isobar excitation [2]. The e.m. current consistent with that potential also couples nucleonic states and states with a Δ-isobar. One-baryon and two-baryon contributions to the e.m. current are taken into account. Results are displayed in two figures. Compared to Ref.[1], both nucleonic and delta currents are calculated within

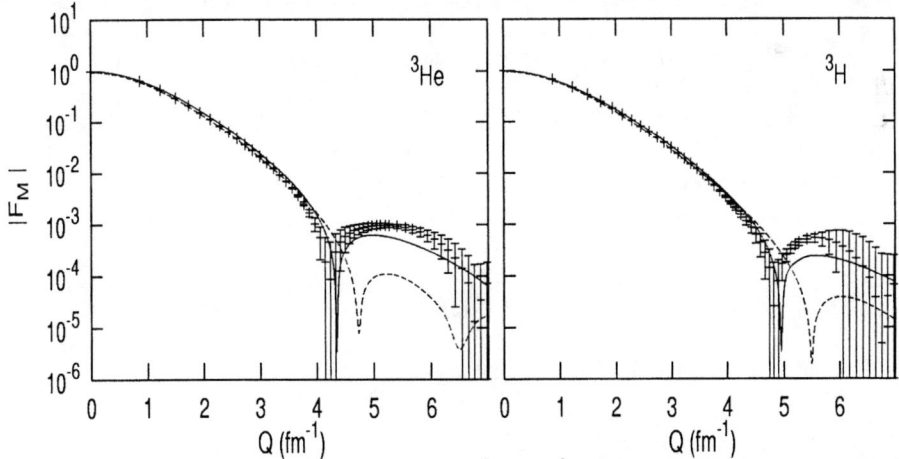

FIGURE 1. Normalized magnetic form factors F_M of ^{3}He and ^{3}H as function of momentum transfer Q. The results for the interaction with Δ-isobar excitation are shown as solid lines, whereas the results for the purely nucleonic reference potential, the Paris potential, are shown as dashed lines. The experimental data are taken from Ref.[4].

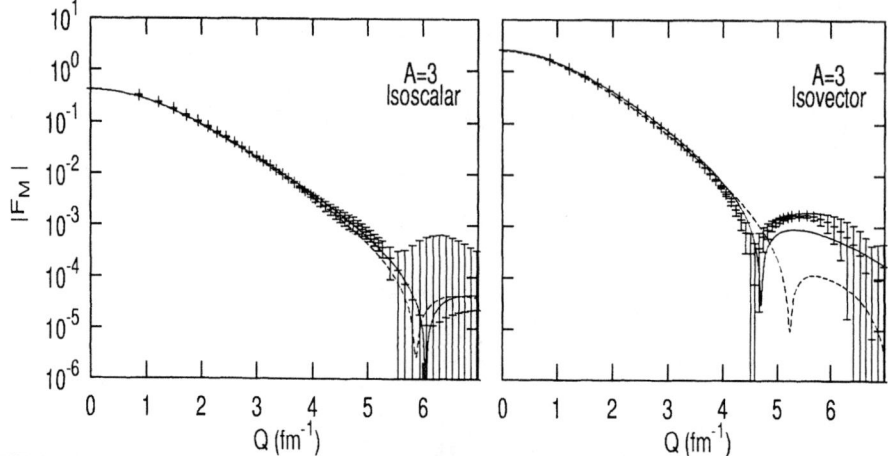

FIGURE 2. Isoscalar and isovector trinucleon magnetic form factors F_M as function of momentum transfer Q. The results for the interaction with Δ-isobar excitation are shown as solid lines, whereas the results for the purely nucleonic reference potential, the Paris potential, are shown as dashed lines. The experimental data are taken from Ref.[4].

the framework of an improved computational technique [3]. Concerning the bound-state wave function, more mesh points and more channels are included in the present calculation, significantly improving the results in and beyond the diffraction minimum technically. Concerning the e.m. current, a sign mistake got detected in the two-baryon transition current from nucleonic to nucleon-Δ states; that contribution is in general very small, but becomes significant when all other contributions cancel, i.e., in the region of the diffraction minimum. Both effects combined, the one arising from the improved bound-state wave function and the other arising from a corrected two-baryon transition current, yield changes in the magnetic form factors at momentum transfer $Q > 4$ fm^{-1}, decreasing the weight of the Δ-isobar for the magnetic form factors somehow, compared with the results of Ref.[1].

L. P. Yuan is supported in part by the DFG grant Sa 247/19-1 and by a graduate-student grant of Lower Saxony, M. Oelsner by the DFG grant Sa 247/20-1/2. J. Adam Jr. is supported by the grant GA CR 202/00/1669. The calculations are performed at Regionales Rechenzentrum für Niedersachen.

REFERENCES

1. Sauer, P.U., and Henning, H., *Few-Body Systems, Suppl.* **7**, 92 (1994).
2. Nemoto, S., Chmielewski, K., Haidenbauer, J., Oryu, S., Sauer, P.U., and Schellingerhout, N.W., *Few-Body Systems* **24**, 213 (1998).
3. Oelsner, M., *PhD Thesis*, University of Hannover, 1999.
4. Amroun, A., *et al.*, *Nucl. Phys.* **A579**, 596 (1994).

HADRON STRUCTURE

Measurement of the electric neutron form factor G_{en} in the exclusive reaction $^3\vec{He}(\vec{e},e'n)$ at momentum transfer $Q^2 = 0.67$ $(GeV/c)^2$

J. Bermuth (for the A1 Collaboration at MAMI)

Inst. f. Kernphysik, J.J. Becherweg 45, 55099 Mainz, Germany

Abstract. The experiment to measure G_{en} in the quasielastic reaction $^3\vec{He}(\vec{e},e'n)$ at four-momentum transfer $Q^2 = 0.67$ $(GeV/c)^2$ has been continued in June 2000. The asymmetry with respect to reversal of the electron helicity has been measured both for perpendicular and for parallel orientation of the ^{3}He spin with respect to $\vec{q}$. From the asymmetry ratio we finally obtained $G_{en} = 0.0468 \pm 0.0064_{syst} \pm 0.0027_{stat}$. In addition, the asymmetry A_y^0 with the target spin normal to the scattering plane and the electrons unpolarized, has been measured in the reaction $^3\vec{He}(e,e'p(n))$. A_y^0 is a direct observable of FSI and MEC effects. It allows to check theoretical predictions. Thereby one gains redundancies in the attempt to derive single particle properties of the free neutrons from measurement on bound neutrons, and one can verify whether ^{3}He can be used as an effective polarized neutron target.

INTRODUCTION

Although the neutron has vanishing electric charge, it has an intrinsic electric charge distribution. This was first observed in the scattering of thermal neutrons on shell electrons of heavy atoms. With the measured mean squared charge radius of $<r_n^2>_{Pb} = -0.1148 \pm 0.0023$ fm^2, this results in a positive slope for G_{en} at $Q^2 \approx 0$ [1]. In order to measure the electric form factor over the full range of momentum transfer one has to investigate the inverse reaction, i.e. scattering of high energy electrons on free neutrons. However, since targets of free neutrons are not available, one is restricted to measurements on bound neutrons in a nucleus. As a consequence one has to use nuclear targets such as ^{2}H or ^{3}He of which the internal structure can be calculated with reasonable accuracy. The uncertainty due to model dependence can be significantly reduced in the quasi-free kinematic process $(e,e'n)$, where the four-momentum is transferred to the single neutron in the nucleus and where the residual nucleus acts as a spectator. Additionally, using both polarized electrons and a polarized target (or alternatively measuring the polarization of the recoil neutron), an interference term between the electric and the dominant magnetic scattering amplitude in the polarization dependent part of the cross section enhances the weak electric scattering amplitude and thus the sensitivity to G_{en} one is looking for. Such experiments were already performed at the Mainz Mikrotron (MAMI) at $(0.15 \leq Q^2/(GeV/c)^2 \leq 0.35)$ using ^{3}He and ^{2}H as targets (see [2] and references therein). Switching the electron helicity changes the sign of the interference term because the magnetic scattering amplitude changes sign, while the electric one does not. For that reason, the interference term can be determined by measuring the asymmetry with respect to electron helicity. In PWIA (plane wave impulse approximation) this asymmetry can be written as

$$A(\theta_S^*) = P_e P_n \frac{aG_e G_m \sin(\theta_S^*)\cos(\phi_S^*) + bG_m^2 \cos(\theta_S^*)}{cG_e^2 + dG_m^2}, \qquad (1)$$

where θ_S^* and ϕ_S^* are the polar and azimuthal angles of the target spin with respect to the momentum transfer $\vec{q}$, respectively. The variables a, b, c, d contain kinematic factors involving the direction of the scattered electron and neutron (see [3]). P_e describes the longitudinal polarization of the electron beam and P_n is the polarization of the bound neutron in the nucleus. The asymmetry $A_\perp$ with target spin perpendicular ($\theta_S^* = 90^o, \phi_S^* = 0^o$ or 180°) to the momentum transfer $\vec{q}$ contains the interference term $G_{en} \cdot G_{mn}$. Thus the ratio $A_\perp/A_\parallel$ with $A_\parallel = A(\theta_S^* = 0^o)$ is proportional to the fraction of G_{en}/G_{mn}. In this ratio, a number of systematic errors cancel. Also, there is no need of an absolute measurement of the polarization product $P_e \cdot P_n$ which drops out to first order.

In addition to the G_{en} measurement, the target asymmetry A_y^0 was measured [5] for an independent proof of the predictions of theoretical calculations. In the reaction $^3\vec{\text{He}}(e,e'n(p))$, where the target spin is perpendicular to the scattering plane ($\theta_S^* = 90^o, \phi_S^* = \pm 90^o$), the asymmetry A_y^0 is defined as the relative change of the cross section with respect to the orientation of the target spin normal to the scattering plane. In PWIA A_y^0 is equal to zero, whereas inclusion of FSI, MEC leads to a finite asymmetry. Thus, if $A_y^0 \neq 0$ can be described by theory, one knows that FSI and MEC are treated properly in the exclusive $^3\vec{\text{He}}(\vec{e},e'n)$ reaction from which G_{en} is extracted.

EXPERIMENTAL SETUP

The here presented value of G_{en} is determined at $\overline{Q}^2 = 0.67$ GeV^{2}c^{-2}. Longitudinal polarized electrons of energy E = 854 MeV were produced using a strained layer GaAs-crystal [6] at the Mainz Mikrotron [7]. At target position the average beam current was i=10 μA. By means of a Møller polarimeter in front of the target a polarization of (82.7 $\pm$ 2.2)% was measured. The angles θ_e, ϕ_e and the momentum of the scattered electrons were reconstructed in the magnetic spectrometer A (see [2]) with solid angle acceptance of 28 msr and a target length acceptance of 5 cm. At spectrometer A the average scattering angle for was $\overline{\theta}_e = 78.6^o$. The nucleons were detected in coincidence with the electrons in four layers of plastic scintillator bars with photomultipliers at both ends, allowing reconstruction of the scattering angles θ_n and ϕ_n of the neutrons. At a distance of 1.7 m from the target and an average scattering angle for the nucleons of $\overline{\theta}_n = 32.2^o$, the detector covered 72 % of the Fermi cone of the neutrons. Protons and neutrons were distinguished in two layers of ΔE-veto detectors placed in front of the E-bars. The nucleon detector was shielded with 10 cm of lead to suppress electromagnetic background. In target direction, the shielding was reduced to 2 cm thickness to allow the detection of protons (e,e'p) and to suppress the (p,n)-conversion process in lead as much as possible. Additionally, collimators were used in the nearby of the entrance and exit windows of the target to further reduce the background.

The ^{3}He gas is spin-polarized by direct optical pumping in its metastable state at pressures around 1 mbar. In order to reach the high densities required for these experiments, the polarized gas is compressed into a target cell by means of a non-magnetic titanium piston compressor [8]. Once the requested pressure of 4 bar is achieved, the cell is disconnected from the ^{3}He polarizer- and compressor-unit and transported to the target chamber in the spectrometer hall. The initial polarization was typically $\approx 50\%$. Long relaxation times are required for a remote type of operation. The target cells made out of quartz glass are spherical at their central part ($V \approx 350$ cm^3) and have two opposing appendices terminated by thin copper foils (total length 20 cm). The inner walls are coated with Cs [4]. Under electron beam conditions the total relaxation time was measured to be 30 – 40 h. Thus the cells were replaced two times per day, whereas an average polarization of $P_{av} = 38\%$ prevails. During the experiment, the target polarization was monitored by NMR techniques and $P_e P_t$ by means of an asymmetry measurement using the spin dependent elastic reaction $^3\vec{\text{He}}(\vec{e}, e')$.

At target position the magnetic spectrometers produce a magnetic stray field of ≈ 2 G with a relative field gradient of 1.5×10^{-2} cm^{-1}. To reduce it significantly, a rectangular box of size $80 \times 60 \times 60$ cm^3 of 2 mm thick μ-metal and iron surrounded the target cell. The holding field of 4 G was produced by three independent pairs of coils mounted inside the box. This allowed us to rotate the target spin in any desired direction. During the experiment a relative field gradient of better than 5×10^{-4} cm^{-1} was achieved.

DATA TAKING AND ANALYSIS

Both reactions $^3\vec{\text{He}}(\vec{e},e'n)$ and $^3\vec{\text{He}}(\vec{e},e'p)$ were measured simultaniously. In a second run the target asymmetry A_y^0 was measured via $^3\text{He}(e,e'n(p))$. The neutron electric form factor was extracted from the ratio $A_\perp / A_\parallel$ as mentioned before. In order to determine the various corrections of the raw asymmetries, a Monte Carlo simulation was developed for the quasielastic scattering process, which uses PWIA and the spectator model (see [2]). Figure 1a shows the results from this simulation which fits to the spherical distribution of the scattered electrons measured with spectrometer A. Included in this simulation is the finite momentum acceptance of the spectrometer A, which cuts the low energy tail of the quasi-elastic peak resulting in a shift of the average $\vec{q}_{av}$-vector relative to the $\vec{q}$ at the top of the quasielastic peak. Further corrections for bremsstrahlung and the missing energy distribution are included. In total this leads to a reduction of G_{en} of 7.4 % relative. Using $G_{mn}(Q^2 = 0.652 \text{ GeV}^2 c^{-2}) = (1.037 \pm 0.012)\mu_n G_D$ from [9] with the magnetic moment μ_n and the dipole form factor G_D we obtain, by combining the results with the result from our previous measurement [2], $G_{en}(Q^2 = 0.67 \text{ GeV}^2 c^{-2}) = 0.0468 \pm 0.0064_{stat} \pm 0.0027_{syst}$. The total systematic error of 5.8 % results from (i) the uncertainty of $\pm 0.1^o$ in the alignement of the magnetic field (ε_{syst} = 4.2 %), (ii) the error in the double polarization ratio $(P_e P_n)_\parallel / (P_e P_n)_\perp$ (ε_{syst} = 2.6 %), (iii) the total error of G_{mn} (ε_{syst} = 1.2 %) and finally (iv) the model dependence (ε_{syst} = 3 %).

In figure 1b our result (o) is shown together with other double polarized coincidence measurements (see [4] and references therein). The hatched region shows the systematic

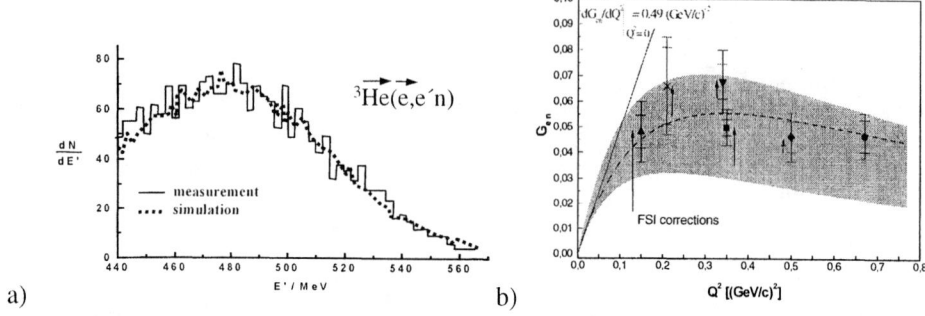

FIGURE 1. a) Energy distribution of the scattered electron in the reaction $^3\vec{\text{He}}(\vec{e},e'n)$. b) The result of this experiment (o) together with the results of other double polarization experiments (see text). The statistic and the sum of statisic and systematic error are shown. Except for (o) all values include the nuclear corrections.

error on G_{en} extracted from the elastic e-d scattering [10], where different nucleon-nucleon-potentials in the calculation of the deuteron-wave function were used. The arrows indicate the theoretical corrections due to FSI and MEC (see [11]) for both ^{3}He- and for ^{2}H-data. The results for the target asymmetry A_y^0 were: $A_y^0(Q^2 = 0.67 \text{ GeV}^2\text{c}^{-2})$ = (0.052 ± 0.031) and $A_y^0(Q^2 = 0.3 \text{ GeV}^2\text{c}^{-2})$ = (0.118 ± 0.027) [5]. As expected it shows that towards higher Q-values FSI and MEC are less important and that corrections on G_{en} are smaller. A theoretical calculation at $\overline{Q}^2$ = 0.67 (GeV/c)^2 as well as further investigations on the structure of ^{3}He are in progress [12]. The experiment was supported by the Deutsche Forschungsgemeinschaft (SFB 443) and the Schweizerische Nationalfonds.

REFERENCES

1. Kopecky, S., et al., *Phys. Rev. C* **56**, 2229 (1997).
2. Rohe, D., et al., *Phys. Rev. Lett.* **83**, 4257 (1999).
3. Becker, J., et al., *Eur. Phys. J. A* **6**, 329 (1999).
4. Bermuth, J., *Ph.D. thesis*, Mainz University, 2001.
5. Merle, P., *Ph.D. thesis*, Mainz University, 2001.
6. Aulenbacher, K., et al., *Nucl. Instr. Methods A* **391**, 498 (1997).
7. Ahrens, J., et al., *Nucl. Phys. News* **2**, 5 (1994).
8. Becker, J., et al., *Nucl. Instr. Methods A* **346**, 45 (1994).
9. Anklin, H., et al., *Phys. Lett. B* **428**, 248 (1998).
10. Platchkov, S., et al., *Nucl. Phys.* **A510**, 740 (1990).
11. Golak, J., et al., *preprint*, nucl-th/0008008, Los Alamos 2000.
12. Heil, W., Distler, M., and Rohe, D., contact persons, *MAMI-Proposal A1/2-00*, Mainz, 2000.

Investigation of Δ medium effects using the $^4He(\gamma, \pi^+ n)$ reaction

D. Branford, for the Edinburgh, Glasgow, Tübingen, Mainz PiP/TOF Collaboration, which is part of the A2 Collaboration at Mainz.

Department of Physics & Astronomy, The University of Edinburgh, Edinburgh, UK

Abstract. A measurement of the quasifree $^4He(\gamma, \pi^+ n)$ reaction was made using unpolarised and linearly polarised tagged photons in conjunction with large solid angle π and n detectors to study predicted Δ medium effects. The measurements are significantly lower than DWIA calculations, which suggests that the amplitude for proton Δ excitation followed by Δ propagation and decay to $\pi^+ + n$ in the nuclear medium is reduced compared to that for free protons. It is concluded that improved calculations are required to quantify Δ medium effects.

INTRODUCTION–Δ MEDIUM EFFECTS

The study of modifications that occur to hadrons in the nuclear medium is important as it provides sensitive tests of QCD in the non–perturbative regime. Measurements have been performed on baryons, mesons and baryon resonances in both normal and the highly compressed nuclear matter produced in relativistic heavy–ion reactions. Results to date are briefly summarised in the table 1.

TABLE 1. Brief summary of nuclear medium effect investigations.

Nucleons	EMC Study	*Swollen nucleons*
	Ratio $\mu G_E^p / G_M^p$	Inconclusive
	Nuclear transparency	Inconclusive
Mesons	Dilepton production in Rel. H.I. reactions	*Maybe ρ, ω masses reduced?*
Resonances	Photon absorption Δ(1232)	*Widths increased Possible effects.* $m_\Delta \propto Q^2$
	$D_{13}(1520)$	Inconclusive
	$S_{11}(1535)$	Inconclusive
	Preformed Δs	Inconclusive

It is evident from these results that only a few experiments have provided conclusive evidence of nuclear medium effects against which to test QCD theories. The main reason for this problem is that the errors that arise in estimating the final state interactions (FSI) are in many cases comparable to the effects under investigation. To overcome this problem and provide new data on nuclear medium effects, we have carried out an investigation of the $^4He(\gamma,\pi^+n)$ reaction in the quasifree π production kinematic regime. The adopted approach was:

1. To study quasifree π^+ production on protons bound in ^4He.
2. Compare to π^+ production on *free protons*.
3. Interpret any observed differences in terms of *nuclear medium effects*.

The quasifree π production reaction was chosen because DWIA calculations [1] for ^{12}C indicated that a small reduction of 3% in the effective mass (m_Δ) of the Δ brought about by the nuclear medium gives large reductions in the differential cross sections (40% at $\theta_\pi \sim 60°$, 20% at $\theta_\pi \sim 120°$). The results also indicated that significant changes can be expected for the polarisation asymmetries $\Sigma = (\sigma_\perp - \sigma_\parallel)/(\sigma_\perp + \sigma_\parallel)$ and these are insensitive to the choice of optical model parameters used to evaluate the FSI. A ^4He target was chosen because it is expected that the medium effects will be increased due to the high central nuclear density and the FSI will be relatively small.

THE EXPERIMENTAL SETUP

The $^4He(\gamma,\pi^+n)$ reaction was studied at MAMI–B, Mainz using tagged photons in the range $E_\gamma = 100-730 MeV$ incident on an 8cm thick liquid ^4He target. Polarised photons were produced using coherent bremsstrahlung from a diamond radiator. The tagging rate was $2 \times 10^8 s^{-1}$ at a resolution of $\Delta E_\gamma = 2MeV$. π^+ particles and neutrons in the ranges $\Theta_\pi = 50°-130°$ and $\Theta_n = 15°-150°$ were detected in a large plastic scintillator hodoscope (PiP) and TOF bars, respectively. Calibration runs were made using a CH_2 target to study the $p(\gamma,\pi^+n)$ reaction. A more detailed description of the setup can be found in Ref. [2].

RESULTS

Figure 1 shows a missing energy plot obtained using the definition $E_m = E_\gamma - E_\pi - E_N - E_R$, where E_π, E_N and E_R are kinetic energies. The results obtained for the $p(\gamma,\pi^+n)$ reaction obtained using the CH_2 target indicated that the overall missing energy resolution was $E_m \sim 7MeV$. The fact that the peak from the $^4He(\gamma,\pi^+n)$ data has a width that is not appreciably greater than $E_m \sim 7MeV$ suggests that the reaction proceeds predominantly by the removal of a 1s proton leaving the residual ^3H nucleus intact and FSI are relatively small. To select events which are primarily of this nature, a cut on the data was made as indicated in Fig. 1.

The efficiency of the detection system was calculated using a GEANT simulation and carrying out internal calibrations. As a check, the absolute cross section results for the $p(\gamma,\pi^+n)$ reaction were compared to results calculated using the prescription of Ref. [3], which is well known to give a good description of all pion production results on nucleons in the $\Delta(1232)$ resonance region. These comparisons indicated that our GEANT calculated efficiency was 16% too high. To correct for this, all our cross section

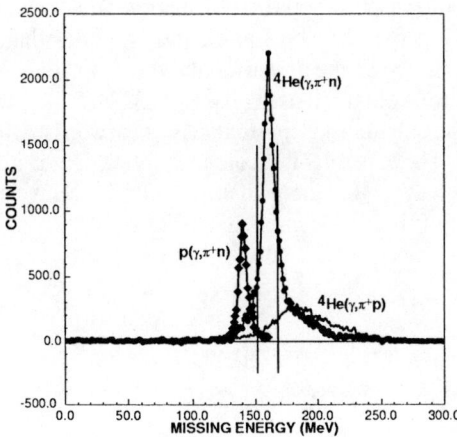

FIGURE 1. Missing energy spectra obtained for the $p(\gamma, \pi^+ n)$, $^4He(\gamma, \pi^+ n)$ and $^4He(\gamma, \pi^+ p)$ reactions.

results were increased by 16%. The polarisation asymmetry results were found, within errors, to be in agreement with published data.

The results for the $^4He(\gamma, \pi^+ n)$ reaction were binned as shown in table 2. The results

TABLE 2. The bins used for the $^4He(\gamma, \pi^+ n)$ data.

Quantity	Range	Bin Size	No. of Bins
E_γ	240–400 MeV	40 MeV	4
E_π	20–180 MeV	10 MeV	16
θ_π	60–120°	15°	4
ϕ_π	(-15)–15°	30°	1
θ_n	10–150°	5°	28
$\phi_n - \phi_\pi$	170–190°	20°	1

were compared to theoretical calculations using the method of Ref. [2], which includes an operator based on the prescription of Ref. [3] for the $p(\gamma, \pi^+ n)$ free proton reaction. The effects of the nuclear medium can be simulated by varying m_Δ, Γ_Δ and the ratio R = (E2)/(M1). FSI are included through DWIA calculations using optical potentials. However, at present, only PWIA calculations are available for the $^4He(\gamma, \pi^+ n)$ reaction. To overcome this problem, we have used calculated $^{12}C(\gamma, \pi^+ n)^{11}B$ results and scaled the FSI in proportion to the nuclear radii r_{He} and r_C as follows:

$$^4He(DWIA) = {}^4He(PWIA) \times \frac{^{12}C(DWIA)}{^{12}C(PWIA)} \times \frac{exp(-r_{He})}{exp(-r_C)}$$

All results were multiplied by a spectroscopic factor of S = 0.8.

Although the shapes of the differential cross section and Σ results are well described by the calculations, all are found to be significantly smaller than the $m_\Delta = 1232 MeV$ DWIA estimates. Figure 2 shows the experimental cross sections integrated over θ_π and E_n compared to calculations obtained using $m_\Delta = 1232 MeV$. It is observed that the data lie significantly below the calculations, particularly at forward angles as anticipated from the ^{12}C calculations [1]. These results indicate that significant Δ nuclear medium effects have been observed. However, it is clear that a quantitative investigation of Δ medium effect in ^4He will require full DWIA calculations for the $^4He(\gamma,\pi^+n)$ reaction.

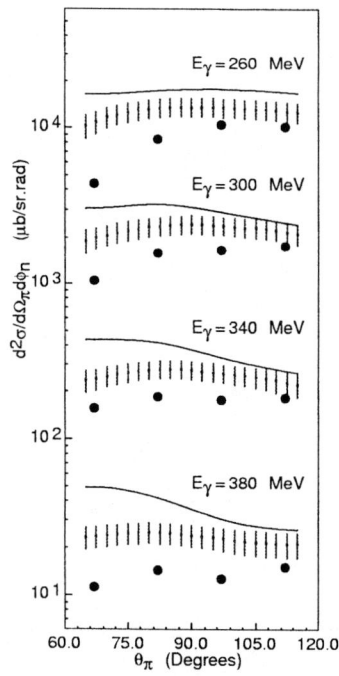

FIGURE 2. Integrated results for the $^4He(\gamma,\pi^+n)$ reaction ($E_\gamma = 340 MeV \times 10$, $300 MeV \times 100$, $260 MeV \times 1000$). Solid lines are PWIA results obtained using $m_\Delta = 1232\ MeV$. The vertical bars are DWIA results including $\pm 15\%$ errors.

REFERENCES

1. Li, X., Wright, L.E., and Bennhold, C., *Phys. Rev. C* **48**, 816 (1993).
2. Branford, D., et al., *Phys. Rev. C* **61**, 014603 (1999).
3. Blomqvist, I., and Laget, J.M., *Nucl. Phys.* **A280**, 405 (1977).

Landau parameters and collective modes in nuclear matter with relativistic non-linear models

J.C. Caillon*, P. Gabinski* and J. Labarsouque*

*Centre d'Etudes Nucléaires de Bordeaux Gradignan, IN2P3, CNRS, Université Bordeaux I,
le Haut Vigneau, 33175 Gradignan Cedex, France*

Abstract. The Landau parameters have been calculated in relativistic non-linear models, on the one hand, from the second derivative of the total energy density with respect to the quasinucleon distribution, on the other, from the longitudinal dielectric function in relativistic RPA with and without the Pauli blocking terms in the polarization operators. The link between the Landau parameters and the collective modes in nuclear matter was shown.

INTRODUCTION

In the last decade, relativistic non-linear models have gained success in the description of nuclear matter and ground state properties of finite nuclei. In the non-linear models, the Lagrangian includes usual linear nucleon-meson couplings (like in the Walecka model [1]) as well as couplings between the mesonic fields. Among many models, the models NL1 [2] contain only σ self-coupling terms in the Lagrangian while TM1 [3] (for medium and heavy nuclei) include in addition an ω self-coupling term and G1 [4] allows all scalar-vector coupling terms up to fourth order.

The non-linear models have been applied to the study of the longitudinal response functions for quasielastic electron scattering [5] and the isoscalar compression modes [6] in the relativistic Random Phase Approximation (RPA). In the calculation of response functions [5], only a class of non-linear diagrams was taken into account in the polarization operators. In the case of isoscalar compression modes, it was shown [6] that the monopole states are not well described with the positive-energy states contribution only, i.e. without the Pauli blocking terms which correct for forbidden transitions between the Dirac sea and the occupied Fermi sea. The questions that now arise are: Are the non-linear diagrams, taken into account in the RPA calculations, sufficient to give a rather good description of the main properties of nuclear matter? Should the Pauli blocking terms be taken into account or not?

In order to try to answer the above questions we consider these relativistic non-linear models in the light of the Landau theory [7]. The Landau parameters can be obtained as usual from the second derivative of the total energy density with respect to the quasinucleon distribution [8] as well as from the longitudinal dielectric function in relativistic RPA with or without the Pauli blocking terms in the polarization operators. By comparing the Landau parameters obtained in these two approaches, we will be able to bring information about the relevance of both non-linear and Pauli blocking terms in the polarization operators. Moreover, we will show the link between the Landau parameters and the collective modes in the space like region like zero sound.

In the next section, we give the expressions of the Landau parameters calculated with the relativistic non-linear models NL1, TM1 and G1. Then, results for the Landau parameters as well as collective modes in nuclear matter are presented. We draw conclusions in the last section.

LANDAU PARAMETERS IN NUCLEAR MATTER

We have found that if the Pauli blocking terms are neglected in the polarization operators, then the Landau parameters F_0 and F_1 are different from those obtained from the total energy density. The Landau parameters are the same only if the Pauli blocking terms are taken into account in the polarization operators. We have found :

$$F_0 = \frac{F_\omega(1+F_\sigma a_S + \frac{\Pi_{\sigma\sigma}^{nl}}{m_\sigma^2}) - F_\sigma(1-v_F^2)(1 - \frac{-\Pi_q^{nl}+\Pi_\eta^{nl}}{m_\omega^2}) + 2\left[F_\sigma F_\omega(1-v_F^2)\right]^{\frac{1}{2}} \frac{\Pi_{\sigma\omega}^{nl}}{m_\sigma m_\omega}}{(1 - \frac{-\Pi_q^{nl}+\Pi_\eta^{nl}}{m_\omega^2})(1+F_\sigma a_S + \frac{\Pi_{\sigma\sigma}^{nl}}{m_\sigma^2}) + \left(\frac{\Pi_{\sigma\omega}^{nl}}{m_\sigma m_\omega}\right)^2}, \quad (1)$$

$$F_1 = \frac{-F_\omega v_F^2}{1+\frac{1}{3}F_\omega v_F^2 + \frac{\Pi_q^{nl}}{m_\omega^2}}. \quad (2)$$

Here, $a_S = \frac{1}{N_F \pi^2}\left[k_F\left(E_F^* + 2\frac{M_N^{*2}}{E_F^*}\right) - 3(M_N^*)^2 \ln\left(\frac{k_F+E_F^*}{M_N^*}\right)\right]$, $\Pi_{\sigma\sigma}^{nl} = 2g_{\sigma 3}\phi_0 + 3g_{\sigma 4}\phi_0^2 - \frac{1}{2}g_{\sigma\omega 4}V_0^2$, $\Pi_q^{nl} = \frac{2}{3}g_{\sigma\omega 3}\phi_0 + g_{\omega 4}V_0^2 + \frac{1}{2}g_{\sigma\omega 4}\phi_0^2$, $\Pi_{\sigma\omega}^{nl} = -\left(\frac{2}{3}g_{\sigma\omega 3} + g_{\sigma\omega 4}\phi_0\right)V_0$, $\Pi_\eta^{nl} = -2g_{\omega 4}V_0^2$, with $M_N^* = M_N - g_{\sigma 1}\phi_0$, $E_F^* = \sqrt{k_F^2 + M_N^{*2}}$, $N_F = \frac{2k_F E_F^*}{\pi^2}$, $v_F = \frac{k_F}{E_F^*}$, $F_\sigma = \frac{g_{\sigma 1}^2 N_F}{m_\sigma^2}$, $F_\omega = \frac{g_{\omega 1}^2 N_F}{m_\omega^2}$. In symmetric nuclear matter, for a static and uniform system in a mean-field theory, the mean fields ϕ_0 and V_0 can be obtained from the Lagrange equations as:

$$\phi_0 = \frac{g_{\sigma 1}\rho_s + \frac{1}{3}g_{\sigma\omega 3}V_0^2}{m_\sigma^2 + g_{\sigma 3}\phi_0 + g_{\sigma 4}\phi_0^2 - \frac{1}{2}g_{\sigma\omega 4}V_0^2}, \quad V_0 = \frac{g_{\omega 1}\rho_B}{m_\omega^2 + \frac{2}{3}g_{\sigma\omega 3}\phi_0 + g_{\omega 4}V_0^2 + \frac{1}{2}g_{\sigma\omega 4}\phi_0^2}.$$

The scalar and baryonic nuclear densities are $\rho_s = \frac{M_N^*}{\pi^2}\left[k_F E_F^* - M_N^{*2}\ln\left(\frac{k_F+E_F^*}{M_N^*}\right)\right]$ and $\rho_B = 2k_F^3/3\pi^2$. The coupling constants ($g_{\sigma 1}$, $g_{\omega 1}$, $g_{\sigma 3}$, $g_{\sigma\omega 3}$, $g_{\sigma 4}$, $g_{\omega 4}$, $g_{\sigma\omega 4}$) and the masses (m_σ, m_ω) depend on the non-linear model used.

COLLECTIVE MODES IN NUCLEAR MATTER

In Fig. 1, we show the link between the Landau parameters and the existence of the zero-sound and instability modes as a function of baryon density for symmetric nuclear

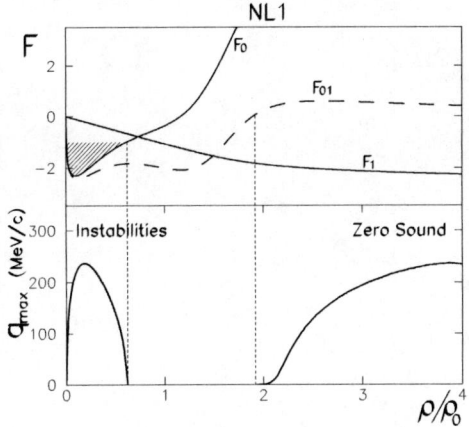

FIGURE 1. The dimensionless Landau parameters F_0 and F_1 (solid curves) are plotted in the upper part, while in the lower part, the highest three-momentum of zero sound and instability modes, q_{max}, is shown, as a function of baryon density, obtained with the NL1 model. The link between the Landau parameters and the existence of the zero-sound mode is shown using the quantity $F_{01} = F_0 + \frac{F_1}{1+\frac{1}{3}F_1}$ (dashed curve).

matter obtained with the non-linear NL1 model. More precisely, in the upper part of Fig. 1 the parameters F_0 and F_1 (solid curves) as well as the quantity F_{01} defined by $F_{01} = F_0 + \frac{F_1}{1+\frac{1}{3}F_1}$ (dashed curve) are plotted, while in the lower part, the momentum at which the modes end up, q_{max}, is shown. It can be observed that, in the low density region, F_0 becomes such as $F_0 < -1$ corresponding to the region of instabilities of nuclear matter. In the same way, the existence of the zero-sound mode is related to the value taken by F_{01} since as it can be seen in Fig.1, zero-sound mode occurs once the condition $F_{01} > 0$ is satisfied. Thus, we can see that the density dependence of the Landau parameters is able to predict at which densities the collective modes in the space-like region will occur.

In Fig. 2, we plotted q_{max} for both zero-sound and instability modes as a function of baryon density for the Walecka, NL1, TM1 and G1 models. The results indicate that whatever the model considered here, the system is stable with respect to density fluctuations at and around the saturation density and exhibits instability in the relatively low density regime. It can also be seen that the zero-sound mode occurs at lower densities for the Walecka model than for more elaborate models including σ self-coupling terms in the Lagrangian like NL1. For other non-linear models, including at least an ω self-coupling term in the Lagrangian (TM1 and G1), the zero sound mode is completely missing.

CONCLUSION

In this work, we have determined the Landau parameters in relativistic non-linear models from the second derivative of the total energy density with respect to the quasinucleon

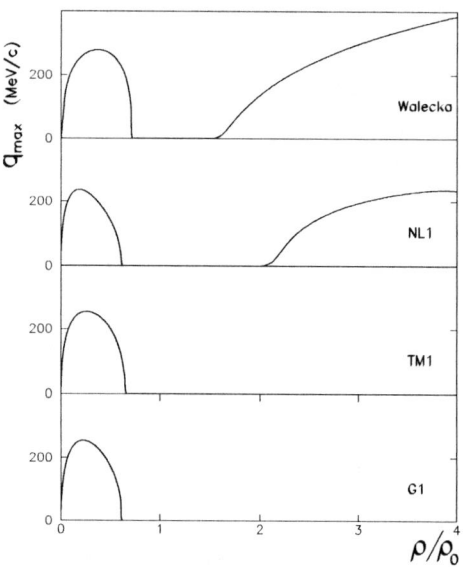

FIGURE 2. Highest three-momentum of both zero-sound and instability modes as a function of the baryon density obtained with the Walecka, NL1, TM1 and G1 models.

distribution as well as from the longitudinal dielectric function in the relativistic RPA. First, the non-linear diagrams that we have taken into account in the polarization operators are the relevant ones since these contributions appear naturally in the calculation of the Landau parameters from the total energy density. Second, the polarization operators including the Pauli blocking terms should be used for all realistic description of processes evaluated in the relativistic RPA. As an application, we have determined the zero-sound and instability modes in relativistic non-linear models. The instabilities of nuclear matter and the zero sound mode occur respectively once the conditions $F_0 < -1$ and $F_{01} > 0$ are satisfied. We have found that the zero-sound mode occurs all the more at lower densities when the model used is more simple.

REFERENCES

1. Serot, B.D., and Walecka, J.D., Advances in Nuclear Physics **16**, 1 (1986).
2. Reinhard, P.G., Rufa, M., Maruhn, J., Greiner, W., and Friedrich, J., *Z. Phys. A* **323**, 13 (1986).
3. Sugahara, Y., and Toki, H., *Nucl. Phys.* **A579**, 557 (1994).
4. Furnstahl, R.J., Serot, B.D., Tang, H.-B., *Nucl. Phys.* **A615**, (1997) 441.
5. Caillon, J.C., Gabinski, P., and Labarsouque, J., *Phys. Rev. C* **63**, 028201 (2001).
6. Zhong-yu Ma, Nguyen Van Giai, Wandelt, A., Vretenar, D., and Ring, P., *Nucl. Phys.* **A686**, 173 (2001).
7. Landau, L.D., *JETP* **3**, 920 (1957); **5**, 101 (1957); **8**, 70 (1959).
8. Matsui, T., *Nucl. Phys.* **A370**, 365 (1981).

Mesons on a transverse lattice

S. Dalley

Centre for Mathematical Sciences, Cambridge CB3 0WA, UK

Abstract. The meson eigenstates of the light-cone Hamiltonian in a coarse transverse lattice gauge theory are investigated. Coupling constants are non-perturbatively renormalised by demanding restoration of space-time symmetries broken by the lattice cutoff. The light-cone wavefunctions of the resulting Hamiltonian are used to compute decay constants, form factors and quark momentum and spin distributions for the pion and rho mesons.

INTRODUCTION

Light-cone quantisation is a Hamiltonian method that treats $x^+ = (x^0 + x^3)/\sqrt{2}$ as canonical time and $\{x^- = (x^0 - x^3)/\sqrt{2}, \mathbf{x} = (x^1, x^2)\}$ as the spatial variables. Lorentz indices μ, ν are split into light-cone indices $\alpha, \beta \in \{+, -\}$ and transverse indices $r, s \in \{1, 2\}$. In transverse lattice gauge theory [1], link variables $M_r(x^+, x^-, \mathbf{x})$ are associated with the link from $\mathbf{x}$ to $\mathbf{x} + a\hat{\mathbf{r}}$ at fixed x^+ and x^-. They physically represent the (averaged) flux between these points. Continuum $SU(N)$ gauge potentials $A_\alpha(x^+, x^-, \mathbf{x})$ and fermions fields $\Psi(x^+, x^-, \mathbf{x})$ are associated to a transverse plane $\mathbf{x} = \text{const.}$. The most general transverse lattice Lagrangian L will consist of all operators invariant under lattice gauge symmetries, Poincaré symmetries manifestly preserved by the lattice cutoff, and renormalisable by dimensional counting with respect to the continuum co-ordinates x^α. In general M can be taken as a general $N{\times}N$ complex matrix, provided it gauge transforms covariantly.

The potential term for L may be expanded in powers of the dynamical fields M, $\Psi^+ = \gamma^0 \gamma^+ \Psi$ (after elimination of non-dynamical fields $A_\pm$ and Ψ^-). Such a power expansion is justified in a region of coupling space where these fields are sufficiently heavy that light-cone wavefunctions of interest converge quickly in parton number. The simplest approximation to the Lagrangian in QCD is then [3, 4]

$$
\begin{aligned}
L = & \sum_\mathbf{x} \int dx^- \sum_{\alpha,\beta=+,-} \sum_{r=1,2} -\frac{1}{2G^2} \text{Tr}\{F^{\alpha\beta} F_{\alpha\beta}\} \\
& + \text{Tr}\{[(\partial_\alpha + iA_\alpha(\mathbf{x})) M_r(\mathbf{x}) - iM_r(\mathbf{x}) A_\alpha(\mathbf{x} + a\hat{\mathbf{r}})][\text{h.c.}]\} \\
& - \mu_b^2 \text{Tr}\{M_r M_r^\dagger\} + i\overline{\Psi}\gamma^\alpha(\partial_\alpha + iA_\alpha)\Psi - \mu_f \overline{\Psi}\Psi \\
& + i\kappa_a \left(\overline{\Psi}(\mathbf{x})\gamma^r M_r(\mathbf{x}) \Psi(\mathbf{x} + a\hat{\mathbf{r}}) - \overline{\Psi}(\mathbf{x})\gamma^r M_r^\dagger(\mathbf{x} - a\hat{\mathbf{r}}) \Psi(\mathbf{x} - a\hat{\mathbf{r}})\right) \\
& + \kappa_s \left(\overline{\Psi}(\mathbf{x}) M_r(\mathbf{x}) \Psi(\mathbf{x} + a\hat{\mathbf{r}}) + \overline{\Psi}(\mathbf{x}) M_r^\dagger(\mathbf{x} - a\hat{\mathbf{r}}) \Psi(\mathbf{x} - a\hat{\mathbf{r}})\right)
\end{aligned}
\quad (1)
$$

where $G, \mu_f, \mu_b, \kappa_a, \kappa_s$ are coupling constants and the $A_- = 0$ gauge is taken.

ONE-LINK APPROXIMATION

We introduce gauge indices $\{i,j \in \{1,2,\cdots N\}\}$ and make a 'one-link' truncation of Fock space. Then a normalised gauge-singlet state of zero transverse momentum $\mathbf{P} = 0$ can be expressed in terms of orthonormal Fock states as

$$|\psi(P^+)\rangle = \frac{1}{\sqrt{\text{Vol}}} \sum_{\mathbf{x}} a^2 \sum_{h,h'} \int_0^1 dx_1 dx_2 \, \delta(x_1 + x_2 - 1)$$

$$\times \left\{ \psi_{hh'}(x_1,x_2) N^{-1/2} b_h^\dagger(x_1,\mathbf{x}) d_{h'}^*(x_2,\mathbf{x}) |0\rangle \right\}$$

$$+ \frac{1}{\sqrt{\text{Vol}}} \sum_{\mathbf{x}} a^2 \sum_{h,h',r} \int_0^1 dx_1 dx_2 dx_3 \, \delta(x_1 + x_2 + x_3 - 1) \qquad (2)$$

$$\times \left\{ \psi_{h(r)h'}(x_1,x_2,x_3) N^{-1} b_h^\dagger(x_1,\mathbf{x}) a_r^\dagger(x_2,\mathbf{x}) d_{h'}^*(x_3,\mathbf{x}+a\hat{r}) |0\rangle \right.$$

$$\left. + \psi_{h(-r)h'}(x_1,x_2,x_3) N^{-1} b_h^\dagger(x_1,\mathbf{x}+a\hat{r}) a_{-r}^\dagger(x_2,\mathbf{x}) d_{h'}^*(x_3,\mathbf{x}) |0\rangle \right\}, \qquad (3)$$

Vol is the volume of transverse space, b and d are quark and anti-quark modes, a_λ are link modes with orientation λ, and $h = \pm$ labels helicity. The boost-invariant momentum fractions $x_1 = k_1^+/P^+$ etc. have been introduced. A state of non-zero transverse momentum is obtained by inserting appropriate phases.

Keeping the only quartic term in P^- that couples to G^2, and all quadratic and cubic interactions, we project $2P^+P^-|\psi(\mathbf{P}=0)\rangle$ onto Fock basis states. This allows us to derive the following set of integral equations for the invariant mass $\mathcal{M}$;

$$\frac{\mathcal{M}^2}{G^2} \psi_{hh'}(x_1,x_2) = \left(\frac{m_f^2}{x_1} + \frac{m_f^2}{x_2}\right) \psi_{hh'}(x_1,x_2) + K(\psi_{hh'}(x_1,x_2))$$

$$+ \frac{(k_a^2 + k_s^2)}{\pi} \left(\frac{1}{x_1}\int_0^{x_1} \frac{dy}{y} + \frac{1}{x_2}\int_0^{x_2}\frac{dy}{y}\right) \psi_{hh'}(x_1,x_2)$$

$$- \sum_\lambda \left\{ \frac{m_f k_s}{2\sqrt{\pi}} \int_0^{x_1} \frac{dy}{\sqrt{y}} \left(\frac{1}{x_1-y}+\frac{1}{x_1}\right) \psi_{h(\lambda)h'}(x_1-y,y,x_2) \right. \qquad (4)$$

$$+ \frac{m_f k_s}{2\sqrt{\pi}} \int_0^{x_2} \frac{dy}{\sqrt{y}} \left(\frac{1}{x_2-y}+\frac{1}{x_2}\right) \psi_{h(\lambda)h'}(x_1,y,x_2-y)$$

$$+ \frac{\text{Sgn}(\lambda) m_f k_a(hi\delta_{|\lambda|1}+\delta_{|\lambda|2})}{2\sqrt{\pi}} \int_0^{x_1} \frac{dy}{\sqrt{y}} \frac{y\psi_{-h(\lambda)h'}(x_1-y,y,x_2)}{(x_1-y)x_1}$$

$$\left. - \frac{\text{Sgn}(\lambda) m_f k_a(h'i\delta_{|\lambda|1}+\delta_{|\lambda|2})}{2\sqrt{\pi}} \int_0^{x_2} \frac{dy}{\sqrt{y}} \frac{y\psi_{h(\lambda)-h'}(x_1,y,x_2-y)}{(x_2-y)x_2} \right\},$$

$$\frac{\mathcal{M}^2}{G^2} \psi_{h(\lambda)h'}(x_1,x_2,x_3) = \left(\frac{m_b^2}{x_2}+\frac{m_f^2}{x_1}+\frac{m_f^2}{x_3}\right) \psi_{h(\lambda)h'}(x_1,x_2,x_3) + K(\psi_{h(\lambda)h'}(x_1,x_2,x_3))$$

$$- \frac{m_f k_s}{2\sqrt{\pi x_2}}\left(\frac{1}{x_1}+\frac{1}{x_1+x_2}\right) \psi_{hh'}(x_1+x_2,x_3)$$

$$-\frac{m_f k_s}{2\sqrt{\pi x_2}}\left(\frac{1}{x_3}+\frac{1}{x_2+x_3}\right)\psi_{hh'}(x_1,x_2+x_3) \qquad (5)$$
$$-\text{Sgn}(\lambda)\frac{(hi\delta_{|\lambda|1}+\delta_{|\lambda|2})m_f k_a}{2\sqrt{\pi x_2}}\left(\frac{x_2\psi_{-hh'}(x_1+x_2,x_3)}{(x_1+x_2)x_1}\right)$$
$$+\text{Sgn}(\lambda)\frac{(h'i\delta_{|\lambda|1}+\delta_{|\lambda|2})m_f k_a}{2\sqrt{\pi x_2}}\left(\frac{x_2\psi_{h-h'}(x_1,x_2+x_3)}{x_3(x_2+x_3)}\right).$$

$$K(\psi_{hh'}(x_1,x_2))=\frac{1}{2\pi}\int_0^1 dy\left\{\frac{\psi_{hh'}(x_1,x_2)-\psi_{hh'}(y,1-y)}{(y-x_1)^2}\right\}, \qquad (6)$$

$$K(\psi_{h(\lambda)h'}(x_1,x_2,x_3)) = \frac{1}{2\pi}\int_0^{x_2+x_3} dy\frac{(x_3+2x_2-y)}{2(x_3-y)^2\sqrt{x_2(x_2+x_3-y)}}\{\psi_{h(\lambda)h'}(x_1,x_2,x_3)$$
$$-\psi_{h(\lambda)h'}(x_1,x_2+x_3-y,y)\}$$
$$+\frac{1}{2\pi x_3}\left(\sqrt{1+\frac{x_3}{x_2}}-1\right)\psi_{h(\lambda)h'}(x_1,x_2,x_3)$$
$$+\frac{1}{2\pi}\int_0^{x_1+x_2} dy\frac{(x_1+2x_2-y)}{2(x_1-y)^2\sqrt{x_2(x_1+x_2-y)}}\{\psi_{h(\lambda)h'}(x_1,x_2,x_3)$$
$$-\psi_{h(\lambda)h'}(y,x_1+x_2-y,x_3)\}$$
$$+\frac{1}{2\pi x_1}\left(\sqrt{1+\frac{x_1}{x_2}}-1\right)\psi_{h(\lambda)h'}(x_1,x_2,x_3). \qquad (7)$$

$\bar{G}=G\sqrt{(N^2-1)/N}$, which has the dimensions of mass, and the following renormalisations have been introduced:

$$\frac{\mu_b}{\bar{G}}\to m_b \; ; \; \frac{\mu_f}{\bar{G}}\to m_f \; ; \; \frac{\kappa_a\sqrt{N}}{\bar{G}}\to k_a \; ; \; \frac{\kappa_s\sqrt{N}}{\bar{G}}\to k_s. \qquad (8)$$

DLCQ [2] is used to discretize longitudinal momentum in these equations, by compactifying x^- on circle of circumference $L = 2\pi K/P^+$, where K is a positive integer, with (anti-)periodic boundary conditions for links (quarks). $\mathcal{M}$ is then a finite matrix for each K. The lattice spacing a is fixed at $a \approx (300\text{MeV})^{-1} \equiv 2/3$ fm, corresponding to $m_b = 0.2$ [5]. This is the maximum quark separation in the one-link approximation. In this case $\bar{G} = 1368$ MeV.

The remaining three-dimensional space of dimensionless couplings $\{m_f, k_a, k_s\}$ was sampled discretely and the eigenvalue problem for $2P^+P^-$ solved for various transverse momenta **P**. A χ^2-test was introduced, at each K, for conditions of relativistic invariance on the π and ρ dispersion and for the conditions that their masses assume the physical values. An unambiguous set of couplings is found at each K that optimises these conditions. Results are then extrapolated in $1/K$.

SOME RESULTS

For the pion distribution amplitude $\phi_\pi(x)$, one finds

$$\psi_{+-}(x, 1-x) = \frac{f_\pi}{2}\sqrt{\frac{\pi}{N}}\phi_\pi(x) \qquad (9)$$

with

$$\phi_\pi(x) = 6x(1-x)\left\{1 + 0.133 C_2^{3/2}(1-2x^2)\right\} . \qquad (10)$$

However, in this one-link approximation the overall normalisation yields $f_\pi = 347(46)$ MeV, compared to the experimental value $f_\pi(\exp.) = 131$ MeV. For the helicity zero component of the rho, one finds a similar shape and $f_{\rho_0} = 315(50)$ MeV, again much more than the experimental value $f_{\rho_0}(\exp.) = 216$ MeV.

The parton distribution function of the pion — the probability for finding a quark carrying certain longitudinal momentum fraction of the hadron — is found to be more peaked at large x than a real pion. The pion elastic electromagnetic form factor $F(-q^2)$ yields a charge radius, given by $r_\pi^2 = 6(\partial F(-q^2)/\partial q^2)_{q^2=0}$, of $r_\pi \sim 0.3$ fm, about half the experimental value. Both these results confirm that the pion wavefunction is being artificially squashed into the valence quark sector with small separations. A large part of the ρ spin is not carried by the quark spins; one finds $s = 0.41(10)$ for the quark spin fraction carried by quarks in a polarized ρ.

It is obviously necessary to do calculations allowing more links in a meson, before quantitative conclusion can be drawn. With a lattice spacing of 2/3 fm, at least two or three links should be allowed, in order to encompass light meson physics.

ACKNOWLEDGMENTS

Work supported by PPARC grant GR/LO3965.

REFERENCES

1. Bardeen, W.A., and Pearson, R.B., *Phys. Rev. D* **14**, 547 (1976).
2. Pauli, H-C. and Brodsky, S.J., *Phys. Rev. D* **32**, 1993 (1985); *ibid* 2001.
3. Dalley, S. *Phys. Rev. D* **64**, 036006 (2001)
4. Burkardt, M., Seal, S.K., hep-ph/0102245.
5. Dalley, S., and van de Sande, B., *Phys. Rev. D* **62**, 014507 (2000).

Heavy quark-antiquark admixture in the proton and its test in exclusive hadronic process

M. Dillig*, G.F. Marranghello†, S.S. Rocha** and C.A.Z. Vasconcellos†

*Institut für Theoretishe Physik III, Universität Erlangen-Nürnberg,
D-91058 Erlangen, Germany
†Instituto de Física, Universidade Federal do Rio Grande do Sul, Bento Gonçalves,
C.P. 15051, Porto Alegre, Brasil
**Universidade do Extremo Sul Catarinense, C.P. 9167, Criciúma, Brasil

Abstract. We investigate the admixture of heavy $s\bar{s}$, $c\bar{c}$ and $b\bar{b}$ flavors to the ground state of the proton. Preliminary perturbative results in a constituent quark model and a covariant approach in the infinite momentum frame are compared with simple estimates based on a meson-exchange model. The test of these virtual $q\bar{q}$ components in the proton induced threshold exclusive meson-production is briefly discussed.

INTRODUCTION

The investigation of the basic structure of hadrons is one of the most important questions in Quantum Chromodynamics (QCD). In the general framework, baryons and mesons are described as $3q$ or $q\bar{q}$ states (with the gluon distribution integrated out). Such a picture allows - given the uncertainties in most QCD inspired quark models - for a qualitative description of hadronic reactions at intermediate scattering energies (at the scale of few GeV).

On the other side, however, a Fock space expansion for baryons yields substantial admixtures to the $3q$ content (Fig. 1)

$$|N> = \alpha|qqq> + \beta|qqq(q\bar{q})> + \gamma|qqqg> +... \qquad (1)$$

from gluonic and $q\bar{q}$ (mesonic) excitations. The increasing number of precision experiments — to quote only the detailed investigation on the spin-content of the proton at HERMES [1] — indicate a substantial component of sea-quark admixtures to the nucleon ground-state. An attractive way to test the presence of $q\bar{q}$ sea quarks in the nucleon is a direct knock out of these components with external probes. Different experiments have been performed in this direction, particularly in electron scattering [2]. These investigations have been recently paralleled by experiments with proton beams in modern hadron facilities like COSY or CELSIUS, where dominantly the exclusive production of heavy mesons near the threshold was (and is) investigated in proton-proton collisions (Fig. 2) [3].

In this note we focus on the admixture of heavy $q\bar{q}$ pairs in the ground state of the proton, i.e. the presence of $s\bar{s}$, $c\bar{c}$ and $b\bar{b}$ pairs. As the energy scale for the excitation of such $q\bar{q}$ pairs for constituent quark excitation energies is $\Delta E \geq 1$ GeV even for a $s\bar{s}$

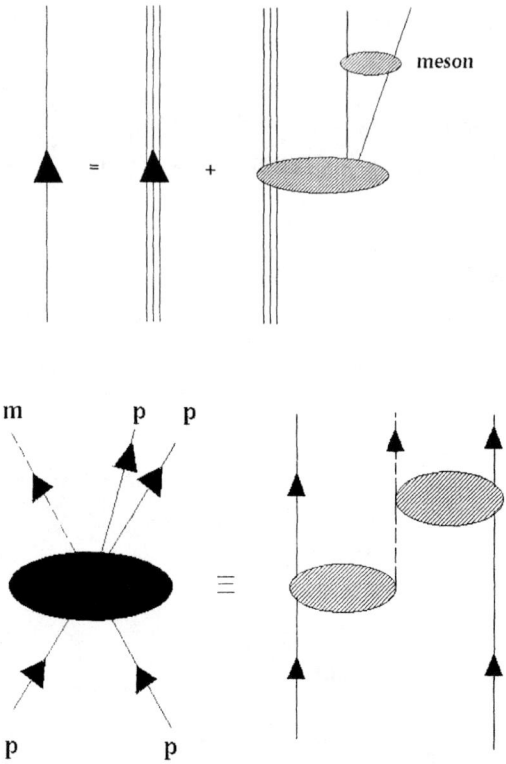

FIGURE 1. Schematic representation of $q\bar{q}$ admixture in the baryonic ground state (a) and of meson-production in pp-collisions (b) (The dashed areas indicate meson or gluon exchange).

pair, it is hoped that, at least a qualitative estimate of such components in baryons can be obtained already in lowest (or low) order perturbation (without diagonalizing the full problem in a given model space). In the following we just proceed along this reasoning.

For a realistic estimate of the $q\bar{q}$ admixture of heavy flavours the appropriate degrees of freedom are of outmost importance. Comparing with current mechanisms for the meson production in the low GeV region the standard approach is based on effective meson - and baryon - degrees of freedom. Such an approach allows for a qualitative description of the overall cross sections; it suffers, however, particularly from the sensitivity of heavy meson exchanges and their continuation off the mass shell [4] (the "off-shellity" is characterized by a typical 4-momentum transfer of $q^2 \leq -1$ (GeV/c)^2 even at threshold). In view of the very short range nature of typically $\leq \frac{1}{4}$ fm of the genuine production vertex, it is tempting to switch over the more fundamental degrees of freedom: these high momentum transfer processes should explicitly probe quark degrees of freedom of the proton (gluonic components are included in the gluon induced exchange qq and pair-creation $q \to qq\bar{q}$ potential). Though also such models contain a variety of phenomenological parameters (as briefly discussed below) they offer the advantage of consistent and coherent exchange mechanism from very light to very heavy $q\bar{q}$ excitations.

For the explicit quark-based description we compare two different approaches: the non-relativistic (constituent) quark model (improved by leading relativistic corrections) and a description on the front form, employing light cone coordinates (equivalently in a transition to the infinite momentum frame - IMF). Beyond the quantitative result in both approaches such a comparison is interesting by itself, as only in the IMF an unambigous identification of the $q\bar{q}$ component is possible (remember that on the light cone $q\bar{q}$ excitations out of the vacuum are strictly suppressed). We briefly summarize the main assumptions and ingredients in our model calculations. In the non-relativistic quark model the 3q-($q\bar{q}$) amplitude in the impulse approximation is schematically given (in momentum space) as

$$|\psi(p,p')>= \frac{< p,p'|V_{q\to qq\bar{q}}|\psi_p >}{E^*(p')+\omega(p,p')-M}\chi_{\text{colour}}, \tag{2}$$

where E and ω denote the energy of the (colored) 3q and $q\bar{q}$ state, respectively. To simplify the formalism we work in this exploratory study in a simple quark-scalar diquark representation for the proton. Then as a typical piece, the colour part of the excited proton state in this colour intermediate state is given in the SU(3) notation as

$$\chi_{\text{colour}} = \left[[(10)\times(01)]^{11}\otimes[(10)\times(01)^{11}]\right]^{00}. \tag{3}$$

With respect to the knock out of the $q\bar{q}$ pair by a (colored) gluon exchange interaction we do not classify the intermediate $3q(q\bar{q})$ state in terms of colour-singlet components, which arise from a rearrangement of the appropriate quark lines, such as a classification with respect to the ΛK^+ component in the proton for example.

In lowest order the $q \to q(q\bar{q})$ transition potential is induced by the one-gluon exchange, yielding the leading spin-colour structure

$$V_{q\to qq\bar{q}}(rij) = \frac{4\pi\alpha}{m_g^2}\left(\frac{\lambda^l\lambda^l}{4}\right)(a\sigma_j\nabla_{ij}+b(\sigma_i\times\sigma_j)\nabla_{ij}+c\sigma_j\nabla_i)\delta(r_{ij})) \tag{4}$$

with the coefficients a, b and c given by the leading terms in the 2-component Pauli expansion of the covariant gluon exchange. Above the gluon exchange is treated with a non-perturbative effective gluon mass m_g and running coupling constant α of the order $\alpha/m_g^2 \approx 1/4$ fm² [5]. In the heavy quark limit (see below), high order corrections to the pair creation potential, arising from the multi-gluon exchange (which may also simulate pomeron exchange) are readily incorporated in a local approximation in the heavy quark mass limit (Fig. 2) (for details we refer to a forthcoming publication [6]).

The proton wave function itself is taken from a resonating group calculation and expanded, for an analytical evaluation, in an harmonic oscillator basis with typical size parameters $a \approx 2/3$ fm (the underlying Hamiltonian contains the one-gluon exchange and a phenomenological harmonic confinement; the baryon (and meson) wave functions are solutions from a Ritz variational principle [7]). On the light cone the $q\bar{q}$ admixture is calculated in a similar spirit. For the derivation of a covariant formalism we start from the manifestly covariant 4-dimensional Bethe-Salpeter equation, which we reduce to a covariant 3-dimensional quasipotential equation by fixing the off-shell propagation of

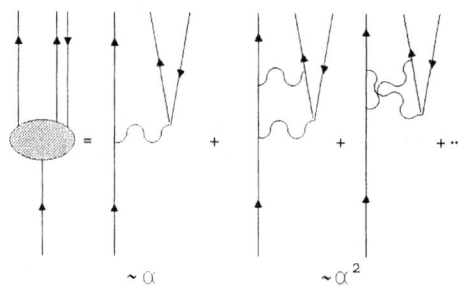

FIGURE 2. Pair creation potential $q \to qq\bar{q}$ in an expansion in α and α^2 (of the running coupling constant; the full line denotes the heavy quark $q\bar{q}$ pair excited).

the $q - q\bar{q}$ state in the proton (we compare the Blankenbecler-Sugar and the Gross limits [8]), which yields schematically for the 2-body Green function

$$G(q,qq) \approx i\pi \frac{\delta(\lambda(q^2 - m_q^2) - (1-\lambda)(Q - m_{qq}^2))}{q^2 - m_q^2 - i\varepsilon} \delta(q + Q - P) \tag{5}$$

for appropriate values of λ; m_{qq} and P are the constituent quark mass and total momentum of the proton, respectively. The bound state light cone wave function is derived for a covariant harmonic confinement with a localized one-gluon exchange (which allows an analytical evaluation of the distribution amplitude in terms of confluent hypergeometric functions [9]). Spin effects are included schematically in a Melosh transformation for a free quark [10].

Typical results show a sizable admixture probability of $q\bar{q}$ pairs in the proton ground state. For $s\bar{s}$ excitation the typical admixture probability is around 5%, depending, however, sensitively on the actual parameters chosen in the calculation, particularly on the high momentum content of the proton in the constituent cluster quark model and in regularization of the quark-gluon vertex (to yeld a bounded Hamiltonian for the resonating group calculation) or the transversal momentum cut off in the calculation in the instant form. Qualitatively, on the percent level, the admixture coefficients found are supported from perturbative one-loop calculations in a meson-exchange model for the $s\bar{s}$-excitation (Fig. 3) with some information on specific mesonic states, such as the $\phi(1020)$ meson, where the coupling to the proton can be estimated from quark counting or via the strong K^+K^- decay channel. The main sensitivity of such an estimate again arises from the regularization of the loop via effective form factors. The idea and the implications of a complex nucleon structure beyond the pure $3q$ content is currently under intensive investigation in modern electron and hadron accelerators. Here an interesting possibility is to study these exotic components by their direct knock-out in predominantly meson production processes. Of particular interest is thereby near threshold exclusive meson production as it dramatically simplifies the structure of the underlying transition amplitude, cutting it down to a single partial wave at threshold. Without entering in a detailed

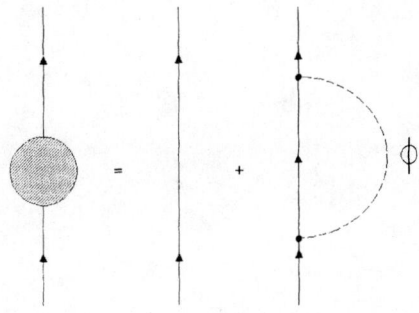

FIGURE 3. Perturbative one-loop ϕ ($q\bar{q}$) excitation in a meson exchange picture. The dashed line denotes the direct $NN\phi$ vertex or its coupling via the dominant $\phi \to KK$ decay channel.

discussion of possible reaction mechanisms, the complexity of the production mechanism itself - the most simple production mode for 1^--vector mesons in the pp collision is a 3-gluon exchange mechanism - and the non-perturbative nature of the hadronization of the $q\bar{q}$ pair towards the mass shell invalidates, at the moment, quantitative conclusions. Here substantial progress both in an experimental and theoretical field is urgently needed; most pressing - and most rewarding - is experimental information on sensitive observables (beyond total cross section) and theoretical progress towards a more realistic fully non-static 4-dimensional evaluations (possibly based on the Bethe-Salpeter equation) of microscopic models.

REFERENCES

1. HERMES Collaboration (Rith, K., for the coll.), in *Deep inelastic scattering and QCD (DIS 98)*, Brussels 1998, p.625.
2. CLAS Collaboration (Burkart, V., for the coll.), *Nucl. Phys.* **A684**, 16 (2001).
3. Grozonha, P., and Kilian, K., *Nucl. Phys.*, **A684**, 130 (2001).
   Bilger, R., et. al., *Nucl. Phys.* **A663**, 1073 (2001).
4. Moskal, P., et. al., *Acta Phys. Polon.* **331**, 2277 (2000).
   Hachner, H., and Heidenbauer, J., *πN Newslett.* **15**, 311 (1999).
5. Kar, S. C., and Gautman, V. P., *Hadronic J.* **19**, 303 (1996).
6. Dillig, M. and Vasconcellos, C. A. Z., preprint Univ. Erlangen, 2001.
7. Hofmann, H. M., et. al., *Few Body Syst. Suppl.* **5**, 471 (1992).
8. Blankenbecler, R., and Sugar, R., *Phys. Rev.* **142**, 1051 (1966).
   Gross, F., *Relativistic Quantum Mechanics and Field Theory*, Wiley& Sons Inc., New York, 1993.
9. Schmitt, M., PhD Thesis, Univ. Erlangen-Nürnberg, 1996.
10. Melosh, H. J., *Phys. Rev. D* **9**, 1095 (1974).

The GDH sum rule and pion photoproduction on the nucleon

Heidi Holvoet (on behalf of the GDH/A2 collaboration)

Dept. Subatomic & Radiation Physics, Ghent University, Proeftuinstraat 86, 9000 Ghent, Belgium

Abstract. Results from the pioneering experiment dedicated to verify the Gerasimov-Drell-Hearn (GDH) sum rule are presented. The measured contribution to the total sum rule up to 800 MeV photon energy is found to be $-(226 \pm 5 \pm 15)$ μb. A significant helicity dependence of the single and double pion photoproduction processes is observed and their contribution to the GDH sum rule in the measured energy range is also discussed.

INTRODUCTION

In the study of the internal structure of the nucleon and of its spin structure, the experimental verification of the Gerasimov-Drell-Hearn sum rule is an important issue in intermediate energy physics today. The Gerasimov-Drell-Hearn sum rule, or GDH sum rule for short, for the nucleon was derived in the middle of the 1960's by Gerasimov [1, 2] and by Drell and Hearn [3]. It is a model independent relation between the dynamical excitation spectrum of the nucleon and its statical properties. Its theoretical derivation relies solely on fundamental physics principles such as causality, flux conservation, gauge invariance and Lorentz (relativistic) invariance. Only one assumption needs to be made, i.e. the so-called *no-subtraction hypothesis* which is related to the convergence of the sum rule integrand. The GDH sum rule for the nucleon reads:

$$\int_{\nu_0}^{\infty} \frac{\sigma_{1/2} - \sigma_{3/2}}{\nu} d\nu = -\frac{2\pi^2 \alpha}{m_N^2} \kappa_N^2.$$

The total photoabsorption helicity cross sections $\sigma_{1/2}$ and $\sigma_{3/2}$ correspond to the absorption of circularly polarised photons by longitudinally polarised nucleons as defined above. The helicity cross section difference is weighted by the inverse of the incoming photon energy ν and integrated from single pion photoproduction threshold ν_0 to infinity. The right-hand side contains the fine structure constant α, the mass of the nucleon m_N, its anomalous magnetic moment κ_N and its spin.

From the very beginning the need for an experimental verification —by measuring the left-hand side— of the GDH sum rule was stressed. Apart from this verification, also the contribution of the various partial photoabsorption processes is wanted with respect to a complete understanding of the sum rule. Results for both these topics, obtained in the pioneering experiment carried out by the GDH collaboration are presented here.

The experiment is divided over two electron accelerator facilities. Photon energies up to 800 MeV and up to 3.2 GeV are obtained at MAMI, Mainz and ELSA, Bonn (both in

Germany), respectively. By the combined energy range a broad and physically important energy range is covered.

Here we concentrate on the results from the Mainz part of the experiment. After a discussion of the experimental apparatus in the next section, results are presented and discussed in detail. The last section summarizes and concludes.

THE GDH EXPERIMENT AT MAMI

Circularly polarised photons are obtained via bremsstrahlung from linearly polarised electrons. The polarised electrons are produced at MAMI from a strained GaAs source [4, 5]. A photon polarisation degree of up to 75% is obtained.

The proton target is a frozen spin butanol (C_4H_9OH) target [6, 7]. At 50 mK a maximum polarisation degree of 87% is reached, with a relaxation time of 200 h.

The central detector is the DAPHNE detector [8]. It has a large angular (94% of 4π) and momentum acceptance. The multiple wire proportional chambers (MWPCs) yield a precise vertex and track reconstruction. A six-layer scintillator telescope surrounds the MWPCs. It yields large efficiencies for charged hadron (protons and charged pions) identification (up to 90%) and reasonable efficiencies for the identification of neutrons and neutral pions (up to 50%). In order to also cover forward angles, the central detector is completed with the forward detectors MIDAS, STAR and FFW. As the forward direction contains a large contribution of electromagnetic background, a N_2-aerogel Čerenkov detector is installed as a veto detector for these unwanted events.

RESULTS

The main observable in the GDH experiment is the total photoabsorption helicity cross section difference $\sigma_{1/2} - \sigma_{3/2}$. It is plotted as function of photon energy in Figure 1(a) [9]. One observes a strong negative peak in the Δ resonance region, below double pion photoproduction threshold, and another, smaller, peak in the second resonance region. Both point towards underlying mechanisms in which the 3/2 intermediate spin state dominates, enhancing the $\sigma_{3/2}$ cross section over $\sigma_{1/2}$. Three model predictions are also shown on the plot: the HDT [10] and SAID multipole analysis [11] and the Unitary Isobar Model (UIM) [12]. The latter contains a contribution from double pion photoproduction and η production, whereas the other two only contain single pion photoproduction. It is obvious from the discrepancy between the data and the SAID curve that double pion photoproduction also yields an important negative contribution. The UIM curve adds an important part by including these extra processes.

The experimental value for the GDH sum rule —from 200 to 800 MeV— is obtained by integrating the cross section difference, weighted by the photon energy. The full integration from 200 up to 800 MeV yields a measured value of $-(226 \pm 5 \pm 15)$ μb, the errors being statistical and systematical, respectively. The measured energy range already renders a large part of the total sum rule. To obtain an idea of the presently obtained agreement with the sum rule value of -204 μb one can complete the mea-

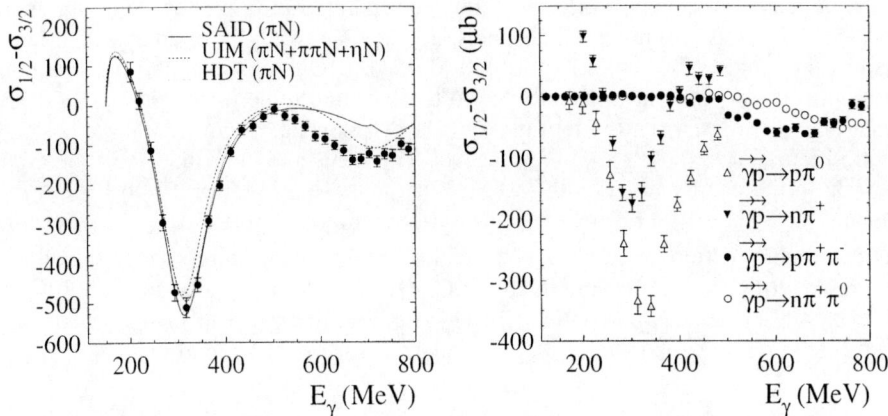

FIGURE 1. (a) Total photoabsorption cross section difference for the proton measured in the GDH experiment, plotted as a function of incoming photon energy. The curves correspond with SAID, HDT and UIM calculations as discussed in the text (b) Cross section difference for the single and double pion channels measured in the GDH experiment.

TABLE 1. Extrapolation of the measured value for the GDH sum rule on the proton and the proton forward spin polarisability by means of the UIM and the Bianchi-Thomas predictions in the unmeasured energy regions.

	E_γ (MeV)	I^p_{GDH} (μb)	I_{γ_0} (10^{-6} fm^4)
GDH experiment	200–800	$-226 \pm 5 \pm 15$	$-187 \pm 8 \pm 13$
UIM [10]	140–200	$+30 \pm 3$	$+104 \pm 14$
UIM [10]	800–1650	-40 ± 19	-3 ± 1
Bianchi-Thomas [13]	> 1650	$+26 \pm 7$	≈ 0
SUM		$-210 \pm 5 \pm 25$	$-86 \pm 8 \pm 19$

sured value with the available model predictions for the unmeasured energy regions. Table 1 gives an overview of this procedure. The UIM model is used for energies below 200 MeV and between 800 and 1650 MeV. The Bianchi-Thomas contribution is added for the higher energy part. In this way a value of $-(210 \pm 5 \pm 15)$ μb is deduced. Within error bars this is in agreement with the GDH sum rule. This is already a good indication for the validity of the sum rule. However, it remains mandatory to continue the experiment towards higher energies to confirm this result. The measurement in Bonn is expected to give a more conclusive answer.

The detector setup with DAPHNE as the central hadron detector is well suited to measure also the partial photoabsorption processes in the energy range from pion photo-production threshold to 800 MEV. For a proton target the active processes are the single and double pion production reactions $\vec{\gamma}\vec{p} \to p\pi^0$, $n\pi^+$, $p\pi^+\pi^-$, $n\pi^+\pi^0$ and $p\pi^0\pi^0$. The study of the helicity dependence of these processes is important in the investigation of the nucleon resonances. Figure 1(b) gives a compilation of the presently available data for the helicity cross section difference for the separated pion photoproduction channels [15]. For the two single pion channels, the data have only been analysed so far in detail in the single pion region, up to 450 MeV [14]. One observes the positive contribution to $\sigma_{1/2} - \sigma_{3/2}$ at threshold due to the E_{0+} multipole in the $n\pi^+$ process. At higher energies the large M_{1+} multipole of the Δ resonance dominates the cross section difference of both single pion processes. The measured contribution to the GDH sum rule in the energy range 200–450 MeV amounts to $-(144 \pm 7)$ μb for the $p\pi^0$ channel and $-(32 \pm 7)$ μb for $n\pi^+$. The cross section difference $\sigma_{1/2} - \sigma_{3/2}$ for the double pion production processes $p\pi^+\pi^-$ and the $n\pi^+\pi^0$ also exhibits a significant negative peak [16]. Their contributions to the GDH sum rule, integrated up to 800 MeV, are $-(25.4 \pm 1 \pm 1.5)$ μb (statistical and systematical error, resp.) from $p\pi^+\pi^-$ and $-(11.3 \pm 0.7 \pm 0.7)$ μb from $n\pi^+\pi^0$. The analysis for $p\pi^0\pi^0$ is ongoing [17].

CONCLUSION

In summary, we have measured the major contribution to the GDH sum rule in the energy range from pion photoproduction threshold to 800 MeV at the MAMI accelerator facility. This result, combined with available model calculations point to a converging GDH integral and the validity of the sum rule. However, only the combination with the experimental data obtained at ELSA will allow to draw more definitive conclusions.

For the single and double pion photoproduction reactions, a strong helicity dependence is observed. We have obtained their relative contributions to the GDH sum rule. Further detailed studies may reveal more information about the properties of the nucleon resonances.

REFERENCES

1. Gerasimov, S. B., *Yad. Fiz.* **2**, 598–602 (1965).
2. Gerasimov, S. B., *Sov. J. Nucl. Phys.* **2**, 430–433 (1966).
3. Drell, S. D., and Hearn, A. C., *Phys. Rev. Lett.* **16**, 908–911 (1966).
4. Herminghaus, H., et al., *Nucl. Instr. Meth.* **138**, 1 (1976).
5. Herminghaus, H., Bau und Betrieb von MAMI, Arbeits- und Ergebnisbericht 1987-1989 (sfb 201), 1989.
6. Bradtke, C., Dutz, H., et al., *Nucl. Instr. Meth.* A **436**, 430–442 (1999).
7. Dutz, H., et al., *Nucl. Instr. Meth.* A **340**, 272–277 (1994).
8. Audit, G., et al., *Nucl. Instr. Meth.* A **301**, 473–481 (1991).
9. GDH-collaboration, Ahrens, J., et al., *Phys. Rev. Lett.* **87**, 022003 (2001).
10. Hanstein, O., Drechsel, D., and Tiator, L., *Nucl. Phys.* **A632**, 561–606 (1998).
11. Arndt, R. A., et al., *Phys. Rev. C* **53**, 430 (1996), (Solution SP00K).
12. Drechsel, D., Hanstein, O., Kamalov, S. S., and Tiator, L., *Nucl. Phys.* **A645**, 145 (1999).
13. Bianchi, N., and Thomas, E., *Phys. Lett. B* **450**, 439–447 (1999).
14. Preobrajenski, I., Ph.D. thesis, Universität Mainz, Germany (2001), in preparation.
15. GDH-collaboration, Ahrens, J., et al., *Phys. Rev. Lett.* **84**, 5950–5954 (2000).
16. Holvoet, H., *Study of the helicity dependence of double pion photoproduction on the proton*, Ph.D. thesis, Ghent University, Belgium (2001).
17. Zapadtka, F., Ph.D. thesis, Universität Göttingen, Germany (2001), in preparation.

Photoproduction of heavy quarkonia

B. Jäger* and W. Schweiger†

*Institute of Theoretical Physics, University of Regensburg, D-93040 Regensburg, Germany
†Institute of Theoretical Physics, University of Graz, Universitätsplatz 5, A-8010 Graz, Austria

Abstract. We investigate the reaction $\gamma p \to V p$, with V denoting a Φ or a J/Ψ meson, within the scope of perturbative QCD, treating the proton as a quark-diquark system. Our predictions extrapolate the existing forward differential cross-section data into the few-GeV momentum-transfer region. In case of the J/Ψ reasonable results are only obtained by properly taking into account its mass in the perturbative calculation of the hard-scattering amplitude.

INTRODUCTION

Within the last few years exclusive photoproduction of vector mesons ($\rho, \omega, \Phi, J/\Psi$) has attained increasing interest which has been stimulated by corresponding experimental efforts at DESY and TJLab. The investigation of high-energy diffractive photoproduction at DESY aims at a better understanding of Pomeron phenomenology in terms of QCD. The 93-031 experiment at TJLab, on the other hand, is situated in the few-GeV (momentum-transfer) region and tries to shed some light on the transition from the non-perturbative (vector-meson dominance) to the perturbative (quark and gluon exchange) production mechanism. For heavy-quarkonium channels the perturbative production mechanism becomes particularly simple. Whereas ρ and ω production can proceed via both, quark and gluon exchange, heavy quarkonia are produced via gluon exchange mainly. Quark exchange is power-suppressed due to the small heavy-quark content of the nucleon.

In the present contribution we report on an attempt to describe photoproduction of Φ and J/Ψ mesons in the perturbative regime by means of a modified version of the hard-scattering approach (HSA), in which the proton is treated as a quark-diquark rather than a three-quark system. This perturbative diquark model accounts effectively for quark-quark correlations inside a baryon and has already successfully been applied to the investigation of various exclusive photon-induced reactions [1, 2, 3, 4, 5, 6]. It provides a consistent description of baryon electromagnetic form factors, Compton scattering off baryons, photoproduction of K mesons, etc., in the range of intermediate momentum transfers ($p_\perp^2 \gtrsim 3$ GeV^2) in the sense that data for these reactions are reproduced with the same set of model parameters.

THE HARD-SCATTERING APPROACH WITH DIQUARKS

Within the HSA an exclusive scattering amplitude M can be written as a convolution integral of a perturbatively calculable hard-scattering amplitude $T_{\{\lambda\}}$ with distribution amplitudes ϕ_H, which contain the bound state dynamics of the involved hadrons H. For the photoproduction reaction $\gamma p \to V p$ this integral takes on the particular form

$$M_{\{\lambda\}}(\tilde{s},\tilde{t}) = \int_0^1 dx_1 dy_1 dz_1 \phi_V^\dagger(z_1) \phi_p^\dagger(y_1) T_{\{\lambda\}}(x_1,y_1,z_1;\tilde{s},\tilde{t}) \phi_p(x_1). \qquad (1)$$

The distribution amplitudes ϕ_H are probability amplitudes for finding the valence-Fock state in the hadron H with its constituents carrying certain fractions x_i, y_i, z_i, $i = 1,2$ of the momentum of their parent hadron. In the diquark approach the valence-Fock state of a baryon is assumed to consist of a quark (q) and a diquark (D). The hard-scattering amplitude $T_{\{\lambda\}}$ is calculated on tree level in collinear approximation. The subscript $\{\lambda\}$ of T and M represents a set of possible photon, proton, and vector-meson helicities. The analytical results are conveniently expressed in terms of Mandelstam variables $\tilde{s}, \tilde{t}$, and $\tilde{u}$, which are obtained by neglecting the proton mass. All hadron masses, however, are fully taken into account in flux and phase-space factors.

The model, as applied in Refs. [1, 2, 3, 4, 6] includes scalar as well as axial-vector diquarks. Feynman rules, the quark-diquark distribution amplitude of the proton, and further details of the diquark model can be found in Ref. [1]. The numerical values of the model parameters for the present work are also adopted from that paper. In order to describe Φ- and J/Ψ-meson photoproduction simultaneously we have modified the Φ-meson distribution amplitude proposed by Benayoun and Chernyak [7] by attaching a flavor dependent exponential factor (cf. Ref. [6]).

Unlike the pure HSA, in which hadron masses are completely neglected in the calculation of the hard-scattering amplitude, we include them in our calculation. This improves the applicability of the model at momentum transfers of only a few GeV, where mass effects may still be important. Our treatment of mass effects parallels the one described in detail in Ref. [6]. In that work an expansion in powers of $(m_H/\sqrt{\tilde{s}})$ at fixed scattering angle has been performed. Only leading order and next-to-leading order terms in the expansion have been kept. Whereas this procedure gives a reasonable description of photoproduction of Φ mesons in the few-GeV momentum-transfer range, it fails in the case of the heavier J/Ψ mesons. Therefore we take the J/Ψ mass fully into account, treat the proton-mass, however, still like in Ref. [6].

RESULTS AND CONCLUSIONS

Photoproduction of Φ mesons in the forward-scattering domain can be well described by simple Pomeron phenomenology [8]. At higher momentum transfers a QCD-inspired version of the Pomeron-exchange model has been proposed by Laget and Mendez-Galain [9], in which the Pomeron is replaced by two (Abelian) gluons. If only those graphs are taken into account in which the two gluons couple to the same quark in the proton, the two-gluon cross section exhibits a characteristic node due to an interference

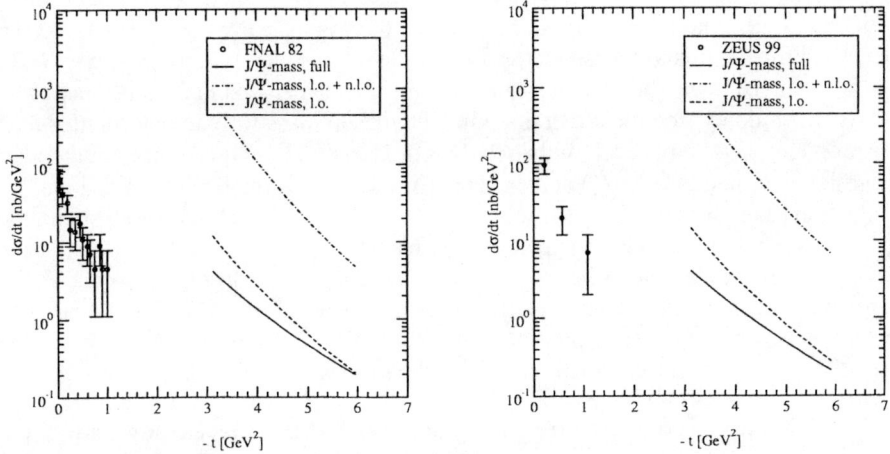

FIGURE 1. Differential cross section for $\gamma p \to J/\Psi p$ versus momentum transfer $-t$ at photon laboratory energies of 150 GeV (left) and 4708 GeV (right). Data are taken from Binkley et al. [12] and Breitweg et al. [13], respectively. Diquark model predictions are shown for different treatments of the J/Ψ mass: full inclusion of the J/Ψ mass (solid), leading (dashed), and next-to-leading order (dash-dotted) of an expansion in $(M_{J/\Psi}/\sqrt{\tilde{s}})$.

of the two different Feynman diagrams contributing to the photoproduction amplitude. By additionally considering diagrams, in which the gluons couple to different quarks in the proton this node is completely washed out [10]. For $|t| \gtrsim 4$ GeV^2 the resulting differential cross section becomes then comparable with the diquark-model prediction [6]. This similarity is not surprising since it is known that the hard-scattering mechanism, i.e. diagrams without loops, in which all hadronic constituents are connected via gluon exchange, becomes dominant in the kinematic range of large t and u.

Difficulties arise if J/Ψ production is considered. Both, Pomeron exchange as well as two-gluon exchange overestimate the differential cross section by at least one order of magnitude. The results are improved by arguing that the coupling of the Pomeron (or the two gluons) to a charmed quark is weaker than to a strange quark. If hadron masses are treated like in Ref. [6] similar difficulties are encountered within the diquark model. With the modified treatment of mass effects in which the meson mass is fully taken into account (for details see, e.g., Ref. [11]), this deficiency is, however, cured.

Figure 1 shows the diquark-model predictions for $d\sigma/dt$ at photon energies of 150 GeV and 4708 GeV, respectively, together with experimental data from FNAL [12] and DESY [13]. Due to the lack of data in the region of large transverse momentum transfers (i.e. large t and u) a direct comparison of our predictions with experiment is not possible, since the perturbatively inspired diquark model is not expected to be applicable in the kinematic region of $p_\perp^2 \lesssim 3$ GeV^2. The diquark-model predictions with the J/Ψ mass fully included, however, extrapolate the low-t data in a reasonable way. By comparing the leading and next-to-leading-order results of the expansion in $(M_{J/\Psi}/\sqrt{\tilde{s}})$ it becomes obvious that such a series expansion converges rather poorly. A closer inspection reveals that the expansion coefficients are angular dependent. The leading-order terms are only dominant for sufficiently large deflections. Towards smaller scattering angles mass-correction terms become increasingly important. For FNAL and HERA energies momentum transfers of a few GeV just mean that we deal with nearly forward scattering and thus we should refrain from an expansion in the J/Ψ mass.

As one would expect, the effect of taking the meson mass fully into account is much less pronounced in Φ production than in J/Ψ production. For equal values of the mass-expansion parameter $(M_V/\sqrt{s})$ and of Mandelstam t the photon laboratory energy for Φ production has to be smaller by about a factor $(M_\Phi/M_{J/\Psi})^2 = 0.11$ and correspondingly the scattering angle has to be larger, so that the leading-order terms in the mass expansion become dominant. But also in Φ production the full inclusion of the Φ mass has a positive effect. It slightly improves the angular dependence of the differential cross-section as compared to the approximate mass treatment of Ref. [6]. A full account of diquark-model predicitions for Φ and J/Ψ photoproduction, which includes also spin observables, can be found in Ref. [11].

To sort out whether perturbative photoproduction mechanisms already start to dominate in the few-GeV momentum-transfer region precision data for $|t| \gtrsim 3$ GeV^2 are doubtlessly needed. A severe test for a perturbative model, like the diquark model, would, however, be its confrontation with data for polarization observables. Hard scattering is closely connected with hadronic helicity conservation. Polarization observables could help to reveal whether the inclusion of constituent-masses and two-quark correlations in terms of diquarks suffices to model higher-twist and non-perturbative effects, or whether other (non-perturbative) mechanisms have to be considered.

ACKNOWLEDGMENTS

B.J. would like to thank the Paul-Urban-Stipendienstiftung for supporting her participation in this conference.

REFERENCES

1. Jakob, R., Kroll, P., Schürmann, M., and Schweiger, W., *Z. Phys.* **A347**, 109-116 (1993).
2. Kroll, P., Pilsner, Th., Schürmann, M., and Schweiger, W., *Phys. Lett.* B **316**, 546-554 (1993).
3. Kroll, P., Schürmann, M., and Guichon, P.A.M., *Nucl. Phys.* **A598**, 435-583 (1996).
4. Berger, C.F., *Exclusive Two-Photon Reactions in the Few-GeV Region*, diploma thesis, Technological University of Graz, Graz (1997).
5. Kroll, P., Schürmann, M., Passek, K., and Schweiger, W., *Phys. Rev. D* **55**, 4315-4328 (1997).
6. Berger, C.F., and Schweiger, W., *Phys. Rev. D* **61**, 114026 (2000).
7. Benayoun, M., and Chernyak, V.L., *Nucl. Phys.* **B329**, 285-313 (1990).
8. Donnachie, A., and Landshoff, P.V., *Nucl. Phys.* **B311**, 509-521 (1989).
9. Laget, J.-M., and Mendez-Galain, R., *Nucl. Phys.* **A581**, 397-428 (1995).
10. Laget, J.-M., *Phys. Lett.* B **489**, 313-318 (2000).
11. Jäger, B., *Photoproduction of Heavy Quarkonia*, diploma thesis, Karl-Franzens University of Graz, Graz (2001).
12. Binkley, M., *et al.*, *Phys. Rev. Lett.* **48**, 73-76 (1982).
13. Breitweg, J., *et al.*, *Eur. Phys. J.* **C14**, 213-238 (2000).

Separation of meson and quark-gluon degrees of freedom in the hadron scattering reactions up to 1 GeV energy region

A.I. Machavariani

Institute für Theoretische Physik der Univesität Tübingen, Tübingen D-72076, Germany

Abstract. The Haag-Nishijima-Zimmermann (HNZ) field theoretical construction of the bound state is applied to the problem of separation of the quark and hadron degrees of freedom in the hadron scattering reactions. We show that the quark-gluon degrees of freedom transform the form of the local hadron field operators in such a way that unified description of hadron interactions on the base of the one-variable vertex meson-hadron functions (like meson-nucleon vertex function $<\mathbf{p}'_N|j_\pi(0)|\mathbf{p}_N>$), becomes impossible.

INTRODUCTION

According to the modern interpretation of the hadron structure, hadrons are treated as quark-gluon clusters. But up to day the almost acceptable quantitative descriptions of the hadron-hadron interaction were obtained in the framework of the simple phenomenological models, where hadrons are considered as the local, structureless fields. Therefore following questions are appearing:

- How can one describe the hadron-hadron scattering reactions as quark cluster (or bound states) interactions in the framework of the consistent quantum field theory and how relates this theory to the corresponding phenomenological models without quark-gluon degrees of freedom?

- What is the characteristic feature of the quark-gluon degrees of freedom in the hadron-hadron interactions up to 1 GeV energy region which cannot be simulated by hadrons as structureless particles?

The first section of my talk is devoted to the first problem. In this section the brief description of the HNZ bound state treatment in the quantum field theory [1] is given. In the second section the effective potentials of the hadron scattering equations with and without quark-gluon degrees of freedom are considered. The difference of the kinematical structure of the corresponding vertex functions which are used as input ingredients by the construction of the above potentials is discussed. The answers to the above questions are presented at the end of the corresponding sections.

CONSTRUCTION OF THE COMPOSITE (BOUND) HADRON STATES FROM THE QUARKS

The unique self-consistent construction of the bound-state fields in the quantum-field theory was presented by HNZ [1] and was developed by Huang and Weldon [2]. In these papers a generalization of the one particle field operator $\Psi(x)$, creation or annihilation operator of this particle $\mathcal{B}_{\text{in(out)}}(\mathbf{p})$ was suggested. This generalization would enable to reformulate the S-matrix theory for the consistent or bound states. The key object in the HNZ formulation is the Heisenberg operator $\mathcal{B}_{\mathbf{p}}(X^0)$ which in the asymptotic region transforms into real particle creation or annihilation operator $\mathcal{B}_{\text{in(out)}}(\mathbf{p}) = \lim_{X^0 \to -\infty(+\infty)} \mathcal{B}_{\mathbf{p}}(X^0)$, where for the fermions $\mathcal{B}_{\mathbf{p}}(X^0) = \int d^3\mathbf{X} \exp(ipX) \bar{u}(\mathbf{p}) \gamma_0 \Upsilon_p(X)$ and operator $\Upsilon_p(X)$ is constructed from the ingredient quark operators $q_i(x_i)$ as

$$\Upsilon_p(X) = \lim_{\rho_{12} \to 0} \lim_{\rho_3 \to 0} T(q_1(x_1)q_2(x_2)q_3(x_3))/\chi_p(X=0,\rho_{12},\rho_3)$$

in the HNZ version, and

$$\Upsilon_p(X) = \int d^4\rho_{12} d^4\rho_3 \chi_p^\dagger(X=0,\rho_{12},\rho_3) T(q_1(x_1)q_2(x_2)q_3(x_3))$$

in the Huang-Weldon [2] version. Here X and ρ_3, ρ_{23} are the c.m. and the two relative Jacobi four-coordinates. The bound state field operator $\Upsilon_p(X)$ is nonlocal, because it depends not only on the coordinate X, but also on the four-momentum of the considering composite particle p. This additional dependence on p is generated by the bound-state wave function $\chi_p(X, \rho_{12}, \rho_3) \equiv \chi_p(x_1, x_2, x_3) = <0|T(q_1(x_1)q_2(x_2)q_3(x_3))|\mathbf{p}>$ which, in principle, is unambiguously determined as the solution of the bound state Bethe-Salpeter equation with the three quark interaction potential. In the same manner one can construct also composite meson fields through the quark-antiquark field operators.

The main property of the composite field operators is that the asymptotic field operators $\mathcal{B}_{\text{in(out)}}(\mathbf{p})$ satisfy the same anticommutation relations as the ordinary local field operators in the conventional quantum field theory

$$[\mathcal{B}_{\text{in(out)}}^\dagger(\mathbf{p}'), \mathcal{B}_{\text{in(out)}}(\mathbf{p})]_+ = \delta(\mathbf{p}' - \mathbf{p}).$$

But the operators $\mathcal{B}_{\mathbf{p}}(X^0)$ do not satisfy the equal-time anticommutation relations. These properties of $\mathcal{B}_{\text{in(out)}}(\mathbf{p})$ allow us to construct the generalised S-matrix for the composite (bound) state field operators [1, 2]. Corresponding construction of the composite field-operators with the three-dimensional field-theoretical wave functions is given in the Ref. [3, 4, 5].

In the asymptotic region one-particle cluster is isolated from the other particles. Therefore the campsite in(out) operator $\mathcal{B}_{\text{in(out)}}(\mathbf{p})$ has the same form as the ordinary fermion operator $b_{\text{in(out)}}(\mathbf{p})$. Moreover, we can identify these asymptotic operators [6]

$$\mathcal{B}_{\text{in(out)}}(\mathbf{p}) \equiv b_{\text{in(out)}}(\mathbf{p}).$$

Otherwise, if we provide $\mathcal{B}_{\text{in(out)}}(\mathbf{p})$ with distinctive features from the $b_{\text{in(out)}}(\mathbf{p})$, we will not be able to reproduce the conventional S-matrix, completeness relations in the ordinary Fock space $\sum_n |n; \text{in(out)}><\text{in(out)}; n| = \mathbf{1}$, etc.

From the identity of the asymptotic hadron states with and without quark degrees of freedom, it follows that the same S-matrix can be obtained in the framework of different theories with different boundary conditions [6]. So in the QED these boundary conditions are given through the Lagrangians and corresponding current operator

$J(x) = (i\gamma_\mu \partial_{X_\mu} - M)\Psi(x)$. Afterwards using the standard technique of the perturbation theory, one can calculate amplitude of the sought reaction. For the hadron reactions up to 1 GeV energy region amplitude may be defined as the solution of the corresponding scattering equations. The effective potentials of these equations are defined through the vertex functions which have the meaning of the input functions. Therefore boundary conditions of the hadron scattering equations are taken into account in the input vertex functions. In the standard field-theory without quarks these vertex functions $<$ out; $\mathbf{p}'_{h'}|J(0)|\mathbf{p}_h;$ in $>$ depend on the one variable (four momentum transfer $t = (p'_h - p_h)^2$) and usually are determined from the dispersion theory or from the phenomenological approximations. In the HNZ formulation of the hadron-hadron scattering with quark degrees of freedom the input vertex functions are more complicated. In these vertices $<$ out; $\mathbf{p}'_{h'}|J_\mathbf{q}(0)|\mathbf{p}_h;$ in $>$ the hadron source operator $J_\mathbf{q}(x) = (i\gamma_\mu \partial_{X_\mu} - M)\Upsilon_q(X)$ introduces an additional dependence on the four-momentum q. Therefore these vertex functions are depending not only on $t = (p'_h - p_h)^2$ but also on the q which is generating by the quark-hadron bound state wave function.

Thus turning to the first question, we want to emphasize that the HNZ approach for the hadrons as quark bound states provides us with the most general formulation of the hadron interaction. This formulation satisfies all first principles of the quantum field theory and the consistent way of construction of the hadron field operators from the quarks is given. These hadron operators are nonlocal, but the HNZ theory on the quark interaction level does not differ from the standard QCD. Therefore, this approach is not in conflict with any effective Lagrangian theory, where quark-gluon degrees of freedom are integrated after the intermediate calculations. But the HNZ formulation is not dependent on the quark-gluon transformation method into hadrons (hadronization methods) and the general equations of the hadron-hadron scattering reactions provide us with additional information about interaction mechanism [5].

THE FIELD-THEORETICAL EQUATIONS WITH AND WITHOUT QUARKS AND THE PHENOMENOLOGICAL TEST OF THE IMPORTANCE OF QUARK DEGREES OF FREEDOM

The field-theoretical equation of the hadron scattering reactions where as input vertex functions are required one variable vertices $<$ out; $\mathbf{p}'_{hadron'}|J(0)|\mathbf{p}_{hadron};$ in $>$, can be derived in the standard way as it was presented in many textbooks. Thus, using the S-matrix reduction formulas and the completness relation of asymptotic fields $\sum_n |n;$ in$($out$)><$ in$($out$);n| = \mathbf{1}$, one can easily derive the field-theoretical spectral decomposition formula [3, 4, 5] which can be considered as the origin of the sought equations. These quadratic-nonlinear, time-ordered and three-dimensional relativistic equations were exactly linearised [3]. The resulting Lippmann-Schwinger-type equations for the two-body scattering amplitude A have the form

$$A = Y + V + (Y + V)G_o A, \qquad (1)$$

where Y is the so called seagull or contact or overlapping term and it can be defined through the equal-time commutator. For example, for the hadron-hadron reaction we have $Y = <$ out; $\mathbf{p}'_{hadron'}|[J(0), b_\mathbf{q}(0)]|\mathbf{p}_{hadron};$ in $>$ and this term contains all contributions from the off mass-shellness of the intermediate particles. G_o denotes the Green function of free particles and V contains all contributions from the on-mass shell particle exchanges. The field-theoretical equations (1) are exactly connected with the all other field-theoretical equations, i.e., we can derive the Bethe-Salpeter equation from the Eq. (1) and vice versa. Therefore, all results obtained in the framework of these time-ordered, three-dimensional equations remain valid in other field-theoretical approaches as well. But unlike to all other field-theoretical equations, it is in the considered formulation that one-variable vertex functions are required by the construction of effective potentials.

According to the HNZ treatment of bound states, equations (1) for the hadron-hadron scattering reactions do not change their form if nucleons and mesons are considered as quark-clusters. In order to transform Eq. (1) for the case of the composite particles, we must replace all local field operators $\Psi(x)$ by the nonlocal operators $\Psi_p(x)$. The off-mass shell intermediate particle exchange potential Y is constructed so [3, 4], that the unitarity condition is satisfied and therefore there are no problems arising from the quark-gluon exchange poles.

At present time by calculation of the every hadron-hadron scattering reaction in the energy region up to 1 GeV different phenomenological vertex functions were used. In particular, the πN vertex functions which were used by description of the NN scattering distinguish strongly from the πN vertex function in the πN scattering reactions [3]. Moreover, even calculation of the S and P partial waves of the elastic πN scattering was performed with different phenomenological vertex functions. In the framework of considered formulation without quarks we have reproduced experimental data for the πN, NN, and γN scattering reactions [3, 7]. On next stage we will try to reproduce the experimental data of multichannel πN, γN, and NN scattering reactions with the same vertex functions.

Keeping in mind, that quark degrees of freedom change only the form of the input vertex functions we can conclude that: if the unified and satisfactory description of the πN, NN, and γN reactions up to 1 GeV energy region is possible with the same one-variable meson-nucleon vertex functions, then quark degrees of freedom do not play an important role in the dynamics of these reactions. Or, in other words, the problem "How important is the quark structure of hadrons in the low and intermediate energy region?" can be reduced to the problem "Is the unified description of the NN, πN, and γN etc. reactions with the identical one-variable phenomenological hadron-hadron vertex functions possible or not?".

REFERENCES

1. Haag, R., *Phys. Rev.* **112**, 669 (1958); Nishijima, K., *Phys. Rev.* **111**, 995 (1958); Zimmermann, W., *Nuovo Cim.* **10**, 598 (1958).
2. Huang, K., and Weldon, H.A., *Phys. Rev. D* **11**, 257 (1975).
3. Machavariani, A.I., *Fiz. Elem. At. Yadra* **24**, 731 (1993).
4. Machavariani, A.I., *Few-Body Phys.* **14**, 59 (1993).
5. Machavariani, A.I., Buchmann, A.J., Faessler, A., and Emelyanenko, G.A., *Ann. of. Phys.* **253**, 149 (1997).
6. Machavariani, A.I., *Acta Phys. Pol.* **31**, 2471 (2000).
7. Machavariani, A.I., and Faessler, A., in preparation.

Single and multi-nucleon antiproton-^{4}He annihilation at rest

P. Montagna (for the OBELIX Collaboration)

Dip. Fisica Nucleare e Teorica, Università di Pavia; INFN Sezione di Pavia, Pavia, Italy

Abstract. The analysis of about $4 \cdot 10^6$ antiproton annihilation events at rest in a NTP ^{4}He gas target, collected with the OBELIX spectrometer at the LEAR accelerator of CERN, is presented. In reaction channels with production of $2\pi^+2\pi^-$, $2\pi^-3\pi^+$, $2\pi^-2\pi^+p$, $2\pi^-\pi^+2p$, $2\pi^-3p$, single and multinucleon annihilations are identified; for the first time some known mesonic and baryonic resonances are observed in $\bar{p}\,^4$He and absolute branching ratios are evaluated.

INTRODUCTION

In general, in the antiproton-nucleus interaction three basic processes may occur:
a) annihilation on a *single* bound nucleon (SNA), which is more probable on the nuclear surface;
b) annihilation on *more* than one nucleon (MNA), which is expected deeply inside the nucleus;
c) same as a) or b), followed by interaction between the annihilation mesons and the residual nuclei (Final State Interaction, FSI).

In case a) only one nucleon is involved in the annihilation, and the remainders behave like spectators; in cases b) or c) additional nucleons are involved. Thus an experimental signature for case a) is the absence of fast nucleons in the final state; for case b) and c) their presence. Note that these last two processes lead to the same final state, and are not distinguishable experimentally.

EXPERIMENTAL ANALYSIS

Helium is a good "laboratory" for the study of the multinucleon annihilation, because it is a very bound system but has few nucleons: so the small number of particles in the final state should make easier the investigation of the annihilation dynamics.

About $\bar{p}\,^4$He annihilation there exists a number of previous studies (see Ref. [1] for a review). In the present work, following a previous one of the Obelix Collaboration [2], we select with higher statistics ($4 \cdot 10^6$ events) several reaction channels and recognize a number of events which undoubtly are annihilations with contribution of more than one nucleon.

The data were collected by means of the Obelix spectrometer (see Ref. [3] for a full description of the experimental layout), which can reconstruct direction and momentum

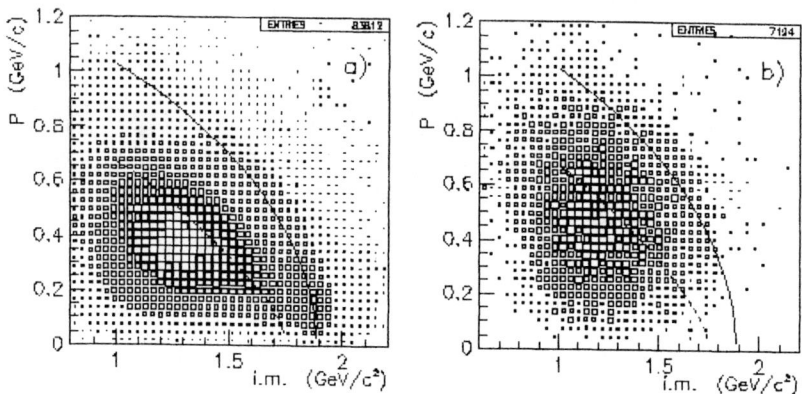

FIGURE 1. $P_{4\pi}$ vs $M_{4\pi}$ distribution for (a) $2\pi^+2\pi^-$ and (b) $2\pi^+2\pi^-p$ events. Solid and dashed lines show the theoretical kinematic relationship (1) respectively for no π^0 and $1\pi^0$ production

of charged pions and protons, respectively with threshold $p_\pi > 80$ MeV/c and $p_p > 300$ MeV/c (such protons in the following are indicated as *fast* protons). The mass identification is performed with two independent methods: dE/dx from spectrometer and β from time-of-flight system.

In order to study SNA and MNA processes in ^{4}He, we considered annihilation events with $2\pi^-$ and 2 or 3 positive particles (π^+ or p) in the final state.

The $2\pi^-2\pi^+$ events, with or without a fast proton, can be distributed in the ($P_{4\pi}, M_{4\pi}$) plane. The relation between total momentum and invariant mass of the 4π system is

$$P_{4\pi} = \sqrt{\frac{(E_0^2 + M_{4\pi}^2 - M_u^2)^2}{4E_0^2} - M_{4\pi}^2} \qquad (1)$$

where $E_0 = (m_{\bar{p}} + m_{^4He} - m_S)$ is the effective annihilation energy (m_S is the mass of the "spectator" nucleons) and M_u is the unmeasured energy of the neutral and slow particles. This functional dependence is shown in Fig. 1, with the relative distribution of $P_{4\pi}$ vs $M_{4\pi}$ of the events with and without fast proton. Note that, at fixed m_S, for constant M_u, $P_{4\pi}$ is a decreasing function of $M_{4\pi}$ and approaches zero for $M_{4\pi} \to (E_0 - M_u)$. So the cluster of events with $P_{4\pi} \simeq 0, M_{4\pi} \simeq 2m_p$ in Fig. 1a indicates exclusive SNA $\bar{p}p \to 2\pi^+2\pi^-$ annihilations, and disappears in Fig. 1b; this difference shows clearly the MNA (or at least SNA + FSI) nature of the annihilations with a detected fast proton.

Figure 1 suggests a simple method to select various reaction sub-channels, by considering that:
- in both cases the events along a band centered on the theoretical line of the relationship (1) are exclusive events ($M_u \simeq 0$) without π^0 production;
- in Fig. 1a, the events with $P_{4\pi} < 300$ MeV/c (in principle $P_{4\pi} \sim 0$, taking into account the Fermi momentum of the nucleons) are dominantly SNA, whereas those with $P_{4\pi} > 300$ MeV/c are dominantly MNA.

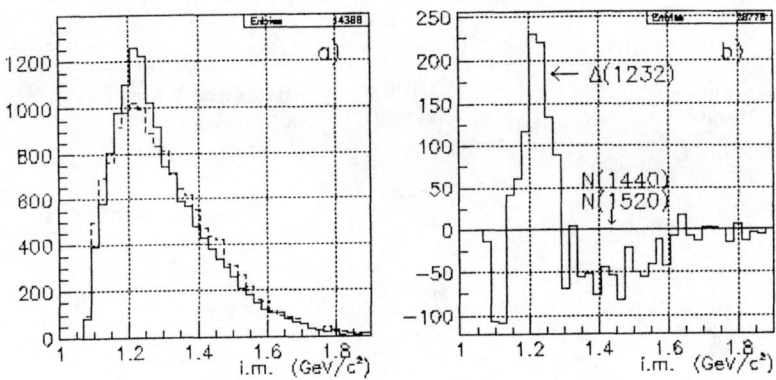

FIGURE 2. (a) Invariant mass distribution of $\pi^+ p$ (solid line) and $\pi^- p$ (dashed line) in $2\pi^+ 2\pi^- p$ events. (b) Difference spectrum between $\pi^+ p$ and $\pi^- p$.

A convincing proof of the goodness of these selection criteria comes from the observation (not shown here) of known mesonic resonances as $\rho^0(770)$, $f_0(980)$ in the invariant masses $\pi^+ \pi^-$: these signals are amplified in 4π events with the $P_{4\pi} < 300$ MeV/c selection, and depleted in the other cases.

In $2\pi^+ 2\pi^- p$ events the production of baryonic resonances is observable. In Figure 2 the difference between the $\pi^+ p$ and $\pi^- p$ distributions clearly shows the presence of $\Delta(1232)$ resonance (in both states Δ^{++} and Δ^0 respectively: note that, due to isospin, $\Delta^{++}/\Delta^0 \simeq 9$). In $\pi^- p$ distribution the $I = 1/2$ $N(1440)$, $N(1520)$ states are visible too.

BRANCHING RATIOS

Using the value $BR_{^4He}(4+5prong) = (29.0 \pm 2.2)\%$ [4] of the absolute branching ratio for all reaction channels with 4 or 5 charged particles in the final state in ^{4}He, measured by a streamer chamber experiment, we evaluated the absolute b.r. for several reaction channels on the basis of the *relative* b.r. to the total number events with 4 and 5 charged particles in the final state (corrected by MC apparatus efficiency). The results are shown in Tab. 1; none of them have been reported before, apart from one ($2\pi^+ 2\pi^- p$ without π^0 production) which confirms the previous value $5 \pm 1 \cdot 10^{-3}$ [2].

Among the multinucleon annihilations, an interesting result concerns the first observation of the reaction $\bar{p}\,^4He \to 2\pi^- 3p$ (136 events). This exclusive reaction is a particular type of the so-called *Pontecorvo reaction*: two-body reactions impossible on free nucleons, hence candidates for MNA annihilations: these reactions, studied in deuterium (see [5] for more details), until now have never been observed in helium.

TABLE 1. Absolute branching ratios. The total errors include the contributions from: a) error on the absolute b.r. from Ref. [4]; b) particle identification; c) kinematic selection of no-π^0 events; d) statistical error.

Final state		B.R. $\cdot 10^{-2}$	Measured n. of events	Total error (%)
$2\pi^+ 2\pi^-$	total	23.62 ± 1.87	83812	7.9
	no π^0	2.752 ± 0.372	9746	13.5
$2\pi^+ 2\pi^- p$	total	3.42 ± 0.34	7194	9.9
	no π^0	0.343 ± 0.065	720	19.1
$2\pi^- 3\pi^+$		1.97 ± 0.32	4278	18.3
$2\pi^- \pi^+ 2p$		0.856 ± 0.096	1862	11.2
$2\pi^- 3p$		0.063 ± 0.010	136	15.2
$2\pi^+ 2\pi^- p$ with Δ^{++} resonance *	total	0.111 ± 0.013	233	11.8
	no π^0	0.0133 ± 0.0028	28	21.4
N^0 resonances	total	0.0694 ± 0.0090	146	12.9
	no π^0	0.0095 ± 0.0023	20	24.4

* $BR(\Delta^0(1232)) = \frac{1}{9} BR(\Delta^{++}(1232))$

FINAL REMARKS

At first glance, all observations are in qualitative agreement with the idea of single nucleon annihilations plus final state interactions. Whether there is also a contribution from direct multinucleon annihilations cannot be deduced simply by a visual inspection of the data. To answer this question theoretical analysis like those made on annihilation in deuterium and ^{3}He are needed and we hope that our data will stimulate such a work.

REFERENCES

1. Bendiscioli, G., and Kharzeev, D., *Riv. Nuovo Cim.* **17**, No.6, (1994).
2. Adamo, A., et al., *Nucl. Phys.* **A569**, 761 (1994).
3. Adamo, A., et al., *Sov. J. Nucl. Phys.* **55**, 1732 (1992).
4. Balestra, F., et al., *Nucl. Phys.* **A465**, 714 (1987).
5. Denisov, O., et al., *Phys. Lett.* B **460**, 248 (1999).

Deuteron and dibaryons in the Skyrme model

E. Norvaišas[*], A. Acus[*], T. Krupovnickas[*] and D.O. Riska[†]

[*]*Institute of Theoretical Physics and Astronomy, Goštauto 12, 2600 Vilnius, Lithuania*
[†]*Helsinki Institute of Physics, 00014 University of Helsinki, Finland*

Abstract. The characteristic feature of the ground state configuration of the Skyrme model description of nuclei is the absence of recognizable individual nucleons. The classical skyrmion with baryon number 2 is axially symmetric, and can be approximated by a simple rational map anzats. Quantized version of biskyrmion then can be identified with dibarion if one employs three collective rotational coordinates. Three independent parameters, however, are not sufficient to identify biskyrmion with deuteron which has different values of spin and isospin. We show that modification of collective coordinate quantization approach by introducing additional three parameters allows us to identify quantized biskyrmion with the deuteron.

INTRODUCTION

The classical Skyrme model solutions with baryon number that is larger than 1 have intriguing geometric structure with striking polyhedral symmetry. The simplest example is the system with baryon number 2, which has axial symmetry. The numerical construction of these multiskyrmion configurations is a demanding task. Employment of the rational map approximation [1] greatly simplifies the study of the quantized modes of the multiskyrmion systems. Such a rational map may be viewed to represent direct formal generalization of Skyrme original hedgehog ansatz for the system with baryon number 1. The deuteron and dibaryons may be viewed as quantized versions of this ground state configuration. The dibaryon case with equal spin and izospin $S = T$ was considered in [3]. The restriction comes from the fact that employed collective coordinate approach included only three independent quantum variables. Here we consider solutions with arbitrary spin S and isospin T. For this generalization to occur, six parameters must be taken into account. The deuteron case with $S = 1$ and $T = 0$ is of particular importance.

THE AXIALLY SYMMETRIC SOLITON WITH B=2

The unitary field $U(\mathbf{r},t) = D^j(\alpha(\mathbf{r},t))$ of the Skyrme model is described by any representation j of $SU(2)$ group and can be expressed in terms of three unconstrained Euler angles $\alpha = (\alpha^1, \alpha^2, \alpha^3)$. The model is defined by the chirally symmetric Lagrangian density

$$\mathcal{L}(U(\mathbf{r},t)) = -\frac{f_\pi^2}{4}\text{Tr}\{R_\mu R^\mu\} + \frac{1}{32e^2}\text{Tr}\{[R_\mu, R_\nu]^2\}, \qquad (1)$$

where the "right" current is defined as $R_\mu = (\partial_\mu U)U^\dagger$ and f_π (the pion decay constant) and e are parameters.

The rational map ansatz for the Skyrme field with $B = 2$ (biskyrmion) in the fundamental representation may be generalized to an arbitrary irreducible representation of $SU(2)$ group

$$\exp[i(\hat{\mathbf{n}} \cdot \sigma) F_R(r)] \Longrightarrow U_R(\mathbf{r}) = \exp[i\hat{n}^a \cdot \hat{J}_a F_R(r)]. \tag{2}$$

Here $F_R(r)$ is a scalar function ("the chiral angle") and $\hat{\mathbf{n}}$ is a unit vector. It's circular components have the following explicit form

$$\begin{aligned}
\hat{n}_{+1} &= -\frac{\sin^2 \vartheta}{\sqrt{2}(1+\cos^2 \vartheta)} e^{2i\varphi}, \\
\hat{n}_0 &= \frac{2\cos \vartheta}{1+\cos^2 \vartheta}, \\
\hat{n}_{-1} &= \frac{\sin^2 \vartheta}{\sqrt{2}(1+\cos^2 \vartheta)} e^{-2i\varphi}.
\end{aligned} \tag{3}$$

Substituting the rational map ansatz 2 into the Lagrangian density reduces it to a simple form, yielding the ordinary differential equation for the chiral angle. The chiral angle solution satisfying the boundary conditions $F_R(0) = \pi$ and $F_R(\infty) = 0$ then can be obtained numerically. The asymptotical behaviour of this numerical solution for $B = 2$ case has the form

$$F_R(\tilde{r}) = C\tilde{r}^{-\frac{1+\sqrt{17}}{2}}. \tag{4}$$

Dimensionless variable $\tilde{r}$ here is defined as $\tilde{r} = ef_\pi r$. The fall off rate of numerical solution is somewhat larger here than in the case of $B = 1$, because the power of $\tilde{r}$ is ≈ -2.56 whereas it was -2 in the case of $B = 1$.

QUANTIZATION OF THE BISKYRMION

The Skyrme model is considered quantum mechanically *ab initio*. For quantization purpose we shall employ two different independent sets of quantum variables $\alpha(t) = (\alpha^1(t), \alpha^2(t), \alpha^3(t))$ and $\beta(t) = (\beta^1(t), \beta^2(t), \beta^3(t))$. The collective rotational coordinates

$$U(F_R, \hat{n}_R, \alpha(t), \beta(t)) = A(\alpha(t)) U_R(F_R, \hat{n}_R) B^\dagger(\beta(t)) \tag{5}$$

are used to separate the variables which depend on the time and spatial coordinates. The coordinates $\alpha(t), \beta(t)$ and corresponding velocities $\dot{\alpha}(t), \dot{\beta}(t)$ form dynamical variables of the theory and satisfy the commutation relations

$$\begin{aligned}
{[\dot{\alpha}^k(t), \alpha^l(t)]} &= -i_R f^{kl}(\alpha, \alpha), & [\dot{\alpha}^k(t), \beta^l(t)] &= -i_R f^{kl}(\alpha, \beta), \\
{[\dot{\beta}^k(t), \alpha^l(t)]} &= -i_R f^{kl}(\beta, \alpha), & [\dot{\beta}^k(t), \beta^l(t)] &= -i_R f^{kl}(\beta, \beta).
\end{aligned} \tag{6}$$

Substitution of (5) into (1) yields the Lagrangian density in terms of the generalized velocities. The generalized momentum operators, which are canonically conjugate to α

and β are defined as usually

$$\pi_k(\alpha) = \frac{\partial L}{\partial \dot\alpha^k}, \quad \pi_k(\beta) = \frac{\partial L}{\partial \dot\beta^k}. \tag{7}$$

These operators satisfy the canonical commutation relations

$$[\pi_k(\alpha), \alpha^l] = -i\delta_{kl}, \quad [\pi_k(\beta), \beta^l] = -i\delta_{kl}, \tag{8}$$

from which the explicit form of the matrices $_Rf^{kl}(\alpha,\alpha), _Rf^{kl}(\alpha,\beta), _Rf^{kl}(\beta,\alpha)$ and $_Rf^{kl}(\beta,\beta)$ can be calculated. It is convenient to introduce two sets of angular momentum operators

$$\begin{aligned}
\hat{J}'_a(\alpha) &= -\frac{i}{\sqrt{2}} \{\pi_k(\alpha), C'^k_{(a)}(\alpha)\}, \\
\hat{J}'_a(\beta) &= -\frac{i}{\sqrt{2}} \{\pi_k(\beta), C'^k_{(a)}(\beta)\},
\end{aligned} \tag{9}$$

where the functions $C'^k_{(a)}(\alpha)$ are defined in [2]. After a long computation, taking into account commutation relations (6) and explicit form of all matrices $_Rf^{kl}$ one obtains the following form of the Lagrangian

$$\begin{aligned}
L \simeq\ &\frac{1}{8}\{\dot\alpha^k, C'^{(a)}_k(\alpha)\}\,_1R_{ab}\{\dot\alpha^\ell, C'^{(b)}_\ell(\alpha)\} + \\
&\frac{1}{8}\{\dot\alpha^k, C'^{(a)}_k(\alpha)\}\,_2R_{ab}\{\dot\beta^\ell, C'^{(b)}_\ell(\beta)\} + \\
&\frac{1}{8}\{\dot\beta^k, C'^{(a)}_k(\beta)\}\,_2R_{ab}\{\dot\alpha^\ell, C'^{(b)}_\ell(\alpha)\} + \\
&\frac{1}{8}\{\dot\beta^k, C'^{(a)}_k(\beta)\}\,_1R_{ab}\{\dot\beta^\ell, C'^{(b)}_\ell(\beta)\} + \cdots .
\end{aligned} \tag{10}$$

Here we have shown only terms quadratic in velocities. The functions $_1R_{ab}$ and $_2R_{ab}$ are closely related to the inverses of $_Rf^{kl}$. The full quantum Lagrangian in terms of angular momentum operators $\hat{J}'_a(\alpha), \hat{J}'_a(\beta)$ has the following simple explicit form

$$L = -M(F_R) - \Delta M(F_R) + \frac{1}{4a_1(F_R)}(\hat{J}'(\alpha) + \hat{J}'(\beta))^2 \\
+ \frac{1}{4}\Big(\frac{1}{a_0(F_R)} - \frac{1}{a_1(F_R)}\Big)(\hat{J}'_0(\alpha) + \hat{J}'_0(\beta))^2. \tag{11}$$

$\Delta M(F_R)$ in the above expression represents the quantum mass correction, which is expressed in terms of integrals of chiral angle function F_R. Because of the axial symmetry of the rational map configuration there are only two different momenta of inertia, which may be defined in the following way

$$\begin{aligned}
a_0(F_R) &= \frac{\pi}{3e^3 f_\pi} \int_0^\infty d\tilde r\, \tilde r^2 \sin^2 F_R \left((12-3\pi)(1+F_R'^2) + 8\frac{\sin^2 F_R}{\tilde r^2}\right), \\
a_1(F_R) &= \frac{\pi}{3\cdot 2e^3 f_\pi} \int_0^\infty d\tilde r\, \tilde r^2 \sin^2 F_R \left(3\pi(1+F_R'^2) + 16\frac{\sin^2 F_R}{\tilde r^2}\right).
\end{aligned} \tag{12}$$

Eigenvectors of the Lagrangian (11) and the corresponding Hamiltonian are necessary to identify deuteron with quantized biskyrmion. It can be easily verified that the normalized eigenstate vectors with fixed spin S and isospin T can be constructed in the following way

$$\left| \begin{matrix} T & S \\ \ell_1 & \ell_2 \\ m_t & m_s \end{matrix} \right\rangle = \sum_{\substack{m_1 \ m_2 \\ m_1' \ m_2'}} \begin{bmatrix} \ell_1 & \ell_2 & T \\ m_1 & m_2 & m_t \end{bmatrix} \begin{bmatrix} \ell_1 & \ell_2 & S \\ m_1' & m_2' & m_s \end{bmatrix}$$

$$\times \frac{\sqrt{(2\ell_1+1)(2\ell_2+1)}}{16\pi^2} D^{\ell_1}_{m_1,m_1'}(\alpha) D^{\ell_2}_{m_2,m_2'}(\beta) |0\rangle. \tag{13}$$

The two D-matrices in the eigenstate vector definition depend on separate generalized dynamical variables and can be considered as different nucleons. Energy expressions can be found for each of these quantum states. The variation of energy expressions yield different integro-differential equations for quantum chiral angle in the deuteron and dibaryon cases, and are under further investigation.

REFERENCES

1. Houghton, C.J., Manton, N.S., and Sutcliffe, P.M., *Nucl. Phys.* **B510**, 507 (1998).
2. Norvaišas, E., and Riska, D.O., *Physica Scripta* **50**, 634 (1994).
3. Krupovnickas, T., Norvaišas, E., and Riska, D.O., *Lith. J. Phys.* **41**, 1 (2001).

Study of charm production in neutrino interactions

O.Sato (CHORUS collaboration)

Furo-cho Chikusa-ku Nagoya, Japan

Abstract. The CHORUS experiment was designed to search for $\nu_\mu \to \nu_\tau$ neutrino oscillation by detecting charged current ν_τ interactions. Thanks to the sub-micron spatial resolution of nuclear emulsion the decay of the short lived τ particle will be recognized in emulsion by the change of its trajectory.

The CHORUS experiment collected about $10^6 \nu_\mu$ CC events during years 1994-1997. Up to now 140,000 ν_μ events have been located and analyzed in the nuclear emulsion target. No candidate events for neutrino oscillation have been found.

The automated emulsion scanning speed increases each year. At this stage, large area volume scanning has become possible. All tracks belonging to an interaction vertex are recognized and measured. This technique is applied not only to the search for neutrino oscillation but also for the recognition of events where charmed particles are produced. Preliminary results on the D^0 production rate in ν_μ CC interactions obtained from a sub-sample of the data are presented here.

THE CHORUS EXPERIMENT

To explore the $\nu_\mu \to \nu_\tau$ oscillation, a total mass of 770 kg of nuclear emulsion has been exposed to an almost pure ν_μ beam ($\bar{E}_\nu \sim 27\text{GeV}$) for 4 years of operations. Electronic detectors downstream of the emulsion target provide kinematic information as well as track and vertex reconstruction [1]. From the analysis of 143,742 charged current and 23,206 neutral current events, CHORUS excludes $\nu_\mu \to \nu_\tau$ oscillation at 90% CL. where $\Delta m^2 > 0.6 eV^2 @ sin^2(2\theta) = 1$ and $sin^2(2\theta) > 6.8 \times 10^{-4}$ for large Δm^2[2].

EVENT LOCATION

Neutrino interactions are reconstructed by a scintillating fiber tracker situated immediately behind the emulsion target. The reconstructed tracks act as prediction for the track search in the most downstream emulsion plate. The resolution for predicted tracks is of the order of $200\mu m$ in position and 2 mrad in slope. Event location in emulsion proceeds from downstream to upstream, using the measurement in each plate as a prediction for the next plate.

Two types of emulsion are used; target sheets and interface sheets. Target sheets are composed of 350 μm layers of emulsion on both sides of a 90 μm plastic base. They are packed together to form a 3cm thick stack of target emulsion, containing 36 plates. Interface sheets employ two thin layers of emulsion on a 800 μm plastic base to yield an angular resolution of the order of 1 mrad. They are placed between the target sheets and the electronic trackers and provide the starting point in the event location. A track is followed up from one plate to the next in the emulsion target until it is no longer found in two consecutive plates. The first of these most likely contains the vertex and is referred to as the 'vertex plate' for the event under study.

In the search for ν_τ charged current events, the vertex location uses only tracks that are possible secondaries from the negatively charged tau lepton. If the event contains a muon, only that track is selected. If no muon exists, all negatively charged tracks with reconstructed momentum in the range from 1 to 20 GeV/c are selected. Since the analysis was focused on neutrino oscillations, the location procedure was not performed for events with a single muon with momentum larger than 30 GeV/c. This kind of events is treated in the currently on going 'Phase II location'.

NETSCAN ANALYSIS

In total 140,000 ν_μ charged current events have been located in emulsion. Within a fiducial volume of $1.5 \times 1.5 \times 6.3 mm^3$ around the position where the muon track stopped, all tracks with slopes less than 400 mrad are scanned by an automatic track recognition device, called the Ultra Track Selector(UTS). The UTS track detection efficiency is above 98% for $\theta < 400$ mrad. It takes 0.3 s per microscope view of $100 \times 100 \mu m^2$ and runs 24 hours per day.

On average 750 tracks are found in the fiducial volume an each plate. The majority of these are muons from beams for other experiments or from neutrino interactions upstream of the CHORUS detector. These background tracks are used in the alignment procedure, achieving a resolution of $0.46 \mu m$ in position and 2 mrad in slope. Only outcoming tracks are retained for physics analysis with two additional selections. One consists in dropping the low momentum(~ 100 MeV) tracks, the other is the request for angular matching with the fiber tracker.

To find decay topologies, the impact parameter is considered for each track with respect to the primary vertex. The cut value varies from 3 μm to 13 μm, depending on the track slope and the vertex depth, defined as distance between the most upstream measurement of the track and the vertex position. If the primary vertex cannot be reconstructed, the minimum distance between tracks is used instead of the impact parameter.

THE RESULTS

Off line analysis for 23,489 charged current events results in 852 selected events. These events have been checked by eye under the microscope to investigate what happens to the selected tracks and finally judge the topology of the decay vertex. A fraction of these events is due to a primary track with an impact parameter measurement in the tail of the distribution, another source of background are through going tracks showing a chance coincidence in the angular matching to the electronic detectors.

The signal for a secondary vertex has been manually confirmed in 471 events, out of the 852 that were selected. The breakdown according to decay multiplicity is shown in

TABLE 1. Charged multiplicity at the decay vertex after human eye confirmation.

Charged decays		kink	trident	5 prong		Neutral decays		vee	4 prong	6 prong	
total=208		98	105	5		total=263		211	51	1	

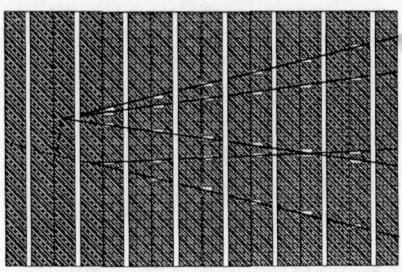

FIGURE 1. Schematic view of reconstructed event by NETSCAN (side view).; UTS picks up tracks at each emulsion sheet surface (ellipses in figure). Event will be analyzed with the impact parameter to a vertex.

table 1. For charged decay candidates, background from so called 'white kink' events is expected to contaminate the 'kink' topology to a non negligible level. The estimation of this background or further analysis of transverse momentum is not yet finished. On the other hand neutral decay candidates are expected to be purely D^0. Monte Carlo simulation gives a total of 6.3 events from K^0 or Λ^0 decays and hadronic interactions in this sample. With an estimated D^0 detection efficiency for already located event of $48.9 \pm 4.0\%$ for vee and $65.5 \pm 6.0\%$ for 4 prong topology, our preliminary result on the D^0 production rate is $\sigma(D^0)/\sigma(CC) = 2.34 \pm 0.15(stat.) \pm 0.17(sys.)$ for $P_\mu < 30$ GeV/c.

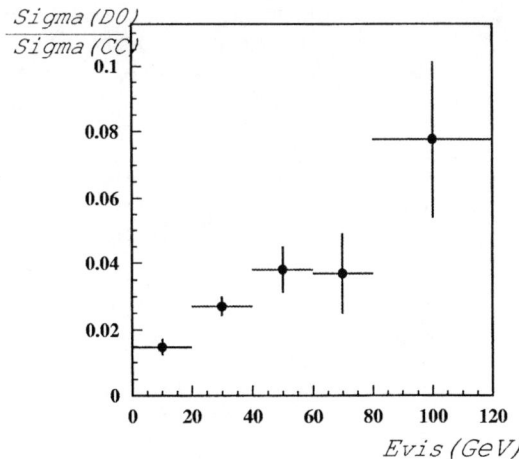

FIGURE 2. Preliminary result on the D^0 production rate as a function of neutrino energy.

TABLE 2. Analyzed statistics in comparison with other neutrino experiments. : The current world average for V_{cd}[PDG] is based on analysis combining data from E531 and CCFR/CDHS.

Emulsion Exp	mean energy(GeV)	number of charm candidates	number of D^0 candidates
CHORUS	27.	471→2800(prospect)	263→1560(prospect)
E531	22.	121	57

Iron etc.Exp	mean energy(GeV)	number of $\mu^+\mu^-$ (ν_μ)	number of $\mu^+\mu^-$ ($\bar{\nu}_\mu$)
CDHS	20.0	9922	2123
CHARM II	23.6	3100	700
NOMAD*	23.6	2714	115
CCFR	140.0	4503	632

* Statistics will increase in near future

SUMMARY AND NEAR FUTURE PROSPECT

The CHORUS experiment has located about 140,000 neutrino charged current events. A subsample of 23,489(17% of located) have been analyzed for the presence of a secondary vertex. Up to now a total of 471 charm candidates has been collected. A preliminary evaluation of the D^0 production rate in neutrino charged current interactions ($\bar{E}_\nu = 27$GeV) has been performed. The value $\sigma(D^0)/\sigma(CC) = 2.34 \pm 0.15(stat.) \pm 0.17(sys.)$ for $P_\mu < 30$ GeV/c is consistent with previous experiments[3][4][5][6][7]. From now on the muon momentum cut will be removed and statistics will increase up to 140,000 charged current events. Assuming the same detection efficiency as in the current sample, one can expect about 2,800 charm candidates in final analysis (in a few years). Table 2 compares the number of analyzed charm events with the statistics in other experiments. We expect the CHORUS experiment to contribute to the study of charmed particle production properties and to V_{cd} measurements.

From the point of view of nuclear physics, CHORUS is trying to find some charmed Hyper nuclei (SUPER fragments). The expected flight length of SUPER fragment is less than $5\mu m$. Nevertheless, they can still be recognized thanks to the sub-micron emulsion resolution. High granularity three-dimensional data is obtained by combining successive CCD images in depth, taken at a spacing below microscope depth of field. For all located events, these data have been taken. The analysis is ongoing and results will be presented in the near future.

REFERENCES

1. Eskut, E., et. al., CHORUS Collaboration, *Phys. Lett. A* **401**, 7 (1997).
2. Eskut, E., et. al., CHORUS Collaboration, *Phys. Lett. B* **497**, 8 (2001).
3. Ushida, N., et. al., E531 Collaboration, *Phys. Lett. B* **206**, 375 (1988).
4. Abromowicz, H., et. al., CDHS Collaboration, *Z. Phys. C* **15**, 19 (1982).
5. Vilain, P., et. al., CHARM II Collaboration, *Eur. Phys. J. C* **11**, 19 (1999).
6. Bazarko, A.O., et. al., CCFR Collaboration, *Z. Phys. C* **65**, 189 (1995).
7. Astier, P., et. al., NOMAD Collaboration, *Phys. Lett. B* **486**, 35 (2000).

Polarizabilities of proton and neutron investigated by Compton scattering on the proton and light nuclei

M. Camen[*], K. Kossert[*], M. Schumacher[*] and F. Wissmann[*]

[*]*Zweites Physikalisches Institut der Universität, Bunsenstraße 7-9, 37073 Göttingen, Germany*

Abstract. In addition to the E2/M1 ratio of the $N \to \Delta$ transition, the electromagnetic polarizabilities and spin-polarizabilities are important structure constants of the nucleon which serve as sensitive tests of chiral perturbation theory and of models of the nucleon. Recently, these quantities have been investigated experimentally at MAMI (Mainz) by high-precision Compton scattering using proton and deuteron targets, where for the latter the method of quasi-free scattering has been applied. Electromagnetic polarizabilities of nucleons bound in nuclei have been measured by nuclear Compton scattering at MAX-lab (Lund) and SAL (Saskatoon).

INTRODUCTION

Compton scattering is an excellent tool for studying the electromagnetic structure of the nucleon. This process supplements on information about photoexcitation of the internal degrees of freedom of the nucleon as contained in the amplitudes for meson photoproduction. In addition, specific two-photon structure constants of the nucleon may be determined as there are the electric and magnetic polarizabilities α and β, respectively, and the spin-polarizabilities γ_0 and γ_π for the forward and backward directions, respectively. Though these structure constants are defined for Compton scattering in the low-energy expansion of the scattering amplitudes, the experimental determination of these polarizabilities requires the understanding of the Compton scattering amplitudes in the first and second resonance regions [1].

Deuteron targets may be used in two complementary approaches, in measurements of coherent Compton scattering by the deuteron [2] and in measurements of quasi-free Compton scattering by the proton or the neutron [3]. Coherent Compton scattering measures the arithmetic average of proton and neutron electromagnetic polarizabilities, whereas the quasi-free scattering process measures the electromagnetic polarizabilities of the recoil nucleon directly.

Compton scattering by spin-saturated nuclei is an excellent tool for studying possible modifications of the electromagnetic polarizabilties in the nuclear medium [4]. In connection with this experimental work careful theoretical investigations of the effects of meson exchange currents have been carried out and found to lead only to minor modifications.

Table 1. Electromagnetic structure parameters of the proton as determined from fits of predictions to Compton scattering data [10] using the SAID-SM99K and MAID2K parameterizations. Errors with index s are statistical+systematic, errors with index m are model errors.

Method		Structure parameter
SAID	EMR(340 MeV)	$= (-1.7 \pm 0.4_s \pm 0.2_m)\%$
	γ_π	$= (-37.1 \pm 0.6_s \pm 3.0_m) \times 10^{-4} \mathrm{fm}^4$
MAID	EMR(340 MeV)	$= (-2.0 \pm 0.4_s \pm 0.2_m)\%$
	γ_π	$= (-40.9 \pm 0.4_s \pm 2.2_m) \times 10^{-4} \mathrm{fm}^4$

COMPTON SCATTERING BY FREE NUCLEONS

Kinematically, the extreme forward ($\theta = 0$) and extreme backward directions ($\theta = \pi$) are the most clearcut cases, corresponding to helicity-nonflip and helicity-flip scattering, respectively. Since in the intermediate state the total spin may be $1/2$ or $3/2$ we may define a spin independent amplitude $f(\omega) = 1/2\{T_{1/2} + T_{3/2}\}$ and a spin dependent amplitude $g(\omega) = 1/2\{T_{1/2} - T_{3/2}\}$ for both cases [1, 5] through the relations

$$T(\theta = 0) = f_0(\omega)\varepsilon' \cdot \varepsilon + g_0(\omega) i\sigma \cdot (\varepsilon' \times \varepsilon)$$
$$T(\theta = \pi) = f_\pi(\omega)\varepsilon' \cdot \varepsilon + g_\pi(\omega) i\sigma \cdot (\varepsilon' \times \varepsilon). \quad (1)$$

The low-energy expansions of the amplitudes $f_0(\omega)$ and $g_0(\omega)$ define the helicity-nonflip electromagnetic polarizability $(\alpha + \beta)$ and forward spin-polarizabilty γ_0, whereas the low-energy expansions of $f_\pi(\omega)$ and $g_\pi(\omega)$ define the helicity-flip electromegnetic polarizability $(\alpha - \beta)$ and the backward spin-polarizability γ_π.

Measurements of total photoabsorption cross sections may be carried out to determine $\alpha + \beta$ and γ_0, using the sum rules

$$\alpha + \beta = \frac{1}{2\pi^2} \int_{\omega_{thr}}^{\infty} \sigma_{tot}(\omega) \frac{d\omega}{\omega^2} \quad \text{and} \quad \gamma_0 = \frac{1}{4\pi^2} \int_{\omega_{thr}}^{\infty} \frac{\sigma_{1/2}(\omega) - \sigma_{3/2}(\omega)}{\omega^3} d\omega, \quad (2)$$

respectively. For the electromagnetic polarizabilities the following results have been obtained [2]

$$\alpha_p + \beta_p = 14.0 \pm 0.3 \qquad \alpha_n + \beta_n = 15.2 \pm 0.5 \quad (3)$$

(in units of $10^{-4} \mathrm{fm}^3$). Very recently, circularly polarized tagged photons and spin-polarized proton targets have been used at MAMI (Mainz) and ELSA (Bonn) to measure spin-dependent total photoabsorption cross sections [6, 7, 8]. In a first place these experiments serve as tests of the Gerasimov-Drell-Hearn (GDH) sum rule [9]. As a by-product these experiments also determine the spin-polarizability for the forward direction

$$\gamma_0 = (-1.87 \pm 0.13) \cdot 10^{-4} \mathrm{fm}^4. \quad (4)$$

It is obvious that the four polarizabilities defined through (1) in principle can genuinely be measured in the forward ($\theta = 0$) and backward ($\theta = \pi$) directions and at low energies. Experiments, however, require intermediate angles and energies where the

Table 2. Status of electromagnetic polarizabilities: The proton values correspond to the new global averages for $\alpha - \beta$ [12]. The neutron values are global averages including the new data from the quasi-free Compton scattering using the SENECA detector [15]. These latter data are still preliminary and may slightly shift in further analyses. The units are 10^{-4} fm^3

$\alpha_p = 12.2 \pm 1.1$	$\beta_p = 1.8 \mp 1.1$	$\alpha_n = 11.6 \pm 2.1$	$\beta_n = 3.6 \mp 2.1$

terms of higher order in ω are not negligible. Therefore, the Compton scattering process has to be investigated in general [10]. In the second resonance region and at backward angles the data mainly serve as a test of the σ-pole ansatz [1] for the asymptotic part of f_π and for the determination of the relevant parameter m_σ of this ansatz. A good fit to the data of the second resonance region is obtained [10] with the SAID-SM99K [11] parameterization of photomeson amplitudes if a parameter of $m_\sigma = 600$ MeV is applied. In the adjustment of predictions to the experimental data of the first resonance region, it appeared reasonable to use $\alpha - \beta$ [12] as an independent input and to base the determination of γ_π and of the E2/M1 ratio on the SAID-SM99K [11] and MAID2K [13] parameterizations, leading to the results labeled SAID and MAID in Table 1. Our result for γ_π is in disagreement with the one of the LEGS group [14] which gave a smaller value of $\gamma_\pi^{\text{LEGS}} = (-27.1 \pm 2.2_{\text{stat+syst}} {}^{+2.8}_{-2.4\text{model}}) \times 10^{-4}$ fm^4. The difference can be traced back to a difference in the measured differential cross sections.

QUASI-FREE AND COHERENT COMPTON SCATTERING BY NUCLEONS BOUND IN THE DEUTERON

Accurate results for the electromagnetic polarizabilities have been obtained for the first time in a recent experiment on quasi-free Compton scattering by the neutron carried out [15] using the large Mainz 48 cm $\oslash$ × 64 cm NaI(Tl) detector and the Göttingen segmented neutron detector SENECA in coincidence. Simultaneously, quasi-free Compton scattering by the proton and quasi-free π^0 photoproduction on the proton and the neutron have been analyzed as tests of the method. A summary of the status of electromagnetic polarizabilities obtained in this way is given in Table 2. The data given for the proton may be considered as stable since little changes to the numbers are expected in the near future. The data given for the neutron take into account the results of all three experiments [15, 16, 17] using the method of quasi-free Compton scattering, and also one experimental result from Coulomb scattering of the neutron [18]. In the latter case the experimental error given by the authors [18] was doubled in order to take into account that the uncertainties of this experiment are believed to be underestimated [19].

Experimental coherent Compton scattering data for the deuteron from Urbana-Champaign [20] and Sascatoon [21] have been analysed by Levchuk and L'vov [2] leading to an unreasonably small value for $\alpha_n - \beta_n$. This problem seems to be solved through recent data measured in Lund [22].

COMPTON SCATTERING BY NUCLEI

^{4}He: Large discrepancies between theory and experiment had been found in earlier work [23, 24] when the calculations were carried on the basis of free-nucleon electromagnetic polarizabilities. Based on these early data, it had to be assumed that in the nuclear medium the magnetic polarizability dominates over the electric one whereas in free space the opposite is true [23, 24]. Our later careful investigation at the tagged-photon facility [25] at MAX-lab (Lund), however, led to a drastic disagreement with the earlier experimental results and to an agreement of the in-medium electromagnetic polarizabilities with the corresponding free-nucleon values.

^{12}C, ^{16}O: Our experiments on ^{12}C and ^{16}O [25, 26] also led to a general agreement between theory and experiment when the free-nucleon electromagnetic polarizabilities were used for the predictions. In contrast to this, a recent experiment on ^{12}C carried out in Saskatoon [27] seems to favour effective electric and magnetic polarizabilities of about equal size, corresponding to a shift in strength from electric to magnetic due to binding. This discrepancy with the Göttingen/Lund result [25, 26] was traced back differences in the experimental differential cross sections.

REFERENCES

1. L'vov, A.I., Petrun'kin, V.A., and Schumacher, M., *Phys. Rev. C* **55**, 359 (1997).
2. Levchuk, M.I., and L'vov, A.I., *Nucl. Phys.* **A674**, 449 (2000).
3. Levchuk, M.I., L'vov, and A.I., Petrun'kin, V.A., *Few-Body Syst.* **16**, 101 (1994).
4. Hütt, M.-Th., L'vov, A.I., Milstein, A.I., and Schumacher, M., *Physics Reports* **323**, 457 (2000).
5. Babusci, B., Giordano, L., L'vov, A.I., Matone, G., and Nathan, A.M., *Phys. Rev. C* **58**, 1013 (1998).
6. Ahrens, J., et al., *Phys. Rev. Lett.* **84**, 5950 (2000).
7. Ahrens, J., et al., *Phys. Rev. Lett.* **87**, 022003 (2001).
8. Holvoet, H., in these proceedings.
9. Gerasimov, S.B., *Sov. J. Nucl. Phys.* **2**, 430 (1966);
   Drell, S.D. and Hearn, A.C., *Phys. Rev. Lett.* **16**, 908 (1966).
10. Galler, G., et al., *Phys. Lett. B* **503**, 245 (2001).
11. Arndt, R.A., Strakovsky, I.I., and Workman, R.L., *Phys. Rev. C* **53**, 430 (1996).
12. Olmos de León, V., et al., *Eur. Phys. J. A* **10**, 207 (2001).
13. Drechsel, D., et al., *Nucl. Phys.* **A645**, 145 (1999).
14. Tonnison, J., et al., *Phys. Rev. Lett.* **80**, 4382 (1998).
15. Kossert, K., Doctoral Thesis, Göttingen 2001 (unpublished).
16. Rose, K.W., et al., *Nucl. Phys.* **A514**, 621 (1990).
17. Kolb, N.R., et al., *Phys. Rev. Lett.* **85**, 1388 (2000).
18. Schmiedmayer, J., et al., *Phys. Rev. Lett.* **66**, 1015 (1991).
19. Wissmann, F., Levchuk, M.I., and Schumacher, M., *Eur. Phys. J. A* **1**, 193 (1998).
20. Lucas, M.A., Ph.D. thesis (1994) University of Illinois at Urbana-Champaign (unpublished).
21. Hornidge, D.L., et al., *Phys. Rev. Lett.* **84**, 2334 (2000).
22. Schröder, B., private communication.
23. Fuhrberg, K., et al., *Nucl. Phys.* **A591**, 1 (1995).
24. Wells, D.P., Ph D thesis, University of Illinois at Urbana-Champaign (1990) (unpublished).
25. Proff, S., et al., *Nucl. Phys.* **A646**, 67 (1999).
26. Häger, D., et al., *Nucl. Phys.* **A595**, 287 (1995).
27. Warkentin, B.J., et al., *Phys. Rev. C* **64**, 014603 (2001).

Two-quark correlations in the hard electromagnetic nucleon form factors

M. Schwärz* and W. Schweiger*

Institute of Theoretical Physics, University of Graz, Universitätsplatz 5, A-8010 Graz, Austria

Abstract. The, so called, "hard-scattering approach" represents a suitable framework for the perturbative treatment of exclusive hadronic processes at large energies and (transverse) momentum transfers. In this context, diquarks can serve as a useful phenomenological concept to model non-perturbative effects which are still observable in the kinematic range accessible by present-day experiments. We outline how a description of baryons as quark-diquark systems can be understood as an effective theory in the sense that the pure quark hard-scattering approach is recovered in the limit of asymptotically large momentum transfers. Our arguments are based on a reformulation of the hard-scattering formalism in terms of quark-diquark degrees of freedom. This reformulation provides the exact form of photon- and gluon-diquark vertices and corresponding vertex functions (diquark form factors) in the limit of asymptotically large momentum transfers – and thus also asymptotic constraints which should be fulfilled by phenomenological quark-diquark models for hard scattering. As an application of this reformulation we present an analysis of the hard electromagnetic nucleon form factors with respect to their quark-diquark content.

It is generally accepted that the, so called, "hard scattering approach" (HSA) gives the correct description of exclusive processes in the limit of asymptotically large (transverse) momentum transfers Q^2 (for an overview on the HSA see, e.g., Ref. [1]). The HSA is based on the factorization of hadronic amplitudes in process dependent, perturbatively calculable hard-scattering amplitudes and process independent distribution amplitudes (DAs) which contain the bound-state dynamics of the hadronic constituents. The corresponding analytical expression for one of the simplest exclusive quantities, the nucleon magnetic form factor, is given by

$$G_M^N(Q^2) = \int_0^1 \left[\prod_{i=1}^3 dx_i \delta\left(1 - \sum_{k=1}^3 x_k\right)\right] \left[\prod_{j=1}^3 dy_j \delta\left(1 - \sum_{l=1}^3 y_l\right)\right]$$
$$\times \phi^{N\dagger}(y_1, y_2, y_3; \tilde{Q}^2) T_H(x_1, \ldots, y_3; Q^2) \phi^N(x_1, x_2, x_3; \tilde{Q}^2). \quad (1)$$

In leading-order perturbation theory the hard-scattering amplitude T_H is calculated on tree level for massless, collinear valence quarks. Two important consequences of this approximation are (fixed-angle) *scaling laws* and *hadronic-helicity conservation*. The latter means, in particular, that no quantitative statement can be made on the Pauli form factors F_2^N within the (conventional) HSA. The distribution amplitude ϕ^N is a probability amplitude for finding the valence Fock state in the nucleon with the quarks carrying the fractions x_i (or y_i) of the nucleon momentum and being collinear within an uncertainty $\tilde{Q}^2$ (which corresponds to the factorization scale). Its dependence on $\tilde{Q}^2$ is given by an evolution equation. The general solution of this evolution equation is known, the integration constants, however, have to be determined by non-perturbative means. In the asymptotic limit $\tilde{Q}^2 \to \infty$ the nucleon DA becomes particularly simple

$$\phi_{as}(x_1, x_2, x_3) \propto x_1 x_2 x_3 \,. \tag{2}$$

Unfortunately it turns out that the leading-order perturbative results for the nucleon magnetic form factors obtained with ϕ_{as} are far away from the experimental data, even at the largest experimentally accessible values of Q^2. Reasonable results can be obtained with asymmetric DAs, like the one proposed by Chernyak et al. [2]

$$\phi_{COZ}(x_1, x_2, x_3) \propto \phi_{as}(x_1, x_2, x_3)(23.814 x_1^2 + 12.978 x_2^2 + 6.174 x_3^2 + 5.88 x_3 - 7.098) \,. \tag{3}$$

This DA fulfills QCD sum-rule constraints on its moments (for $\tilde{Q}^2 \approx 1$ GeV^2). It has, however, been objected that such an asymmetric DA leads to a situation which is problematic for a perturbative calculation [3]: If the momentum of a hadron is very unequally shared between the constituents the hard subprocesses are dominated by rather small gluon virtualities.

It is, nevertheless, not necessary to completely give up the HSA when working in the few-GeV momentum-transfer region. After having observed that strongly asymmetric DAs are required to reproduce experimental data it is, of course, tempting to associate this asymmetry with diquark clustering. This idea has been pursued in a series of papers [4, 5, 6, 7, 8], in which a systematic study of various (photon-induced) exclusive reactions has been carried out within a HSA-based phenomenological quark-diquark model of baryons. The dynamics of the scalar and vector diquarks occurring within this model is determined by the gauge-boson diquark vertices and corresponding form factors. The form factors account for the composite nature of diquarks and have been chosen in such a way that the scaling behaviour of the pure quark HSA is recovered in the limit of asymptotically large momentum transfers. What one gains by modelling two-quark clusters by means of diquarks is that the gluons on the quark-diquark level are, on the average, harder than on the pure quark level. The kinematic range in which a perturbative treatment is well justified is thus extended to smaller (overall) momentum transfers on the quark-diquark level.

A more formal justification of diquarks can be achieved by observing that the diquark model – and not only its scaling behaviour – should evolve into the pure quark HSA in the limit of asymptotically large momentum transfers. This suggests a reformulation of the pure quark HSA in terms of quarks and diquarks. Two obvious restrictions on this reformulation are that the leading order hard-scattering amplitude on quark-diquark level should only consist of tree graphs (like in the pure quark HSA), and that the result of the reformulation should not depend on whether quarks 1 and 2, 1 and 3, or 2 and 3 are grouped to a diquark. For the case of baryon form factors it has explicitly been shown that such a reformulation is possible [9]. This can be seen by splitting the calculation of the hard-scattering amplitude on the pure quark level into two steps, namely first the calculation of two-quark subgraphs with the two quarks being in a particular spin-flavour state and afterwards the calculation of the full graphs on quark-diquark level with the vertices and vertex functions obtained for the two-quark subgraphs. Correspondingly, the baryon DA has to be decomposed into terms which belong to certain spin-flavour states of the two-quark system (i, j) consisting of quarks i and j. The key observation is then

TABLE 1. Decomposition of the proton magnetic form factor into diquark contributions for the asymptotic proton DA (Eq. (2)) and the proton DA proposed by Chernyak et al. (Eq. (3)), respectively. $S^I[q_1q_2]$ and $V_h^I[q_1q_2]$ denote scalar and vector diquarks (with isospin I and helicity h), respectively, consisting of quarks q_1 and q_2. The various diquark contributions are further decomposed into a 3- and 4-point part, $G_M^{3,p}$ and $G_M^{4,p}$, depending on whether one or two gauge bosons couple to the diquark. The constants $C_{as} = 4.646 \times 10^{-2}$ and $C_{COZ} = 1.266$ are chosen in such a way that the largest contribution, i.e. the $S^0[ud] \to S^0[ud]$ transition, becomes 1.

Transition	Asymptotic DA			COZ DA		
	$C_{as}^{-1}Q^4G_M^{3,p}$	$C_{as}^{-1}Q^4G_M^{4,p}$	Sum	$C_{COZ}^{-1}Q^4G_M^{3,p}$	$C_{COZ}^{-1}Q^4G_M^{4,p}$	Sum
$V_0^1[ud] \to V_0^1[ud]$	0.111	0.000	0.111	0.016	0.003	0.019
$V_0^1[ud] \leftrightarrow V_0^0[ud]$	0	0	0	0	0.004	0.004
$V_0^1[ud] \leftrightarrow S^1[ud]$	0	0	0	0	0	0
$V_0^1[ud] \leftrightarrow S^0[ud]$	0	-0.500	-0.500	0	-0.063	-0.063
$V_0^0[ud] \to V_0^0[ud]$	0	0	0	0.001	0	0.001
$V_0^0[ud] \leftrightarrow S^1[ud]$	0	0	0	0	-0.002	-0.002
$V_0^0[ud] \leftrightarrow S^0[ud]$	0	0	0	0	-0.011	-0.011
$S^1[ud] \to S^1[ud]$	0	0	0	≈ 0	≈ 0	0.001
$S^1[ud] \leftrightarrow S^0[ud]$	0	0	0	0	-0.002	-0.002
$S^0[ud] \to S^0[ud]$	1.000	0	1.000	1.000	-0.005	0.995
$V_1^1[ud] \to V_1^1[ud]$	0	-0.056	-0.056	0	-0.002	-0.002
$V_1^1[ud] \leftrightarrow V_1^0[ud]$	0	0	0	0	0.036	0.036
$V_1^0[ud] \to V_1^0[ud]$	0	0	0	0	0.007	0.007
$V_0^1[uu] \to V_0^1[uu]$	-0.111	0	-0.111	-0.016	0.027	0.012
$V_0^1[uu] \leftrightarrow S^1[uu]$	0	0	0	0	-0.003	-0.003
$S^1[uu] \to S^1[uu]$	0	0	0	≈ 0	0.005	0.004
$V_1^1[uu] \to V_1^1[uu]$	0	-0.444	-0.444	0	-0.013	-0.013
	Total		0	Total		0.985

that the groupings $(1,2)$, $(1,3)$, and $(2,3)$ are connected by an appropriate interchange of the momentum fractions which means that the integration variables in the convolution integral, Eq. (1), are only renamed. Thus the convolution integral, Eq. (1), is independent on whether quarks 1 and 2, 1 and 3, or 2 and 3 are grouped to a diquark (of certain spin and flavour). By summing furthermore all the tree graphs on quark-diquark level over the groupings $(1,2)$, $(1,3)$, and $(2,3)$ every graph is counted twice on quark level. It thus suffices to calculate all the tree graphs on quark-diquark level for a particular (but arbitrary) grouping (i, j) and to multiply the result with a statistical factor $(3/2)$ to embrace all the tree graphs on quark level.

Without explicitly calculating the (sometimes rather complicated) vertex structure of the gauge-boson diquark vertices this reformulation can already be used to analyze which diquark states are (dis)favoured by a particular baryon distribution amplitude. The diquark contents of the proton magnetic form factor for the asymptotic DA, Eq. (2), and the COZ-DA, Eq. (3), are compared in Tab. 1. In both cases the most important contribution comes from the $S^0[ud]$ diquark. In case of the asymptotic DA this contribution is, however, completely compensated by contributions from vector diquarks and also transitions between scalar and vector diquarks. The asymmetry of the COZ-DA, on the other hand, leads to a strong suppression of all the other contributions in favour of the $S^0[ud]$ diquark. A similar observation can also be made for the neutron magnetic form factor [9]. This suggests that a nucleon at intermediate momentum transfers appears approximately like a quark-scalar-diquark system.

Up to this point the term "diquark" has just referred to a spin-flavour state of two quarks in a baryon. Diquarks start to play a dynamical role if one assumes that two quarks in a particular spin-flavour state become strongly correlated at moderately large momentum transfers such that they cannot be resolved but appear rather as a quasi-elementary particle. This should be reflected by the corresponding diquark form factors which acquire a value of $O(1)$ at small enough momentum transfers when the diquark appears nearly pointlike. If all the correlations between the valence quarks of a baryon are absorbed into the diquark form factors – which means that all deviations from the asymptotic form of the baryon DA are shifted to the quark-quark DA of the diquark – the perturbative calculation of the diquark form factors may become problematic, since the exchanged gluon becomes relatively soft. The way out is to consider the diquark form factors (at intermediate momentum transfers) as non-perturbative quantities which have to be parameterized phenomenologically. In order to finally arrive at a diquark model for hard exclusive scattering which can be understood as an effective model in the sense that it reproduces the results of the pure quark HSA in the limit of asymptotically large momentum transfers one should then take the gauge-boson diquark vertices resulting from the reformulation of the pure quark HSA and use the results for the corresponding vertex functions as asymptotic constraints for the parameterization of the diquark form factors. A corresponding program is presently carried out.

REFERENCES

1. Brodsky, S.J., and Lepage, G.P., "Exclusive Processes in Quantum Chromodynamics", in *Perturbative Quantum Chromodynamics*, edited by A.H. Mueller, World Scientific, Singapore, 1989, pp. 93-240.
2. Chernyak, V.L., Ogloblin, A.A., and Zhitnitsky, I.R., *Sov. J. Nucl. Phys.* **48**, 536-541 (1988).
3. Isgur, N., and Llewellyn-Smith, C.H., *Nucl. Phys.* **B317**, 526-572 (1989).
4. Jakob, R., Kroll, P., Schürmann, M., and Schweiger, W., *Z. Phys. A* **347**, 109-116 (1993).
5. Kroll, P., Pilsner, Th., Schürmann, M., and Schweiger, W., *Phys. Lett. B* **316**, 546-554 (1993).
6. Kroll, P., Schürmann, M., and Guichon, P.A.M., *Nucl. Phys.* **A598**, 435-583 (1996).
7. Kroll, P., Schürmann, M., Passek, K., and Schweiger, W., *Phys. Rev. D* **55**, 4315-4328 (1997).
8. Berger, C.F., and Schweiger, W., *Phys. Rev. D*, **61**, 114026 (2000).
9. Schwärz, M., *Mass Effects and Two-Quark Correlations in the Perturbative Electromagnetic Nucleon Form Factors*, diploma thesis, Karl-Franzens University of Graz, Graz (2001).

The GDH-experiment at ELSA

Thorsten Speckner (for the GDH-collaboration)

Physikalisches Institut IV, Universität Erlangen-Nürnberg, Erwin Rommel Str. 1, 91058 Erlangen, Germany

Abstract. The GDH-collaboration performs the first experimental verification of the fundamental GDH sum rule at the electron accelerators MAMI and ELSA between 0.14 and 3.1 GeV. The helicity dependent total cross section of circularly polarized real photons on longitudinally polarized nucleons is measured. The setup at ELSA and preliminary results for the proton are presented here.

MOTIVATION

The GDH sum rule has been derived in 1966 by Gerasimov [1], Drell and Hearn [2]. It connects well known static properties of the nucleon like the anomalous magnetic moment κ and the mass m (precision $< 10^{-6}$) to dynamic observables of the nucleon. As dynamic observables, the total photo absorption cross sections σ of real circularly polarized photons on longitudinally polarized nucleons in the helicity states 3/2 (spins parallel) and 1/2 (spins antiparallel) are regarded. Its difference is weighted by the inverse of the photon energy and integrated up to infinity:

$$\int_0^\infty d\nu \frac{\sigma_{3/2}(\nu) - \sigma_{1/2}(\nu)}{\nu} = \frac{2\pi^2\alpha}{m^2}\kappa^2 \qquad (1)$$

The dispersion theoretic derivation assumes only very fundamental physical principles like causality, unitarity, Lorentz invariance and gauge invariance. The only non-fundamental assumption of the derivation is the so-called *No-Subtraction-Hypothesis*. The experimental verification of the sum rule is up to the present knowledge most likely a check of the validity of this hypothesis. Additionally, the measured cross section difference delivers constraints to multipole analyses in the resonance region and gives information about the validity of Regge models in the high energy regime.

SETUP

To cover a wide photon energy range the experiment is split up into two parts. At MAMI the resonance region from pion threshold up to 0.8 GeV is covered. Data taking on the proton was finished in 1998 and results have recently been published [3, 4]. The MAMI part is discussed by H. Holvoet in these proceedings [5]. In the following, we focus on the high energy part (0.7-3.1 GeV) of the measurement carried out at ELSA in Bonn.

Circularly polarized photons

Longitudinally polarized electrons, which are produced by a GaAs crystal, are accelerated in ELSA up to 3.2 GeV and are extracted to the experiment with a duty cycle of up to 95%. Since ELSA is a storage type accelerator, depolarizing resonances occur which have to be overcome [6, 7]. Therefore, dependent on energy, a longitudinal electron polarization between 50% and 73% is available at the experiment. The maximum extracted beam current is 2 nA. The electron beam helicity is randomly reversed at the source every few seconds to give access to the helicity states $3/2$ and $1/2$.

The circularly polarized photons are produced via bremsstrahlung radiation in a thin metal foil from the longitudinal polarized electrons. The helicity transfer of the bremsstrahlung process is well known and depends only on the portion of the transferred energy from the electron to the emitted photon [8]. The tagging system covers an energy range of 68-96% and therefore five primary electron energy settings are necessary to cover the photon energy range from 0.7 up to 3.1 GeV.

The photon beam is defined in its emittance and divergence by an active collimator system [9]. By the collimation of a high energy photon beam, both charged and neutral low energy particles are produced. While electrons and positrons can be removed from the beam by a magnetic field, secondary photons can reach the hadron target. They carry less energy than determined by the tagging system and can cause background events in the detector. Therefore the collimation process itself is used to identify and to veto such secondary photons.

The electron polarization is permanently monitored by the GDH-Møller-Polarimeter, which is situated in the primary electron beam behind the tagging system. Møller-scattering occurs on polarized electrons provided by a magnetized Vacoflux foil. Both scattered electrons are momentum separated by a dipole magnet and detected in coincidence in an arrangement of 14 leadglass detectors. This two arm spectrometer has an acceptance of $\Theta_{CMS} = [65^o, 115^o]$ and allows fast determination of the electron beam polarization with almost no background contamination.

Longitudinally polarized target

The actively collimated photon beam impinges on the longitudinally polarized nucleons, which are provided by a frozen spin target [11]. Beads of butanol (C_4H_9OH) are used as target material. Since C and O are spinless particles, only the H is polarized. Using dynamic nuclear polarization, a maximum value of about 85% is achieved for the proton polarization. A newly developed horizontal $^3He/^4He$ dilution cryostat keeps the target temperature at about 60 mK. By using an internal super conducting holding coil, which provides a holding field of 0.4 T, decay times of up to 200h are achieved. The spin orientation was reversed from time to time to allow for systematic studies. Due to its horizontal alignment the cryostat can be inserted into the GDH-Detector from upstream side, resulting in a minimal dead solid angle in the center of mass system.

GDH-Detector

In the ELSA energy range the photo absorption process leads to multi particle final states which are hard to detect individually. In addition, some of these partial channels are even unknown. To avoid systematic uncertainties arising from undetected final states, the total photo absorption cross section measurement is done inclusively.

The concept of the **GDH-Detector** [10] is to detect at least one reaction product from all possible hadronic final states with almost complete acceptance concerning solid angle and efficiency. These conditions are fulfilled by an arrangement of hadron detection modules, which surround the target with a solid angle of $99.6\% \cdot 4\pi$. Each hadron detection module consists of a first scintillator plate for the detection of charged hadrons and a following lead-scintillator sandwich for the detection of decay photons. The detection efficiency for the hadronic final states is higher than 99%. An important part of the detector concept is the suppression of electromagnetic background caused by pair production and compton scattering off electrons. Due to the Lorentzboost this background is emitted into far forward directions and can be suppressed by means of a threshold CO_2-Čerenkov detector which covers a solid angle of $0^o \leq \theta \leq 15^o$.

RESULTS

The complete setup has been tested extensively with unpolarized nuclei. We obtained unprecedented data quality for solid state targets like Be and C. Additionally, we extracted the hydrogen cross section from the measured cross sections of the solid state C- and $[CH_2]_n$-targets by the difference method. Several test measurements have been done with polarized beam and polarized target to exclude possible sources of fake asymmetries.

The doubly polarized data presented here was obtained during the data taking period in January 2001 at electron energies of 1.0 GeV and 1.4 GeV. Figure 1 shows $\sigma_{3/2} - \sigma_{1/2}$ obtained at ELSA together with the MAMI data up to 0.8 GeV [4]. The data sets from the different accelerators and the different primary electron energy settings match each other excellently. The 2nd and 3rd resonance are clearly visible and even much more pronounced than in the unpolarized cross section. The measured difference is positive up to 1.4 GeV. The agreement of the data with models like the unitary isobar model MAID [13, 14] is not sufficient, esp. in the 2nd and 3rd resonance.

Calculating the running GDH integral in dependence of the upper integration bound one clearly sees (Fig. 2) that the measured integral is above the predicted sum rule value for the proton of 205 μbarn. On the other hand, there exists a Regge-based analysis of deep-inelastic-scattering data by Bianchi and Thomas [12]. The obtained parameterization can be extrapolated to $Q^2 = 0$. This gives a negative sign for $\sigma_{3/2} - \sigma_{1/2}$ in the Regge regime. The negative contribution to the sum rule would improve the agreement between experiment and sum rule prediction. This shows that the question of a sign change is of prime importance. Analysis of the data up to 3.1 GeV will also clarify the validity of the Regge prediction in this energy regime and finally give an answer to the question wether the GDH sum rule holds or not.

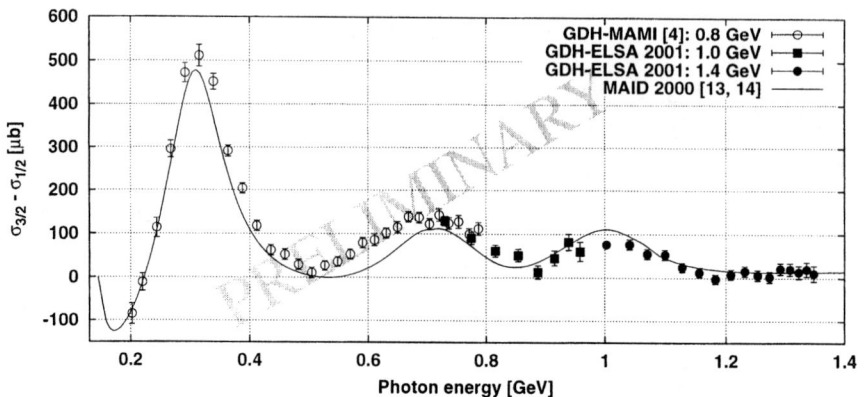

FIGURE 1. Doubly polarized cross-section difference off the proton.

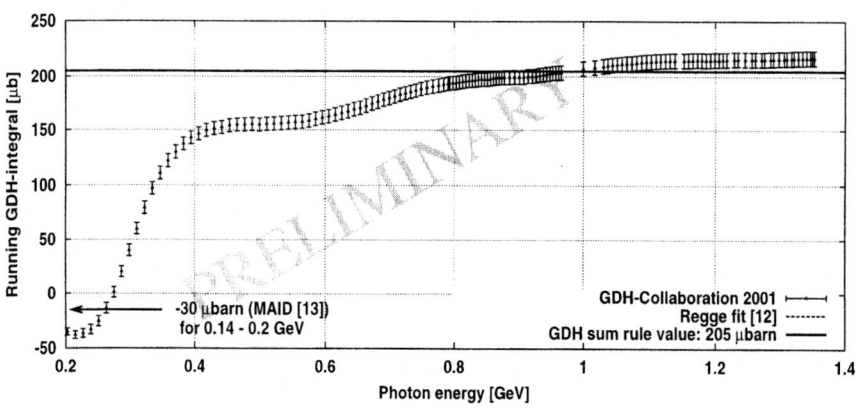

FIGURE 2. Running GDH-integral up to 1.4 GeV. Error bars indicate statistical error only.

REFERENCES

1. Gerasimov, S.B., *Sov. J. Nucl. Phys.* **2**, 430-433 (1966).
2. Drell, S.D., and Hearn, A.C., *Phys. Rev. Lett.* **16**, 908-911 (1966).
3. GDH-collaboration: Ahrens, J., et al., *Phys. Rev. Lett.* **84**, 5950-5953 (2000).
4. GDH-collaboration: Ahrens, J., et al., *Phys. Rev. Lett.* **87**, 022003 (2001).
5. Holvoet, H., in these proceedings.
6. Nakamura, S., et al., *Nucl. Inst. Meth. A* **411**, 93-106 (1998).
7. Hoffmann, M., et al., *SPIN 2000 - Proc., Osaka, Japan*, AIP, New York, 2001, pp. 756-760.
8. Olsen, H., and Maximon, L.C., *Phys. Rev.* **114**, 887-904 (1959).
9. Zeitler, G., et al., *Nucl. Inst. Meth. A* **459**, 6-15 (2001).
10. Helbing, K., et al., subm. *Nucl. Inst. Meth. A*, (2001).
11. Bradtke, C., et al., *Nucl. Inst. Meth. A* **436**, 430-442 (1999).
12. Bianchi, N., and Thomas, E., *Phys. Lett. B.* **450**, 439-447 (1999).
13. Drechsel, D., et al., *Nucl. Phys. A.* **645**, 145-174 (1999).
14. Drechsel, D., et al., *Phys. Rev. D* **63**, 114010 (2001).

Critical behavior of the characteristics of hadron-nucleus and nucleus-nucleus interactions depending on the centrality degree of collisions

M. K. Suleymanov[1,2], M. Šumbera[1], A. S. Vodopianov[2], I. Zborovský[1]

1) Nuclear Physics Institute ASCR, CZ-250 68 Řež near Prague, Czech Republic
2) LHE JINR, RU-141 980 Dubna, Moscow Region, Russia

Abstract. Correlation between secondary particles produced in high energy hadron-nucleus and nucleus-nucleus collisions are studied using variable Q characterizing centrality of the collision or the disintegration degree of participating nuclei. The results indicate that there is a critical value $Q = Q^*$ at which the change in the reaction dynamics takes place. A formation of percolation clusters was suggested as one of the reasons of the critical behavior of the event characteristics near Q^*. An anomalous peak in the angular distributions of slow protons produced in $\pi^- + {}^{12}C$ interactions at momenta of 40 GeV/c and 5 GeV/c with $Q^* = 4$ provides an additional support for the existence of such critical behavior.

Information on the properties of nuclear matter at high baryon and energy density can be extracted by studying the dependence of the characteristics of high energy hadron-nucleus and nucleus-nucleus collisions on the centrality of collision Q. The variable Q can be related e.g. to the number of participant nucleons or to the energy of the spectator matter measured via zero degree calorimetry. From the point of view of the reaction dynamics variation of the characteristics of these collisions with variable Q can occur either continuously or discontinuously. We consider the latter a manifestation of some critical phenomena.

At present there are many experimental results demonstrating existence of discontinuous change with Q of some event characteristics (see Fig. 1 from [1]). Moreover, for $\pi^- + {}^{12}C$ interactions at 40 GeV/c [2] regime change reveals itself as a turnover in the distribution of the variable Q itself. Since for such a small system as $\pi^- + {}^{12}C$ statistical description seems to be inadequate we propose to use the percolation approach instead. The percolation approach has been applied to describe processes of nuclear fragmentation [3] and central nuclear collisions [4] as critical phenomena. Similarly, we assume that dynamics of high energy hadron-nucleus and nucleus-nucleus is determined by the formation of percolation clusters. The collective behavior emerges automatically without need to resort to the concepts of equilibrium or non-equilibrium statistical theory.

The idea that the dynamics of high energy nuclear reactions is governed by the formation of percolation clusters acquires wider applicability if we try to use it in order to explain irregularities observed in the angular distributions of slow protons from most violent $\pi^- + {}^{12}C$ collisions. Events with the total disintegration of the carbon nuclei at incident momentum of 40 GeV/c reproduced from [5] are displayed in Fig. 2. The angular distribution from the same reaction but at the lower incident momentum of 5 GeV/c was studied in Ref. [6] (not shown here). It reveals the same irregularity. It is very attractive to consider these "peaks" in the angular spectra of slow protons to be related to the formation of percolation clusters.

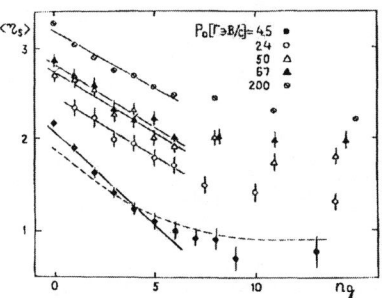

FIGURE 1. Average values of pseudorapidity $\eta_s = -\log(\tan(\Theta/2))$ for s-particles depending on the number of g-particles for pEm-reactions [1]. The dashed line corresponds to the cascade evaporation model calculation.

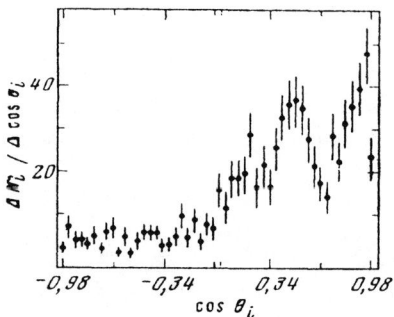

FIGURE 2. Angular distributions of slow protons from cental $\pi^- + {}^{12}C$ collisions.

REFERENCES

1. Vokál, S., Šumbera, M., *Yad. Fiz.* **39**, 1474 (1984).
1. Abdinov, O. B., et al., *JINR Rapid Communication* **1 [75]**, Dubna, 1996.
2. Desbois, J., *Nucl. Phys.* **A466**, 724 (1987) : Nemeth, J., et al., *Z. Phys. A* **325**, 347 (1986); Leray, S., et al., *Nucl. Phys.* **A511**, 414 (1990); Santiago, A.J., and Chung, K.C., *J. Phys.G: Nucl. Part. Phys.* **16**, 1483 (1990).
3. Campi, X., and Desbois, J., "Percolation picture of nucleus break-up.", *preprint IPNOTH 85-9*, Orsay, 1985; Bauer, W., et al., *Nucl.Phys.* **452** 699 (1986); Botvina, A.S., and Lanin, L.V., *Yad. Fiz.* **55**, 688 (1992).
5. Anoshin, A..I., et al., *Yad. Fiz.* **33**, 164 (1981).
6. Abdinnov, O.B., et al., *JINR Preprint 1-80-859*, Dubna, 1980.

Virtual Compton scattering: First results from JLab

R. Van de Vyver (for the JLab-Hall A/VCS Collaboration)

Dept. Subatomic & Radiation Physics - RUG, 9000 Gent, Belgium

Abstract. The Virtual Compton Scattering process off the proton has been studied at Q^2-values of 1.0 and 1.9 (GeV/c)^2 using the Hall A facilities of JLab. Data were taken below and above the pion production threshold as well as in the resonance region. Preliminary results, especially in relation with the Generalised Polarisabilities, will be presented.

INTRODUCTION

The Virtual Compton Scattering (VCS) process has the potential for becoming an adequate probe for the study of the internal structure of the nucleon. Its interest lies in the fact that it allows access to the 6 new observables called Generalised Polarisabilities (GPs) [1, 2], which are extensions of the electromagnetic polarisabilities $\bar{\alpha}$ and $\bar{\beta}$ that have been studied with Real Compton Scattering (RCS). In practise, VCS off the proton is accessible through the photon electro-production reaction $ep \to e'p'\gamma$, which is the coherent sum of two processes, the Bethe-Heitler (BH) and the true VCS one. This means that the real photon γ (with q' the modulus of its cm three-momentum) can be emitted either by the incoming or the outgoing electron (BH contribution) or in the VCS process on the proton itself. In this VCS reaction the photon is produced by the proton as a charged point particle (Born term) or by the excited nucleon (Non-Born term); the latter can be parametrized in terms of these GPs. The first two contributions, Bethe-Heitler and Born ($BH+B$), can be accurately calculated within QED with the use of the electromagnetic form factors of the proton. The $ep \to e'p'\gamma$ reaction cross section can then be conveniently arranged as

$$d^5\sigma_{exp} = d^5\sigma_{BH+B} + [\mathcal{M}_o - \mathcal{M}_o^{BH+B}] * [\Phi q'_{cm}] + O(q'_{cm})$$

wherein $d^5\sigma_{BH+B}$ represents the [Bethe-Heitler + Born] cross section which has terms $O(1/q',1,q'...)$ and $[\mathcal{M}_o - \mathcal{M}_o^{BH+B}] * [\Phi q'_{cm}]$ is the lowest order Non-Born contribution to the cross section while $[\Phi q'_{cm}]$ is a phase space factor; the term $[\mathcal{M}_o - \mathcal{M}_o^{BH+B}]$ contains a linear combination of GPs. It is precisely from this term that two new structure functions can be deduced: $P_{LL} - \frac{1}{\varepsilon}P_{TT}$ and P_{LT} (unpolarised experiment at fixed Q^2 and ε, the latter quantity being the virtual photon's polarisation), which themselves consist of linear combinations of five (out of six independent) GPs [3].
From an analysis of the data below the pion production threshold, first such results were obtained at MAMI/Mainz [4, 5, 6] at $Q^2 = 0.33$ (GeV/c)^2. Meanwhile, there has been a new theoretical development by Pasquini et al. [7], allowing an interpretation of the measured cross section data up to the Δ-resonance region. This is the more interesting

as a new VCS experiment (E93050) has been performed at the 4 GeV *cw* electron accelerator of JLab (USA) for which first, preliminary results will be presented here. With the aim to deepen our knowledge about these GPs, measurements were performed at $Q^2 = 1.0$ and 1.9 (GeV/c)^2; additionally, it was also the intention to search the cross section for so-called missing resonances.

EXPERIMENTAL METHOD AND DATA ANALYSIS

A measurement of the VCS reaction cross section [8] puts stringent requirements on the quality of the experimental equipment used, such as the duty factor of the accelerator, the energy spread of the electron beam, the presence of high-resolution magnetic spectrometers, etc. Those conditions could be met by the facilities available in Hall A at JLab [9] and using a cryogenic liquid hydrogen target. The identification of the VCS signal in the $e + p \rightarrow e' + p' + x$ reaction (with $x = \gamma$ or $\pi°$) relies on a coincidence between the scattered electron and the recoil proton while the third particle is reconstructed from a missing mass separation technique. For the determination of the absolute cross section values, the accurate knowledge of the solid angles subtended by the spectrometers is required; these were estimated using an extensive Monte Carlo simulation [10] which generates events starting from the $BH + B$ cross section behaviour and incorporates all resolution deteriorating effects. The procedure was checked by a determination of the elastic scattering cross section at $Q^2 = 1.0$ (GeV/c)^2 and agreement within 3% was found with the theoretical prediction based on the Bosted parametrisation [11] complemented with the results from a recent measurement at TJNAF [12] of the ratio of $\mu G_E^p(Q^2)/G_M^p(Q^2)$.

PRELIMINARY RESULTS - DISCUSSION

In the present stage of the analysis, preliminary results are available for all kinematical settings but only those at $Q^2 = 1.0$ (GeV/c)^2 will be discussed here. Figure 1 shows the values determined in the leptonic plane as a function of $\theta_{cm}^{\gamma^*\gamma}$ for $q' = 45, 75$ and 105 MeV (solid points), the full curve representing the $BH + B$ cross section. A rather strong effect on the cross section caused by the polarisability effect is observed around 30°. An iteration procedure for the solid angle simulation in which these polarisabilities are taken into account, makes the effect also to become visible around $-200°$, as expected and indicating the importance of a meticulous analysis. This further shows that the magnitude of the deduced structure functions $P_{LL} - \frac{1}{\varepsilon}P_{TT}$ and P_{LT} is of the order of only $1/10$ [13] of that at $Q^2 = 0.33$ (GeV/c)^2 [4].
At $Q^2 = 1.9$ (GeV/c)^2 [14] similar effects are being observed, be it that the deviation from the $BH + B$ cross section becomes apparent only above the pion production threshold giving support to the results from the recently developed DR approach [7].
In the resonance region no experimental data exist as yet for the magnitude of or the structure in the cross section for the VCS process. In the present measurement, this reaction was studied at at $Q^2 = 1.0$ (GeV/c)^2 and for a total centre-of-mass energy W

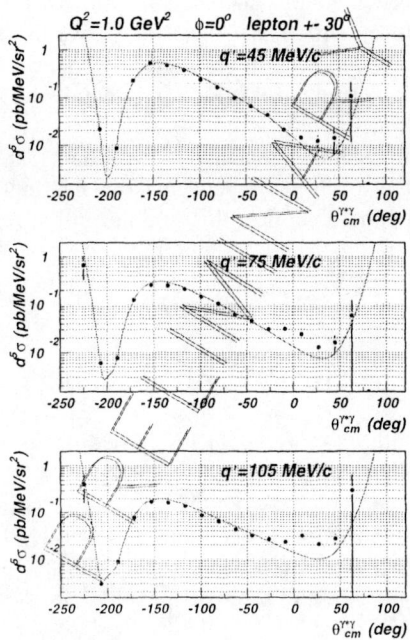

FIGURE 1. Absolute cross sections for the $ep \to e'p'\gamma$ process as a function of $\theta_{cm}^{\gamma^*\gamma}$ and for different values of q'; the solid line represents the $BH + B$ cross section

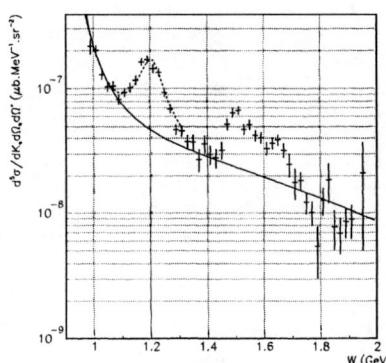

FIGURE 2. Deduced cross sections in the resonance region for $Q^2 = 1.0$ (GeV/c)^2, $\cos\theta_{cm}^{\gamma^*\gamma} = -0.975$ and $\phi = 180°$; the solid curve represents the $BH + B$ contribution while also the prediction from the DR formalism in the Δ-region is plotted (dashed line).

ranging from 0.95 to 2.0 GeV [15, 16] with, among others, the aim to possibly identify any of the so-called missing resonances. The experimental result for $\phi=180°$ is depicted in Fig. 2. The solid curve shows the contribution to the cross section from the $BH + B$ process while the dashed curve is the result from a calculation in the Δ-resonance using the DR formalism [7]. It is evident that, also beyond this Δ-resonance, a net cross section contribution with pronounced structure is observed above the $BH + B$ value. Although the present data do not show any sign of "missing" resonances, the agreement in the Δ-resonance with the DR prediction is encouraging especially as there is no normalisation between theory and experiment.

As a conclusion it seems that a first pass analysis of the data from the E93050 experiment at JLab already suggests the strong Q^2-dependence of the structure functions $P_{LL} - \frac{1}{\varepsilon}P_{TT}$ and P_{LT}; further analysis should confirm this finding. In the resonance region, one observes a pronounced structure in the cross section for the $ep \to e'p'\gamma$ reaction; its detailed interpretation, however, awaits further theoretical developments.

ACKNOWLEDGMENTS

We are grateful to the authors of Ref. [7] for providing us with the results of their calculations prior to publication. This work was supported by DOE, NSF by contract DE-AC05-84ER40150 under which the Southeastern Universities Research Association (SURA) operates the Thomas Jefferson National Accelerator Facility for DOE, by the French CEA, the Université Blaise Pascal de Clermont-Ferrand and the CNRS/IN2P3 (France), the FWO-Flanders (Belgium), the BOF-Gent University (Belgium) and by the European Commission ERB FMRX-CT96-0008.

REFERENCES

1. Guicho, P.A.M., Liu, G.Q., and Thomas, A.W., *Nucl. Phys.* **A591**, 606 (1995).
2. Drechsel, D., Knöchlein, G., Metz, A., and Scherer, S., *Phys. Rev. C* **55**, 424 (1997); Drechsel, D., et al., *Phys. Rev. C* **57**, 941 (1998).
3. Guichon, P.A.M., and Vanderhaeghen, M., *Progress in Particle and Nuclear Physics* **41**, 125 (1998).
4. Roche, J., et al., *Phys. Rev. Lett.* **85**, 708 (2000).
5. d'Hose, N., et al., Proceedings of the Conference on Perspectives in Hadronic Physics (12-16 May 1997, ICTP, Trieste, Italy), Eds. S. Boffi, C. Ciofi degli Atti and M. Giannini, World Scientific Publ. Co. (Singapore), ISBN 981-02-3321-3 (1998) p. 47-58.
6. Van Hoorebeke, L., et al., Proceedings of the Conference on Perspectives in Hadronic Physics (10-14 May 1999, ICTP, Trieste, Italy), Eds. S. Boffi, C. Ciofi degli Atti and M. Giannini, World Scientific Publ. Co. (Singapore), ISBN 981-02-4110-0 (2000) p. 295-304.
7. Pasquini, B. et al., hep-ph/0102335 (February 2001).
8. Bertin, P.Y., Guichon, P.A.M., and Hyde-Wright, C., CEBAF Experiment-93050 (1993).
9. Web page of TJNAF Hall A, http://hallaweb.jlab.org (2000).
10. Van Hoorebeke, L., private communication.
11. Bosted, P.E., *Phys. Rev. C* **51**, 409 (1995).
12. Jones, M.K., et al., *Phys. Rev. Lett.* **84**, 1398 (2000).
13. Degrande, N., PhD thesis (2001), Gent University, Gent, Belgium - *unpublished*.
14. Jaminion, S., PhD thesis (2000), Université Blaise Pascal, Clermont-Ferrand, France - DU1259, EDSF:303.
15. Todor, L., PhD thesis (2000), Old Dominium University, Norfolk, USA - *unpublished*.
16. Laveissière, G., PhD thesis (*in preparation*), Université Blaise Pascal, Clermont-Ferrand, France.

PHYSICS OF STRANGENESS

Non-mesonic weak decay of Λ-hypernuclei and nuclear structure

M.F. Aristizabal*, H.C. Wu* and W.A. Ponce*

Departamento de Fisica, Universidad de Antioquia, Medellin, Colombia

Abstract. This work studies the nuclear structure effects in non-mesonic weak decay of the p-shell Λ-hypernuclei in the framework of symmetry model SU(4) ⊗ SU(3). A comparison between symmetry model and single particle shell model gives an estimate of the range of nuclear structure effects.

INTRODUCTION

The study of the hypernuclear weak decay [1] includes two topics: the $\Lambda N \to NN$ process and the nuclear structure effects. While previous studies have been focused on the $\Lambda N \to NN$ process, investigation of the nuclear structure effects is relatively rare. [2]

In dealing with nuclear medium, pioneering works used nuclear matter. Later, studies based on finite nuclei [3] brought about improvements. A recent work by Parreño et. al. [4] employed spectroscopic factor to account for the shell structure.

In this work we investigate the nuclear structure effects of non-mesonic decay based on nuclear model wave functions. The shell model has been approved good in explaining data for the p- and sd-shell nuclei. However, the big model space often causes technical difficulties. The symmetry model based on group SU(4) is a good approximation to the shell model in the description of nuclear structure [5]. The advantage of a symmetry model is that its limiting case can facilitate analytical calculation, therefore, it makes easier to give predictions on systematic behavior of hypernuclei in the entire shell region.

A long standing puzzle in the study of the hypernuclear weak decay is a big discrepancy between calculated and experimental values of the neutron-induced-to-proton-induced ratio Γ_n/Γ_p (referred to as "n-p ratio"). We hope to find the reason for this discrepancy by analyzing the nuclear structure effects. In this work we restrict ourselves to the one pion exchange (OPE) mechanism.

FORMALISM

In practice the model basis we use is SU(4) ⊗ SU(3), where SU(3) refers to the orbital symmetry [6]. The limiting case of the symmetry SU(4) ⊗ SU(3) provides a reasonable approximation of the nuclear wave function.

The initial state is consisting of a hyperon Λ and a nuclear core with A-1 nucleons. For the calculation of decay rates, a nucleon is separated from the nuclear core and then recoupled to Λ. The crucial elements in this recoupling procedure are the coefficients of fractional parentage (cfp), which are calculated following group theoretical techniques [5, 6]. For example, the cfp of separating a p-shell nucleon reads as follows,

$$\langle [f^{A-1}]\alpha_1 L_1 \beta_1 S_1 T_1 J_1 \{ |[C2,1]; j_N \rangle = \left\{ \frac{\mathcal{N}_{[1]}\mathcal{N}_{[f^{A-2}]}}{\mathcal{N}_{[f^{A-1}]}} \right\}^{\frac{1}{2}} \left\{ \begin{array}{ccc} 1 & L_2 & L_1 \\ \frac{1}{2} & S_2 & S_1 \\ j_N & J_2 & J_1 \end{array} \right\} \quad (1)$$

$$[\hat{L}_1 \hat{S}_1 \hat{j}_N \hat{J}_2]^{\frac{1}{2}} \langle [\tilde{1}]1; [\tilde{f}^{A-2}]\alpha_2 L_2 || [\tilde{f}^{A-1}]\alpha_1 L_1 \rangle \langle [1]\tfrac{1}{2}\tfrac{1}{2}; [f^{A-2}]\beta_2 S_2 T_2 ||[f^{A-1}]\beta_1 S_1 T_1 \rangle,$$

where $[f^{A-i}]\alpha_i L_i \beta_i S_i T_i J_i$ ($i = 1, 2$) represent the SU(4) irrep and the spin-isospin of p-shell nucleons in the nuclear core ($i = 1$) and in the residue with A-2 nucleons ($i = 2$). $\mathcal{N}_{[f]}$ is a dimension of the irrep $[f]$ of a symmetric group. $\langle [\tilde{1}]1; [\tilde{f}^{A-2}]\alpha_2 L_2 || [\tilde{f}^{A-1}]\alpha_1 L_1 \rangle$ and $\langle [1]\tfrac{1}{2}\tfrac{1}{2}; [f^{A-2}]\beta_2 S_2 T_2 ||[f^{A-1}]\beta_1 S_1 T_1 \rangle$ are the isoscalar factors of SU(3) and SU(4) groups, respectively. $[C2, 1]$ stands for the residue state $([f^{A-2}]\alpha_2 L_2 \beta_2 S_2 T_2) J_2$.

In order to study the nuclear structure effects, we express decay rates as follows:

$$\Gamma(m_{t_N}, J) = \sum_{i=1}^{11} \mathcal{R}_i(m_{t_N}, J) \mathcal{T}_i, \quad (2)$$

where $\mathcal{R}_i$ are analytical factors purely related to the nuclear structure, $\mathcal{T}_i$ are integrals purely related to the $\Lambda N \to NN$ process, $m_{t_N} = -\frac{1}{2}(\frac{1}{2})$ corresponds to the neutron (proton) induced decay, J is the total angular momentum of hypernucleus, and $i = 1 - 11$ correspond to eleven sets of initial ΛN state and final NN state (see Table 1).

RESULTS ON $^{12}_{\Lambda}$C DECAY AND DISCUSSION

The formalism is applied to the non-mesonic weak decay of $^{12}_{\Lambda}$C. The single particle orbits are taken the same as of the nucleus ^{11}C. The following results are preliminary:
A. *Nuclear structure effects:* Table 1 lists a typical set of the factors of nuclear structure ($\mathcal{R}_i$) and the factors of two-body process ($\mathcal{T}_i$). The nuclear structure factors are given for the proton-induced and neutron-induced decay (all with $J = 1$). Numbers in parenthesis of column 5 are the values of $\mathcal{T}_i \cdot \mathcal{R}_i$; column 2 is the shell from which a nucleon is separated, and column 3 is the type of interaction. Columns 7 and 8 are the corresponding states of ΛN and N pair. From Table 1 the total decay rate and the n-p ratio can be calculated as $\Gamma_{NM}/\Gamma_{free} = 2.82$ and $\Gamma_n/\Gamma_p = 0.096$, respectively. It is easy to see that the smallness of the n-p ratio is caused by two factors: the contribution of a tensorial interaction is dominant and the neutron is prohibited by Pauli principle from participating in one of the strongest tensorial process, i.e. $\mathcal{T}_3$. The total decay rate depends on the properties of single particle orbits which have important effects. For example, Fig. 1 (solid line) gives the dependence of the total decay rate and n-p ratio on B which is the sum of the binding energies of nucleon and hyperon. When B increases, the total decay rate decreases, which is due to the reduction of phase space caused by higher B.

TABLE 1. Factors of the process $\Lambda N \to NN$ and nuclear structure.

i	shell	V	$\mathcal{T}_i$	$\mathcal{R}_i(n,1)$	$\mathcal{R}_i(p,1)$	l_R	l_r	S	l_p	l_t	$S'T'$
1	p	c	0.00006	9.1668 (0.00057)	18.000 (0.00116)	1	0	0 1	1	0	01 10
2	p	c	0.00001	1.9815 (0.00001)	46.9386 (0.00028)	0	1	0 1	0	1	00 11
3	p	t	0.13414	0.0000 (0.0)	12.9445 (1.73640)	1	0	1	1	2	10
4	p	t	0.11633	0.5195 (0.06043)	0.3158 (0.03674)	0	1	1	0	1	11
5	p	t	0.03342	0.2450 (0.00819)	0.2792 (0.00933)	0	1	1	0	3	11
6	p	pv	0.07092	1.1698 (0.08296)	2.9177 (0.20692)	1	0	1 0,1	1	1	00 11
7	p	pv	0.01334	0.0681 (0.00009)	2.2639 (0.03020)	0	1	1 0,1	0	2	01 10
8	p	pv	0.00429	0.1250 (0.00054)	2.4722 (0.01061)	0	1	1 0,1	0	0	01 10
9	s	c	0.00012	4.5000 (0.00056)	8.9998 (0.00112)	0	0	0 1	0	0	01 10
10	s	t	0.05190	0.0000 (0.0)	6.7499 (0.35035)	0	0	1	0	2	10
11	s	pv	0.12405	0.7500 (0.09304)	1.5000 (0.18607)	0	0	1 0,1	0	1	00 11
SUM				(0.24649)	(2.56917)						

The above results are based on a SU(4) ⊗ SU(3) wave function, which uses the L-S coupling. We also calculate the total decay rate and n-p ratio in the framework of the single particle shell model (SPSM), where the p-shell nucleons are all in $j = 3/2$ orbits (j-j coupling). The results of SPSM are also given in Fig. 1 (solid line with marks '+'). The dependence of the decay rate and n-p ratio on B are similar to the predictions of the symmetry model, but there is a difference of about 20% in the value of n-p ratio, which shows the importance of nuclear structure. While the SU(4) model and SPSM are two extremes of nuclear structure, a realistic case, such as the one used in [4], lies between these two limits [7]. The difference of the n-p ratio between two extremes gives an estimate of the range of the nuclear structure effects. Neither extreme of nuclear structure is able to reproduce the data of the n-p ratio ($\sim 1.$), therefore, one can conclude that the study in nuclear structure alone is not enough to solve the discrepancy of the n-p ratio.

B. *SU(4) wave function vs. spectroscopic factor:* The decay rate can be expressed schematically as follows

$$\Gamma = \alpha \sum_{[C2,l]} |\beta\, \text{cfp}([C2,l],\tfrac{1}{2}) + \gamma\, \text{cfp}([C2,l],\tfrac{3}{2})|^2, \quad (3)$$

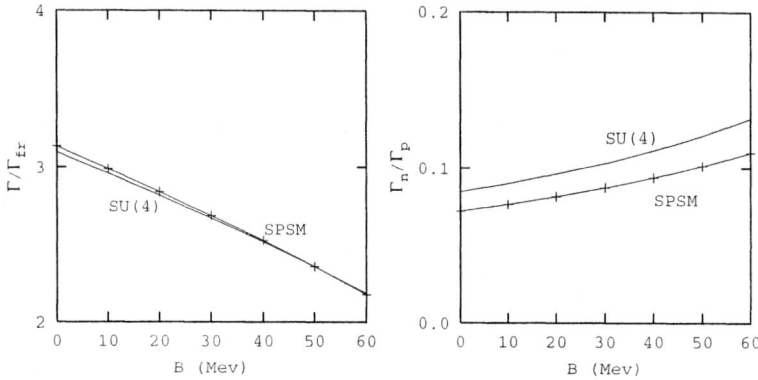

FIGURE 1. Dependence of decay rates on binding energy.

where α, β and γ are quantities not depending on $[C2, l]$. When spectroscopic factors are employed, only the squared term of cfp is used and the interference terms are neglected, which will cause a difference in the decay rate of about 10%. However, when the symmetry of $SU(4) \otimes SU(3)$ is broken, configuration mixing is needed. This mixing will bring in many more interference terms which will cancel each other to some extent. Therefore the differences between the results of a nuclear wave function and of spectroscopic factors will be smaller.

C. *Short-range correlations and form factor:* The effects of the short range correlations and the form factor are also studied. Our results are similar to previous studies [1, 4]. Finally we note that the $\Delta I = \frac{1}{2}$ rule is enforced as was done by other authors [1, 4]. However, there lacks a theoretical foundation for this rule, neither from the aspects of hadrons nor from the aspects of quarks. It is interesting to study the possible contribution from the $\Delta I = \frac{3}{2}$ channel in a phenomenological way. This study is currently underway.

ACKNOWLEDGMENTS

The authors thank the University of Antioquia for financial support.

REFERENCES

1. Cohen, J., *Prog. Part. Nucl. Phys.* **25**, 139 (1990);
   Oset, E., and Ramos, A., *Prog. Part. Nucl. Phys.* **41**, 191 (1998).
2. Dubach, J. F., Feldeman, G. B., Holstein, B. R., and de la Torre, L., *Ann. Phys.* **249**, 146 (1996).
3. Cheung, C. Y., Heddle, D. P., and Kisslinger, L. S., *Phys. Rev. C* **27**, 335 (1983).
4. A. Parreno, A. Ramos and C. Bennhold, *Phys. Rev. C* **56**, 339 (1997).
5. Wigner, E. P., *Phys. Rev.* **51**, 106 (1937); Vogel, P., and Ormand, W. E., *Phys. Rev C* **47**, 623 (1993); Hecht, K. T., and Pang, S. C., *J. Math. Phys.* **10**, 1571 (1969).
6. Elliott, J. P., *Proc. Royal Soc. A* **245**, 128 & 562 (1958); Vergados, J. D., *Nucl. Phys.* **A111**, 681 (1968).
7. Cohen, S., and Kurath, D., *Nucl. Phys.* **A101**, 1 (1967).

Low-energy K^- optical potentials: deep or shallow?

A. Cieplý[*], E. Friedman[†], A. Gal[†] and J. Mareš[*]

[*]Nuclear Physics Institute, 25068 Řež, Czech Republic
[†]Racah Institute of Physics, The Hebrew University, Jerusalem 91904, Israel

Abstract. The K^- optical potential in the nuclear medium is evaluated self consistently from a free-space K^-N t matrix constructed within a coupled-channel chiral approach. The fit of model parameters gives a good description of the low-energy data *plus* the available K^- atomic data. The resulting optical potential is relatively 'shallow' in contradiction to the potentials obtained from phenomenological analysis. The calculated (K^-_{stop},π) hypernuclear production rates are very sensitive to the details of kaonic bound state wave function. The (K^-_{stop},π) reaction could thus serve as a suitable tool to distinguish between shallow and deep K^- optical potentials.

INTRODUCTION

Calculations existing to date for the $\bar{K}$ nucleus interaction at threshold essentially give two different predictions for the depth of the K^- nucleus potential at nuclear matter density. The phenomenological density dependent (DD) [1] and the relativistic mean field (RMF) [2] models, which describe the kaonic atom data very well, produce deep attractive potentials. In contrast, chirally inspired models of the $\bar{K}N$ interaction [3, 4, 5] yield relatively shallow attractive potentials while giving worse description of the kaonic atoms. Recently, Baca et al. [6] improved significantly the fit to the atomic data by adding to the optical potential of Ref. [5] a phenomenological '$t\rho$' term. However, this improvement was achieved at the cost of losing the direct connection of the optical potential to the chirally inspired microscopic model of the $\bar{K}N$ interaction.

In this contribution based on our recent work [7] we demonstrate that one can find such parameters of the chirally motivated microscopic model of the $\bar{K}N$ interaction that the low-energy $\bar{K}N$ data plus the K^- atomic data are fitted *simultaneously* without introducing any additional 'non-chiral' terms. The quality of the K^- atomic fit provided by our optical potential is superior to other chirally motivated approaches. Furthermore, we demonstrate that the outcome of K^- initiated reactions at low energy is sensitive to the wavefunction of the kaon inside the nucleus, where different optical potentials produce noticeably different wavefunctions. As an example, we discuss the (K^-_{stop},π) reaction into specific hypernuclear states.

CHIRAL $\bar{K}N$ AMPLITUDE

We follow the chirally motivated model developed by Weise and collaborators [3, 4] for the $\bar{K}N$ scattering and reactions near threshold. The $\Lambda(1405)$ resonance is generated dynamically by solving coupled Lippmann-Schwinger equations for the meson-baryon t-matrix. The channels included in the model are: K^-p, $\bar{K}^0 n$, $\pi^0\Lambda$, $\pi^+\Sigma^-$, $\pi^0\Sigma^0$, $\pi^-\Sigma^+$. The $\bar{K}N$ interaction is treated selfconsistently [8]. This means that the K^- optical potential constructed from the elementary $\bar{K}N$ amplitudes enters the in-medium propagator through the kaon selfenergy correction. We included only the kaon and nucleon selfenergies in $\bar{K}N$ sector and neglected the selfenergy corrections in the pion-hyperon channels, for simplicity.

The model was used [7] in χ^2 fits to K^--atomic data and to representative low energy K^-p data. We started from the original parametrization of Refs. [3, 4] and refitted the meson-baryon coupling constant f_π and the four inverse range parameters α_i that characterize the range of meson-baryon interaction in the included channels. All parameters turned out to be close to their initial values resulting in a moderate modification of the original model. The details of our fitting procedure can be found in Ref. [7].

The fit to kaonic atoms data includes 65 data points throughout the periodic table. Our results are summarized in Table 1 where we also show the results of similar fits performed with other K^- nuclear optical potentials. The degree of success of those potentials is characterized by the values of χ^2 per data point. We also show the depths of the optical potentials V_R (real) and V_I (imaginary) at nuclear density $\rho_0 = 0.17$ fm^{-3}. The first three columns show values obtained for the present model in three different cases: 'no Π' - no medium effects beyond Pauli blocking are included; '$\Pi_{\bar{K}}$' - the self-consistent calculation including the $\bar{K}$ selfenergy; and '$\Pi_K + \Pi_N$' - the kaon and nucleon selfenergies are included. In order to improve the fit to the data for the chiral optical potentials by Ramos and Oset (RO) [5] and by Schaffner-Bielich et al. (SKE) [9], we followed the approach of Ref. [6] and added a phenomenological '$t\rho$' term to the potential. Although our model gives clearly superior description of the data when compared with the other two chiral models, the phenomenological DD model [1] and the combined DD+RMF model [2] give even better values of χ^2/N.

At zero energy the optical potential is related to the isospin averaged (effective) $\bar{K}N$ threshold scattering amplitude a_{eff} by the standard formula

$$V_{\mathrm{opt}}^{\bar{K}} = -(4\pi/2\mu_{KN})\, a_{\mathrm{eff}}(\rho)\, \rho\ . \qquad (1)$$

Figure 1 shows the density dependence of a_{eff} for the three cases discussed above. The change of the sign of Re a_{eff} from negative to positive corresponds to the transition from a repulsive free-space interaction to an attractive one in the nuclear medium. Taking the K^- selfenergy into account generally leads to a weaker density dependence of the threshold scattering amplitude. This indicates that the selfconsistent K^- optical potential

TABLE 1. Fits to the K^- atomic data (see text for notation).

model	no Π	$\Pi_{\bar{K}}$	$\Pi_{\bar{K}}+\Pi_N$	RO	RO $+t\rho$	SKE	SKE $+t\rho$	DD	DD + RMF
χ^2/N	16.5	6.72	2.24	4.62	2.73	12.7	2.46	1.28	1.40
V_R(MeV)	-117	-71	-55	-44	-58	-34	-159	-190	-185
V_I(MeV)	-67	-85	-60	-54	-23	-62	-95	-55	-60

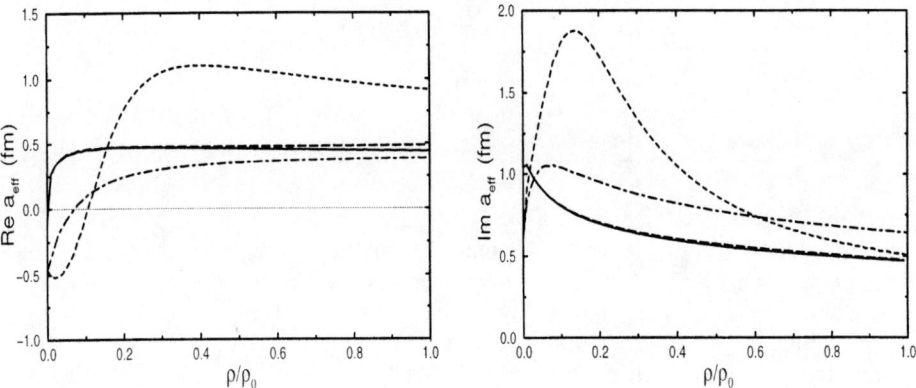

FIGURE 1. Real (left) and imaginary (right) parts of a_{eff} as function of density ρ/ρ_0, calculated in the 'no Π' (short-dashed line), '$\Pi_{\bar{K}}$' (dot-dashed line) and '$\Pi_{\bar{K}} + \Pi_N$' (solid line) regimes.

is well approximated by a '$t\rho$' form (where t =const.) over a wide range of densities. A genuine ρ dependence of t appears only at very low densities.

The long-dashed line which is hard to distinguish from the solid line in Figure 1 visualizes the results obtained when we added the pion and hyperon selfenergies to the propagator in non-kaonic channels. This result fully justifies the simplification used in Ref. [7] (and mentioned above) but it contradicts the observations of Ref. [5]. Although we used only a simple local effective pion potential in this work it is hard to believe that a more sophisticated form of the pion selfenergy [5] can make such difference. We have no clear understanding of the discrepancy at the moment.

STOPPED K^- REACTIONS

Various models of K^- optical potential at threshold can be tested in (K^-_{stop}, π) reactions to specific Λ hypernuclear states. We have studied the sensitivity of the capture rates to the choice of the K^- nucleus optical potential within the distorted wave impulse approximation (DWIA) [10, 7]. Here we limit the discussion to the following K^- capture-at-rest reactions on ^{12}C:

$$K^- + {}^{12}\text{C} \longrightarrow \pi^- + {}^{12}_{\Lambda}\text{C}, \quad K^- + {}^{12}\text{C} \longrightarrow \pi^0 + {}^{12}_{\Lambda}\text{B}. \tag{2}$$

Our results are summarized in Table 2. Different optical potentials, ordered according to their central depth, were tested. The 'chiral' potential corresponds to the relatively shallow potential of the present work and the deep potential is represented by the density dependent 'DD' approach [1]. The 'effective' potential was obtained as a best-fit solution for the standard form $V_{\text{opt}} = t_{\text{eff}}\rho$. As these potentials were obtained within *global* fits to the available K^--atomic data, we add another potential of the $t\rho$ form 'fixed' to fit exclusively the carbon data.

TABLE 2. Calculated capture rates on ^{12}C per stopped K^- (in units of 10^{-3}) to the summed $p_N \to s_\Lambda$ 1^- excitations in $^{12}_\Lambda$C and $^{12}_\Lambda$B for various optical potentials.

final $^A_\Lambda Z$	chiral	effective	fixed	DD
$^{12}_\Lambda$C	0.231	0.169	0.089	0.063
$^{12}_\Lambda$B	0.119	0.087	0.046	0.032

The calculated capture rates per K^-, are shown for the production of the 1^- hypernuclear ground states off ^{12}C. It is clear that the deeper the K^- optical potential is, the lower the calculated rate becomes. This pattern is caused by the strong-interaction bound D state generated by all but the 'chiral' potentials. The atomic wavefunction acquires then *extra* nodes within the nucleus, thus causing substantial cancelations in the DWIA amplitude. All the calculated rates shown in Table 2 are much lower than the measured values [11]. Unfortunately, some uncertainties inherent in the calculation (pion distortion, adopted value for in-medium branching ratio) make it difficult to compare the present results with experiment.

CONCLUSIONS

- Only minor modifications of the parameters involved in the chirally motivated model for the $\bar{K}N$ amplitude are necessary to get good agreement with the atomic data while maintaining the good description of low-energy free space K^-p data. No phenomenological '$t\rho$' term is required in our analysis.
- In our model, the pion and hyperon selfenergies have marginal impact on the $\bar{K}N$ amplitude.
- The optical potential constructed from the 'chiral' $\bar{K}N$ amplitude does not yield as strong attraction as phenomenological potentials do.
- Λ-hypernuclear ground-state production rates calculated for the (K^-_{stop}, π) reactions on carbon vary by more than a factor of 3 for the optical potentials adopted in our work.

REFERENCES

1. Friedman, E., Gal, A., and Batty, C.J., *Nucl. Phys.* **A579**, 518 (1994).
2. Friedman, E., Gal, A., Mareš, J., and Cieplý, A., *Phys. Rev. C* **60**, 024314 (1999).
3. Kaiser, N., Siegel, P.B., and Weise, W., *Nucl. Phys.* **A594**, 325 (1995).
4. Waas, T., Kaiser, N., and Weise, W., *Phys. Lett. B* **365**, 12 (1996); **379**, 34 (1996).
5. Ramos, A., and Oset, E., *Nucl. Phys.* **A671**, 481 (2000).
6. Baca, A., García-Recio, C., and Nieves, J., *Nucl. Phys.* **A673**, 335 (2000).
7. Cieplý, A., Friedman, E., Gal, A., and Mareš, J., in print *Nucl. Phys.* **A**, preprint nucl-th/0104087.
8. Lutz, M., *Phys. Lett. B* **426**, 12 (1998).
9. Schaffner-Bielich, J., Koch, V., and Effenberger, M., *Nucl. Phys.* **A669**, 153 (2000).
10. Gal, A., and Klieb, L., *Phys. Rev. C* **34**, 956 (1986).
11. H. Tamura, R.S. Hayano, H. Outa and T. Yamazaki, *Prog. Theor. Phys. Suppl.* **117**, 1 (1994).

Few-body $\Lambda\Lambda$ hypernuclei and the onset of stability for $\Lambda\Xi$ hypernuclei

I.N. Filikhin and A. Gal

Racah Institute of Physics, The Hebrew University, Jerusalem 91904, Israel

Abstract. New few-body calculations of light $\Lambda\Lambda$ hypernuclei are reviewed in the wake of recent reports on experimental candidates for $^{4}_{\Lambda\Lambda}$H and $^{6}_{\Lambda\Lambda}$He. The self consistency of the $\Lambda\Lambda$ hypernuclear world data is discussed and the implications for the strength of the $\Lambda\Lambda$ interaction are pointed out. For stranger systems, novel few-body calculations of light $\Lambda\Xi$ hypernuclei, using the strong $\Lambda\Xi$ interactions due to the SU(3) extension of the Nijmegen soft-core NSC97 model, are reported and the onset of stability for these hypernuclei is predicted.

INTRODUCTION

Very little is known experimentally on strangeness $S = -2$ hypernuclear systems, and virtually nothing about systems with higher strangeness content. Multistrange hadronic matter in finite systems and in bulk is predicted on general grounds to be stable, up to strangeness violating weak decays (see Ref. [1] for a recent review). Hyperons are believed to contribute macroscopically to the composition of neutron-star matter [2], which could even be made exclusively out of hyperons (see Ref. [3] for the more speculative suggestion of hyperstars). The study of multistrange systems can provide rather stringent tests of microscopic models for the baryon-baryon interaction, for which there is hardly any two-body data beyond the abundant NN scattering data and the scarce and poor YN low-energy data. Over the years, the Nijmegen group has constructed a number of one-boson-exchange (OBE) models for the baryon-baryon interaction (NN, $\Lambda N - \Sigma N$, and $\Xi N - \Lambda\Lambda - \Sigma\Sigma - \Lambda\Sigma$) using SU(3)$_{\text{flavor}}$ symmetry to relate baryon-baryon-meson coupling constants and phenomenological hard or soft cores at short distances (for a recent review, see Ref. [4]). In addition, the Jülich group has constructed OBE models for the YN interaction along the lines of the Bonn Model for the NN interaction using SU(6) symmetry to relate coupling constants and short-range form factors [5]. The SU(6) quark model has also been used to derive baryon-baryon interactions within a unified framework of a $(3q) - (3q)$ resonating group method, augmented by a few effective meson exchange potentials of scalar and pseudoscalar meson nonets directly coupled to quarks [6].

Until recently only three candidates existed for $\Lambda\Lambda$ hypernuclei which fit events seen in emulsion experiments [7, 8, 9]. The $\Lambda\Lambda$ binding energies deduced from these events indicated that the $\Lambda\Lambda$ interaction is strongly attractive in the $^1S_0$ channel [10, 11], although it had been realized [12, 13] that the binding energies of the two older events, $^{6}_{\Lambda\Lambda}$He and $^{10}_{\Lambda\Lambda}$Be, are inconsistent with each other. This outlook is perhaps undergoing an important change following the very recent report from the KEK hybrid-emulsion experiment E373 on a new candidate [14] for $^{\ \ 6}_{\Lambda\Lambda}$He, with binding energy substantially smaller than that deduced from the older event [8]. Furthermore, there are also indications from the AGS experiment E906 for the production of light $\Lambda\Lambda$ hypernuclei [15], perhaps as light even as $^{\ 4}_{\Lambda\Lambda}$H, in the (K^-,K^+) reaction on ^{9}Be.

In this talk we review ongoing three-body and four-body calculations of light $\Lambda\Lambda$ hypernuclei, mostly using simple $\Lambda\Lambda$ interaction potentials which simulate the low-energy s-wave scattering parameters produced by the Nijmegen OBE models. Our own work is compared to other recent calculations [16, 17]. The purpose of these calculations is twofold: to check the self consistency of the data, particularly for $^{\ \ 6}_{\Lambda\Lambda}$He and $^{\ 10}_{\Lambda\Lambda}$Be [18] which are treated here as clusters of α's and Λ's; and to find out which of the recent Nijmegen Soft-Core (NSC97) models [19, 20], or the Extended Soft-Core (ESC00) interaction model [4], is the most appropriate one for describing well these $\Lambda\Lambda$ hypernuclei.

A novel piece of work which we report here preliminarily concerns few-body calculations of multistrange hypernuclei consisting, in addition to Λ's, also of a (doubly strange $S = -2$) Ξ hyperon. Schaffner et al. [21] were the first ones to observe that Ξ hyperons would become particle stable against the strong decay $\Xi N \to \Lambda\Lambda$ if a sufficient number of bound Λ's Pauli blocked this decay mode. Subsequent calculations [22, 23] established the generality and robustness of this feature across the periodic table. The lightest system of this kind was argued to be $^{\ \ 7}_{\Xi\Lambda\Lambda}$He, where the two bound Λ hyperons Pauli block any further conversion into the $1s$ shell. Whereas $^{\ \ 7}_{\Xi\Lambda\Lambda}$He is still awaiting a proper four-body $\Xi\Lambda\Lambda\alpha$ calculation, we have studied the possibility of stabilizing a Ξ hyperon in the isodoublet $^{\ 6}_{\Xi\Lambda}$H - $^{\ 6}_{\Xi\Lambda}$He hypernuclei due to the particularly strong $\Xi\Lambda$ attraction in the SU(3) extended Nijmegen Soft Core NSC97 models [20]. The three-body $\Xi\Lambda\alpha$ system is also required as input for doing the four-body $\Xi\Lambda\Lambda\alpha$ calculation.

METHODOLOGY

Faddeev-Yakubovsky equations

The bound states of the three-body systems considered in this review are obtained by solving the differential Faddeev equations [24]

$$(H_0 + V_\beta^c + V_\beta^s - E)U_\beta = -V_\beta^s \sum_{\gamma=1, \gamma\neq\beta}^{3} U_\gamma , \qquad (1)$$

where V_β^c and V_β^s are Coulomb and short-range pairwise interactions respectively in the channel β, H_0 is the kinetic energy operator, E is the total energy and the wavefunction of the three-body system is given as a sum over the three Faddeev components U_β, $\beta=1,2,3$, corresponding to the two-body breakup channels.

When two constituents of the three-body system are identical, for example Λ hyperons in $^{\ 5}_{\Lambda\Lambda}$H ($t\Lambda\Lambda$) or $^{\ 6}_{\Lambda\Lambda}$He ($\alpha\Lambda\Lambda$) generically of the form $C\Lambda\Lambda$ (C = core), the coupled set of Faddeev equations simplifies as follows:

$$(H_0^u + V_{\Lambda\Lambda} - E)U = -V_{\Lambda\Lambda}(W - P_{23}W), \quad (H_0^w + V_{\Lambda C} - E)W = -V_{\Lambda C}(U - P_{23}W), \qquad (2)$$

where P_{23} is the permutation operator for the Λ hyperons (particles 2,3). For $^9_\Lambda$Be, here considered as a $\Lambda\alpha\alpha$ system, a similar simplification occurs, but where the minus sign in Eqs. (2) arising from the Fermi-Dirac statistics for the Λ's is replaced by a plus sign owing to the Bose-Einstein statistics for the α's.

Four-particle wavefunctions Ψ are decomposed, in the Faddeev-Yakubovsky method [24], into components which are in a one to one correspondence with all chains of partitions. The chains consist of two-cluster partitions a_2 (e.g., $(ijk)l$ or $(ij)(kl)$) and three-cluster partitions a_3 (e.g., $(ij)kl$) obeying the relation $a_3 \subset a_2$. The latter means that the partition a_3 can be obtained from the partition a_2 by splitting up one subsystem. It is easy to see that there exist generally 18 chains of partitions for a four-particle system. The components $\Psi_{a_3 a_2}$ obey the Yakubovsky equations [25]

$$(H_0 + V_{a_3} - E)\Psi_{a_3 a_2} + V_{a_3} \sum_{(c_3 \neq a_3) \subset a_2} \Psi_{c_3 a_2} = -V_{a_3} \sum_{d_2 \neq a_2} \sum_{(d_3 \neq a_3) \subset a_2} \Psi_{d_3 d_2} \ . \quad (3)$$

Here H_0 is the kinetic-energy operator and V_{a_3} stands for the two-body potential acting within the two-particle subsystem of a partition a_3 (e.g., $V_{a_3} = V_{ij}$ with $a_3 = (ij)kl$).

The s-wave Faddeev-Yakubovsky equations are solved using the cluster reduction method (CRM) which was developed by Yakovlev and Filikhin [26] and has been recently applied [27] to calculate bound states and low-energy scattering for systems of three and four nucleons. In this method, the Faddeev-Yakubovsky components are decomposed in terms of the eigenfunctions of the Hamiltonians of the two- or three-particle subsystems. Due to the projection onto elements of an orthogonal basis, one obtains a set of equations with effective interactions corresponding to the relative motion of the various clusters. A fairly small number of terms, usually less than 10, is sufficient to generate a stable and precise numerical solution. The CRM has been recently used by one of us [28] to study $^9_\Lambda$Be and $^6_{\Lambda\Lambda}$He in terms of the three-cluster systems $\Lambda\alpha\alpha$ and $\alpha\Lambda\Lambda$, respectively.

Potentials

The hyperon-hyperon (YY) interaction potentials in the $^1S_0$ channel which are used as input to the Faddeev equations are of the three-range Gaussian form

$$V_{YY'} = \sum_i^3 v^{(i)}(r) \exp(-r^2/\beta_i^2) \ , \quad (4)$$

following the work of Hiyama et al. [16] where a $\Lambda\Lambda$ potential of this form was fitted to the Nijmegen model D (ND) hard-core interaction [29] assuming the same hard core for the NN and $\Lambda\Lambda$ potentials in the $^1S_0$ channel. We have renormalized the strength of the medium-range attractive component ($i=2$) of this potential such that it yields values for the scattering length and for the effective range as close to values prescribed by us. In Table 1 we list some low-energy parameters produced by the soft-core NSC97 model YY potentials [20] and the corresponding values derived by fitting potentials of the form (4). The $\Lambda\Lambda$ and $\Lambda\Xi$ fitted potentials in the $^1S_0$ channel are shown in Fig. 1.

TABLE 1. $^1S_0$ scattering lengths a and effective ranges r in fm, for $\Lambda\Lambda$, $\Lambda\Xi^0$ and $\Lambda\Xi^-$ hyperons in different potential models

Potential	Model	$a_{\Lambda\Lambda}$	$r_{\Lambda\Lambda}$	$a_{\Lambda\Xi^0}$	$r_{\Lambda\Xi^0}$	$a_{\Lambda\Xi^-}$	$r_{\Lambda\Xi^-}$
Eq. (4)	b	-0.38	15.2	-1.14	4.91	-1.18	4.78
	e	-0.50	10.6	-2.65	3.07	-2.75	3.03
NSC97 [20]	b	-0.38	10.24	-1.14	3.80	-1.18	3.84
	e	-0.50	9.11	-2.65	2.89	-2.75	2.88

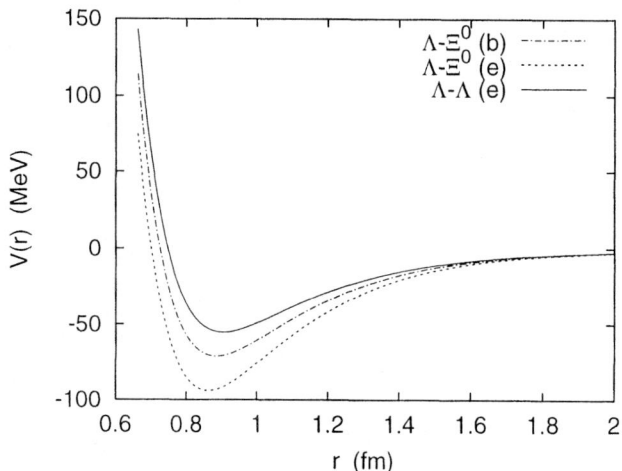

FIGURE 1. Selected YY potentials, simulating NSC97 interaction models [20].

We note that the $\Lambda\Xi$ interaction is rather strong, but that the $\Lambda\Lambda$ interaction is quite weak and is considerably weaker than the $\Lambda\Xi$ interaction for the same type of NSC97 model (here e). In the following, we will demonstrate our calculational results mostly for version e, which according to Ref. [19] is close to describing well the spin dependence of the ΛN interaction in Λ hypernuclei. The other 'good' version f yields very close results.

The $\alpha\alpha$ short-range interaction, and the $\Lambda\alpha$ and $\Xi\alpha$ interactions, are given in terms of a two-range Gaussian (Isle) potential

$$V_{Y\alpha} = V_{\rm rep}^{(Y)} \exp(-r^2/\beta_{\rm rep}^2) - V_{\rm att}^{(Y)} \exp(-r^2/\beta_{\rm att}^2) \ . \quad (5)$$

Here the superscript Y stands also for α. For the $\alpha\alpha$ short-range interaction potential we used the s-wave component of the Ali-Bodmer potential [30]. The $\Lambda\alpha$ interaction parameters were taken from Ref. [31] where the binding energy and mesonic weak decay of $^5_\Lambda$He were studied. Similar potentials were constructed by us for the $\Lambda-^3$H and $\Lambda-^3$He singlet and triplet interactions by fitting to the known binding energies of $^4_\Lambda$H and $^4_\Lambda$He, respectively. The calculated binding energies are shown in Table 2, in comparison with the experimentally known values.

TABLE 2. Binding energies (B_Λ) of Λ hypernuclei in MeV

B_Λ	$^4_\Lambda H(0^+)$	$^4_\Lambda H(1^+)$	$^4_\Lambda He(0^+)$	$^4_\Lambda He(1^+)$	$^5_\Lambda He$	$^9_\Lambda Be$
calc.	2.06	1.04	2.36	1.24	3.09	6.67
exp.*	2.04±0.04^a	1.00±0.04^b	2.39±0.03^a	1.24±0.04^b	3.12±0.02^a	6.71±0.04^a

* ^{a}Ref. [32] ^{b}Ref. [33]

For constructing an Isle-type potential for the $\Xi\alpha$ interaction, we had to extrapolate from the Woods-Saxon potential depth of about 14 - 16 MeV for a Ξ^- in ^{11}B, as determined recently by studying the excitation spectrum in the (K^-, K^+) reaction on ^{12}C [34, 35]. Since the ^4He central density is about 50% larger than for ^{11}B, we used a depth parameter of $V_0 = 24$ MeV for the $\Xi\alpha$ Woods-Saxon potential

$$V_{\Xi\alpha}(r) = -V_0(1+\exp(r-R)/a)^{-1}, \qquad (6)$$

where $R = r_0 A^{1/3}$, with $r_0 = 1.1$ fm and $a = 0.65$ fm for $A = 4$. The value of the central depth of this potential is $V_{\Xi\alpha}(0) = 22.5$ MeV. This $\Xi\alpha$ potential binds $\Xi^0\alpha$ and $\Xi^-\alpha$ by $B_\Xi = 2.09, 3.64$ MeV respectively. The depth $V_{att}^{(\Xi)}$ of the attractive component of the Isle potential (5) was then adjusted so as to reproduce these binding energies. We note that these $\Xi\alpha$ bound systems, in all of the calculations reported below, come out particle unstable against the strong decays $^5_\Xi He \to ^4_\Lambda He + \Lambda$, $^5_\Xi H \to ^4_\Lambda H + \Lambda$.

RESULTS AND DISCUSSION

We first applied the $\alpha\alpha$ and $\Lambda\alpha$ potentials, specified by Eq. (5) and the subsequent text, to the solution of the Faddeev equations for the $\alpha\alpha\Lambda$ system. The calculated ground-state binding energy of $^9_\Lambda Be$ in this three-body model is given in Table 2 and is in excellent agreement with the measured B_Λ value, with no need for renormalization or for introducing three-body interactions. Having thus gained confidence in the appropriateness of the $\alpha\alpha$ and $\Lambda\alpha$ s-wave potentials, we then applied these potentials to the solution of the Faddeev-Yakubovsky equations for several $\Lambda\Lambda$ hypernuclei, using different $\Lambda\Lambda$ interactions generically of the form (4).

$\Lambda\Lambda$ hypernuclei

The ground-state energies $E_{\Lambda\Lambda}$ obtained by solving the s-wave three-body ($\alpha\Lambda\Lambda$) Faddeev equations for $^6_{\Lambda\Lambda} He$ and the s-wave four-body ($\alpha\alpha\Lambda\Lambda$) Yakubovsky equations for $^{10}_{\Lambda\Lambda} Be$ are given in Table 3 for $\Lambda\Lambda$ potentials ordered according to their degree of attraction. The corresponding $\Lambda\Lambda$ binding energies $B_{\Lambda\Lambda}$ are given by $B_{\Lambda\Lambda} = -E_{\Lambda\Lambda}$ for $^6_{\Lambda\Lambda} He$ and $B_{\Lambda\Lambda} = -(E_{\Lambda\Lambda} + 0.1 \text{ MeV})$ for $^{10}_{\Lambda\Lambda} Be$. The strongest $\Lambda\Lambda$ attraction is provided by the potential of the uppermost row, simulating the very recent ESC00 model [4] which was largely motivated by wishing to get a $B_{\Lambda\Lambda}$ value for $^6_{\Lambda\Lambda} He$ as close to that reported by Prowse [8]; indeed our calculation reproduces it. A significantly smaller $B_{\Lambda\Lambda}$ value

TABLE 3. Ground-state energies of $\Lambda\Lambda$ hypernuclei ($E_{\Lambda\Lambda}$ in MeV) with respect to the breakup threshold of the free clusters, for $\Lambda\Lambda$ potentials (4) specified by scattering length a and effective range r (in fm)

Model	α-Λ pot.	$^5_\Lambda$He	$^9_\Lambda$Be	$a_{\Lambda\Lambda}$	$r_{\Lambda\Lambda}$	$^{\ \ 6}_{\Lambda\Lambda}$He	$^{\ 10}_{\Lambda\Lambda}$Be
ESC00	Isle	-3.09	-6.58	-10.6	2.23	-10.7	-19.4
ND				-2.81	2.95	-9.10	-17.7
				-0.77	2.92	-7.70	-16.4
				-0.31	3.12	-6.98	-15.6
NSC97e				-0.50	10.6	-6.82	-15.4
NSC97b				-0.38	15.2	-6.60	-15.2
				0.0	–	-6.27	-14.8
[16](ND)	ND	-3.12	-6.67	-2.80	2.81	-9.34	-17.15
exp.*		-3.12 ±0.02[a]	-6.62 ±0.04[a]	–	–	-10.9±0.6[b] -7.25±0.19$^{+0.18}_{-0.11}$[d]	-17.6±0.4[c] (-14.5±0.4)[c]

* [a]Ref. [32] [b]Ref. [8] [c]Ref. [18] [d]Ref. [14] [e]assuming $^{\ 10}_{\Lambda\Lambda}$Be → π^- + p + $^9_\Lambda$Be*

is obtained for our simulation of model ND which, however, reproduces well the $B_{\Lambda\Lambda}$ value reported for $^{\ 10}_{\Lambda\Lambda}$Be [18]. Down the list, the NSC97 models give yet smaller $B_{\Lambda\Lambda}$ values, which for $^{\ \ 6}_{\Lambda\Lambda}$He are close to the very recent experimental report [14].

It has been shown in previous cluster calculations [12, 13] that the calculated $B_{\Lambda\Lambda}$ values for $^{\ \ 6}_{\Lambda\Lambda}$He and for $^{\ 10}_{\Lambda\Lambda}$Be are correlated nearly linearly with each other. Our calculations also indicate such a correlation, as demonstrated in Fig. 2 by the dotted line. This line precludes any joint theoretical explanation of the $^{\ \ 6}_{\Lambda\Lambda}$He and $^{\ 10}_{\Lambda\Lambda}$Be experimental candidates listed in Table 3. If $B_{\Lambda\Lambda}(^{\ 10}_{\Lambda\Lambda}Be) = 17.7 \pm 0.4$ MeV [7, 18], then the theoretically implied $B_{\Lambda\Lambda}$ value for $^{\ \ 6}_{\Lambda\Lambda}$He is about 9.1 ± 0.4 MeV, considerably below the value reported by Prowse [8] but considerably above the value reported very recently by Takahashi et al. [14]. These calculated $B_{\Lambda\Lambda}$ values were obtained using a $\Lambda\Lambda$ potential of the form (4) identical to the one (also marked ND) used in the cluster calculation [16] which is listed separately in the table. The two calculations essentially agree with each other for $^{\ \ 6}_{\Lambda\Lambda}$He, whereas for $^{\ 10}_{\Lambda\Lambda}$Be our calculation provides about 0.5 MeV more binding. If $B_{\Lambda\Lambda}(^{\ 10}_{\Lambda\Lambda}Be) = 14.6 \pm 0.4$ MeV, on the assumption that the π^- weak decay of $^{\ 10}_{\Lambda\Lambda}$Be ground-state occurred to the first excited doublet levels of $^9_\Lambda$Be at 3.1 MeV [36], then the theoretically implied $B_{\Lambda\Lambda}$ value for $^{\ \ 6}_{\Lambda\Lambda}$He is 6.1 ± 0.4 MeV, significantly below the value reported by either one of the $^{\ \ 6}_{\Lambda\Lambda}$He observations.

For $V_{\Lambda\Lambda} = 0$, the lower-left point on the dotted line in Fig. 2 corresponds to approximately zero incremental binding energy $\Delta B_{\Lambda\Lambda}$ for $^{\ \ 6}_{\Lambda\Lambda}$He, where

$$\Delta B_{\Lambda\Lambda}(^{\ \ A}_{\Lambda\Lambda}Z) = B_{\Lambda\Lambda}(^{\ \ A}_{\Lambda\Lambda}Z) - 2B_\Lambda(^{(A-1)}_{\ \ \ \Lambda}Z) \ . \tag{7}$$

This is also explicitly listed in Table 4 and is easy to understand owing to the rigidity of the α core. However, the corresponding $\Delta B_{\Lambda\Lambda}$ value for $^{\ 10}_{\Lambda\Lambda}$Be is fairly substantial, about 1.5 MeV, reflecting a basic difference between the four-body $\alpha\alpha\Lambda\Lambda$ calculation and any three-body approximation in terms of a nuclear core and two Λ's as in $^{\ \ 6}_{\Lambda\Lambda}$He. To demonstrate this point we show by the dot-dash line in Fig. 2 the results of a three-

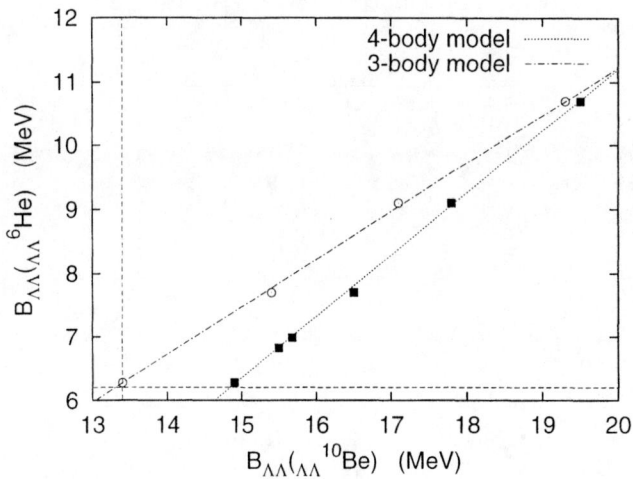

FIGURE 2. Calculated binding energies ($B_{\Lambda\Lambda}$ in MeV) for $^{\,\,6}_{\Lambda\Lambda}$He in a three-body $\alpha\Lambda\Lambda$ model, and for $^{10}_{\Lambda\Lambda}$Be in a four-body $\alpha\alpha\Lambda\Lambda$ model and in a three-body ^{8}Be $\Lambda\Lambda$ model.

body calculation for $^{10}_{\Lambda\Lambda}$Be in which the ^{8}Be core is not assigned an $\alpha\alpha$ structure. In this calculation, the Λ–^{8}Be potential was chosen in a Woods-Saxon form, where the depth (for a given geometry) was fixed by requiring the ground-state binding energy of the Λ–^{8}Be system to be given by the measured $B_{\Lambda}(^{9}_{\Lambda}Be)$ value. The dependence on the geometry proved mild. This three-body calculation gives smaller $B_{\Lambda\Lambda}$ values for $^{10}_{\Lambda\Lambda}$Be than the four-body calculation does. The difference between the two calculations is particularly pronounced as the $\Lambda\Lambda$ interaction is weakened down to zero. When this occurs, the $\alpha\alpha$ correlations which are absent in the three-body calculation, and which are built in within the Yakubovsky equations of the four-body calculation, become dominantly important. These correlations are responsible for the 1.5 MeV binding-energy difference in the limit of $V_{\Lambda\Lambda} \to 0$. Other calculations [12, 13] which pointed out the linear relationship discussed above found smaller values, not exceeding 0.5 MeV, in this limit.

The results of Table 3 for $^{\,\,6}_{\Lambda\Lambda}$He and $^{10}_{\Lambda\Lambda}$Be are shown again in Table 4, in terms of the incremental binding energy $\Delta B_{\Lambda\Lambda}$ (7), together with results for the $A = 5$ isodoublet hypernuclei $^{\,\,5}_{\Lambda\Lambda}$H and $^{\,\,5}_{\Lambda\Lambda}$He which are considered as three-cluster systems ^{3}H$\Lambda\Lambda$ and ^{3}He$\Lambda\Lambda$ respectively. These $\Lambda\Lambda$ hypernuclei are particle stable for all the $\Lambda\Lambda$ attractive potentials used in the present calculation. As the $\Lambda\Lambda$ interaction strength goes to zero, $\Delta B_{\Lambda\Lambda}(A = 5)$ approaches approximately zero, similarly to $\Delta B_{\Lambda\Lambda}(A = 6)$. We recall that both of the $A = 5, 6$ $\Lambda\Lambda$ hypernuclear systems are treated here as three-body clusters. A nearly linear correlation appears between $\Delta B_{\Lambda\Lambda}(A = 6)$ and $\Delta B_{\Lambda\Lambda}(A = 5)$, as shown in Fig. 3. Judging by the slope of the straight lines prevailing over most of the interval shown in the figure, the $\Lambda\Lambda$ interaction is more effective in binding $^{\,\,6}_{\Lambda\Lambda}$He than binding either one of the $A = 5$ $\Lambda\Lambda$ hypernuclei. For this range of mass values, the heavier the nuclear core is - the larger $\Delta B_{\Lambda\Lambda}$ is, implying that no saturation is yet reached. This

TABLE 4. Incremental binding energies ($\Delta B_{\Lambda\Lambda}$ in MeV) for $\Lambda\Lambda$ potentials (4) specified by the scattering length a and effective range r (in fm)

Model	$a_{\Lambda\Lambda}$	$r_{\Lambda\Lambda}$	$^{5}_{\Lambda\Lambda}$H *	$^{5}_{\Lambda\Lambda}$He	$^{6}_{\Lambda\Lambda}$He	$^{10}_{\Lambda\Lambda}$Be
ESC00	-10.6	2.23	3.46	3.68	4.51	6.1
ND	-2.81	2.95	2.11	2.27	2.91	4.4
	-0.77	2.92	1.02	1.13	1.51	3.1
	-0.31	3.12	0.53	0.59	0.79	2.3
NSC97e	-0.50	10.6	0.50	0.55	0.63	2.1
NSC97b	-0.38	15.2	0.37	0.40	0.41	1.9
	0	–	0.11	0.12	0.08	1.5
[16](ND)	-2.80	2.81	–	–	3.10	3.74
[17](ND)	-2.80	2.81	2.8	2.7	4.3	–
exp.[†]					4.7 ± 0.6^{a} $1.01\pm0.20^{+0.18}_{-0.11}{}^{c}$	4.3 ± 0.4^{b} $(1.2\pm0.4)^{d}$

* $\Delta B_{\Lambda\Lambda}(A=5)$ is relative to the $(2J+1)$ average of the ($^{4}_{\Lambda}$H, $^{4}_{\Lambda}$He) levels of Table 2
[†] [a]Ref. [8] [b]Ref. [18] [c]Ref. [14] [d]assuming $^{10}_{\Lambda\Lambda}$Be $\rightarrow \pi^{-} + p + {}^{9}_{\Lambda}$Be*

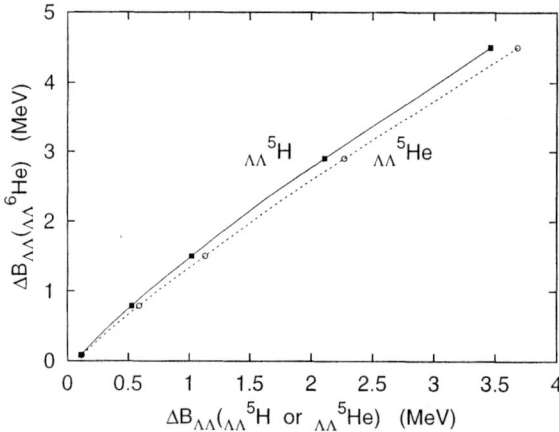

FIGURE 3. Incremental binding energies ($\Delta B_{\Lambda\Lambda}$ in MeV) calculated for $A = 5, 6$ in a three-body model.

conclusion holds also for the three-body models of $^{10}_{\Lambda\Lambda}$Be and $^{6}_{\Lambda\Lambda}$He shown in Fig. 2 (for heavier nuclear cores $\Delta B_{\Lambda\Lambda}$ should start occasionally decreasing slowly to zero with A). Returning to Fig. 3 we note that $\Delta B_{\Lambda\Lambda}$ for $^{5}_{\Lambda\Lambda}$He is slightly larger than for $^{5}_{\Lambda\Lambda}$H, owing to the tighter binding of the core Λ hypernucleus $^{4}_{\Lambda}$He as compared to $^{4}_{\Lambda}$H.

We have already compared our calculations, discussing Table 3, with those of Ref. [16]. In Table 4 we also added a comparison with the calculations of Ref. [17]. Their $\Delta B_{\Lambda\Lambda}$ values are incredibly higher everywhere than ours, and this holds also for their ND-type calculation of $^{5}_{\Lambda}$He.

TABLE 5. $\Lambda\Xi$ and $\Lambda\Lambda$ binding energies, in MeV, for versions of model NSC97 [20]

NSC97	$B_{\Lambda\Xi^-}(^{6}_{\Lambda\Xi}\text{H})$	$B_{\Lambda\Lambda}(^{5}_{\Lambda\Lambda}\text{H})$	$B_{\Lambda\Xi^0}(^{6}_{\Lambda\Xi}\text{He})$	$B_{\Lambda\Lambda}(^{5}_{\Lambda\Lambda}\text{He})$
model b	7.80	2.79	6.53	3.27
model e	8.88	2.92	7.63	3.42

$\Lambda\Xi$ hypernuclei

Recall the strongly attractive $^1S_0$ $\Lambda\Xi$ potentials, plotted in Fig. 1, which simulate two of the interactions of model NSC97. Along with the phenomenological $\Lambda\alpha$ and $\Xi\alpha$ Isle-type potentials (5), we have all the ingredients necessary for doing a Faddeev calculation of the $\alpha\Lambda\Xi$ three-body system. The results of this calculation for the $\Lambda\Xi$ binding energies of the isodoublet hypernuclei $^{6}_{\Lambda\Xi}\text{H}$ ($\alpha\Lambda\Xi^-$) and $^{6}_{\Lambda\Xi}\text{He}$ ($\alpha\Lambda\Xi^0$) are given in Table 5, where the $\Lambda\Lambda$ binding energies for $^{5}_{\Lambda\Lambda}\text{H}$ and $^{5}_{\Lambda\Lambda}\text{He}$ are also listed. These latter $B_{\Lambda\Lambda}$ values determine the location of the lowest particle-stability thresholds relevant for the stability of the $\alpha\Lambda\Xi$ systems, namely whether or not these $S=-3$ systems are particle stable with respect to the strong decays

$$^{6}_{\Lambda\Xi}\text{H} \to {}^{5}_{\Lambda\Lambda}\text{H} + \Lambda \ , \quad {}^{6}_{\Lambda\Xi}\text{He} \to {}^{5}_{\Lambda\Lambda}\text{He} + \Lambda \ . \tag{8}$$

The appropriate level schemes are shown in Fig. 4 on the next page for models NSC97b and NSC97e. It is seen that $^{6}_{\Lambda\Xi}\text{He}$ is particle stable in these NSC97 models. The mirror hypernucleus $^{6}_{\Lambda\Xi}\text{H}$ is unstable because the Ξ^- hyperon is heavier by 6.5 MeV than Ξ^0.

Our prediction for the stability of $^{6}_{\Lambda\Xi}\text{He}$ would remain in effect, particularly for model NSC97e, even if the binding energy of $^{5}_{\Lambda\Lambda}\text{He}$ is increased by a fraction of MeV, using Fig. 3 to scale it with the recently reported $^{6}_{\Lambda\Lambda}\text{He}$ binding energy [14]. This is the first result of its kind, predicting $^{6}_{\Lambda\Xi}\text{He}$ as the lightest particle-stable $S=-3$ hypernucleus, and the lightest and least strange particle-stable hypernucleus in which a Ξ hyperon is bound.

SUMMARY

In this work we studied light multistrange hypernuclear systems which may be described in terms of few-body clusters and treated by solving the three-body Faddeev and four-body Faddeev-Yakubovsky equations. For $^{10}_{\Lambda\Lambda}\text{Be}$, the Faddeev-Yakubovsky solution of the $\alpha\alpha\Lambda\Lambda$ problem given here is the first one of its kind. It yields more binding for $^{10}_{\Lambda\Lambda}\text{Be}$ than that produced by most of the previous calculations, where comparison is meaningful. It confirms, if not aggravates, the incompatibility of the 'old' experimental determination of the binding energy of $^{6}_{\Lambda\Lambda}\text{He}$ [8] with that of $^{10}_{\Lambda\Lambda}\text{Be}$ [7]. The 'new' experimental determination of the binding energy of $^{6}_{\Lambda\Lambda}\text{He}$ [14] is found to be even more incompatible with that of $^{10}_{\Lambda\Lambda}\text{Be}$. Assuming that the determination of $B_{\Lambda\Lambda}(^{10}_{\Lambda\Lambda}\text{Be})$ was plagued by an unobserved γ deexcitation involving either $^{10}_{\Lambda\Lambda}\text{Be}^*$ or $^{9}_{\Lambda}\text{Be}^*$, somewhere along the sequence of tracks observed in the emulsion event carefully reanalyzed in Ref.

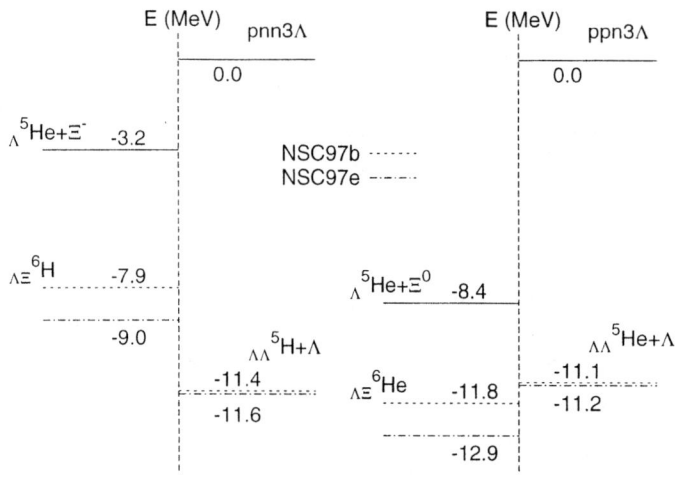

FIGURE 4. Calculated level scheme of $^{\ \ 6}_{\Lambda\Xi}\text{H}$ and $^{\ \ 6}_{\Lambda\Xi}\text{He}$ hypernuclei.

[18], does not resolve this incompatibility. Adding $^{\ 13}_{\Lambda\Lambda}\text{B}$ [9] as input does not alleviate it either, since the possibility of unobserved γ deexcitation cannot be dismissed also for this species, while on the theoretical side the analysis of $^{\ 13}_{\Lambda\Lambda}\text{B}$ in terms of a few-body cluster is more dubious than for the lighter $\Lambda\Lambda$ species.

Discarding past history of this emulsion experimentation for $\Lambda\Lambda$ hypernuclear events identified as heavier than $^{\ 6}_{\Lambda\Lambda}\text{He}$, because of the ambiguities mentioned here, one remains with the very recent report from the KEK E373 experiment [14] which claims to have identified uniquely $^{\ 6}_{\Lambda\Lambda}\text{He}$, with $\Delta B_{\Lambda\Lambda} \sim 1$ MeV. No particle-stable excited states are possible for this species or for its Λ hypernuclear core $^{5}_{\Lambda}\text{He}$, so this event - if confirmed - should be taken as the most directly relevant constraint on the $\Lambda\Lambda$ interaction.

Moreover, $^{\ 6}_{\Lambda\Lambda}\text{He}$ is also ideally suited for three-body cluster calculations such as the Faddeev equations here solved for the $\alpha\Lambda\Lambda$ system. Using s-wave soft-core $\Lambda\Lambda$ potentials that simulate several of the Nijmegen $\Lambda\Lambda$ interaction models, we have shown that model NSC97 is the only one capable of coming close to the observed binding, short by about 0.5 MeV of the new value [14]. In fact, we estimate the theoretical uncertainty of our Faddeev calculation for $^{\ 6}_{\Lambda\Lambda}\text{He}$ as bounded by 0.5 MeV, and such that the precisely calculated binding energy is *larger* by a fraction of this bound than the $\Delta B_{\Lambda\Lambda}$ values shown in Table 4. Taking into account such possible corrections would bring our calculated $\Delta B_{\Lambda\Lambda}$ values to within the error bars of the reported $\Delta B_{\Lambda\Lambda}$ value. There are two possible origins for this theoretical uncertainty, one is the restriction to s-waves in the partial-wave expansion of the Faddeev equations, excluding higher ℓ values; the other one is ignoring the off-diagonal $\Lambda\Lambda - \Xi N$ interaction which admixes Ξ components into the $^{\ 6}_{\Lambda\Lambda}\text{He}$ wavefunction. Both effects have been tested in several previous calculations and found small. For example, a recent work by Yamada and Nakamoto [37] using model ND finds an increase of 0.4 MeV in the calculated $B_{\Lambda\Lambda}(^{6}_{\Lambda\Lambda}\text{He})$ value due to a 0.3% (probability) Ξ component.

If model NSC97, as our few-body calculations suggest, is indeed the right SU(3) extrapolation from fits to NN and YN data, then it is unlikely that $^{4}_{\Lambda\Lambda}\text{H}$ is particle stable. The subtleties involved in estimating whether or not this species provides for the onset of $\Lambda\Lambda$ binding in nuclei were clearly demonstrated in Ref. [38]. Its $\Delta B_{\Lambda\Lambda}$, at any rate, should not exceed 0.5 MeV for considerably stronger $\Lambda\Lambda$ interactions [17]. If the AGS E906 events conjectured in Ref. [15] as evidence for $^{4}_{\Lambda\Lambda}\text{H}$ are confirmed as such in a future extension of this experiment, this four-body system $pn\Lambda\Lambda$ would play as a fundamental role for studying theoretically the hyperon-hyperon forces as $^{3}_{\Lambda}\text{H}$ ($pn\Lambda$) has played for studying theoretically the hyperon-nucleon forces [39]. In fact, the published data does not rule out $^{5}_{\Lambda\Lambda}\text{H}$ as the newly discovered species in this experiment [40], and it would seem natural that the $A = 5$ $\Lambda\Lambda$ hypernuclei mark the onset of $\Lambda\Lambda$ hypernuclear stability.

Finally, if model NSC97 is indeed the right model, particularly its e version which has been tuned up to agree with the ΛN spin doublet splittings in light Λ hypernuclei [19], then the onset of Ξ-nuclear stability is likely to be at $^{6}_{\Lambda\Xi}\text{He}$, as the calculations reported here suggest.

ACKNOWLEDGMENTS

This work is partially supported by the trilateral DFG grant GR 243/51-2. I.N.F. is also partly supported by the Russian Ministry of Education grant E00-3.1-133.

REFERENCES

1. Schaffner-Bielich, J., and Gal, A., *Phys. Rev. C* **62**, 034311 (2000).
2. Glendenning, N.K., *Ap. J.* **293**, 470 (1985); Balberg, S., and Gal, A., *Nucl. Phys.* **A625**, 435 (1997).
3. Schaffner-Bielich, J., Hanauske, M., Stocker, H., and Greiner, W., astro-ph/0005490.
4. Rijken, Th. A., *Nucl. Phys.* **A691**, 322 (2001).
5. Reuber, A., Holinde, K., and Speth, J., *Nucl. Phys.* **A570**, 543 (1994).
6. Fujiwara, Y., Nakamoto, C., and Suzuki, Y., *Phys. Rev. C* **54**, 2180 (1996).
7. Danysz, M., et al., *Nucl. Phys.* **49**, 121 (1963).
8. Prowse, R.J., *Phys. Rev. Lett.* **17**, 782 (1966).
9. Aoki, S., et al., *Prog. Theor. Phys.* **85**, 1287 (1991).
10. Dover, C.B., Millener, D.J., Gal, A., and Davis, D.H., *Phys. Rev. C* **44**, 1905 (1991).
11. Yamamoto, Y., Takaki, H., and Ikeda, K., *Prog. Theor. Phys.* **86**, 867 (1991).
12. Bodmer, A.R., Usmani, Q.N., and Carlson, J., *Nucl. Phys.* **A422**, 510 (1984).
13. Wang, X.C., Takaki, H., and Bando, H., *Prog. Theor. Phys.* **76**, 865 (1986).
14. Takahashi, H., et al., *Phys. Rev. Lett.* (in press, 2001).
15. Ahn, J.K., et al., *Phys. Rev. Lett.* **87**, 132504 (2001).
16. Hiyama, E., Kamimura, M., Motoba, T., Yamada, T., and Yamamoto, Y., *Prog. Theor. Phys.* **97**, 881 (1997).
17. Nemura, H., Suzuki, Y., Fujiwara, Y., and Nakamoto, C., *Prog. Theor. Phys.* **103**, 929 (2000).
18. Dalitz, R.H., Davis, D.H., Fowler, P.H., Montwill, A., Pniewski, J., and Zakrzewski, J.A., *Proc. Roy. Soc. London A* **426**, 1 (1989).
19. Rijken, Th.A, Stoks, V.G.J., and Yamamoto, Y., *Phys. Rev. C* **59**, 21 (1999).
20. Stoks, V.G.J., and Rijken, Th.A., *Phys. Rev. C* **59**, 3009 (1999).

21. Schaffner, J., Greiner, C., and Stoecker, H., *Phys. Rev. C* **46**, 322 (1992).
22. Schaffner, J., Dover, C.B., Gal, A., Greiner, C., and Stoecker, H., *Phys. Rev. Lett.* **71**, 1328 (1993).
23. Schaffner, J., Dover, C.B., Gal, A., Greiner, C., Millener, D.J., and Stoecker, H., *Ann. Phys. [NY]* **235**, 35 (1994).
24. Merkuriev, S.P., and Faddeev, L.D., *Quantum Scattering Theory for Systems of Several Particles*, Kluwer Academic Press, Dodrecht, 1993.
25. Merkuriev, S.P., Yakovlev, S.L., and Gignoux, C., *Nucl. Phys.* **A431**, 125 (1984).
26. Yakovlev, S.L., and Filikhin, I.N., *Phys. At. Nucl.* **58**, 754 (1995); *ibid.* **60**, 1794 (1997).
27. Filikhin, I.N., and Yakovlev, S.L., *Phys. At. Nucl.* **62**, 1588 (1999); *ibid.* **63**, 63 (2000); *ibid.* **63**, 79 (2000).
28. Filikhin, I.N., and Yakovlev, S.L., *Phys. At. Nucl.* **63**, 336 (2000).
29. Nagels, M.M., Rijken, Th.A., and de Swart, J.J., *Phys. Rev. D* **12**, 744 (1975); *ibid.* **15**, 2547 (1977).
30. Ali, S., and Bodmer, A.R., *Nucl. Phys.* **80**, 99 (1966).
31. Kurihara, Y., Akaishi, Y., and Tanaka, H., *Phys. Rev. C* **31**, 971 (1985); see also *Prog. Theor. Phys.* **71**, 561 (1984) for a justification of the Isle potential.
32. Davis, D.H., and Pniewski, J., *J. Contemp. Phys.* **27**, 91 (1986); see also Davis, D.H., in *LAMPF Workshop on (π, K) Physics*, edited by B.F. Gibson, W.R. Gibbs and M.B. Johnson, AIP Conference Proceedings 224, New York, 1991, pp. 38-48.
33. Bedjidian, M., *et al.*, *Phys. Lett. B* **83**, 252 (1979).
34. Fukuda, T., *et al.*, *Phys. Rev. C* **58**, 1306 (1998).
35. Khaustov, P., *et al.*, *Phys. Rev. C* **61**, 054603 (2000).
36. May, M., *et al.*, *Phys. Rev. Lett.* **51**, 2085 (1983).
37. Yamada, T., and Nakamoto, C., *Phys. Rev. C* **62**, 034319 (2000).
38. Nakaichi-Maeda, S., and Akaishi, Y., *Prog. Theor. Phys.* **84**, 1025 (1990).
39. Miyagawa, K., Kamada, H., Gloeckle, W., Yamamoto, H., Mart, T., and Bennhold, C., *Few-Body Syst. Suppl.* **12**, 324 (2000).
40. Gal, A., in preparation.

Optical potentials in kaonic atoms

Carmen García-Recio*, Juan Nieves*, Eulogio Oset† and Angels Ramos**

Departamento Física Moderna, University of Granada, E-18071 Granada, Spain
†*Departamento de Física Teórica and IFIC, Centro Mixto Universidad de Valencia-CSIC, Paterna, Aptdo. Correos 22085, 46071 Valencia, Spain*
**Departament d'Estructura i Constituents de la Matèria, Universitat de Barcelona, 08028 Barcelona, Spain*

Abstract. A microscopic optical potential based on a chiral model is used as a starting point for studying kaonic atoms levels. We add to this potential a phenomenological part fitted to the experimentally known shifts and widths of kaonic levels. This fitted potential is used to predict deeply bound atomic levels, as well as nuclear levels. Comparison with the predictions of other optical models found in the literature is done. The effects on the kaonic atoms levels of certain known non-local contributions to the optical potential are also analyzed.

THE SELFCONSISTENT K^- SELFENERGY OF RAMOS-OSET

The problem of kaonic atoms has regained interest recently. First, due to the new perspective that the use of chiral Lagrangians has brought into the problem [1]. Second, because of the need to obtain accurately the kaon selfenergy in a nuclear medium, in view of the possibility to get kaon condensates in neutron-proton stars. Third, the interpretation of the enhancement of the K^- yields in heavy ion reactions relies on the value of the K^- selfenergy in the nuclear medium.

The dominance of the s–wave in the elementary $\bar{K}N$ interaction has been the justification for using traditionally s–wave $\bar{K}$ nucleus optical potentials [2, 3], by means of which good agreement with data can be obtained. In the work of Ramos-Oset [1], the s–wave self-energy $\Pi^s_{\bar{K}}$ of the K^- meson in nuclear matter is calculated in a selfconsistent microscopic approach, using an in medium effective $\bar{K}N$ interaction, t^{eff}_{ij}, obtained from the lowest-order meson-baryon chiral lagrangian V_{ij} by solving the coupled-channel Bethe-Salpeter equation for the meson-baryon sector with strangeness $S = -1$. For our calculations of shifts and widths in kaonic atoms, we take as starting point this local theoretical potential, $V^{(1)}_{\text{opt}}$ (dashed line in Fig. 1). We also use an improved version of this potential, $V^{(1\Sigma^*)}_{\text{opt}}$ (shown in Fig. 1 with solid line), which includes the Σ^*h–excitation in the selfconsistent calculation of the s–wave K^- optical potential.

CALCULATION OF SHIFTS AND WIDTHS OF KAONIC ATOMS

Purely s–wave potential. By using the s–wave potentials (1) and ($1\Sigma^*$) we calculate the shifts and widths corresponding to 63 experimental data (see Table 1) obtaining

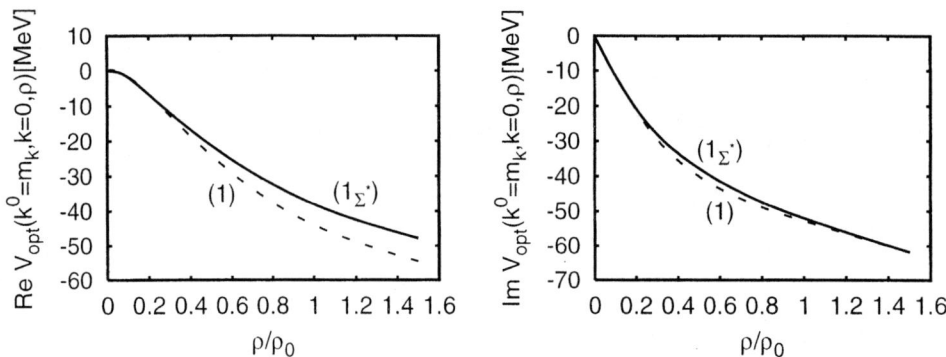

FIGURE 1. K^- optical potentials versus nuclear density.

values of χ^2 per data of 3.76 and 2.89, respectively. The agreement with the data is quite satisfactory if one takes into account that the potentials are purely theoretical without any free parameter.

p–wave. The lowest order p–wave optical potential [4] includes the lowest order p–wave chiral lagrangian plus the contribution of the Λh, Σh and $\Sigma^* h$ excitations. Using the s–wave potential $V_{opt}^{(1)}$ plus this p–wave part to solve the Klein-Gordon equation, we find the results of row (2) in Table 1, with $\chi^2/N = 4.00$. We observe that the change in the shifts and widths due to the p–wave are much smaller than the experimental uncertainties.

s–wave induced non-local effects. The s–wave optical potential of a K^- in a nuclear medium of density ρ depends on the K^- energy ω and momentum $\vec{k}$. For kaonic atoms, the potentials (1) and (1_{Σ^*}) were evaluated at threshold ($\omega = m_K, \vec{k} = 0$). Now, we consider the corrections to the optical potential due to the explicit ω and $\vec{k}$ dependence. Expanding the K^- selfenergy at first order around threshold:

$$\Pi(\omega, \vec{k}, \rho) = 2\omega V_{opt} = \Pi(m_K, 0, \rho) + b(\rho)\vec{k}^2 + c(\rho)(\omega - m_K). \quad (1)$$

The momentum $\vec{k}$ is not defined for a K^- bound in an atom and instead it becomes an operator. Having evaluated the selfenergy in nuclear matter, it is not well defined which kind of $\vec{\nabla}\vec{\nabla}$ operator will correspond to the factor $\vec{k}^2$. So different realizations for the operator $\vec{k}^2$ has been considered, see rows (3a) and (3b) in Table 1. The effect of considering the $c(\omega - m_K)$ term is shown in row (4) of the table. We observe that the effect of any of these non local terms on shifts and widths are smaller that the error bars of the experimental data. For more details on non-local effects see ref. [4].

TABLE 1. Shifts and widths of representative kaonic atom levels in eV obtained for different potentials. Also χ^2 per number of data, $N = 63$, is shown. Row (1) correspond to the local potential of ref. [1]. Rows (2) to (4) correspond to different non-local additions to the dominant piece (1):
  (2) Only p–wave non-local effects due to the coupling of K^-N to Λ, Σ and Σ^*, are added.
  (3) Only lowest order in momentum $b\vec{q}^2$ non-local terms of the s–wave potential, see eq. (1), are included in two different ways: (3a) $-\vec{\nabla}b\vec{\nabla}$, (3b) $-\vec{\nabla}b\vec{\nabla} - 0.5(\Delta b)$.
  (4) Only energy dependent $c(\omega - \mu)$ "non-local" effects of eq. (1) are added.
The results of row (1_{Σ^*}) are obtained from a theoretical s–wave optical potential, like in row (1), but including Σ^*h excitations. Potentials (1m), ($1_{\Sigma^*}b_0$) and ($1_{\Sigma^*}B_0$) are best–fit local potentials.

	χ^2/N	$^{10}_{5}B$		$^{27}_{13}Al$		$^{63}_{29}Cu$		$^{112}_{48}Cd$		$^{238}_{92}U$	
		$-\varepsilon_{2p}$	Γ_{2p}	$-\varepsilon_{3d}$	Γ_{3d}	$-\varepsilon_{4f}$	Γ_{4f}	$-\varepsilon_{5g}$	Γ_{5g}	$-\varepsilon_{7i}$	Γ_{7i}
(1)	3.76	217	551	109	368	384	1121	528	1437	330	1090
(2)	4.00	213	542	110	362	392	1110	543	1420	350	1076
(3a)	3.20	211	565	102	397	361	1229	494	1588	302	1291
(3b)	4.00	234	564	118	383	415	1172	568	1515	357	1196
(4)	3.69	217	552	110	371	388	1141	534	1465	337	1166
(1_{Σ^*})	2.89	208	575	105	398	373	1219	512	1550	320	1201
(1m)	1.52	159	742	69	438	335	1290	490	1610	270	1100
($1_{\Sigma^*}b_0$)	1.30	156	722	68	449	309	1368	453	1732	255	1241
($1_{\Sigma^*}B_0$)	1.41	148	673	70	438	308	1386	448	1781	279	1297
Exp	-	208 ±35	810 ±100	80 ±13	443 ±22	370 ±47	1370 ±170	400 ±100	2010 ±440	260 ±400	1500 ±750

FITS TO ATOMIC DATA

By adding a fitted correction δV^{fit} to the previous theoretical optical potentials, we achieve theoretically founded phenomenological potentials that describe the experimental data quite satisfactorily. We consider two different forms for the fitted part:
$2\mu\, \delta V^{\text{fit}}_{b_0}(r) = -4\pi\,(1 + \mu/M_N)\,\rho(r)\,\delta b_0$, with one complex parameter δb_0, and
$2\mu\, \delta V^{\text{fit}}_{B_0}(r) = -4\pi\,(1 + \mu/M_N)\,\rho(r)\,(\rho(r)/\rho_0)^{1/3}\,(i\,\delta\text{Im}B_0)$, with one real parameter $\delta\text{Im}B_0$. The quantity μ is the K^--nucleus reduced mass. The phenomenological potential (1m) is obtained from (1) plus $\delta V^{\text{fit}}_{b_0}$ with $\delta b_0 = (0.078 - i0.25)$ fm. The potential ($1_{\Sigma^*}b_0$), from (1_{Σ^*}) plus $\delta V^{\text{fit}}_{b_0}$ with $\delta b_0 = (0.0750 - i0.200)$ fm. And the potential ($1_{\Sigma^*}B_0$), from (1_{Σ^*}) plus $\delta V^{\text{fit}}_{B_0}$ with $\delta\text{Im}B_0 = -0.260$ fm. They provide good fits with χ^2 per data of about 1.4, see rows (1m), ($1_{\Sigma^*}b_0$) and ($1_{\Sigma^*}B_0$) of Table 1.

Predictions: deeply–bound atomic levels. The fitted potential (1m), described above, is used to predict binding energies and widths of deeply bound atomic states, not yet observed. They are shown, for ^{208}Pb, in Fig. 2. See ref. [3] for other nuclei. One can see that the levels, including the widths, do not overlap for a given angular momentum.

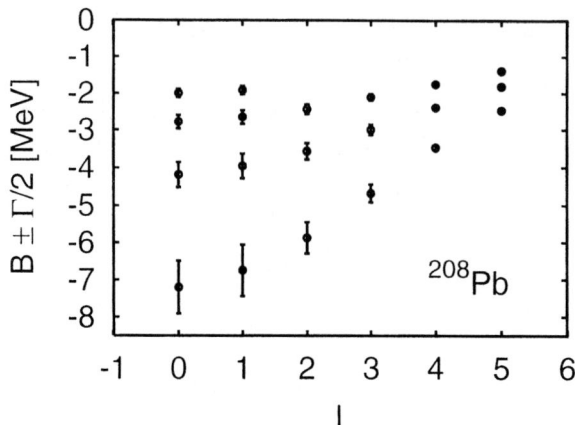

FIGURE 2. Binding energies B of deeply bound atomic levels in ^{208}Pb versus angular momentum l. The error bar stands for the full width Γ of each level. They have been computed using the $V_{opt}^{(Im)}$ potential.

CONCLUSIONS

- The selfconsistent microscopic approach based on the chiral lagrangian (1_{Σ^*}) is quite good: $\chi^2/N = 2.9$, for 63 atomic data, with no free parameter in the model.
- Non-local effects, associated to proper p–wave contributions or to momentum and energy dependence of the s–wave selfenergy, are negligible at this stage, because their effect on shifts and widths are smaller than the current uncertainties of data.
- An improved fitted potential, ($1_{\Sigma^*} b_0$), provides a very good fit with $\chi^2/63 = 1.3$.
- Deeply-bound kaonic levels are narrow and separable, so subjected to experimental observation via nuclear reactions. The widths of these levels are sensitive to different potentials by about 20%.

ACKNOWLEDGMENTS

This work was partially supported by DGICYT contract PB98-1367 and Junta de Andalucía under grant FQM 0225

REFERENCES

1. Ramos, A., and Oset, E., *Nucl. Phys.* **A671**, 481 (2000), and references therein.
2. Friedman, E., and Gal, A., *Phys. Lett. B* **459**, 43 (1999).
3. Baca, A., García-Recio, C., and Nieves, J., *Nucl. Phys.* **A673**, 335 (2000).
4. García-Recio, C., Nieves, J., Oset, E., and Ramos, A., nucl-th/0012075.

A meson exchange model for the YN interaction

J. Haidenbauer*, W. Melnitchouk[†] and J. Speth*

Forschungszentrum Jülich, IKP, D-52425 Jülich, Germany
[†]*Jefferson Lab, 12000 Jefferson Avenue, Newport News, VA 23606, USA*

Abstract. We present a new model for the hyperon-nucleon (ΛN, ΣN) interaction, derived within the meson exchange framework. The model incorporates the standard one boson exchange contributions of the lowest pseudoscalar and vector meson multiplets with coupling constants fixed by SU(6) symmetry relations. In addition — as the main feature of the new model — the exchange of two correlated pions or kaons, both in the scalar-isoscalar (σ) and vector-isovector (ρ) channels, is included.

INTRODUCTION

The hyperon-nucleon (YN) interaction is an ideal testing ground for studying the importance of SU(3) flavor symmetry breaking in hadronic systems. Existing meson exchange models of the YN force usually assume SU(3) flavor symmetry for the hadronic coupling constants, and in some cases [1, 2] even the SU(6) symmetry of the quark model. The symmetry requirements provide relations between couplings of mesons of a given multiplet to the baryon current, which greatly reduce the number of free model parameters. Specifically, coupling constants at the strange vertices are connected to nucleon-nucleon-meson coupling constants, which in turn are constrained by the wealth of empirical information on NN scattering. Essentially all YN interaction models can reproduce the existing YN scattering data, so that at present the assumption of SU(3) symmetry for the coupling constants cannot be ruled out by experiment.

One should note, however, that the various models differ dramatically in their treatment of the scalar-isoscalar meson sector, which describes the baryon-baryon interaction at intermediate ranges. For example, in the Nijmegen models [3, 4] this interaction is generated by the exchange of a genuine scalar meson SU(3) nonet. The Tübingen model [5], on the other hand, which is essentially a constituent quark model supplemented by π and σ exchange at intermediate and short ranges, treats the σ meson as an SU(3) singlet.

In the YN models of the Jülich group [1, 2] the σ (with a mass of 550 MeV) is viewed as arising from correlated $\pi\pi$ exchange. A rough estimate for the ratios of the σ-coupling strengths in the various channels can then be obtained from the relevant pion couplings. In practice, however, in the Jülich YN models, which start from the Bonn NN potential, the coupling constants of the fictitious σ meson at the strange vertices ($\Lambda\Lambda\sigma$, $\Sigma\Sigma\sigma$) are free parameters — a rather unsatisfactory feature of the models.

These problems can be overcome by an explicit evaluation of correlated $\pi\pi$ exchange in the various baryon-baryon channels. A corresponding calculation was already done for the NN case in Ref. [6]. The starting point there was a field theoretic model for both the $N\bar{N} \to \pi\pi$ Born amplitudes and the $\pi\pi$ and $K\bar{K}$ elastic scattering [7]. With the help of unitarity and dispersion relations, the amplitude for the correlated $\pi\pi$ exchange in the NN interaction was computed, showing characteristic discrepancies with the σ and ρ exchange in the (full) Bonn potential.

In a recent study [8] the Jülich group presented a microscopic derivation of correlated $\pi\pi$ exchange in various baryon-baryon (BB') channels with strangeness $S = 0, -1$ and -2. The $K\bar{K}$ channel was treated on an equal footing with the $\pi\pi$ channel in order to reliably determine the influence of $K\bar{K}$ correlations in the relevant t-channels. In this approach one can replace the phenomenological σ and ρ exchanges in the Bonn NN [9] and Jülich YN [1] models by correlated processes, and eliminate undetermined parameters such as the $BB'\sigma$ coupling constants. As a first application of the full model [10] for correlated $\pi\pi$ and $K\bar{K}$ exchange, we present here new results for YN cross sections for various YN channels.

Alternative approaches to describing baryon-baryon interactions using effective field theory, based on chiral power counting schemes, have been applied to the NN interaction. However, at present a quantitative description of NN scattering within a consistent power counting scheme is still problematic [11]. Furthermore, it is not clear that such a description can be applied in the strangeness sector, where the expansion parameter, m_K/m_N, may no longer be small enough to allow an accurate low order truncation. In addition, contact terms, which parameterize the intermediate and short range interaction, do not fulfill any SU(3) relations, and cannot be fixed by currently available YN data. At present, therefore, to obtain a quantitative description of YN scattering data one is forced towards a more traditional approach, such as that adopted here.

POTENTIAL FROM CORRELATED $\pi\pi + K\bar{K}$ EXCHANGE

Let us briefly describe the dynamical model [6, 8] for correlated two-pion and two-kaon exchange in the baryon-baryon interaction, both in the scalar-isoscalar (σ) and vector-isovector (ρ) channels. The contribution of correlated $\pi\pi$ and $K\bar{K}$ exchange is derived from the amplitudes for the transition of a baryon-antibaryon state ($B\bar{B}'$) to a $\pi\pi$ or $K\bar{K}$ state in the pseudophysical region by applying dispersion theory and unitarity. For the $B\bar{B}' \to \pi\pi, K\bar{K}$ amplitudes a microscopic model is constructed, which is based on the hadron exchange picture.

The Born terms include contributions from baryon exchange as well as ρ-pole diagrams (cf. Ref. [12]). The correlations between the two pseudoscalar mesons are taken into account by means of a coupled channel ($\pi\pi$, $K\bar{K}$) model [7, 12] generated from s- and t-channel meson exchange Born terms. This model describes the empirical $\pi\pi$ phase shifts over a large energy range from threshold up to 1.3 GeV. The parameters of the $B\bar{B}' \to \pi\pi, K\bar{K}$ model, which are interrelated through SU(3) symmetry, are determined by fitting to the quasiempirical $N\bar{N}' \to \pi\pi$ amplitudes in the pseudophysical region, $t \leq 4m_\pi^2$ [8], obtained by analytic continuation of the empirical πN and $\pi\pi$ data.

From the $B\bar{B}' \to \pi\pi$ helicity amplitudes one can calculate the corresponding spectral functions (see Ref. [8] for details), which are then inserted into dispersion integrals to obtain the (on-shell) baryon-baryon interaction:

$$V^{(0^+,1^-)}_{B'_1,B'_2;B_1,B_2}(t) \propto \int_{4m_\pi^2}^{\infty} dt' \frac{\rho^{(0^+,1^-)}_{B'_1,B'_2;B_1,B_2}(t')}{t'-t}, \quad t < 0. \tag{1}$$

Note that the spectral functions characterize both the strength and range of the interaction. Clearly, for the exchange of an infinitely narrow meson the spectral function becomes a δ-function at the appropriate mass.

RESULTS AND DISCUSSION

As shown by Reuber et al. [8], in the σ channel the strength of the correlated $\pi\pi$ and $K\bar{K}$ exchange decreases with the strangeness of the baryon-baryon channels becoming more negative. For example, in the hyperon-nucleon systems (ΛN, ΣN) the scalar-isoscalar part of the correlated exchanges is about a factor of 2 weaker than in the NN channel, and, in particular, is also weaker than the phenomenological σ meson exchange used in the original Jülich YN model [1]. Accordingly, we expect that the microscopic model with correlated $\pi\pi$ exchange will lead to a YN interaction which is less attractive.

Besides replacing the conventional σ and ρ exchanges by correlated $\pi\pi$ and $K\bar{K}$ exchange, there are in addition several new ingredients in the present YN model. First of all, we now take into account contributions from $a_0(980)$ exchange. The a_0 meson is present in the original Bonn NN potential [9] and for consistency should also be included in the YN model. Secondly, we consider the exchange of a strange scalar meson, the κ, with mass ~ 1000 MeV. Let us emphasize, however, that these particles are not viewed as being members of a scalar meson SU(3) multiplet, but rather as representations of strong meson-meson correlations in the scalar–isovector ($\pi\eta$–$K\bar{K}$) [12] and scalar–isospin-1/2 (πK) channels, respectively. In principle, their contributions can also be evaluated along the lines of Ref. [8], however, for simplicity in the present model they are effectively parameterized by one boson exchange diagrams with the appropriate quantum numbers. In any case, these phenomenological pieces are of rather short range, and do not modify the long range part of the YN interaction, which is determined solely by SU(6) constraints (for the pseudoscalar and vector mesons) and by correlated $\pi\pi$ and $K\bar{K}$ exchange.

In Figure 1 we compare the integrated cross sections for the new YN potential (solid curves) with the $YN \to Y'N$ scattering data for various channels. The agreement between the predictions and the data is clearly excellent in all channels. Also shown are the predictions from the original Jülich YN model A [1] (dashed curves). The main qualitative differences between the two models appear in the $\Lambda p \to \Lambda p$ channel, for which the Jülich model [1] (with standard σ and ρ exchange) predicts a broad shoulder at $p_{lab} \approx$ 350 MeV/c. This structure, which is not supported by the available experimental evidence, is due to a bound state in the $^1S_0$ partial wave of the ΣN channel. It is not present in the new model. The agreement in the other channels is equally good, if not better, for the new model. Further results and more details about the model can be found in Ref. [13].

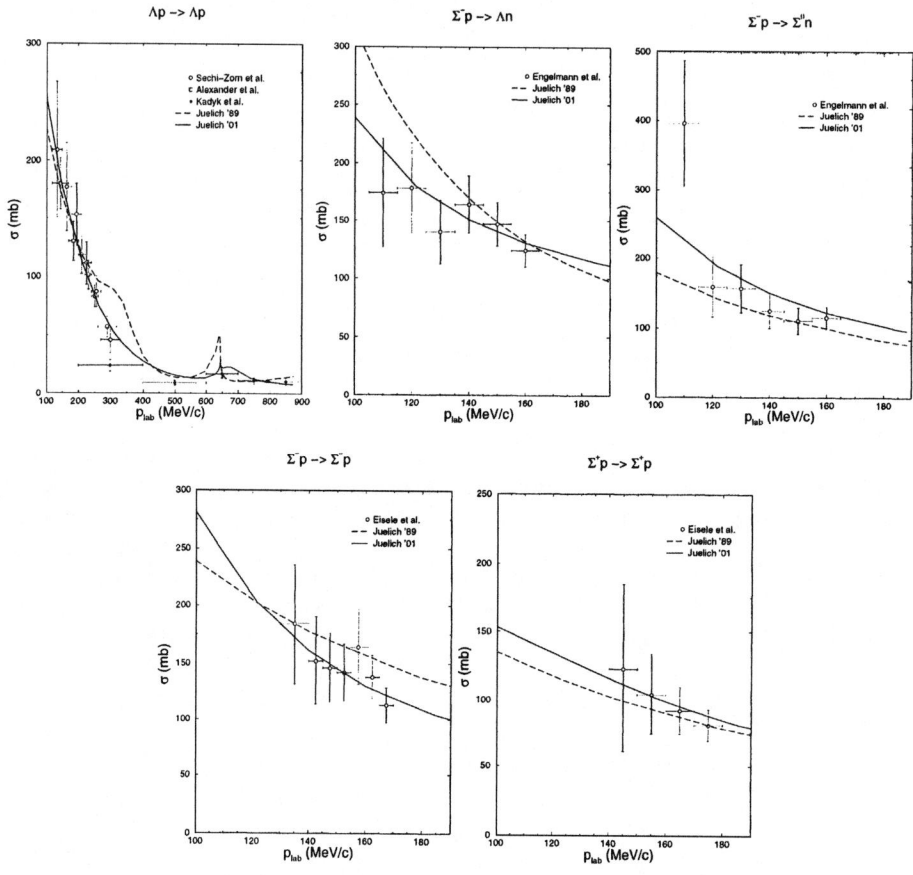

FIGURE 1. Cross sections for YN scattering. The solid lines are results of the new YN model, based on correlated $\pi\pi$ and $K\bar{K}$ exchange, while the dashed are results of the Jülich YN model A [1].

REFERENCES

1. Holzenkamp, B., Holinde, K., and Speth, J., *Nucl. Phys.* **A500**, 485 (1989).
2. Reuber, A., Holinde, K., and Speth, J., *Nucl. Phys.* **A570**, 543 (1994).
3. Maessen, P.M.M., Rijken, T.A., and de Swart, J.J., *Phys. Rev. C* **40**, 2226 (1989).
4. Rijken, Th.A., Stoks, V.G.J., and Yamamoto, Y., *Phys. Rev. C* **59**, 21 (1999).
5. Straub, U., *et al.*, *Nucl. Phys.* **A483**, 686 (1988); *Nucl. Phys.* **A508**, 385c (1990).
6. Kim, H.-C., Durso, J.W., and Holinde, K., *Phys. Rev. C* **49**, 2355 (1994).
7. Lohse, D., Durso, J.W., Holinde, K., and Speth, J., *Nucl. Phys.* **A516**, 513 (1990).
8. Reuber, A., Holinde, K., Kim, H.-C., and Speth, J., *Nucl. Phys.* **A608**, 243 (1996).
9. Machleidt, R., Holinde, K., and Elster, Ch., *Phys. Rep.* **149**, 1 (1987).
10. Haidenbauer, J., Melnitchouk, W., and Speth, J., *Nucl. Phys.* **A663**, 549 (2000).
11. Entem, D.R., and Machleidt, R., `nucl-th/0107057`.
12. Janssen, G., Pearce, B.C., Holinde, K., and Speth, J., *Phys. Rev. D* **52**, 2690 (1995).
13. Haidenbauer, J., Melnitchouk, W., and Speth, J., in preparation.

Four-body calclations of $^4_\Lambda H$ and $^4_\Lambda He$ with realistic YN and NN interactions

E. Hiyama[*], M. Kamimura[†], T. Motoba[**], T. Yamada[‡] and Y. Yamamoto[§]

[*]*Institute of Particle and Nuclear Studies, High Energy Accelerator Research Organization (KEK), Tsukuba, 305-0801, Japan*
[†]*Department of Physics, Kyushu University, Fukuoka 812-8581,Japan*
[**]*Laboratory of Physics, Osaka Electro-Comm. University, Neyagawa 572-8530, Japan*
[‡]*Laboratory of Physics, Kanto Gakuin University, Yokohama 236-8501, Japan*
[§]*Physics Section, Tsuru University, Tsuru, Yamanashi 402-8555, Japan*

Abstract. In order to discuss the important role of the $\Lambda - \Sigma$ conversion, the four-body calculations for $^4_\Lambda He$ and $^4_\Lambda H$ with high accuracy have been performed in the framework of the variational method with Jacobian-coordinate Gaussian-basis functions. All the rearrangement channel of both $NNN\Lambda$ and $NNN\Sigma$ are explicitly taken into account for the first time using realistic NN and YN interactions.

INTRODUCTION

One of the major tasks of hypernuclear physics is to investigate what influence a Λ particle receives from many-nucleon systems. For example, in the free space, $\Lambda \to N + \pi$ process is dominat. On the other hand, if a Λ particle is surrounded by the many-nucleon systems, this process is strongly suppressed by the Pauli brocking. Then, in many-nucleon systems, $\Lambda N \to NN$ process is dominant. Another example is that the Λ particle can be converted into Σ particle by the $\Lambda N - \Sigma N$ coupling of a YN interaction. The study of the role of Σ particle in Λ hypernuclei is one of the most interesting subjects in hypernuclear physics. For this study, $^4_\Lambda He$ and $^4_\Lambda H$ are useful systems. because both of the spin-doublet states have been observed.

The main purpose of this work is, first, to solve four-body problem of $^4_\Lambda He$ and $^4_\Lambda H$ by taking account of $NNN\Lambda(\Sigma)$ channels explicitly with the use of realistic NN and YN interactions and, secondly, to clarify the role of the $\Lambda N - \Sigma N$ coupling in the $A = 4$ hypernuclei quantitatively. As a first step before going to the use of sophisticated OBE models, we employ here the $\Lambda N - \Sigma N$ coupled potential with central, spin-orbit and tensor terms [5] which simulates the scattering phase shifts given by NSC97f. We employ the AV8 potential as the NN interaction.

RESULTS

In order to solve the four-body problem precisely, we have employed the coupled-rearrangement-channel variational method with the use of Jacobian-coordinate Gaussian basis function [1, 2]. This method has been successfully applied to the bound states of

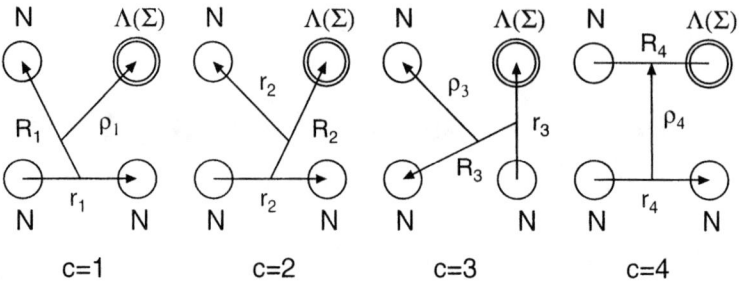

FIGURE 1. Jacobian coordinates for all the rearrangement channels of $3N + \Lambda(\Sigma)$ of $^4_\Lambda$H and $^4_\Lambda$He. The three nucleons have to be antisymmetrized.

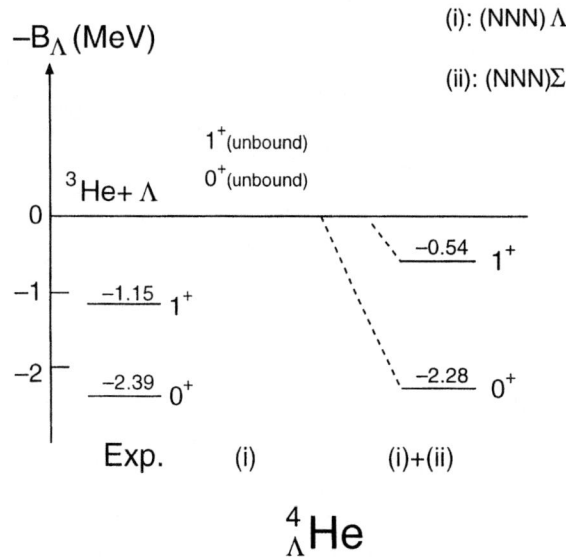

FIGURE 2. Calculated energy levels of $^4_\Lambda$He. The channels successively included are (i) $(NNN)\Lambda$, (ii) $(NNN)\Sigma$. Energy is measured from the ^{3}He+Λ threshold.

various three- and four-body systems [1-4]. Jacobian coordinates of $A = 4$ hypernuclei are illustrated in Fig.1.

All the calculations have been performed both for $^4_\Lambda$He and $^4_\Lambda$H. Calculated B_Λ of the 0^+ ground state and the 1^+ excited state of $^4_\Lambda$He are illustrated in Fig. 2 together with the observed values. In the case when only the $NNN\Lambda$ channel is considered, both of the two states are unbound. When the $(NNN)\Sigma$ channel is included, the 0^+ and 1^+ states become bound. The calculated probabilities of the $NNN\Sigma$-channel admixture are 2.08% and 1.03% for the 0^+ and 1^+ states, respectively, in $^4_\Lambda$He. Although these values are small, the Σ-channel components turn out to play an essential role in the binding mechanism of the $A = 4$ hypernuclei. The calculated binding energy of the 0^+ state al-

TABLE 1. Calculated energies of the 0^+ and 1^+ states of $^4_\Lambda$He and $^4_\Lambda$H. The energies E are measured from the $NNN\Lambda$ four-body breakup threshold. As for the ^{3}He (^{3}H) nucleus, the calculated binding energy is $-7.12\,(-7.77)$ MeV. $\bar{r}_{A-B}$ denotes the r.m.s. distances between particles A and B, while $\bar{r}_A$ stands for the r.m.s. radius of particle A measured from the c.m. of $3N$.

	$^4_\Lambda$He		$^4_\Lambda$H	
J	0^+	1^+	0^+	1^+
E (MeV)	-9.40	-7.66	-10.10	-8.33
E^{exp} (MeV)	-10.11	-8.87	-10.52	-9.53
B_Λ (MeV)	2.28	0.54	2.33	0.59
B_Λ^{exp} (MeV)	2.39	1.15	2.04	1.05
$\bar{r}_{N-N}$ (fm)	2.86	3.03	2.83	2.99
$\bar{r}_{\Lambda-N}$ (fm)	3.77	5.74	3.75	5.70
$\bar{r}_{\Sigma N}$ (fm)	2.24	2.48	2.23	2.46
$\bar{r}_N$ (fm)	1.65	1.75	1.64	1.73
$\bar{r}_\Lambda$ (fm)	3.39	5.47	3.37	5.43
$\bar{r}_\Sigma$ (fm)	1.67	1.81	1.66	1.80

most reproduces the observed binding energy, while the 1^+ state is less bound by 0.6 MeV and hence the $0^+ - 1^+$ splitting is larger than the observed splitting. The calculated B_Λ of $^4_\Lambda$H is similar to that of $^4_\Lambda$He as shown in Table I. The calculated value of $B_\Lambda(^4_\Lambda\mathrm{He}(0^+)) - B_\Lambda(^4_\Lambda\mathrm{H}(0^+)) = -0.05$ MeV is different from the experimental one, $+0.35$ MeV, although the Coulomb potential between charged particles $(p, \Sigma^\pm)$ is included. This difference should be attributed to the charge-symmetry-breaking component which is not included in our adopted YN interaction.

It is of interest to see the spatial location of the N, Λ and Σ components in the $A=4$ hypernuclei. We calculated the correlation functions (two-body densities) of the NN, ΛN and ΣN pairs (Fig.3) and estimated the r.m.s distance between the pair particles $(\bar{r}_{NN}, \bar{r}_{\Lambda N}, \bar{r}_{\Sigma N})$ as well as the r.m.s. radii $(\bar{r}_N, \bar{r}_\Lambda, \bar{r}_\Sigma)$ of N, Λ and Σ measured from the c.m. of $3N$ (Table I). The NN correlation function in $^4_\Lambda$He exhibits almost the same shape as that in the ^{3}He nucleus, indicating that the dynamical change due to the Λ participation is small. The ΛN correlation function is of larger range and flatter than the NN one, because the strength of the ΛN interaction is significantly smaller than the NN case. The ΣN correlation function is much shorter-ranged than the ΛN one due to the large virtual excitation energy (80 MeV) of $\Lambda \to \Sigma$. These features are verified by the r.m.s. distances $(\bar{r}_{NN}, \bar{r}_{\Lambda N}, \bar{r}_{\Sigma N})$ listed in Table I. Furthermore, it is interesting to see the following feature of the Σ-admixture in Fig. 3: In spite of the totally small probability of the Σ-mixing (2 %), the ΣN component at short distances is not so small in comparison with the ΛN ones. This enhanced short-ranged component of the Σ-mixing is expected to be reflected in the non-mesic decay of $\Lambda N \to NN$.

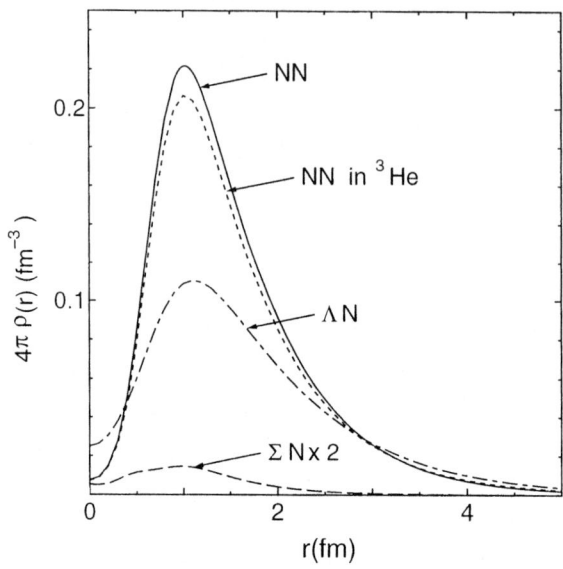

FIGURE 3. Correlation functions (two-body densities) of the NN, ΛN and ΣN pairs in the 0^+ state of $^4_\Lambda$He together with that for the NN pair in ^{3}He. In order to see the behavior of this function clearly, the correlation function of ΣN pair has been multiplied by factor 2.

SUMMARY

We have developed a calculational method of the four-body bound-state problem so that it becomes possible to make precise four-body calculations of $^4_\Lambda$H and $^4_\Lambda$He taking both the $NNN\Lambda$ and $NNN\Sigma$ channels explicitly with the use of realistic NN and YN interactions. As a result, we succeeded in making clear the role of $\Lambda - \Sigma$ conversion and deriving the amount of the Σ-mixing in $A = 4$ hypernuclei quantitatively. However, the YN interaction employed here is not sufficient to reproduce the binding energy of the excited state of 1^+, although those of ground states of $^3_\Lambda$H, $^4_\Lambda$H and $^4_\Lambda$He are in good agreement with the experimental values. It is a future problem to explore the feature of the $\Lambda - \Sigma$ conversion in Λ hypernuclei with the use of more reasonable YN interactions through systematic study of structure for the few-body hypernuclear systems.

REFERENCES

1. Kamimura, M., *Phys. Rev. A* **38**, 621 (1988).
2. Kameyama, H., Kamimura, M., and Fukushima, Y., *Phys. Rev. C* **40**, 974 (1989).
3. Hiyama, E., and Kamimura, M., *Nucl. Phys.* **A588**, 35c (1995).
4. Hiyama, E., Kamimura, M., Motoba, T., Yamada, T., and Yamamoto, Y., *Phys. Rev. C* **53**, 2075 (1996).
5. Shinmura, S., (private communication).

How does nucleus shrink by participation of Λ hyperon?

E. Hiyama*, M. Kamimura†, T. Motoba**, T. Yamada‡ and Y. Yamamoto§

*Institute of Particle and Nuclear Studies, High Energy Accelerator Research Organization (KEK), Tsukuba, 305-0801, Japan
†Department of Physics, Kyushu University, Fukuoka 812-8581,Japan
**Laboratory of Physics, Osaka Electro-Comm. University, Neyagawa 572-8530, Japan
‡Laboratory of Physics, Kanto Gakuin University, Yokohama 236-8501, Japan
§Physics Section, Tsuru University, Tsuru, Yamanashi 402-8555, Japan

Abstract. We investigated shrinkage of the core nucleus in a hypernucleus by participation of a Λ hyperon. Three- and four-body calculations of $^{7}_{\Lambda}$Li, $^{13}_{\Lambda}$C and $^{4}_{\Lambda}$He were performed. In $^{7}_{\Lambda}$Li, sizable shrinkage of the core nucleus ^{6}Li was predicted and compared with a recent experimental result. In $^{13}_{\Lambda}$C, it is shown that the nuclear shrinkage by Λ particle much depends on the character of the core nuclear states. An effect of the $\Lambda - \Sigma$ conversion on the shrinkage is discussed in the case of $^{4}_{\Lambda}$He.

INTRODUCTION

Purposes of hypernuclear physics may be stated as i) study of YN and YY interactions from the structure calculations, ii) study of the influence of the Λ particle to the surrounding nuclear many-body system, and iii) study of the influence of the surrounding nuclear many-body system to the Λ particle. Among a variety of interesting subjects related to the second item, we shall discuss the question "How does nucleus shrink by participation of the Λ hyperon?"

Nucleus is a very hard body and the nucleon density is almost independent of mass number. The inner core is hardly compressed by adding nucleons to the surface region. But, what will happen if a Λ particle is added to the core? Since the Λ does not feel the Pauli blocking by nucleons, it can reach deep inside. It might be possible that the Λ particle pulls the surrounding nucleons and thus compresses the nuclear size to a certain extent. But, how to observe the shrinkage of the nuclear size by the Λ participation? In other words, how to measure the size of a hypernucleus for the first time?

More than fifteen years ago, Bando, Motoba and Ikeda pointed out that the effect of the nuclear size shrinkage would appear in the reduction of $B(E2)$ value in the E2 transition in a hypernucleus compared with that in the corresponding core-nucleus transition [1]; note that the $B(E2)$ is proportional to r^4.

In this paper, we shall discuss the nuclear shrinkage in hypernuclei $^{7}_{\Lambda}$Li, $^{13}_{\Lambda}$C and $^{4}_{\Lambda}$H($^{4}_{\Lambda}$He).

NUCLEAR SHRINKAGE IN $^{7}_{\Lambda}$LI

A few years ago, the present authors proposed to experimentalists to measure the specific $B(E2; 5/2^+_1 \to 1/2^+_1)$ strength in $^{7}_{\Lambda}$Li and compare it with the empirical $B(E2; 3^+_1 \to 1^+_1)$ value of the corresponding transition in the core nucleus, ^{6}Li. We suggested that this measurement of $E2$ transition of $^{7}_{\Lambda}$Li provide a unique opportunity to derive the hypernuclear size [2]. However, it should be noted that one cannot conclude straightforwardly the size-shrinkage from the reduction of the $B(E2)$ value since the $B(E2)$ operator is composed of not only the r^2 term but also the $Y_{2\mu}(\theta, \phi)$ terms.

In Ref. [2], with the use of three-body model of $\alpha + p + n$ for ^{6}Li and $^{5}_{\Lambda}$He $+ p + n$ for $^{7}_{\Lambda}$Li, we examined this problem and found that the addition of the Λ particle does not change the internal motion of the np pair but contracts only the core-(np) relative motion (see Fig.4 of [2]). Then, the expectation value of the angle part of the $E2$ operator, $Y_{2\mu}(\theta, \phi)$, is not affected by the contraction of the core-(np) distance. Therefore, we concluded that one can use the relation

$$\frac{\bar{R}_{core-pn}(^{7}_{\Lambda}\text{Li})}{\bar{R}_{core-pn}(^{6}\text{Li})} = \left[\frac{B(E2; 5/2^+ \to 1/2^+)}{\frac{7}{9}B(E2; 3^+ \to 1^+)} \right]^{\frac{1}{4}},$$

which gives the degree of the shrinkage in the distance $\bar{R}_{core-pn}(^{7}_{\Lambda}\text{Li})$. Within the framework of $^{5}_{\Lambda}$He $+ p + n$ for $^{7}_{\Lambda}$Li, we predicted $B(E2; 5/2^+_1 \to 1/2^+_1) = 2.42\ e^2\text{fm}^4$ and therefore reduction in $\bar{R}_{core-pn}$ due to the Λ participation will be 25 %. Recently, with the use of a more sophisticated model, the four-body model of $\alpha + \Lambda + n + p$ for $^{7}_{\Lambda}$Li, we recalculated the $B(E2)$ value and obtained $B(E2; 5/2^+_1 \to 1/2^+_1) = 2.85\ e^2\text{fm}^4$ which means that the core nuclear size will be reduced by 22 %.

The first observation of the hypernuclear $B(E2)$ strength was recently performed in the KEK-E419 experiment for $B(E2; 5/2^+_1 \to 1/2^+_1)$ and obtained $B(E2; 5/2^+_1 \to 1/2^+_1) = 3.6 \pm 0.5\ e^2\text{fm}^4$ [3]. Therefore, this experimental data shows that hypernuclear size is reduced by 19 % compared with the core nucleus, which is consistent with our prediction. This confirms, for the first time, the shrinkage of the nuclear size induced by the gluelike character of the Λ hyperon.

NUCLEAR SHRINKAGE IN $^{13}_{\Lambda}$C

One may ask the next question "Does every nucleus shrink when Λ is added?" Our answer is No. It depends on the property of the nuclear states. Namely, i) shrinkage will be drastic in well-developed clustering states (loosely coupled systems) and ii) shrinkage will be negligible in shell-model states (compactly bound systems). This was precisely studied for ^{12}C in our previous work [4, 5] with the use of the $3\alpha + \Lambda$ four-body model. It is known that the 0^+_1 state is a shell-model state while the 0^+_2 state at $E_x = 7.65$ MeV is a well-developed clustering state and that both states are simultaneously well described by the microscopic 3α cluster model. We calculated the r.m.s. distances between two α clusters, $\bar{r}_{\alpha-\alpha}$ in the $^{13}_{\Lambda}$C and ^{12}C. In the second $1/2^+$ state of $^{13}_{\Lambda}$C which

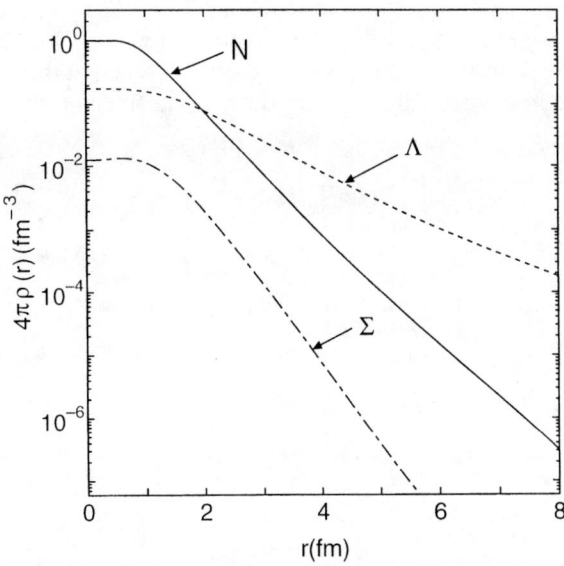

FIGURE 1. Calculated one-body densities of N, Λ and Σ particles in the 0^+ state of $^4_\Lambda$He. Volume integrals of the densities are 1.0, 0.98 and 0.02 for N, Λ and Σ particles, respectively.

is dominantly composed of the $0_2^+(^{12}C)$ and the Λ, a sizable contraction of the $\alpha-\alpha$ distance, $\bar{r}_{\alpha-\alpha} = 4.5$ fm, is seen compared to $\bar{r}_{\alpha-\alpha} = 6.3$ fm in the 0_2^+ state of ^{12}C. This contraction is more vividly seen in Fig.4(a) of [4] which illustrates the density of the $\alpha-\alpha$ relative motion as a function of the $\alpha-\alpha$ distance in the cases of both $^{13}_\Lambda C$ and ^{12}C.

On the other hand, in the $1/2^+$ ground state of $^{13}_\Lambda C$ which is dominantly composed of the $0_1^+(^{12}C)$ and the Λ, we have $\bar{r}_{\alpha-\alpha} = 2.9$ fm which represents little shrinkage from $\bar{r}_{\alpha-\alpha} = 3.0$ fm in the ground 0_1^+ state of ^{12}C. This is visually given in Fig.4(b) of [4] in contrast to Fig.4(a), which reasonably means that shell-like states are hard to contract by the addition of a Λ particle.

NUCLEAR SHRINKAGE IN $^4_\Lambda$HE

Now, one may ask the third question "What is the effect of the $\Lambda-\Sigma$ conversion on the shrinkage in Λ-hypernuclei?" The answer was already given in our recent work on the structure of $^4_\Lambda$He ($^4_\Lambda$H) [6] in which a precise four-body calculation was performed for the first time in the presence of realistic YN and NN interactions including the $NNN\Lambda$ and $NNN\Sigma$ channels explicitly. In Fig. 1, we illustrate the one-body densities of a single nucleon and Λ and Σ hyperons for the 0^+ state of $^4_\Lambda$He. We notice that the Λ particle is located much outside the core nucleons and therefore the dynamical change of the core nucleus due to the Λ-particle partition is small. The nucleon r.m.s. radius is $\bar{r}_N = 1.65$ fm which is shrinked by 8 % from the corresponding $\bar{r}_N = 1.79$ fm in ^{3}He. On the other hand, the Σ hyperon comes close to the nucleons and therefore generates large dynamical contraction of the core nucleus in the $NNN\Sigma$-channel space; $\bar{r}_N = 1.49$ fm is obtained with the Σ-channel amplitude only and the reduction rate amounts to 17 %.

REFERENCES

1. Motoba, T., Bandō, H., and Ikeda, K., *Prog. Theor. Phys.* **70**, 189 (1983); Motoba, T., Bandō, H., Ikeda, K., and Yamada, T., *Prog. Theor. Phys. Suppl.* **81**, 42 (1985).
2. Hiyama, E., Kamimura, M., Miyazaki, K., and Motoba, T., *Phys. Rev. C* **59**, 2351 (1999).
3. Tanida, K., *et al.*, *Phys. Rev. Lett.* **86**, 1982 (2001).
4. Hiyama, E., Kamimura, M., Motoba, T., Yamada, T., and Yamamoto, Y., *Prog. Theor. Phys.* **97**, 881 (1997).
5. Hiyama, E., Kamimura, M., Motoba, T., Yamada, T., and Yamamoto, Y., *Phys. Rev. Lett.* **85**, 270 (2000).
6. Hiyama, E., Kamimura, M., Motoba, T., Yamada, T., and Yamamoto, Y., *Phys. Rev. C* **64**, (2001), in press.

Single particle properties of Λ hypernuclei in the density dependent relativistic hadron field theory

Christoph Keil* and Horst Lenske*

Theoretische Physik Uni Gießen, Heinrich-Buff-Ring 16, 35392 Gießen, Germany

Abstract. The density dependent relativistic hadron field theory is used to describe single particle properties of Λ hypernuclei. The discussion focuses on the spin-orbit systematics in the relativistic mean-field formalism by discussing general effects of the Λ continuum threshold and the delocalization of the Λ wave function. Theoretical predictions for the hypernuclear Auger effect are presented.

INTRODUCTION

Due to the experimental difficulties in pursuing hyperonic scattering experiments the modeling of hypernuclei is crucial for investigating interactions in the baryon octet. For this reason the development and application of microscopic models linking in-medium and free interactions is necessary. The extension of the density dependent relativistic hadron field theory (DDRH) [1] into the strangeness sector and to hypernuclei is such a microscopic link. Besides an appropriate microscopic description it is essential to chose suitable systems in which dynamic many-body effects play a minor role in order to get more direct access to generic hyperon interactions. In this respect we present our studies concerned with *weakly bound Λ states* in hypernuclei. It is found that deeply bound Λ states are fairly safe from subtle dynamical effects – though caution needs to be taken there in other respects. Such deeply bound Λ states can ideally be found in heavy hypernuclei. For the investigation of heavy hypernuclei there is, besides the standard spectroscopic methods, also the possibility of the spectroscopy of *Auger neutrons* [2] as planned in a recently proposed experiment at JLab [3]. We present Auger transition rates for $^{209}_{\Lambda}$Pb calculated within our self-consistent model and discuss the spectroscopic content of the spectra.

THE DDRH MODEL

The DDRH model [1] is a relativistic Lagrangian field theory with baryonic and mesonic degrees of freedom. In-medium interactions are derived from a Dirac-Brueckner (DB) calculation using the free space Bonn A interaction. The DB self-energies are mapped onto DDRH self-energies leading to vertex functionals that depend on Lorentz scalars of the baryonic field operators. The Lagrange density of DDRH is given by $\mathcal{L} = \mathcal{L}_\text{B} + \mathcal{L}_\text{M} + \mathcal{L}_\text{int}$. $\mathcal{L}_\text{B}$ and $\mathcal{L}_\text{M}$ denote the free baryonic and mesonic Lagrange densities, whereas the interaction Lagrangian is given by:

$$\mathcal{L}_{int} = \overline{\Psi}_F \hat{\Gamma}_\sigma(\overline{\Psi}_F, \Psi_F) \Psi_F \sigma - \overline{\Psi}_F \hat{\Gamma}_\omega(\overline{\Psi}_F, \Psi_F) \gamma_\mu \Psi_F \omega^\mu \quad (1)$$
$$- \frac{1}{2} \overline{\Psi}_F \hat{\vec{\Gamma}}_\rho(\overline{\Psi}_F, \Psi_F) \gamma_\mu \Psi_F \vec{\rho}^\mu - e \overline{\Psi}_F \hat{Q} \gamma_\mu \Psi_F A^\mu$$

Due to the functional structure of the vertices $\Gamma(\overline{\Psi}_F, \Psi_F)$ the baryonic equations of motion contain additional *rearrangement self-energies* which arise due to the variation of $\mathcal{L}_{int}$ with respect to Ψ. These rearrangement self-energies account for static polarization effects in the baryonic medium, assuring the thermodynamical consistency of the theory and Lorentz covariance of the field equations [1].

SINGLE PARTICLE PROPERTIES OF Λ HYPERNUCLEI

Weakly bound Λ states

In this section we point out the subtleties of extracting information on generic hyperon interactions from weakly bound Λ systems on two examples: (1) the influence of the close continuum threshold on the structure of Λ spin-orbit (s.o.) doublets and (2) the influence of the delocalization of the Λ wave function resulting from the weak binding of the Λ. These effects are *independent of the specific interaction* and, in general, contribute as an apparent quenching of spin-orbit strengths.

Continuum threshold level compression

From weakly bound systems, e.g. neutron-rich dripline nuclei [4], it is known that a s.o. doublet which approaches its continuum threshold gets compressed before actually becoming unbound. This is due to the so called avoided level crossing, reflecting the fact that a state which becomes unbound has to cross other states lying lower in the continuum before it reappears as a resonance. In a mean-field description this situation is in particular found in the light p-shell hypernuclei. The effect is investigated in a model study by varying the $\omega\Lambda\Lambda$ coupling in $^{16}_\Lambda$O artificially such that the centroid energy $E_{centroid}$ of the 1p s.o. doublet approaches the continuum threshold (for notation see Fig. 1). The resulting s.o. splitting ΔE depends sensitively on the location of $E_{centroid}$ relative to the continuum threshold. Therefore a clean extraction of the generic s.o. interaction from light hypernuclei is hardly to achieve because the threshold compression will always be superimposed on the generic s.o. interaction. However, around the $^{40}_\Lambda$Ca region the Λ 1p doublet has moved far enough into the bound region such that the threshold effect ceases to be important.

Wave function delocalization

Another effect which diminishes the influence of the s.o. interaction on the observed splitting by an additional reduction is the delocalization of the Λ wave function due to the shallow Λ potential (e.g. [5]). This comes about because the splitting is proportional

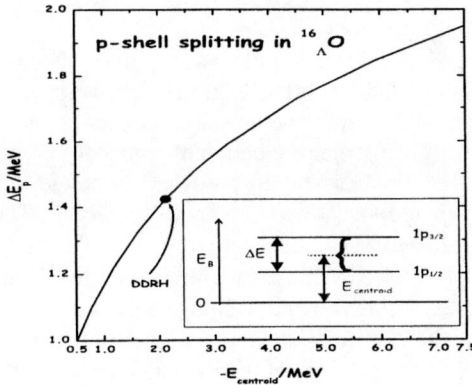

FIGURE 1. The 1p Λ s.o. splitting ΔE is plotted versus the centroid energy $E_{centroid}$. The insert defines the notation used.

to the overlap integral between the single particle density of the considered state and the s.o. potential which is well localized in the nuclear surface. To determine the importance of this delocalization we define the following measure for the fraction of the Λ and neutron wave function $F_{\Lambda,n}$ residing inside the nuclear rms radius:

$$N_{\Lambda,n}(R) = N_o \int_0^R dr\, r^2\, |F_{\Lambda,n}(r)|^2 \qquad (2)$$

where N_o is chosen such that $N_{\Lambda,n}(R) \to 1$ for $R \to \infty$ and $R = \sqrt{\langle r^2 \rangle}$. Results for different nuclei are shown in Table 1. It is obvious that this effect is especially pronounced in light hypernuclei and in the weakly bound orbits in the heavy mass region.

We conclude from these investigations, that hyperon interactions still can be determined provided that both the centroid energy and the relative splitting of a doublet are considered. In this respect the new γ spectroscopic results, reporting an almost vanishing s.o. splitting in e.g. $^{13}_{\Lambda}$C [6], point to an extremely weak effective Λ s.o. strength but for decisive conclusions on the generic interaction more strongly bound heavy hypernuclei should be investigated.

TABLE 1. Localization coefficients $N_{\Lambda,n}(r_{rms})$ as defined in eq. (2).

	$^{40}_{\Lambda}$Ca		$^{51}_{\Lambda}$V		$^{89}_{\Lambda}$Y		$^{208}_{\Lambda}$Pb	
	N_Λ	N_n	N_Λ	N_n	N_Λ	N_n	N_Λ	N_n
$1p_{3/2}$	0.51	0.71	0.59	0.75	0.68	0.81	0.79	0.88
$1p_{1/2}$	0.50	0.71	0.60	0.76	0.69	0.82	0.80	0.86
$1d_{5/2}$	0.23	0.47	0.33	0.54	0.47	0.64	0.66	0.80
$1d_{3/2}$	0.17	0.46	0.30	0.54	0.49	0.67	0.69	0.84
$1f_{7/2}$	–	–	–	–	0.27	0.45	0.52	0.72
$1f_{5/2}$	–	–	–	–	0.26	0.47	0.55	0.77

Hypernuclear Auger effect

Auger rates for the capture of a Λ and the de-excitation of heavy hypernuclei are calculated within the DDRH model [1]. Different to a previous attempt [2] our approach treats the binding mean field, wave functions and interactions in a self-consistent manner. The process is described by the production of a neutron-particle–neutron-hole state due to the de-excitation of the Λ, where the particle is unbound. A schematic illustration of the process for the 1g shell is shown in the insert of Fig. 2. The details of the calculation will be given elsewhere. Fig. 2 shows the partial life times for the Λ transitions starting from the 1g shell plotted against the kinetic energy of the emitted neutron. The manifestation of the s.o. splitting is marked. Due to the complexity of the full spectrum it will be important to tag the Λ initial state in the measurements of Auger spectra. Further analysis of the global spectral features of the hypernuclear Auger effect is in progress.

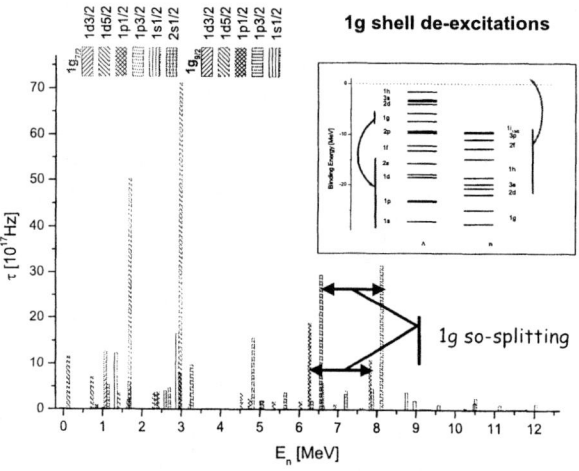

FIGURE 2. The partial life times of the 1g Λ shell in $^{209}_{\Lambda}$Pb due to Auger neutron emission are plotted versus the kinetic energy of the outgoing neutron. The insert shows the Λ and neutron level schemes for this transition where the involved states are marked.

ACKNOWLEDGMENTS

This work has been partially supported by DFG grant Le439/4-3 and the European Graduate School "Complex Systems of Hadrons and Nuclei, Gießen–Copenhagen".

REFERENCES

1. Keil, C.M., Hofmann F., and Lenske, H., *Phys. Rev. C* **61**, 064309 (2000); nucl-th/9911014.
2. Likar, A., Rosina, M., and Povh, B., *Z. Phys. A* **324**, 35 (1986).
3. Margaryan, A., *et al.*, JLab letter of intent, 2000.
4. Lalazissis, G.A., Vretenar, D., Poschl, W., and Ring, P., *Nucl. Phys. A* **632**, 363 (1998).
5. Rijken, T.A., Stoks, V.G., and Yamamoto, Y., *Phys. Rev. C* **59**, 21 (1999); nucl-th/9807082.
6. Ajimura, S., *et al.*, *Phys. Rev. Lett.* **86**, 4255 (2001).

Near threshold K^+K^- meson-pair production in proton-proton collisions

A. Khoukaz[*], C. Quentmeier[*], H.-H. Adam[*], J. T. Balewski[1,†],
A. Budzanowski[†], D. Grzonka[**], L. Jarczyk[‡], K. Kilian[**], P. Kowina[§],
N. Lang[*], T. Lister[*], P. Moskal[‡], W. Oelert[**], R. Santo[*], G. Schepers[**],
T. Sefzick[**], S. Sewerin[**], M. Siemaszko[§], J. Smyrski[‡], A. Strzałkowski[‡],
M. Wolke[**], P. Wüstner[**] and W. Zipper[§]

[*] *Institut für Kernphysik, Westfälische Wilhelms-Universität, D-48149 Münster, Germany*
[†] *Institute of Nuclear Physics, PL-31-342 Cracow, Poland*
[**] *IKP and ZEL, Forschungszentrum Jülich, D-52425 Jülich, Germany*
[‡] *Institute of Physics, Jagellonian University, PL-30-059 Cracow, Poland*
[§] *Institute of Physics, University of Silesia, PL-40-007 Katowice, Poland*

Abstract. The near threshold total cross section and angular distributions of K^+K^- pair production via the reaction $pp \longrightarrow ppK^+K^-$ have been studied at an excess energy of $Q = 17$ MeV using the COSY-11 facility at the cooler synchrotron COSY. The obtained cross section as well as an upper limit at an excess energy of $Q = 3$ MeV represent the first measurements on the K^+K^- production in the region of small excess energies where production via the channel $pp \longrightarrow pp\Phi \longrightarrow ppK^+K^-$ is energetically forbidden.

INTRODUCTION

Studies on the kaon-pair production are stimulated by the continuing discussion on the nature of the scalar resonances $f_0(980)$ and $a_0(980)$, which have been interpreted as conventional $q\bar{q}$ states [1], qq-$\bar{q}\bar{q}$ states [2] or as $K\bar{K}$ molecules [3, 4]. Furthermore, exclusive K^- production data are also of special interest in the context of sub-threshold kaon production experiments in nucleus-nucleus interactions, which are expected to probe the antikaon properties at high baryon density.

At the cooler synchrotron COSY [5] near threshold measurements on the reaction $pp \to ppK^+K^-$ have been performed at the internal beam facility COSY-11 [6], using a hydrogen cluster target [7] in front of a C-shaped COSY-dipole magnet, acting as a magnetic spectrometer. Tracks of positively charged particles, detected in a set of two drift chambers, are traced back through the magnetic field to the interaction point, leading to a momentum determination. The velocities of these particles are accessible by a time-of-flight path behind the drift chambers, consisting of scintillation hodoscopes in a distance of ~ 9.3 m. By measuring the momentum and the velocity, a particle identification of the positively charged ejectiles is possible and the four momentum

---

[1] Present address: Indiana University Cyclotron Facility, Bloomington, Indiana 47408, USA

vectors can be completely reconstructed. This yields a full event reconstruction for the reaction type $pp \to ppK^+X$ ($X = K^-$) by detecting both outgoing protons as well as the K^+ meson and identifying the X-particle using the missing mass method. Additionally, a silicon pad detector allows to detect the hit position of outgoing K^- mesons.

RESULTS

The reaction channel $pp \longrightarrow ppK^+K^-$ has been studied using the COSY-11 spectrometer at incident proton beam momenta of p_{beam} = 3.311 GeV/c and 3.356 GeV/c, corresponding to excess energies of Q = 3 MeV and 17 MeV [8]. At the higher excess energy an unambiguous detection of events from the ppK^+K^- reaction is possible, leading to a total number of $N = 61^{+0}_{-5}$ accumulated K^+K^- events. Due to the high precision of the experimental facility a K^- missing mass resolution of FWHM $\sim$ 2 MeV/c^2 was achieved.

The accessibility of the four-momentum vectors of all ejectiles from ppK^+K^- events allows to study angular distributions of the particles or systems of particles. Monte-Carlo simulations on the free $pp \to ppK^+K^-$ reaction, considering the pp FSI and the Coulomb interaction, have been performed to determine the acceptance of the detection

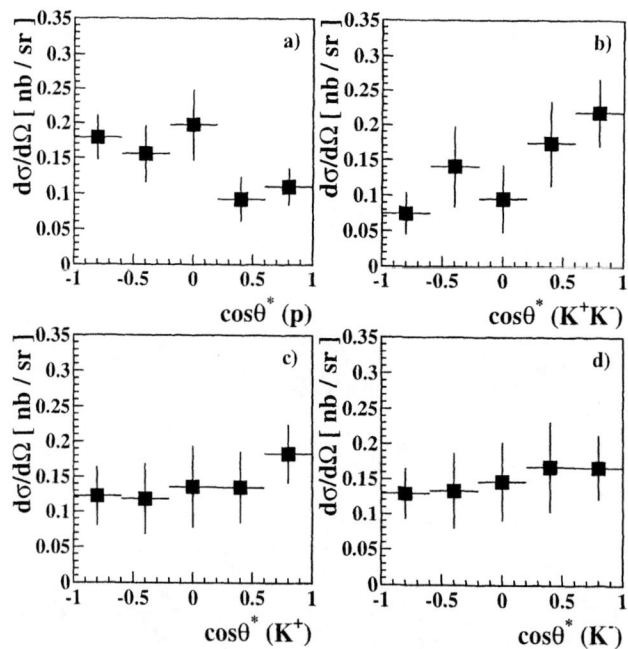

FIGURE 1. Angular distributions in the overall CMS relative to the beam direction of the extracted ppK^+K^- events for both outgoing protons (a), the K^+K^- system (b), the K^+ mesons (c) and the K^- mesons (d).

system in order to obtain acceptance corrected kinematical distributions. The overall detection efficiencies for events from the non-resonant K^+K^- production, requiring the detection of both protons and the K^+ meson, were determined to be $\varepsilon(3 \text{ MeV}) = (6.4 \cdot 10^{-2})^{+43\%}_{-28\%}$ and $\varepsilon(17 \text{ MeV}) = (7.4 \cdot 10^{-3})^{+10\%}_{-13\%}$. These quantities take into account the kaon decay, detection and track reconstruction efficiencies as well as the influence of the error in the absolute excess energies, which are known with a precision of $\Delta Q = 1$ MeV, caused by the uncertainty in the determination of the absolute COSY beam momentum ($\Delta p/p = 10^{-3}$). In Fig. 1 the angular distributions in the center of mass system relative to the beam direction for the extracted ppK^+K^- events are shown for both outgoing protons (a), the K^+K^- system (b), the K^+ mesons (c) and the K^- mesons (d). Within the statistical errors the measured distributions of the protons and the kaons show no significant deviation from an isotropic emission.

The luminosity was determined by comparing the differential counting rates of elastically scattered protons with data recorded by the EDDA collaboration [9]. The integrated luminosities were extracted to be $\int L dt = 0.84$ pb^{-1} ± 1% (stat.) ± 5% (syst.) at $Q = 3$ MeV and $\int L dt = 4.50$ pb^{-1} ± 1% (stat.) ± 5% (syst.) at $Q = 17$ MeV, corresponding to a mean luminosity of $L = 2 \cdot 10^{30}$ cm^{-2} s^{-1}.

In Fig. 2 the present result at $Q = 17$ MeV (filled symbol) and a data point from the DISTO collaboration [10], excluding the contribution from the Φ, are plotted as function of the excess energy. These data represent the available world data for the K^+K^- production via the reaction channel $pp \to ppK^+K^-$ in the near threshold region (threshold: $p_{\text{beam}} = 3.30175$ GeV/c).

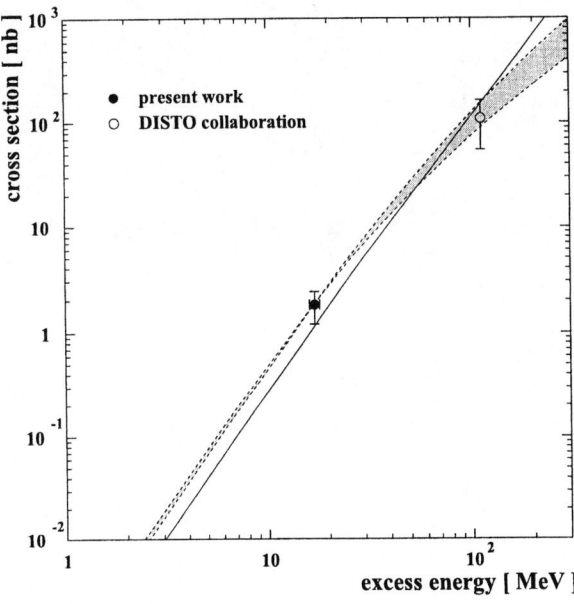

FIGURE 2. Total cross sections for the free K^+K^- pair production in proton-proton collisions. The curves are described in the text.

The total cross section at an excess energy of $Q = 17 \pm 1$ MeV was determined to be $\sigma = (1.80 \pm 0.27^{+0.28}_{-0.35})$ nb, including statistical and systematical errors [8]. The upper limit at $Q = 3$ MeV was determined to be $\sigma < 0.16$ nb on the basis of a confidence level of 95% [8].

Figure 2 shows parametrizations of the K^+K^- cross sections assuming different production processes. The solid line, representing a fit to the data points on the basis of a four-body S-wave phase space expectation including the proton-proton final state interaction (FSI) [8], describes the data points adequately within the error bars. Therefore, the total cross section data points are consistent with a description based on the free ppK^+K^- production with no distinct effects of higher partial waves or strong K^+K^- final state interactions. Although not suggested by our previously discussed results, one can calculate the cross section for the $pp \to ppf_0(980) \to ppK^+K^-$ channel, leading to a value of $\sigma(pp \to ppf_0 \to ppK^+K^-) = 1.84 \pm 0.29^{+0.25}_{-0.33}$ nb. The dashed lines present three-body phase space calculations for the ppK^+K^- final state via the excitation of the broad f_0 resonance including the pp FSI and normalized to this $\sigma(pp \to ppf_0 \to ppK^+K^-)$. Here we assumed the $f_0(980)$ to be a Breit-Wigner distribution with a mass of $m = 980$ MeV/c^2. The effect of the large uncertainty about the width of the $f_0(980)$ resonance ($\Gamma = 40$ MeV/c^2 to 100 MeV/c^2 [11]) is indicated by the dashed area. Nevertheless, within the error bars also this description is consistent with the measured data. Consequently, the two data points are in agreement with both the assumption of a non-resonant as well as a resonant production via the f_0, always neglecting effects of higher partial waves.

REFERENCES

1. Morgan, D., and Pennington, M. R., *Phys. Review D* **48**, 1185 (1993).
2. Jaffe, R., *Phys. Review D* **15**, 267 (1977).
3. Weinstein, J., and Isgur, N., *Phys. Review D* **41**, 2236 (1990).
4. Lohse, D., et al., *Nucl. Physics A* **516**, 513 (1990).
5. Bechstedt, U., et al., *Nucl. Instr. & Methods B* **113**, 26 (1996); Maier, R., *Nucl. Instr. & Methods A* **390**, 1 (1997).
6. Brauksiepe, S., et al., *Nucl. Instr. & Methods A* **376**, 397 (1996).
7. Dombrowski, H., et al., *Nucl. Instr. & Methods A* **386**, 228 (1997).
8. Quentmeier, C., et al., accepted for publication in *Phys. Letters B*; doctoral thesis, Westfälische Wilhelms-Universität Münster, Germany, 2001.
9. Albers, D., et al., *Phys. Review Letters* **78**, 1652 (1997).
10. Balestra, F., et al., *Phys. Letters B* **468**, 7 (1999).
11. Review of Particle Physics, Particle Data Group, *Eur. Phys. J. C* **15**, 1-878 (2000).

Composition and properties of multi-strange hypernuclear systems

D. E. Lanskoy

Institute of Nuclear Physics, Moscow State University, 119899 Moscow, Russia

Abstract. Conditions of stability of multi-strange nuclear matter consisting of nucleons, Λ, and Ξ hyperons are analysed in the nonrelativistic effective potential approach and the relativistic mean field theory. Nucleon and Ξ drip lines confining the bound state region are deduced. Possibility of simultaneous conversion of all Ξ hyperons in finite and infinite systems is discussed.

STABILITY AND COMPOSITION OF INFINITE HYPERNUCLEAR MATTER

For a long time, it was believed that multi-strange hypernuclei stable with respect to the strong interaction consist of nucleons and Λ hyperons. It was shown later [1-4] that these hypernuclei contain generally also Ξ hyperons since conversion $\Xi N \longrightarrow \Lambda\Lambda$ becomes Pauli blocked when the number of Λ's is sufficiently high. We study conditions of stability and properties of multi-strange infinite matter at the whole range of strangeness s and electric charge q per baryon.

Stability of the matter with respect to the strong interaction imposes the following conditions [5]: the saturation condition; negativity of the chemical potentials of all the baryonic species (preventing single-particle emission); and the chemical equilibrium conditions with respect to the $\Xi N \Longleftrightarrow \Lambda\Lambda$ conversion:

$$\mu_p + \mu_{\Xi^-} - 2\mu_\Lambda + m_p + m_{\Xi^-} - 2m_\Lambda = 0, \qquad (1)$$

$$\mu_n + \mu_{\Xi^0} - 2\mu_\Lambda + m_n + m_{\Xi^0} - 2m_\Lambda = 0, \qquad (2)$$

where μ_i is the chemical potential of baryon i.

We adopt two dynamical models: the nonrelativistic effective potential approach with Skyrme potentials and the relativistic mean field theory. Choice of potentials in the nonrelativistic approach is described elsewhere [5]. In the relativistic approach, we use parameter set TM1 [6] extended for multi-strange systems in Ref. [7] and already employed in hypernuclear matter calculations [8].

In Fig. 1, we present bound state regions of the hypernuclear matter in the $s-q$ plane. Since exact isospin symmetry is implied, only a half of a bound state region lying below the zero isospin line $s = (1-q)/2$ is depicted. The bound state region is confined by n and Ξ^- drip lines corresponding to zero μ_n and μ_{Ξ^-} (and also by not shown p and Ξ^0 drip lines). Besides a small $pn\Lambda$ domain and the main domain of $pn\Lambda\Xi^0\Xi^-$ systems, the bound state region includes also a domain of neutron-rich $pn\Lambda\Xi^-$ matter (note that Ξ^- hyperons appear in neutron-rich matter at smaller s than in isosymmetric matter). In

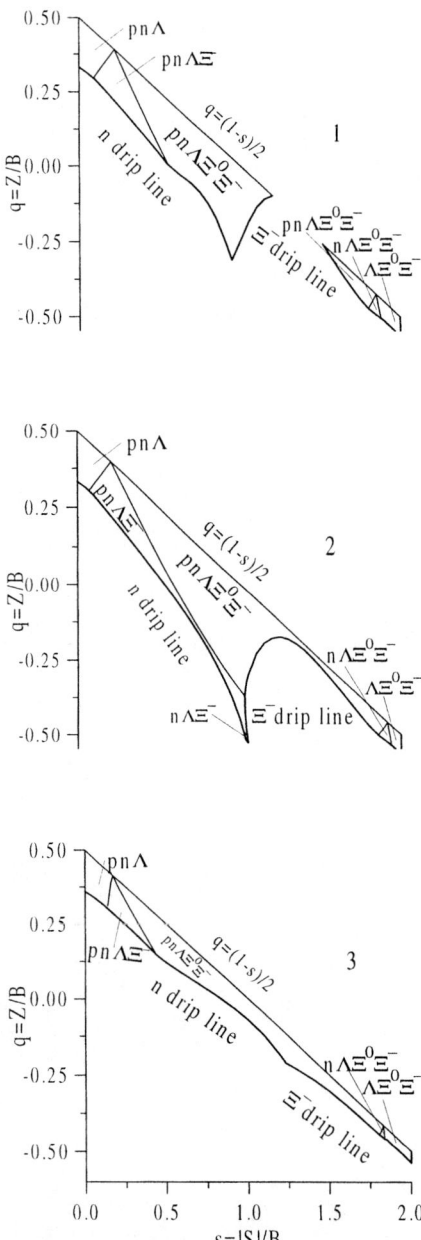

FIGURE 1. Bound state region in the nonrelativistic (upper and middle parts) and the relativistic (lower part) approaches. Upper (middle) part is calculated with ΞN potential SΞN1 (SΞN2) from Ref. [9]. s and q are strangeness and charge per baryon.

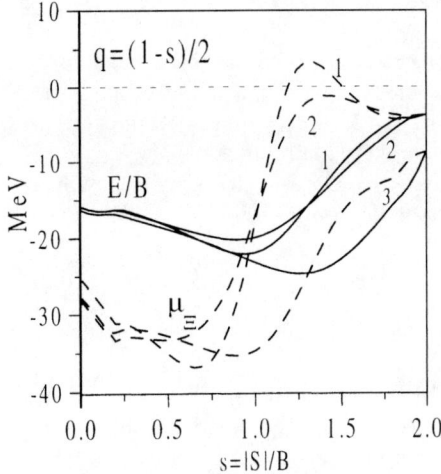
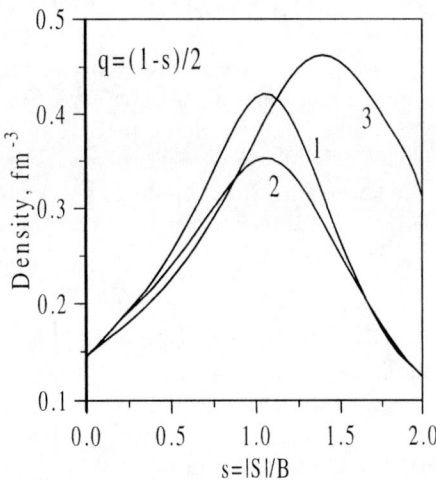

FIGURE 2. Energy per baryon (solid curves) and Ξ^- chemical potential (dashed curves) in isosymmetric hypernuclear matter. Numbering of the curves corresponds to numbering of the parts of Fig. 1.

FIGURE 3. Saturation density of isosymmetric hypernuclear matter. Numbering of the curves corresponds to numbering of the parts of Fig. 1.

all the cases, pure hyperonic ($\Lambda\Xi^0\Xi^-$) as well as $n\Lambda\Xi^0\Xi^-$ systems are also predicted at large s (however, the results at s close to 2 are rather ambiguous due to uncertainties in hyperon-hyperon interactions).

The only difference between the upper and the middle parts of Fig. 1 is ΞN potential used. Potential SΞN1 is not weaker than SΞN2 one [9] as seen in Fig. 2, where energy per baryon for the symmetric matter is displayed. However, it gives higher saturation densities (Fig. 3), and, therefore, higher baryonic Fermi momenta. As a result, the region of positivity of μ_{Ξ^-} is wider and reaches also the isospin symmetry line (see dashed curves in Fig. 2).

The bound state region in the relativistic mean-field calculation incorporates only nearly symmetric matter, since the TM1 set includes strong $NN\rho$ and $\Xi\Xi\rho$ couplings. The region becomes much wider if one decreases $\Xi\Xi\rho$ coupling or, alternatively, inserts $\Xi\Xi\delta$ coupling (where δ is the scalar isovector meson). At $s > 1$, the matter is more dense and bound stronger in the relativistic approach.

The calculations are made with relatively strong $\Lambda\Lambda$ interactions as suggested from known data on $\Lambda\Lambda$ hypernuclei. New KEK experiment [10] casts doubt on these data. We repeat the nonrelativistic calculation reducing the $\Lambda\Lambda$ potential by a factor of five (and keeping other hyperonic potentials the same). The drip lines confining the bound state region are almost unchanged, though the boundaries between the domains of the different compositions alter. It is not so surprizing, since the n and Ξ^- drip lines are not affected by the $\Lambda\Lambda$ interaction directly.

MANY-BODY CONVERSION OF Ξ HYPERONS

Here, we point out a new channel of the strong decay, which can be crucial for the stability of multi-strange hypernuclei in some cases. For a specific example, let us consider the multi-strange α particle analog $(2p2n2\Lambda 2\Xi^0 2\Xi^-)$ suggested in [1]. It was shown [1] that the $\Xi N \longrightarrow \Lambda\Lambda$ conversion is most likely Pauli blocked.

However, if the total binding energy of the hypernucleus $B < 2(m_{\Xi^-} + m_p - 2m_\Lambda) + 2(m_{\Xi^0} + m_n - 2m_\Lambda) \approx 103$ MeV then the simultaneous conversion

$$(2p2n2\Lambda 2\Xi^0 2\Xi^-) \longrightarrow 10\Lambda \qquad (3)$$

is allowed. If $80 < B < 103$ MeV, then the channel (3) with ten final particles in continuum is the unique channel of the strong decay of this hypernucleus.

Many-body (simultaneous) conversion can be important in some light hypernuclei with a large spacing between single-particle levels. On the other hand, it can be proved that in uniform infinite saturated matter stable with respect to single-baryon emission and two-body conversion $\Xi N \Longleftrightarrow \Lambda\Lambda$, the many-body conversion is strictly forbidden. For instance, let us consider the case $z_p > z_{\Xi^-}$, $z_n > z_{\Xi^0}$, where z_i is the fraction of baryon i. Then the energy per baryon of the saturated and chemically equilibrated matter can be represented as follows:

$$\begin{aligned}E/B &= \mu_p(z_p - z_{\Xi^-}) + \mu_n(z_n - z_{\Xi^0}) + \mu_\Lambda(z_\Lambda + 2z_{\Xi^0} + 2z_{\Xi^-}) \\ &\quad - z_{\Xi^-}(m_p + m_{\Xi^-} - 2m_\Lambda) - z_{\Xi^0}(m_n + m_{\Xi^0} - 2m_\Lambda).\end{aligned} \qquad (4)$$

The second row is the possible energy release per baryon from the conversion of all the Ξ hyperons with the opposite sign. Each term in the first row is also negative since $\mu_i < 0$ for all i. Thus, the binding energy per baryon is greater than the possible energy release, therefore the simultaneous conversion is impossible.

REFERENCES

1. Schaffner, J., Greiner, C., and Stöcker, H., *Phys. Rev. C* **46**, 322-329 (1992).
2. Schaffner, J. et al., *Phys. Rev. Lett.* **71**, 1328-1331 (1993).
3. Dover, C.B., and Gal, A., *Nucl. Phys.* **A560**, 559-585 (1993).
4. Schaffner, J. et al., *Ann. Phys.* **235**, 35-76 (1994).
5. Lanskoy, D.E., "Structure of multi-strange nuclear matter," in *Strangeness Nuclear Physics*, edited by Il-T. Cheon et al., World Scientific, Singapore, 2000, pp. 347-354.
6. Sugahara, Y., and Toki, H., *Nucl. Phys.* **A579**, 557-572 (1994).
7. Schaffner, J., and Mishustin, I.N., *Phys. Rev. C* **53**, 1416-1429 (1996).
8. Schaffner, J., and Gal, A., *Phys. Rev. C* **62**, 034311 (2000).
9. Lanskoy, D.E., *Few-Body Syst. Suppl.* **9**, 277-280 (1995).
10. Nakazawa, K., in these proceedings.

KN phase shifts in a model with a spin-orbit interaction

S. Lemaire[*], J. Labarsouque[*] and B. Silvestre-Brac[†]

[*]*Centre d'Etudes Nucléaires de Bordeaux Gradignan, IN2P3, CNRS, Université Bordeaux I, le Haut Vigneau, 33175 Gradignan Cedex, France*
[†]*Institut des Sciences Nucléaires, IN2P3, CNRS, Université Joseph Fourier, 53 Av. des Martyrs, 38026 Grenoble Cedex, France*

Abstract. The I=1 and I=0 kaon-nucleon s, p, d, f, g-waves phase shifts have been calculated in a non relativistic quark potential model using the resonating group method (RGM). The interquark potential includes gluon exchanges with a spin-orbit interaction. This force was determined to reproduce as well as possible the meson and baryon spectra. The same force is employed for the cluster and intercluster dynamics and the relative *KN* wave function is calculated without any approximation. While some channels are correctly described, the theory is still unable to explain others.

INTRODUCTION

In this work we investigate *KN* elastic scattering, an important tool to understand high density regions of nuclei [1, 2]. Our calculation essentially rely on three requirements which aim at reducing as much as possible the approximations. First, the same quark-quark interaction is used to build the K and N wave functions and to generate the dynamics of the *KN* interaction in order to ensure a better consistency between the dynamics inside the clusters and the dynamics between the clusters; second, the quark-quark interaction must reproduce as well as possible the meson and baryon spectra; third, the relative *KN* wave function is let completely free to take the form that dynamics wants without any presupposed parametrization.
The goal of the present work aims at calculating s, p, d, f, g-waves *KN* elastic phase shifts using the resonating group method approach (RGM) in a non relativistic constituent quark model. Following the previous result [3], we suppose that the effect of the relativistic kinetic energy for quarks remains negligible for higher waves. We start with the input ingredients for building the potential without detailing the clusters wave-functions since it has already been done in [4, 5]. Next we show the fundamental equation governing the scattering. Then, results are discussed and last, conclusions are drawn.

INTERQUARK POTENTIAL

We take the AL1 potential due to B. Silvestre-Brac and C. Semay [6] and we add to this AL1 potential a phenomenological spin-orbit term. The adopted potential is of the form

$$V_{ij}(r) = -\frac{3}{16}\vec{\lambda}_i.\vec{\lambda}_j \left[V^{(c)}(r) + \vec{\sigma}_i.\vec{\sigma}_j V_{ij}^{(\sigma)}(r) + \vec{l}_{ij}.\vec{s}_{ij} V_{ij}^{(ls)}(r)\right]. \quad (1)$$

In this expression, $\vec{l}_{ij}$ and $\vec{s}_{ij}$ are the relative angular momentum and spin of the pair of quarks i and j. The form we adopt for the spin-orbit term is $V_{ij}^{ls}(r) = V_{1_{ij}} e^{-r^2/r_{1_{ij}}^2}$. The form of the central and hyperfine part of this interquark potential is given in [4, 5].
The parameters entering the AL1 potential were determined to give a good description of meson and baryon spectra. The parameters of the LS term were adjusted to reproduce at best the LS multiplets.

SCATTERING EQUATION

The trial wave-function we take for the KN system reads as (in the center of mass frame of the system):

$$\Psi_{NK}(1,2,3,4,5,\vec{x},\vec{y},\vec{z},\vec{R}) = \int dr h(r) \mathcal{A} \phi(1,2,3,4,5,\vec{x},\vec{y},\vec{z},\vec{R},r), \quad (2)$$

$$\phi(1,2,3,4,5,\vec{x},\vec{y},\vec{z},\vec{R},r) = \left\{[\Psi_N(1,2,3,\vec{x},\vec{y})\Psi_K(4,5,\vec{z})]^{C,S=\frac{1}{2},I=0,1} \frac{\delta(R-r)}{r} Y_L(\hat{R})\right\}^{J,M}. \quad (3)$$

The cluster wave functions are $\Psi_N(1,2,3,\vec{x},\vec{y})$ and $\Psi_K(4,5,\vec{z})$; the coordinate $\vec{R}$ corresponds to the relative position between K and N and a variational principle is applied to the relative wave-function $h(r)$.
Then s, p, d, f, g phase shifts can be calculated from the solution of a Hill-Wheeler equation (see [3]):

$$\frac{d^2 h(r)}{dr^2} - \frac{L(L+1)}{r^2} h(r) + k^2 h(r) + \int_0^\infty dr' [\chi_{NL}(k^2, r, r')] \frac{h(r')}{r'} = 0, \quad (4)$$

where $|\vec{k}|$ is the relative impulse of both colliding particles and χ_{NL} is the non local kernel.

Interquark interaction and masses of kaon and nucleon

All the parameters entering in the interquark AL1 potential were determined by B. Silvestre-Brac and C. Semay [6] to reproduce as well as possible the baryon and meson spectra. The spin-orbit term in the gluon interaction enables us to raise the degeneracy of states for a given total angular momentum J of the hadron spectra.
We develop the space-functions of kaon and nucleon up to two gaussians. The masses we obtain for one gaussian are $M_K = 0.532$ GeV, $M_N = 1.055$ GeV and for two gaussians $M_K = 0.493$ GeV, $M_N = 1.038$ GeV.

S, P, D, F, G NON RELATIVISTIC K-N PHASE SHIFTS

The s, p, d, f, g-waves KN phase-shifts are calculated and compared with experimental data [7]. We show in Figs. 1, 2 only the p and d waves. We can note that neither our calculations in this constituent quark model (Fig.1) nor calculations done in all the previous works are able to reproduce correctly the P_{13} wave. Although the P_{11} channel is correctly reproduced, the absolute value of the calculated phase shifts of channels with isospin $I = 0$ remain much too small. For higher waves, the absolute value of the phase shifts are underestimated and their values remain too small as compared to experiment, except for the D waves with isospin $I = 1$. The D_{13} wave is particularly well reproduced. However we are aware that a good description of high angular momentum phase shifts cannot be obtained without taking into account meson exchange mechanisms. This study is a first approach to s, p, d, f, g-waves phase shifts of KN scattering. It points out that the spin-orbit part of the one gluon exchange interquark potential is necessary but not sufficient enough to explain the $L \neq 0$ KN scattering process.

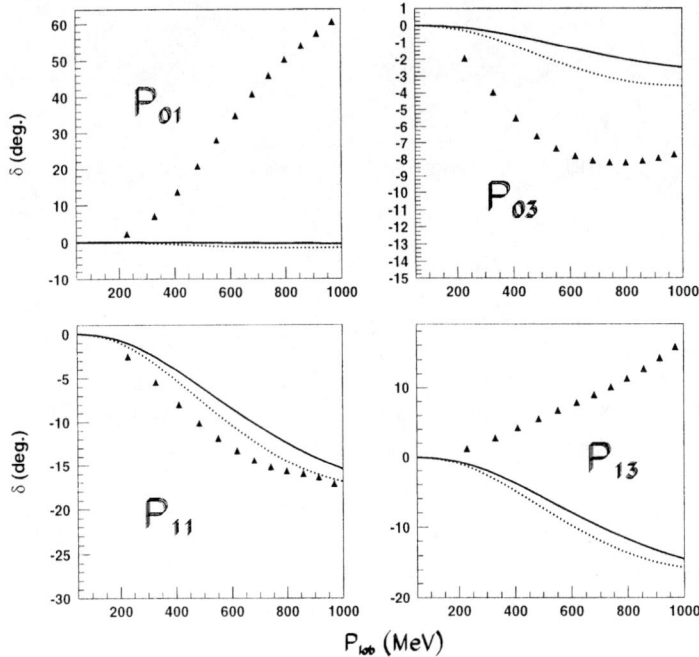

FIGURE 1. KN p-wave phase shifts as a function of laboratory momentum for 1 gaussian (solid line) and 2 gaussians (dotted line). Experimental data are taken from [7]. The first subscript refers to the isospin quantum number and the second one to twice the total spin of the channel.

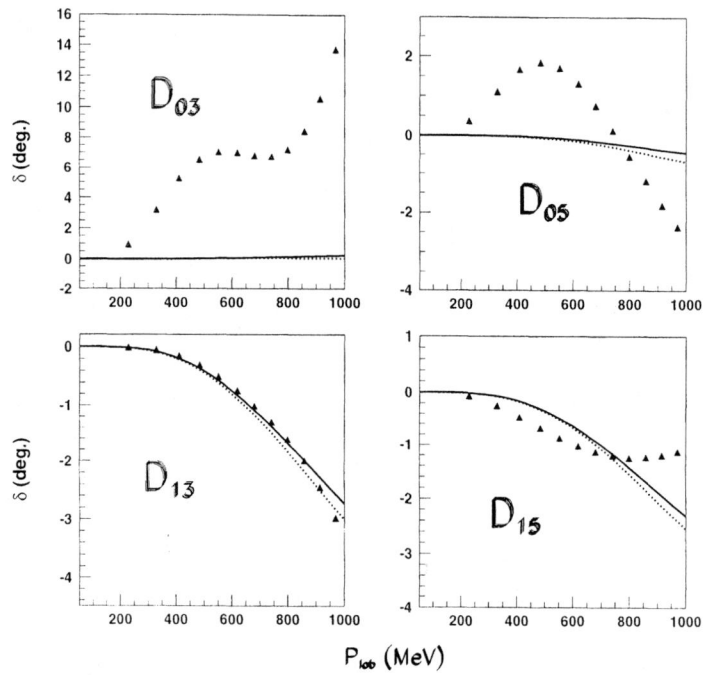

FIGURE 2. KN d-wave phase shifts. Same notation as in Fig. 1.

CONCLUSION

We have studied s, p, d, f, g-wave phase shifts for elastic KN scattering in a model based on RGM calculations with quark degrees of freedom. We used, for the force between quarks, a potential including a spin-orbit interaction which gives quite good description of meson and baryon spectra. The P_{11}, D_{13}, D_{15} waves are well reproduced in this quark constituent model whereas the others are rather poorly reproduced. The difference between experimental data and calculations for s, p, d, f, g wave phase shifts tells us that probably some physical ingredients are still missing in our approach. In fact we know that the influence of meson exchange between quarks is far from negligible for L-waves phase shifts.

REFERENCES

1. Coker, W. R., Lumpe, J. D., and Ray, L., *Phys. Rev. C* **31**, 1412 (1985).
2. Abgrall, Y., Belaidi, R., and Labarsouque, J., *Nucl. Phys.* **A462**, 781 (1987).
3. Lemaire, S., Labarsouque, J., and Silvestre-Brac, B., to be published in *Nucl. Phys. A*.
4. Silvestre-Brac, B., Leandri, J., and Labarsouque, J., *Nucl. Phys.* **A589**, 585 (1995).
5. Silvestre-Brac, B., Labarsouque, J., and Leandri, J., *Nucl. Phys.* **A613**, 342 (1997).
6. Silvestre-Brac, B., and Semay, C., Int. Report, ISN93-69, 1993.
7. Hyslop, J. S., Arndt, R. A., Roper, L. D., and Workman, R. L., *Phys. Rev. D* **46**, 961 (1992).

Progress of the Nuclotron accelerator and the hypernuclear program

A.I.Golokhvastov *, S.A.Khorozov*, J. Lukstins*, A. Parfenov* and N.E.Vasyukhin *

*Joint Institute for Nuclear Research, RU-141980 Dubna, Moscow Region, Russia

Abstract. A revised and extended hypernuclear research program for the Nuclotron at the Laboratory of High Energies (JINR, Dubna) is presented. Experimental problems and parameters of the high resolution part of the spectrometer are discussed considering experiments with two alpha particles from ^{8}Be decays.

INTRODUCTION

In the last few years, the Nuclotron [1], a new superconducting accelerator, has been developed step by step to reach the designed specifications. The ultimate aim of this machine is 12 GeV energy for protons but the accelerator will be generally used as a source of medium energy (6 GeV per nucleon) ion beams. The top aim is to accelerate and extract the uranium beam. However, the main stream of the approved research program [2] for the next 3 – 5 years includes experiments with light and medium ion beams as well as polarized deuteron beams. Tritium and neutron beams are expected to be also produced. There was a large delay between the first accelerated beam in 1994 and the first successful test of the beam extraction system and the first experiment in the extracted beam in March 2000, that was a crucial moment in the Nuclotron progress. For example, the beam extraction efficiency was increased up to 60-70% in this run. At the end of 2000, a helium refrigerator was installed to overcome the run duration limit of ten days due to low power of the liquid nitrogen factory. Now a Nuclotron run can be prolonged for any time necessary for experiments. So, the Nuclotron can be nominated now as an actually working machine.

It should be noted that for the hypernuclear research program the main improvement of the accelerator properties is a significant gain of the run time efficiency. While the Synchrophasotron short beam spill of 0.5 s was repeated every 8-10 s, at the Nuclotron the beam spill can be larger than 10 s with quite a short interval (less than 3 s), required to accelerate particles. In other words, the efficiency of any accelerator run will be increased by a factor of more than 16 for hypernuclear experiments in comparison with the previous experiments at the Synchrophasotron. Another gain of expected statistics in hypernuclear experiments will be obtained by using proportional chambers (or maybe, by quicker detectors) instead of the streamer chamber. Estimates have shown that the data collection rate can be increased by a factor of 200-300 in comparison with our previous experiments.

Though the Nuclotron beam energy and run time were limited up to now, we have got some experience from run by run progress. Indeed, the energy of the extracted beam is 1.5-3.0 GeV in July 2001 (while hypernuclear experiments can be carried out at 4-6 GeV), but it will be increased up to 6 GeV in 2002. Thus, it is high time to revise the hypernuclear research program [3, 4] for the new accelerator. In particular, we discuss possible hypernuclear experiments to reach two main goals: mesonic decays of lightest hypernuclei and (generally nonmesonic) decays of medium hypernuclei (details presented and discussed at this Conference by L. Majling). In the case of the lightest hypernuclei, we propose to measure the binding energy of the lightest hypernucleus, hypertriton, with an accuracy significantly better than that measured up to now and to measure the lifetimes of the lightest hypernuclei with an error of 2-3 %.

The measurement of the lifetime and production cross section follows our first experiments [5, 6] with statistics increased by a factor of 100. The main advantage of the approach is an opportunity to measure the lifetimes practically without systematic errors because the lifetimes are calculated by using the measured distribution of the decay points. While decay vertices can be located with a high precision, the only source of small systematic errors can be due to the uncertainty of the momentum spectra of the hypernuclear samples.

In our previous experiments the first reliable measurement of hypernuclei production in the interaction of relativistic ions was carried out, and the production cross sections [6] were obtained in a qualitative agreement with the predictions of the coalescence model [7, 8]. However, we have no data considering the production energy dependence. Therefore, experiments with $^4_\Lambda$H at different Nuclotron energies will be performed to check the predictions of the theory [9].

BEAM OF $^3_\Lambda$H HYPERNUCLEI

A quite new approach is an idea to use an intensive beam (hundreds of hypernuclei per day) of loosely bound hypernuclei $^3_\Lambda$H to investigate their Coulomb dissociation in order to estimate binding energy of the nuclei within small errors. Indeed, calculations show [10, 11] that the Coulomb dissociation of $^3_\Lambda$H should be very sensitive to the value of the binding energy. The binding energy of the Λ-particle in the nucleus, B_Λ, is one of the basic characteristics of hypernuclei, and the measurement of B_Λ for $^3_\Lambda$H is most informative because this hypernucleus is the simplest one for theoretical calculations [12]. The current experimental value [13] of the hypertritium binding energy, $B_\Lambda = 0.13 \pm 0.05$, provides very loose constrains for theoretical models.

Estimates have shown that the 10 % accuracy of measuring the dissociation cross section at the Nuclotron is quite realistic. This level of errors is good enough for the task because the Coulomb dissociation cross section on a U target changes from a few barns to 60 barns if the binding energy of Λ in the hypertriton drops from 0.15 MeV to 0.01 MeV. The possibility to estimate the binding energy using this method is especially attractive in case of a very low value of the binding energy.

Considering the investigation of hypernuclear Coulomb dissociation, one should take into account that a number of nuclear processes will contribute to the total cross section which should be measured or estimated at least. Thus, the extraction of the binding energy from the measured cross sections will be somewhat model dependent. Basing on recent theoretical calculations [11], we expect that the dependence of various parameters in the models is not strong. Moreover, the nuclear part of the total cross section varies in a narrow range for reasonable binding energy limits, in contrast to the contribution of the Coulomb dissociation.

Different absorbers will be used in an experiment, that will allow one to investigate the model itself and, perhaps, to provide some details of Λd and ΛN interactions [11] because the model calculations have been already performed for different absorbers. It should be also mentioned that the nuclear part of the total cross section can be measured experimentally because the Coulomb dissociation is negligible ($Z^{1.92}$ dependence) in the case of light absorbers.

In experiments with mesonic decays the trigger tested in previous experiments will be used, while the estimates of efficiency and data collection rates are based on our experience. The trigger Čerenkov counters measure the charge (Z) of the hypernucleus and the charge of the daughter nucleus which is equal to Z+1 for the mesonic decay. Beam nuclei fragments can not simulate this increase of the charge and the trigger can be tuned to any background rejection level.

NONMESONIC DECAYS

The trigger measuring charge of hypernuclei and decay products can be used in the case of nonmesonic decays just as for the decays when the negative pion is emitted. However, two particles – a residual nucleus and a proton – should be registered and measured. A similar charge signal response in the trigger detectors can be produced by beam fragments - that is why additional restrictions and cuts should be used since the background problems will be more serious in this case [14].

In this context, a very surprising and original idea is to search for two α particles from ^{8}Be decays. Four p-shell hypernuclei decay via the chain ending with an immediate decay $^8\text{Be} \to \alpha + \alpha$. A set of interesting data can be obtained [15] if the production and decay of these nuclei are observed:

$$^{8}_{\Lambda}\text{Li} \to \pi^- + {}^8\text{Be}, \qquad ^{9}_{\Lambda}\text{Be} \to \pi^- + p + {}^8\text{Be},$$

$$^{10}_{\Lambda}\text{Be} \to n + n + {}^8\text{Be}, \qquad ^{10}_{\Lambda}\text{B} \to n + p + {}^8\text{Be}.$$

The task seems to be very difficult. If ^{8}Be decays from the ground state, then the back to back momentum of two alpha-particles is very low while the longitudinal momentum is above 20 GeV/c. This means that both alphas hit the detectors practically at the same point and there is no hope to specify the event by using proportional chambers. There is no sense to move the tracking detector at a larger distance beyond the decay volume because this will not allow to measure the location of the decay point within the required accuracy and the identification of the hypernuclear decay fails as well as for chambers detecting two alphas as one cluster.

Fortunately, ^{8}Be often decays from the excited states and a significant part of the events can be triggered and registered by high resolution devices. Thus, silicon microstrips (or a scintillating fiber tracker) are suggested for the central part of the spectrometer. A set of programs for Monte Carlo calculations was elaborated and calculations have shown that a number of the lost hypernuclear events can be reduced to 30-60% level. The exact value depends strongly on assumptions of the required accuracy for the vertex location. Anyway, it is clear that the high resolution detector can partially solve this problem. The size of the high resolution part of the detector does not exceed 6 cm x 6 cm. The same calculations have shown that for the first task (mesonic decays of hydrogen hypernuclei) proportional chambers are adequate and not more than 5-15% of events are expected to be lost due to one cluster for two particles or an ambiguous vertex location.

A more complicated problem for experiments with ^{8}Be is the expected enormous number of background triggers due to the beam nuclei fragmentation in the edge of the trigger counters. This mechanism can not be suppressed to rather low values. The first estimates, based on the analysis of the properties of a test Čerenkov counter, have shown that one can hope to reduce the background trigger rate to the level when approximately 50% efficiency for hypernuclear events can be expected. This estimate is very optimistic and does not include some additional sources of the background. On the other hand, the counter was tested at 2 GeV and one can have better parameters at 6 GeV energy. Anyway, the upgrade of detectors and trigger is promising for this type of hypernuclear experiments.

ACKNOWLEDGMENTS

This work has been supported in part by RFBR grant 99-02-17655.

REFERENCES

1. Issinsky, I., et al., in Proc. Particle Accelerator Conf. (Vancouver, 1997), IEEE, Vancouver, V. 1, p. 181, 1997.
2. Joint Institute for Nuclear Research. Research Program of the Laboratory of High Energies, JINR, Dubna, 1999.
3. Avramenko, S.A., et al., *JINR Rapid Comm.* **5 [68]**, 14 (1994).
4. Avramenko, S.A., et al., *Nucl. Phys.* **A585**, 91c (1995).
5. Abdurakhimov, A.U., et al., *Nuovo Cim. C* **102**, 645 (1989).
6. Avramenko, S., et al., *Nucl. Phys.* **A547**, 95c (1992).
7. Wakai, M., et al., *Phys. Rev. C* **38**, 748, (1988).
8. Bandō, H., et al., *Int. J. Mod. Phys. A* **21**, 4021, (1990).
9. Wakai, M., *Nucl. Phys.* **A547**, 89c (1992).
10. Lyuboshitz, V.L., *Yad. Fiz.* **51**, 1013 (1990).
11. Evlanov, M.V., et al., *Nucl. Phys.* **A632**, 624 (1998).
12. Miyagawa, K., and Glöckle, W., *Nucl. Phys.* **A585**, 169 (1995).
13. Davis, D.H., and Pniewski, J., *Contemp. Phys.* **27**, 91 (1986).
14. Bartke, J., et al., *J. Phys. G*, **25**, 429 (1999).
15. Majling, L., and Batusov, Yu., Proc. HYP2000, *Nucl. Phys.* **A**, in press.

$^{10}_{\Lambda}$Be and $^{10}_{\Lambda}$B hypernuclei: A clue to some puzzles in nonmesonic weak decay

L. Majling*, A. Parreño†, A. Margaryan** and L. Tang ‡

*Nuclear Physics Institute, Academy of Sciences, CZ-25068 Řež, Czech Republic
†Dept. d'Estructura i Constituents de la Matèria, U. B., Diagonal 647, E-08028 Barcelona, Spain
**Yerevan Physics Institute, 2 Alikhanian Br. Str., Yerevan, 375036, Armenia
‡T. Jefferson National Accelerator Facility, Newport News, VA 23606, U.S.A

Abstract. We demonstrate how the nuclear structure aspects of the problem, an often unwelcome detail of the calculations attempting to understand basic two-body interactions, can be used to pick out components of the effective weak Hamiltonian.

INTRODUCTION

The field of weak decay of Λ-hypernuclei has experienced an impressive progress in the last few years. We mention here only the new experimental results [1], and calculations [2] and [3]. For more information we refer to [4]. The main problem concerning the weak decay of Λ-hypernuclei is the disagreement between the theoretical and experimental values for the ratio Γ_n/Γ_p between the neutron- and proton-induced widths. The theoretical calculations underestimate the central data points for all considered hypernuclei, although the large experimental error bars do not permit any definitive conclusion. The data are quite limited and not precise since it is difficult to detect the products of the non mesonic decays, especially for the neutron-induced one. In order to solve the Γ_n/Γ_p puzzle, many attempts have been made up to now, but without success. Among these we recall the introduction of mesons heavier than the pion in the ΛN transition potential; the role of two-nucleon stimulated decay; the description of the short range baryon-baryon interaction in terms of quark degrees of freedom.

METHOD

The present experimental resolution for the detection of the outgoing nucleons does not allow to identify the final state of the residual nucleus in the process $^A_\Lambda Z \rightarrow {}^{A-2}Z + nn$ and $^A_\Lambda Z \rightarrow {}^{A-2}(Z-1) + np$. As a consequence, the measurements supply decay rates averaged over several nuclear final states.

TABLE 1. Spectroscopic factors extracted from the experimental results for $^9\text{Be}(p,d)^8\text{Be}$ reaction [7] are compared with predictions of various calculations based on the intermediate coupling shell model [8]

		^{8}Be	[6]		calculated		measured
J_i^π	T_i	E_i	Γ_i, (keV)	decay	E_i	S_i^n	S_i^n
0^+	0	0.00	6.8 eV	α	0.00	0.550	0.67±0.14
2^+	0	3.04	1500	α	3.09	0.755	1.49±0.23
4^+	0	11.4	~3500	α	10.30	0.000	no fit
2^+	1 (0)	16.63	108	α,γ	16.76	0.505	} 1.14±0.19
2^+	0 (1)	16.92	74	α,γ	16.89	0.430	
1^+	1	17.64	11	p,γ	17.59	0.223	0.27±0.05

We focus our attention on one peculiar case [5]. It is well known that removing one nucleon from ^{9}Be or ^{9}B results in ^{8}Be* [6]:

$$^9\text{Be} \rightarrow p + (^8\text{Li} \xrightarrow{\beta^-} {}^8\text{Be}^*); \quad ^9\text{Be} \rightarrow n + {}^8\text{Be}^*;$$
$$^9\text{B} \rightarrow n + (^8\text{B} \xrightarrow{\beta^+} {}^8\text{Be}^*); \quad ^9\text{B} \rightarrow p + {}^8\text{Be}^*.$$

Prevailing part of the final states of the residual nuclei ultimately decays into α-α channels (see Table 1). Through this unique process it would be possible to identify the final states. (These events are recognized easily as 'hammer tracks' in the emulsion.)

So, due to these specific properties of the core nuclei ^{9}Be and ^{9}B, it may be possible to measure the partial decay widths Γ_i^τ for the $^{10}_\Lambda$Be and $^{10}_\Lambda$B hypernuclei.

PARTIAL DECAY WIDTHS

The nonmesonic decay rate Γ_{nm} can be written as [9]

$$\Gamma_{nm} = \sum_\tau \Gamma^\tau = \sum_\tau \sum_i \Gamma_i^\tau$$

($\tau = n$ (neutron) or p (proton)), where the partial width Γ_i^τ is

$$\Gamma_i^\tau = |\langle \Psi^{A-2}(\{i\}) \otimes \psi^{NN}(JT) | V_{weak} | [\Psi^{A-1}(\{c\}) \otimes \psi^\Lambda(\tfrac{1}{2})]^J \rangle|^2$$

(the shorthand notation $\{i\} \equiv E_i, J_i, T_i, \tau_i$ and $\{c\} \equiv E_c, J_c, T_c, \tau_c$ is used).

It is possible to factorize this expression as:

$$\Gamma_i^\tau = \sum_{SJ} G_J^2(\{c\},\{i\},\tau LSJ) \cdot w_{\ell\tau}^{SJ},$$

with

$$w_{\ell\tau}^{SJ} = |\sum_{L'S'} \langle l_1 l_2 : L'S'JT | V_{weak} | \tau\ell\, s_\Lambda : L = \ell SJ \rangle|^2,$$

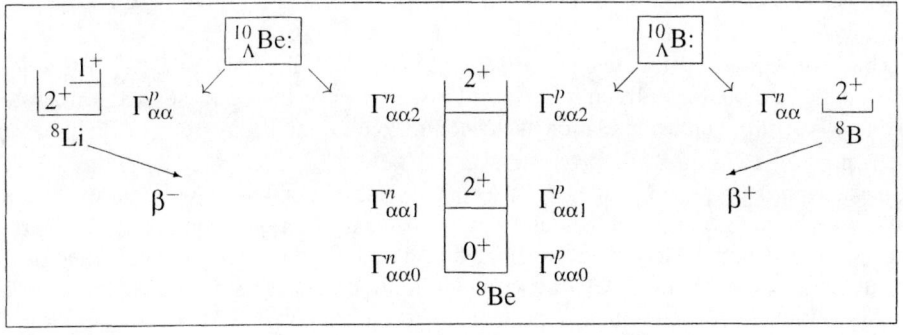

FIGURE 1. Notation of the partial widths $\Gamma^{\tau}_{\alpha\alpha\, i}$

the **spin-dependent part** of the weak interaction matrix element and $G_{\mathcal{J}}(\{c\},\{i\},\tau LSJ)$, the $N\Lambda$ **pair** fractional parentage coefficient, defined through

$$[\Psi^{A-1}(\{c\})\otimes\psi^{\Lambda}(\tfrac{1}{2})]^{\mathcal{J}} \equiv |\ell^k(E_c J_c T_c)\otimes s_\Lambda : \mathcal{J}\rangle =$$
$$= \sum_{i,\tau LSJ} G_{\mathcal{J}}(\{c\},\{i\},\tau LSJ) \cdot |\ell^{k-1}(E_i J_i T_i)\otimes \tau\ell\, s_\Lambda(LSJ) : \mathcal{J}\rangle,$$

$$G_{\mathcal{J}}(\{c\},\{i\},\tau LSJ) = \sqrt{k}\cdot(T_i t_i \tfrac{1}{2}\tau\,|\,T_c t_c)\cdot$$
$$\sum_j g_i^c(\ell j)\cdot U(J_i j \mathcal{J}\tfrac{1}{2} : J_c J)\cdot U(\ell \tfrac{1}{2} J \tfrac{1}{2} : jS).$$

In Fig. 1. the relevant states of A = 8 isotopes are displayed. The similarity of the structure of the $\Gamma^n_{\alpha\alpha\,i}$ ($^{10}_\Lambda$Be) and $\Gamma^p_{\alpha\alpha\,i}$ ($^{10}_\Lambda$B) is clearly seen.

RESULTS

Table 2 demonstrates that these partial widths (related to ^8Z excited states) are various combinations of four matrix elements (eight for different τ), hence their study offers a unique possibility to determine all needed matrix elements of the weak interaction [9] and can shed some light to the resolution of the Γ^n/Γ^p puzzle [10].

TABLE 2. Partial $\Gamma^{\tau}_{\alpha\alpha\,i}(^{10}_\Lambda\text{Be})$ and total $\Gamma^{\tau}_{tot}(^{10}_\Lambda\text{Be})$ decay widths

$\Gamma^n_{\alpha\alpha 2}$	=	$0.096\, w^{01}_{1n} + 0.520\, w^{11}_{1n} + 0.439\, w^{12}_{1n}$
$\Gamma^n_{\alpha\alpha 1}$	=	$0.388\, w^{01}_{1n} + 0.051\, w^{11}_{1n} + 0.408\, w^{12}_{1n}$
$\Gamma^n_{\alpha\alpha 0}$	=	$0.412\, w^{01}_{1n} + 0.206\, w^{11}_{1n}$
Γ^n_{tot}	=	$1.187\, w^{01}_{1n} + 0.822\, w^{11}_{1n} + 1.248\, w^{12}_{1n} + 0.117\, w^{10}_{1n} + 0.354\, w^{00}_{0n} + 1.271\, w^{11}_{0n}$
Γ^p_{tot}	=	$0.569\, w^{01}_{1p} + 0.535\, w^{11}_{1p} + 0.981\, w^{12}_{1p} + 0.165\, w^{10}_{1p} + 0.437\, w^{00}_{0p} + 1.313\, w^{11}_{0p}$
$\Gamma^p_{\alpha\alpha}$	=	$0.441\, w^{01}_{1p} + 0.491\, w^{11}_{1p} + 0.548\, w^{12}_{1p} + 0.157\, w^{10}_{1p}$

NONMESONIC WEAK DECAY STUDY OF $^{10}_{\Lambda}$BE AT CEBAF

Recently, the first hypernuclear spectroscopy experiment, the $^{12}C(e,e'K^+)^{12}_{\Lambda}B$ reaction, was performed successfully at TJNAF [11]. We propose to use these electron-photon beams to carry out an investigation of the nonmesonic weak decay of the $^{10}_{\Lambda}$Be hypernucleus produced via the (γ, K^+) reaction on ^{10}B target. For the detection of the delayed α-particles it is proposed to use large acceptance (2π sr) detector based on low-pressure multiwire proportional chambers and low-pressure **multistep** chambers [12]. During the past years such devices were developed for a time-zero fission fragment detector for the direct measurement of heavy hypernuclei lifetimes at Jefferson Lab [13]. By means of this low-pressure technique it is possible to detect protons (0.1 MeV $< E_p < 5$ MeV), deuterons (0.2 MeV $< E_d < 10$ MeV) and α particles (0.4 MeV $< E_\alpha < 100$ MeV) as well as more heavy particles.

ACKNOWLEDGMENTS

We are grateful to Yu. Batusov, T. Motoba and A. Ramos for stimulating discussions. This work has been supported by GACR, grant No. 202/00/1667.

REFERENCES

1. Park, H., et al., *Phys. Rev. C* **61**, 054004 (2000).
2. Alberico, W., et al., *Phys. Rev. C* **61** 044314 (2000).
3. Jido, D., Oset, E., and Palomar, J. E., nucl-th/0101051.
4. Parreño, A., and Ramos, A., nucl-th/0104080.
5. Majling, L., and Batusov, Yu., Proc. Int. Conf. HYP2000: *Nucl. Phys. A*, in press.
6. Ajzenberg - Selove, F., *Nucl. Phys.* **A490**, 1 – 226 (1988).
7. Schoonover, J.L., Li, T.Y., and Mark, S.K., *Nucl. Phys.* **A176**, 567 – 576 (1971).
8. Barker, F.C., *Nucl. Phys.* **83**, 418 – 448 (1968).
9. Parreño, A., Ramos, A., and Bennhold, C., *Phys. Rev. C* **56**, 339 – 364 (1997).
10. Oset, E., and Ramos, A., *Progr. Part. Nucl. Phys.*, **41**, 191 – 253 (1998).
11. Hungerford, E., Proc. Int. Conf. HYP2000: *Nucl. Phys. A*, in press.
12. Assamagan, K., et al., *Nucl. Instr. and Meth. A* **426**, 405 – 419 (1999).
13. Tang, L., "Direct Measurement of the Lifetime of Heavy Hypernuclei at CEBAF Using Low Pressure MWPC Technique", in *Strangeness Nuclear Physics*, edited by Il-Tong Cheon et al., Proc. APCTP Workshop SNP'99, World Scientific, Singapore, 2000, pp. 393 – 400.

$^{10}_{\Lambda}$Be and $^{10}_{\Lambda}$B hypernuclei on the NUCLOTRON

J. Lukstins[*], V. Nikitin[*], A. Parfenov[*], L. Majling[†], D. Chren[**], B. Sopko[**] and J. Bartke[‡]

[*]Joint Institute for Nuclear Research, RU-141980 Dubna, Moscow Region, Russia
[†]Nuclear Physics Institute, Academy of Sciences, CZ-25068 Řež, Czech Republic
[**]Czech Technical University, Technická 4, CZ-16607 Prague, Czech Republic
[‡]H. Niewodniczański Institute of Nuclear Physics, PL-30055 Kraków, Poland

Abstract. Recently, the Nuclotron accelerator at the Laboratory of High Energy has supplied the first extracted beam of medium energy ions. This opens up possibilities to perform hypernuclear experiments. Dedicated detector set up, including as main parts a silicon strip vertex detector, Cherenkov detectors and a magnetic spectrometer, opens possibilities to continue experiments performed on the Dubna synchrophasotron ion beams and, at the same time, to test a new method of identification of some hypernuclei using the $\alpha\alpha$ decay of the ^{8}Be residual nucleus as a trigger.

INTRODUCTION

The production of hypernuclei in relativistic heavy ion collisions has several advantages: relativistic time dilatation gives the flight path of the order of 10 cm, which would allow measurement of their lifetime and the resolution of various decay channels; hypernuclear decay products thrown into small laboratory solid angle; wide variety of hypernuclear weak-interaction strangeness-changing decays can be studied.

Until now, only two experiments of this kind have been performed. In 1975 the group of Arizona looked for hypernuclei produced by the 2.1A GeV ^{16}O beam at the LBL Bevalac [1]. At the end of the 80-ies hypernuclear experiments were performed on the Dubna synchrophasotron ion beams (^{3}He, ^{4}He, ^{6}Li). The production cross sections of $^{4}_{\Lambda}$H and $^{3}_{\Lambda}$H as well as the lifetime of $^{4}_{\Lambda}$H were measured [2].

The calculation in the coalescence scheme was performed [3] to pertinently describe the results [2]. It makes possible to verify the model of hypernuclear production in relativistic heavy ion collisions and to limit the parameters entering it by using values for cross sections already measured. It pointed at hypernuclei which were produced abundantly in the process employed, but not analyzed. The production of kaons and especially of pions in high energy nuclear collisions is large, thus the **secondary yields** of hypernuclei through (K,π) and (π,K) reactions may be quite appreciable. The theoretical results exhibit a pronounced enhancement at $^{5}_{\Lambda}$He and $^{9}_{\Lambda}$Be. In experiment [2] the mesonic weak decay used as a signature caused severe limitation on hypernuclei examined. The $\alpha\alpha$ decay of the ^{8}Be residual nucleus could be used as a trigger for production of $^{8}_{\Lambda}$Li ($^{8}_{\Lambda}$Li $\to \pi^{-}\,^{8}$Be), $^{9}_{\Lambda}$Be ($^{9}_{\Lambda}$Be $\to \pi^{-}p\,^{8}$Be), $^{10}_{\Lambda}$Be ($^{10}_{\Lambda}$Be $\to nn\,^{8}$Be) and $^{10}_{\Lambda}$B ($^{10}_{\Lambda}$B $\to np\,^{8}$Be) hypernuclei in relativistic heavy ion collisions.

NUCLOTRON AND DETECTOR SET-UP

In the last few years, the Nuclotron, a new superconducting accelerator, has been developed step by step to reach the designed specifications. Protons will be accelerated up to 12 GeV by this machine, but the accelerator will generally be used as a source of medium energy ion beams.

It is a proper time to revise the research program for the new accelerator. In particular, remind of possible hypernuclear experiments in which we see two main goals: mesonic decays of the lightest hypernuclei and nonmesonic decays of medium hypernuclei [4].

The trigger measuring charge of hypernuclei and decay products can be used in the case of nonmesonic decays just as for decays when negative pion is emitted. The trigger tuned to search for two α particles from ^{8}Be decays is an original idea [5]. Four p-shell hypernuclei decay via the chain ending with an immediate decay ^{8}Be $\rightarrow \alpha\alpha$. A set of interesting data can be obtained if the production and decay of these nuclei are observed [6]. At the first sight, the task is very difficult. If ^{8}Be decays from the ground state, the back to back momentum of two α-particles is very low while the longitudinal momentum is above 20 GeV/c. This means that both α's hit the detectors at the same point and there is no practical way to specify the event. Fortunately, ^{8}Be decays mainly from the **excited states** in this case and significant part of events can be triggered and registered by high resolution devices. Thus, silicon microstrips are suggested for the trigger and a scintillating fibre tracker for the central part of the spectrometer.

The experimental identification of hypernuclei can be based on the decay of these nuclei in the vacuum volume behind the target in events, when Λ-hyperons produced in interactions of beam nuclei with the target can be absorbed by relativistic fragments in an inclusive process [7]. The velocities of these hypernuclei are close to those of projectile nuclei.

Registration and analysis of hypernucleus decays can be performed with the experimental set-up SPHERE (4π geometry spectrometer with an analyzing magnet SP-40 for measurements of momenta and charges of all charged particles) [8], which is already in operation at the Laboratory of High Energy JINR.

ACKNOWLEDGMENTS

This work has been supported by GA ASCR, grant No. 1048703.

REFERENCES

1. Nield, ,K., et al., *Phys. Rev. C* **13**, 1263 – 1266 (1976).
2. Avramenko, S., et al., *Nucl. Phys.* **A547**, 95c – 100c (1992).
3. Bandō, H., et al., *Nucl. Phys.* **A501**, 900 – 914 (1989).
4. Lukstins, J., Proc. Int. Conf. HYP2000. *Nucl. Phys. A*, in press.
5. Bartke, J., et al., *in* Proc. XV. Int. Sem. on High Energy Physics Problems, JINR, Dubna, Sept. 2000.
6. Majling, L., and Batusov, Yu., Proc. Int. Conf. HYP2000: *Nucl. Phys. A*, in press.
7. Bartke, J., et al., *J. Phys. G* **25**, 429-435 (1999).
8. Mikhalev, D.P., Nikitin, V.A., Okonov, E.O., and Parfenov, A.N., *JINR P1-95-549*, Dubna, 1995.

Electromagnetic K^+ production on the deuteron and hyperon-nucleon interactions

K. Miyagawa*, T. Mart[§], C. Bennhold[•], and W. Glöckle[¶]

*Department of Applied Physics, Okayama University of Science, 1-1 Ridai-cho, Okayama 700, Japan
[§] Jurusan Fisika, FMIPA, Universitas Indonesia, Depok 16424, Indonesia
[•] Center for Nuclear Studies, The George Washington University, Washington, D.C. 20052
[¶] Institut für Theoretische Physik II, Ruhr-Universität Bochum, D-44780 Bochum, Germany

Abstract. Electro- and photo production processes of K^+ on the deuteron are investigated theoretically and various observables including beam and recoil polarizations are predicted. Modern hyperon-nucleon forces as well as an updated production operator on the nucleon are used. Sizable effects of the hyperon-nucleon final state interaction are seen around the Λ and Σ thresholds.

INTRODUCTION

Recently rigorous calculations for light hypernuclei have been performed [1, 2], which give us a valuable piece of information on low-energy properties of the hyperon-nucleon interaction. However, no clear understanding of the hyperon-nucleon interaction around the Σ threshold has emerged. Electro- and photoproduction processes of the kaon on light nuclei offer a unique possibility for studying the hyperon-nucleon interaction in the continuum, especially near the Σ threshold. Thus we analyze inclusive $d(\gamma, K^+)$ and exclusive $d(\gamma, K^+\Lambda(\Sigma))$ processes and make predictions on various observables including beam and recoil polarizations. We also present a preliminary result of the electro-production process $d(e,e'K^+)$. Our main aim is to investigate the coupled $\Lambda N - \Sigma N$ interaction in the final states, and the modern hyperon-nucleon interaction of the Nijmegen group [3], NSC97f and NSC89 are used. Those interactions have been found to give a reasonable binding energy for the hypertriton. Another aim of this analysis is to obtain the information about the elementary cross sections on the neutron, such as $\gamma + n \to K^+ + \Sigma^-$, which is allowed to access in kinematic regions where final-state interaction effects are small. An updated amplitude [4] for the elementary processes is used.

PHOTO- AND ELECTRO PRODUCTION

The predictions of the inclusive $d(\gamma, K^+)$ cross sections are given as a function of lab momentum P_K in Fig. 1. The incident photon energy is 1.3 GeV, while the outgoing kaon angle is fixed to 3 degrees. The two pronounced peaks around $P_K = 943$ and 811 MeV/c are due to the quasifree scattering between photon and one of the nucleons. The results with the final state YN interaction NSC97f are compared to the results with NSC89 and to the PWIA results. Sizable FSI effects are seen above the Λ threshold for both of the YN interactions, but only the NSC97f interaction shows a prominent enhancement around the Σ threshold.

For the same E_γ and θ_K, the exclusive $d(\gamma, K^+\Lambda)n$ cross sections and double polarization observables C_z at $P_K = 870$ MeV/c are shown in Fig. 2. While the values for C_z in PWIA are almost 100%, the FSI results show dramatic deviations. In Fig. 3, the cross sections and the Λ recoil polarization P_y are depicted with E_γ and P_K kept the same, but with $\theta_K = 17°$. The cross sections take much larger values than in the case of $\theta_K = 3°$ at the forward Λ angles. The Λ polarizations along the y-axis P_y with the FSI clearly differ from the Λ polarizations in PWIA.

We also present preliminary results for the electroproduction process $d(e, e'K^+)$ in Fig. 4. As in the case of the photoproduction, YN FSI effects are seen near both Λ and

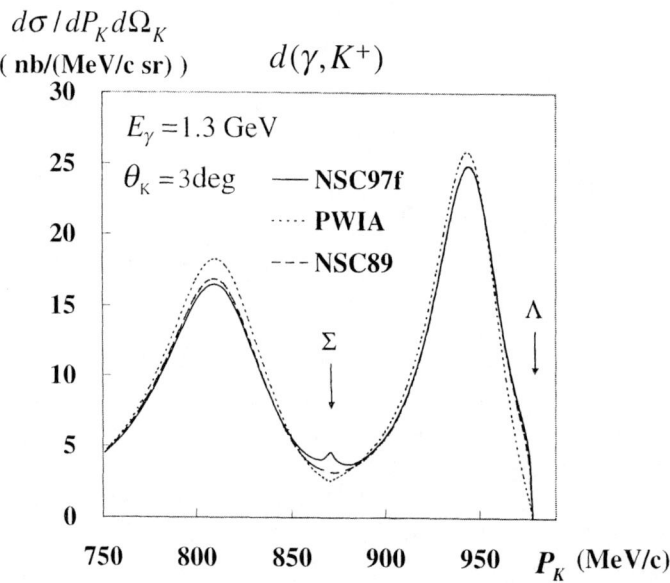

FIGURE 1. Inclusive $d(\gamma, K^+)$ cross section as a function of kaon lab momentum. The $K^+\Lambda N$ and $K^+\Sigma N$ thresholds are indicated by the arrows.

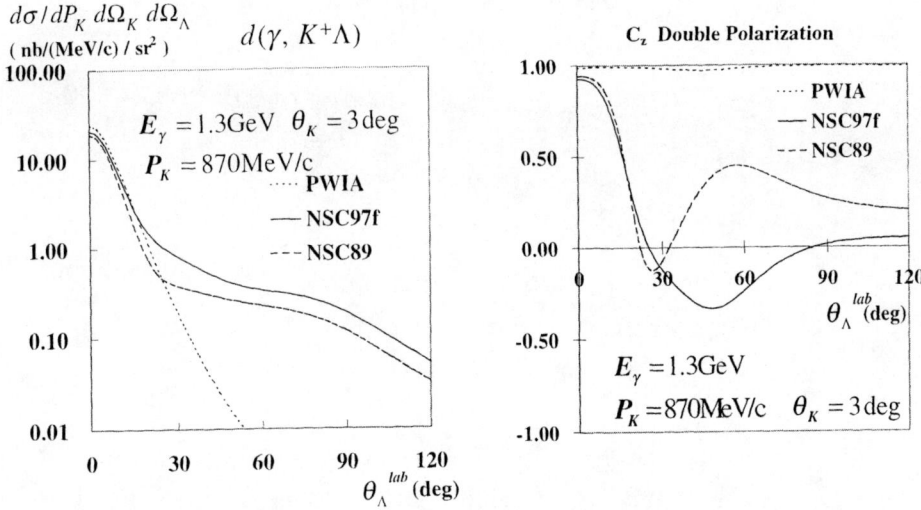

FIGURE 2. Exclusive $d(\gamma, K^+\Lambda)$ cross section and the double polarization C_Z for lab momentum $P_K = 870$ MeV/c and polar angle $\theta_K = 3$ degrees

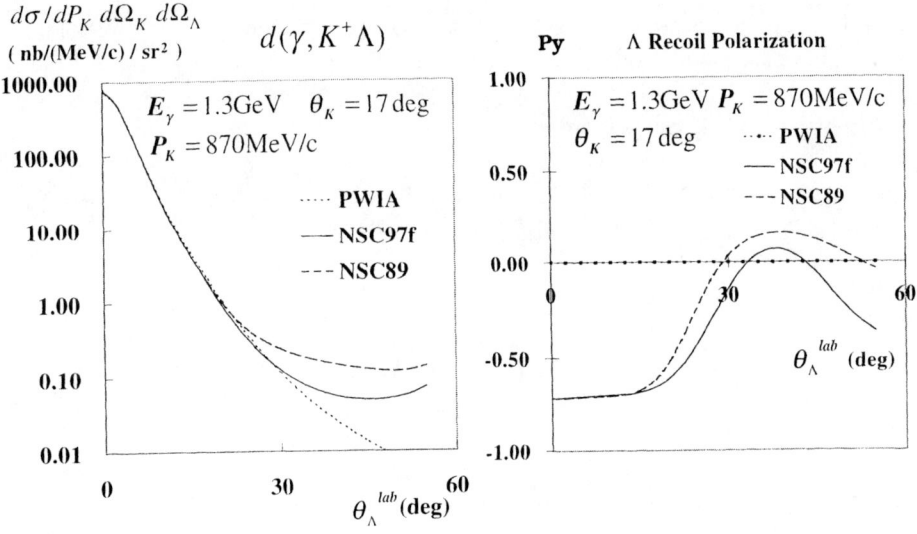

FIGURE 3. Exclusive $d(\gamma, K^+\Lambda)$ cross section and the Λ recoil polarization P_y for lab momentum $P_K = 870$ MeV/c and polar angle $\theta_K = 17$ degrees.

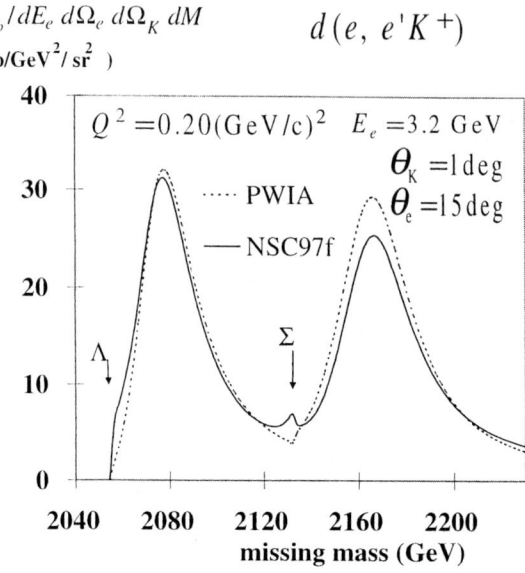

FIGURE 4. Missing mass spectrum for the reaction $d(e,e'K^+)$. The Λ and Σ thresholds are indicated by the arrows.

Σ thresholds. However, the FSI has effects in a wide range above the Σ threshold. A prominent enhancement around the Σ threshold is not a simple threshold effect but is caused by a YN t-matrix pole in the complex momentum plane [5].

In summary, we investigate cross sections and polarization observables for the K^+ photoproduction on the deuteron. Especially, in the polarization observables, we find large hyperon-nucleon FSI effects. Also in the electroproduction process, cross sections show a prominent enhancement around the Σ threshold. A systematic analysis for a wide range of kinematics for both photo- and electroproduction processes is under way.

REFERENCES

1. Miyagawa, K., Kamada, H., Glöckle, W., Yamamura, H., Mart, T., and Bennhold, C., *Few-Body Systems Suppl.* **12**, 324 (2000); nucl-th/0002035.
2. Nogga, A., Ph. D. Thesis, Ruhr-Universität Bochum, 2001 (unpublished).
3. Rijken, Th.A., Stoks, V.G.J., and Yamamoto, Y., *Phys. Rev. C* **59**, 21 (1999), and references therein .
4. Lee, C.F.X., Mart, T., Bennhold, C., Haberzettl, H., and Wright, L.E., nucl-th/9907119; Bennhold, C., Haberzettl, H., and Mart, T., nucl-th/9909022.
5. Miyagawa, K., and Yamamura, Y., *Phys. Rev. C* **60**, 024003 (1999); nucl-th/9904002.

$\Lambda\Lambda$ interaction indicated by "Lambpha" ($^{6}_{\Lambda\Lambda}$He double hypernucleus)

K.Nakazawa (for the KEK-PS E373 collaboration)

Phys. Dept. of Gifu Univ., Gifu 501-1193, Japan

Abstract. We have carried out a hybrid-emulsion experiment E373 at KEK to study a double strangeness nuclear system. By 11% data analysis of all, we uniquely identified the formation and the decay sequence of a $^{6}_{\Lambda\Lambda}$He double hypernucleus as; $\Xi^- + ^{12}C \rightarrow ^{6}_{\Lambda\Lambda}He + ^4He + t$, $^{6}_{\Lambda\Lambda}He \rightarrow ^{5}_{\Lambda}He + \pi^- + p$. The $\Lambda\Lambda$ interaction energy has been confirmed to be 1.01 ± 0.20 (measured) $^{+0.18}_{-0.11}$ (syst.) MeV using the most probable Ξ^- binding energy of 0.13 MeV which would be the level energy of ^{12}C atomic 3D state. This value demonstrates the $\Lambda\Lambda$ interaction is attractive but not so large as 4~5 MeV which has been believed for forty years.

OVERVIEW OF THE EXPERIMENT

Study of doubly strangeness ($S = -2$) system is important to understand the baryon-baryon interaction under the flavor $SU(3)$ symmetry. In the $S = -2$ state, theoretically, the baryon-baryon interaction can be attractive at short distance, compared with the repulsive core of the nucleon-nucleon interaction. Therefore, the doubly strangeness nuclei can be quite different from the ordinary nuclei or Λ hypernuclei. The H dibaryon of a stable 6-quark state has been proposed using color-magnetic interaction [1], and the strange quark matter is an extreme example in this field. However, we know very little about $S = -2$ systems experimentally. Theoretical predictions of the $\Lambda\Lambda$ interaction energy, $\Delta B_{\Lambda\Lambda}$, are in a wide range from -6 MeV to 10 MeV [2].

About forty years ago, a $^{10}_{\Lambda\Lambda}$Be event was presented by Danysz et al. in expected four events of Ξ^- capture at rest [3]. In 1966, a $^{6}_{\Lambda\Lambda}$He event by Prowse[4] was reported, however only a sketch of the event is left and it has not been studied independently by other scientists. The $\Delta B_{\Lambda\Lambda}$ value has been favored to be 4 – 5 MeV based on the results of those events. By the hybrid-emulsion experiment of E176 at KEK in 1991, we have successfully confirmed the existence of a double hypernucleus among $\sim$ 80 events of Ξ^- capture at rest in nuclear emulsion [5]. The nuclide in the event was interpreted as $^{10}_{\Lambda\Lambda}$Be or $^{13}_{\Lambda\Lambda}$B with $\Delta B_{\Lambda\Lambda}$ of $-4.8^{+0.7}_{-0.8}$ MeV (repulsive interaction) or $+4.9\pm0.8$ MeV (attractive), respectively. By the events of Danysz et al. and E176, $\Delta B_{\Lambda\Lambda}$ with 4 – 5 MeV may be only an upper limit because there remained possibility for the production and/or decay of the excited state of the nucleus.

In 1995, the E373 experiment with a hybrid-emulsion method at KEK started to confirm the findings of E176 with better accuracy, to decide the $\Lambda\Lambda$ interaction, and to provide new discoveries about double strangeness nuclei [6]. In E373, the quasi-free $'p'(K^-, K^+)\Xi^-$ reactions using 1.66 GeV/c K^- beam were induced in the diamond

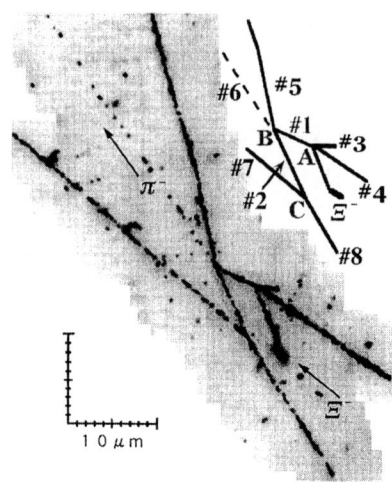

FIGURE 1. NAGARA event showing production and decay of "Lambpha" observed in E373. A Ξ^- hyperon was captured at the point A, and the double hypernucleus (#1) and two stable nuclei were emitted. At the point B, track #1 decayed into #2 (single-Λ hypernucleus), #5 and #6. Track #6 was identified as a π^- meson by its characteristic topology at the end point in the emulsion. Track #7 and #8 are decay daughters of #2 from the point C. Although track #7 was escaping from the emulsion stack, its end point was located in the downstream SCIFI block, and it was recognized not to be π^- by its illumination.

(^{12}C) target located upstream of the emulsion. Some of the Ξ^- hyperons were stopped in the emulsion stack. The high precision tracking detector, micro fiber-bundle tracker [7], was placed between the target and the emulsion to guide the Ξ^- tracks into the first emulsion plate. The automated scanning system was also successfully developed for Ξ^- track tracing in the emulsion to reduce the time for emulsion analysis [8]. In E373, we can obtain 10^3 stopping Ξ^- events which is ten times more than presented by E176. Although we have finished analysis of only 11 % of the data, an event showing the decay topology of the twin single Λ hypernuclei [9] and an $^{6}_{\Lambda\Lambda}$He double hypernucleus event [10] were successfully detected. In this letter, we will discuss the latter event of a beautiful double hypernucleus and the $\Lambda\Lambda$ interaction and make comparison with previous results.

OBSERVATION OF "LAMBPHA" : $^{6}_{\Lambda\Lambda}$HE DOUBLE HYPERNUCLEUS

The event of a double hypernucleus is very clearly recognized as shown in Fig. 1. This event found in Gifu is named "*NAGARA*", which is the river originating in Gifu, Japan. Coplanarities by charged tracks except for the Ξ^- hyperon at the point A and B were well established within measurement errors.

At the point C, track #7 and #8 show large visible energy with their long ranges. Then we could easily reject the case of a mesonic decay of single-Λ hypernuclei at C.

To satisfy the Q-value of a non-mesonic decay of single ones, we found that #7 and #8 would have single unit of charge, therefore track #2 has to be a $_\Lambda$He nucleus.

At the decay point B of track #1, three charged particles (#2, #5 and #6) were emitted. Possible decay modes of track #1 were kinematically checked, track #2 was assigned to be the $_\Lambda$He hypernucleus. After the reconstruction, there remained 16 decay modes with $\Delta B_{\Lambda\Lambda} > -20$ MeV including the cases of the neutron(s) emission from the point B. All of them are present in the decay of the He double-Λ hypernucleus.

At the production point A of track #1, the kinematical analysis of possible production modes of the double hypernucleus was made by assuming the Ξ^- hyperon was captured by a light nucleus in the emulsion, i.e. ^{12}C, ^{14}N or ^{16}O. We set $\Delta B_{\Lambda\Lambda} < 20$ MeV to get possible modes. As a result, ten production modes were accepted with the case of the neutron(s) emission.

$\Delta B_{\Lambda\Lambda}$ and the binding energy of two Λ hyperons, $B_{\Lambda\Lambda}$, obtained from both the formation and the decay point were compared. Finally, the interpretation of the nucleus was uniquely established as:

$$\Xi^- + {}^{12}C \rightarrow {}^{\ \ 6}_{\Lambda\Lambda}He + {}^4He + t,$$
$$^{\ \ 6}_{\Lambda\Lambda}He \rightarrow {}^5_\Lambda He + \pi^- + p.$$

The production and/or decay in the excited state of $^{\ \ 6}_{\Lambda\Lambda}$He and/or $^5_\Lambda$He are not possible. The vertex coplanarities at both the point A and B support the interpretation without neutron emission. The finding of $^{\ \ 6}_{\Lambda\Lambda}$He is very important because it is a closed-shell nucleus of p, n and Λ. Bando et al. [11] expected that it would give us information on the structure of multi-strangeness cluster system, and proposed the name "Lambpha" for $^{\ \ 6}_{\Lambda\Lambda}$He by analogy with the α particle. We obtained the values of $B_{\Lambda\Lambda}$ and $\Delta B_{\Lambda\Lambda}$ to be 7.25 ± 0.19 (measured) $^{+0.18}_{-0.11}$ (syst.) MeV and 1.01 ± 0.20 (measured) $^{+0.18}_{-0.11}$ (syst.) MeV, respectively, using the most probable Ξ^- binding energy of 0.13 MeV in ^{12}C 3D atomic state. By the obtained $B_{\Lambda\Lambda}$, the lower mass limit of the H dibaryon would be 2223.7 MeV at a 90 % confidence level.

$\Lambda\Lambda$ interaction ; comparison with the past experimental results

It is very interesting to check the consistency of the results from previous events. They are shown in Fig.2 with two events obtained by E373. Except for the $^{\ \ 6}_{\Lambda\Lambda}$He events, the interpretation is not unique due to the production cases of the double-Λ or the daughter single-Λ hypernucleus in the excited states, and therefore $^{\ \ 10}_{\Lambda\Lambda}$Be($^9_\Lambda$Be*), $^{\ \ 13}_{\Lambda\Lambda}$B($^{13}_\Lambda$C*) and ($^{\ \ 10}_{\Lambda\Lambda}$Be*) are listed. We calculated the event of E176 from the production topology. The event is plotted as $^{\ \ 10}_{\Lambda\Lambda}$Be on the right in the area of E176. From the figure, $\Delta B_{\Lambda\Lambda}$ of $4-5$ MeV seems to be still supported as a possibility by three or four events. However, due to larger reliability of the NAGARA event than that of Prowse [4], it is clear that our results are consistent, within the errors, with $\Delta B_{\Lambda\Lambda} \approx 1$ MeV. Also, the disagreement between our $\Delta B_{\Lambda\Lambda}$ ($^{\ \ 6}_{\Lambda\Lambda}$He) and that reported in Ref. [4] confirms the doubts about the authenticity of the event in [4].

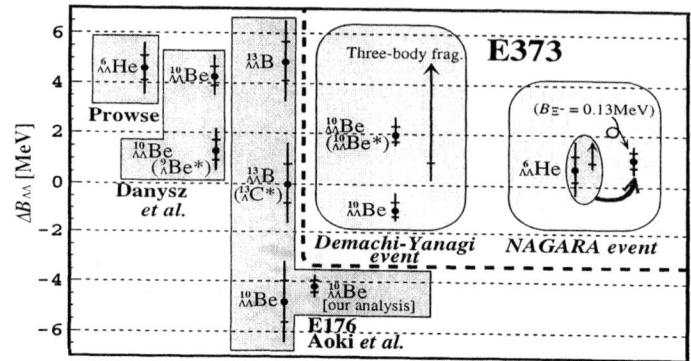

FIGURE 2. The $\Delta B_{\Lambda\Lambda}$ obtained from the past three events and the E373 events.

CONCLUSION

A hybrid-emulsion experiment E373 at KEK has been successfully carried out to study doubly strangeness nuclear system using 1.66 GeV/c K^- beam. By the analysis of 11% of the data, "Lambpha" ($^{6}_{\Lambda\Lambda}$He double hypernucleus) was observed. The production and decay of the "Lambpha" was uniquely interpreted as: $\Xi^- + ^{12}C \rightarrow\ ^{6}_{\Lambda\Lambda}$He+^{4}He+$t$, $^{6}_{\Lambda\Lambda}$He $\rightarrow\ ^{5}_{\Lambda}$He+$\pi^-$+$p$. There are no modes in the excited states. By use of the most probable Ξ^- binding energy of 0.13 MeV which is the calculated level of ^{12}C atomic $3D$ state, kinematical analysis of the event gives us information about the $\Lambda\Lambda$ interaction energy, $\Delta B_{\Lambda\Lambda}$, and the binding energy of two Λ, $B_{\Lambda\Lambda}$, as 1.01 ± 0.20 (measured) $^{+0.18}_{-0.11}$ (syst.) MeV and 7.25 ± 0.19 (measured) $^{+0.18}_{-0.11}$ (syst.) MeV, respectively. Previous experimental results, except for the old $^{6}_{\Lambda\Lambda}$He, are consistent with the $\Delta B_{\Lambda\Lambda} \approx 1$ MeV. Our obtained value of $\Delta B_{\Lambda\Lambda}$ confirmed the attractive $\Lambda\Lambda$ interaction, but very weak rather than 4 − 5 MeV.

REFERENCES

1. Jaffe, R.L., *Phys. Rev. Lett.* **38**, 195 (1977).
2. Bando, H., *Prog. Theor. Phys.* **67**, 699 (1977); Yamamoto, Y., et al., *Prog. Theor. Phys.* **86**, 867 (1991); Himeno, H., et al., *Prog. Theor. Phys.* **89**, 109 (1993); Schaffner, J., et al., *Ann. Phys.* **235**, 35 (1994); Yamamoto, Y., et al., *Prog. Theor. Phys. Suppl.* **117**, 361 (1995); Car, S.B., et al., *Nucl. Phys.* **A625**, 143 (1997); Yamada, T., and Nakamoto, C., *Phys. Rev. C* **62**, 034319 (2000).
3. Danysz, M., et al., *Phys. Rev. Lett.* **11**, 29 (1963); *Nucl. Phys.* **49**, 121 (1963); Dalitz, R.H., et al., *Proc. R. Soc. Lond. A* **426**, 1 (1963).
4. Prowse, D.J., *Phys. Rev. Lett.* **17**, 782 (1966).
5. Aoki, S., et al., *Prog. Theor. Phys.* **85**, 1287 (1991).
6. Nakazawa, K., KEK proposal E373, 1995; Nakazawa, K., *Nucl. Phys.* **A585**, 75c (1995); ibid. **A639**, 345c (1998).
7. Ichikawa, A., et al., *Nucl. Instr. Meth. A* **417**, 220 (1998).
8. Ichikawa, A., et al., to be submitted to *Nucl. Instr. Meth. A*.
9. Ichikawa, A., et al., *Phys. Lett. B* **500**, 37 (2001).
10. Takahashi, H., et al., subm. *Phys. Rev. Lett.*
11. Bando, H., Ikeda, K., and Motoba, T., *Prog. Theor. Phys.* **66**, 1344 (1981); ibid. **67**, 508 (1982).

K^+-meson production in subthreshold pA collisions with ANKE [1]

M. Nekipelov (for the ANKE collaboration)[2]

Institut für Kernphysik, Forschungszentrum Jülich, 52425 Jülich, Germany

Abstract. The production of K^+-mesons in proton-nucleus interactions at projectile energies far below and above the free nucleon-nucleon threshold of 1.58 GeV has been investigated using the ANKE spectrometer at COSY-Jülich. Double differential cross sections of K^+ production have been obtained for diamond, copper and gold strip targets and beam energies $T_p = 1.0...2.3$ GeV. At $T_p = 1.0$ GeV the complete momentum spectrum of kaons emitted at forward angles ($\theta \leq 12°$) has been measured. The analysis of the target-mass dependence of the cross sections does not allow unambiguous conclusions about the underlying reaction mechanisms. Therefore, further measurements at forward emission angles, in particular of coincident K^+p and K^+d pairs, are needed.

A central topic of hadron physics is the influence of the nuclear medium on elementary processes. The investigation of positively charged K^+-meson production is particularly well suited since this meson is relatively heavy so that its production requires strong medium effects. Due to the strangeness conservation, kaons with relatively low momenta which are considered in this work, are not absorbed in nuclear matter after their creation and can carry away the information about the production mechanism. Also quasi-elastic scattering of K^+ mesons in nuclei is substantially suppressed leading to a large mean free path of $\lambda_{K^+} \sim 5$ fm. In this context the production of K^+-mesons in proton-nucleus interactions at projectile energies far below and above the free nucleon-nucleon threshold of 1.58 GeV has been investigated using the large acceptance ANKE spectrometer [1] at an internal target position of the accelerator COSY-Jülich [2].

The spectrometer, including its detectors and the data-acquisition system (DAQ), is optimized to study K^+ spectra at different beam energies. The main experimental challenge is that at low energies the background (of mostly pions and protons) is approximately factor of 10^6 more intense than the signal (kaons). The detection system of ANKE allows to suppress this huge background and to detect rare K^+ mesons in the momentum range $p \approx 200 - 500$ MeV/c. Combining hardware and software criteria such as time-of-flight, information from the wire chambers and delayed signals from the K^+ decay gives us a unique possibility to identify kaons. The detectors and corresponding data-analysis procedures for K^+ identification are described in detail in [3].

The measured double differential cross sections for K^+ production in pC interactions

---

[1] Work is partially supported by BMBF (WTZ grant RUS-685-99), A.v. Humboldt foundation, RFBR (00-02 17808), RMS (FNP 125.03), US CRDF (RP2-2247), INTAS (2000-587)
[2] For a complete collaboration list see: http://www.fz-juelich.de/ikp/anke

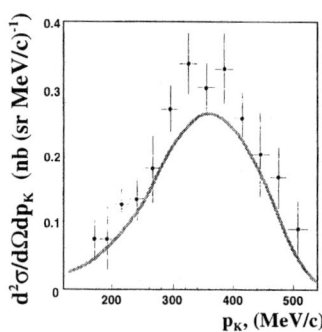

FIGURE 1. Double differential K^+ production cross section for the $p(1.0\ \text{GeV})^{12}\text{C} \to K^+X$ reaction as a function of the K^+ momentum. The solid line is a result of a calculation within the folding model assuming the two-step production mechanism (*E.Ya.Paryev, private communication*).

at the lowest beam energy are shown in Fig. 1. This is the first complete coverage of the momentum spectrum at deep subthreshold energies. The cross sections were calculated by a normalization to known pion-production cross sections measured in similar kinematical conditions [5, 6].

In addition to the direct production on a single target nucleon the K^+ mesons can also be produced in two-step processes with an intermediate pion production, i.e a $pN_1 \to \pi X$ reaction, followed by $\pi N_2 \to K^+X$ on a second target nucleon. Depending on the beam energy, the K^+ production may be due to both the direct and two-step reaction mechanisms. At low beam energies the two-step process is energetically favorable since the intrinsic nucleon motion can be utilized twice. Based on a folding calculation of the two-step kaon production a reasonable description of the measured cross section was obtained (see Fig. 1).

The analysis of the target-mass dependence aims to deduce the exponent α by fitting the cross sections measured with different target nuclei with an A^α dependence. We have analyzed available data on the K^+ production in pA collisions at beam energies $T_p \leq 2.9$ GeV [7]. One expects that the K^+ production cross section is proportional to $\simeq A^{0.7}$ if the kaons are dominantly produced via the direct production mechanism. Since two nucleons are needed for the two-step kaon production, stronger A dependence for the two-step production as compared with the direct kaon production is expected.

Total K^+-production cross sections σ_{tot} in proton-nucleus collisions at $T_p \leq 1.0$ GeV for Be, C, Cu, Sn and Pb as target nuclei have been measured by Koptev et al. [8] at the Petersburg Nuclear Physics Institute (PNPI). The data can well be described by a constant value $\alpha = 1.04 \pm 0.01$ independent of the beam energy T_p. The strong A dependence of the total K^+ production cross section has been interpreted [8-10] as an indication for the dominance of the non-direct K^+-meson production in pA collisions at energies far below the free nucleon-nucleon threshold.

Kaon production induced by 2.1 GeV protons (i.e. above the free NN threshold) on NaF and Pb targets has been studied by Schnetzer et al. [11] at the Lawrence Berkley Laboratory (LBL). The mass dependence is rather weak ($\alpha = 0.56 \pm 0.05$) for $\theta_K = 15°$, and increases to larger angles ($\alpha = 0.88 \pm 0.08$ for $\theta_K = 80°$). This might be explained

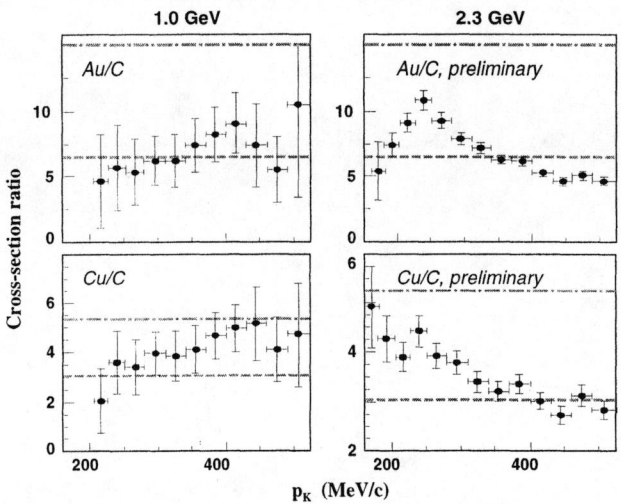

FIGURE 2. Ratios of the K^+ production cross sections for Cu/C (lower figures) and Au/C (upper figures) as a function of the kaon momentum for the proton beam energy of 1.0 GeV (left) and 2.3 GeV (right). Lines illustrate the ratios if the cross sections scale like $A^{2/3}$ (dashed lines) and A^1 (dashed-dotted lines).

by the fact that the contribution of the two-step mechanism becomes stronger with increasing production angle. The data measured by other groups [12, 13] in the beam energy range 1.2...1.5 GeV give a value of the exponent, which is very close to the expected $\alpha \approx 0.7$ indicating the dominance of the one-step kaon production.

Fig. 2 shows the target-mass dependence of the K^+ lab-momentum evaluated from the ANKE data for two different energies. At the highest investigated energy of 2.3 GeV the average cross-section ratios are in a good agreement with an $A^{0.7}$ dependence. One observes a strong momentum dependence, possibly resulting from rescattering of the produced kaons in the nuclear environment.

A fit to the lowest beam energy data with a constant value gives $\alpha = 0.74 \pm 0.05$. A comparison with the values for α obtained from the total cross sections at the same beam energy [8] shows that the ANKE data have much softer A dependence. The only difference between the PNPI and the ANKE measurements is that at ANKE the K^+ mesons were detected inside the forward cone $0° \leq \theta_K \leq 12°$, whereas at PNPI angular-integrated spectra were obtained. Therefore, the very soft A dependence may be attributed to some special features of K^+ production at forward angles. It can be speculated that at 1.0 GeV, α increases with increasing K^+-emission angle similar to the behavior that has been deduced from the LBL data. It could also be that the two-step production combined with rescattering effects of the produced kaons may result in a weaker A dependence.

The analysis of the cross sections as well as their target-mass dependence does not yet allow an unambiguous conclusion about the underlying reaction mechanisms. Therefore, it was proposed to investigate the production K^+ mesons correlated with protons and

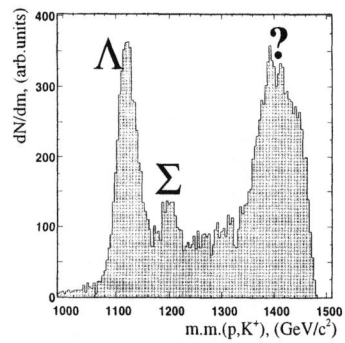

FIGURE 3. K^+p missing mass spectrum for the reaction $p(2.65\,\text{GeV})p \to pK^+Y$.

deuterons from one-step or two-step production mechanisms. The K^+d final state can only be observed in the case of the two-step reaction, while K^+p pairs come both from the two-step and direct production mechanisms. In the latter case an analysis of more than two correlated particles will provide a recognition of the πN or pN channels [14].

The possibility of such a correlation experiment was tested during the last ANKE beam-time devoted to the a_0^+ production in pp collisions at $T_p = 2.65$ GeV [15]. The missing mass spectrum of the K^+p final state is shown in Fig. 3. One can clearly see two peaks close to the Λ and Σ^+ masses. In addition, the broad peak on the right-hand side of the distribution can be attributed to the production of higher resonances like $\Sigma(1380)$ and $\Lambda(1405)$ [16]. Additional measurements at lower energy (1.2 GeV) should give us an answer whether we can directly measure the contributions of different mechanisms.

REFERENCES

1. Barsov, S., et al., *Nucl. Instr. Methods Phys. Res. A* **462**, 364 (2001).
2. Maier, R., *Nucl. Instr. Methods Phys. Res. A* **390**, 1 (1997).
3. Büscher, M., et al., *Nucl. Instr. Methods Phys. Res. A* (in print).
4. Koptev, V., et al., *Phys. Rev. Lett.* **87**, 022301 (2001).
5. Papp, J., et al., *Phys. Rev. Lett.* **34**, 601 (1975).
6. Abaev, V.V., et al., *J. Phys. G* **14**, 903 (1988).
7. Büscher, M., et al., subm. *Phys. Rev. C*; nucl-ex/0107011.
8. Koptev, V.P., et al., *JETP* **67**, 2177 (1988).
9. Cassing, W., et al., *Phys. Lett. B* **238**, 25 (1990).
10. Sibirtsev, A.A., and Büscher, M., *Z. Phys. A* **347**, 191 (1994).
11. Schnetzer, S., et al., *Phys. Rev. C* **40**, 640 (1989).
12. Debowski, M., et al., *Z. Phys. A* **356**, 313 (1996).
13. Badalà, A., et al., *Phys. Rev. Lett.* **80**, 4863 (1998).
14. Koptev, V., and Rudy, Z., *"Forward K^+ production in pA collisions"*, beam-time request for COSY proposal #70 (2001); available via http://www.fz-juelich.de/ikp/anke/.
15. Büscher, M., et al., *"Study of a_0^+-mesons at ANKE"*, beam-time request for COSY proposal #55 (2000); available via http://www.fz-juelich.de/ikp/anke/.
16. Kravchenko, P., Proc. 2nd ANKE workshop on *"Strangeness production in pp and pA interactions at ANKE"*, June 21/22 2001, Gatchina, Russia; *Berichte des Forschungszentrums Jülich*, to be published.

Three-, four- and five-body calculations of s-shell hypernuclei with realistic interactions

H. Nemura*, Y. Suzuki† and Y. Akaishi*

*Institute of Particle and Nuclear Studies, KEK, Tsukuba 305-0801, Japan
†Department of Physics, Niigata University, Niigata 950-2181, Japan

Abstract. We perform stochastic variational calculations for s-shell hypernuclear systems, $^3_\Lambda$H, $^4_\Lambda$H, $^4_\Lambda$H* and $^5_\Lambda$He with realistic NN and YN interactions including explicitly Σ degrees of freedom. The G3RS potential is used for the NN interaction. We test four sets of the YN interactions; D2, SC97e(S), SC97f(S) and SC89(S). The Λ-separation energies and the probabilities in Σ-channel are successfully calculated for all the s-shell hypernuclei. The bound state solution of $^5_\Lambda$He is the first result obtained by the five-body calculation with realistic interactions. The largest probability of Σ-component is found in the ground state of four-body system. With change of the YN interaction from D2 to SC97e(S), SC97f(S) and SC89(S), the Σ-probabilities increase except for the case of $^4_\Lambda$H using the D2. It is found that the $B_\Lambda \left(^3_\Lambda H\right)$ grows with increasing P_Σ while $B_\Lambda \left(^4_\Lambda H^*\right)$ and $B_\Lambda \left(^5_\Lambda He\right)$ decrease. The present result implies that the $\Lambda - \Sigma$ coupling plays an important role in understanding the binding mechanism of s-shell hypernuclei. The results obtained by using the SC97e(S) are in relatively good agreement with experimental data. The most important contribution in the case of the SC97e(S) is the strong spin-triplet tensor $\Lambda - \Sigma$ coupling. The D2 is also a reasonable phenomenological interaction which reproduces well the experimental binding energies.

INTRODUCTION

Few-body calculations for $A = 3 - 5$ s-shell hypernuclei are important not only to study the structure of the hypernuclei but also to pin down the characteristic feature of the hyperon-nucleon(YN) interaction (particularly in *even*-parity states) since no reliable phase shift analysis has been available yet. Dalitz *et al.* [1] determined a central ΛN potential which reproduces the separation energies of both $^3_\Lambda$H and $^4_\Lambda$H. This phenomenological interaction, however, is not consistent with the separation energy of $^5_\Lambda$He. The experimental separation energy $B_\Lambda \left(^5_\Lambda He\right) = 3.12 \pm 0.02 \text{MeV}$ is smaller by about 2 MeV than the calculated value. The suppression of tensor forces [2] or of the $\Lambda N - \Sigma N$ coupling was discussed to be a possible mechanism to resolve the anomalously small binding of $^5_\Lambda$He. The recent study by Brueckner-Hartree-Fock calculation [3] showed the importance of the *coherent* $\Lambda - \Sigma$ *coupling* in resolving the anomaly. Motivated by these, we make stochastic variational A-body calculations for s-shell hypernuclei, $^3_\Lambda$H, $^4_\Lambda$H, $^4_\Lambda$H* and $^5_\Lambda$He with realistic NN and YN interactions including explicitly Σ degrees of freedom.

FORMALISM

We take the G3RS potential [4] for the NN interaction and the D2, SC97e(S), SC97f(S) and SC89(S) potentials [3] for the YN interaction. The D2 potential does not contain any non-central components, while the others have the non-central (tensor and spin-orbit) components. We omit small nonstatic correction terms (($\mathbf{L} \cdot \mathbf{S})^2$ and $\mathbf{L}^2$ terms) in the G3RS NN interaction. The SC97e(S), SC97f(S) and SC89(S) interactions have Gaussian form factors where parameters are set to reproduce the low-energy S matrix of corresponding original Nijmegen YN interactions. The D2 interaction is phase-equivalent to the Nijmegen model D [5]. We use only the even-parity states in these interactions.

The binding energies of various systems are calculated in a complete A-body treatment. Since the Σ degrees of freedom are taken into account, and the interactions contain the tensor operator, the variational trial function must be flexible enough to search the accurate eigenenergies. The trial function is given by a combination of basis functions:

$$\Psi_{JMTM_T} = \sum_{k=1}^{N} c_k \varphi_k, \quad \text{with} \quad \varphi_k = \mathcal{A}\{G(\mathbf{x};A_k)[\theta_{L_k}(\mathbf{x};u_k,K_k) \times \chi_{S_k}]_{JM} \eta_{kTM_T}\}. \quad (1)$$

Here $\mathcal{A}$ is an antisymmetrizer acting on nucleons. For the spin χ_{S_k} and the isospin η_{kTM_T} functions, all possible configurations are taken into account. The abbreviation $\mathbf{x} = (\mathbf{x}_1, \cdots, \mathbf{x}_{A-1})$ is a set of relative coordinates and u_k has $(A-1)$ real numbers, $u_k = (u_{k:1}, \cdots, u_{k:A-1})$. For the spatial part, the basis function is constructed by the correlated Gaussian (CG) multiplied by the orbital angular momentum part expressed by the global vector representation (GVR). The CG, $G(\mathbf{x};A_k)$, and the GVR of $\theta_L(\mathbf{x})$ are defined by

$$G(\mathbf{x};A_k) = \exp\left\{-\frac{1}{2}\sum_{i<j}^{A} \alpha_{kij}(\mathbf{r}_i - \mathbf{r}_j)^2\right\} = \exp\left\{-\frac{1}{2}\sum_{i,j=1}^{A-1} A_{kij}\mathbf{x}_i \cdot \mathbf{x}_j\right\}, \text{ and} \quad (2)$$

$$\theta_{L_k}(\mathbf{x};u_k,K_k) = v_k^{2K_k}\mathcal{Y}_{L_k}(\mathbf{v}_k) = v_k^{2K_k+L_k}Y_{L_k}(\hat{\mathbf{v}}_k), \quad \text{with} \quad \mathbf{v}_k = \sum_{i=1}^{A-1} u_{k:i}\mathbf{x}_i. \quad (3)$$

The stochastic variational method (SVM) with the above CG basis produces accurate solutions. Ref. [6] shows the details and the recent progresses of SVM.

RESULTS

Table 1 lists the results of the calculations. For the SC89(S) interaction, no bound states have been obtained for $^4_\Lambda \text{H}^*$ and $^5_\Lambda \text{He}$. This result is consistent with the Variational Monte Carlo calculation [7]. The D2, SC97e(S) and SC97f(S) interactions give reasonable B_Λ values for both $^3_\Lambda \text{H}$ and $^4_\Lambda \text{H}$. The D2 interaction gives reasonable B_Λ value for $^4_\Lambda \text{H}^*$ as well and the smallest discrepancy of $B_\Lambda \left(^5_\Lambda \text{He}\right)$ between the experiment and the calculation. These results are also in good agreement with the calculations within the Brueckner-Hartree-Fock method [3].

TABLE 1. Λ separation energies of $A = 3-5$ Λ-hypernuclei (in MeV).

YN	$B_\Lambda \left(^3_\Lambda H\right)$	$B_\Lambda \left(^4_\Lambda H\right)$	$B_\Lambda \left(^4_\Lambda H^*\right)$	$B_\Lambda \left(^5_\Lambda He\right)$
D2	0.05	2.1	1.1	2.7
SC97e(S)	0.09	2.0	0.8	1.8
SC97f(S)	0.18	2.1	0.6	1.4
SC89(S)	0.27	1.7	-	-
Expt.	0.13 ± 0.05	2.04 ± 0.04	1.00 ± 0.04	3.12 ± 0.02

The SC97e(S) and SC97f(S) interactions, which contain the tensor and spin-orbit components, give bound states for both $^4_\Lambda H^*$ and $^5_\Lambda He$. The calculated values of $B_\Lambda \left(^4_\Lambda H^*\right)$ and $B_\Lambda \left(^5_\Lambda He\right)$ are smaller than the experimental values. This is the *first* calculation which has produced the bound ground state of $^5_\Lambda He$ using the realistic NN and YN interactions.

Table 2 lists the Σ-probability P_Σ for $^3_\Lambda H$, $^4_\Lambda H$, $^4_\Lambda H^*$ and $^5_\Lambda He$. The P_Σ value grows with the increasing number of particles except for $^4_\Lambda H$. The largest P_Σ value is seen in $^4_\Lambda H$, which reflects the strong enhancement due to the coherent $\Lambda - \Sigma$ coupling [3]. This tendency is independent of the choice of the YN interactions.

Based on the present study, it is possible to search for such a YN interaction that reproduces the experimental B_Λ values of all $A = 3-5$ s-shell hypernuclei. The D2 interaction can be such a *phenomenological* candidate which contains the $\Lambda N - \Sigma N$ coupling of only the central-type. On the realistic level, the SC97e(S) could be closest to the reasonable interaction. On purpose to seek more reasonable YN interaction, it is efficient to see the relation of the energy levels with the Σ-probabilities as the YN interaction changes. Table 2 shows an interesting tendency: The P_Σ's grow with the change from D2 to SC97e(S), SC97f(S) and SC89(S), with the exception of the $P_\Sigma \left(^4_\Lambda H\right)$ in the case of the D2. The exception is probably due to unusual strong enhancement by the *pure* central $\Lambda - \Sigma$ coupling. The separation energy $B_\Lambda \left(^3_\Lambda H\right)$ increases with the change of YN interaction (i.e. increasing the P_Σ) while the $B_\Lambda \left(^4_\Lambda H^*\right)$ and the $B_\Lambda \left(^5_\Lambda He\right)$ decrease. This result implies an important remark: The key to understand the binding mechanism of s-shell hypernuclei and to resolve the anomalously small binding of $^5_\Lambda He$ is the $\Lambda - \Sigma$ coupling. The important difference of D2 from the phenomenological interaction by Dalitz *et al.* [1] which are both central is the inclusion of explicit Σ degrees of freedom.

The tensor interaction is also important. In fact, in the case of realistic interaction as SC97e(S), the dominant contribution to the P_Σ is the tensor term of the $\Lambda - \Sigma$ coupling.

TABLE 2. Probabilities of finding a Σ particle in $A = 3-5$ Λ-hypernuclei.

YN	$P_\Sigma \left(^3_\Lambda H\right)$	$P_\Sigma \left(^4_\Lambda H\right)$	$P_\Sigma \left(^4_\Lambda H^*\right)$	$P_\Sigma \left(^5_\Lambda He\right)$
D2	0.0014	0.0190	0.0041	0.0056
SC97e(S)	0.0016	0.0144	0.0090	0.0138
SC97f(S)	0.0023	0.0180	0.0104	0.0162
SC89(S)	0.0049	0.0252	-	-

The probabilities of the dominant state in Σ-channel, $P_\Sigma(L,S,S_c)$, are:

$$P_\Sigma(2,\tfrac{3}{2},1) = 0.0009 \quad \text{for } {}^3_\Lambda\text{H}, \qquad P_\Sigma(2,2,\tfrac{3}{2}) = 0.0082 \quad \text{for } {}^4_\Lambda\text{H},$$
$$P_\Sigma(2,1,\tfrac{1}{2}) = 0.0052 \quad \text{for } {}^4_\Lambda\text{H}^*, \qquad P_\Sigma(2,\tfrac{3}{2},1) = 0.0122 \quad \text{for } {}^5_\Lambda\text{He}, \quad (4)$$

where L, S and S_c are the total orbital angular momentum, the total spin and the core nucleus spin, respectively. The D-state dominance and the additive relation of the spin part $S = S_c + \tfrac{1}{2}$ imply the strong spin-triplet tensor $\Lambda - \Sigma$ coupling.

CONCLUSIONS

The stochastic variational calculations has been performed for A-body system ($A \leq 5$). It is the *first time* that the bound state solution of ${}^5_\Lambda$He is obtained by the five-body calculation with the realistic interactions including Σ degrees of freedom.

The SC97e(S) interaction seems to be most appropriate among the YN interactions employed in this study. According to the study using Faddeev method [8], the SC89 interaction reproduces the experimental $B_\Lambda \left({}^3_\Lambda\text{H}\right)$ value. The SC89 interaction, however, did not pass the test in reproducing the experimental $B_\Lambda \left({}^4_\Lambda\text{H}^*\right)$ and $B_\Lambda \left({}^5_\Lambda\text{He}\right)$ as was also shown by Variational Monte Carlo method [7]. In comparison with the SC89(S), the SC97e(S) has a little weaker spin-singlet strength and has relatively stronger spin-triplet strength with a weaker Σ transition. As a result, the difference between the ${}^1S_0$ and ${}^3S_1$ scattering phase shifts becomes small.

The D2 is another solution for an effective YN interaction in which the effect of the tensor component is renormalized suitably in the central part of the YN interaction. The D2 interaction is expected to be useful interaction for studying the structure of light hypernuclei.

ACKNOWLEDGMENTS

We are thankful to Y. Fujiwara for useful discussions. One of the authors (H. N.) would like to thank K. Miyagawa, H. Kamada, A. Nogga and K. Varga for helpful discussions and JSPS Research Fellowships for Young Scientists.

REFERENCES

1. Dalitz, R. H., Herndon, R. C., and Tang, Y. C., *Nucl. Phys.* **B47**, 109-137 (1972).
2. Shinmura, S., Akaishi, Y., and Tanaka, H., *Prog. Theor. Phys.* **71**, 546-560 (1984).
3. Akaishi, Y., Harada, T., Shinmura, S., and Khin Swe Myint, *Phys. Rev. Lett.* **84**, 3539-3541 (2000).
4. Tamagaki, R., *Prog. Theor. Phys.* **39**, 91-107 (1968).
5. Shinmura, S., private communication.
6. Suzuki, Y., and Varga, K., *Stochastic Variational Approach to Quantum-Mechanical Few-Body Problems*, Lecture Notes in Physics, Vol. 54, Springer-Verlag, Berlin Heidelberg, 1998.
7. Carlson, J. A., "Light Hypernuclei and the Hyperon-Nucleon Interaction," in *LAMPF Workshop on (π, K) Physics*, edited by B. F. Gibson et al., AIP Conf. Proc. 224, New York, 1991, pp.198-210.
8. Miyagawa, K., Kamada, H., Glöckle, W., and Stoks, V., *Phys. Rev. C* **51**, 2905-2913 (1995).

Non-mesonic decay of Λ-hypernuclei and the Γ_n/Γ_p ratio

J. E. Palomar

Departamento de Física Teórica and IFIC, Centro Mixto Universidad de Valencia-CSIC, Institutos de Investigación de Paterna, Apdo. correos 22085, 46071, Valencia, Spain

Abstract. We take an approach to the Λ nonmesonic weak decay in nuclei based on the exchange of mesons. The one pion and one kaon exchange are considered, together with the exchange of two pions, either correlated, leading to an important scalar-isoscalar exchange (σ-like exchange), or uncorrelated (box diagrams). A drastic reduction of the OPE results for the Γ_n/Γ_p ratio is obtained and the new results are compatible with all present experiments within errors. The absolute rates obtained for different nuclei are also in fair agreement with experiment.

INTRODUCTION

The problem of the Γ_n/Γ_p ratio is the most persistent problem related to the non-mesonic decay of Λ-hypernuclei. The problem lies in the large discrepancy between the theoretical ratio provided by the OPE (one pion exchange) model (around 1/8) and the experiment. Experimentally one has results for $^5_\Lambda$He from [1] with a ratio 0.93±0.5 and for $^{12}_\Lambda$C with ratios $1.33^{+1.12}_{-0.81}$ [1], $1.87^{+0.91}_{-1.59}$ [2] and 0.70±0.30, 0.52±0.16 [3]. More recent results for $^{12}_\Lambda$C are still quoted as preliminary [4, 5] but also range in values around unity with large errors. Certainly the OPE model is too simple, but the different attempts improving on the model have not been succesful. Only recently some works are giving values for the ratio that are within the error bands of some of the experiments [6, 7, 8]. Here I report on the recent work [9], where in addition to one pion exchange we have considered kaon and ω exchange, and correlated as well as uncorrelated two-pion exchange. We do not consider here the effects of the exchange of more particles like ρ, K^*, η, since they have been seen to be not relevant [10].

FORMALISM AND ONE PION EXCHANGE

The decay of the Λ in nuclear matter was investigated in [11] with the propagator approach which provides a unified picture of different decay channels of the Λ. The decay width of the Λ is calculated in infinite nuclear matter, and is extended to finite nuclei with the local density approximation. We will use here the same technique.

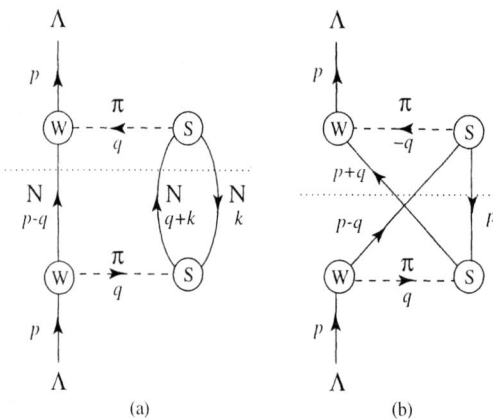

FIGURE 1. Lowest order Λ self-energy. The non-mesonic width comes from the imaginary part when the intermediate states cut by a horizontal line are placed on shell.

For the effective $\pi \Lambda N$ weak interaction we use the Lagrangian:

$$\mathcal{L}_{\Lambda \pi N} = iG\mu^2 \bar{\psi}_N [A + B\gamma_5] \vec{\tau} \cdot \vec{\phi}_\pi \psi_\Lambda + \text{h.c.} \tag{1}$$

where μ denotes the pion mass, G is the weak coupling and we implement the $\Delta I = 1/2$ rule by assuming that the Λ behaves as a $I = 1/2$, $I_z = -1/2$ state in the isospin space. The A and B constants are determined by the parity violating and the parity conserving amplitudes of the non-leptonic decay of the free Λ: $A = 1.06$, $B\mu/(2M_N) = -0.527$.

In order to evaluate the Λ decay width, Γ, in a nuclear medium due to a certain $\Lambda N \to NN$ transition, we start with the calculation of the self-energy in the medium, Σ, shown in Fig. 1, and then we take its imaginary part:

$$\Gamma = -2\text{Im}\Sigma \tag{2}$$

In addition we take into account the ph and Δh excitations to all orders in the sense of the RPA, and also short range correlations are introduced as done in [11]. We show our results for $^{12}_\Lambda\text{C}$ from OPE in Table 1. From the results of this table one gets the Γ_n/Γ_p ratio of $\sim 1/8$ and a too large non-mesonic width. As we can see there is a large longitudinal p-wave contribution in the case of the proton, which is the responsible of these bad results. OPE is not enough, and therefore more processes have to be considered.

ONE KAON EXCHANGE AND TWO-PION EXCHANGE

The Λ non-mesonic decay with one K exchange takes place through the diagrams shown in Fig. 1, substituting the pions by kaons in the figure. The inclusion of K exchange is

TABLE 1. OPE contributions to Γ_p and Γ_n: s-wave, longitudinal p-wave, transverse p-wave and interference between longitudinal and transverse p-wave.

	S	P_L	P_T	$P_{int.LT}$	Total
$\Gamma_p/\Gamma_\Lambda^{free}$	0.177	0.684	0.012	0.082	0.956
$\Gamma_n/\Gamma_\Lambda^{free}$	0.089	0.049	0.003	-0.021	0.119

then straightforward, once the weak KNN and strong $K\Lambda N$ vertices are fixed. The strong $K\Lambda N$ vertex is given by chiral Lagrangians, and for the weak NNK vertex we have used the results of reference [12], where for the parity violating part the amplitude is assumed to behave as the 6th component of the $SU(3)$ generators, and for the parity conserving part the pole model is used. The calculation gives an interference between kaon and pion of -0.629 in units of the free Λ decay width in the longitudinal p-wave in the proton case. This contribution is most welcome since it kills the large longitudinal p-wave contribution from OPE in the proton case (see table 1) and helps us to get better values for the ratio and the total width.

We include also two-pion exchange, both correlated and uncorrelated (box diagrams). The correlated exchange accounts for the σ exchange, since in [13] it was found that the σ meson is dynamically generated by the in-flight two pion interaction when summing up the s-wave t-matrix of the $\pi\pi$ scattering to all orders using the Bethe-Salpeter equation. The former picture of the σ meson was used to describe its role in the NN interaction in ref [14], getting opposite sign to the conventional contributions taking only the exchange of a σ particle. We have followed an analogous model to the one of the aforementioned reference, finding an interesting important cancelation between correlated and uncorrelated exchange at the relevant momentum of ~ 400 MeV.

Finally we have also considered ω exchange, taking a strong $NN\omega$ coupling that allows us to describe the Argonne v_{14} potential. For the weak coupling we use the one given in [15]. This ω contribution is small but still helps to obtain a better agreement with the experiment for the total widths.

RESULTS AND CONCLUSIONS

In Table 2 we show the results for $^{12}_\Lambda C$ separating the different contributions. There we can see how the introduction of K exchange reduces the proton rate by a factor of two and increases the neutron rate also by a factor of two, thus increasing the Γ_n/Γ_p ratio in a factor of four and reducing also the total rate from OPE. The effect of two-pion and ω exchange is not so important, due in part to the cancellation between correlated and uncorrelated two-pion exchange contributions.

Finally, in Table 3 we present our results for Γ_p, Γ_n and the Γ_n/Γ_p ratio for different nuclei. Here, to get the total width, we have added to the non-mesonic width the mesonic one taken from [16] and the $2p2h$ induced one calculated in [17].

As we can see, our results for the ratio are considerably improved with respect to the OPE ones and compatible with most of the present experiments. The total rates obtained

TABLE 2. Decay rates obtained when considering the different contributions.

	π	$\pi+K$	$\pi+K+2\pi$	$\pi+K+2\pi+\omega$
$\Gamma_p/\Gamma_\Lambda^{\text{free}}$	0.956	0.522	0.571	0.504
$\Gamma_n/\Gamma_\Lambda^{\text{free}}$	0.119	0.273	0.308	0.265

TABLE 3. Decay rates and the Γ_n/Γ_p ratio for different hypernuclei.

Nucleus	$\Gamma_p/\Gamma_\Lambda^{\text{free}}$	$\Gamma_n/\Gamma_\Lambda^{\text{free}}$	$(\Gamma_p+\Gamma_n)/\Gamma_\Lambda^{\text{free}}$	Γ_n/Γ_p	$\Gamma_{\text{tot}}/\Gamma_\Lambda^{\text{free}}$
$^{12}_\Lambda$C	0.504	0.265	0.769	0.53	1.289
$^{28}_\Lambda$Si	0.665	0.351	1.016	0.53	1.386
$^{40}_\Lambda$Ca	0.694	0.366	1.060	0.53	1.390
$^{56}_\Lambda$Fe	0.754	0.398	1.152	0.53	1.452
$^{89}_\Lambda$Y	0.785	0.414	1.199	0.53	1.499
$^{139}_\Lambda$La	0.765	0.403	1.168	0.53	1.468
$^{208}_\Lambda$Pb	0.847	0.446	1.293	0.53	1.593

also agree with experimental data.

In conclusion, we have found that the contribution of K exchange is essential to reduce the total decay rate from the OPE results and simultaneously increase the value of the Γ_n/Γ_p ratio from values around 0.12 for the OPE to values around 0.52. The fact that the correlated two-pion exchange contribution has opposite sign to the uncorrelated one is also helpful. The results obtained are in agreement with most of the experiments.

REFERENCES

1. Szymanski, J.J., et al., *Phys. Rev. C* **43**, 849 (1991).
2. Noumi, H., et al., *Phys. Rev. C* **52**, 2936 (1995).
3. Montwill, A., et al., *Nucl. Phys.* **A234**, 413 (1974).
4. Outa, H., et al., *Nucl. Phys.* **A670**, 281 (2000).
5. Hashimoto, O., private communication.
6. Sasaki, K., Inoue, T., and Oka, M., *Nucl. Phys.* **A669**, 331 (2000); [nucl-th/9906036]; erratum *Nucl. Phys.* **A678**, 455 (2000).
7. Oka, M., talk given at Hyper2000 conference, Torino, October 2000.
8. Parreño, A., and Ramos, A., nucl-th/0104080.
9. Jido, D., Oset, E., and Palomar, J.E., to appear in *Nucl. Phys.* **A**; nucl-th/0101051.
10. Shmatikov, M., *Nucl. Phys* **A580**, 538 (1994).
11. Oset, E., and Salcedo, L.L., *Nucl. Phys.* **A443**, 704 (1985).
12. Dubach, J.F., Feldman, G.B., Holstein, B.R., and de la Torre, L., *Annals Phys.* **249**, 146 (1996).
13. Oller, J.A., and Oset, E., *Nucl. Phys* **A620**, 438 (1997); erratum *Nucl. Phys* **A652**, 407 (1999).
14. Oset, E., Toki, H., Mizobe, M., and Takahashi, T.T., *Prog. Theor. Phys.* **103**, 351 (2000).
15. Parreño, A., Ramos, A., and Bennhold, C., *Phys. Rev. C* **56**, 339 (1997).
16. Nieves, J., and Oset, E., *Phys. Rev. C* **47**, 1478 (1993).
17. Ramos, A., Oset, E. and Salcedo, L.L., *Phys. Rev. C* **50**, 2314 (1994).

Pion and kaon vector form factors and some applications

J. E. Palomar

Departamento de Física Teórica and IFIC, Centro Mixto Universidad de Valencia-CSIC, Institutos de Investigación de Paterna, Apdo. correos 22085, 46071, Valencia, Spain

Abstract. The pion and kaon coupled-channel vector form factors are described by making use of the resonance chiral Lagrangian results together with a suitable unitarization method in order to take care of the final state interactions[1]. A very good reproduction of experimental data is accomplished for the vector form factors up to $\sqrt{s} \leq 1.2$ GeV. We then apply these form factors to calculate the width of τ decay to two pions and to two kaons and also to evaluate the contribution of these mesons to the anomalous magnetic moment of the muon.

UNITARIZATION

Using an appropriate unitarization method we take into account the final state interaction corrections to the tree level amplitudes calculated from lowest order χPT [2] and from the inclusion of explicit resonance fields in a chiral symmetry fashion as given in [3].

We will work in the isospin limit, with $|\pi\pi\rangle$ and $|K\bar{K}\rangle$ states (and the ρ resonance) in the $I = 1$ channel, and we will use matrix notation, labeling pions with 1 and kaons with 2. In the $I = 0$ case we only have kaons (and the ω and ϕ resonances). Starting from the unitarity of the S-matrix and the introduction of the electromagnetic meson form factor $F_{MM'}(s)$:

$$\langle \gamma(q)|T|M(p)M'(p')\rangle = e\varepsilon_\mu(p-p')^\mu F_{MM'}(s) \qquad (1)$$

with $q^2 = s$, e the modulus of the electron charge and ε_μ the photon polarization vector, we can arrive at the expression:

$$F^I(s) = [1 + \tilde{Q}(s)^{-1} \cdot K^I(s) \cdot \tilde{Q}(s) \cdot g^I(s)]^{-1} \cdot R^I(s) \qquad (2)$$

where $\tilde{Q}_{ij}(s) = p_i(s)\delta_{ij}$ and $K^I(s)$ is the matrix collecting the tree level scattering amplitudes[2] between definite $\pi\pi$ and $K\bar{K}$ isospin states. $F^I(s)$ is the column matrix

---

[1] For a more detailed version of the description of the form factors see [1].
[2] Derived from lowest order χPT plus s-channel vector resonance exchange contributions [2] corresponding to the transition $i \to j$.

$F^I(s)_i = F_i^I(s)$; $R^I(s)$ is a vector made up by functions without any cut and $g^I(s)$ is the diagonal matrix given by the loop with two meson propagators (see ref. [4]). The expression for the P-wave amplitudes $T^I(s)$ that we have used to go from eq.(1) to eq.(2) is the one provided by the N/D method adapted to the chiral framework (see ref. [4]).

In the large N_c limit loop physics is suppressed. Thus, in this limit $F^I(s) = R^I_{N_c^{\text{leading}}}(s) = F_t^I(s)$, where $F_t^I(s)$ is the tree level form factor[3]. We can write:

$$F^I(s) = [1 + \tilde{Q}(s)^{-1} \cdot K^I(s) \cdot \tilde{Q}(s) \cdot g^I(s)]^{-1} \cdot [F_t^I(s) + R^I_{\text{subleading}}(s)] \quad (3)$$

with $R^I_{\text{subleading}}(s)$ being $O(N_c^{-1})$. If we require that the vector form factor of eq. (3) vanishes when $s \to \infty$ we find that the subleading part of $R^I(s)$, which at first can be a polynomial, must be a constant. One can fix the values of the d_i^I parameters appearing in the $g_i^I(s)$ functions (see ref [4]) and $R^I_{\text{subleading}}$ by matching our results to those of one loop χPT, and the bare masses of the resonances by the requirements that the moduli of the scattering amplitudes have a maximum for $\sqrt{s} = M_{\text{resonance}}^{\text{physical}}$.

RESULTS

If we compare our results with the experiment we see that we can describe in a very precise way the pion and kaon vector form factors and the P-wave $\pi\pi$ phase shifts up to $s = 1.44$ GeV^2 (even for higher energies in the case of the phase shifts). In the low energy region we have very good agreement with two loop χPT. The resonance regions are also well reproduced. We have also applied these form factors to calculate the partial width of the decay of a τ lepton going to two pions and to two kaons, and also the contributions of these mesons to the anomalous magnetic moment of the muon. We get good results for the pion case, while for the kaon case the results are not so good since the value of the form factor at energies higher than 1.2 GeV plays a non-negligible role.

REFERENCES

1. Oller, J. A., Oset, E., and Palomar, J. E., *Phys. Rev. D* **63**, 114009 (2001), [hep-ph/0011096].
2. Gasser,J., and Leutwyler, H., *Nucl. Phys.* **B250**, 465, 517, 539 (1985).
3. Ecker, G., Gasser, J., Pich, A. and de Rafael, E., *Nucl. Phys.* **B321**, 311 (1989).
4. Oller, J. A., and Oset, E., *Phys. Rev. D* **60**, 074023 (1999).

---

[3] The tree level form factors are also evaluated using the lowest order χPT Lagrangian [2] plus the chiral resonance Lagrangian [3]

Recent results on the nonmesonic weak decay of hypernuclei within a one-meson-exchange model

A. Parreño [1*], A. Ramos* and C. Bennhold[†]

Departament d'Estructura i Constituents de la Matèria, Universitat de Barcelona, Diagonal 647, E-08028 Barcelona, Spain
[†]*Center for Nuclear Studies and Department of Physics, The George Washington University, Washington DC, 20052, USA*

Abstract. We update [1] the results presented in Ref. [2] for the nonmesonic decay (NMD) of $^{12}_\Lambda C$ and $^5_\Lambda He$. We pay special attention to the role played by Final State Interractions (FSI) in the decay observables. We follow a One-Meson-Exchange (OME) model which includes the exchange of the π, ρ, K, K^*, η and ω mesons. We also present recent predictions for different observables concerning the decay of the doubly strange $^6_{\Lambda\Lambda} He$ hypernucleus [3].

INTRODUCTION

The Λ particle, the lightest among the hyperons, decays in free space into nucleons and pions, with a lifetime of $\approx 2.632 \times 10^{-10}$ sec. This (mesonic) decay mode is dominant for the very light s-shell hypernuclei but, as the number of nucleons increases, it gets suppressed due to the low momentum of the outgoing nucleon, which gets Pauli blocked. Therefore, for hypernuclei with $A \simeq 5$ or larger, the decay is assumed to proceed mainly via the two-body reaction $\Lambda N \to NN$, the so-called nonmesonic decay (NMD) mode. Since at present the availability of stable hyperon beams is very limited, this NMD mode is the only source of information on the $|\Delta S| = 1$ hyperon-nucleon (YN) interaction. Further, the decay of doubly-strange hypernuclei, as the decay of $^6_{\Lambda\Lambda} He$ studied here, provides additional information due to novel hyperon-induced decay mechanisms, namely $\Lambda\Lambda \to \Lambda N$ and $\Lambda\Lambda \to \Sigma N$.

Our framework includes the exchange of the pseudoscalar π, η, K octet for the long-range part while parametrizing the short-range part through the exchange of the vector ρ, ω and K^* mesons. Realistic baryon-baryon forces for the $S = 0, -1$ and -2 sectors [4] are used to account for the strong interaction in the initial and final states.

FORMALISM

Analytic expressions of the total and partial decay rates, as well as of the Parity Violating (PV) asymmetry for the decay of single- and double-Λ hypernuclei can be found in Refs.

---

[1] and *IFAE, Universitat Autònoma de Barcelona, E-08193 Bellaterra, Barcelona, Spain.*

[1, 3]. Details on how to derive the transition potential and its final form can be found also there. Only some basic aspects of the formalism are going to be outlined here.

The transition potential is derived by performing a nonrelativistic reduction of the Feynman diagram associated to the exchange of the meson under consideration. In analogy to One-Boson-Exchange (OBE) based models of the strong interaction, the present formalism includes not only the exchange of the long-ranged pion, but also more massive mesons which account for shorter distances. Those are the ρ, K, K^*, η and ω mesons. In order to account for finite size effects, we include a monopole form factor at each vertex, where the value of the cut-off depends on the meson.

As it is well known, one of the sources of uncertainty in OBE models comes from the coupling constants between baryons and mesons (BBM). In the strong sector the different interaction models use SU(3) in order to obtain the BBM couplings that are not constrained experimentally. Recently, the Nijmegen group has made available new baryon-baryon interactions in the strangeness $S = 0, -1, -2, -3$ and -4 sectors [4], where the $S = -2 \rightarrow -4$ versions are SU(3) extensions of the models in the $S = 0$ and -1 sectors, which are fitted to experimental data. The authors of Ref. [4] give six different models, which fit the available NN and YN scattering data equally well but are characterized by different values of the magnetic vector $F/(F+D)$ ratio, ranging from 0.4447 (model NSC97a) to 0.3647 (model NSC97f).

In the weak sector, only the decay of the Λ and Σ hyperons into nucleons and pions can be experimentally observed. For the other mesons, $SU_w(6)$ represents a convenient tool to obtain the PV amplitudes, while for the Parity Conserving (PC) ones, we use a pole model [2, 5] with only baryon pole resonances.

In order to take into account the effects of the strong interaction between the baryons, correlated wave functions are obtained from a G-matrix calculation for the initial ΛN and $\Lambda\Lambda$ states. Our treatment of FSI is restricted [1] to the study of the mutual influence between the two emitted baryons. To include the effect of these FSI in the decay process, we obtain a scattering BB wave function from a Lippmann-Schwinger (T-matrix) equation using the NSC97f potential model of Ref. [4]. In the next section, we will also show the results obtained with other more simplistic approaches. These approaches include, for instance, the absence of FSI or the use of a phenomenological correlation function multiplying the uncorrelated wave function.

RESULTS

We present updated results for the weak nonmesonic decay of hypernuclei to the light of the new Nijmegen baryon-baryon potentials [4]. These strong interaction models influence the weak decay mechanism, not only through the coupling constants and form factors at the strong vertex involved in the two-body reaction, $\Lambda N \rightarrow NN$ (and $\Lambda\Lambda \rightarrow YN$ in $\Lambda\Lambda$-hypernuclei, where Y denotes a hyperon in the final state), but also via the PC piece of the weak vertex, obtained from a pole model, as well as from the corresponding correlated wave functions for the initial ΛN and final NN states.

Table 1 shows our estimations for the decay observables (in units of the free Λ decay rate, $\Gamma_\Lambda = 3.8 \times 10^9 \text{ s}^{-1}$) of $^5_\Lambda\text{He}$ and $^{12}_\Lambda\text{C}$. Those numbers have been obtained working

TABLE 1. Weak decay observables for $^5_\Lambda$He and $^{12}_\Lambda$C. The strong NSC97a (left column) and NSC97f (right column) potential models were used [4]. For the final NN wave function we used the solution of a T-matrix equation with either NSC97a or NSC97f.

	Γ_{nm}		Γ_n/Γ_p		Γ_p		A_p	
	a	f	a	f	a	f	a	f
$^5_\Lambda$He	0.425	0.317	0.343	0.457	0.317	0.218	−0.675	−0.682
EXP:	0.41 ± 0.14[9]		0.93 ± 0.55[9]		0.21 ± 0.07[9]		0.24 ± 0.22 [7]	
$^{12}_\Lambda$C	0.726	0.554	0.288	0.341	0.564	0.413	0.358	0.367
EXP:	1.14 ± 0.08[8]		$1.33^{+1.12}_{-0.81}$[9]		$0.31^{+0.18}_{-0.11}$[10]		0.05 ± 0.53*	
	0.89 ± 0.15 ± 0.03[10]		$1.87 \pm 0.59^{+0.32}_{-1.00}$[10]					
	1.14 ± 0.2[9]		0.70 ± 0.3[11]					
			0.52 ± 0.16[11]					

* This number was obtained by dividing the experimental asymmetry, $\mathcal{A} = -0.01 \pm 0.10$[6], by a polarization of $P_y = -0.19$.

consistently within each of the strong models of the Nijmegen group. The new results for the nonmesonic rates compare favourably with the present experimental data. The n/p ratio has increased with respect to our previous works and it now lies practically within the lower side of the error band. The asymmetry for $^{12}_\Lambda$C is also compatible with experiment [6] but that for $^5_\Lambda$He disagrees strongly from the recent experimental observation [7]. The latter work finds a small and positive value for the elementary asymmetry parameter a_Λ in $^5_\Lambda$He, while that for $^{12}_\Lambda$C is large and negative. Our meson-exchange model does not explain the present experimental differences and understanding this issue is one of the current challenges, both experimental and theoretical, in the study of the weak decay of hypernuclei.

We found a tremendous influence on the weak decay observables from the way FSI are considered, especially in the case of total and partial decay rates. In Table 2 we compare the results obtained by using different approaches to implement FSI. A phenomenological implementation of FSI effects, $f_{\text{phen}} = 1 - j_0(q_c r)$ with $q_c = 3.93$ fm^{-1}, or not including them at all, gives rise to decay rates that differ by more than a factor of two, and to a neutron-to-proton ratio about 20% larger from what is obtained with the more realistic calculation that uses the proper NN scattering wave function. The K-matrix solution represents an approximation which is only appropriate for standing waves, i.e. non-propagating solutions, as is the case in the nuclear medium. The differences observed in the decay rates and the n/p ratio are much larger than the uncertainties tied to the different strong interaction models commented above. Therefore, accurate calculations of the nonmesonic weak decay of hypernuclei demand a proper treatment of FSI effects through the solution of a T-matrix using realistic NN interactions.

Predictions for the decay observables of $^6_{\Lambda\Lambda}$He are shown in Table 3. The $\Lambda N \to NN$ rate is found to be more than twice as large as in $^5_\Lambda$He due to the increased binding of the second Λ hyperon. [2] The total hyperon-induced rate is 4% of the total nonmesonic

---

[2] We have to note here that this number could change in the light of the new data presented by Prof. Nakazawa during this conference: $B(^6_{\Lambda\Lambda}\text{He}) = 6.93 \pm 0.54$ MeV and $\Delta B(^6_{\Lambda\Lambda}\text{He}) = 0.69 \pm 0.54$ MeV.

TABLE 2. Weak decay observables for $^5_\Lambda$He using different approaches to FSI. The NSC97f model was used.

$^5_\Lambda$He	Γ_{nm}	Γ_n/Γ_p	Γ_p	A_p
T	0.317	0.457	0.218	-0.682
K	0.475	0.471	0.323	-0.650
$f_{\text{phen}}(r)$	0.766	0.619	0.473	-0.671
no FSI	0.721	0.614	0.447	-0.654

TABLE 3. Partial weak decay rates for $^6_{\Lambda\Lambda}$He.

$\Lambda n \to nn$	0.30	$\Lambda\Lambda \to \Lambda n$	3.6×10^{-2}
$\Lambda p \to np$	0.66	$\Lambda\Lambda \to \Sigma^0 n$	1.3×10^{-3}
$\Lambda N \to NN$	0.96	$\Lambda\Lambda \to \Sigma^- p$	2.6×10^{-3}
Γ_n/Γ_p	0.46	$\Lambda\Lambda \to YN$	4.0×10^{-2}

rate, and it is dominated by the $\Lambda\Lambda \to \Lambda n$ mode, which allows direct access to exotic vertices like $\Lambda\Lambda K$, unencumbered by the usually dominant pion exchange. Indeed, one-loop log corrected χPT results[3] modify the $\Lambda\Lambda \to \Lambda n$ by 50% while changing the $\Lambda N \to NN$ only at the 15% level, demonstrating the power of this weak mechanism to test χPT in the weak SU(3) sector. With a free Λ in the final state this new mode should be distinguishable from the usual nucleon-induced decay channels.

ACKNOWLEDGMENTS

This work has been partially supported by the U.S. Dept. of Energy under Grant No. DE-FG03-00-ER41132, by the DGICYT (Spain) under contract PB98-1247, by the Generalitat de Catalunya project SGR2000-24, and by the EEC-TMR Program EURODAPHNE under contract CT98-0169.

REFERENCES

1. Parreño, A., and Ramos, A., nucl-th/0104080.
2. Parreño, A., Ramos, A., and Bennhold, C., *Phys. Rev. C* **56**, 339 (1997).
3. Parreño, A., Ramos, A., and Bennhold, C., nucl-th/0106054.
4. Stoks, V. G. J., and Rijken, Th. A., *Phys. Rev. C* **59**, 3009 (1999); Rijken, Th. A., Stoks, V. G. J., and Yamamoto, Y., *Phys. Rev. C* **59**, 21 (1999).
5. Dubach, J. F., Feldman, G. B., Holstein, B. R., de la Torre, L., *Ann. Phys. (N.Y.)* **249**, 146 (1996); de la Torre, L., *Ph.D. Thesis, Univ. of Massachusetts*, 1982.
6. Ajimura, S., et al., *Phys. Lett. B* **282**, 293 (1992).
7. Ajimura, S., et al., *Phys. Rev. Lett.* **18**, 4052 (2000).
8. Bhang, H., et al., *Phys. Rev. Lett.* **81**, 4321 (1998).
9. Szymanski, J. J., et al., *Phys. Rev. C* **43**, 849 (1991).
10. Noumi, H., et al., *Phys. Rev. C* **52**, 2936 (1995).
11. Montwill, A., et al., *Nucl. Phys.* **A234**, 413 (1974).

Finite temperature effects on the $\bar{K}$ optical potential

L.Tolós*, A.Polls* and A.Ramos*

Departament d'Estructura i Constituents de la Matèria, Barcelona 08028, Spain

Abstract. By solving the Bethe-Goldstone equation, we have obtained the $\bar{K}$ optical potential from the $\bar{K}N$ effective interaction in nuclear matter at $T = 0$. We have extended the model by incorporating finite temperature effects in order to adapt our calculations to the experimental conditions in heavy-ion collisions. In the rank of densities $(0-2\rho_0)$, the finite temperature $\bar{K}$ optical potential shows a smooth behaviour if we compare it to the $T=0$ outcome. Our model has also been applied to the study of the ratio between K^+ and K^- produced at GSI with T around 70 MeV. Our results point at the necessity of introducing an attractive $\bar{K}$ optical potential.

THE MODEL AT $T = 0$

The $\bar{K}$ optical potential in nuclear matter is computed using a $\bar{K}N$ effective interaction, G-matrix, derived microscopically from a meson-exchange potential [1]. The G-matrix is given by the solution of the Bethe-Goldstone equation

$$G(w) = V + V \frac{Q}{w - H_0 + i\eta} G(w), \qquad (1)$$

which takes into account the different channels ($\bar{K}N$, $\pi\Sigma$, $\pi\Lambda$), which are coupled by the strong interaction.

At lowest order in the Brueckner-Hartree-Fock (BHF) theory, the single-particle potential for antikaons $U_{\bar{K}}$ at $T=0$ is given by

$$U_{\bar{K}}(k, E_{\bar{K}}^{qp}) = \sum_{N \leq F} \langle \bar{K}N \mid G_{\bar{K}N \rightarrow \bar{K}N}(w = E_N^{qp} + E_{\bar{K}}^{qp}) \mid \bar{K}N \rangle, \qquad (2)$$

in which $E_{\bar{K}}^{qp}$ is self-consistently determined.

In order to introduce in-medium effects on the intermediate states of the G-matrix, we have employed mean field single-particle propagators for the baryons. The single-particle potential for nucleons has been derived from a $\sigma - \omega$ model, while for Λ and Σ, we have followed the parameterization of Ref. [2]. Finally, the pion is dressed with the momentum and energy dependent self-energy of Ref. [3], that incorporates the coupling to particle-hole, Δ-hole and two particle-two hole excitations.

Figure 1 (left) shows the real and imaginary part of $U_{\bar{K}}$ as a function of $k_{\bar{K}}$ at $T=0$ for $\rho = \rho_0$ including or not the dressing of pions. The long-dashed line has been calculated determining the complex $\bar{K}$ single particle energy self-consistently. However, the pion self-energy in the intermediate states is not included. This procedure is similar to the one presented in Ref. [4] with the inclusion here of a $\sigma - \omega$ model for the nucleon dressing. The presence of the $\Lambda(1405)$ resonance, dynamically generated in our model, introduces

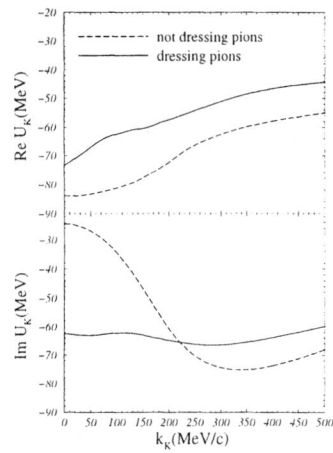 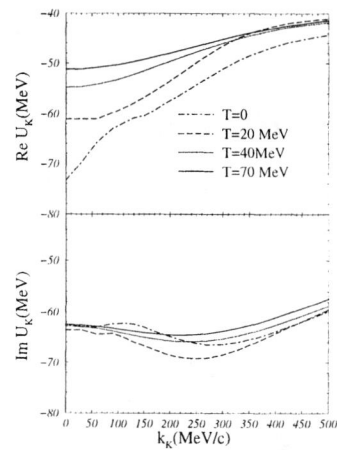

FIGURE 1. $U_{\bar{K}}$ as a function of $k_{\bar{K}}$ for $T = 0$, including or not the dressing of pions (left), and for different T (right), at $\rho = 0.17$ fm^{-3}.

a strong energy dependence on the $\bar{K}N$ amplitude which governs, in turn, the behaviour of $U_{\bar{K}}$. The Pauli blocking on the intermediate nucleons has a repulsive effect on the resonance, produced by having moved the threshold of intermediate states to higher energies. The self-consistent incorporation of the $\bar{K}$ properties on the $\bar{K}N$ G-matrix moves the resonance back in energy, but it dilutes as the density increases. If we now include the dressing of pions (solid line), we obtain a less attractive real part, going from -84 MeV to -73 MeV at $k = 0$ MeV, and a smoother behaviour for the imaginary part. This can be understood by looking again to the $\bar{K}N$ amplitude, specially for $L = 0$. Due to the additional dressing of the pions, the resonance dissolves even faster loosing structure, although it shows a tendency to move to higher energies.

TEMPERATURE EFFECTS

The introduction of temperature in the calculation affects the Pauli blocking of the intermediate nucleon states in the G-matrix. In addition, the $\bar{K}$ optical potential is calculated according to

$$U_{\bar{K}}(k, E_{\bar{K}}^{qp}) = \int d^3 k\, n(k,T)\, \langle \bar{K}N \mid G_{\bar{K}N \to \bar{K}N}(w,T) \mid \bar{K}N \rangle , \qquad (3)$$

where $n(k,T)$ is the nucleon momentum distribution at finite temperature. By looking at the right hand side on Fig.1, one observes that, as the temperature increases, $U_{\bar{K}}$ is less attractive and shows a smoother behaviour in momentum. At finite T, the sum over momenta in Eq. (3) is extended over the corresponding Fermi distribution, and higher nucleon momentum translates into a weaker interaction. Moreover, the effective interaction has also changed. However, this last effect is less important. At $T = 70$ MeV, the momentum dependence is very smooth, reducing in a factor of three the difference between low and high momentum. However, $U_{\bar{K}}$ is still attractive. This fact can be used to understand the enhancement of the observed ratio K^-/K^+ in heavy-ion collisions at GSI energies.

RATIO K^-/K^+

Heavy-ion collisions provide a unique possibility to create a dense and hot nuclear system to study the in-medium properties of hadrons, like in-medium effects on the $\bar{K}$. The measured ratio of particles K^-/K^+ can be understood assuming an in-medium attractive potential for the K^-. However, the increase of this ratio could also be explained from the in-medium enhanced production of antikaons through Σ hyperons [5].

In direct nucleon-nucleon collisions, the production of K^+ and K^- is governed by quite distinct thresholds (for $NN \rightarrow K^+\Lambda N$, the threshold is at 1.58 GeV and for $NN \rightarrow K^+K^-NN$ is 2.5 GeV), and the K^+ multiplicity exceeds the K^- one by 1-2 orders at a given energy above thresholds. However, this large difference disappears for nucleus-nucleus collisions (C + C, Ni + Ni [6, 7]), where the data nearly fall on the same curve. Furthermore, we observe that this ratio is approximately constant for C + C, Ni + Ni and Au + Au [6, 7, 8], although absorption of K^- via $K^-N \rightarrow Y\pi$ is supposed to be higher for heavier nuclei [9]. It turns out that both effects can be interpreted by assuming an in-medium attractive $U_{\bar{K}}$ [10, 11].

Analysis of K^+/K^- in Ni + Ni at GSI energies [6, 7] gives a ratio around 30. In Reference [12], it was shown that this ratio can be explained in terms of a *thermal model*, i.e., the final state can be described as a *hadronic gas* in chemical equilibrium at a given temperature. Imposing exact strangeness conservation, one obtains,

$$\frac{K^+}{K^-} = \frac{g_{K^+}V^2 \int \frac{d^3p}{(2\pi)^3} e^{\frac{-E_{K^+}}{T}} \left(g_{K^-} \int \frac{d^3p}{(2\pi)^3} e^{\frac{-E_{K^-}}{T}} + g_\Lambda \int \frac{d^3p}{(2\pi)^3} e^{\frac{-E_\Lambda + \mu_B}{T}} + g_\Sigma \int \frac{d^3p}{(2\pi)^3} e^{\frac{-E_\Sigma + \mu_B}{T}} \right)}{g_{K^-}V^2 \int \frac{d^3p}{(2\pi)^3} e^{\frac{-E_{K^-}}{T}} \left(g_{K^+} \int \frac{d^3p}{(2\pi)^3} e^{\frac{-E_{K^+}}{T}} \right)},$$

where g_α are the spin-isospin degeneracies and V is the interaction volume. The above expression shows that the presence of K^+ must be compensated by Λ, Σ and K^- to conserve strangeness equal to zero, while K^- can only be compensated by K^+.

It is found that the data [13] can be explained using a $T = 70 \pm 10$ MeV and $\mu_B = 720 \pm 20$ MeV. G. Brown [14] introduced the notion of "broad-band equilibration" in heavy-ion processes at GSI energies. In this interpretation, due to the compensation

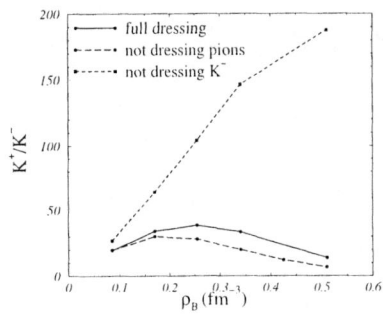 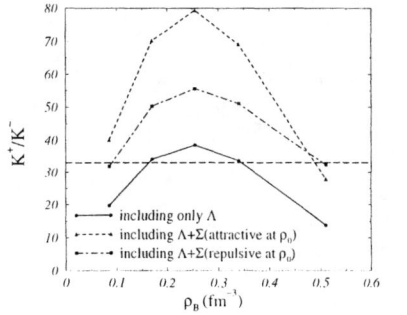

FIGURE 2. K^+/K^- as a function of density at $T = 70$ MeV. Left: including or not $U_{\bar{K}}$. Right: included $U_{\bar{K}}$, including or not Σ.

between the increase in μ_B as the density increases and the density dependence of $U_{\bar{K}}$, the ratio is constant over a large range of densities, not only for $\rho \sim \frac{1}{4}\rho_0$, but also up to $2\rho_0$.

In Figure 2 (left), we show the importance of dressing K^-. Plotting the ratio K^+/K^- as function of density, it can be seen that the effect of including $U_{\bar{K}}$ reduces the ratio that, otherwise, would increase steadily (short-dashed line). It is also seen that dressing the pions induces appreciable differences in the calculations (long-dashed line).

Once the full dressing is included, Fig.2 (right hand) shows the sensitivity of the ratio to the chemical potential, and hence to density, as well as the importance of introducing the Σ in the balance equation. The "broad-band equilibration" is hardly reproduced when the Σ is taken into account, and one obtains large differences depending on the character (attractive or repulsive) of the Σ optical potential.

CONCLUSIONS

Performing a microscopic calculation of the Brueckner-Hartree-Fock type, we conclude that dressing pions in the calculation of $U_{\bar{K}}$ introduces significant differences, around 13% at zero momentum. Moreover, the introduction of temperature gives rise to a smoother behaviour in momentum, being attractive even at momentum as large as 500 MeV, relevant in heavy-ion collisions. Heavy-ion collisions are the perfect scenario to test in-medium $\bar{K}$ properties, specially in understanding the ratio K^-/K^+. The attempts to reproduce this ratio require an attractive $U_{\bar{K}}$. The ratio shows a measurable density dependence. The "broad-band equilibration" is far from being reproduced if hyperons like Σ are taken into account, showing large differences with regard to only including Λ hyperons.

ACKNOWLEDGMENTS

We are very grateful to Dr. Jürgen Schaffner-Bielich for the useful discussions that have made possible this work and L.T. wishes to acknowledge BNL for the hospitality during her stay.

REFERENCES

1. Müller-Groeling, A., Holinde, K., and Speth, J., *Nucl. Phys.* **A513**, 557 (1990).
2. Balberg, S., and Gal, A., *Nucl. Phys.* **A625**, 435 (1997).
3. Oset, E., and Ramos, A., *Nucl. Phys.* **A671**, 481 (2000).
4. Tolós, L., Ramos, A., Polls, A., and Kuo, T.T.S., *Nucl. Phys.* **A690**, 547 (2001).
5. Schaffner-Bielich, J., Koch, V., and Effenberg, M., *Nucl. Phys.* **A669**, 153 (2000).
6. Barth, R., et al., *Phys. Rev. Lett.* **78**, 4007 (1997); Laue, F., et al., *Phys. Rev. Lett.* **82**, 1640 (1999).
7. Menzel, M., et al., *Phys. Lett. B* **495** (2000) 26.
8. Förster, A., Ph.D Thesis, TU Darmstad.
9. Senger, P., *Nucl. Phys.* **A685**, 312 (2001).
10. Cassing, W., Bratkovskaya, E.L., Mosel, U., Teis, S., and Sibirtsev, A., *Nucl. Phys.* **A614**, 415 (1997); Bratkovskaya, E.L., Cassing, W., and Mosel, U., *Nucl. Phys.* **A622**, 593 (1997).
11. Li, G.Q., Lee, C.-H., and Brown, G., *Nucl. Phys.* **A625**, 372 (1997); *ibid*, *Phys. Rev. Lett.* **79**, 5214 (1997).
12. Cleymans, J., Oeschler, H., and Redlich, K., *Phys. Rev. C* **59**, 1663 (1999).
13. Cleymans, J., Elliot, D., Keränen, A., and Suhonen, E., *Phys. Rev. C* **57**, 3319 (1998).
14. Brown, G., Rho, M., and Song, C., *Nucl. Phys.* **A690**, 184 (2001).

Phenomenological value of the $\pi\Lambda\Sigma$ coupling

Sławomir Wycech* and Benoit Loiseau[†]

Sołtan Institute for Nuclear Studies, 00-681 Warsaw, Poland
[†]*Laboratoire de Physique Nucléaire et de Hautes Énergies, 75252 Paris CEDEX 05, France*

Abstract. Level widths in Σ-hyperonic atoms are due to the $\Sigma N \to \Lambda N$ conversion. In high angular momentum states it is dominated by the one-pion exchange. A joint analysis of the scattering and atomic data allows to obtain the pion-hyperon coupling constant $f^2_{\pi\Lambda\Sigma}/4\pi = 0.048 \pm 0.005$ (statistic) ± 0.004 (systematic) or $G^2_{\pi\Lambda\Sigma}/4\pi = 13.3 \pm 1.4$ (statistic) ± 1.1 (systematic).

INTRODUCTION

Hyperonic atoms allow to test the long range part of hadron-nucleon interactions, the one-pion exchange (OPE) [1]. Proper conditions are found in high angular momentum atomic states. There, the centrifugal barrier prevents hyperons to get close to the nucleus and the OPE effect is enhanced over those of two-pion or heavier meson exchanges. The lifetimes of Σ atoms are determined by the rate of transition

$$\Sigma^- + p \to \Lambda + n \qquad (1)$$

occurring at the nuclear surface. In general the Σ inelastic scattering [2, 3] and atomic [4, 5, 6] data are old and uncertain, however a recent measurement of Powers *et al.* [6] provides the precise width of 17(3) eV for the Σ^-Pb atom in $n = 10, l = 9$ state.

The inelastic as well as the elastic hyperon interactions, have been described by meson-exchange models [7, 8, 9, 10]. The pionic coupling $f^2_{\pi\Lambda\Sigma}/4\pi$ is either assumed or fitted to the scattering data. In particular : $f^2_{\pi\Lambda\Sigma}/4\pi = 0.041$ has been used by Nijmegen group [7, 8] while the Jülich group [10] obtains $f^2_{\pi\Lambda\Sigma}/4\pi = 0.034$ and/or 0.028. On the other hand, the high l atomic states are an analog of free ΣN interactions in high partial waves. The OPE effects are dominant and short-range forces are strongly suppressed. It turns out that the atomic data favor larger values of the pion hyperon coupling.

THE WIDTHS OF ATOMIC Σ^- LEVELS

The Σ^- hyperon, bound into atomic orbits, cascades down to states where nuclear interactions convert it into Λ. Some of these orbits have been found by the X-ray data to be circular states of high l [4, 5, 6]. The centrifugal barrier forces the conversion to occur at an extreme nuclear surface. In some "upper" states the distances involved are as large as twice the nuclear radius. This allows quasi-free scattering, single-particle picture of the nucleus and quasi free description of final states.

The $\Sigma - \Lambda$ conversion mechanism presented here follows calculations done in Ref. [11]. The atomic level width is given by an optical model expression

$$\Gamma/2 = -\frac{2\pi}{\mu_{\Sigma N}} \int d\mathbf{R} \int d\mathbf{u}\, \rho(\mathbf{R}-\mathbf{u})\, Im\, t_{\Sigma N}(\mathbf{u})\, |\Psi_\Sigma(\mathbf{R})|^2, \quad (2)$$

where $\Psi_\Sigma(\mathbf{R})$ is the atomic wave function, $\mu_{\Sigma N}$ is the ΣN reduced mass, $\rho(\mathbf{R})$ is the proton density and $t_{\Sigma N}$ is the elastic scattering matrix. Since $\Psi_\Sigma \sim R^l$ and l is large, the nuclear and atomic densities in Eq.(2) are well separated and it is only the longest range components of $Im\, t_{\Sigma N}(\mathbf{u})$ that matter. A microscopic approach that discloses the strength and range involved in $Im\, t_{\Sigma N}$ is needed. This quantity is related by unitarity to the inelastic reactions $\Sigma^- p \to \Lambda n, \Sigma^0 n$. In the atomic conditions : the $\Sigma^0 n$ channel is blocked and the unitarity is affected by the nucleus. The latter determines the energy balance generating the nucleon binding and recoil energy. For OPE the unitarity gives

$$Im\, t_{\Sigma N}(u) = \frac{(\sqrt{2} f_{\pi NN} f_{\pi \Lambda \Sigma})^2}{(m_\pi^2 4\pi)^2} \frac{\mu_{N\Lambda}}{2\pi} q \left(\frac{1}{3} m_\pi^{*4} + \frac{2}{3} q^4\right) F_{\text{fold}}(u) \quad (3)$$

where F_{fold} describes the range of atomic-nuclear folding

$$F_{\text{fold}}(u) = \int d\mathbf{w} \frac{d\Omega_q}{4\pi} \frac{\exp(-m^*|\mathbf{u}+\mathbf{w}/2|)}{|\mathbf{u}+\mathbf{w}/2|} \frac{\exp(-m^*|\mathbf{u}-\mathbf{w}/2|)}{|\mathbf{u}-\mathbf{w}/2|} J(w) \exp(i\mathbf{q}.\mathbf{w}). \quad (4)$$

The main bulk of $F_{\text{fold}}(u)$ is given by $16\pi \exp(-2u\sqrt{m^{*2}+q^2})/\sqrt{m^{*2}+q^2}$. The range is related both to the mass of the exchanged meson and to the momentum transfer, as expected from the uncertainty principle. To obtain Eq.(3) few manipulations are required

1) The OPE for the $\Sigma p - \Lambda n$ conversion is introduced into the nuclear transition matrix. It involves some energy transfer changing the pion mass m_π into m_π^* [11].

2) The spins of initial and intermediate baryons are averaged and summed. These generate the 1/3 and 2/3 weights in Eq.(3).

3) The Λ, n emission probabilities are calculated and summed over the final states. For an isolated Σp pair this procedure would produce the absorptive amplitude $Im\, t_{\Sigma N}$. Here, this summation extends over free Λ, n states and single proton-hole nuclear states. It generates $Im\, t_{\Sigma N}$ which in the nuclear case is folded over nuclear and atomic wave functions. The wave functions arise in the form of mixed, single particle density matrices. Next, the mixed densities are reduced into densities. This approximation jointly with off-shell corrections, nucleon recoil in the intermediate states, and nucleon form factor, introduce a corrective factor $J(w)$ which is very close to unity.

The energy conservation fixes the final state relative momentum q which is also the momentum carried by the π meson. Its central value is about 260 MeV but some 10 % changes are due to binding and recoil energies.

RESULTS

An attempt to describe the atomic level widths with the simple OPE mechanism is summarised in Table 1. One obtains the best fit value $f_{\pi \Lambda \Sigma}^2/4\pi = 0.050(6)$ with $\chi^2/\text{data} = 2.38/4$.

TABLE 1. Upper level widths, Γ(eV), in circular orbits. l is the level angular momentum.

Atom	Ref.	l	$\Gamma_{experiment}$(eV)	$\Gamma_{calculated}$(eV)
^{27}Al	[5]	4	0.24(6)	0.21
^{28}Si	[5]	4	0.41(10)	0.51
^{40}Ca	[4]	5	0.41(22)	0.17
^{208}Pb	[6]	9	17(3)	15

In fact, the fitted parameter is $f_{\pi NN} f_{\pi\Lambda\Sigma}$, and here an average value of $f^2_{\pi NN}/4\pi = 0.08$ was used. The error of $f_{\pi\Lambda\Sigma}$ given above is due to the experimental uncertainties. In addition, there exist uncertainties of the nuclear structure and simplifications in the conversion mechanism which induce systematic errors. To obtain these we analyze the Σp inelastic scattering data and discuss the basic nuclear parameters.

The nuclear parameters of importance are the energy excess determined by the proton binding energies E_B and nuclear recoil energy E_{rec}. The related uncertainty is removed by a remarkable numerical stability of the atomic level width with respect to changes of $E_B + E_{rec}$. It is related to the form of the pion coupling in Eq. (3). For an increased (decreased) q one finds a decrease (increase) in the atomic-nucleus overlap integral which is perfectly balanced by the increase (decrease) of $q \times (q^4)$ in Eq. (3). The energy release uncertainty yields a 0.001 uncertainty in the coupling constant. Nuclear factors of prime importance are the proton density distributions $\rho(r)$. These are based on muonic atoms and electron scattering data. The transition from charge to proton densities is done by relating few low moments of these densities. The proton density is not unique as there exists a number of experimental charge densities. To estimate the uncertainty a "statistics" over available four to five densities per nucleus is performed. For the $f^2_{\pi\Lambda\Sigma}$ the corresponding uncertainty is 0.003.

In addition to OPE, other short-range interactions are known to contribute and, in principle, one needs a full interaction model. However the data on the YN interactions are sparse and the models [7] to [10] differ in their content and their results. Here, we take a phenomenological attitude. A plausible form of the background, motivated by these theoretical models, is assumed. The choice for additional amplitudes is based on several observations :

(1) One expects the scattering $t(r)$ matrix to be of the form $t(\mathbf{r}) = V(\mathbf{r}) + V(\mathbf{r}) \int d\mathbf{r}' \ G(\mathbf{r} - \mathbf{r}') \ t(\mathbf{r}')$ where $t(\mathbf{r}) = t_\pi(\mathbf{r}) + t_b(\mathbf{r})$. The second term of this equation is due to multiple scattering effects. The "background" amplitude of interest may be due to the short range part of potential $V(r)$ and/or to the multiple scattering contribution. What is significant is that the range involved in both these terms is expected to be at least shorter than that of the two-pion exchange amplitude.

(2) The low momentum (140 – 170 MeV/c) inelastic $\Sigma^- p$ cross section of Engelman [2] shows no structure and no enhancement at low energies that would indicate a large scattering length in this system. However, to be on a safe side, we fit the Massachusetts data [3] taken at higher momenta (200 – 580 MeV/c, 8 data points) to be more independent of the initial state interactions.

We consider several models with different background. The results of the best fits to the atomic and scattering data are summarized in Fig.1. The models are: – A) OPE approximation is used to describe both the atomic widths and the reaction cross sections. – B) The two pion effects are simulated by Yukawa potential of the $1/(2m_\pi)$ range and f_b^2 coupling. Two cases, spin triplet dominance and tensorial force are tried. – C) A tensor component [11], the range of which is given by the ρ mass of 770 MeV. – D) From the SU(3) symmetry one expects a strong coupling with K meson as found in Refs. [7, 10] and a $\pi + K$ model is tried. – E) The ρ meson exchange amplitude is added. – F) The $\rho + K$ meson exchange amplitude is added. This is a three parameter fit but only a marginal improvement has been obtained. The results show that the best fit value of $f_{\pi\Lambda\Sigma}^2$ depends rather weakly on the uncertain short range interaction mechanism. The systematic error due to the unprecise knowledge of the short range interaction is found to be 0.003.

Averaging these best fits gives a value of

$$f_{\pi\Lambda\Sigma}^2/4\pi = 0.048 \pm 0.005 \text{ (statistic)} \pm 0.004 \text{ (systematic)}. \tag{5}$$

This corresponds to $G_{\pi\Lambda\Sigma}^2/4\pi = f_{\pi\Lambda\Sigma}^2[(M_\Lambda + M_\Sigma)/m_\pi]^2/4\pi = 13.3 \pm 1.4$ (statistic) $\pm$ 1.1 (systematic). In order to reduce the large statistical part, both new measurements of the Σ atomic widths and free $\Sigma^- p \to \Lambda n$ scattering data would be welcome.

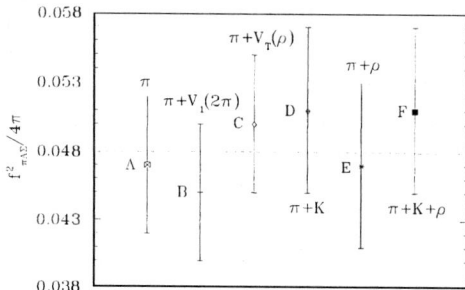

FIGURE 1. Results of best fit to the atomic and scattering data with different background amplitudes. The errors include statistics and systematics.

Support from KBN grant 5P 03B 04521 is acknowledged.

REFERENCES

1. Bongart, K., and Pilkuhn, H., *Phys. Letters* **76B**, 32 (1978).
2. Engelman, R., et al., *Phys. Letters* **21**, 586 (1966).
3. Stephen, D., Ph.D. Thesis, Univ. of Massachusetts, 1970.
4. Backenstoss, G., et al., *Z. Phys. A* **273**, 137 (1975).
5. Batty, C.J., et al., *Phys. Letters* **74B**, 27 (1978).
6. Powers, R.J. et al., *Phys. Rev. C* **47**, 1263 (1993).
7. Maessen, P.M.M., Rijken, Th.A., and de Swart, J.J., *Phys. Rev. C* **40**, 2226 (1989).
8. Stocks V.G.J., and Rijken, Th.A., *Phys. Rev. C* **59**, 3009 (1999).
9. Rijken, Th.A., Stocks, V.G.J., and Yamamoto, Y., *Phys. Rev. C* **59**, 21 (1999).
10. Reuber, A., Holinde, K., and Speth, J., *Nucl. Phys.* **A570**, 543 (1994).
11. Loiseau, B., and Wycech, S., *Phys. Rev. C* **63**, 034003 (2001).

MESON DYNAMICS

Microscopic description of η-photoproduction on light nuclei

V.B. Belyaev*, N.V. Shevchenko*, S.A. Rakityansky†, W. Sandhas** and S. Sofianos†

*Joint Institute for Nuclear Research, Dubna, 141980, Russia
†Physics Dept., University of South Africa, P.O. Box 392, Pretoria 0003, South Africa
**Physikalisches Institut, Universität Bonn, D-53115 Bonn, Germany

Abstract. A microscopic four-body description of near-threshold coherent photoproduction of the η meson on the (3N)-nuclei is given. The photoproduction cross-section is calculated using the Finite Rank Approximation (FRA) of the nuclear Hamiltonian. The results indicate that the final state interaction of the η meson with the residual nucleus plays an important role in the photoproduction process. Sensitivity of the results to the choice of the ηN T-matrix is investigated. The importance of obeying the condition of ηN unitarity is demonstrated.

The high level of reliability of the modern few-body theory provides the means for making conclusions about underlying two-body interactions from experimental data on relevant few-body processes. The purpose of this report is to present the results of few-body calculations concerning the coherent η-photoproduction on the tritium and ^{3}He targets,

$$^3H(\gamma, \eta)^3H \quad \text{and} \quad ^3He(\gamma, \eta)^3He \, , \qquad (1)$$

from which some conclusions about the T-matrices describing the elastic ηN scattering and the photoproduction process $N(\gamma, \eta)N$ could be made. To the best of our knowledge, no experimental data on such coherent reactions has been published yet.

To describe the few-body dynamics of this reaction, we employ the method based on the Finite-Rank Approximation (FRA) [1] of the nuclear Hamiltonian H_A. This approximation consists in neglecting the continuous spectrum in the spectral expansion

$$H_A = \mathcal{E}_0|\psi_0\rangle\langle\psi_0| + \text{continuum} \qquad (2)$$

of this Hamiltonian. Physically, this means that we exclude the processes of (virtual) excitations of the nucleus during its interaction with the η meson. It is clear that the stronger the nucleus is bound, the smaller is the contribution from such processes to the elastic ηA scattering. By comparing with the results of the exact Faddeev calculations, it was shown[2] that even for ηd scattering (with weakest nuclear binding) the FRA method works reasonably well, which implies that we can obtain sufficiently accurate results applying this method to η^{3}H and η^{3}He scattering.

To include a photon into the FRA formalism, we follow the same procedure as in Ref.[3] where the coherent η-photoproduction on deuteron was treated in the framework of the exact AGS equations and the photon was introduced by considering the ηN and

γN states as two different channels of the same system. This implies that the T-operator describing the ηN interaction, should be replaced by 2×2 matrix, viz.

$$t_{\eta N} \to \begin{pmatrix} t^{\gamma\gamma} & t^{\gamma\eta} \\ t^{\eta\gamma} & t^{\eta\eta} \end{pmatrix}, \qquad (3)$$

where $t^{\gamma\gamma}$ describes the Compton scattering, $t^{\eta\gamma}$ the photoproduction process, and $t^{\eta\eta}$ the elastic ηN scattering. All calculations was performed in the first order on electromagnetic interaction.

The problem of constructing ηN potential or directly the corresponding T-matrix $t^{\eta\eta}$ has no unique solution since the only experimental information we have consists of the two complex numbers, namely, position of the S_{11}-resonance pole $E_0 - i\Gamma/2$ and the ηN scattering length $a_{\eta N}$. In the present work, we use three different versions of $t^{\eta\eta}$ all of which have the same separable form

$$t^{\eta\eta}(k',k;z) = g(k')\tau(z)g(k), \qquad (4)$$

with the same formfactors $g(k) = (k^2 + \alpha^2)^{-1}$ (where $\alpha = 3.316\,\text{fm}^{-1}$, see Ref. [4]) but with different versions of the propagator $\tau(z)$. The version I is motivated by the dominance of the S_{11} resonance at the near-threshold energies and has simple Breit-Wigner form

$$\tau(z) = \frac{\lambda}{z - E_0 + i\Gamma/2}, \qquad (5)$$

which guaranties that the $t^{\eta\eta}$ has the resonance pole at $z = E_0 - i\Gamma/2$ (with $E_0 = 1535\,\text{MeV} - (m_N + m_\eta)$ and $\Gamma = 150\,\text{MeV}$ see Ref. [5]). In this version of $\tau(z)$ the strength parameter λ is chosen to reproduce the ηN scattering length $a_{\eta N} = (0.55 + i0.30)\,\text{fm}$.

An alternative way (version II) of constructing the two-body T-matrix $t^{\eta\eta}$ is to solve the corresponding Lippmann-Shwinger equation with an appropriate separable potential having the same form-factors $g(k)$. However, a one-term separable T-matrix obtained in this way, does not have a pole at $z = E_0 - i\Gamma/2$. To recover the resonance behaviour in this case, we use the trick suggested in Ref. [6], namely, we use a potential with an energy-dependent strength, which resulted in

$$\tau(z) = -\frac{4\pi\alpha^3}{\mu_{\eta N}} \cdot \frac{\Lambda(\zeta - z) + C\zeta}{\zeta - z - [\Lambda(\zeta - z) + C\zeta]/(1 - i\sqrt{2z\mu_{\eta N}}/\alpha)^2}, \qquad (6)$$

where the constants Λ, C, and ζ are chosen in such way that the corresponding scattering amplitude reproduces the same (as for version I) scattering length $a_{\eta N}$ and has a pole at $z = E_0 - i\Gamma/2$. Moreover, it is consistent with the condition of the two-body unitarity because it obeys the Lippmann-Schwinger equation.

The version III has the same functional form as version I, but different value of λ which is now fixed using the condition of the ηN unitarity, namely, $(1 - 2\pi i t^{\eta\eta})(1 - 2\pi i t^{\eta\eta})^+ = 1$. The resulting $t^{\eta\eta}$ gives $a_{\eta N} = (0.76 + i0.61)\,\text{fm}$.

Therefore all the three versions of $t^{\eta\eta}$ have a pole at $z = E_0 - i\Gamma/2$, the first two of them reproduce the same $a_{\eta N}$, the versions II and III are consistent with the unitarity condition but give different $a_{\eta N}$.

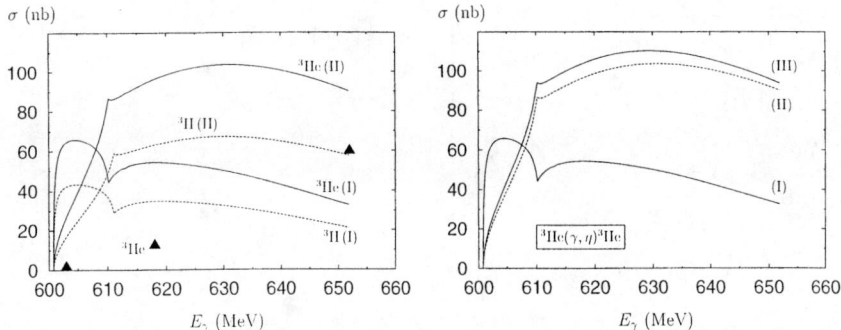

FIGURE 1. Total photoproduction cross-section on ^{3}He and ^{3}H nuclei for different $t^{\eta\eta}$-matrices. Triangles are the results for ^{3}He taken from Ref.[11].

In constructing the photoabsorption (production) T-matrix $t^{\gamma\eta}$, we use the corresponding on-shell T-matrix $t^{\gamma\eta}_{on}(E)$ of Ref. [7] and extend it off the energy shell,

$$t^{\gamma\eta}_{off}(k',k;E) = \frac{\kappa^2 + E^2}{\kappa^2 + k'^2} \, t^{\gamma\eta}_{on}(E) \, \frac{\alpha^2 + 2\mu_{\eta N}E}{\alpha^2 + k^2} \,, \qquad (7)$$

using Yamaguchi form-factors which disappear (go to unity) on the energy shell with κ being a parameter. Varying κ in our calculations, we found that the dependence of the photoproduction cross-sections on the choice of this parameter is rather weak, and we simply put $\kappa = \alpha$. It is known that $t^{\gamma\eta}$ is different for neutron and proton. In this work we assume that they have the same functional form (7) and differ by a constant factor, $t^{\gamma\eta}_n = A t^{\gamma\eta}_p$. Multipole analysis [8] gives for this factor the estimate: $A = -0.84 \pm 0.15$.

To obtain the nuclear wave function ψ_0 (which is needed for the expansion (2) of H_A), we solve the few-body equations of the Integro-Differential Equation Approach (IDEA) [9] with the Malfliet–Tjon potential [10].

Figures 1 and 2 show the results of our calculations for the total (integrated over the directions of the outgoing meson) cross-section σ of the coherent processes (1). The calculations were done for two nuclear targets, namely, ^{3}H and ^{3}He. The curves corresponding to the three versions of $t^{\eta\eta}$ are denoted respectively as (I), (II), and (III).

As is seen in Fig.1 (left plot), the two versions of $t^{\eta\eta}$, (I) and (II), give significantly different results despite the fact that both of them reproduce the same $a_{\eta N}$ and the S_{11} resonance. This indicates that the scattering of the η meson on the nucleons (final state interaction) is very important in the description of the photoproduction process. This conclusion becomes even more substantiated when our curves are compared to the corresponding points (triangles) calculated for the ^{3}He target in Ref. [11] where the final state interaction was treated using an optical potential of the first order. It is well-known that the first-order optical theory is not adequate at the energies near resonances. This is the reason why the calculations of Ref. [11] underestimate σ near the threshold.

Another conclusion, following from the fact that the curves (I) in Fig.1 (left plot) are significantly different from the corresponding curves (II), is that the two-body unitarity is important as well. To clarify this statement, we compare (see Fig.1, right plot) three

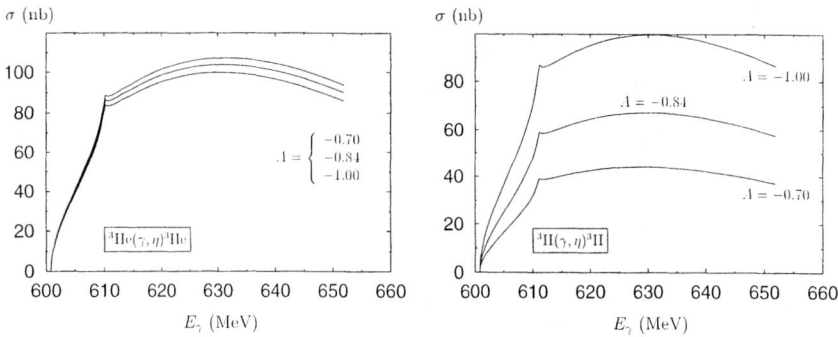

FIGURE 2. Total photoproduction cross-section on ^{3}He (left plot) and ^{3}H nuclei (right plot) for 3 different values of parameter A.

curves corresponding to the three choices of $\tau(z)$ in (4). Surprisingly, the curves (II) and (III) almost coincide despite the fact that they correspond to different $a_{\eta N}$ while both obey the two-body unitarity condition.

Fig.2 shows the dependence of σ on the choice of the parameter A for the cases of ^{3}He and ^{3}H target (left and right plots respectively). An interesting observation here is that the cross-section for η photoproduction is more sensitive to this parameter when tritium rather than on ^{3}He target is used. This means that between these two nuclei, the tritium is a preferable candidate for a possible experimental determination of the ratio A.

Authors would like to thank Division for Scientific Affair of NATO for support (grant CRG LG 970110) and DFG-RFFR for financial assistance (grant 436 RUS 113/425/1). One of authors (V.B.B.) willing to thank Physikalishes Institut of Bonn University for hospitality.

REFERENCES

1. Belyaev, V., *Lectures on the Theory of Few-Body systems*, Springer Verlag, Heidelberg, 1990.
2. Shevchenko, N.V., Rakityansky, S.A., Sofianos, S.A., Belyaev, V.B., and Sandhas, W., *Phys. Rev. C* **58**, R3055 (1998).
3. Shevchenko, N.V., Belyaev, V.B., Rakityansky, S.A., Sandhas, W., and Sofianos, S.A., *Nucl. Phys.* **A689**, 383 (2001).
4. Bennhold, C. and Tanabe, H., *Nucl. Phys.* **A530**, 625 (1991).
5. Particle Data Group, *Phys. Rev. D* **50**, 1173 (1994).
6. Deloff, A., *Phys. Rev. C* **61**, 024004 (2000).
7. Green, A.M., and Wycech, S., *Phys. Rev. C* **60**, 035208 (1999).
8. Mukhopadhyay, N.C., Zhang, J.F., Benmerouche, M., *Phys. Lett. B* **364**, 1 (1995).
9. Fabre de la Ripelle, M., Fiedeldey, H., and Sofianos S.A., *Phys. Rev. C* **38**, 449 (1988); Oehm, W., Sofianos, S.A., Fiedeldey, H., and Fabre de la Ripelle, M., *Phys. Rev. C* **44**, 81 (1991).
10. Malfliet, R.A., and Tjon, J.A., *Nucl. Phys.* **A127**, 161 (1969); *Ann. Phys. (N.Y.)* **61**, 425 (1970).
11. Tiator, L., Bennhold, C., Kamalov, S.S., *Nucl. Phys.* **A580**, 455 (1994) and private communication.

Meson photoproduction on the proton at GRAAL

J.-P. Bocquet[1]

for the GRAAL collaboration:

O. Bartalini[2], V. Bellini[3], M. Castoldi[2], P. Corvisiero[2], A. D'Angelo[2],
J.-P. Didelez[4], R. Di Salvo[2], G. Gervino[5], F. Ghio[6], B. Girolami[6],
M. Guidal[4], E. Hourany[4], V. Kouznetsov[7], R. Kunne[4], A. Lapik[7],
P. Levi Sandri[8], A. Lleres[1], D. Moricciani[2], V. Nedorezov[7], L.Nicoletti[1],
D. Rebreyend[1], F. Renard[1], N.V. Rudnev[9], M. Sanzone[10], C. Schaerf[2],
M. L. Sperduto[4], C.M. Sutera[4], A. Turinge[9], and A. Zucchiatti[10].

[1]*IN2P3, Institut des Sciences Nucléaires, 38026 Grenoble, France*
[2]*INFN sezione di Roma II and Università di Roma "TorVergata", 00133, Italy*
[3]*INFN sezione di Catania and Università di Catania, 95100, Italy*
[4]*IN2P3, Institut de Physique Nucléaire, 91406 Orsay, France*
[5]*INFN sezione di Torino and Università di Torino, 10125*
[6]*INFN sezione di Roma I and Istituto Superiore di Sanità, Roma, 00161, Italy*
[7]*Institute for Nuclear Research, 117312 Moscow, Russia*
[8]*INFN Laboratori Nazionali di Frascati, 00044, Italy*
[9]*Institute of Theoretical and Experimental Physics, 117259 Moscow, Russia*
[10]*INFN sezione di Genova and Università di Genova, 16146, Italy*

Abstract. The GRAAL (Grenoble Anneau Accélérateur Laser) facility provides a polarized and tagged photon beam (from 500 to 1500 MeV) obtained by Compton scattering of laser light on the electron beam of ESRF (European Synchrotron Radiation Facility) in Grenoble. Mesons photoproduction off the nucleon allows to investigate the excited states of the nucleon. The $p(\gamma, \eta p)$ channel has been analysed and the cross-sections are given up to 1100 MeV.

INTRODUCTION

The production of mesons in photo-reactions is sensitive to the underlying dynamics of quarks and the measurement of cross-sections and polarization observables provide information on the production mechanism through contributing resonances. Most of the theoretical work on the nucleon excitation spectrum has been done with quark models which contain three massive constituent quarks and predict a resonance spectrum much more rich [1] than observed experimentally (only about 30% of states). Since the majority of experimental data come from $N(\pi, \pi N)$ reactions it has been suggested [2] by quark models that the missing states couple strongly to other channels, such as $p(\gamma, \eta p)$.

With the GRAAL photon beam many channels can be accessed simultaneously, some of them have been completely studied (spin observables and cross sections) [3, 4, 5] and some others are still under study. Recent analysis of p(γ,ηp) cross-section and beam asymmetries suggest the possibility of new a resonance contributing to these channels.

EXPERIMENTAL SET-UP AND EVENT SELECTION

We have used the GRAAL polarized and tagged photon beam obtained by the backscattering of laser light on the high energy electrons circulating in the 6.04 GeV storage ring of the ESRF (European Synchrotron Radiation Facility) in Grenoble. In the measurements presented here, the visible line of an Argon laser (514 nm) has been used to produce a gamma-ray energy spectrum extending from the lowest tagged energy of 500 MeV, to the maximum energy of 1100 MeV. The tagging detector provides an energy resolution of 16 MeV (FWHM) which is limited by the emittance and the energy spread of the electron beam. The linear polarization of the beam varies from 0.98 at the maximum photon energy to 0.60 at the η threshold (700 MeV).

The 4π detector, for the detection of neutral and charged particles, consists of a cylindrical central part, that provides high energy resolution for photons and protons and a forward detector, giving time of flight (TOF) and angular information for charged particles and neutrons. The particles emitted into the central part at angles between 25^0 and 155^0 with respect to the beam axis, pass through two coaxial cylindrical wire chambers, a barrel made of 32 plastic scintillators, that provides ΔE information for particle identification, and the BGO crystal ball made of 480 crystals, 21 radiation lengths each. It has cylindrical symmetry around the beam axis and the energy resolution for γ ray detection is 3% at 1 GeV.

The particles emitted in the forward direction at angles less than 25^0 pass two plane wire chambers that provide tracking angular resolution of 0.5^0 and a double wall of plastic scintillators covering an area of 3×3 m^2 and located 3 m away from the target. This detector gives ΔE information and a measurement of the time of flight (TOF) with a resolution of 600 ps (FWHM), with an angular resolution of 2^0. It is followed by a calorimeter consisting of 16 vertical modules (lead/scintillator sandwiches) covering the same area as the double plastic wall and providing an angular resolution of 3^0, a TOF resolution of 600 ps, but a poor energy resolution for photons.

Complete detection of the reaction products is required in the event analysis. The photons from neutral meson decays ($\pi^\circ \rightarrow 2\gamma$, $\eta \rightarrow 2\gamma$ and $\eta \rightarrow 3\pi^0 \rightarrow 6\gamma$) are identified in the BGO calorimeter, while the recoil proton is tracked in wire chambers and characterized by time of flight (ToF) and dE/dx measurements.

The cross section normalization takes into account the target thickness, the photon beam intensity, the detection efficiency and the branching ratios of the η-meson decays taken from [6] ($\eta \rightarrow 2\gamma$: 39.21$\pm$0.34 %, $\eta \rightarrow 3\pi^0$: 32.2$\pm$0.4 %). The photon intensity is monitored by thin plastic scintillators located between the target and a total absorption calorimeter (Spacal) which serves as a beam dump. Both detectors are in coincidence with the tagging system and their ToF spectra are measured for correction of accidentals. The low efficiency of the thin monitor ($\simeq$ 2.7%) prevents pile-up effects at high photon rate and is measured by comparison with the Spacal rate at low flux.

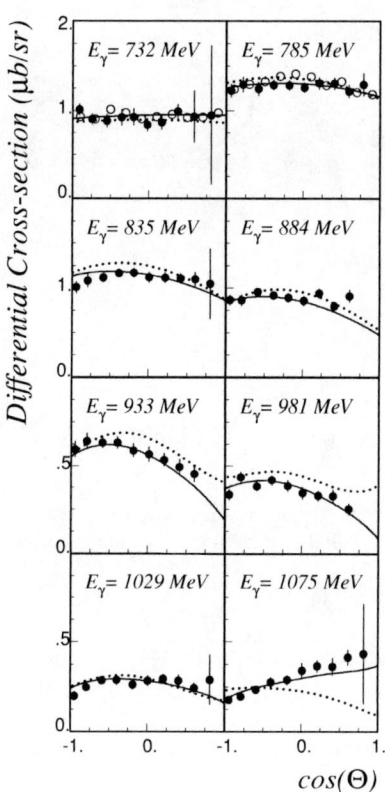

FIGURE 1. Differential cross section for the p(γ,ηp) reaction. The experimental data from Mainz are plotted for the two low energy bins (open circles). The models B (dotted line) and C (full line) of B. Saghai are shown (see text for details).

The differential cross section, plotted in Fig. 1 for a sample of photon energies as a function of $\cos\theta$ where θ is the η C.M. polar angle, was measured every 17 MeV between the threshold and 1100 MeV (233 points). For the two lowest energies of the figure, data from Mainz are also shown and illustrate the good agreement between both experiments.

The data can be compared with a quark model of Li and Saghai, [7, 8] which includes all known resonances. An interesting property of this approach is that it directly links data to the quark model, hence to quantities such as mixing angles of resonance configurations [9]. In their most recent analysis [10], the authors have shown that to reproduce correctly the data they must include a new S_{11} resonance around 1700 MeV. This new resonance, included in the C model, improves significantly the agreement with the experimental data, as compared to the B model (not including the resonance).

The total cross section has been obtained by integration of the differential cross section, using a polynomial fit in $\cos\theta$ to extrapolate to the unmeasured region ($\simeq 10\%$). The result from the threshold to 1100 MeV is plotted in Fig. 2. One can notice the

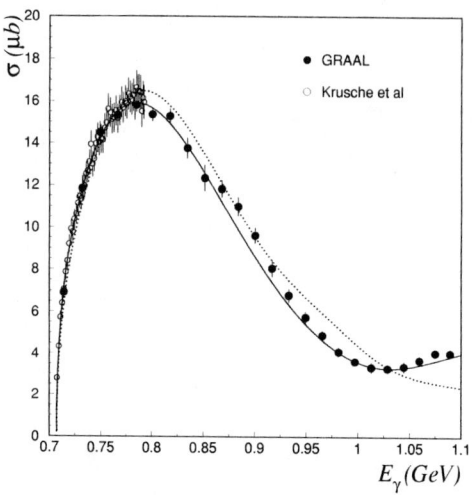

FIGURE 2. Total cross section for the p(γ,ηp) reaction. The experimental data from Mainz are plotted for the two low energy bins (open circles). The models B (dotted line) and C (full line) of B. Saghai are shown (see text for details).

nice agreement with previous measurements obtained at Mainz [11] except for a small discrepancy around 800 MeV. The model C here again shows a better agreement with the experimental data.

In summary, we have measured the differential cross section for the reaction p(γ,ηp) from the threshold to 1100 MeV photon lab. energy, completing our previous measurement of the beam asymmetry Σ over the same energy range. Below 900 MeV, the reaction mechanism is well understood and these data will contribute to a precise determination of the dominant $S_{11}(1535)$ parameters. Above this energy, S-wave dominance vanishes and the data show a rapid transition to a new regime with a clear P-wave contribution. We are looking forward to definitive analyses that will allow the extraction of as yet unknown couplings of baryon resonances to the η meson.

REFERENCES

1. For a recent review, see Vrana, T.P., Dytman, S.A., and Lee, T.S.H., *Phys. Rep.* **328**, (2000) and references therein.
2. Capstick, S., and Roberts, W., *Phys. Rev. D* **47**, 1994 (1993); *Phys. Rev. D* **49**, 4570 (1994).
3. Ajaka, J., et al., *Phys. Rev. Lett.* **81**, 1797 (1998).
4. Ajaka, J., et al., *Phys. Lett. B* **475**, 372 (2000).
5. Renard, F., Thesis, Univ. J. Fourier (Grenoble, 1999), available upon request.
6. Groom, D.E., et al. (Particle Data Group), *Eur. Phys. J. C* **15**, 1 (2000).
7. Li, Z., and Saghai, B., *Nucl. Phys.* **A644**, 345 (1998).
8. Saghai, B., private communication.
9. Saghai, B., N*2000 Workshop Proceedings (to appear in World Scientific), JLab (2000).
10. Saghai, B., and Li, Z., nucl-th/0104084, submitted to *Eur. Phys. J.*.
11. Krusche, B., et al., *Phys. Rev. Lett.* **74**, 3736 (1995).

Exclusive measurement of the $pp \to pp\pi^-\pi^+$ reaction close to threshold[1]

W. Brodowski*, R. Bilger*, H. Calén†, H. Clement*, C. Ekström†, K. Fransson**, S. Häggström‡, B. Höistad‡, J. Johanson‡, A. Johansson‡, T. Johansson‡, K. Kilian§, S. Kullander‡, A. Kupść†, P. Marciniewski‡, A. Mörtsell‡, B. Morosov¶, W. Oelert§, J. Pätzold*, R.J.M.Y. Ruber‡, M. Schepkin‖, J. Stepaniak††, A. Sukhanov¶, P. Sundberg‡, A. Turowiecki‡‡, G.J. Wagner*, Z. Wilhelmi‡‡, J. Zabierowski§§, A. Zernov¶ and J. Zlomanczuk‡

*Physikalisches Institut der Universität Tübingen, Morgenstelle 14, D-72076 Tübingen, Germany
†The Svedberg Laboratory, S-751 21 Uppsala, Sweden
**Department of Radiation Sciences, Uppsal University, S-751 21 Uppsala, Sweden
‡Department of Radiation Sciences, Uppsala University, S-751 21 Uppsala, Sweden
§IKP-Forschungszentrum Jülich GmbH, D-52425 Jülich, Germany
¶Joint Institute for Nuclear Research Dubna, 101000 Moscow, Russia
‖Institute for Theoretical and Experimental Physics, 117218 Moscow, Russia
††Soltan Institute for Nuclear Studies, PL-00681 Warsaw, Poland
‡‡Institute of Experimental Physics, Warsaw University, PL-0061 Warsaw, Poland
§§Soltan Institute for Nuclear Studies, PL-90137 Lódz, Poland

Abstract. The two-pion production in proton-proton collisions in vacuum has been measured for the first time exclusively in the energy range $T_p = 650 - 750$ MeV with the WASA/PROMICE detector at CELSIUS. The differential cross sections reveal $pp \to pp^*(1440) \to pp\sigma \to pp(\pi^-\pi^+)_{I=I=0}$ to be the dominant process with a significant contribution from the path $pp \to pp^*(1440) \to p\Delta\pi \to pp(\pi^-\pi^+)_{I=I=0}$. Hence this reaction is very well suited for studying the still poorly known Roper resonance and may also provide an ideal σ source to study effects of the chiral restauration in the nuclear medium.

The reaction $NN \to NN\pi\pi$ is the basic hadronic two-pion production process, be it in NN collisions in vacuum or in the nuclear medium. This reaction offers a variety of aspects concerning the dynamics of the total system as well as that of its subsystems $\pi\pi, NN, \pi N, \pi\pi N$ and πNN. Close to threshold the reaction is expected to be dominated by the excitation of $N^*(1440)$ in one of the participating nucleons, since single Δ excitation leads to the emission of a single pion only. Hence the lowest order to which the Δ mechanism contributes is the $\Delta\Delta$ excitation, which is expected to be dominant at higher energies only.

Therefore this reaction offers the possibility to study the still not very well known Roper resonance in more detail. This is particularly interesting, if its deexcitation pro-

---

[1] supported by BMBF (06 TU 886, 06 TU 987) and DFG (Graduiertenkolleg)

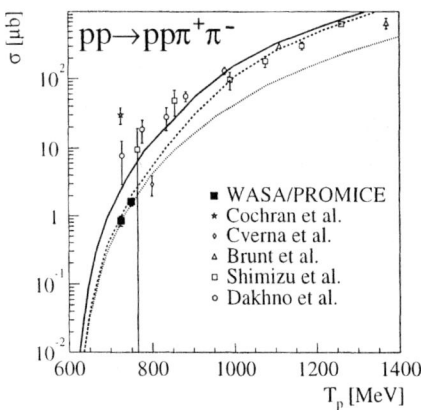

FIGURE 1. Energy dependence of the total cross section. Results from this work are shown by solid squares, open symbols denote Refs. [4] (asteriks), [5] (diamonds), [8] (triangles), [6] (squares) and [7] (circles). The dotted line shows the pure phase space behavior normalized arbitrarily, solid and dashed lines represent theoretical predictions [13] with and without pp FSI, respectively.

ceeds via the emission of the σ-meson, the chiral partner of the pion.

In fact, the σ meson which appears as a tightly correlated pion pair in the scalar-isoscalar channel $(\pi\pi)_{I=J=0}$, is expected to exhibit pronounced effects of chiral restauration when propagating in nuclear matter. According to chiral perturbation theory the σ strength function should get concentrated near the $\pi\pi$ threshold the more the higher the density of the surrounding nuclear medium [1, 2]. This is contrary, e.g., to the situation of the ρ meson, whose width is expected to increase drastically in the nuclear medium [3]. Hence for the quest of chiral restauration the nucleon-included two-pion production can serve as an ideal σ production source in the nuclear medium, if it proceeds via the process $NN \to NN^*(1440) \to NN\sigma \to NN(\pi\pi)_{I=J=0}$.

The invariant mass spectra $M_{pp\pi^-}$ and $M_{pp\pi^+}$ are of special interest with regard to dibaryonic resonances. In particular, if such resonances were NN-decoupled, then they would need to decay into $NN\pi$ with a narrow width, in case their masses are not far above the $NN\pi$ threshold. Hence they should be detectable in $NN \to NN\pi\pi$, if their production cross section is not too low.

All previous data on this reaction below 1 GeV stem from inclusive magnetic spectrometer measurements [4, 5] or low-statistics bubble chamber experiments [6, 7], which partly also have been inclusive [7]. Hence we have carried out high-statistics exclusive measurements of the $pp \to pp\pi^-\pi^+$ reaction at the CELSIUS storage ring using the PROMICE/WASA detector setup with clusterjet H_2 target [9]. Protons and pions have been registered in the forward detector, which covers the polar angle range $4° \leq \Theta \leq 21°$. Protons have been identified by the $\Delta E - E$ method, positive pions in addition by their delayed pulse from subsequent muon decay. The thus identified $pp\pi^+$ events yield very clean spectra. From the measured four-momenta of $pp\pi^+$ the full $pp\pi^+\pi^-$ events have been reconstructed by kinematical fits with one overconstraint (1C-fit). Detector acceptance and efficiencies have been deduced from Monte-Carlo (MC) simulations of the

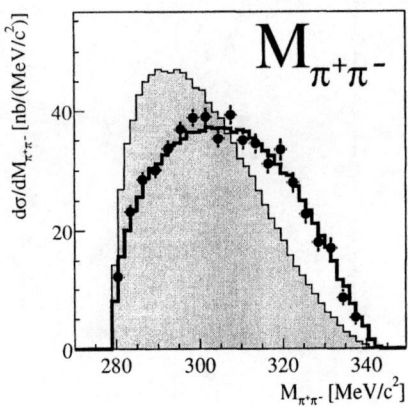

FIGURE 2. Distribution of the invariant mass $M_{\pi^+\pi^-}$. The shaded area shows the pure phase space. The solid curve is a calculation assuming the coherent superposition of the routes $pp \to pp^*(1440) \to pp\sigma \to pp(\pi\pi)_{I=l=0}$ and $pp \to pp^*(1440) \to p\Delta\pi \to pp(\pi\pi)_{I=l=0}$.

detector setup, which have been checked against single pion production data taken simultaneously as well as separately. The absolute normalization has been obtained by simultaneous measurements of elastic scattering and its comparison to literature data [10].

Figure 1 shows total cross sections obtained in this experiment [11, 12] in comparison with previous results for $pp \to pp\pi^-\pi^+$ [4, 5, 6, 7, 8]. In the overlap region our results are approximately one order of magnitude below the previous data obtained by bubble chamber measurements on deuterium. They are, however, in tentative agreement with the LAMPF magnetic spectrometer datum at $T_p = 800$ MeV [4]. Solid and dashed lines in Fig. 1 show calculations of the Valencia group [13] with and without nucleon-nucleon final-state interaction (FSI), respectively. The dotted line is a pure phase space calculation normalized arbitrarily to our data points.

The differential cross sections reveal the dynamics of the $\pi\pi$-production process. From the invariant pp mass distribution M_{pp} we find a pp FSI as expected from NN phase shifts, however, much weaker than assumed in the Valencia group calculations [12]. The concave shape of the proton angular distribution reveals heavy meson exchange as the dominating process. The pion angular distributions are flat, i.e. the pions are produced predominantly in s-waves. This is in favor of $pp \to pp^*(1440) \to pp\sigma \to pp(\pi\pi)_{I=l=0}$ being the dominant process. The $\Delta\Delta$ process involves angular momenta other than $l = 0$ and hence leads to wildly different angular distributions. Aside from the decay of Roper resonance by emission of a σ meson, also the branch $p^*(1440) \to \Delta\pi \to p(\pi\pi)_{I=l=0}$ plays a significant role. This is seen by the data for $M_{\pi^+\pi^-}$ (Fig. 2) as well as the distribution of the opening angle between the two pions (Fig. 3), which are far from being described by phase space. This process via the Δ excitation goes to leading order like $\mathbf{k}_1 \cdot \mathbf{k}_2$, which in turn enters directly into the observables shown in Figs. 2 and 3. In order to obtain a good description also for these observables we need a 20% admixture in amplitude by the route $pp^*(1440) \to p\Delta\pi \to pp(\pi\pi)_{I=l=0}$. The solid

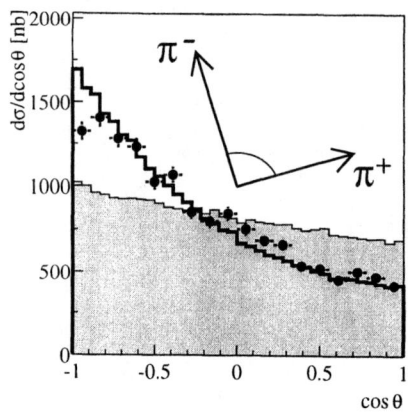

FIGURE 3. Same as Fig. 2, but for the distribution of the opening angle between the two pions

lines in Figs. 2 and 3 show the coherent superposition of both routes. This calculation gives a very good description of all differential data.

The first exclusive data of solid statistics for the $pp \to pp\pi^+\pi^-$ reaction reveal this reaction to be driven by the excitation of the Roper resonance, at least at energies not too far from threshold. This provides a powerful tool to investigate the properties of this resonance, which can hardly be excited by electromagnetic probes. Further on, since in both reaction routes the pion pairs are produced in the scalar-isoscalar channel, this reaction offers an ideal σ source in the nuclear medium to investigate there effects of partial restauration of chiral symmetry in the σ channel [1, 2].

REFERENCES

1. Hatsuda, T., Kunihiro, T., and Shimizu, H., *Phys. Rev. Lett.* **82**, 2840 (1999).
2. Aouissat, Z., et al., *Phys. Rev. C* **61**, 012202 (1999).
3. Klingl, F., Kaiser, N., and Weise, W., *Nucl. Phys.* **A624**, 527 (1997).
4. Cverna, F.H., et al., *Phys. Rev. C* **23**, 1698 (1981).
5. Cochran, D.R.F., et al., *Phys. Rev. D* **6**, 3085 (1972).
6. Shimizu, F., et al., *Nucl. Phys.* **A386**, 571 (1982).
7. Dakhno, L.G., et al., *Sov. J. Nucl. Phys.* **37**, 540 (1983).
8. Brunt, D.C., et al., *Phys. Rev.* **187**, 1856 (1969).
9. Calen, H., et al., *Nucl. Instr. Meth. A* **379**, 57 (1996).
10. Arndt, R., et al., *Phys. Rev. C* **56**, 635 (1997); program package SAID.
11. Johanson, J., doctoral thesis, Uppsala University 2000 and to be published.
12. Brodowski, W., doctoral thesis, Univ. Tübingen 2001
    http://www.pit.physik.uni-tuebingen.de/forschung.html.
13. Alvarez-Ruso, L., Oset, E., and Hernández, E., *Nucl. Phys.* **A633**, 519 (1998) and private communication.

Determination of πN scattering lengths from pionic hydrogen and pionic deuterium x-ray data

A. Deloff [1]

Soltan Institute for Nuclear Studies, Warsaw, Poland

Abstract. The πN s-wave scattering lengths have been inferred from a joint analysis of the pionic hydrogen and the pionic deuterium x-ray data using a non-relativistic approach in which the πN interaction is simulated by a short-ranged potential.

The determination of low-energy pion-nucleon (πN) parameters has been the focus of much theoretical and experimental efforts. The s-wave πN scattering lengths (a_{2I} for total isospin I=1/2,3/2, or they isoscalar and isovector combinations b_0 and b_1, respectively) are of particular importance serving as testing ground for various theoretical considerations. In recent years major advances have been made in the experimental and theoretical investigation of the πN system. With the advent of meson factories (LAMPF, PSI and TRIUMF) and the corresponding influx of the new high accuracy πN scattering data considerable progress has been achieved in the πN phase shift analyses [1, 2, 3] providing means to examine even such subtleties as isospin symmetry breaking effects [2, 4, 5]. Recently, the πN scattering experiments have been complemented by high quality pionic x-ray measurements performed, both on pionic hydrogen [6, 7] and on pionic deuterium [8] and these data constitute an independent source of information on the low-energy πN scattering parameters. On the theoretical side, the physical quantities bearing on the low-energy πN interaction have now become accessible to calculations [9] conducted within quantum chromodynamics (QCD). Since QCD is known to be highly non-perturbative at low energies, its low-energy implementation has been based instead on a chiral perturbation theory in which the effective Lagrangian is expanded in increasing powers of derivatives in meson fields and quark masses. This approach in practice involves a Taylor expansion in the meson four-momenta and therefore it may be expected that the lower the energy is, the more accurate are the predictions. In this context, the precise knowledge of the experimental values of the low-energy πN scattering parameters is essential for further development of the theory.

The purpose of this work (for a more complete account cf. [10]) is to extract the s-wave πN scattering lengths using exclusively the pionic hydrogen and pionic deuterium x-ray data. The key reason for proceeding along this route is that the low-energy regime can be thereby investigated without recourse to scattering data and there is no danger that the low-energy parameters have been largely determined by the data at high energies.

---

[1] Supported by KBN grant 5P03B04521

The measurement of x-ray transitions in pionic hydrogen provides the values of the 1s level shift ε and the level width Γ and these two quantities are related to the complex π^-p scattering length $a_{\pi p}$ by the Deser-Trueman formula [11]

$$\varepsilon + i\tfrac{1}{2}\Gamma = 2\alpha^3\mu^2 a_{\pi p}, \qquad (1)$$

where α is the fine structure constant and μ is π^-p reduced mass. Assuming that the underlying πN interaction is isospin invariant, the scattering lengths a_{2I} could be inferred from (1) alone. The pionic hydrogen system may be envisaged as a two-channel problem involving the $\pi^0 n$ and $\pi^- p$ states (channel 1 and 2, respectively). Introducing a two-channel K-matrix, isospin invariance can be invoked to pin down its elements at the single unsplit threshold

$$K = \begin{pmatrix} \tfrac{1}{3}a_1 + \tfrac{2}{3}a_3 & \tfrac{\sqrt{2}}{3}(a_3 - a_1) \\ \tfrac{\sqrt{2}}{3}(a_3 - a_1) & \tfrac{2}{3}a_1 + \tfrac{1}{3}a_3 \end{pmatrix}$$

and the complex π^-p scattering length takes the form

$$a_{\pi p} = K_{22} + ip_t K_{12}^2/(1 - ip_t K_{11}), \qquad (2)$$

where p_t is the momentum in the $\pi^0 n$ channel evaluated at the $\pi^- p$ threshold. Using (2) in (1), results in two algebraic equations for a_1 and a_2 that can be solved analytically (as γn channel is left out we replace Γ by $P\Gamma/(1+P)$ in (1) where P is Panofsky ratio). The b_0 and b_1 scattering lengths obtained this way are shown in Fig. 1 where the error bars reflect the experimental uncertainty on ε, Γ and the Panofsky ratio [13].

In order to find out what the deuteron data can teach us about πN scattering lengths, we calculated the πd scattering length by solving the appropriate three-body πNN problem. This task was accomplished, both within the static approximation, and also by using the Faddeev formalism. We have assumed that the πN forces are of a very short range and this supposition follows from a particle exchange picture: there is no sufficiently light particle presently known that might be capable of generating forces whose range would exceed 0.3-0.4 fm (which roughly corresponds to a vector meson exchange). In this situation a logical point of departure is the zero-range limit. In this case the same static formula for $a_{\pi d}$ can be derived from: *(i)* a set of boundary conditions; *(ii)* a static solution of Faddeev equations, and *(iii)* a summation of Feynman diagrams [12]. The final result is

$$a_{\pi d} = \frac{2}{1 + m/2M}\left\langle \frac{\tilde{b}_0 + (\tilde{b}_0 + \tilde{b}_1)(\tilde{b}_0 - 2\tilde{b}_1)/r}{1 - \tilde{b}_1/r - (\tilde{b}_0 + \tilde{b}_1)(\tilde{b}_0 - 2\tilde{b}_1)/r^2}\right\rangle, \qquad (3)$$

where m and M are the pion and the nucleon masses, respectively, and $\tilde{b}_j = (1 + m/M)b_j$. The expectation value is with respect of the deuteron wave function (r is n-p separation). Static solution in coordinate space (3) is very appealing and helps to develop an intuitive picture of how the individual πN amplitudes contribute to build up the πd scattering length. By solving numerically the Faddeev equations we show that the accuracy of (3) is comparable with the present experimental uncertainty on $a_{\pi d}$

TABLE 1. πd scattering length obtained from the static model and from a Faddeev calculation in the zero-range model for different b_0 and b_1 (all in $10^{-2}/m_\pi$ units).

b_0/b_1		-9.47	-9.05	-8.63
-0.65	static	-3.89	-3.69	-3.49
	Faddeev	-3.97	-3.76	-3.55
	ditto with Δ	-3.59	-3.37	-3.16
-0.22	static	-2.99	-2.78	-2.58
	Faddeev	-3.07	-2.85	-2.65
	ditto with Δ	-2.68	-2.46	-2.25
0.21	static	-2.08	-1.87	-1.68
	Faddeev	-2.16	-1.95	-1.74
	ditto with Δ	-1.76	-1.54	-1.34
	experiment *		-2.65±0.05	

* from [8] subtracting Coulomb correction

(cf. table 1). The requirement that the πd scattering lengths obtained as a solution of the Faddeev equations be in agreement with experiment imposes a constraint on a_{2I}. Such agreement can be obtained only when the input πN scattering lengths belong to a relatively small subset of values that are consistent with pionic hydrogen data. The πN scattering lengths that belong to this subset simultaneously satisfy the constraints imposed by the pionic hydrogen and pionic deuterium data (they are depicted as a black strip in Fig. 1). Subsequently, we lifted the zero-range limitation representing the πN interaction by Yamaguchi potential with inverse range parameter β. The pionic hydrogen bound state problem was solved afresh and for an assigned β, the appropriate bound state condition was used to pin down the s-wave πN potentials and by using them in the Faddeev equations we calculated the πd scattering length. The latter quantity was found

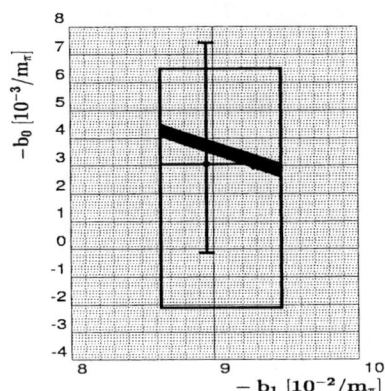

FIGURE 1. Constrains on the isoscalar and isovector scattering lengths imposed by pionic deuterium data: the black strip corresponds the one standard deviation region. The rectangle represents the values inferred in [6] from pionic hydrogen data, whereas the cross shows the Deser-Trueman formula result.

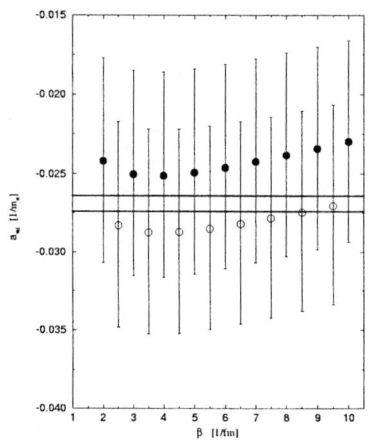

FIGURE 2. πd scattering length vs. the inverse range parameter β of the πN potential. Full (open) circles correspond to a Faddeev calculation with (without) p-wave πN interaction. The area between horizontal lines represents one standard deviation region.

to be almost independent upon β (cf. Fig. 2) but was rather sensitive to the a_{2I} values used as input in the Faddeev equations. The above result, supporting the zero-range approach ($\beta \to \infty$), could be explained by the fact that the range of the πN interaction that was considered physically justified was small in comparison with the deuteron size. Our analysis of the pionic hydrogen is consistent with that of [6]. This is a direct consequence of the fact that Deser-Trueman formula (1), that depends neither upon the shape of the πN potential nor upon its range, provides such a good approximation that we can make considerable progress in deducing the πN scattering lengths without committing ourselves in great deal to the nature of the πN dynamics. Although the lack of knowledge of the range of the πN interaction is responsible for some uncertainty in the deduced πN scattering lengths but this uncertainty is rather small, at the level of 1%. The main source of error is still the experimental uncertainty in the pionic hydrogen data.

REFERENCES

1. Arndt, R.A., *et al.*, http://said.phys.vt.edu/analysis/pin_analysis.html.
2. Gibbs, W.R., Li Ai and Kaufmann, W.B., *Phys. Rev.* **C 57**, 784 (1998).
3. Gashi, A., *et al.* hep-ph/0009081.
4. Gibbs, W.R., Li Ai and Kaufmann, W.B., *Phys. Rev. Lett.* **74**, 3740 (1995).
5. Matsinos, E., *Phys. Rev.* **C 56**, 3014 (1997).
6. Sigg, D., *et al.*, *Nucl. Phys.* **A609**, 310 (1996).
7. Schröder, H.C., *et al.*, *Phys. Lett.* **B 469**, 25 (1999).
8. Hauser, P., *et al.*, *Phys. Rev.* **C 58**, R1869 (1998).
9. Fettes, N., Meissner, U.-G., and Steininger, S., *Nucl. Phys.* **A640**, 199 (1998).
10. Deloff, A., nucl-th/0104067
11. Deser, S., *et al.*, *Phys. Rev.* **96**, 774 (1954); Trueman, T.L., *Nucl. Phys.* **26**, 57 (1961).
12. Kolybasov, V.M., and Kudryavtsev, A.E., *Sov. Phys. JETP* **36**, 18 (1973).
13. Spuller, J., *et al.*, *Phys. Lett.* **B 67**, 479 (1977).

η–photoproduction with SAPHIR at ELSA

J. Ernst (for the SAPHIR collaboration)

Institut für Strahlen- und Kernphysik der Universität Bonn, Nussallee 14-16, D 53115 Bonn

Abstract. At the electron stretcher ring ELSA in Bonn the photoproduction of mesons off the proton is investigated with tagged photons from threshold up to 2.8 GeV. The main components of the SAPHIR detector are sketched together with the methods of data analysis. Preliminary results from a rather complete data set from η–production up to $E_\gamma = 1.8$ GeV are presented.

INTRODUCTION

Meson photoproduction beyond single–pion production off the proton is a powerful method to search for the many "missing baryon resonances" predicted by constituent quark models (see e.g. [1]), and to clarify the role of other underlying reaction mechanisms. "Missing resonances" should have a notable coupling strength to the photon and should couple weakly to $N\pi$ but stronger to one or more of $\Delta\pi$, $N\eta$, $N\rho$, $N\omega$, $N\Phi$, $N\eta'$, ΛK and ΣK channels. Hence, photoproduction experiments have a rather unexploited discovery potential for "Missing Resonances" as pointed out by Capstick & Roberts [2]. Especially, investigating $p\gamma \to N\eta$ has high selectivity for $I=\frac{1}{2}$ resonances since isospin conservation rules out Δ-excitations in this channel. The electron stretcher assembly ELSA at Bonn is able to provide tagged photons up to energies of about 3 GeV. Investigations on the photoproduction of the η–meson being the next-heavier non–strange meson beyond the pion, have started in recent years showing a dominant production from excitation and decay of the $S_{11}(1535)$ resonance [3, 4, 5, 6, 7]. Here, we give a short status report on the ongoing analyses of η–photoproduction off the proton with the charged particle momentum resolving 4π–detector SAPHIR.

THE SAPHIR DETECTOR ARRANGEMENT AT ELSA

The ELSA stretcher ring at Bonn University provides continuous electron beams up to 3.2 GeV (for details see [8]). The 4π–spectrometer SAPHIR allows to track outgoing charged particles from tagged photon induced reactions. It is especially suited for particles from threshold reactions leaving the target under forward angles. The main components of the SAPHIR setup [9] are shown in Fig. 1. The tagging facility TOPAS II [10] provides tagged photons of $31\% \leq E_\gamma \leq 94.4\%$ of E_e. The decelerated electrons are analysed by the tagging magnet. Here, 14 plastic scintillators release fast timing signals while the 352 wires of two proportional chambers on top allow a finer mesh of 349 photon energy channels with a resolving power of 0.028 to 2.6%. The liquid H_2 target is placed in the center of the C–shaped magnet operated at 0.6 Tesla. It is surrounded by 14

Spectrometer Arrangement for PHoton Induced Reactions

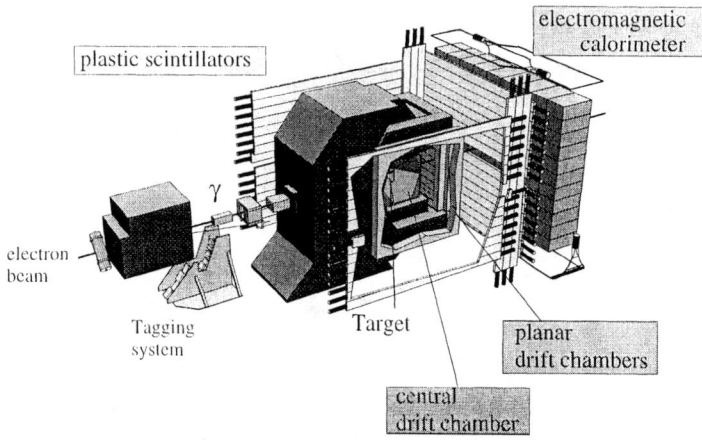

FIGURE 1. The SAPHIR setup. (Planar drift chambers and electromagnetic calorimeter were not used).

circular layers of hexagonal drift cells forming the central drift chamber. Six layers are alternately bent to the vertical by $\pm 5°$ providing vertical information of the particle trajectories. The measurement of drift–times from the central drift chamber together with the known strength of the magnetic field in each cell allows to reconstruct the momenta of the charged ejectiles. The time–of–flight (TOF) of at least 2 particles hitting the wall of scintillator bars behind the drift chambers determines their velocities and finally their masses taking the deduced momenta into account. The central horizontal bar in the TOF wall is left out for avoiding the huge background from magnetically spread out e^+e^- pairs. Details on the performance of the drift chambers, the TOF system, the total flux measuring γ–veto detector behind the TOF wall (not shown in Fig. 1) and on the methods of data analysis may be found in [7, 9, 10, 11].

METHOD OF ANALYSIS

In several beam–times data on $p\gamma \to p\eta$ were taken at $E_e = 1.6$ and 2.8 GeV. The data acquisition system recorded events obeying the following trigger conditions: a time–start signal from at least one tagger scintillator followed by at least two TOF stop–hits from the TOF wall, in anti-coincidence with the γ–veto detector. True hadronic events were sorted out offline. The $p\gamma \to p\eta$ reaction is investigated by 3–prong events from the decay chain $p\eta \to p\pi^+\pi^-\pi^0/\gamma$ $(23.0\pm0.4\%)/(4.75\pm0.11\%)$ [12]. The 4–momentum coordinates of the unobserved π^0 are determined from a kinematical fit taking into account the 4–momenta of the initial state and the measured 3–momenta of the 3 charged final state particles. The mixed–in contribution of the second neutral decay possibility is corrected for by the subsequent Monte Carlo (MC) simulation of both decay channels. The data analysis is not straightforward since it requires the full MC simulation of the reaction and its detection with the rather complex SAPHIR setup

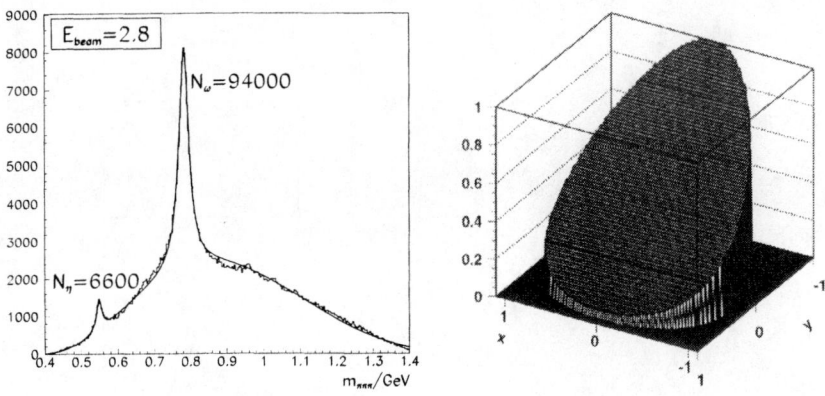

FIGURE 2. At left: Invariant $\pi^+\pi^-\pi^0$ mass spectrum taken at $E_e = 2.8$ GeV. At right: Modified Dalitz plot of charged $\eta \to \pi^+\pi^-\pi^0$ decay with $x = \sqrt{3}(T_+ - T_-)/Q$, $y = 3T_0/Q - 1$ and $Q = T_+ + T_- + T_0$.

including the highly inhomogeneous magnetic field. For the central drift chamber it was found that the drift–time resolution strongly depends on the velocity of particles since ionisation clusters are not uniformly distributed and are wider and wider separated with increasing β-values. These and other effects may lead to a distorted trajectory of one of the three charged particles. Then, the 3 prongs do not appear to stem from the same vertex, and the event is discarded. The amount of true η–events had to be separated from a high background due to $\Delta^0\pi^+$ production and other reactions. In addition, the analysis proved that the tagger deteriorated during the runs due to a gradual damage of the cathode being an Al coated plastic foil, especially at the locus of high electron flux related to low energy γ-quanta. This explains why in toto the η–production observed was relatively low at $E_e = 2.8$ GeV (cf. left part of Fig. 2). In the MC calculations it was tried to account for all mentioned effects [12] best to our knowledge. Finally, at each energy the results were normalized with a factor gained from comparing the total 3-prong yield with respective cross sections of the literature [3]. The MC simulations took also into account that the dominating $\eta \to \pi^+\pi^-\pi^0$ decay is not governed by phasespace but follows the distribution [13] displayed in Fig. 2 at right.

RESULTS AND DISCUSSION

The extracted $p\gamma \to p\eta$ cross sections from a first preliminary analysis were already presented in [7]. Below $E_\gamma = 900$ MeV within errors they well agree with the published results from Refs. [4, 5, 6]. Here, we show the corresponding angular distributions above $E_\gamma = 900$ MeV (Fig. 3). At 950 MeV, where the $S_{11}(1535)$ resonance still dominates, the distribution rather favours backward angles while with increasing energy more and more forward production shows up indicating possibly strong ρ- and ω–exchange in the t–channel. Regge–model inspired calculations [14] are under way.

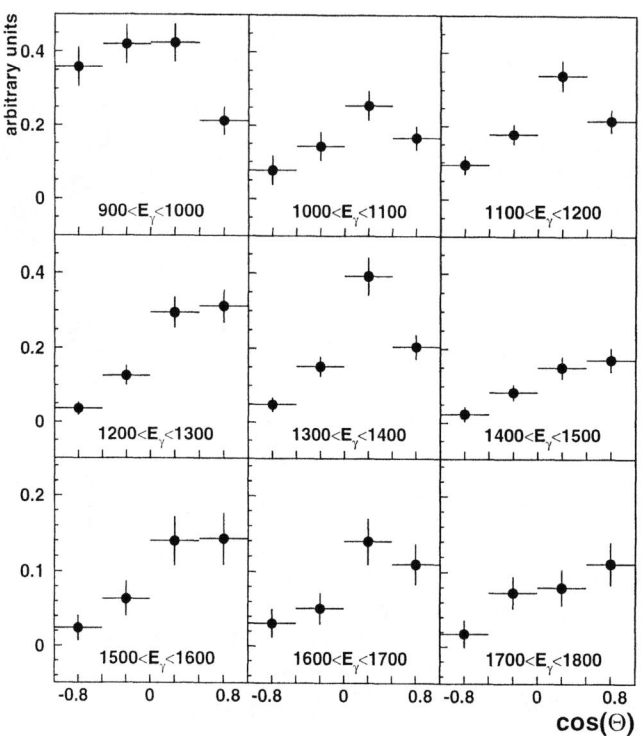

FIGURE 3. Measured differential cross sections of $p\gamma \to p\eta$ in a.u.

REFERENCES

1. Metsch, B.C., *Nucl. Phys.* **A675**, 161c (2000).
2. Capstick, S., and Roberts, W., *Phys. Rev. D* **47**, 1994 (1993), and dito *D* **49**, 4570 (1994).
3. AHHM Collaboration, *Nucl. Phys. B* **108**, 45 (1976).
4. Wilhelm, M., *Elektroproduktion von η–Mesonen am Proton im Bereich der $N(1535)S_{11}$–Resonanz*, Ph.D. thesis, Bonn University, Nussallee 12, D 53115 Bonn (1993), Bonn–IR–93–43 (1993).
5. Krusche, B., et al., *Phys. Rev. Letters* **74**, 3736 (1995).
6. Rebreyend, D., *Nucl. Phys.* **A663**, 436c (2000), and preprint arXiv:hep-ex/0011098 30 Nov 2000.
7. Barth, J., et al., The SAPHIR Collaboration, *Nucl. Phys. A*, contribution to the VII Int. Conf. on Hypernuclear and Strange Particle Physics, Torino, Italy, Oct. 23-27, 2000, in press.
8. Nakamura, S., et al., *Nucl. Instr. Methods A* **411**, 93 (1998), and refs. therein.
9. Schwille, W. J., et al., *Nucl. Instr. Methods A* **344**, 470 (1994).
10. Link, J., *Untersuchung der Photoproduktion von η'–Mesonen mit dem SAPHIR–Detektor*, Ph.D. thesis, Bonn University, Bonn, Nussallee 14-16, D 53115 (2000).
11. Plötzke, R., *Phys. Letters B* **444**, 555 (1998).
12. van Pee, H., *Photoproduction of η–mesons with SAPHIR*, Ph.D. thesis, Bonn University, Nussallee 14-16, D 53115 Bonn (2001), tentative title of thesis in progress, and private communication.
13. Abele, A., et al., *Phys. Letters B* **417**, 197 (1998).
14. Guidal, M., Laget, J.-M. and Vanderhaeghen, M., *Nucl. Phys.* **A627**, 645 (1997), and *Phys. Letters B* **400**, 6 (1997).

The ITEP study of inclusive pion double charge exchange: Experiment and interpretation

B. M. Abramov*, L. Alvarez-Ruso [†], Yu. A. Borodin*, S. A. Bulychjov*, I. A. Dukhovskoi*, A. B. Kaidalov*, A. I. Khanov*, A. P. Krutenkova *, V. V. Kulikov*, M. A. Matsuk*, I. A. Radkevich*, A. I. Sutormin*, E. N. Turdakina* and M. J. Vicente Vacas [†]

*State Research Center "Institute of Theoretical and Experimental Physics" Moscow, 117218, Russia
[†]University of Valencia, Burjassot, Spain

Abstract. The inclusive pion double charge exchange (DCX) on light nuclei at intermediate energies $T_0 = 0.59, 0.75$ and 1.1 GeV was studied recently in the ITEP experiment on $\pi^- A$ interactions. The differential cross sections of these processes decrease with energy rather weakly and exceed the theoretical prediction within the conventional sequential single charge exchange mechanism with neutral pion in the intermediate state (i.e. Glauber elastic rescattering) at the highest energy by about a half an order of magnitude. Our updated analysis of the data on ^{16}O (as well as new outgoing pion momentum spectra on ^{16}O and $^{6,7}Li$ at 0.59 GeV and some auxiliary measurements) confirms our evidence for this anomalous DCX energy behavior. Thus other mechanisms of DCX are needed to explain the observed energy dependence. The theoretical analysis showed that the Glauber inelastic rescatterings (IR) mediated by intermediate states of several (mainly two) pions appeared to give an important contribution to the inclusive DCX cross section at energies $T_0 > 0.6$ GeV. To estimate the IR contribution in the framework of Gribov-Glauber approach to DCX, the OPE model which is known to give a good description of $\pi^- p \to \pi^+ \pi^- n$ reaction above ~ 2 GeV/c is used. It is shown that IR with 2π intermediate state dominate the pion DCX in a wide energy range above ~ 1 GeV.

INTRODUCTION

In the general scheme of the pion double charge exchange (DCX) on nuclei, this reaction appears as a two-step transition on two like nucleons in the nuclear medium. In the conventional mechanism the pion propagates in the matter undergoing twice the sequential single charge exchange (SSCX). The ITEP experimental and theoretical study showed that the inclusive pion DCX at GeV energies has an interesting feature. Above 0.6 GeV the measured cross section of this process [1,2] does not fall sharply as it is expected in the SSCX mechanism [3] but decreases relatively slowly, that is interpreted qualitatively within Gribov formalism as the dominance in the DCX [4] of inelastic Glauber rescatterings.

In this report we present some additional experimental information on inclusive DCX spectra for light nuclei and on A dependence of the cross section at 0.59 GeV emphasizing the importance of the nuclear medium effects for quantitative understanding of the DCX processes. We give an updated results for the reaction on ^{16}O and also for quantitative estimation of the DCX energy dependence up to 4 GeV made in the OPE model for the mechanism of the inelastic Glauber rescatterings.

THE ITEP EXPERIMENT

We studied pion DCX in the inclusive (not exclusive) reaction $\pi^- + A \to \pi^+ + X$ at energies higher than used at meson factories. To avoid the contribution of an additional pion production we chose a special kinematical region, $\Delta T \equiv T_0 - T < 140$ MeV, where T is the kinetic energy of the outgoing pion.

The experiment was performed at the 3-m magnet spectrometer (see, e.g., [5]) in a pion beam of the ITEP 10-GeV proton synchrotron at $T_0 = 0.59$, 0.75 and 1.1 GeV. Forward going pions were identified by TOF system. The Cherenkov counters were used to suppress the background of positrons. Nuclear targets (^{6}Li, ^{7}Li, C, Al, Cu, In, Ta, Bi, H_2O and D_2O) were placed near the center of the magnet so that beam and outgoing pions were registered in spark chambers and their momenta were measured in the magnetic field of the spectrometer. The momentum resolution $\Delta p/p$ of the spectrometer was $\sim 1\%$. The calibration measurements of the elastic backward $\pi^- p$ scattering for $-0.99 \leq \cos\vartheta \leq -0.86$ on H_2O target gave the cross section that agrees with that of Ref. [6] within the statistical errors (less than 10%) for all energies studied.

THE DCX SPECTRUM AND A DEPENDENCE.

The ΔT dependence of the doubly differential cross section, $d^2\sigma/d\Omega dT$, on ^{16}O at $T_0 = 0.59$ GeV was obtained (Fig. 1a) as the mean value of the cross sections on H_2O and D_2O targets. As it is seen the cross section grows monotonously in the range ΔT from 0.03 to 0.25 GeV even at the threshold (see arrow) of the additional pion production process. The analogous ΔT dependences of the DCX spectra on ^{6}Li and ^{7}Li are given in [7]. The solid-line histogram in Fig.1a presents the DCX contribution to the cross section of the reaction $\pi^{-16}O \to \pi^+ X$, calculated for the SSCX mechanism in the framework of the model [3] which was intended for the description of all the pion-nucleus interaction processes above 0.4 GeV. In these calculations Fermi motion of nucleons, Pauli blocking and absorption were taken into consideration. It appeared that the theoretical calculation overestimates the measured cross section for $\Delta T < 140$ MeV. The similar overestimation is seen also for the inclusive spectra [8] at lower energies $T_0 = 0.4 - 0.5$ GeV for the reaction $\pi^{+16}O \to \pi^- X$ (see [7]).

It is worth to mention here, that forward cross sections of exclusive reactions $\pi^{+14}C \to \pi^{-14}O$ and $\pi^{+18}O \to \pi^{-18}Ne$, which were calculated within Glauber theory without free parameters [9], are also appeared to be higher than the measured ones at $T_0 = 300 - 525$ MeV [10]. The authors of [9] succeeded in reaching the agreement of their calculations with the data of [10] after they took into consideration a medium polarization, which leads to the renormalization of charge exchange πN amplitude. An account of this effect in the cascade model [3] substantially decreases the value of the DCX cross section for the reaction on ^{16}O (dashed histogram in Fig.1a) as well as the theoretical cross sections at $T_0 = 0.4 - 0.5$ GeV mentioned above .

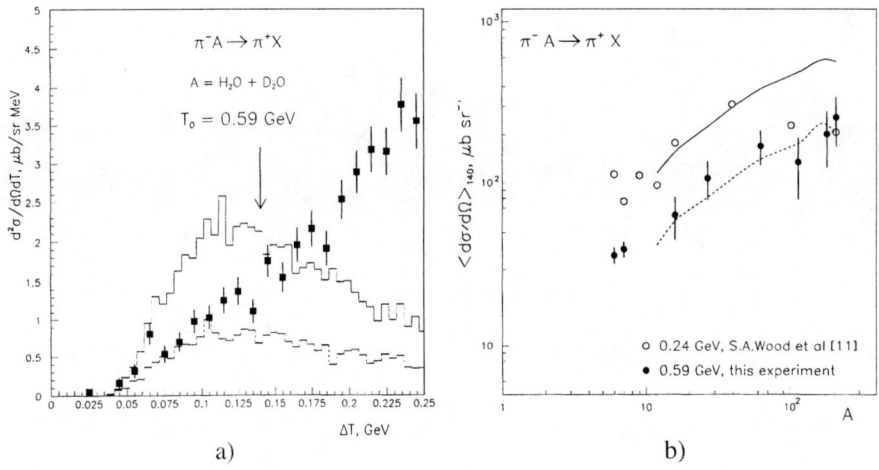

FIGURE 1. (a) The ΔT spectrum at $T_0 = 0.59$ GeV. (b) The A dependence of the inclusive DCX cross section. Theoretical curves (see text) are for $T_0 = 0.59$ GeV.

The effect of such a renomalization is shown also for the A dependence of the DCX cross section integrated over the region $0 \leq \Delta T \leq 140$ MeV (Fig. 1b). Here, dashed and solid curves stand for the SSCX calculations with and without taking into account the medium polarization. It appears that the effect changes only the normalization, not the shape of the curve.

ENERGY DEPENDENCE OF INCLUSIVE DCX.

Recently it was shown [4] in the relativistic Gribov formalism approach to DCX that inelastic rescatterings (IR) with two (and more) pions in the intermediate states give a dominant contribution to the process of DCX at energies above ~ 1 GeV. In [12] an elaborated approach is applied to study in detail the amplitudes of the $\pi \to 2\pi$ transition on nucleon which govern the IR contribution with the two-pion intermediate states. In this case the one pion exchange (OPE) model is used for the calculation of the transition amplitude. As the result, the forward inclusive pion DCX cross sections are predicted for energies above ~ 1 GeV, where IR prevail over the conventional SSCX mechanism (see Fig. 2). The dotted and solid curves in this figure are for the OPE model calculation of the IR mechanism [12] within two approaches, respectively. The lower and upper dashed curves stand for the SSCX mechanism with and without core polarization, respectively. Results of future experiments are expected to be approximately in the region between solid and dotted curves, approaching the solid one at higher energies.

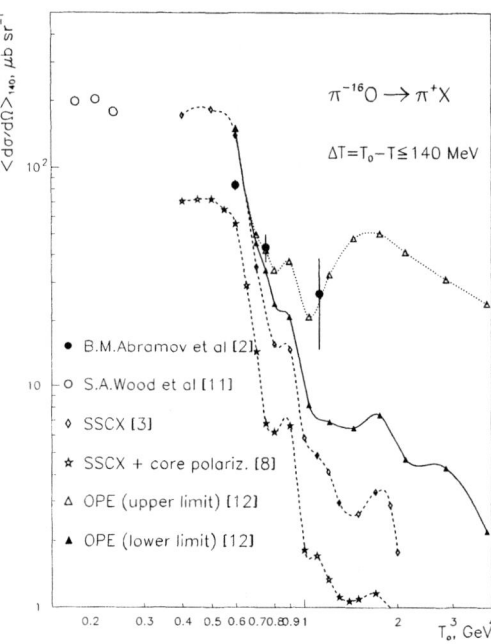

FIGURE 2. The energy dependence of the partly integrated inclusive DCX cross section.

ACKNOWLEDGMENTS

This work was partly supported by Grants RFBI 98-02-17179 and 00-15-96545.

REFERENCES

1. Abramov, B.M., et al., *Yad. Fiz.* **59**, 399-402 (1996); English translation: *Phys. Atomic Nucl.* **59**, 376-380 (1996).
2. Abramov, B.M., et al., *Few–Body Systems Suppl.* **9**, 237-240 (1995).
3. Vicente Vacas, M.J., Khankhasayev, M.Kh., and Mashnik, S.G., nucl-th/9412023.
4. Kaidalov, A.B., and Krutenkova, A.P., *Yad. Phys.* **60**, 1334-1339 (1997); English translation: *Phys. Atomic Nucl.* **60**, 1206-1211 (1997).
5. Abramov, B.M., et al., *Nucl. Phys.* **A372**, 301-316, (1981).
6. Arndt, R.A., *Phys. Rev. C* **52**, 2120-2130 (1995).
7. Burleson, R.G., in *Pion Nucleus Double Charge Exchange*, edited by W.R. Gibbs and M.J. Leitch. (World Sci., Singapore, 1990), p.79.
8. Abramov, B.M., et al., *Yad. Fiz.* **65**, (2002); English translation: *Phys. Atomic Nucl.* **65**, (2002).
9. Oset, E., and Strottman, D., *Phys. Rev. Lett.* **70**, 146-149 (1993); Oset, E., Strottman, D., Toki, H., and Navarro, J., *Phys. Rev. C* **48**, 2395-2402 (1993).
10. Williams, A.L., et al., *Phys. Lett. B* **216**, 11-14 (1989).
11. Wood, S.A., et al., *Phys. Rev. C* **46**, 1903-1921 (1992).
12. Kaidalov, A.B., and Krutenkova, A.P., *J. Phys. G* **27**, 893-911 (2001).

Search for medium effects in backward pion deuteron quasielastic scattering on ^{6}Li

B. M. Abramov*, Yu. A. Borodin*, S. A. Bulychjov*, I. A. Dukhovskoi*,
A. I. Khanov*, A. P. Krutenkova *, V. V. Kulikov*, M. A. Matsuk*,
I. A. Radkevich* and E. N. Turdakina*

Institute of Theoretical and Experimental Physics, Moscow, 117218, Russia

Abstract. The first results of the ITEP experiment on a backward pion-deuteron quasielastic scattering are presented. The experiment was done on the 3-m magnet spectrometer on a pion beam of the ITEP 10-GeV proton synchrotron. The measurements were made with a set of nuclei from ^{6}Li to ^{209}Bi at three negative pion momenta 0.72, 0.88, and 1.28 GeV/c. It is the first experiment on the backward quasielastic πd scattering in full kinematics. Here the preliminary result for ^{6}Li is presented. For the reaction ^{6}Li$(\pi^-,\pi^-d)^4$He the parameters of quasideuteron cluster in ^{6}Li have been determined which are in a good agreement with the measurements on a proton beam.

It is widely accepted that at high momentum transfer the pion-deuteron elastic scattering is dominated by the contribution of six-quark bag configuration in a deuteron. Relative admixture of this configuration can be a function of nuclear density. So it is interesting to compare the elastic πd scattering and quasielastic πd scattering in ^{6}Li, which has a well known α–d cluster structure. We have performed the first measurement of quasielastic deuteron knockout by pions, ^{6}Li$(\pi^-, d\pi^-)^4$He (1) at a high momentum transfer at beam momenta 0.72 –1.28 GeV/c.

The experiment was performed at the 3-m magnet spectrometer (see, e.g., [1]) on a pion beam of the ITEP 10-GeV proton synchrotron. Nuclear targets were placed near the center of the $3\times1\times0.5$ m^3 dipole magnet. One half of the magnet was used as a forward going deuteron spectrometer. Another half was used as a backward scattered pion (as well as a beam pion) spectrometer. Deuterons were identified by TOF with a 1.5 m^2 scintillator wall placed at 6 m from the target. All particles were registered in spark chambers and their momenta were measured in the magnetic field of the spectrometer. The resolution in the excitation energy of the rest nucleus and in the internal momentum of a quasideuteron cluster was 9.5 MeV and 17 MeV/c, respectively.

The events of the reaction (1) were selected using the cut on the value of $E_{miss} = T_0 - T_\pi - T_d - T_\alpha$, where T is the kinetic energy, and indices 0, π, d, and α stand for the initial and final pions, deuteron, and residual nucleus, respectively. The E_{miss} distributions (for $p_0 = 0.72$ GeV/c, see Fig. 1a) were fitted analogously to [2] by two Gaussians with the experimental resolution and a smooth tail. The first peak is for the reaction (1), the second one effectively describes the contribution of dd and 2p2n final states.

The Gaussian fit to the quasideuteron Fermi-momentum, $\vec{p}_F = \vec{p}_0 - \vec{p}_d - \vec{p}_\pi$, distributions, of the form $\exp[-(p_F/\kappa)^2]p_F^2$, gives $\kappa = 64 \pm 4$ MeV/c for the range -0.020

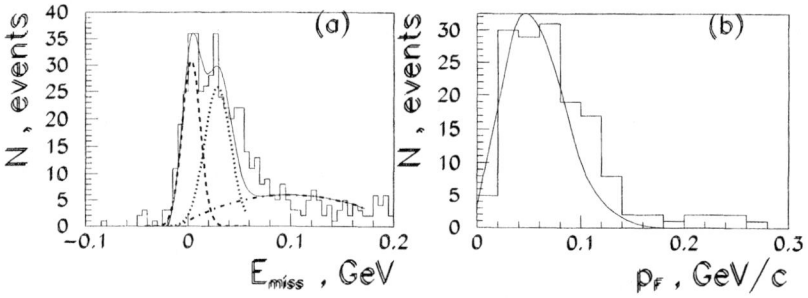

FIGURE 1. (a) E_{miss} distribution, (b) Fermi-momentum distribution for $-0.020 \leq E_{miss} \leq 0.015$ GeV.

TABLE 1. The effective numbers of deuterons.

T_0, GeV	Reaction	κ, MeV/c	n_d(PWIA)	Ref.
0.59	^{6}Li(p,dp)^{4}He	73 ±1.6	0.78 ± 0.10	[4]
0.67	^{6}Li(p,dp)^{4}He	51.5 ±2.5	0.83 ± 0.08	[2]
0.59	^{6}Li(π,dπ)^{4}He	64 ±4	0.75 ± 0.07	this exp.
0.75	^{6}Li(π,dπ)^{4}He		0.77 ± 0.15	this exp.
1.15	^{6}Li(π,dπ)^{4}He		1.26 ± 0.44	this exp.

$\leq E_{miss} \leq 0.015$ GeV (see Fig. 1b). In the plane-wave impulse approximation (PWIA) a ratio of the measured cross section of the reaction (1) to the elastic scattering cross section on a free deuteron [3] gives the effective numbers of deuterons n_d in ^{6}Li (see Table 1) which are in a good agreement with the measurements on proton beams.

Summarizing, we note the following results.
(i) For the first time the cross sections for the reaction ^{6}Li(π^-,π^-d)^{4}He was measured.
(ii) The parameters of quasideuteron cluster in ^{6}Li were determined in PWIA: $\kappa = 64 \pm 4$ MeV/c and $n_d = 0.75 \pm 0.07$, which are in a good agreement with the measurements on proton beams.
(iii) The n_d values were estimated at $p_0 = 0.72$, 0.88 and 1.28 GeV/c. An energy dependence of n_d was not seen in the studied range of q^2: $1.4 \leq q^2 \leq 2.5$ (GeV/c)^2. A small cross section has precluded from performing the measurements at the most interesting region of $q^2 \geq 4$ (GeV/c)^2, where the six quark component dominates.
This work was partly supported by Grants RFBI 00-02-17163 and 00-15-96545.

REFERENCES

1. Abramov, B.M., et al., *Nucl. Phys.* **A372**, 301-316, (1981).
2. Albrecht, D., et al., *Nucl. Phys.* **A338**, 477-494, (1980).
3. Keller, R., et al., *Phys. Rev. D* **11**, 2389-2399, (1975).
4. Kitching, P., et al., *Phys. Rev. C* **11**, 420-425, (1975).

Vector meson production with linearly polarised photons at Jefferson Lab

K. Livingston

Dept of Physics & Astronomy, University of Glasgow, Glasgow G12 8QQ, Scotland

Abstract. A Coherent Bremsstrahlung facility for the production of tagged, linearly polarised photons has recently been installed in Hall B at Jefferson Lab. The first production running with this facility will be in June/July 2001 when the g8 programme of experiments on vector meson production will take data. We will measure the decay products from the vector meson resulting from the reaction $\vec{\gamma}p \to VN$ in the energy range of $1.1 \leq E_\gamma \leq 2.2$ GeV. Here, $V = (\rho, \omega, \text{or } \phi)$. In a practical sense, we will be measuring the angular momenta, and thus the spin density matrix elements $\rho_{\alpha\beta}^j$, which determine the angular distribution of the daughter spin-0 mesons that decay from the parent vector meson. With linearly-polarised photons, one has access to six more independent spin density matrix elements than can be obtained in an unpolarised vector meson photoproduction experiment. By using the CEBAF Large Acceptance Spectrometer (CLAS) and the photon tagger, we will accurately measure the decay angular distribution (and hence the spin density matrix elements) as functions of the scattering angle and the incident photon energy with high precision.

Several new and upgraded devices will be commissioned prior to production running. A report on the results of the commissioning and the status of the running experiment will be presented.

EXPERIMENTAL SETUP

In the coherent bremsstrahlung technique an electron beam incident on a diamond radiator can be used to produce tagged, linearly polarised photons. The energy and plane of polarisation of the photons can be selected by adjusting orientation of the crystal relative to the electron beam [1]. The g8 programme of experiments on vector meson production was the first experiment to use this technique at Jefferson Lab. The

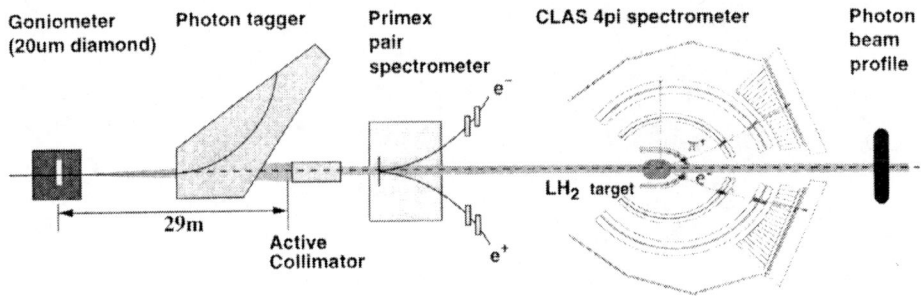

FIGURE 1. The experimental setup.

experimental setup is shown in figure 1. With the exception of the CLAS detector itself all other devices shown are new, or were substantially upgraded for g8 running, and were commissioned at the beginning of the run period. A brief description of the components is given below.

Diamond radiator. Several techniques have been developed by the Glasgow Group for selecting crystals, including examination with a petrographic microscope and measurement of rocking curves at a synchrotron light facility. The rocking curve measures the width of the Bragg peak, and gives a very good indication of the quality of the diamond. Figure 2 shows the result of these measurements made on a 100 μm synthetic crystal. The fact that the width of the Bragg peak is close to the theoretical value indicates that the crystal has a very low mosaic spread. On the basis of this, the crystal was ground down to less than 20 μm for use in the experiment.

Goniometer. This was provided and installed by the group from George Washington University. It is capable of orienting the crystal to a precision of better than 10 μrad.

Tagging spectrometer. The focal plane of tagging spectrometer has 384 plastic scintillators (e-counters) which provide signals to scalers and TDCs [2]. A major upgrade of the focal plane took place prior to running, since a high quality, low noise photon spectrum is essential to determine the degree of linear polarisation. The spectrum derived from the scalers gives rapid feedback on the uncollimated photon spectrum while the spectrum derived from the TDC (started by a downstream detector) shows the photon spectrum beyond the collimator.

Active collimator. Tight collimation can strongly enhance the ration of coherent / incoherent bremsstrahlung, and hence increase the polarisation. A 2 mm diameter collimator was used. This also had 4 embedded scintillators at the front end to provide an asymmetry measurement to monitor the position and stability of the beam.

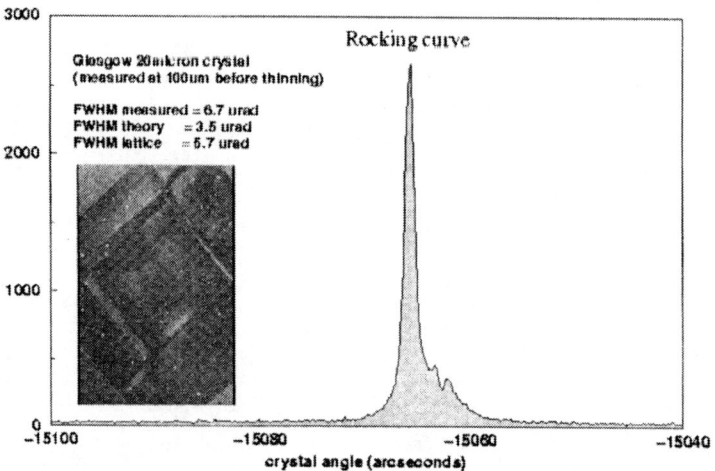

FIGURE 2. Measurements made on the crystal.

Pair spectrometer. The pair spectrometer has been installed for the Primex experiment and is used for photon normalisation [3].

RESULTS OF COMMISSIONING

To produce polarised photons 2 GeV from a 6 GeV electron beam it is necessary to have an angle of approximately 1 μrad between the beam and the crystal lattice. Hence, a major part of setting up a coherent bremsstrahlung facility is the initial alignment of the crystal, where its default orientation relative to the beam is measured. The process used at Jlab is an extension of that developed by Lohman et al. [4], where a series of scans are taken. A scan consists of a sequence of small angular movements of the crystal and the corresponding accumulation of a photon energy spectrum. Figure 3 shows the result of a scan where a series of sinusoidal steps in horizontal and vertical rotation axes are used to set the [100] crystal axis at 60 mrad from its default position and sweep it through a 360° cone on the azimuthal axis.

The ridges on the plot show the energy of the coherent peak changing as the angle between the beam and crystal changes. As can be seen in the figure, the orientation of the beam relative to the crystal can be found by fitting a template (shown with dashed lines) consisting of 8 lines separated by 45°. These lines must coincide with the points at which the ridges touch the inner circle (corresponding to the lowest tagged photon energy). The offset, in θ_v, θ_h coordinates, between the center of the template and the centre of the circle gives the angular offset between the beam and the default crystal position, and the angle of the template gives the default azimuthal orientation of the crystal.

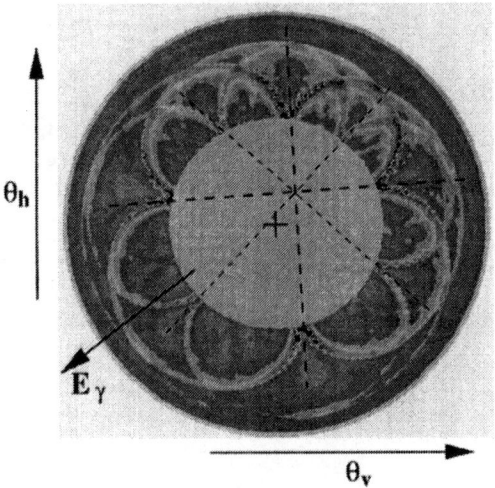

FIGURE 3. Simulated scan illustrating the method of aligning the crystal.

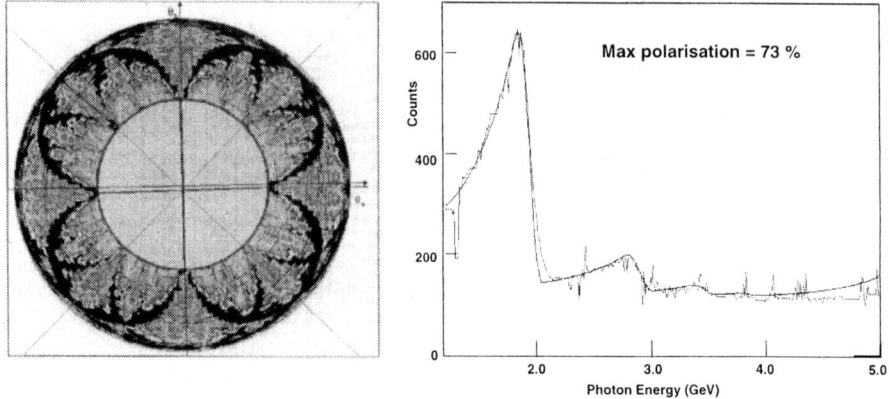

FIGURE 4. a) Scan with the crystal almost perfectly aligned. b) Coherent bremsstrahlung spectrum obtained online.

Figure 4a shows a scan taken during the alignment. There is almost a 4-fold symmetry, which shows that the crystal is aligned almost perfectly with the beam. This allowed the coherent peak to be positioned at any desired photon energy. Figure 4b shows the photon spectrum with the coherent peak set at 2 GeV, and a fit carried out using a coherent bremsstrahlung code [5], giving a maximum polarisation in the peak of 73%. However, this data was obtained from free running scalers on the tagger e-counters, and as such does not show the effect of collimation. With the effect of the collimator included in the calculation, the predicted maximum polarisation is increased to 84%. This was the anticipated degree of polarisation for the the production running, which began at the time of presentation of this paper.

ACKNOWLEDGMENTS

Thanks to all members of the g8 run group, and to all those from the CLAS collaboration, GWU and Catholic University, Washington who helped get the system up and running.

REFERENCES

1. Timm, U., *Fortschritte der Physik* **17**, u765 (1969).
2. Sober, D., et al., *Nucl. Inst. and Meth. A* **440**, 263 (2000).
3. Dale, D., et al., *Proposal E-99-014*, URL http://www.jlab.org/exp_prog/generated/~apphallb.html.
4. Lohman, D., Schumacher, M., et al., *Nucl. Inst. and Meth. A* **343**, 494 (1994).
5. Natter, A., *ANB Code*, URL http://www.pit.physik.uni-tuebingen.de/~natter/software/brems-anb.html.

Nucleon and meson effective masses in the relativistic mean field theory

Ryszard Mańka* and Ilona Bednarek*

*Institute of Physics, Silesian University, Uniwersytecka 4, 40-007 Katowice, Poland.

Abstract. Nucleon and meson effective masses in the nonlinear Relativistic Mean Field theory (RMF) introducing a nonlinear $\omega - \rho$ and σ coupling motivated by the Quark Meson Coupling model (QMC) is explored.

In the quark meson coupling model [1] nucleon properties are modified by the meson coupling directly to the quarks and not to the nucleons (paper by Li et.al. [2]). The starting point in this paper is the construction of the effective Lagrangian motivated by the QMC model. The scalar σ, isoscalar-vector ω and isovector-vector ρ mesons are denoted by φ, ω_μ, b_μ^a, respectively. The Lagrange density function for this model has the following form

$$\mathcal{L} = \tfrac{1}{2}\partial_\mu\varphi\partial^\mu\varphi - U(\varphi) - \tfrac{1}{4}\Omega_{\mu\nu}\Omega^{\mu\nu} + \tfrac{1}{2}(M_\omega - g_{\omega\sigma}\varphi)^2\omega_\mu\omega^\mu + \tfrac{1}{4}c_3(\omega_\mu\omega^\mu)^2$$
$$- \tfrac{1}{4}R^a_{\mu\nu}R^{a\mu\nu} + \tfrac{1}{2}(M_\rho - g_{\rho\sigma}\varphi)^2 b^a_\mu b^{a\mu} + \Lambda_v(g_\rho g_\omega)^2(\omega_\mu\omega^\mu)(b^a_\mu b^{a\mu})$$
$$- \tfrac{1}{4}F_{\mu\nu}F^{\mu\nu} + i\overline{\psi}\gamma^\mu D_\mu\psi - \overline{\psi}(M - g_{N\sigma}\varphi)\psi \,. \tag{1}$$

The parameters entering the Lagrangian function (1) are the coupling constants c_3, g_ω, g_ρ, $g_{N\sigma}$ for meson fields and the self-interacting coupling constants g_2 and g_3. The coupling constants $g_{\omega\sigma}$ and $g_{\rho\sigma}$ are taken from the QMC model. The modification of the density dependence of the ρ mean field can lead to the change of the neutron-skin thickness [3]. This effect is observed in the Parity Radius Experiment (PREX) at the Jefferson Laboratory which aims to measure the neutron radius in ^{208}Pb via parity violating electron scattering.

In this paper the variational method based on the Feynman-Bogolubov inequality [4] is incorporated (see more details in [5]) and the mesons effective masses are calculated. The meson effective masses as functions of the baryon density n_B are presented in Fig. 2. This figure provides a comparison of the meson effective masses obtained for different values of the coupling constants. Straight dotted lines show the ω and ρ masses in the simplest model with no additional nonlinear terms. When this model is extended by the nonlinear vector-isoscalar and vector-isovector and the vector-scalar interactions the effective meson masses are obtained. As baryon number increases the ω meson effective mass increases and then become smaller. The ρ meson effective mass increases with the increasing baryon number.

The additional nonlinear meson interaction can be explained in terms of effective meson masses that are modified in the nuclear medium. The additional nonlinear couplings

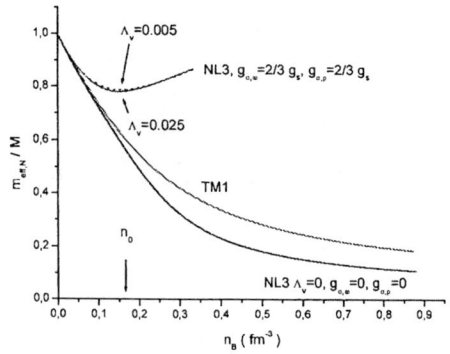

FIGURE 1. Nucleon effective mass $m_{eff,N} = M - g_{N\sigma}\sigma$ as function of the baryon density n_B (fm^{-3}).

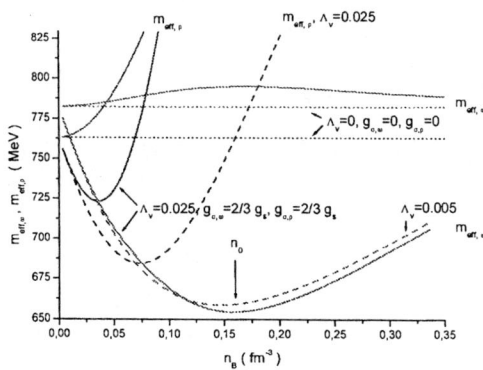

FIGURE 2. The mesons ω and ρ effective mass as function of the baryon density n_B (fm^{-3}).

Λ_V and $g_{\omega\sigma}$ and $g_{\rho\sigma}$ change the density dependence of the ω and ρ fields, whereas Λ_V itself influences the density dependence of the symmetry energy.

REFERENCES

1. Guichon, P.A.M., *Phys. Lett. B* **200**, 235 (1988);
   Guichon, P.A.M., Saito, K., Rodionov, E., Thomas, A.W., *Nucl. Phys.* **A601**, 349 (1996).
2. Li, G.Q., Ko, C.M., Brown, G.E., *Nucl. Phys.* **A606**, 568 (1996).
3. Horowitz, C.J., and Piekarewicz ,J., *Phys. Rev. Lett.* **86**, 5647 (2001); *Nucl. Phys.* **A640**, 281 (1998); Saito, K., Tsushima, K., Thomas, A.W., *Phys. Rev. C* **55**, 2637 (1997);
4. Mańka, R., Kuczyński, J., and Vitiello, G.V., *Nucl. Phys.* **B276**, 533 (1986);
   Mańka, R., and Vittiello, G.V., *Annals of Physics* **199**, 61 (1990).
5. Mańka, R., Bednarek, I., and Przybyła, G., *Phys. Rev. C* **62**, 015802 (2000).

Compton scattering at GRAAL

D. Moricciani*, O. Bartalini[†], V. Bellini**, J.-P. Bocquet[‡], M. Capogni*,
M. Castoldi[§], Annalisa D'Angelo*, Annelisa d'Angelo[†], J. P. Didelez[¶],
R. Di Salvo*, A. Fantini*, G. Gervino[∥], F. Ghio[††], B. Girolami[††],
A. Giusa**, M. Guidal[¶], E. Hourany[¶], V. Kouznetsov[‡‡], A. Lapik[‡‡], P. Levi
Sandri[†], A. Lleres[‡], V. Nedorezov[‡‡], L. Nicoletti*, C. Randieri**,
D. Rebreyend[‡], F. Renard[‡], N. V. Rudnev[§§], C. Schaerf*, M. L. Sperduto**,
C.M. Sutera[¶¶], A. Turinge***, A.Zabrodin[‡‡] and A. Zucchiatti[§]

*INFN, Sezione di Roma 2,Italy
[†]INFN, Laboratori Nazionali di Frascati, Italy
**INFN, Laboratori Nazionali del Sud, Italy
[‡]Institut des Sciences Nuclèaire, IN2P3, Grenoble, France
[§]INFN, Sezione di Genova, Italy
[¶]Institut de Physique Nuclèaire, IN2P3, Orsay, France
[∥]INFN, Sezione di Torino, Italy
[††]INFN, Sezione di Roma 1, Italy
[‡‡]Institute for Nuclear Research, Moscow, Russia
[§§]Institute of Theoretical and Experimental Physics, Moscow, Russia
[¶¶]INFN, Sezione di Catania
***L. Kurchatov Institute of Atomic Energy, Moscow, Russia

Abstract. Preliminary results on the Compton scattering on the proton are presented in energy region around 1 GeV. Data are from the GRAAL facility: where a polarised $\vec{\gamma}$–beam in the energy range of 500-1600 MeV is combined with a 4π detector having a very good energy resolution for photons.

INTRODUCTION

Probing of the nucleon with polarised tagged photons provides important information on the spectrum of baryon resonances. Measurements of polarisation observables in photon-induced reactions, in particular in meson photoproduction, are essential for comparison with recent theoretical models [1, 2, 3, 4].

The backward Compton scattering of laser light on the high energy electrons circulating in storage rings provides beams of gamma-rays useful for the study of photonuclear reactions [5]. After the successful operation of the Ladon beam in Frascati [6], several back-scattered beams have become available for nuclear and particle physics research.

The main characteristics of these beams are the low background of photons of lower energies and the high degree of polarisation. Instead of the ($\approx 1/k$) spectrum of bremsstrahlung beams, Ladon beams have a quasi-flat spectrum with the maximum intensity at the highest energy. In the scattering of very relativistic electrons, helicity is a good quantum number and therefore there is little transfer of angular momentum from the electron to the photon: at the highest photon energy, the polarisation of the gamma-ray is very close to that of the laser light and can be modified by changing the polarisation of the light with optical tools. In this way it is easy to rotate the linear polarisation of the beam, to change from linear to circular polarisation and to flip the photon spin.

EXPERIMENTAL APPARATUS AND RESULTS

The Graal facility at the ESRF in Grenoble combines a polarised gamma-ray beam with a detection apparatus that covers the entire solid angle with two little holes in the front and the back for the passage of the beam. It consists of a BGO ball made of 480 crystals (21 radiation lengths each) that covers a θ angle between 25° and 155° plus wire chambers, thin plastic scintillators and a shower wall in the forward direction ($\theta < 25°$) for the detection of gamma-rays and neutrons (Figure 1).

With a maximum energy of ≈ 1.6 GeV the Graal beam allows the study of all baryon resonances with masses above the Delta and up to ≈ 2 GeV.

We report some preliminary results on the measurements of the beam polarisation asymmetries in the Compton scattering on the proton in the angular range where both proton and photon are detected inside the BGO (central region). The main background is represented by the π^0 photo-production when one of the two photons from the π^0 decay is not detected. The π^0 decay, which contribute as background for Compton, could be symmetric or asymmetric in energy. In the first case the two photons could produce two overlapping clusters in the BGO; in order to optimize the cluster reconstruction different statistical methods have been developed [7]. In the second case, where a low energy photon of π^0 decay can be lost, a comparison between data and Monte Carlo simulation can increase the reconstruction efficiency.

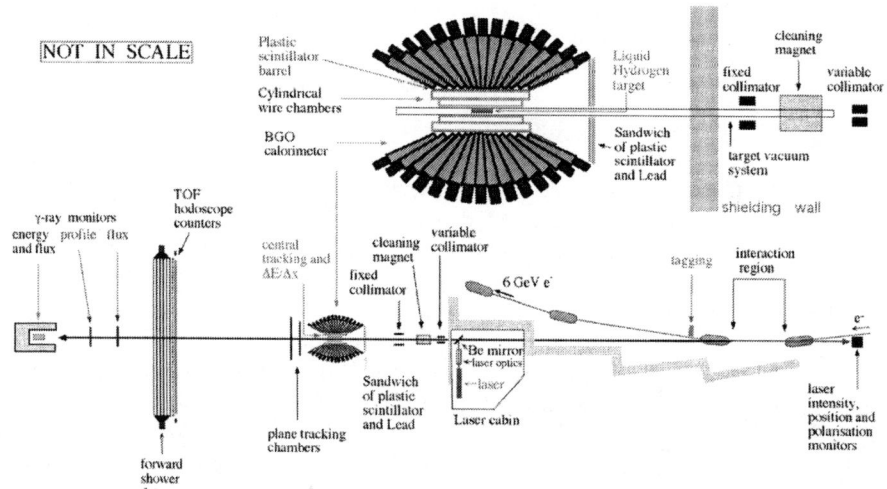

FIGURE 1. The Graal experimental apparatus. Not in scale.

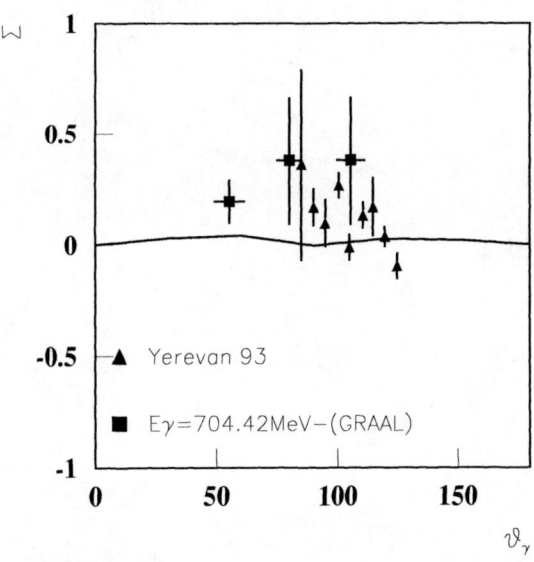

FIGURE 2. The Compton Scattering Asymmetry.

A kinematical fit allows the identification of Compton reaction and the π^0 remaining background. Preliminary results have been extracted for the beam asymmetry Σ in Compton scattering, that are shown in Figure 2. Graal data (solid squares) for $E_\gamma = 705$ MeV are in good agreement with Yerevan data [8] (solid triangles) for $E_\gamma = 750$ MeV. On the other side the comparison of data with the theoretical model from [9] is not yet satisfying. The observed disagreement demands for further theoretical investigation.

REFERENCES

1. Arndt, R., Strakovsky, I., and Workman, R., *Bull. Am. Phys. Soc.* **45**, 34 (2000); Arndt, R., Strakovsky, I., Workman, R., *Phys. Rev. C* **53**, 430 (1996). The SAID SP01, WI00, and SM95 solutions and the single pion photoproduction data base are aviable via telnet gwdac.phys.gwu.edu, user said .
2. Saghai, B., and Li, Z., DAPNIA/SPhN-01-04 Submitted to *Eur. Phys. J. A*.
3. Tiator, L., *et al.*, *Phys. Rev. C* **60**, 35210 (1999).
4. Feuster, T., and Mosel, U., *Phys. Rev. C* **59**, 460 (1999).
5. Babusci, D., *et al.*, *La Rivista del Nuovo Cimento* **19**, 5 (1996).
6. Casano L., *et al.*, *Laser and Unconventional Optics Journal* **55**, 3 (1975).
7. Zucchiatti, A., *et al.*, *Nucl. Inst. and Meth. A* **425**, 536 (1999).
8. Adamian, F.V., *et al.*, *J. Phys. G* **19**, L193, (1993).
9. L'vov, A., *et al.*, *Phys. Rev. C* **55**, 359 (1997).

Some features of the pd → ppnπ⁰ reaction at 1.037 GeV

J. Stepaniak*, H. Calén†, L. Gustafsson**, B. Höistad**, M. Jacewicz**,
A. Johansson†, T. Johansson†, A. Kupść†, S. Kullander**,
R.J.M.Y. Ruber**, C. Ekström†, K. Fransson**, J. Złomańczuk**,
A. Turowiecki‡, Z. Wilhelmi‡, J. Zabierowski§, J. Greiff¶, I. Koch¶,
B. Morosov‖, R. Bilger††, W. Brodowski††, H. Clement,†† and B. Shwartz‡‡

*Institute for Nuclear Studies, PL-00681 Warsaw, Poland
†The Svedberg Laboratory, S-75121 Uppsala, Sweden
**Department of Radiation Sciences, Uppsala University, S-75121 Uppsala, Sweden
‡Institute of Experimental Physics, Warsaw University, PL-00681 Warsaw, Poland
§Institute for Nuclear Studies, PL-90137 Lodz, Poland
¶Institut für Experimentalphysik, Universität Hamburg, D-22761 Hamburg, Germany
‖Joint Institute for Nuclear Research, Dubna, 101000 Moscow, Russia
††Physikalisches Institut, Universität Tübingen, D-72076 Tübingen, Germany
‡‡Budker Institute of Nuclear Physics, Novosibirsk 630090, Russia

Abstract. Preliminary results of the measurement of pd → ppnγγ reaction at around 1 GeV are presented. The reaction was measured at the CELSIUS storage ring using the WASA/PROMICE set-up. The quasi free mechanism was found to be dominant with $\Delta(1232)$ excitation in the proton-pion system. The signal at the position of Roper resonance was observed in the proton-π^0 invariant mass. A class of "non-spectator" events was observed with small proton-proton invariant mass.

INTRODUCTION

The aim of the measurement of pion production in proton-deuteron collisions was:

- to investigate to what extent the pion production in the three nucleon system is a superposition of quasi-free interactions involving two nucleons,
- to search for the near threshold formation of Roper resonance, which is possible in the proton-deuteron system at the energy considered, but below the threshold in free proton-nucleon interactions,
- to find out, how important is the charge exchange process $p + d \rightarrow (pp) + \Delta$ when the protons are emitted with small relative momentum, being possibly in a $^1S_0$ state.

The information on neutral pion production from proton-deuteron interactions is scarce at the energy considered here. The quasi-free mechanism is believed to dominate. This belief is supported by the charge pion production studies on deuteron.

The mechanism of the neutral pion production in the free proton-proton scattering is not fully understood not only near the threshold, but in the resonance region as well. The

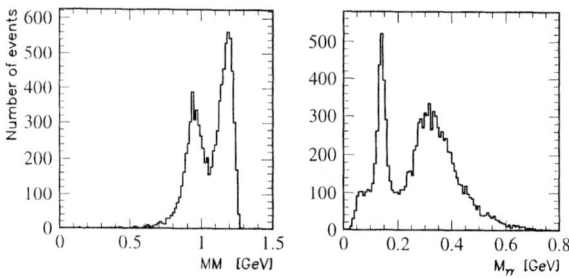

FIGURE 1. (a)(left)Total ppγγ missing mass MM; (b)(right) the γγ invariant mass distribution.

pp → ppπ^0 reaction was measured at 800 MeV at Saturne [7] and by Andreev et al. [6] in the energy range 600-900 MeV. The shape of various differential distributions was found to be consistent with the predictions of one pion exchange model, but the predicted total cross section was found to be lower than the measured one by a factor of about two. Such discrepancy was also observed by Konig and Kroll [8]. They compared data from older experiments to the refined version of peripheral model (with all the pion-nucleon partial waves taken into account).

EXPERIMENTAL PROCEDURE

The experiment was carried out using WASA/PROMICE set-up at CELSIUS accelerator. The proton beam at an energy of 1037 MeV interacted with deuterons in a cluster jet target. Details of the experimental set-up are to be found in Ref. [2].

A sample of about 130000 events of pd → ppγγ + X reaction with two identified protons has been selected. The photons were observed in two arrays of CsI(Na) scintillator at angles between 30 and 90 degrees with respect to the beam and the protons energies and angles were reconstructed in the set of drift chambers and the plastic range hodoscope as well as in the CsI arrays preceded by two layers of plastic DE counters.

To give an idea about the resolution in the neutral pion mass the two photon invariant mass distributions are shown in Fig. 1. The bump at the higher invariant masses is mainly due to the events with two neutral pion production when only two photons, out of four, are detected.

The protons with energies lower than about 35 MeV are stopped before they reach the active part of the detector. It is equivalent to 258 MeV/c momentum. This cut rejects majority of events with spectator protons.

The selection of events with one pion production was done by using a cut on the missing mass in p + p → p + p + γ + γ + MM process. The observed MM distribution is shown in Fig. 1a. An enhancement at the neutron mass is clearly seen. The region of high MM can be attributed to the two neutral pions production. The tail of events with one pion production with secondary interaction of final proton(s) in the material of the detector can also contribute to the high MM region.

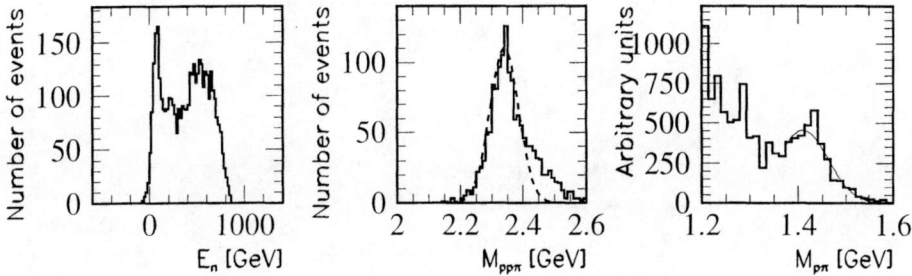

FIGURE 2. (a) Neutron energy E_n calculated from missing energy; (b) invariant mass of $pp\pi^0$ system for events with slow neutrons, dashed curve shows the results of the Monte Carlo calculation of the quasi-free process; (c) the $p\pi$ mass in the same class of events, corrected for the detector acceptance.

RESULTS

A working sample of the reaction $p+d \rightarrow p+p+\pi^0+n$ has been selected by using the criterion $0.8 < Missing\ Mass(pp\gamma\gamma) < 1.05$ GeV and by requirement of the two photon mass to be consistent with the π^0 mass. The acceptance of the detector favours the fast pion production events and excludes the majority of events with spectator protons.

The differential cross sections in various variables were compared to the model calculation based on quasi-free delta production and the four body phase-space. Large portion of the selected sample can be described by the quasi-free proton-proton production with a spectator neutron.

In the distribution of the kinetic energy of the unseen neutron two regions can be distinguished (Fig. 2a). One with slow neutron and the other, corresponding to the charge exchange process $p+n \rightarrow n+p$ or $p+n \rightarrow \Delta^0 + p$. Figure 2b shows the proton-proton-π^0 invariant mass plot for the events with additional cut on neutron energy E_n < 200 MeV. The position of the maximum coincides with the CMS energy 2.34 GeV available in the nucleon-nucleon system with the target nucleon at rest. The dashed curve shows the result of a Monte Carlo calculation for the quasi-free process with the target nucleon. It was assumed that the target nucleon moves with a Fermi momentum calculated according to Paris deuteron wave function and the pion is produced via the decay of delta isobar.

The curve does not describe the tail of the distribution in the high mass region. The plot of proton-pion invariant mass for the events from the region $M_{pp\pi^0} > 2.4$ GeV shows the possible presence of Roper resonance at 1440 MeV (Fig. 2c).

In the proton-proton invariant mass distribution an enhancement is observed in the region of the small invariant mass of the protons with respect to the phase-space Monte Carlo calculation corrected for the detector acceptance. A natural explanation is the final state interaction in the $^1S_0$ diproton and the peripheral nature of charge-exchange process. In Figure 3 the energy transfer from deuteron to the low mass proton pair is shown. Events in the large energy transfer region are correlated with small neutron energy, which excludes the process $p+d \rightarrow (pp)+\Delta^0$, since the delta isobar should

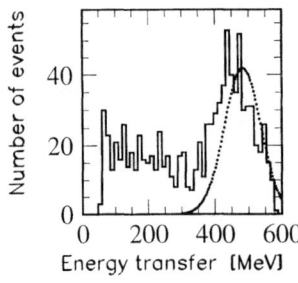 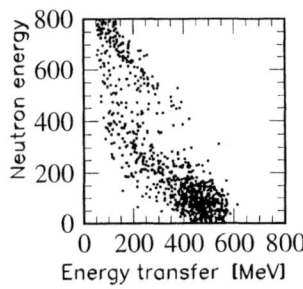

FIGURE 3. (a)(left) Energy transfer from deuteron to the proton pair for events with invariant mass of the two protons M_{pp} less than 1.9 GeV, dotted curve shows the result of the Monte Carlo calculation for the quasi-free pion production on the proton target; (b)(right) plot of neutron energy versus the energy transfer ($M_{pp} < 1.9$ GeV).

be fast in this case. Such a mechanism was discussed in Refs. [4] and [3]. In inclusive experiments with only proton pair measured one could not distinguish that mechanism from the $p + p + n_s \rightarrow pp + \pi^0 + n_s$ process. In our case it is clearly possible.

A further, higher statistics study with determination of the absolute cross sections is in progress.

ACKNOWLEDGMENTS

We thank D. Reistad and the operating team of the CELSIUS accelerator for excellent beam and target support. This work was partly supported by Polish KBN grant number 5 P03B 094 20.

REFERENCES

1. Gaarde, C., *Nucl. Phys.* **A478**, 475c (1988).
2. Calen, H., et al., *Nucl. Inst. Meth. Phys. Res. A* **379**, 57 (1996).
3. Mundigl, S., and Weise, W., *Phys. Rev. C* **39**, 710 (1989).
4. Bugg, D.V., and Wilkin, C., *Nucl. Phys.* **A467**, 575 (1987).
5. Brunt, et al., *Phys. Rev.* **187**, 1856 (1969).
6. Andreev, V.P., et al., *Phys. Rev. C* **50**, 15 (1994).
7. Comptour, C., et al., *Nucl. Phys.* **A579**, 369 (1994).
8. König, A., and Kroll, P., *Nucl. Phys.* **A356**, 345 (1981).

On radiative muon capture in hydrogen

E. Truhlík* and F.C. Khanna[†]

*Institute of Nuclear Physics, Academy of Sciences of the Czech Republic, CZ–250 68 Řež n. Prague, Czechia
[†]Theoretical Physics Institute, Department of Physics, University of Alberta, Edmonton, Alberta, Canada, T6G 2J1 and TRIUMF, 4004 Wesbrook Mall, Vancouver, BC, Canada, V6T 2A3

Abstract. We analyze the radiative capture of the negative muon in hydrogen using amplitudes derived within the chiral Lagrangian approach. Besides the leading order terms, we extract from these amplitudes the corrections up the order $O(1/M^2)$ (M is the nucleon mass). We estimate also the $\Delta(1232)$ excitation effects and processes described by an anomalous Lagrangian. Employing the off-shell parameters of the model obtained from the analysis of the pion photoproduction allows us to explain two times more of the discrepancy between the PCAC value g_P^{PCAC} of the induced pseudoscalar constant g_P and of g_P extracted from the recent TRIUMF experiment, than the standard approach can explain. By varying these parameters independently, the model can describe the high energy part of the experimental photon spectrum reasonably well for the values of $g_P \approx g_P^{PCAC}$. Our results for capture rates agree with the latest calculations within 10%.

INTRODUCTION

The least known of the four form factors, $g_V(q^2)$, $g_M(q^2)$, $g_A(q^2)$ and $g_P(q^2)$, which enter the charged weak interaction of the nucleon with a lepton [1], is the induced pseudoscalar form factor, $g_P(q^2)$. Elementary calculations lead to

$$g_P(q^2) = 2Mg_A m_l \Delta_F^\pi(q^2), \quad g_A \equiv g_A(0) = -1.267. \tag{1}$$

where $\Delta_F^\pi(q^2)$ is the pion propagator and m_l is the lepton mass.

The best way to search for the effect of the form factor $g_P(q^2)$ is the muon capture. In the elementary process of the ordinary muon capture (OMC),

$$\mu^- + p \longrightarrow \nu_\mu + n, \tag{2}$$

according to Eq. (1), the value of the induced pseudoscalar form factor g_P is

$$g_P^{OMC}(p) \equiv g_P(q^2 = 0.877 m_\mu^2) = \frac{2M m_\mu}{0.877 m_\mu^2 + m_\pi^2} g_A = 6.78 g_A = -8.59, \tag{3}$$

which is close to the value of g_P^{PCAC} predicted by PCAC.

Another attractive tool to extract the value of g_P^{PCAC} is the radiative muon capture (RMC),

$$\mu^- + p \longrightarrow \nu_\mu + \gamma + n, \tag{4}$$

The theory of the RMC was elaborated by many authors (see Refs. [2]–[9] and references therein).

The nuclear Hamiltonian suitable for use in nuclear physics calculations was provided by Rood and Tolhoek [3]. Recently, this amplitude was produced [9] from a chiral Lagrangian of the $N\pi\rho\omega a_1$ system. It satisfies the corresponding continuity equations and the consistency condition exactly.

A set of the relativistic Feynman diagrams was used by Fearing [6] to calculate the photon energy spectrum for the reaction (4). This activity was extended by Beder and Fearing [8] to include the Δ excitation processes. A recent comparison of the TRIUMF experiment [10] with the Beder–Fearing calculations provided a value of g_P which is enhanced by $\approx 50\%$ in comparison with the g_P^{PCAC}.

Here we report briefly on our results[1] obtained for the reaction (4) by making the non-relativistic reduction of the amplitudes [9]. Besides, we included the Δ isobar using again the formalism of chiral Lagrangians which we extend by adopting results of Ref. [13]. Then the resulting $N\Delta\pi\rho a_1$ Lagrangian consists of three terms and is characterised by three couplings and four arbitrary parameters A, X, Y, Z. The parameters X, Y, Z, which reflect the off–shell ambiguity of the massive spin 3/2 field were found [13] by analyzing the data on pion photoproduction. With only the X, Y, Z–independent part included, one obtains [8] about 7% effect from the Δ excitation to the photon spectrum.

We have also analyzed the contribution due to amplitudes constructed from an anomalous Lagrangian of the $\pi\rho\omega a_1$ system [15]. We have found that the influence of this contribution on the photon energy spectrum is not significant.

CAPTURE RATES

We now present the results for the singlet (Λ_s) and triplet (Λ_t) capture rates and for the capture rate Λ_T for the mixture of muonic states relevant to the TRIUMF experiment [10].

A) Δ isobar is included, the parameters used are $f^2_{\pi N\Delta}/4\pi = 0.371$, $G_1 = 2.525$, $Y = Z = -0.5$. The choice of Y and Z corresponds to Δ on–shell.

$$\Lambda_s = 3.4 \times 10^{-3} s^{-1}, \quad \Lambda_t = 101.8 \times 10^{-3} s^{-1}, \quad \Lambda_T = 31.9 \times 10^{-3} s^{-1}. \quad (5)$$

The contribution of the Δ excitation to Λ_t is $\approx 2\%$.

Very recent calculations [11] yield $\Lambda_s = (2.90 - 3.10) \times 10^{-3} s^{-1}$ and $\Lambda_t = (112 - 114) \times 10^{-3} s^{-1}$. Having in mind that Λ_s results as the difference of two large and almost equal numbers, the agreement between our value of Λ_s and the BHM value can be considered as satisfactory. However, the difference of $\approx 10-12\%$ between the triplet capture rates is too large. A half of this discrepancy can be understood by checking the integration over the phase volume [12]. But the source of the remaining difference of $\approx 5-7\%$ between the results for the Λ_t is not clear.

B) The Δ isobar is not included, the value $g_P = 1.5 g_P^{PCAC}$. The rates are

$$\Lambda_s = 8.3 \times 10^{-3} s^{-1}, \quad \Lambda_t = 115.4 \times 10^{-3} s^{-1}, \quad \Lambda_T = 39.4 \times 10^{-3} s^{-1}. \quad (6)$$

---

[1] For more details see [12].

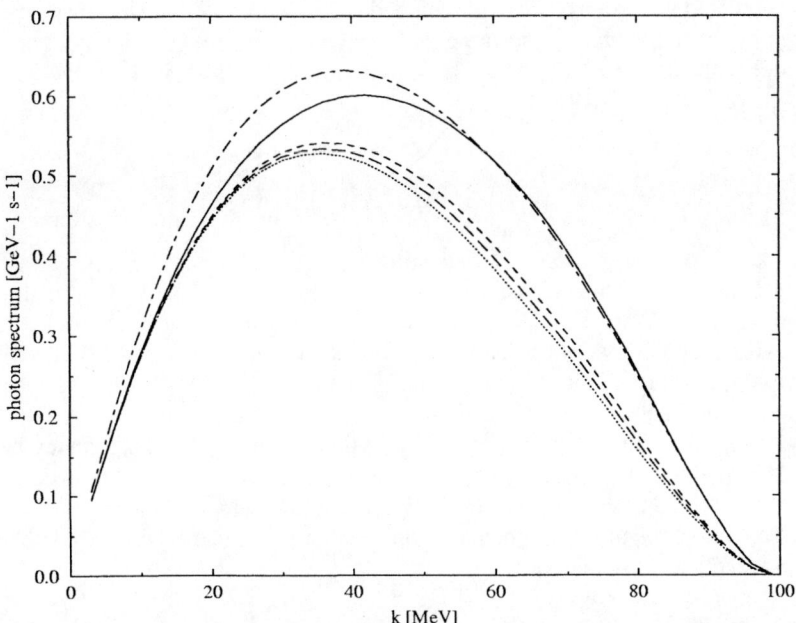

FIGURE 1. Influence of the Δ isobar parameters on the photon spectra corresponding to the mixture of muonic states in TRIUMF experiment [10]. For the explanation of the curves see text.

The strong dependence of the capture rates on g_P has already been known to Opat [2].
C) The Δ isobar is included as in A), but $Y = 1.75$ and $Z = -1.99$,

$$\Lambda_s = 6.0 \times 10^{-3}\,s^{-1},\ \Lambda_t = 115.8 \times 10^{-3}\,s^{-1},\ \Lambda_T = 37.8 \times 10^{-3}\,s^{-1}. \tag{7}$$

These capture rates are close to those calculated in the case B) for $g_P = 1.5 g_P^{PCAC}$. In order to achieve the same enhancement in the rates, one should put the Δ isobar somewhat more off–shell than it is needed in pion photoproduction processes. On the other hand, the value of the off–shell parameters X, Y, Z depend strongly on whether the pion production amplitude is unitarized or not and on the method of unitarization [13].

PHOTON SPECTRA

In Fig. 1, we show how the photon spectra, relevant to the mixture of the muonic states for the TRIUMF experiment, depend on the parameters of our model. The dotted curve corresponds to the model without the Δ isobar included and $g_P = g_P^{PCAC}$, the long–dashed curve is the photon spectrum for the Δ isobar on–shell, the dashed curve is calculated for the Δ isobar off–shell, the parameters $Y = 1.75, Z = -0.8$ are at the boundary of the region allowed by the photoproduction data [13]. In this case, two times more discrepancy is explained in comparison with the long–dashed curve. The dependence of the photon spectrum on the change in g_P is illustrated by the dash–dotted curve, calculated without the Δ isobar included and $g_P = 1.5 g_P^{PCAC}$. This curve is closely followed by the solid curve up to the photon energies $\approx 60 MeV$ which corresponds to the model with the Δ isobar off-shell, the parameter $Y = 1.75$ is the same as for the dashed curve, but $Z = -1.99$.

SUMMARY

In this paper, we have presented the capture rates and photon energy spectra for the RMC in hydrogen, calculated using an effective Hamiltonian derived from the chiral Lagrangian of the $N\Delta\pi\rho\omega a_1$ system. For the resonant part of our Lagrangian we have extended our model by adopting results of the model developed in Ref. [13]. If we restrict ourselves to the values of the parameters of the model extracted [14] from the data on the pion production by the electromagnetic interaction off the nucleon, the effect due to the Δ isobar excitation correction can explain two times more of the discrepancy between the PCAC value of the g_P and of g_P extracted from the experiment [10], than can explain the approach with the Δ isobar on-shell. If one is allowed to vary the parameters of the model independently, the experimental photon spectrum can be described for the values of the induced pseudoscalar constant, $g_P \approx g_P^{PCAC}$, reasonably well.

ACKNOWLEDGMENTS

This work is supported by the grant GA ČR 202/00/1669. The research of F. C. K. is supported in part by NSERCC.

REFERENCES

1. Blin-Stoyle, R.J., *Fundamental Interactions and the Nucleus*, North-Holland/American Elsevier, London/New-York, 1973.
2. Opat, G.I., *Phys. Rev.* **134**, B428 (1964).
3. Rood, H.P.C. and Tolhoek, H.A., *Nucl. Phys.* **A70**, 658 (1965).
4. Adler, S.L. and Dothan, Y., *Phys. Rev.* **151**, 1267 (1966); **ibid**, **164**, 2062 (1967) (erratum).
5. Christillin, P. and Servadio, S., *Nuovo Cim.* **42A**, 165 (1977).
6. Fearing, H.W., *Phys. Rev.* C **21**, 1951 (1980).
7. Gmitro, M. and Truöl, P., *Advances in Nucl. Phys.* **18**, 241 (1987).
8. Beder, D.S. and Fearing, H.W., *Phys. Rev.* D **35**, 2130 (1987); **ibid**, **39**, 3493 (1989).
9. Smejkal, J., Truhlík, E. and Khanna, F.C., *Few-Body Systems* **26**, 175 (1999).
10. Wright, D.H. et al., *Phys. Rev.* C **57**, 373 (1998).
11. Bernard, V., Hemmert, T.R. and Meißner, U.G., *Nucl. Phys.* **A686**, 290 (2001).
12. Truhlík, E. and Khanna, F.C., On radiative muon capture in hydrogen, nucl-th/0102005.
13. Davidson, R., Mukhopadhyay, N.C. and Wittman, R., *Phys. Rev.* D **43**, 71 (1991).
14. Benmerrouche, M., Davidson, R.M. and Mukhopadhyay, N.C., *Phys. Rev.* C **39**, 2339 (1989).
15. Truhlík, E., Smejkal, J. and Khanna, F.C., *Nucl. Phys.* **A689**, 741 (2001).

Pion electroproduction amplitude at threshold and nucleon weak axial form factors

E. Truhlík

Institute of Nuclear Physics, Academy of Sciences of the Czech Republic, CZ-250 68 Řež, Czechia

Abstract. We analyze amplitudes for the pion electroproduction on proton derived from Lagrangians based on the local chiral $SU(2) \times SU(2)$ symmetries. We show that such amplitudes do contain information on the nucleon weak axial form factor F_A in both soft and hard pion regimes, which is in contrast with the recent Haberzettl's claim that the pion electroproduction at threshold cannot be used to extract any information regarding F_A. We also show that these amplitudes do not contain the induced pseudoscalar form factor G_P.

INTRODUCTION

The low energy theorems for the pion production by an external electroweak interaction were formulated after the development of the current algebra and PCAC [1],[2]. According to them, the amplitude $M_\lambda^{nj}(q,k)$ for the production of a soft pion $\pi^n(q)$ off the nucleon by a vector–isovector current $\hat{J}_\lambda^j(k)$ is written as

$$f_\pi M_\lambda^{nj}(q,k) \xrightarrow{q \to 0} iq_\mu \left\langle p' \Big| \int d^4 y e^{-iqy} T\left(\hat{J}_{5\mu}^n(y)\hat{J}_\lambda^j(0)\right) \Big| p \right\rangle + \varepsilon^{njm} \langle p'|\hat{J}_{5\lambda}^m(0)|p\rangle, \quad (1)$$

where the matrix element of the nucleon weak axial current is [3]

$$\langle p'|\hat{J}_{5\lambda}^m(0)|p\rangle = i\bar{u}(p')\left[g_A F_A(q_1^2)\gamma_\lambda \gamma_5 - 2ig f_\pi \Delta_F^\pi(q_1^2)q_{1\lambda}\gamma_5\right]\frac{\tau^m}{2}u(p), \quad (2)$$

$q_1 = p' - p = k - q$ and $g_A = 1.267$.

Starting from Eq.(1), a "master formula" for the amplitude $M_\lambda^{nj}(q,k)$ can be derived [4]. For this purpose, the contribution to the divergence due to the coupling of the axial current to the external nucleon lines (Fig. 1a and 1b) can be extracted from the current–current amplitudes.

Besides, it was shown [5] that the evaluation of the contribution to the divergence of the current–current amplitudes using the current algebra and PCAC due to the process of Fig. 1c with $B = \pi$ yields two pieces: one is exactly the pion pole production term and the other one cancels the induced pseudoscalar term. Then the soft pion production amplitude corresponds to Fig. 2 and it is

$$M_\lambda^{nj}(q,k) = \frac{g}{2M}\bar{u}(p')\left[\slashed{q}\tau^n S_F(P)\hat{J}_\lambda^j(k) + \hat{J}_\lambda^j(k)S_F(Q)\slashed{q}\tau^n\right]u(p)$$
$$+ \frac{(k-2q)_\lambda}{(k-q)^2 + m_\pi^2}F_\pi(k^2)\varepsilon^{njm}g\,\bar{u}(p')\gamma_5\tau^m u(p) + i\frac{g_A}{2f_\pi}F_A(k^2)\varepsilon^{njm}\bar{u}(p')\gamma_\lambda\gamma_5\tau^m u(p). \quad (3)$$

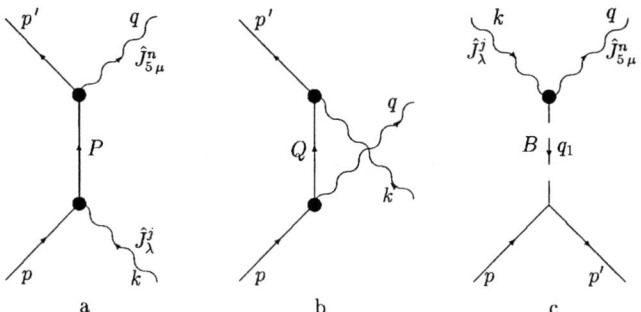

FIGURE 1. The current–current amplitudes contributing to the first term on the right hand side of Eq.(1). Graphs a,b – the nucleon terms; graph c – the contact terms of the $B = \pi$ or a_1 range.

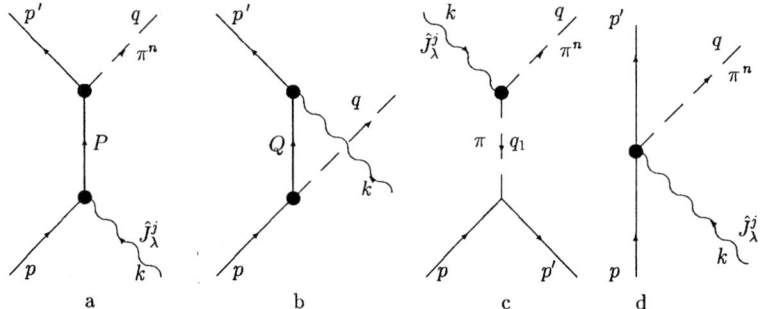

FIGURE 2. The soft pion production amplitude according to Eq.(3). Graphs a,b – the nucleon Born terms; graph c – the pion pole term; graph d – the Kroll–Ruderman term.

It is clear that besides the nucleon Born terms, the soft pion amplitude Eq. (3) contains the Kroll–Ruderman (contact) term including the nucleon weak axial form factor, and the pion pole production term.

The relevance of the nucleon weak axial form factor to the pion electroproduction off the nucleon has been questioned recently by Haberzettl in Ref. [6], where he concluded that the pion electroproduction processes at threshold cannot be used to extract any information regarding the nucleon weak axial form factor. The pion production amplitude [7],[8] was rederived [6] using an analogue of Eq. (1) valid beyond the soft pion limit. Neglecting possible Schwinger terms, the amplitude is

$$M_\lambda^{nj}(q,k) = \frac{q^2 + m_\pi^2}{f_\pi m_\pi^2} \left[iq_\mu \left\langle p' \left| \int d^4y\, e^{-iqy} T\left(\hat{j}_{5\mu}^n(y) \hat{j}_\lambda^j(0) \right) \right| p \right\rangle + \varepsilon^{njm} \left\langle p' \left| \hat{j}_{5\lambda}^m(0) \right| p \right\rangle \right]. \quad (4)$$

After making our own study of the problem and comparing it with the earlier results [6], we came to the conclusion that the problem is with a misinterpretation of the results by Haberzettl [6]. Here we present a short account of the results, referring for the details to Ref. [9].

PION ELECTROPRODUCTION AMPLITUDE

Our study of the pion electroproduction amplitude is based on chiral Lagrangians possessing hidden local chiral symmetry $SU(2) \times SU(2)$ [3],[10]. The pion electroproduction amplitude was derived in the same way as it was done in [6]: first constructing the current–current amplitudes and then calculating the divergence.

According to Fig. 1, the current–current amplitudes consist of three sets of terms: of the nucleon Born terms and of the pion- and a_1 pole terms.

Let us start with the nucleon Born terms represented by the amplitude $T^{B,nj}_{\mu\lambda}$. As a result of calculations we obtain

$$iq_\mu T^{B,nj}_{\mu\lambda} = f_\pi m_\pi^2 \Delta_F^\pi(q^2) M^{nj}_{B,\lambda}, \tag{5}$$

where $M^{nj}_{B,\lambda}$ is the nucleon Born pion electroproduction amplitude for the pseudovector πNN coupling.

The current–current amplitude $T^{pp,nj}_{\mu\lambda}$ due to the pion exchange in Fig. 1c contributes into the divergence as

$$iq_\mu T^{pp,nj}_{\mu\lambda} = f_\pi m_\pi^2 \Delta_F^\pi(q^2) M^{nj}_{pp,\lambda} - f_\pi q_{1\lambda} \Delta_F^\pi(q_1^2) \varepsilon^{njm} g \Gamma_5^m(p',p), \tag{6}$$

where $\Gamma_5^m(p',p) = \bar{u}(p')\gamma_5\tau^m u(p)$. Using this result and the corresponding equation for the matrix element of the nucleon weak axial current, we find that the induced pseudoscalar form factor disappears from the pion electroproduction amplitude and the pion pole production amplitude $M^{nj}_{pp,\lambda}$ appears instead.

The contribution from the a_1 pole current–current amplitude, $T^{a_1 p,nj}_{\mu\lambda}$, can be calculated analogously. The result is

$$iq_\mu T^{a_1 p,nj}_{\mu\lambda}(x) = f_\pi m_\pi^2 \Delta_F^\pi(q^2) \sum_{x=a,b} M^{nj}_{a_1 p,\lambda}(x) - ig_A m_\rho^2 \Delta_{\lambda\zeta}^{a_1}(q_1) \varepsilon^{njm} \Gamma_{5\zeta}^m(p',p) \tag{7}$$

where $\Gamma_{5\zeta}^m(p',p) = \bar{u}(p')\gamma_\zeta\gamma_5\tau^m u(p)$ and

$$M^{nj}_{a_1 p,\lambda}(a) = i\frac{g_A}{2f_\pi} m_{a_1}^2 \left\{ \Delta_{\nu\zeta}^{a_1}(q_1) \Delta_F^\rho(k^2) [k_\nu q_\lambda - (k\cdot q)\delta_{\nu\lambda}] \right.$$
$$\left. + \frac{1}{2}\Delta_{\nu\lambda}^\rho(k) \Delta_F^{a_1}(q_1^2) [q_\zeta q_{1\eta} - (q\cdot q_1)\delta_{\zeta\nu}] \right\} \varepsilon^{njm} \Gamma_{5\zeta}^m(p',p), \tag{8}$$

$$M^{nj}_{a_1 p,\lambda}(b) = i\frac{g_A}{2f_\pi} m_{a_1}^2 \Delta_{\lambda\zeta}^{a_1}(q_1) \varepsilon^{njm} \Gamma_{5\zeta}^m(p',p). \tag{9}$$

Substituting this result into the equation for the pion electroproduction amplitude one finds that the first term in the matrix element of the axial current is cancelled, too.

Summing up the partial results Eqs. (5), (6) and Eq. (7), we obtain the resulting pion electroproduction amplitude,

$$M^{nj}_\lambda(q,k) = M^{nj}_{B,\lambda} + M^{nj}_{pp,\lambda} + \sum_{x=a,b} M^{nj}_{a_1 p,\lambda}(x). \tag{10}$$

The amplitudes $M^{nj}_{B,\lambda}$ and $M^{nj}_{pp,\lambda}$ are the well–known nucleon Born and pion pole production terms, respectively. So only the contribution of the amplitudes $M^{nj}_{a_1p,\lambda}(a)$ and $M^{nj}_{a_1p,\lambda}(b)$ remains to be considered. In the soft pion limit, the amplitude $M^{nj}_{a_1p,\lambda}(a) \sim O(kq)$, while the amplitude $M^{nj}_{a_1p,\lambda}(b)$ restores the contact term retaining the nucleon weak axial form factor F_A. It follows that this model is consistent with the prediction of the current algebra and PCAC.

DISCUSSION AND SUMMARY

Our Lagrangian, reflecting the hidden local symmetry, helped us to construct the current–current amplitudes of Fig. 1. Subsequent calculation of the divergence of these amplitudes and the use of Eq. (4) led us to observe that the pion and a_1 exchanges in Fig. 1c do contribute non–negligibly even in the soft pion limit. The contribution proceeds in such a way that the matrix element of the axial current [the second term on the right hand side of Eq. (4)] is eliminated and the pion pole and a_1 pole pion production terms appear. Moreover, one of the a_1 pole pion production amplitudes is nothing but the Kroll–Ruderman term containing again the nucleon weak axial form factor $F_A(k^2)$ in the soft pion limit. Actually, any model, respecting the local chiral symmetry, should provide at threshold the same result. So the claim [6] that the pion electroproduction processes at threshold cannot be used to extract any information regarding the nucleon weak axial form factor should be considered as precipitous.

It follows from our results that the pion electroproduction amplitude does not contain the induced pseudoscalar part of the nucleon weak axial current either for soft or hard pions, which is at variance with [11]. It also means that the measurement of this quantity [12] is a misconception.

This work is supported by the grant GA ČR 202/00/1669.

REFERENCES

1. Adler, S.L. and Dashen, R., *Current Algebras*, W.A. Benjamin, New York, 1968.
2. De Alfaro, V., Fubini, S., Furlan, G. and Rossetti, C., *Currents in Hadron Physics*, North–Holland/American Elsevier, Amsterdam–London/New York, 1973.
3. Smejkal, J., Truhlík, E. and Khanna, F.C., *Few–Body Systems* **26**, 175 (1999).
4. Adler, S.L., in: Proc. of the 6th Hawaii Topical Conf. in Particle Physics, eds. Dobson, D.N. *et al.*, University of Hawaii at Manoa, 1975, p. 5.
5. Ivanov, E.A. and Truhlík, E., *Fiz. Elem. Chastits At. Yadra* **12**, 492 (1981) [*Sov. J. Part. Nucl.* **12**, 198 (1981)].
6. Haberzettl, H., *Phys. Rev. Lett.* **85**, 3576 (2000).
7. Haberzettl, H., *Phys. Rev.* C **56**, 2041 (1977).
8. Haberzettl, H., *Phys. Rev.* C **62**, 034605 (2000).
9. Truhlík, E. On the pion electroproduction amplitude, *Phys. Rev.* C, in press; nucl–th/0105038.
10. Smejkal, J., Truhlík, E. and Göller, H., *Nucl. Phys.* **A 624**, 655 (1997).
11. Guichon, P.A.M, Comment about pion electro–production and the axial form factors, hep–ph/0012126.
12. Choi, S. *et al.*, *Phys. Rev. Lett.* **71**, 3927 (1993).

Measurement of the $p + {}^6Li \to {}^7Be + \eta$ reaction near threshold energies

Marcel Uličný[*] (for the GEM Collaboration)
[http://ikpgem02.ikp.kfa-juelich.de/gem]

Forschungszentrum Jülich, 52425 Jülich, Germany

Abstract. The $p + {}^6Li \to {}^7Be + \eta$ reaction near the kinematical threshold has been studied within the GEM Collaboration, following the previous line of η production and η bound states studies on light nuclei. The proton beam of 1297 MeV/c, from the COSY, Forschungszentrum Jülich, Germany, has been used to test the experimental setup. For the registration of the recoiled nuclei, the BigKarl Spectrometer operates as a full solid angle analyzer. The results of the test run as well as further detector improvements are discussed.

INTRODUCTION

Our aim is the investigation of the η production on light nuclei in order to study the reaction mechanism as well as a possible strong η-nucleus interaction in a final state. For the present investigation we chose the reaction

$$p + {}^6Li \to {}^7Be + \eta \qquad (1)$$

at 1297 MeV/c, close to its absolute threshold of 1273.17 MeV/c. The threshold is far below the threshold of the nucleon-nucleon collision. On the other hand, the momentum transfer in the CMS is 1015.74 MeV/c, which is far above the Fermi momentum of the struck nucleon. Hence the reaction has to proceed in several steps and/or on a nucleon cluster.

For the present reaction there was only one earlier measurement [1] done at somewhat different energy, aimed at measuring the η's via their two photon decay. Only 8 events were detected in total out of which 3 were estimated to be associated with the background.

The above number includes the sum of ${}^7Be$ states up to 10 MeV excitation energy.

---

[*] On leave from University of P.J. Šafárik in Košice, Slovakia [http://www.upjs.sk]

EXPERIMENTAL SETUP

To measure the reaction we used BigKarl magnetic spectrometer which works in a focusing mode in the horizontal plane and dispersive mode in the vertical direction.

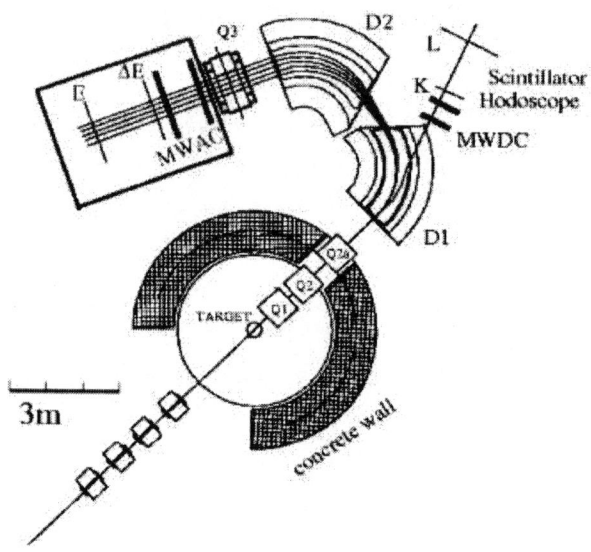

FIGURE 1. The BigKarl magnetic spectrometer.

FIGURE 2. Focal plane vacuum detection system installation.

At the so called focal area a vacuum detection system (Fig.2) was built to measure position, energy and Time of Flight (TOF) observables of the particle passing thru. The system consists of a set of two layers of plastic scintillator (BC 408 from Bicron) paddles of which there was a signal readout on each side by phototubes. This gave us an opportunity, in combination with the BigKarl Spectrometer, to gather complete kinematics information about the particles involved in the reaction studied.

RESULTS OF THE TEST MEASUREMENT

We performed standard Monte Carlo (MC) simulations using the GEANT modeling package of CERN. We could estimate the topological displacement of groups of particles with respect to the energy loss and TOF (Fig.3).

Within the feasibility study of the experiment we could nicely reproduce the spectra predicted by simulations and achieve ≈6% difference effect between the production spectra taken on the full target and the empty target measurement.

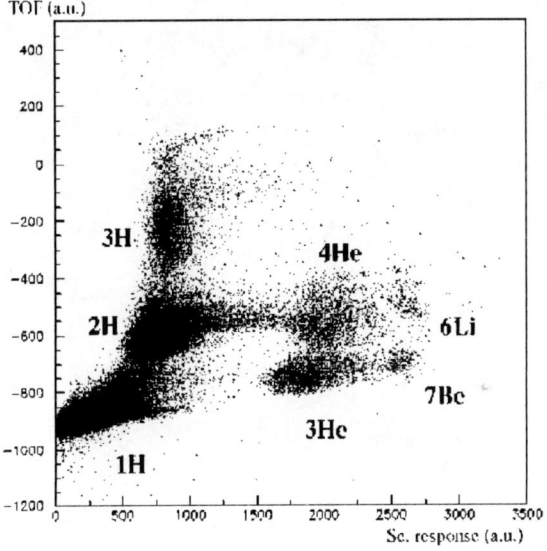

FIGURE 3. Particle identification for the (1) production runs(shown in a.u.).

FURTHER IMPROVEMENTS

To fully exploit the spectrometer precision we decided to improve the spatial resolution of the detector by adding two packs of Multiwire Avalanche Chambers (MWAC), constructed to work in low pressure environment (Fig.4).

They should improve our horizontal position resolution by a factor of at least 100 (resolution ~1mm).

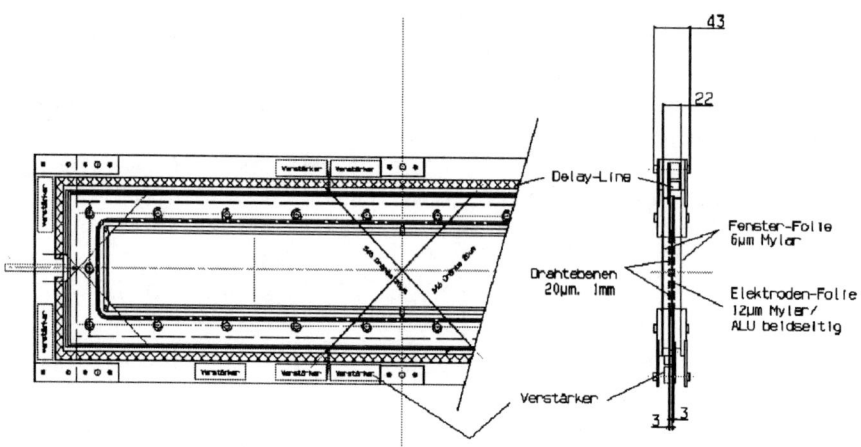

FIGURE 4. The Multiwire Avalanche Chambers.

FUTURE PLANS

After a short test run planned to the end of the year 2001 we would like to apply for further three weeks of beamtime to collect sufficient statistics for the measured reaction since the estimated total cross-section is rather small (few nanobarns).
The detection system mentioned above is of course suitable for studies of various kind in future. Particularly experiments involving light and semi-heavy nucleons in the exit channel, with short air penetration lengths, requiring high spatial resolution capabilities are preferred.

ACKNOWLEDGMENTS

The Forschungszentrum Jülich supported this work by Contract No. FFE 41397018.

REFERENCES

1. Scomparin, E., et al., *J. Phys. G* **19**, L51 (1993).

The ηN scattering length and ηd final states

Sławomir Wycech[*] and Anthony M. Green [†]

[*]Sołtan Institute for Nuclear Studies, 00-681 Warsaw, Poland
[†]Department of Physical Sciences and Helsinki Institute of Physics, University of Helsinki,
FIN-00014 Helsinki, Finland

Abstract. The K-matrix used to describe the coupled πN, ηN, γN, $\pi\pi N$ systems favours large $0.7 - 1.05$ fm ηN scattering lengths. These generate a sharp cusp, due to a virtual state in the elastic ηd amplitude at the threshold. However, its effect on final state interactions in the $pn \to \eta d$ reaction is shown to be moderate.

THE K-MATRIX FOR ηN INTERACTIONS

One can check the two body ηN interaction models with few body η- physics. At low energies these interactions are dominated by the $N(1540)$ resonance. To account for this dominance one parametrises the K matrix as

$$K_{i,j} = \frac{g_i g_j}{E_o - E} + B_{i,j} . \tag{1}$$

The pole term represents the resonance coupled to channel states by g_i. A background matrix $B_{i,j}$ due to additional interactions is necessary to describe the data. In the ηN channel $B_{\eta\eta}$ is not well known on a phenomenological level and is not understood by theory. This K matrix for channels πN, ηN, γN, $\pi\pi N$ was found in ref. [1]. The best solution yields the "cusp" in Fig. 1, due to the resonance and attractive $B_{\eta\eta}$. The scattering length is $a_{\eta N} = 0.75(4) + i0.27(3)$ fm. Similar values are obtained by other methods [2], but the actual number is not settled. Another slightly less likely solution is allowed with a small "chiral" $a_{\eta N} = 0.2 + i0.3$ fm [3]. It corresponds to a broad resonance, repulsive $B_{\eta\eta}$, and it resembles the finding in the η Helium system [4].

The η few-nucleon systems may test the ηN interaction. The $np \to \eta d$ reaction is the simplest available possibility [5].

THE η-DEUTERON SYSTEM

The ηN interaction, due to the $N(1540)$, is attractive. This mechanism by itself cannot produce bound states, unless it is supplemented by a large B. On the other hand, quasibound or virtual states may exist in the ηd system. To find these the summation of a multiple scattering series given in ref. [6] is used. The "cusp" amplitude generates length $A_{\eta d} = 2.5 + i2.8$ fm corresponding to a "virtual" ηd state. An order of magnitude enhancement is found in the ηd elastic scattering close to the threshold. With Re $a_{\eta N}$

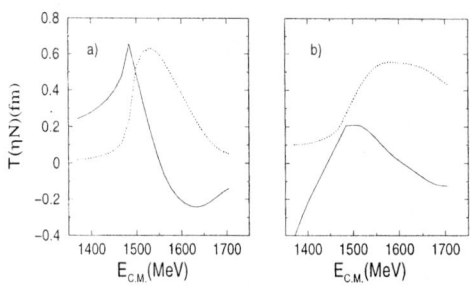

FIGURE 1. Two K-matrix fits for the elastic ηN scattering amplitude. A "cusp" solution on the left [1] and a "chiral" solution on the right [3].

as large as 1 fm one enters a critical region of Re $A_{\eta d} \approx 0$ which signals the ηd system close to binding. Genuine quasi-bound singularity is formed with Re $a_{\eta N} > 1.2$ fm. That is beyond the limits allowed by our K-matrix model. Elastic cusps are are not observable, yet. The observed $np \to \eta d$ amplitude indicates a rather smooth behaviour [5]. It is shown in Fig. 2 and compared with calculations. The threshold cusp is moderate. This reflects the necessity for the NN pair to propagate from the small η formation region to the large deuteron region. Scattering lengths in the range of 0.6-0.8 fm are allowed.

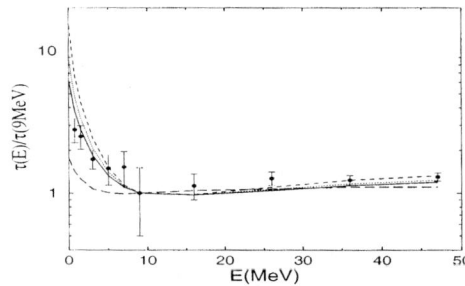

FIGURE 2. The reduced cross section $\tau = \sigma(pn \to \eta d)/p_\eta(E)$ from Ref. [5] arbitrarily normalised. The theoretical equivalent is calculated for Re $a_{\eta N} = 0.75$ fm (solid line), 0.87 fm (dotted line), 1.05 fm (dashed line) and 0.21 fm (long-dashed line) [7].

Support from KBN grant 5P 03B 04521 is acknowledged.

REFERENCES

1. Green, A.M., and Wycech, S., *Phys. Rev* **C55**, 2167R (1997).
2. Batinic, M., Dadic, I., Slaus, I., Svarc, A., Nefkens, B.M.K., and Lee, T.-S.H., *Physica Scripta* **58**, 15 (1998).
3. Arndt, R.A., Green, A.M., Workman, R.L., and Wycech, S., *Phys. Rev C* **58**, 3636 (1998).
4. Wilkin, C., *Phys. Rev C* **47**, 938R (1993).
5. Calén, H., et al., *Phys. Rev. Lett.* **80**, 2069 (1998); **79**, 2642 (1997).
6. Green, A.M., Niskanen, J.A., and Wycech, S., *Phys. Rev. C* **54**, 1970 (1996).
7. Wycech, S., and Green, A.M., *Nucl. Phys.* **A663**, 529c (2000), and submitted to *Phys. Rev. C.*

APPENDIX

8th International Conference
Mesons and Light Nuclei
July 2nd-6th, Prague

Conference Program

Sunday, July 1st

    13:00-20:00   *Registration* (hotel Krystal)

Monday, July 2nd

    8:45- 9:00   *Conference opening*

    9:00-10:30   *Morning session I (F.L.Gross)*
        45' H.Griesshammer(Munich): An introduction to few-nucleon systems in effective field theory
        45' L.E.Marcucci(Pisa): Electro-weak structure of few-body nuclei

    10:30-11:00   *Coffee break*

    11:00-12:30   *Morning session II (D. Phillips)*
        45' E.Epelbaum(Jülich): Chiral dynamics in few-nucleon systems
        45' H.O.Meyer(IUCF,Bloomington): The search for a three-nucleon force in polarization experiments with a storage rings

    12:30-14:30   *Lunch break*

    14:30-16:10   *Parallel sessions I*
    16:10-16:40   *Coffee break*
    16:40-18:00   *Parallel sessions II*

    19:00   *Dinner*

Tuesday, July 3rd

- 9:00-10:30   *Morning session I (A.Rinat)*
  - 45' T.Motoba(Osaka/Seattle): Recent topics in hypernuclear spectroscopy
  - 45' K.Imai(Kyoto): Recent progress of spectroscopy of light hypernuclei
- 10:30-11:00   *Coffee break*

- 11:00-12:25   *Morning session II (T.Motoba)*
  - 45' L.Tang(Hampton/JLab): First experiment on spectroscopy of Λ-hypernuclei by electroproduction at JLab
  - 20' A.Gal(Jerusalem): Few-body $\Lambda\Lambda$ hypernuclei and the onset of stability of $\Lambda\Xi$ hypernuclei
  - 20' A.Parreño(Barcelona): Recent results on the non-mesonic weak decay of hypernuclei within a OME model

- 12:25-14:20   *Lunch break*

- 14:20-15:40   *Parallel sessions I*
- 15:40-16:10   *Coffee break*
- 16:10-17:30   *Parallel sessions II*

- 17:45   *Dinner*
- 20:00   *Concert*

Wednesday, July 4th

- 9:00-10:30   *Morning session I (A.Gal)*
  - 45' V.Vento(Valencia): Quark degrees of freedom in hadronic systems
  - 45' V.Burkert(JLab): Nucleon structure in the resonance region
- 10:30-11:00   *Coffee break*

- 11:00-12:05   *Morning session II (V.Vento)*
  - 45' G. van der Steenhoven(NIKHEF): Hard exclusive processes at HERMES
  - 20' R. Van de Vyver(Ghent): Virtual Compton Scattering: First Results from JLab

- 12:05-13:30   *Lunch break*

- 13:30-18:00   *Sightseeing Tours*

- 19:00   *Dinner*

Thursday, July 5th

9:00-10:05		*Morning session I (A.Stadler)*
	45'	F.L.Gross(W&M/JLab): Deuteron- recent theoretical and experimental results
	20'	S.Širca(MIT): Recent measurements with the out-of-plane spectrometer system at MIT-Bates
10:05-10:25		*Coffee break*
10:25-11:30		*Morning session II (W.Klink)*
	45'	J.-F.Mathiot(Clermont-Ferrand): Recent progress in light front dynamics
	20'	A.Siepe(Bonn): Neutron-proton and neutron-neutron quasi-free scattering in the n+d breakup reaction at 26 MeV
11:30-12:30		*Lunch break*
13:00-14:25		*Afternoon session I (J.F.Mathiot)*
	45'	W.Klink(Iowa): Point form analysis of nucleon and deuteron form factors
	20'	M.Birse(Manchester): A renormalisation group approach to two-body scattering with long-range forces
	20'	J.Gegelia(Ferrara): Doublet channel neutron-deuteron scattering in leading order EFT
14:25-15:00		*Coffee break*
15:00-16:45		*Afternoon session II (C.Schaerf)*
	45'	E.Fragiacomo(Trieste): Pion pairs in nuclear matter
	20'	A.Rinat(Rehovot): Neutron structure function from inclusive scattering on nuclei
	20'	Th.Speckner(Erlangen): The GDH Experiment at ELSA
	20'	K.Livingston(Glasgow): Vector meson production with linearly polarised photons at Jefferson Lab
17:00-19:00		*Poster Session (D.Phillips)*
	40'	Brief presentations of posters
19:00		*Conference Dinner*

Friday, July 6th

9:00-10:05 *Morning session I (M.Birse)*
- 45' D.Phillips(Athens,Ohio): Probing the effectiveness: Chiral perturbation theory calculations of low-energy electromagnetic reactions on deuterium
- 20' J.Smolík(Prague): The DIRAC Experiment at CERN

10:05-10:25 *Coffee break*

10:25-11:30 *Morning session II (E.Epelbaum)*
- 45' R.Timmermans(KVI Groningen): Chiral symmetry and the NN interaction
- 20' R.Vinh Mau(Paris): Nuclear forces and quark degrees of freedom

11:30-12:30 *Lunch break*

13:00-14:25 *Afternoon session I (J.Zlomanczuk)*
- 45' T.Peña(IST Lisbon): Probing the ηN interaction with η production reactions
- 20' J.Ernst(Bonn): η-photoproduction with SAPHIR at ELSA
- 20' J.-P.Bocquet(Grenoble): Meson photoproduction on the proton at GRAAL

14:25-15:00 *Coffee break*

15:00-16:25 *Afternoon session II (H.O.Meyer)*
- 45' J.Zlomańczuk(Uppsala): Pion production in pN collisions at close to threshold energies
- 20' H.Clement(Tübingen): Exclusive measurements of the $p+ p \rightarrow p+ p+ \pi^+ + \pi^-$ reaction close to threshold
- 20' M.Nekipelov(Jülich): K^+-production in subthreshold pA Collisions with ANKE

16:25-16:45 *Coffee break*

16:45-17:25 *Afternoon session III (R.Timmermans)*
- 20' M.Schwamb(Mainz): The two-nucleon system in the Δ-resonance region including full meson retardation
- 20' J.Haidenbauer(Jülich): A meson-exchange model for the YN interaction

17:30-17:45 *Conference closing (F.L.Gross)*

19:00 *Dinner*

Parallel Sessions
MONDAY, 14:30-16:10

Session Ia Chairperson - R.Timmermans

14:30-14:50   J.McGovern(Manchester): Compton scattering from the
                                proton at NLO in the chiral expansion
14:50-15:10   A.Moalem(Beer-Sheva): Effective chiral theory for
                                pseudoscalar and vector mesons
15:10-15:30   M.D.Cozma(Groningen): pp bremsstrahlung and
                                low energy NN interaction
15:30-15:50   S.Dalley(Cambridge): Mesons on a transverse Lattice
15:50-16:10   F.Fernandez(Salamanca): Quark model study of
                                the NN*(1440) components of the deuteron

Session Ib Chairperson - M.T.Peña

14:30-14:50   V.B.Belyaev(Dubna): Microscopic description of
                                η-photoproduction on light nuclei
14:50-15:10   D.Moricciani(Roma): Preliminary results on strangeness
                                photoproduction and Compton scattering
                                on the proton at GRAAL
15:10-15:30   A.Deloff(Warsaw): Determination of πN scattering lengths
                          from pionic hydrogen and pionic deuterium X-ray data
15:30-15:50   D.Cabrera-Urbán(Valencia): The ρ meson in nuclear matter
                                - a chiral unitary approach
15:50-16:10   M.Uličný(Jülich): Measurement of the p + ^{6}Li --> ^{7}Be + η
                                reaction near threshold energies

Session Ic Chairperson - A.Parreño

14:30-14:50   Ch.Keil(Giessen): Single particle properties of
                                Λ hypernuclei in the density dependent
                                relativistic hadron field theory
14:50-15:10   E.Hiyama(KEK): Four-body calculations of $_\Lambda^4$H and $_\Lambda^4$He with
                                realistic YN and NN interactions
15:10-15:30   M.Kamimura(Kyushu): How does nucleus shrink by
                                participation of a Λ hyperon?
15:30-15:50   K.Nakazawa(Gifu): Λ-Λ interaction indicated by "Lambpha"
                                (^{6}He$_{\Lambda\Lambda}$ double hypernucleus)
15:50-16:10   H.Nemura(Tsukuba): Three-, four- and five-body
                                calculations of s-shell hypernuclei
                                with realistic interactions

MONDAY, 16:40-18:20

Session IIa Chairperson - W.Schweiger

16:40-17:00 V.Karmanov(LPI,Moscow): Stability of bound states in the light-front Yukawa model
17:00-17:20 B.Juliá-Díaz(Salamanca): Quark model study of the triton bound state
17:20-17:40 V.Kolybasov(LPI,Moscow): New polarization test for Δ-isobar admixture in ^{3}He
17:40-18:00 S.Levin(Stockholm/St.Petersburg): Asymptotical structure of the three-body Coulomb Green's function for the case of two charged particles
18:00-18:20 L.Yuan(Hannover) : Triton two-body photo disintegration with Δ-isobar excitation

Session IIb Chairperson - L.Tang

16:40-17:00 J.Bermuth(Mainz): Measurement of the electric neutron form factor G_E^n in the exclusive reaction ^{3}He(e,e'n) at momentum transfer $Q^2=0.67(GeV/c)^2$
17:00-17:20 D.Branford(Edinburgh): Investigation of Δ medium effects using the ^{4}He(γ,π^+n) reaction
17:20-17:40 M.Schumacher(Göttingen): Polarizabilities of proton and neutron investigated by Compton scattering on the proton and light nuclei
17:40-18:00 P.Montagna(Pavia): Single and multi-nucleon antiproton-^{4}He annihilation at rest
18:00-18:20 A.Krutenkova(ITEP,Moscow): The ITEP study of inclusive pion double charge exchange: Experiment and interpretation

Session IIc Chairperson - M.Dillig

16:40-17:00 S.Wycech(Warsaw): Phenomenological value of the $\pi\Lambda\Sigma$ coupling
17:00-17:20 J.Stepaniak(Warsaw): Some features of the pd--> ppnπ^0 reaction at 1.037 GeV
17:20-17:40 E.Truhlík(Řež): On radiative muon capture in hydrogen
17:40-18:00 J.E.Palomar-Burdeus(Valencia): Non-mesonic decay of Λ hypernuclei and the Γ_n/Γ_p ratio
18:00-18:20 J.Smejkal(Řež): Electromagnetic isoscalar $\rho\pi\gamma$ exchange current and the anomalous action

TUESDAY, 14:20-15:40

Session IIIa Chairperson - J.Haidenbauer

14:20-14:40 W.Schweiger(Graz): Two-quark correlations in the hard electromagnetic nucleon form factors
14:40-15:00 Th.Melde(Graz): Meson dynamics and the resulting "3-Nucleon-Force" diagrams: Results from a simplified test case
15:00-15:20 R.F.Wagenbrunn(Graz): Covariant electromagnetic and axial form factors of the nucleon from a chiral quark model
15:20-15:40 E.Norvaišas(Vilnius): Deuteron and dibaryons in the Skyrme model

Session IIIb Chairperson - V.Belyaev

14:20-14:40 J.Lukstins(Dubna): Progress of the nuclotron accelerator and the hypernuclear program
14:40-15:00 A.Khoukaz(Munster): Near threshold K^+K^- meson-pair production in proton-proton collisions
15:00-15:20 H.Holvoet(Ghent): The GDH sum rules and pion photoproduction on the nucleon
15:20-15:40 N.Matsuda(Hiroshima): Phase-shift analysis for proton-helium scattering at 4.0-200 MeV

Session IIIc Chairperson - R.Timmermans

14:20-14:40 C.García-Recio(Granada): Optical potentials in kaonic atoms
14:40-15:00 A.Cieplý(Řež): Low-energy K^- optical potentials: deep or shallow?
15:00-15:20 L.Tolós(Barcelona): Finite temperature effects on the anti-K optical potential
15:20-15:40 M.F.Aristizabal-Vargas(Medellin): Non-mesonic weak decay of Λ-hypernuclei and nuclear structure

TUESDAY, 16:10-17:30

Session IVa Chairperson - R.Van de Vyver

16:10-16:30 B.Jäger(Graz): Photoproduction of heavy quarkonia
16:30-16:50 J.-Ch.Caillon(Gradignan): Landau parameters and collective modes in nuclear matter with relativistic non-linear models
16:50-17:10 S.Lemaire(Gradignan): KN phase shifts in a model with a spin orbit interaction

Session IVb Chairperson - S.Wycech

16:10-16:30 M.Dillig(Erlangen): Heavy quark-antiquark admixture in the proton and its test in exclusive hadronic processes
16:30-16:50 O.Sato(Nagoya): Study of charm production in neutrino interactions
16:50-17:10 A.Machavariani(Tübingen/Dubna): Separation of the quark and meson degrees of freedom in the relativistic field-theoretical equations

Session IVc Chairperson - J.Lukstins

16:10-16:30 K.Miyagawa(Okayama): Electromagnetic K^+ production on the deuteron and hyperon-nucleon interaction
16:30-16:50 D.Lanskoy(Dubna): Composition and properties of multi-strange hypernuclear systems
16:50-17:10 L.Majling(Řež): $_\Lambda^{10}$Be and $_\Lambda^{10}$B Hypernuclei: A clue to some puzzles in non-mesonic weak decay

Poster Session
Thursday, 17:00-19:00

17:00-17:40 Short 3min presentation of the posters

E.Gedalin, A.Moalem, L.Razdolskaya(Beer-Sheva):
Pseudoscalar meson mixing in effective field theory

E.Gedalin, A.Moalem, L.Razdolskaya(Beer-Sheva):
On convergence of the ChPT HFF expansion for one loop
contribution to meson production in NN collisions

L.Yuan(Hannover):
Trinucleon Magnetic form factors

M.Suleymanov(Řež/Dubna):
Critical behavior of the characteristics of hadron-nucleus and
nucleus-nucleus interactions depending on the centrality degree
of collisions

A.Machavariani(Tübingen/Dubna):
The field-theoretical description of the $\gamma p \rightarrow p' \pi' \gamma'$
reaction in the Δ resonance region and determination of the
magnetic moment of the Δ^+ resonance.

E.Truhlík(Řež):
Pion electroproduction amplitude at threshold and
nucleon weak axial form factors

I.Bednarek, R.Mańka(Katowice):
Nucleon and meson effective masses in the relativistic
mean field theory

A.Krutenkova(ITEP,Moscow):
Search for medium effects in backward pion deuteron
quasielastic scattering on ^6Li

S.Wycech(Warsaw):
The ηA scattering length ηd final states

L.Majling(Řež)
$_\Lambda^{10}Be$ and $_\Lambda^{10}B$ hypernuclei on the nuclotron

J.E.Palomar-Burdeus(Valencia):
Pion and kaon vector form factors and some applications

A.Mutygullina(Kazan):
a) Quark-gloun retardation effects and an anomalous off-shell
   behavior of the two-nucleon amplitudes
b) Retardation effects from quark confinement on low-energy
   hadron dynamics

LIST OF PARTICIPANTS

Jiří Adam
Dept. of Theor. Physics
Nuclear Physics Institute
250 68 Řež near Prague
CZECH REPUBLIC
adam@ujf.cas.cz

Maria F. Aristizabal Vargas
Physics Department
University of Antioquia
Calle 67
53108 Medellin
COLOMBIA
mafe@pegasus.udea.edu.co

Adriana Banu
GSI
Planckstrasse 1
64291 Darmstadt
GERMANY
banu@gsi.de

Ilona Bednarek
Institute of Physics
University of Silesia
Uniwersytecka 4
40007 Katowice
POLAND
bednarek@us.edu.pl

Vladimir B. Belyaev
Bogoliubov Lab. of Theor. Physics
JINR
141980 Dubna
RUSSIA
belyaev@thsun1.jinr.ru

Giorgio Bendiscioli
Dip. di Fisica Nucleare e Teorica
Università di Pavia
via Bassi 6
27100 Pavia
ITALY
bendiscioli@pv.infn.it

Jörg Bermuth
Institut für Physik
J.J. Becherweg 45
55099 Mainz
GERMANY
bermuth@kph.uni-mainz.de

Michael Birse
Dept. of Phys. and Astronomy
University of Manchester
M13 9PL Manchester
UNITED KINGDOM
mike.birse@man.ac.uk

Jean-Paul Bocquet
Institut des Sciences Nucleaires
53, Avenue des Martyrs
38026 Grenoble
FRANCE
bocquet@isn.in2p3.fr

Derek Branford
Dept. of Physics and Astronomy
Edinburgh University
King's Buildings
Mayfield Road
EH9 3JZ Edinburgh
UNITED KINGDOM
db@np.ph.ed.ac.uk

Volker Burkert
Jefferson Laboratory
12000 Jefferson Ave.
Newport News
VA 23606
U.S.A.
burkert@jlab.org

Petr Bydžovský
Dept. of Theor. Physics
Nuclear Physics Institute
250 68 Řež near Prague
CZECH REPUBLIC
bydz@ujf.cas.cz

Daniel Cabrera Urbán
Dept. of Theoretical Physics
Universidad de Valencia
Apartado Oficial 22085
46071 Valencia
SPAIN
daniel.cabrera@ific.uv.es

Jean-Christophe Caillon
Centre d'Etudes Nucleaires
Uni. Bordeaux I, IN2P3
Le Haut Vigneau
33175 Gradignan Cedex
FRANCE
caillon@cenbg.in2p3.fr

Tim Van Cauteren
Ghent University
Proeftuinstraat 86
9000 Ghent
BELGIUM
tim.vancauteren@rug.ac.be

Miroslav Červený
Dept. of Theor. Physics
Nuclear Physics Institute
250 68 Řež near Prague
CZECH REPUBLIC
cerny@ujf.cas.cz

Aleš Cieplý
Dept. of Theor. Physics
Nuclear Physics Institute
250 68 Řež near Prague
CZECH REPUBLIC
cieply@ujf.cas.cz

Heinz Clement
Physikalisches Institut
Universität Tübingen
Auf der Morgenstelle 14
72076 Tübingen
GERMANY
clement@pit.physik.uni-tuebingen.de

Mircea Dan Cozma
Kernfysisch Versneller Inst.
Zernikelaan 25
9747 AA Groningen
THE NETHERLANDS
cozma@kvi.nl

Simon Dalley
Cambridge University
Wilberforce Road
CB3 0WA Cambridge
UNITED KINGDOM
sd214@damtp.cam.ac.uk

Andrzej Deloff
Soltan Inst. for Nuclear Studies
Hoza 69
00-681 Warsaw
POLAND
deloff@fuw.edu.pl

Manfred Dillig
Universität Erlangen-Nürnberg
Staudstrasse 7
91058 Erlangen
GERMANY
mdillig@theorie3.physik.uni-erlangen.de

Jan Dobeš
Dept. of Theor. Physics
Nuclear Physics Institute
250 68 Řež near Prague
CZECH REPUBLIC
dobes@ujf.cas.cz

Dolors Eiras
Universidad de Barcelona
Av. Diagonal, 647
08028 Barcelona
SPAIN
dolors@ecm.ub.es

Evgeny Epelbaum
Forschungszentrum Jülich GmbH
Leo-Brandt Strasse 1
52425 Jülich
GERMANY
e.epelbaum@fz-juelich.de

Jürgen Ernst
Universität Bonn
Nussallee 14-16
53115 Bonn
GERMANY
ernst@iskp.uni-bonn.de

Francisco Fernandez
Universidad de Salamanca
Plaza de la Merced s/n
37008 Salamanca
SPAIN
fdz@gugu.usal.es

Enrico Fragiacomo
Università di Trieste
Strada Costiera 11
34127 Trieste
ITALY
enrico.fragiacomo@ts.infn.it

Avraham Gal
Racah Inst. of Physics
The Hebrew University
Jerusalem 91904
ISRAEL
avragal@vms.huji.ac.il

Carmen Garcia-Recio
Departamento Física Moderna
Universidad de Granada
18071 Granada
SPAIN
g_recio@ugr.es

Edward Gedalin
Department of Physics
Ben-Gurion Uni. of the Negev
P.O.B. 653
84105 Beer-Sheva
ISRAEL
gedal@bgumail.bgu.ac.il

Jambul Gegelia
INFN, Sezione di Ferrara
via Paradiso 12
44100 Ferrara
ITALY
gegelia@fe.infn.it

Harald W. Griesshammer
Inst. für Theoretische Physik
Technische Uni. München
James Franck Strasse
85747 Garching
GERMANY
hgrie@physik.tu-muenchen.de

Franz Gross
Jefferson Laboratory
12000 Jefferson Ave.
VA 23606 Newport News
U.S.A.
gross@jlab.org

Johann Haidenbauer
Forschungszentrum Jülich GmbH
Leo-Brandt Strasse 1
52425 Jülich
GERMANY
j.haidenbauer@fz-juelich.de

Emiko Hiyama
KEK
Oho 1-1
305-0801 Tsukuba
JAPAN
hiyama@post.kek.jp

Heidi Holvoet
Ghent University
Proeftuinstraat 86
9000 Ghent
BELGIUM
heidi@inwfsun1.rug.ac.be

Jiří Hošek
Dept. of Theor. Physics
Nuclear Physics Institute
250 68 Řež near Prague
CZECH REPUBLIC
hosek@ujf.cas.cz

Herbert Hübel
Universität Bonn
Nussallee 14-16
53115 Bonn
GERMANY
hubel@iskp.uni-bonn.de

Ken'ichi Imai
Department of Physics
Kyoto University
Kyoto 606
JAPAN
imai@nh.scphys.kyoto-u.ac.jp

Barbara Jäger
Inst. für Theoretische Physik
Universität Graz
Universitätsplatz 5
8010 Graz
AUSTRIA
barbara.jaeger@uni-graz.at

Bruno Juliá-Diaz
Universidad de Salamanca
Plaza de la Merced s/n
37008 Salamanca
SPAIN
fdz@gugu.usal.es

Masayasu Kamimura
Kyushu University
Hakozaki 6-10-1
812-8581 Fukuoa
JAPAN
kami2scp@mbox.nc.kyushu-u.ac.jp

Vladimir A. Karmanov
Lebedev Physical Institute
Leninsky Prospekt 53
117924 Moscow
RUSSIA
karmanov@sci.lebedev.ru

Christoph Keil
Universität Giessen
Heinrich-Buff-Ring 16
35392 Giessen
GERMANY
christoph.m.keil@theo.physik.uni-giessen.de

Alfons Khoukaz
Institut für Kernphysik
Wilhelm-Klemm-Str. 9
48149 Münster
GERMANY
khoukaz@ikp.uni-muenster.de

William H. Klink
Dept. Physics and Astronomy
University of Iowa
Iowa City
52242 Iowa
U.S.A.
william-klink@uiowa.edu

Victor M. Kolybasov
Lebedev Physical Institute
Leninsky prospect 53
117924 Moscow
RUSSIA
kolybasov@sci.lebedev.ru

Anna P. Krutenkova
ITEP
B.Cheremushkinskaya 25
117218 Moscow
RUSSIA
krutenkova@vxitep.itep.ru

Dmitry E. Lanskoy
Inst. of Nuclear Physics
Moscow State University
119899 Moscow
RUSSIA
lanskoy@npi.msu.su

Sébastien Lemaire
Centre d'Etudes Nucleaires
Uni. Bordeaux I, IN2P3
Le Haut Vigneau
33175 Gradignan Cedex
FRANCE
lemaire@cenbg.in2p3.fr

Sergey Levin
Dept. of Atomic and Molecular Phys.
Stockholm University
Vanadisvagen 9
11385 Stockholm
SWEDEN
levin@physto.se

Kenneth Livingston
Dept. of Phys. and Astronomy
Univeristy of Glasgow
G12 8QQ Glasgow
UNITED KINGDOM
k.livingston@physics.gla.ac.uk

Juris Lukstins
Laboratory of High Energies
JINR
141980 Dubna
RUSSIA
juris@sunhe.jinr.ru

Lubomír Majling
Dept. of Theor. Physics
Nuclear Physics Institute
250 68 Řež near Prague
CZECH REPUBLIC
majling@ujf.cas.cz

Alexander Matschavariani
Bogoliubov Lab. of Theor. Physics
JINR
141980 Dubna
RUSSIA
alexander.matschavariani@uni-tuebingen.de

Veronica Malafaia
Centro de Fisica das Inter. Fund.
Instituto Superior Tecnico
Avenida Rovisco Pais
1049-001 Lisbon
PORTUGAL
vmalafaia@mail.telepac.pt

Ryszard Manka
University of Silesia
Institute of Physics
Universytecka 4
40007 Katowice
POLAND
manka@us.edu.pl

Laura Elisa Marcucci
Dipartimento di Fisica
Università di Pisa
Via Buonarroti, 2
56127 Pisa
ITALY
marcucci@df.unipi.it

Jiří Mareš
Dept. of Theor. Physics
Nuclear Physics Institute
250 68 Řež near Prague
CZECH REPUBLIC
mares@ujf.cas.cz

Jean-Francois Mathiot
Lab. de Physique Corpusculaire
Universite Blaise-Pascal
63177 Aubiere Cedex
FRANCE
mathiot@in2p3.fr

Masanori Matsuda
Division of Material Science
Faculty of Integrated Arts and Sci.
Hiroshima University
Kagamiyama 1-7-1
Higashi-Hiroshima 739-8521
JAPAN
masa@hiroshima-u.ac.jp

Judith McGovern
Department of Physics
University of Manchester
Brunswick St
M13 9PL Manchester
UNITED KINGDOM
judith.mcgovern@man.ac.uk

Thomas Melde
Inst. für Theoretische Physik
Universität Graz
Universitätsplatz 5
8010 Graz
AUSTRIA
thomas.melde@kfunigraz.ac.at

Hans-Otto Meyer
Indiana Uni. Cooler Facility
Milo B. Sampson Lane
Bloomington, IN 47405
U.S.A.
meyer@iucf.indiana.edu

Kazuya Miyagawa
Dept. of Applied Physics
Okayama University of Science
1-1 Ridai-cho
700 Okayama
JAPAN
miyagawa@dap.ous.ac.jp

Amnon Moalem
Department of Physics
Ben-Gurion Uni. of the Negev
Hanesiim
84105 Beer-Sheva
ISRAEL
moalem@bgumail.bgu.ac.il

Paolo Montagna
INFN, Sezione di Pavia
Via A. Bassi, 6
27100 Pavia
ITALY
Paolo.Montagna@pv.infn.it

Dario Moricciani
INFN, Sezione di Roma2
Via della Ricerca Scientifica 1
00133 Roma
ITALY
Dario.Moricciani@roma2.infn.it

Toshio Motoba
Physics Department
Brookhaven National Laboratory
P.O.Box 5000
NY 11973-5000 Upton
U.S.A.
motoba@isc.osakac.ac.jp

Aigul Mutygullina
Department of Physics
Kazan State University
Kremlevskaya, 18
420008 Kazan
RUSSIA
Aigul.Mutygullina@ksu.ru

Juan E. Palomar Burdeus
Dept. of Theoretical Physics-IFIC
Universidad de Valencia
Apartado Oficial 22085
46071 Valencia
SPAIN
palomar@condor3.ific.uv.es

Kazuma Nakazawa
Physics Dept. of Gifu Uni.
1-1 Yanagido
501-1193 Gifu
JAPAN
nakazawa@cc.gifu-u.ac.jp

Assumpta Parreño
Dept. ECM, Facultat de Fisica
Universidad de Barcelona
Diagonal 647
08028 Barcelona
SPAIN
assum@ecm.ub.es

Mikhail Nekipelov
Institut für Kernphysik
Forschungszentrum Jülich
Leo-Brandt Strasse 1
52425 Jülich
GERMANY
m.nekipelov@fz-juelich.de

M. Teresa Peña
Physics Department, IST
Av. Rovisco Pais
1049-001 Lisboa
PORTUGAL
teresa@gtae3.ist.utl.pt

Hidekatsu Nemura
Inst. of Part. and Nucl. Studies
KEK
Oho 1-1
305-0801 Tsukuba
JAPAN
nemura@post.kek.jp

Teake Penninga
Kernfysisch Versneller Instituut
Zernikelaan 25
9747 AA Groningen
THE NETHERLANDS
Penninga@kvi.nl

Egidijus Norvaišas
Inst. of Theor. Phys. and Astronomy
A.Goštauto 12
2600 Vilnius
LITHUANIA
norvaisas@itpa.lt

Daniel Phillips
Dept. of Phys. and Astronomy
Ohio University
OH 45701 Athens
U. S. A.
phillips@phy.ohiou.edu

Henk Polinder
Inst. for Theoretical Physic
University of Nijmegen
Toernooiveld 1
Nijmegen
THE NETHERLANDS
polinder@sci.kun.nl

Lubov Razdolskaya
Department of Physics
Ben-Gurion Univ. of the Negev
P.O.B. 653
84105 Beer-Sheva
ISRAEL
ljuba@bgumail.bgu.ac.il

Avraham Rinat
Dept. of Particle Physics
Weizmann Institute of Science
76100 Rehovot
ISRAEL
fnrinat@wicc.weizmann.ac.il

Osamu Sato
Department of Physics
Nagoya University
Furo-cho Chikusa-ku
464-8602 Nagoya
JAPAN
sato@flab.phys.nagoya-u.ac.jp

Vladimír Šauli
Dept. of Theor. Physics
Nuclear Physics Institute
250 68 Řež near Prague
CZECH REPUBLIC
sauli@ujf.cas.cz

Carlo Schaerf
Dipartimento di Fisica
Università di Roma "Tor Vergata"
via della Ricerca Scientifica 1
00133 Roma
ITALY
schaerf@roma2.infn.it

Martin Schumacher
II. Physikalisches Institut
Bunsenstrasse 7-9
37073 Göttingen
GERMANY
schumacher@physik2.uni-goettingen.de

Michael Schwamb
Institut für Kernphysik
Johannes-Gutenberg Universität
J.J.-Becherweg 45
55099 Mainz
GERMANY
schwamb@kph.uni-mainz.de

Wolfgang Schweiger
Inst. für Theoretische Physik
Universität Graz
Universitätsplatz 5
8010 Graz
AUSTRIA
wolfgang.schweiger@uni-graz.at

Andre Siepe
ISKP
Universität Bonn
Nussallee 14-16
53115 Bonn
GERMANY
siepe@iskp.uni-bonn.de

Simon Širca
MIT Lab for Nuclear Science
77 Massachusetts Avenue
02141 Cambridge, MA
U.S.A.
sirca@mitlns.mit.edu

Jaroslav Smejkal
Dept. of Theor. Physics
Nuclear Physics Institute
250 68 Řež near Prague
CZECH REPUBLIC
smejkal@ujf.cas.cz

Jan Smolík
Institute of Physics
Na Slovance 2
18000 Praha
CZECH REPUBLIC
smolik@fzu.cz

Miloslav Sotona
Dept. of Theor. Physics
Nuclear Physics Institute
250 68 Řež near Prague
CZECH REPUBLIC
sotona@ujf.cas.cz

Thorsten Speckner
Physikalisches Institut IV
Uni. of Erlangen-Nürnberg
Erwin-Rommel-Str.1
91058 Erlangen
GERMANY
Thorsten.Speckner@physik.uni-erlangen.de

Alfred Stadler
CFNUL
Av. Prof. Gama Pinto, 2
1649-003 Lisboa
PORTUGAL
stadler@cii.fc.ul.pt

Petr Štecher
Dept. of Theor. Physics
Nuclear Physics Institute
250 68 Řež near Prague
CZECH REPUBLIC
stecher@ujf.cas.cz

Gerard van der Steenhoven
NIKHEF
P.O. Box 41882
1009 DB Amsterdam
THE NETHERLANDS
gerard@nikhef.nl

Joanna Stepaniak
Institute for Nuclear Studies
Hoza 69
00681 Warsaw
POLAND
joste@fuw.edu.pl

Mais Suleymanov
Dept. of Theor. Physics
Nuclear Physics Institute
250 68 Řež near Prague
CZECH REPUBLIC
mais@sunhe.jinr.ru

Liguang Tang
Jefferson Laboratory
12000 Jefferson Ave.
Newport News
VA 23606
U.S.A.
tangl@jlab.org

Rob Timmermans
Theory Group
Kernfysisch Versneller Instituut
Zernikelaan 25
9747 AA Groningen
THE NETHERLANDS
timmermans@kvi.nl

Laura Tolós
Departament ECM
Universidad de Barcelona
Avda. Diagonal, 647
08028 Barcelona
SPAIN
laura@ecm.ub.es

Emil Truhlík
Dept. of Theor. Physics
Nuclear Physics Institute
250 68 Řež near Prague
CZECH REPUBLIC
truhlik@ujf.cas.cz

Marcel Uličný
Forschungszentrum Jülich GmbH
Leo-Brandt Strasse 1
52425 Jülich
GERMANY
m.ulicny@fz-juelich.de

Vicente Vento
Universidad de Valencia
C/ Dr. Moliner, 50
46100 Burjassot (Valencia)
SPAIN
Vicente.Vento@uv.es

Robert Vinh Mau
LPNHE
Universite Paris VI
4 Place Jussieu Tour 12-13 E3 case 127
75252 Paris Cedex 05
FRANCE
rvinhmau@in2p3.fr

Robert Van de Vyver
Dept. Subatomic and Radiation Phys.
Ghent University
Proeftuinstraat 86
9000 Ghent
BELGIUM
robert@inwfsun1.rug.ac.be

Robert F. Wagenbrunn
Institut für Theoretische Physik
Universität Graz
Universitätsplatz 5
8010 Graz
AUSTRIA
Robert.Wagenbrunn@uni-graz.at

Wolfram von Witsch
Inst. für Strahlen- und Kernphysik
Universität Bonn
Nussallee 14-16
53115 Bonn
GERMANY
vwitsch@iskp.uni-bonn.de

Slawomir Wycech
Soltan Inst. for Nuclear Studies
Hoza 69
00-681 Warsaw
POLAND
wycech@fuw.edu.pl

Avivi I. Yavin
Tel Aviv University
Ramat Aviv
69978 Tel Aviv
ISRAEL
avivi1@012.net.il

Luping Yuan
Inst. für Theoretische Physik
Universität Hannover
Appelstrasse 2
30167 Hannover
GERMANY
yuan@itp.uni-hannover.de

Imrich Zborovský
Dept. of Theor. Physics
Nuclear Physics Institute
250 68 Řež near Prague
CZECH REPUBLIC
zborovsky@ujf.cas.cz

Jozef Zlomańczuk
Dept. of Radiation Sciences
Uppsala University
Box 535
751 21 Uppsala
SWEDEN
Jozef.Zlomanczuk@tsl.uu.se

AUTHOR INDEX

A

Abramov, B. M., 515, 519
Acus, A., 369
Adam, H.-H., 437
Adam, Jr., J., 323, 327
Ahmidouch, A., 173
Akaishi, Y., 471
Alvarez-Ruso, L., 515
Ambrozewicz, P., 173
Androic, D., 173
Angelescu, T., 173
Arenhövel, H., 299
Aristizabal, M. F., 397
Asaturyan, R., 173
Avery, S., 173

B

Baker, O. K., 173
Balewski, J. T., 437
Barford, T., 229
Bartalini, O., 499, 527
Bartke, J., 457
Bednarek, I., 525
Bellini, V., 499, 527
Belyaev, V. B., 495
Bennhold, C., 459, 481
Bermuth, J., 331
Bertovic, I., 173
Bilger, R., 211, 503, 531
Birse, M. C., 229
Blankleider, B., 233
Bocquet, J.-P., 499, 527
Boffi, S., 319
Borodin, Y. A., 515, 519
Branford, D., 335
Breuer, H., 173
Brodowski, W., 211, 503, 531
Budzanowski, A., 437
Bulychjov, S. A., 515, 519
Burkert, V. D., 3

C

Cabrera, D., 237
Caillon, J. C., 339
Calén, H., 211, 503, 531
Camen, M., 377
Canton, L., 287
Capogni, M., 527
Carbonell, J., 271
Carlini, R., 173
Castoldi, M., 499, 527
Cha, J., 173
Chmielewski, K., 323, 327
Chren, D., 457
Chrien, R., 173
Christy, M., 173
Cieplý, A., 401
Clement, H., 211, 503, 531
Cole, L., 173
Corvisiero, P., 499
Cozma, M. D., 257

D

Dalley, S., 343
Danagoulian, S., 173
D'Angelo, A., 499, 527
d'Angelo, A., 527
Dehnhard, D., 173
Deloff, A., 507
de Melo, J. P. B. C., 315
Demetriou, P., 315
Didelez, J.-P., 499, 527
Dillig, M., 347
Di Salvo, R., 499, 527
Dukhovskoi, I. A., 515, 519

E

Ekström, C., 211, 503, 531
Elaasar, M., 173
Elander, N., 279
Elster, C., 17
Empl, A., 173
Ent, R., 173

Entem, D. R., 263
Epelbaum, E., 17
Ernst, J., 511

F

Fäldt, G., 211
Fantini, A., 527
Fenker, H., 173
Fernández, F., 263, 267
Filikhin, I. N., 405
Fonseca, A. C., 323
Fragiacomo, E., 29
Fransson, K., 211, 503, 531
Friedman, E., 401
Fujii, Y., 173
Furic, M., 173

G

Gabinski, P., 339
Gainutdinov, R. K., 291, 295
Gal, A., 401, 405
Gan, L., 173
García-Recio, C., 417
Garrow, K., 173
Gasparian, A., 173
Gedalin, E., 241, 245, 247
Gegelia, J., 233
Gervino, G., 499, 527
Ghio, F., 499, 527
Gilman, R., 55
Girolami, B., 499, 527
Giusa, A., 527
Glöckle, W., 17, 459
Glozman, L. Y., 319
Golokhvastov, A. I., 449
Green, A. M., 547
Greiff, J., 531
Grießhammer, H. W., 41
Gross, F., 55
Grzonka, D., 437
Gueye, P., 173
Guidal, M., 499, 527
Gustafsson, L., 211, 531

H

Häggström, S., 503
Haidenbauer, J., 267, 421
Harvey, M., 173
Hashimoto, O., 173
Henning, H., 327
Hinton, W., 173
Hiyama, E., 425, 429
Höistad, B., 211, 503, 531
Holvoet, H., 353
Hourany, E., 499, 527
Hu, B., 173
Huhn, V., 303
Hungerford, E., 173

I

Imai, K., 69

J

Jacewicz, M., 531
Jackson, C., 173
Jäger, B., 357
Jarczyk, L., 437
Johanson, J., 211, 503
Johansson, A., 211, 503, 531
Johansson, T., 211, 503, 531
Johnston, K., 173
Juengst, H., 173
Juliá-Díaz, B., 263, 267

K

Kaidalov, A. B., 515
Kamada, H., 17
Kamimura, M., 425, 429
Karmanov, V. A., 271
Keil, C., 433
Keppel, C., 173
Khanna, F. C., 311, 535
Khanov, A. I., 515, 519
Khorozov, S. A., 449
Khoukaz, A., 437
Kilian, K., 211, 437, 503
Klink, W., 319

Klink, W. H., 79
Koch, I., 531
Kolybasov, V. M., 275
Kossert, K., 377
Kouznetsov, V., 499, 527
Kowina, P., 437
Krupovnickas, T., 369
Krutenkova, A. P., 515, 519
Kulikov, V. V., 515, 519
Kullander, S., 211, 503, 531
Kunne, R., 499
Kupść, A., 211, 503, 531

L

Labarsouque, J., 339, 445
Lacombe, M., 315
Lan, K., 173
Lang, N., 437
Lanskoy, D. E., 441
Lapik, A., 499, 527
Lemaire, S., 445
Lenske, H., 433
Levin, S. B., 279
Liang, Y., 173
Likhachev, V. P., 173
Limkaisang, V., 283
Lister, T., 437
Liu, J. H., 173
Livingston, K., 521
Lleres, A., 499, 527
Loiseau, B., 315, 489
Lukstins, J., 449, 457

M

Machavariani, A. I., 361
Mack, D., 173
Maeda, K., 173
Majling, L., 453, 457
Mangin-Brinet, M., 271
Mańka, R., 525
Marciniewski, P., 503
Marcucci, L. E., 89
Mareš, J., 401
Margaryan, A., 173, 453
Markowitz, P., 173
Marranghello, G. F., 347

Mart, T., 459
Martoff, J., 173
Mathiot, J.-F., 101
Matsuda, M., 283
Matsuk, M. A., 515, 519
McGovern, J. A., 249
Meißner, U.-G., 17
Melde, T., 287
Melnitchouk, W., 421
Meyer, H. O., 113
Miyagawa, K., 459
Miyoshi, T., 173
Mkrtchyan, H., 173
Moalem, A., 241, 245, 247
Montagna, P., 365
Moricciani, D., 499, 527
Morosov, B., 211, 503, 531
Mörtsell, A., 503
Moskal, P., 437
Motoba, T., 125, 425, 429
Mutygullina, A. A., 291, 295

N

Nagata, J., 283
Nakazawa, K., 463
Nedorezov, V., 499, 527
Nekipelov, M., 467
Nemoto, S., 327
Nemura, H., 471
Nicoletti, L., 499, 527
Nieves, J., 417
Nikitin, V., 457
Nogga, A., 17
Norvaišas, E., 369

O

Oelert, W., 211, 437, 503
Oelsner, M., 323, 327
Oset, E., 417

P

Palomar, J. E., 475, 479
Parfenov, A., 449, 457
Parreño, A., 453, 481

Pätzold, J., 503
Peña, M. T., 137
Petkovic, T., 173
Phillips, D. R., 149
Plessas, W., 319
Polls, A., 485
Ponce, W. A., 397

Q

Quentmeier, C., 437

R

Radici, M., 319
Radkevich, I. A., 515, 519
Rakityansky, S. A., 495
Ramos, A., 417, 481, 485
Randieri, C., 527
Razdolskaya, L., 241, 245, 247
Rebreyend, D., 499, 527
Reinhold, J., 173
Renard, F., 499, 527
Riska, D. O., 369
Rocha, S. S., 347
Roche, J., 173
Ruber, R. J. M. Y., 211, 503, 531
Rudnev, N. V., 499, 527

S

Sandhas, W., 495
Sandri, P. Levi, 499, 527
Santo, R., 437
Sanzone, M., 499
Sarsour, M., 173
Sato, O., 373
Sato, Y., 173
Sauer, P. U., 323, 327
Sawafta, R., 173
Schaerf, C., 499, 527
Schepers, G., 437
Schepkin, M., 503
Scholten, O., 257
Schumacher, M., 377
Schwamb, M., 299
Schwärz, M., 381

Schweiger, W., 357, 381
Sefzick, T., 437
Semay, C., 315
Sewerin, S., 437
Shevchenko, N. V., 495
Shwartz, B., 531
Siemaszko, M., 437
Siepe, A., 303
Silvestre-Brac, B., 445
Simicevic, N., 173
Širca, S., 303
Smejkal, J., 311
Smith, G., 173
Smolik, J., 251
Smyrski, J., 437
Sofianos, S., 495
Sopko, B., 457
Speckner, T., 385
Sperduto, M. L., 499, 527
Speth, J., 421
Stepaniak, J., 211, 503, 531
Stepanyan, S., 173
Strzałkowski, A., 437
Sukhanov, A., 211, 503
Suleymanov, M. K., 389
Sundberg, P., 211, 503
Sutera, C. M., 499, 527
Sutormin, A. I., 515
Suzuki, Y., 471
Svenne, J. P., 287

T

Tadevosyan, V., 173
Takahashi, T., 173
Tamura, H., 173
Tang, L., 173, 453
Tanida, K., 173
Timmermans, R. G. E., 187, 257
Tjon, J. A., 257
Tolós, L., 485
Truhlík, E., 311, 535, 539
Turdakina, E. N., 515, 519
Turinge, A., 499, 527
Turowiecki, A., 211, 503, 531

U

Ukai, M., 173
Uličný, M., 543
Uzzle, A., 173

V

Valcarce, A., 263, 267
van der Steenhoven, G., 161
Van de Vyver, R., 391
Vasconcellos, C. A. Z., 347
Vasyukhin, N. E., 449
Vento, V., 199
Vicente Vacas, M. J., 515
Vinh Mau, R., 315
Vodopianov, A. S., 389
von Witsch, W., 303
Vulcan, W., 173

W

Wagenbrunn, R. F., 319
Wagner, G. J., 211, 503
Wätzold, L., 303
Weber, C., 303
Wells, S., 173
Wilhelmi, Z., 211, 503, 531
Wilkin, C., 211
Wissmann, F., 377
Witała, H., 17, 303
Wolke, M., 437
Wood, S., 173
Wu, H. C., 397
Wüstner, P., 437
Wycech, S., 489, 547

X

Xu, G., 173

Y

Yakovlev, S. L., 279
Yamada, T., 425, 429
Yamaguchi, Y., 173
Yamamoto, Y., 425, 429
Yan, C., 173
Yoshino, H., 283
Yuan, L., 173
Yuan, L. P., 323, 327

Z

Zabierowski, J., 211, 503, 531
Zabrodin, A., 527
Zborovský, I., 389
Zernov, A., 211, 503
Zhu, X., 173
Zipper, W., 437
Złomańczuk, J., 211, 503, 531
Zucchiatti, A., 499, 527